Hubertus Buchstein
Enduring Enmity

Political Science | Volume 140

Hubertus Buchstein, born in 1959, professor of political theory at Universität Greifswald, Germany. He has obtained his doctorate at Freie Universität Berlin and has taught at the New School for Social Research in New York. His research focusses on modern democratic theory, the history of political ideas, the use of lotteries in political decision making, and Critical Theory.

Hubertus Buchstein

Enduring Enmity

The Story of Otto Kirchheimer and Carl Schmitt

Translated by Sandra H. Lustig

[transcript]

Bibliographic information published by the Deutsche Nationalbibliothek
The Deutsche Nationalbibliothek lists this publication in the Deutsche Nationalbibliografie; detailed bibliographic data are available in the Internet at https://dnb.dnb.de /

First published in 2024 by transcript Verlag, Bielefeld

Cover layout: Maria Arndt, Bielefeld
Copy-editing: Linda Jayne Turner
Printed by: Majuskel Medienproduktion GmbH, Wetzlar
https://doi.org/10.14361/9783839464700
Print-ISBN: 978-3-8376-6470-6
PDF-ISBN: 978-3-8394-6470-0
ISSN of series: 2702-9050
eISSN of series: 2702-9069

Printed on permanent acid-free text paper.

Contents

Postwar Democracies

Conclusion

Appendix

Acknowledgments

Many scholars on both sides of the Atlantic generously shared their knowledge in discussions with me during my research for this book, but one person stands out for her tireless work helping me write and complete it. For four years, Sandra H. Lustig was an invaluable collaborator on this book. First and foremost, I would like to express my gratitude for the outstanding quality of her work and for her dedication to this project, doing so much more than "simply" translate the manuscript, most of which I wrote in German, into English. Her persistent questioning and critical comments compelled me to express myself more clearly, made me aware of numerous inaccuracies and errors, and resulted in substantial changes to the end product. Following her advice, I decided that the story was best told in chronological order and restructured the book. Sandra then worked through the entire manuscript again with an editor's eye for detail and for the big picture, for content and for language. In particular, she prompted me to clarify Schmitt's antisemitism, take a closer look at Kirchheimer's and Schmitt's correspondence after 1945, revisit the order of events in the conflict over George Schwab's doctorate, and reexamine the correspondence between Kirchheimer and Schmitt after World War II.

Most importantly, Sandra drew on her in-depth understanding of Nazi German to develop and successfully apply an approach to translating it into accurate, comprehensible, and readable English. She researched what the Nazis meant by their own terminology and compiled a Glossary to provide the English reader with extensive information to better illuminate the text. Even most native German speakers today do not appreciate how the Nazis changed the German language to further their own political and propagandistic goals. The majority of English-speaking readers are unlikely to be aware of this linguistic issue at all. Sandra then provided brief explanations of pertinent terms throughout the book, enabling readers to understand Schmitt within the historical, linguistic, and intellectual context. I advise readers to consult her Translator's Preface for an incisive explanation of Nazi German and her Glossary for clarification of individual words and terms. Her research stands out—at least to the best of my knowledge—as unique in the vast secondary literature on Carl Schmitt and should guide scholars to navigate the problem of distinguishing the meaning of his words. Translators tackling Schmitt's numerous works from the Nazi period—most of which have not been translated—may well consider adopting her approach to rendering the precise meaning of his writing into English.

Next, I would like to thank Linda Jayne Turner. She did a wonderful job copyediting the entire manuscript and communicating with me about a number of inconsistencies she found. She made the language more consistent throughout and encouraged me to provide explanations for English-speaking readers in some places and to make cuts in others. She also crosschecked the names of everyone mentioned in the book against the Index of Names and ensured that they were all included and introduced at the first mention if appropriate. Her meticulous work eliminated my original errors and inaccuracies from the final publication. I will miss the insightful discussions about my manuscript with both Linda and Sandra. Yet any remaining errors or ambiguities are due solely to my intransigence in responding to their questions and comments.

The impetus to write this book came from the reactions to a lecture about Kirchheimer and Schmitt I gave at the New School for Social Research in New York in February 2019 during my Theodor Heuss Professorship. The controversial and intense discussion that followed my talk became the starting point for exploring the relationship between the two of them in more detail. In the early phase of the work for this book, I benefited greatly from the critical comments from my colleagues at the New School in the academic year 2018/19, namely Andrew Arato, Richard J. Bernstein (†), Federico Finchelstein, Carlos Forment, Andreas Kalyvas, William Milberg, and James E. Miller. Additional early encouragement to write this book came from Jean Louise Cohen, Gerhard Göhler, Bob Jessop, Christina Lafont, Herfried Münkler, Claus Offe, Kari Palonen, Hartmut Rosa, Thomas Saretzki, Rainer Schmalz-Bruns (†), and Nadia Urbinati.

The support from the many members of the staff at the archives listed in the bibliography was invaluable. In particular, my thanks go to Jodi Boyle and Brian Keough for assisting me in searching through the papers of Otto Kirchheimer and John H. Herz in the German Intellectual Émigré Collection at the State University of New York in Albany, to Jenny Swadosh at the archive of the New School for Social Research in New York, to Dirk Braunstein at the archive of Institut für Sozialforschung in Frankfurt, and to Gerd Giesler of the Carl-Schmitt-Gesellschaft e.V.

While working on the book I received invaluable suggestions and/or material from Andreas Anter, Harald Bluhm, Rainer Erd, Jürgen Habermas, Dirk Jörke, Bernd Ladwig, Simone Ladwig-Winters, Detlef Lehnert, Marcus Llanque, Reinhard Mehring, Volker Neumann, Peter Niesen, Kerstin Pohl, Michael Sauthoff, Frank Schale, William E. Scheuerman, Alfons Söllner, Grit Straßenberger, and Jürgen Tröger. With regard to the title of the book, Jodi Dean, Frank Fischer, Chad A. Goldberg, and Axel Müller all gave insightful input. In addition, I owe special thanks to Douglas G. Morris and Rieke Trimçev for their thorough critical reviews of earlier versions of the introduction and conclusion.

At the University of Greifswald, I had the wonderful opportunity to draw on the expertise of my editorial team of Otto Kirchheimer's Collected Works, namely Henning Hochstein, Lisa Klingsporn, Moritz Langfeldt, and Merete Peetz. Thomas Stamm-Kuhlmann and Katrin Rhein provided valuable assistance in deciphering a number of handwritten letters and notes from the archives by Schmitt, Kirchheimer, and Rudolf Smend. Lena Else Droese, Fabian Fleßner, Steffi Krohn, Anna Luise Munsky, and Finja-

Pauline Schöbel collaborated in the search for literature and additional source material. I also had the privilege to receive critical feedback from the participants of the Greifswald Colloquium for Political Theory, namely Sara Gebh, Eva Hausteiner, Niels Hegewisch, Tobias Müller, Ronny Rhode, Eno Trimçev, Rieke Trimçev, Milos Rodatos, and Mounir Zahran. I owe special thanks to Fabian Fleßner for preparing the manuscript for publication and for creating the Index of Names. I would like to thank Annika Linnemann at transcript publishing for supporting my project and her endless patience in the face of the significantly delayed submission of my manuscript.

I had the opportunity to present parts of the book in lectures or at conferences hosted at the New School for Social Research, the University of Augsburg, the Freie Universität Berlin, the University of Bonn, the Technical University of Darmstadt, the University of Halle, the University of Hamburg, the University of Helsinki, the University of Jena, and New York University. Some earlier versions of sections of the book have been previously published in articles in the following journals: *Antisemitism Studies, Constellations, Finnish Yearbook of Political Thought, Leviathan, New German Critique, Redescriptions, WestEnd—Neue Zeitschrift für Sozialforschung, Zeitschrift für Parlamentsfragen, Zeitschrift für Politik,* and *Zeitschrift für Politische Theorie.*

Writing this book was only possible thanks to a fellowship from the Volkswagen Foundation in the academic year 2021/22. It is ironic that this foundation sponsored the project since neither Schmitt nor Kirchheimer had a driver's license. Both, however, did deal with the German automotive industry, including the Volkswagen company, in their oeuvre: whereas for Schmitt, cars and tanks exemplified the modern age of accelerated mobility, Kirchheimer examined the German steel industry as a supporter and profiteer of Nazism and the Volkswagen company in particular as a creation of the Nazi regime. Being able to step back from all my teaching and administrative duties at my department thanks to the generous Volkswagen Foundation fellowship provided the invaluable opportunity for me to focus fully on the research for this book.

Finally, my deepest gratitude goes to Otto Kirchheimer's daughter Hanna Kirchheimer-Grossman (Arlington, VA) and his son Peter Kirchheimer (New York City) for granting me access to documents and photos in their family archive. I am even more grateful for the stimulating discussions with them and their three children about the life and work of their father and grandfather. It is touching for me to see that Otto Kirchheimer's turbulent life story, his writings and his intellectual manner have found a permanent place in the memories of the younger generation of the Kirchheimer family like Rebecca Kirchheimer, too.

I dedicate this book to the memory of my father, Walter Buchstein (1905–1977), and to the countless stories he told me about how he had experienced social and political life as a young man in 1920s and 1930s Germany.

Translator's Preface

By Sandra H. Lustig[1]

The spirit of a language is most clearly revealed in its untranslatable words.
(Marie von Ebner-Eschenbach)

When I began working on this book, I soon realized that translating Carl Schmitt poses greater challenges than translating most legal and political theorists. This is particularly true of his oeuvre from 1933 on and the versions of his older works which he revised during the Nazi period; only a few of his publications from that time are available in English translation. Viewing Schmitt's work within the societal context in which he produced it is essential for comprehending it fully. And as the societal context changed, so did his writing.

As is typical of totalitarian and authoritarian regimes, propaganda was crucial to the Nazis' pervasive efforts to gain and retain the support of broad swaths of the population for the regime itself and all its actions. Nazi propaganda permeated virtually everything written and spoken in Germany during that time, from publishing and the media to teaching at schools to scholarly production, all of which were controlled by the state. Language was key to making all this succeed. The German language was changed in various ways under the Nazis, resulting in what is commonly called Nazi German, as will be explained in more detail below. If readers of Nazi German texts today aim to understand their precise meaning, they must be aware of these linguistic changes.

In this translator's preface, I explain some of the most important features of Nazi German to help English-speaking readers of Schmitt's writing at least from the Nazi period develop such an awareness, to explain how I have dealt with the problems of translating Nazi German in this book, and to encourage scholars to question whether it might be worth rethinking their interpretations of Schmitt's oeuvre against this background. With his masterful command of the German language, there can be no doubt that he

1 I would like to thank Adam Blauhut, Hubertus Buchstein, Susanne Elfferding, and Rebecca Garron for reading drafts of this translator's preface, discussing countless linguistic nuances with me, and helping me tease out exactly what I wanted to say.

used it very deliberately, and at least during the Nazi period, he wrote in Nazi German. Incidentally, Otto Kirchheimer did not write in Nazi German; his writing is clear and comprehensible. If he used untranslatable words on occasion—and German, like every other language, has some, such as *Heimat* and *Weltschmerz*—then it was because they express concepts particular to German culture or law, not to convey Nazi ideology in an obfuscating way.

Nazi German also used specific elements of style and rhetoric,[2] but analyzing them is beyond the scope of this preface; the focus here and in the Glossary is on individual words and terms.

Nazi German

Nazi German has been described as "a manipulative and aggressive language based on a vocabulary of emotionalism, radicalization, deception, defamation, and brutalization. [...] [T]he Nazis were able to shape not only the communicative process but also the psychological and sociopolitical thinking of most Germans" (Mieder 2002, xvi). Nazi German worked in various ways. Among a number of linguistic techniques, it gave existing words new meanings, e.g., *gesund* (no longer only: healthy, but now also: corresponding to the norm of the NSDAP; see Glossary). It invented new words and terms, e.g., *Ehrenkreuz der Deutschen Mutter* (literally: Cross of Honor of the German Mother, a medal awarded to "worthy mothers of German *Blut*" (see Glossary) with at least four, later three, children (Schmitz-Berning 2007, 164–165). Nazi German used old Germanic terms to evoke "sentiments of mystical blood ties and emotions to make Germans believe in the transcendental German nation (*Volk*)" (Doerr 2002, 29) and an archaic mythology of Germanness that readers today may have encountered in Wagnerian operas. For instance, *Bund der Rechtswahrer* (Alliance of Preservers/Defenders of the Law; see Glossary) has a more august ring to it than *Vereinigung der Rechtsanwälte* (Association of Lawyers), as it was previously called.

While Nazi German was sometimes unreservedly clear, e.g., *Die Juden sind unser Unglück* (The Jews are our misfortune/ruin), it could also give words "deliberately equivocal meanings" (Paechter et al. 1944, 110). "The ambiguity of this para-logical language serves not only to deceive the enemy but also as an instrument of social control" (Paechter et al. 1944, 8). Kirchheimer wrote of "the brutal, cynical reality of the new legal system behind the fog of Nazi phrases" (see Chapter 8, p. 209)—in reference to Schmitt helping to create and cultivate that fog. The title of the *Gesetz zur Wiederherstellung des Berufsbeamtentums* (literally: Law for the Restoration of the Public Civil Service) illustrates how Nazi German "obscure[d] and sanitize[d] the reality to which it refer[red]" (Morris 2002, xi). It states the need for the civil service to be restored, implying that it had been damaged or destroyed; the solution to that alleged problem was to expel political opponents, Jews, and all others not considered Aryan from the civil service (see Chapter 7). Without even mentioning the words "Jew" or "Jewish"—by design—this law robbed

2 See, for example, the chapter "The Spirit and Structure of Nazi Language" (Paechter et al. 1944, 5–15).

all Jewish civil servants of their livelihoods. Nazi German often used code or allusion to hint at what was not expressed explicitly, for example, the word *Sonderbehandlung* (literally: special treatment; in fact: execution or mass killing of Jews or opponents of the regime, Schmitz-Berning 2007, 584–587). At times, the Nazi regime relied "more on the execution of implied policies than on outspoken orders" (Kirchheimer/Herz 1945a, 467, see Chapter 13, p. 335). While a target audience clued in to this form of communication understood the general thrust, this obscurantism created fear and terror because it left people in the dark about the exact details of a message.

Therefore, nobody, and certainly not translators, should take the words in Nazi German texts at face value; depending on the author, target audience, purpose, and general context, the meaning to be conveyed by individual words may be very different. Even scholars whose mother tongue is not German but who have become fluent in it "might have difficulty in understanding the at-times-absurd, impenetrable, and euphemistic language of the Nazis" (Mieder 2002, xvii). As a matter of fact, fully grasping much of Nazi German is difficult even for native German speakers today (although extreme right-wing parties do use some code words and other features of Nazi German, and they resonate with their audience).

The goal of much of Nazi German was not simply to describe things and concepts. It was to conjure up strong emotions such as loyalty to the Nazi system or hatred of certain groups of people. "For what, after all, was the ultimate purpose and eventual success of all these overblown emotions? Emotion was not itself the be-all and end-all, it was only a means to an end, a step in a particular direction. Emotion had to suppress the intellect and itself surrender to a state of numbing dullness without any freedom of will or feeling; how else would one have got hold of the necessary crowd of executioners and torturers?" (Klemperer 1947, 252). Rose, too, argues that such "emotionality [...] completely flooded the German listeners' critical defenses and appealed directly to their whole soul and being—a mood of ecstatic joy reinforced by the sense of excitement and dynamism that was conjured up by the host of new modern jargon words that Nazi-Deutsch invented. When it came to words connected to Jews, of course, the emotional resonance was of very long standing in German language and culture, and even today German words such as "Jude" (Jew) bear an intensity of revulsion and reaction that is present in no other European language" (2002, vii). And Nazi German could easily draw on the "long tradition of antisemitic vocabulary and phrases" (Mieder 2002, xv) in the German language.

One of the main and most enduring emotions conjured incessantly by Nazi German was the feeling of "us versus them." Nazi ideology was based on the notion that human beings were *not* all created equal. On the contrary, some groups of people were considered superior to others; some were therefore to rule over others, who were to be subjugated or even exterminated. Every person was born as a member of a particular group, and this determined how a person was seen, what opportunities they should or should not have, and what their "rightful" place in society was to be. All this was immutable: once a member of a group, always a member of that group. (The impacts of this notion extended to criminal law; see Chapter 8, p. 205, regarding *Willensstrafrecht* and *Täterstrafrecht*.)

Such groups were defined according to pseudo-biological notions of *Art, Blut, Gemeinschaft, Rasse,* and *Volk*. Translated as biological terms, their meanings are as follows: *Art* = species; *Blut* = blood; *Gemeinschaft* = biocenosis or community; *Rasse* = race;

Volk = a group of animals of the same species (e.g., *Bienenvolk* = bee colony). Nazi German did not use these terms in a strictly biological sense but shifted them to the sociopolitical sphere. For instance, *Volk* referred to a biologistically defined people: a *Volk* was defined not by citizenship or culture or language, but by an allegedly shared biology. I encourage readers to peruse the Glossary at this point.

As explained in the Glossary, *Art*, *Blut*, *Gemeinschaft*, *Rasse*, and *Volk* were defined in reference to each other: *Art*, *Gemeinschaft*, and *Volk* were defined by *Blut* and *Rasse*; *Rasse* and *Volk* were based on the notion of *Gemeinschaft*; *Blut* was defined by *Rasse*, or the two terms were considered synonymous; and each term had its own nuances. Since their definitions overlap, it is often difficult, if not impossible, to discern precise differences in meaning.

Nazi German used *Art*, *Blut*, *Rasse*, or *Volk* to form more than one hundred compound words, many of which also included terms such as *fremd* (foreign, alien), *gleich* (same, similar), or *eigen* (own, one's own), which themselves indicated belonging or non-belonging. Dozens more words were formed by adding a prefix or suffix. Some examples: "*Aufartung*: enhancement of the *Art* by means of measures of *Rassehygiene*" (Schmitz-Berning 2007, 73); "*artfremd*: … alien, foreign (to the essential nature of German culture)" (Neuburger 1944, 8); "*volksbewusst* … conscious of national entity [used by the Nazis as synonym of rassenbewusst (race-conscious)]" (ibid., 117); "*volksfremd* … alien to the spirit of the German people" (ibid., 117); "*Volksschädling*, … people's parasite [used in connection with persons violating measures of national urgency]" (ibid., 118). Referring to the use of *Volk* as a prefix to countless words, which intensified their emotional charge by connecting them to the people included in the *Volk* while implicitly excluding the others—think *Volkswagen*—Victor Klemperer commented: "The term '*Volk*' … is now as customary in spoken and written language as salt at table, everything is spiced with a soupçon of *Volk*" (1947, 30).

Alone or in compounds, these words were laden with emotion; those referring to the groups favored by the Nazis were intended to convey positive feelings (belonging, superiority, pride, beauty, flawlessness, etc.); those referring to the others negative ones (othering, exclusion, inferiority, disgust, ugliness, etc.). These terms express belonging versus non-belonging, inclusion versus exclusion, us versus them, and they imply domination versus subjugation, persecution, and extermination—even without explicitly stating as much. They are inherently antisemitic code words as they exclude Jews, most without mentioning "Jew" or "Jewish."

Translating Nazi German

Translation is a process of decoding and encoding. Translators must begin with the words on the page, but translation is not simply a mechanical process of transposing sequences of words from one language to another. The translator has to understand (decode) the meaning intended by the author in detail within the societal context of the text in general, reading between the lines where necessary, and then express (encode) that meaning in a way that is both accurate and comprehensible to the target audience.

Throughout the process, translators must be aware of their own biases to avoid distorting the meaning of the text.

Translating works written during the Nazi period for a twenty-first-century audience means transferring a text across time, space, and sociocultural contexts. It entails bridging the linguistic gap between Nazi German and English; the political, legal, and cultural gaps between the dictatorship of Nazi Germany and present-day democratic societies; and a leap in time of more than three-quarters of a century.

Much of Schmitt's writing is on legal and political theory, which by definition both rely heavily on specific, specialized language. Any legal translation has to deal with the challenge of transposing terms, concepts, institutions, laws, procedures, etc. in all their intricacies from the language of one legal system to the language of another; in the case of German to English, this also involves taking the step from a system based on Roman law to one based on common law. In many cases, there is no legal concept in the target language that corresponds exactly to that in the source language, so no corresponding term exists, either. Similar problems arise in political theory translations. A few examples: the functions of political parties and the role of the president during the Weimar Republic do not correspond precisely to those of the US, and the meaning of terms such as state, parliamentarism, political representation, and political leadership differs to a certain degree in the traditions of German and American political language. Kirchheimer's difficulties in finding proper translations for *Rechtsstaat* and *politische Justiz* exemplify this general problem. Concerning texts on legal and political theory of the Nazi period in particular, it should be borne in mind that the system of Nazi rule employed the Nazi linguistic project—which was designed to obfuscate its violence and brutality, as explained above—to express that ideology in laws and other legal measures. Understanding such texts requires readers to be familiar with Nazi German.

Ambiguous language is notoriously difficult to translate. Translators may attempt to use equally ambiguous words and figures of speech in the target language, but they often do not exist. Another option is for the translator to interpret the ambiguous language, decide which meaning was intended, and then render that meaning in the target language. Since multiple interpretations are possible, however, translators may legitimately be accused of incorrectly translating the passage in question, or of bias, irrespective of which interpretation they decide on. This is all the more true when it comes to translating language that relies on readers grasping the author's intended connotations.

When translating antisemitic language in particular, one is faced with the challenge that several features of Nazi German come together: code words, emotional charge, exclusion of Jews, "the constant fluctuation in German antisemitic language between the metaphorical and the physical meaning of words that in English are concrete" (Rose 2002, viii), and language that is ambiguous at times, enabling plausible deniability and perhaps chosen for precisely that reason. "If anti-Semitic sentiment is uttered overtly, one can discuss it by simply pointing to explicitly anti-Semitic passages. But if it is uttered in code and by allusion [...], one must first interpret the statements to uncover their meaning before arguing against their anti-Semitic content, and whoever made such statements can deny that that interpretation is correct" (Lustig 2006, 208).

Translating Nazi German makes both parts of the process of decoding and encoding more difficult than usual. The first step is to identify the many code words and terms whose meanings had been changed. If the translator fails to do so and instead translates the meaning usually given to a particular word today, then the translation will be inaccurate and the specific meaning intended by the Nazis will be lost. After identifying a word as code, the translator must crack the code to understand the precise meaning of the term, both its denotation and its connotations, at its particular place in the text. Simply referring to standard German dictionaries will not help here as they do not include Nazi German code, for example for *Art* and *Gemeinschaft* (see Glossary). Once the translator has ascertained the meaning in Nazi German, the final step is to express that meaning in a way that is accurate, comprehensible to an English-speaking readership, and if possible, readable. The main challenge here is that, in many cases, no words exist in English that correspond to the Nazi German terms, as explained above. Many words that might appear to be translations are in fact not quite accurate or even plain wrong. For instance, the Nazi concept of *Volk* encompassed emotionally charged and exclusionary notions of nation, *Rasse*, purity, bonds to the soil, not being cosmopolitan (itself a code word), and so on (see Glossary). Yet a US reader encountering its common translation "people" today may well associate concepts such as "people's republic"—which are entirely different. As Rose has argued, direct translations of terms used in Nazi propaganda, such as "'patriotism,' 'rebirth,' 'mobilization,' 'order,' 'dignity,' 'national community,' [fail to] convey [their] powerful emotional resonance—almost religious in its intensity" (2002, vii) and leave English-speaking audiences to wonder why Nazi propaganda films such as Leni Riefenstahl's *Triumph of the Will* had such an overpowering impact on the German audience of the day.

Now, Schmitt not only wrote in Nazi German at times, he also had his own way of using words, as discussed in Chapter 6: He considered them devices to be employed when struggling for the authority to interpret terms and concepts. The body of terms he defined in his own way became a Schmittian lexicon, as it were, for instance, his concepts of dictatorship and democracy in his *Constitutional Theory*. Schmitt scholar Reinhard Mehring even chose *Kriegstechniker des Begriffs* [Military technologist of terms and concepts] (2014b) as the title of his biographical studies on Schmitt. This practice of Schmitt's in fact predated the Nazi period: in an analysis published in 1931 of Schmitt's *Constitutional Theory*, political philosopher Eric Voegelin correctly pointed out that Schmitt's political intentions were behind his unconventional definitions and that he did not express these explicitly (see Chapter 6, p. 149). Translators are therefore confronted with the additional task of teasing out exactly which meaning Schmitt intended with each of his terms within the context of his argument in a particular text written at a particular point in time and in a particular political situation—and then rendering that meaning in the translation. It would be incorrect to simply translate such terms as they are usually used today; the reader must be able to discern that the term as used by Schmitt has a different meaning than usual, and that Schmitt changed its meaning with a specific purpose in mind.

There are various ways to deal with these translation issues. Some publications simply use the Nazi German words, often in italics, in the English text without further explanation, for example, *völkisch*. In my view, this does not do justice to the needs of English-speaking readers who likely do not understand the Nazi German terms, let alone the intricacies of meaning. Another option is to use direct English translations in quotation marks, for example, "Aryan," again without explanation. English-speaking readers unfamiliar with Nazi German terminology and ideology may wonder what the quotation marks are supposed to signify and what the words actually mean.

Another alternative is to use common English terms as supposed equivalents of Nazi German ones. This approach often does not put across the proper meaning of the terms and instead yields an incorrect translation that may be skewed, sanitized, and/or trivialized. For instance, translating *Gleichschaltung* as "coordination" or "synchronization" sanitizes the term by eliminating the brutality and the severe consequences of the process of forcibly aligning organizations with the Nazi system and dismissing Jews, leftists, and others from their jobs (see Glossary). Some authors forge English terms as translations of neologisms which the Nazis coined to denote things and concepts that were particular to the Nazi system. For example, they use "people's unit, national storm,"[3] or "People's Assault"[4] for *Volkssturm*—a militia of poorly equipped civilian boys and men drafted by the Nazi regime in a last-ditch effort to defend the fatherland—but such coinages mean nothing in themselves without further explanation and certainly do not convey the proper meaning.

As a matter of principle, I hesitate to critique translations without knowing the conditions under which the translator worked. Translation is often piecework, making translators work against the clock, which may result in inadvertent errors. In addition, every translation involves an interpretation of the source text, even if it is not particularly ambiguous, and various translators may legitimately interpret the source text differently and select different words to express different meanings in the target language.

Translators of Schmitt in particular who lack knowledge of Nazi German and the Schmittian lexicon may tone down these features inherent to his thinking and writing. As a result, scholars working with such erroneous—and possibly biased and sanitized—translations of Schmitt's work may well misunderstand or fail to understand exactly what he wrote, construct flawed or misguided theories about Schmitt and his oeuvre, or even put words in his mouth. Reviewing published translations of Schmitt's writings for accuracy and biases would be an interesting task, albeit one far beyond the scope of this book.

Another issue relating to potentially biased and sanitized receptions of Schmitt is the fact that many of his works, scholarly and otherwise, from the Nazi period have not been translated into English. Since scholars unable to read German and Nazi German texts cannot access this part of Schmitt's oeuvre, they may not be aware of the depth of Schmitt's own antisemitism and of his support of the Nazi regime and its ideology.

Some mistranslations of terms coined by Schmitt seem to have wended their way into usage in the literature. Jeffrey Seitzer translated *dilatorischer Formelkompromiß* as "dilatory formal compromise" in *Constitutional Theory* (Schmitt 1928b, 85). However, the *Formel* in *Formelkompromiss* has nothing to do with whether the compromise is formal or informal. (That would be *formaler Kompromiss* or *informeller Kompromiss*.) A *Formelkompromiss* is a formulaic compromise, *Formel* being the formula used in this type of compromise: the wording that the parties can agree on, even if the policy problem remains unresolved.

3 Robert Michael and Karin Doerr (2002, 426).

4 United States Holocaust Memorial Museum, Holocaust Encyclopedia, accessed 5 March 2024, https://encyclopedia.ushmm.org/content/en/article/indoctrinating-youth.

German grammar permits two interpretations of the term: the adjective *dilatorisch* may refer either to *Formel* or to *Kompromiss*, so it could be correctly translated either as "compromise on a dilatory formula" or as "dilatory formulaic compromise." Kirchheimer in "Weimar—and What Then? An Analysis of a Constitution" (1930e) refers to Schmitt's *dilatorischer Formelkompromiss*, and the translation by Arnold H. Heidenheimer, reviewed by John H. Herz, uses the first option; following discussions with Hubertus Buchstein, I decided on the second.

In his translator's note to *Land and Sea* (2015), Samuel Garrett Zeitlin explains that Schmitt "explicitly contrasted his notion of *Völkerrecht* with prevailing notions of 'international law,' offering the latter term in quotation marks in English in his German-language texts" (ix) from 1937 to 1943, and that he used "law of peoples" for Schmitt's writing from that period "to allow readers to interpret Schmitt's notion of this term as something potentially quite different from the notions of international law prevalent in the 1940s or those notions of international law prevalent today" (ibid., ix–x). Yet such a clear demarcation may be debatable because there were also continuities in Schmitt's thinking on international law over time; in any case, this differentiation is not relevant in the context of this book, so Hubertus Buchstein and I opted for the common translation "international law."

In *Behemoth* (1944), Franz L. Neumann used translations of Nazi terms that I would avoid: blood for *Blut*, community for *Gemeinschaft*, folk for *Volk*, Leader for *Führer*, large space for *Großraum*, and leadership principle for *Führerprinzip*. Leadership principle (which has no particular meaning in English; after all, some form of leadership is practiced in virtually every organization and institution) would be a translation of *Führungsprinzip*, not of *Führerprinzip*. I am reluctant to criticize Neumann for this. After all, he wrote this massive and seminal book in a very short time to analyze and explain the Nazi system while it was in power and World War II was raging; the first edition was published in 1942, predating what were probably the first works explaining Nazi terms in English (Paechter et al. 1944 and Neuburger 1944) by two years; he wrote in a language that was not his native tongue; it is unclear how familiar the book's copy editors were with Nazi German terminology; and expecting Neumann and his colleagues to solve these complex translation problems under the given circumstances would be unrealistic. I would apply the same considerations to Kirchheimer's translation of *gesundes Volksempfinden* as "sound feelings of the people" (1940b, 448) and his use of "leadership principle" as the translation of *Führerprinzip* in the title of "Leadership Principle and Criminal Responsibility" (Kirchheimer/Herz 1945).

* * *

Aiming for this book translation to be accurate, comprehensible, and readable, I have taken the following approach. Some Nazi German terms simply require a brief explanation, so I retain them in the text and provide a gloss, sometimes in a footnote, for readability. I have also compiled a Glossary with definitions and explanations of key Nazi German terms that occur in this book. Particularly enlightening sources either present the Nazis' own definitions (Schmitz-Berning 2007) or were prepared at the time to explain Nazi German terms in English (Paechter et al. 1944 and Neuburger 1944); in some cases, I added a few words of my own. The purpose of the Glossary is to explain the meaning or meanings of these terms, including their connotations and emotional charge. The Glossary shows that these terms cannot simply be translated into individual English words, but rather require considerable explanation.[5] Incorporating such explanations in the body of the text would make it unreadable and would distract from the argument being made, so I retain the Nazi German terms there and provide a brief gloss as well as a reference to the Glossary where appropriate. I encourage readers to study the Glossary with a view to fathoming the Nazi mindset and ideology as a backdrop for understanding Schmitt's writing from that time.

Some terms are explained, a few specific translation problems are discussed, and critiques of published translations are offered in various places in the book, for example *Lage* in Chapter 1; *Artgleichheit, Gleichartigkeit,* and *Homogenität* in Chapters 4 and 10; *Mißgeschick* and *das Elementare* in Chapter 9; mistranslations and omissions in Schwab's translations in Chapter 10; and *Raum, Großraum,* and other compounds of *Raum* in Chapter 12.

* * *

Readers and translators of Schmitt should be aware of the intricacies of the German language, the purposes and linguistic techniques of Nazi German, and Schmitt's practice of creating his own specific definitions of terms, in short: they should understand that one cannot take his writing at face value. The fact that terms in his writing may be untranslatable because no corresponding terms exist in English does not mean that their meanings are opaque or even incomprehensible; rather, it means that the process of translation—decoding the precise meaning in (Nazi) German and then rendering that meaning in English—is more complex than usual. Much work remains to be done on reexamining Schmitt's works at least from the Nazi period for his use of Nazi German, translating more of them into English, reviewing existing translations for accuracy in this respect, and potentially reassessing the meaning and implications of his oeuvre. To conclude with just a single example: what does it mean for receptions of Schmitt if we know that he wrote not only of *konkretes Ordnungsdenken*, but sometimes of *gesundes, konkretes Ordnungsdenken*, that is, not only of concrete-order thinking, but also of concrete-order thinking corresponding to the norm of the NSDAP? I leave this for scholars to ponder.

5 Indeed, there is an entire monograph devoted to a single term: *Sonderbehandlung*, see Wulf (1963).

Chapter 1:
Introduction: Refuting the Legends

"Are you coming as a friend or as an enemy?," Carl Schmitt reportedly asked Kirchheimer when he unexpectedly arrived on his doorstep in Plettenberg in November 1949 (see Söllner 1996, 114). The question is an apt allusion to Schmitt's *Concept of the Political* in which he had famously defined the political by making the distinction between friend and enemy and had singled out the deadly conflict between enemies as the basic category of the political process. In other words, the emphasis in his formulation was exclusively on the concept of the enemy. However, there is more to this anecdote than a witty allusion to Schmitt's famous essay. Alfons Söllner told me in a recent conversation that he does not recall his source for the anecdote, that he probably heard it in the 1980s, and that the most likely sources were either his conversations with George Schwab, a great admirer of Schmitt's, or with John H. Herz, a close friend of Kirchheimer's. Söllner raised doubts about its accuracy and also called it a cleverly contrived allegory.[1]

The speculative status of this anecdote points to the uncertainty of what we know about the relationship between Kirchheimer and Schmitt. Both Carl Schmitt (1888–1985) and Otto Kirchheimer (1905–1965) are paradigmatic figures in the history of political thought of the twentieth century: Schmitt as the most original, dazzling, and controversial German constitutional law teacher and Kirchheimer, who received his doctorate under Schmitt, as a member of the Frankfurt School in exile. Their journeys through life intertwined repeatedly between 1926 and 1965, whereby their roles evolved and they were in frequent contact, both directly and indirectly. And, tellingly, for Schmitt's part, their relationship even extended beyond Kirchheimer's death. This book explores the personal, political, and theoretical dimensions of the relationship between these two thinkers from opposite political camps in times of tremendous political upheaval. I will describe the cross-fertilization of their thinking as well as Kirchheimer's learning process that led him far away from Schmitt's concepts and theories. This book also challenges the feel-good interpretation in the secondary literature of their alleged friendship. Conversely, I will shed light on the different phases and various constellations of an enduring enmity.

1 Alfons Söllner in conversations with the author on 21 April 2021 and 10 May 2023.

Carl Schmitt is the more prominent of the two authors. International interest in his work has become practically a "tsunami" (Bernstein 2011, 403) and has culminated in the *Oxford Handbook of Carl Schmitt* (see Meierhenrich and Simons 2016a).[2] Schmitt was a masterful wordsmith who wrote in different styles, from dry legal opinion to forceful polemical essays to literary works. Many of his books and articles contain veiled allusions and messages for the initiated. He was aware of the changing styles of political thought and liked to play with them. Schmitt chose his words carefully and his writings are filled with subtle linguistic twists. His scholarly apparatus reveals some, but not all, of his sources. The thrill of deciphering the arcana in his writing has undoubtedly contributed to the ongoing reception of his work.

Otto Kirchheimer is well known among legal scholars, political scientists, and historians for his work in multiple fields: as a critical analyst of the Weimar Constitution and the Nazi legal system, as a member of the Frankfurt School in exile, as an inspirer of Foucault's critical criminology, as a member of the legal team that prepared the Nuremberg Trials, as a European politics researcher in the US State Department, as the inventor of the concept of the "catch-all party" in comparative political science, and as the author of the seminal book *Political Justice*.[3] His rich oeuvre reflects both the range of his political experiences and his evolution as a legal and political theorist in the Weimar Republic and the early Nazi era, in his exile in France and the US, and during the founding and establishment of the two new German states after 1945. In the vast secondary literature, however, his connection to Schmitt overshadows all these facets of his oeuvre.

1. Repeated visits and friendship after World War II?

In the literature about both Schmitt and critical theory, Kirchheimer has been assigned the role of the first "left-Schmittian," someone who started to borrow intensively from Schmitt earlier and with considerably stronger effects than other authors of the Frankfurt School such as Walter Benjamin and Franz L. Neumann. Consequently, the name Otto Kirchheimer has become associated with the beginning of a genealogy of authors from the left who rely on concepts and theories adopted from Schmitt. Seyla Benhabib, for instance, names Kirchheimer along with contemporary theorists such as Ernesto Laclau and Chantal Mouffe as the first author for whom "Schmitt is the *éminence grise* to whom one turns when the liberal-democratic project is in deep crisis" (Benhabib 2012, 689). Kirchheimer seems to fit this exceptional role perfectly: in this view, not only was he an outstanding left-wing doctoral student of Schmitt's in Bonn in 1926–28 who made extensive use of Schmitt's thinking in his own writing during the Weimar Republic but

---

2 The best and (regularly updated) bibliography on the literature about Schmitt can be found on the website of the Carl-Schmitt-Gesellschaft e.V., accessed 2 January 2024, https://www.carl-schmitt.de/forschung/literatur-zu-carl-schmitt/#tab1.

3 See Jay (1973), Söllner (1982), Tribe (1987), Luthardt and Söllner (1989), Scheuerman (1994), Wiggershaus (1995), Schale (2006), Ooyen and Schale (2011), Schale, Klingsporn, and Buchstein (2018), Simard (2020), Klingsporn (2023), and Simard (2023).

he was also in personal contact with him again after 1945, regardless of Schmitt's participation in the establishment of the Nazi regime and his support for this regime until the end of the war.

A number of authors from the extensive secondary literature on both Schmitt and Kirchheimer have constructed a positive description of their "friendship" (Neumann 1981, 239) on this basis. The editors of Schmitt's diaries call Kirchheimer Schmitt's "favorite student" and use this characterization as proof "that Schmitt got along very well with many Jews" (Tielke and Giesler 2020, 51). When reading authors of various political convictions, we get the impression that Kirchheimer and Schmitt shared an almost lifelong understanding of legal and political theory that bridged their well-known political differences. In his foreword to the American reissue of *The Concept of the Political*, Tracy B. Strong refers to the late Kirchheimer as Schmitt's "colleague and friend" (Strong 1996, ix). The late Reinhart Koselleck mentioned the "good friendship"[4] between the two of them in an interview. Martin Tielke referred to the relationship between Kirchheimer and Schmitt up until the late 1950s as a "friendship." [5] And the editor of the journal *Telos*, Gary S. Ulmen, who became a proponent of Schmitt's in the 1980s, was referring to Kirchheimer when he stated: "There has been an important Jewish reception of Schmitt [...] let's face it: Critical Theory makes strange bedfellows."[6] Continuing in this vein, John McCormick states that Kirchheimer was among the German émigrés whose work was still influenced by Schmitt, but "chose to acknowledge him as little as possible" (McCormick 1998, 849) because of his political affiliation with the Nazi regime.

Various authors have claimed that Kirchheimer visited Schmitt at his home in Plettenberg several times after World War II and tried to stay in close intellectual exchange with him. This view can be found in most scholarly contributions that mention the relationship between Schmitt and Kirchheimer. The editor of Schmitt's diaries, Martin Tielke, states that it was Kirchheimer who initiated contact with Schmitt after 1945 (see Tielke 2018, xxvii). Schmitt's best biographer, Reinhard Mehring, writes about Kirchheimer visiting Schmitt "repeatedly" after the war. In his view, Kirchheimer even played the active role in the relationship as he writes that, ultimately, the "efforts of Otto Kirchheimer to have a renewed relationship to his old doctoral advisor from the period in Bonn failed" (Mehring 2014a, 432).[7] Helmut Quaritsch and George Schwab go one step further and sardonically use Kirchheimer's allegedly multiple visits to Schmitt's home against a "mentally unbalanced" (see Quaritsch 1995, 72) Kirchheimer as evidence of his purported opportunism (see Schwab 1988a, 80–82). Joseph W. Bendersky speaks of Kirchheimer's "return" (Bendersky 2016, 137) to Schmitt, which allegedly began with visits in 1947. Rolf Wiggershaus writes in his seminal history of the Frankfurt School that "Kirchheimer visited Schmitt on several occasions" (Wiggershaus 1995, 470). In his brilliant book about Schmitt's personal networks after 1945, Dirk van Laak mentions "several visits" (van Laak 1993, 135) of Kirchheimer's to Plettenberg after 1945. To the

4 Koselleck in an interview with Claus Peppel in 1994 quoted in Schmitt and Koselleck (2019, 377).

5 Tielke in his editorial notes in Schmitt and Schnur (2023, 211).

6 Cited in Zwarg (2017, 368).

7 This version can be found even in the new revised edition of his excellent biography of Schmitt, see Mehring (2022a, 473).

leftist legal scholar Jürgen Seifert, the numerous contacts between the two after 1945 were an impressive testament to the fact that "rejection of positions did not necessarily have to mean personal enmity" (Seifert 1985, 199).

According to the existing literature, the relationship between the two of them seems to have been a kind of personal friendship despite their political differences. The—as I will prove in this book, erroneous—claim that Kirchheimer visited Schmitt repeatedly after 1945 at his home in Plettenberg seems to fuel this kitschy legend. As a matter of fact, Kirchheimer only visited Schmitt once, in November 1949, and they only met in person one more time after that, in Cologne in June 1953.

Jürgen Seifert's retrospective statement fits perfectly with Schmitt's vocabulary in *The Concept of the Political*. The enemy is "solely the public enemy" (Schmitt 1932a, 28); they are part of a collectivity of people fighting against another collectivity. Schmitt's definition leaves room for positive private relationships between two individuals from different fighting collectivities. As a matter of fact, Schmitt emphasized that "the political enemy does not need to be morally evil or aesthetically ugly" (Schmitt 1932a, 27). This is Schmitt's understanding of the civilizing aspect of enmity: enemies may treat each other with respect on the personal level; indeed, they may even like each other. Kirchheimer—at least in his Weimar writing—had a similar idea about the separation of personal and political enmity, albeit arguing from his left-wing political position. He used the German term *Feind* in several publications between 1928 and 1932. The enemy was the class enemy whose position was defined by class membership.[8] In the Marxist tradition, the class enemy is a character mask behind which there might be a person one may get along well with on a personal level. In both Schmitt's and Kirchheimer's understanding, enemies can become close friends—and then can tragically, but necessarily, turn away from each other after the outbreak of an existential fight between the two collectivities they belong to.

I have listed the sources that claim the contacts between Kirchheimer and Schmitt were friendly for a long time in order to illustrate the importance of the biographical dimension in assessing their relationship. Kirchheimer became a candidate for filling the role of a political enemy as well as a personal friend of Schmitt's not only because of his writing, but also because of the various authors' assumptions about the personal relationship between the two. As a matter of fact, as I will demonstrate in Chapter 15, when Kirchheimer stopped by Schmitt's house in Plettenberg in 1949, his intention was to show Schmitt that he, who in 1933 had been one of those whom Schmitt had wanted to see driven out of Germany as an enemy once and for all, as evidenced in his Nazi writing, had survived—in a dual sense, both as a Jew and as a leftist.

8 See Kirchheimer (1928a, 13—translated as "arch foe"), (1929b, 183), (1930a, 327), (1930e, 39), and (1932a, 62). Between 1933 and 1945, Kirchheimer used the term *Feind* when he paraphrased Nazi documents.

2. Grasping the *Lage*: Two theorists of concrete situations

It is difficult to receive Schmitt's work impartially. Nor is it easy to insert his work into current international debates about legal and political theory. This results in much cherry-picking from Schmitt's oeuvre as is currently the case particularly in China, with the Nouvelle Droite in France, and in Russia with its aggressive geopolitical agenda leading to the military attack on Ukraine.[9] There are various reasons to raise doubts about the dominant interpretative lines in the literature on Schmitt in the English-speaking academic world. Some interpretations of his works are so vague and peppered so strongly with literary associations that it is difficult to recognize Schmitt, the eminently political writer, in them. Others are so far detached from Schmitt's theoretical and political impulses that using him as a reference point becomes almost superfluous. The fixation in this literature on some of Schmitt's obscure ideas, for example, his eschatology, needs to be overcome by recontextualizing his work in the political constellations of its time (see Finchelstein 2022, 96–100). These shortcomings certainly have something to do with the fact that not all relevant sources are available in English. Important works by Schmitt, particularly from the Nazi era, have not yet been translated. In addition, some translations into English blur important linguistic nuances, Schmitt's choice of words, and the shifting meaning of some of his terms. These factors make Schmitt's writing in German accessible only to scholars with an extraordinary command of the language, including the specific use of language of the Nazi period. I would like to employ the concrete understanding of legal and political theory that is fortunately shared by both Schmitt and Kirchheimer to counter the prevailing cherry-picking and abstract readings.

Kirchheimer broadly characterized his interpretation of the debates on Weimar constitutional law as "sociological" (Kirchheimer 1933e, 500) analysis. He claimed to view existent legal and political institutions in their particular "social function" (Kirchheimer 1928b, 162) and postulated certain changes in the social function of institutions (such as parliaments, interest groups, the judicial institutions, or property rights) as the starting point for his political critique. In a number of his publications during the Weimar period, Kirchheimer referred to the German metaphor *Lage*. Within the theoretical framework of Marxism, *Lage* had a military and territorial meaning. The Prussian military expert Carl von Clausewitz was the tradition-forming author for using *Lage* as a synonym for the antagonistic positions in a battle. The line of left-wing uses of *Lage* started most prominently with Friedrich Engels's book *Die Lage der arbeitenden Klasse in England* (1844).[10] In the early 1920s, *Lage* had become a crucial term used frequently by Lenin, Georg Lukács, Austro-Marxists, and reformist social democrats alike. *Lage* was the overarching descriptive term for the analysis of the concrete positions in the struggle between the social classes (*Klassenlage*). Following the Marxist tradition, Kirchheimer made extensive use

9 See de Benoist (2007), Auer (2015), Lilla (2016), Marchal and Shaw (2017), and Lewis (2020).

10 The authorized translation by Florence Kelley appeared in 1885 titled *The Condition of the Working Class in England*. Engels is well known for his admiration of the Prussian military and Clausewitz in particular. In the translation, he nevertheless renounced the militant semantic content that he chose to use for his German readers.

of this term in his Weimar writing.[11] Whereas the young Kirchheimer described his approach using this metaphor stemming from the world of the military, he later chose a different metaphor instead: it was a pacified metaphor from the social practice of craftspeople. In 1964, in one of the few statements in which he explicitly addressed his own methodological approach, he called it a "handicraft to decipher government systems in full activity, to diagnose them or to substitute better ones in his mind."[12] Such a craft, however, "increases the operational risk" of errors due to its specific reference to concrete political situations. Nevertheless, he claimed he didn't "want to wait until the obituary of a political system [was] due" but instead sought to intervene in current political constellations with his studies. Even though Kirchheimer switched to a different metaphor, he still kept the original idea of *Lage*.

Schmitt also repeatedly emphasized the close connection between his scholarly work and concrete political constellations, albeit in a different way. Clausewitz's term *Lage* had become popular in the vocabulary of the Prussian military and also inspired a number of authors on the extreme right such as Ernst Jünger and Gottfried Benn during the Weimar Republic. Schmitt also frequently used the phrase "*Analyse der Lage*" (analysis of the *Lage*) or simply stated "*Das ist die Lage!*" (That is the *Lage!*) throughout his long career.[13] For Schmitt, the metaphor *Lage* always had a double meaning. It simultaneously designates the aspect of a situation of being bound and its potential for change. It is therefore not to be confused with the arbitrary, the freely available, the merely opportune. Every *Lage* requires a decision. The standard English translation "concrete situation analysis" disregards the military and territorial semantic component of the term, and Schmitt's polemical meaning gets lost at this point.[14] Schmitt thought that legal and political theory was a polemical practice, which gave every academic debate the character of a political struggle. In his view, all political and legal theories emerge out of concrete political battles and disputes. Anyone who denies such a close connection to concrete constellations in political battles, he wrote in 1930, is simply using the "specifically political

11 See Kirchheimer (1929b, 182, 185), (1929c, 193, 195), (1930i, 220, 237), and (1932e, 370). The posthumous translation of *Weimar—and What Then?* by John H. Paasche uses either "position," (Kirchheimer 1930e, 44) or "political situation" (62). In his late German work, the term *Lage* can only rarely be found (see Kirchheimer 1957c, 380).

12 This and the following two quotes are in Kirchheimer (1964, 501).

13 The term is part of the title of his critique of parliamentarism in *Die geistesgeschichtliche Lage des heutigen Parlamentarismus* (Schmitt 1926e). It can also be found in his *Verfassungslehre* (Schmitt 1928c, 69) and in the title of a programmatic lecture in the final phase of the Nazi era, "Die Lage der europäischen Rechtswissenschaft (1943/44)" (see Schmitt 1950b). After the war, he used it in *Der Nomos der Erde* (see Schmitt 1950e, 54) as well as in his foreword to the 1963 German edition of *Der Begriff des Politischen* (see Schmitt 1963b, 12).

14 Ellen Kennedy chose the title *The Crisis of Parliamentary Democracy* (see Schmitt 1923a) for her English translation of Schmitt's *Die geistesgeschichtliche Lage des heutigen Parlamentarismus*. Her translation of the term *Lage* in this book is "intellectual circumstances" (Schmitt 1926a, 1); she claims that her departure from the literal translation of the book does "capture the spirit" (Kennedy 2000, x) of Schmitt. Gary S. Ulmen in his translation of *The Nomos of the Earth* gets rid of the term altogether (Schmitt 1950d, 86). Jeffrey Seitzer comes closer in the articulation of the territorial and military components of the term by using "position" in his translation of *Constitutional Theory* (Schmitt 1928b, 119).

trick of presenting oneself as apolitical and the opponent as political." He complained about "too much methodology and not enough method" in German legal and political thought (Schmitt 1930c, 165 and 175). Four years later, he coined the term "concrete-order thinking" (Schmitt 1934h, 225) to describe his own methodological approach in opposition to legal positivism. In an interview five years after Kirchheimer's death, Schmitt explained his scientific practice as follows: "I have a method that is peculiar to me: to let the phenomena approach me, to wait and to think from the concrete material, so to speak." (Schmitt 1970, 11) As discussed in more detail in Chapter 6, Schmitt's *konkrete Lageanalyse* (concrete analysis of the *Lage*) is a method that claims to situate every theoretical debate within the context of a political battle. With reference to Schmitt, *Lageanalyse* has become a key concept in particular among right-wing authors and Schmitt enthusiasts in Germany and it remains so to this day.[15]

The methodological commitments of Kirchheimer and Schmitt, expressed in metaphors from different social practices, have two epistemological axioms in common. First, the axiom of inevitable situatedness: no legal or political theories will be able to evade their dependence on the specifics of particular sociohistorical, cultural, and geographical contexts and social power relations; and neither should they. Second, the axiom of interventionism: the inevitable relatedness of theory and praxis becomes a productive virtue by making active use of legal and political theorizing as an instrument of political intervention. Despite many other fundamental differences between Schmitt's approach and Kirchheimer's neo-Marxist critical theory, they agree on these two—however vague—epistemological axioms.

3. Through the lens of the other

The methodological approach of *Lagenalyse* shared by the two authors is also my methodological key for reconstructing their story in this book. Their numerous dialogues, disagreements, and repeated confrontations can only be understood in the context of the changing political situation. Thus, the overarching questions addressed in this book are as follows: What was the relationship between these two political thinkers from opposite political camps and how did it change in its personal, political, and theoretical dimensions over the course of time? Our understanding of Schmitt shifts if viewed through the lens of Kirchheimer's analyses and commentaries—and vice versa. This line of inquiry results in three sets of more specific questions.

First, on the level of their legal and political theories, how significant and inspiring were Schmitt's theories, categories, and concepts for Kirchheimer's work—and vice versa? What was the explicit dialogue between the two like? Can traces of implicit dialogues be identified? What subject areas did their receptions of each other cover and what, if any, is the meaning of what they overlooked or chose to disregard? In what kind of modalities did these receptions take place—from direct adoption and integration into their own theoretical framework to suggestions, critiques, or even instrumentalization?

15 See Willms (1982), Arndt (1985), Oberlercher (1993), Sander (1993), Schneider (1993), and Maschke (2011, 22). For a critical review, see Priester (2015).

Are there any surprising thematic overlaps even in the absence of a direct reception of the other? Does Kirchheimer's reading of Schmitt contribute to a better understanding of his work—and vice versa? Or are the lenses—one or both—distortive in a way that they serve the interests of the author peering through them?

Second, on the personal level, how did their personal relationship develop and change—on the one hand, the Jewish and socialist student and, on the other, the Catholic and right-wing extremist professor seventeen years his senior? Were there any role changes and shifts in the balance of power between them during the four politically turbulent decades in multiple political systems? Are there any indications that their personal relationship influenced the theoretical substance of their work? Or, conversely, are there any indications that changes in the theoretical substance of their work influenced their personal relationship?

And third, to a minor degree, on the level of contemporary debates in political theory, does the controversy between the two point to any hidden treasures in legal and political theory that are worth being unearthed? Is there anything to be learned from Kirchheimer's grappling with Schmitt's work for a new perspective in the debate with the protagonists of today's left-Schmittianism?

These three levels of exploration cannot be treated separately. The polemical practices in the writing of both authors can only be adequately understood within their constantly and dramatically changing political contexts. I therefore reconstruct the multifaceted relationship between Kirchheimer and Schmitt in chronological order: the years of the Weimar Republic between 1926 and January 1933 (Chapters 2 to 6); the Nazi period in Germany and Kirchheimer's exile in France and the US (Chapters 7 to 13); and the postwar years until Kirchheimer's untimely death in 1965 (Chapters 14 to 17). Thus, the book combines a double political biography with the discussion of systematic questions of legal and political theory. The theoretical, political, and personal links between Schmitt and Kirchheimer illuminate crucial points in the recent history of political ideas, in German and European contemporary history, and in transatlantic intellectual history as well as the role of German exiles in the American academic system, the subject of antisemitism, and German-Jewish relations in the twentieth century.

The personal, political, and theoretical dimensions of the relationship between Kirchheimer and Schmitt are inextricably linked. Part of the story told in this book is the self-serving manner in which Schmitt himself contributed to the legend that the relationship between the two men after 1945—with the exception of a conflict in 1961/62—had remained essentially friendly. The book is not only about the cross-fertilization of their thinking, but it will also challenge the feel-good interpretation of their alleged friendship and reveal their numerous conflicts—and the different phases and different constellations of an enduring enmity.

4. Enduring enmity in changing *Lagen*

Between 1926 and 1965, significant changes occurred to both Schmitt and Kirchheimer's individual *Lagen*, and these changes affected their relationship in considerable ways.

During these almost forty years, their enduring enmity went through four distinctly different phases.

The *first phase* starts with Kirchheimer's decision to follow the advice of his mentor at Berlin University, the legal scholar Rudolf Smend, to move to Bonn and study with Schmitt. His decision was inspired by the motivation to learn as much as possible from Schmitt to further his left-wing politics, in particular about his theory of dictatorship and his criticism of parliamentary democracy. During the time they shared in Bonn, their enmity was a kind of abstract relationship to each other, on the level of political standpoints only. As will be demonstrated in Chapter 2, Kirchheimer and Schmitt did not view each other merely as sympathizing with rival political camps but as representatives of political forces that saw each other as political enemies, ready and willing to fight the other side until they achieved irreversible victory. To Kirchheimer, Schmitt was a militant right-wing ideologist of the bourgeois class enemy—a kind of "Lenin of the bourgeoisie."[16] The two respected each other and their political views, although Schmitt's acknowledgment of Kirchheimer was accompanied by a certain condescension. At this point in their relationship, their political enmity was compatible with a seemingly friendly relationship on the personal level and Schmitt's patronage of his doctoral student.

Kirchheimer was completely satisfied with the opportunity to study with Schmitt. He found it very stimulating to attend his seminars, read his books and articles, and discuss them with him in person. Schmitt, for his part, learned from his doctoral student about ongoing theoretical and strategic debates among intellectuals of the socialist left. Kirchheimer's dissertation gave him evidence for his belief that not only revolutionary communists but even reformist social democratic Marxists saw the existing Weimar Republic as merely a transition period toward a better socialist future. Such a view seemed to confirm Schmitt's expectation of future political instability and also reinforced his political counterprogram: the search for an authoritarian political model in order to build a strong and stable state. In his citations, he soon made Kirchheimer a key witness for his conviction of the uncompromising enmity of the left toward the existing bourgeois state.

After the two met again in Berlin in 1928, their relationship quickly shifted from the former teacher-student constellation. These years in Berlin became a time of rapidly growing success for both of them. Schmitt easily succeeded on the academic and political stages and Kirchheimer soon gained a certain public notoriety as a harsh critical voice on the left wing of the Social Democratic Party (SPD). Their personal contact remained good between 1928 and 1931. As in Bonn, they went for walks and met frequently, occasionally with their families, as will be described in Chapters 3 and 5. In addition to their personal communication, they exchanged manuscripts and offprints so that both were able to quote from the other's as yet unpublished writing. Nevertheless, neither Kirchheimer nor Schmitt showed any intention of approaching the other to look for common ground that could lead to compromises in their political aspirations or theoretical reflections. They continued to see themselves as representatives of political forces engaged in tough battles, although still hoping to learn from the enemy-other in order to further their own political ambitions.

16 John H. Herz in a discussion at the Kirchheimer symposium in Berlin on 11 November 1985.

With the dramatically changing political *Lage* in the upcoming crisis of the Weimar Republic in 1932, Kirchheimer and Schmitt's intellectual disagreements lost their sterility and transformed into a *second phase*. As will be shown in Chapters 5 and 6, an increasing number of political conflicts between them became public. Kirchheimer still attended Schmitt's seminars at the Handelshochschule in Berlin and introduced him to other younger socialists. Schmitt, in turn, supported Kirchheimer in his (unsuccessful) applications for academic scholarships. For his *habilitation*,[17] he aspired to at the University of Berlin; however, Kirchheimer turned to his other mentor Rudolf Smend and not to Schmitt. Whereas their contact remained cordial on the personal level, Schmitt was careful to conceal from Kirchheimer his active role in the futile efforts in late 1932 to install a presidential dictatorship, albeit not yet under Hitler's chancellorship.

Kirchheimer contradicted Schmitt in his publications on almost every important topic: Schmitt's theory of parliamentarism and democracy; the role of the president in the constitutional order of the republic; the *Rechtsstaat* and property rights; the role of political parties in a modern democracy; international law; the legitimate limits of legal constitutional changes; political myths; and Italian fascism as a political alternative to the current system. By contrast, until the end of the Weimar Republic, Schmitt made no critical comments about Kirchheimer in his publications but treated him as a representative of a radical socialist left that wanted to overcome the constitutional order. In doing so, he ignored Kirchheimer's turn toward the defense of the republic and even misquoted him in his influential book *Legality and Legitimacy*.

Their conflict escalated and turned—at least from Schmitt's side—into a level of enmity *ad personam* just as the Weimar Republic suffered its final blows. Schmitt wanted to prevent the revitalization of the Weimar Republic and conjured the specter of civil war or a state of trade unions that would become a socialist republic. To Schmitt, with his Catholic background and upbringing, Kirchheimer's way of thinking was synonymous with agnostic socialism. Seen through Kirchheimer's lens, Schmitt was drifting toward an authoritarian economic liberalism that intended to eliminate the central social and democratic elements of the Weimar Constitution. At this crucial political moment, Kirchheimer co-authored an essay which reads like an incisive review of his previous critical debates with Schmitt. It also included a general attack against him on the methodological level. When the two discussed their political, theoretical, and methodological disagreements in person at Schmitt's home in November 1932, the conflict escalated to a new level from Schmitt's side. Afterwards, he noted "*scheußlich, dieser Jude*" ("vile, this Jew")[18] in his diary, referring to Kirchheimer. The entry indicates that at this point in time, Schmitt's antisemitic sentiments were no longer distinguishable from his substantive differences with Kirchheimer.

Once Hitler had been installed in power, the new *Lage* transformed Schmitt and Kirchheimer's relationship into a *third phase*. Hitler's inauguration as the new Chancellor of the Reich took both Kirchheimer and Schmitt by surprise. And they both misinterpreted the new chancellorship initially, albeit for different reasons. Kirchheimer's

17 A postdoctoral qualification in many European countries, including Germany, required in order to become a full professor.

18 Carl Schmitt, diary entry of 6 November 1932 (Schmitt 2010, 231).

underestimation of the Nazis was to some extent a logical result of his theories. Just like many German Marxists of the day, he interpreted Italian fascism as a phenomenon that could prevail only in industrially backward societies. What had distinguished his analytical acuity in the years 1930 to 1932—his description, inspired by Marxism, of the societal functions of state and politics—no longer helped him come to terms with the new *Lage*. To Kirchheimer, the main risk to the parliamentary republic stemmed from a bureaucracy that had taken on a life of its own with a presidential dictatorship at the top—in other words, precisely what Schmitt had declared to be his political ideal before 1933. So ironically, it was presumably the fact that Kirchheimer knew Schmitt's way of thinking very well that contributed to him temporarily losing sight of the danger of a successful Nazi mass movement. He was among the majority of leftists of that time who perceived only minor differences between the authoritarian governments of Chancellors Franz von Papen and Kurt von Schleicher on the one hand and Adolf Hitler's regime on the other and who underestimated the residual protective function of bureaucratic state institutions of the former presidential dictatorship.

After Hitler was declared Chancellor on 30 January 1933, Schmitt was not sure whether the new regime would be able to stay in power. It took him a few weeks to grasp the new political *Lage* and the brutal energy of the new regime. He still decided to support it and quickly became its most influential legal theorist, whereas Kirchheimer saw no other option but to flee to France after being released from prison. He had been detained for a few days in May 1933 for political reasons. From then on, there was a new asymmetry to their enmity. Schmitt no longer dealt specifically with Kirchheimer but simply included him in the group of all those Germans who were in detention or had been forced to emigrate. He called all of them enemies of the Reich and even accompanied this label with threats of violence, stating that enmity toward the Jewish émigrés was part of an existential life-or-death struggle for the German *Volk* (people/nation in a racial sense, of common blood and with a common destiny; see Glossary).

Schmitt and Kirchheimer agreed that the new regime was a response to the civil war-like conditions of 1932. In November 1932, Schmitt had already spoken of the inevitability of civil war in Germany. By supporting Hitler, he linked the Nazi regime with the suggestion of a civil order rescued from a dangerous civil war situation. Hitler appeared to be the one preventing civil war and permanently overcoming it. Kirchheimer, on the other hand, described the fact that Hitler's party had been able to stabilize its power position not as preventing civil war but as the uncompromising first victory of one civil war party over the others. He was convinced the measures taken by the Nazi regime were a continuation of this civil war with the additional instruments available to state agencies. Kirchheimer observed Schmitt's many activities in detail from his exile in Paris and London and reported on them in journals and magazines. He was the first to call Schmitt "*the* theorist of the Nazi Constitution" (Kirchheimer 1933c, 533). A few months later, émigré journalist Waldemar Gurian referred to him as the "crown jurist of the Third Reich."[19] Gurian's label was more striking and was immediately used polemically by Schmitt's opponents outside of Germany as well as by his Nazi competitors. Whereas Gurian's term captures the essence of a political lawyer for a regime who becomes dependent on his

19 Waldemar Gurian in the émigré journal *Deutsche Briefe*, 26 October 1934.

superior, the *Führer* (see Glossary), and serves his master for the sake of ambition or careerism, Kirchheimer's more laconic characterization left the question of Schmitt's personal motives open. But by emphasizing Schmitt's role as the leading legal theorist of the Nazi regime, he also set the tone for the Frankfurt School philosopher Herbert Marcuse's first study of Nazi political thought (see Marcuse 1934).

In the summer of 1935, the enmity between Kirchheimer and Schmitt escalated to a new level. This time, it was Kirchheimer who intensified the dispute. As will be reconstructed in more detail in Chapter 8, Kirchheimer chose a new and direct tactic aimed directly at Schmitt. Using the pseudonym Dr. Hermann Seitz, he wrote the booklet *State Structure and Law in the Third Reich* for the resistance in Germany; the title alluded to one of Schmitt's most widely distributed booklets in Nazi Germany. To boost distribution, the booklet's cover design, color, and typesetting were designed to make it appear to be part of a Nazi book series edited by Schmitt. The illegal booklet was distributed to a few thousand lawyers across Germany. Schmitt was infuriated when he found out about it. He instantly assumed Kirchheimer was the author and demanded that the Nazi authorities crack down on everyone involved in its production. If the Gestapo had caught Kirchheimer, he would likely have been interned in a concentration camp or worse.

Even though Paris still was a safe place for Kirchheimer, he rightfully expected a German military attack on neighboring countries in the near future and was determined to leave France for the US. He was eventually able to do so in the fall of 1936. Schmitt's militant antisemitism, which will be analyzed in detail in Chapter 10, reached its pinnacle in the fall of 1936 when he claimed that the deadly poison of Jewry and Judaism had for decades permeated the German state and German academia unhindered. To Schmitt, Jews—like his former student Kirchheimer—had become the public enemy *par excellence*. There can be no beating around the bush: it was precisely this kind of domestic declaration that Jews were the enemy that made the extermination of the Jews in the Holocaust possible and that made the Holocaust unique as a crime against humanity.

The same year, Schmitt himself got caught up in the machinery of the Nazi system. As will be described in depth in Chapter 9, the surviving files from the *Reichsführer* of the *Schutzstaffel* (SS; see Glossary) also mentioned his former contacts with Kirchheimer. However, Schmitt's fall from grace was not a case of persecution of a supposed enemy or opponent of the regime. It was an initiative by other Nazi jurists to limit his leading role. The best way to understand Schmitt's fall is through the analytical lens of Kirchheimer's writing in exile about the political system of Nazi Germany. Whereas Schmitt in his numerous written works and speeches had admired the Nazi regime for overcoming the pluralism of the Weimar Republic and creating a tripartite structure of unity of the German state, Kirchheimer countered that no such unity existed. Contrary to the official ideology, he argued that the Nazi state had never become a homogeneous entity but was instead a polycracy. It was a system based on compromises between five major social groups—the Nazi party, the army, industrial and financial capital, the agrarian Junkers, and the state bureaucracy—that were constantly struggling for influence against each other. The party hierarchies below the level of the *Führer* were regrouped time and again. Thus, the position of every individual in the system was subject to sudden shifts. This analytical approach permits us to identify the reasons for the activities of the *Reichsführer* of the SS against Schmitt as being founded less in his person and rather in the complex internal struggles between competing groups within the Nazi system. Schmitt, the triumphant Nazi theorist of tripartite state unity, had become caught in the clutches of the polycratic power structure that Kirchheimer had described in his writing in exile.

Only after another year had passed, in late 1937, did both Schmitt and Kirchheimer find themselves in a new *Lage*. Kirchheimer had succeeded in moving to the United States and was working for the Institute of Social Research (ISR) in New York under the leadership of Max Horkheimer. In November 1938, the Nazi government revoked his German citizenship and he and his daughter Hanna became stateless. Shortly afterwards, the University of Bonn revoked his doctoral degree of 1928 (it took the university until November 2023 to give it back eighty-five years later).[20] Meanwhile Schmitt had found his way back into the top ranks of Nazi jurists by throwing himself into international law as the main subject of his theoretical research. Kirchheimer commented on this new twist in Schmitt's new works soon after their publication, concluding that Germany would attack neighboring countries within a short space of time. After Germany had started the war, both men contributed to the fight against the enemy on the other side: Schmitt in his writing and lectures for Nazi Germany and Kirchheimer at the Office of Strategic Services (OSS) on the side of the US. His research at the OSS included analyzing the political mood in Germany, determining the Reich's military capabilities, and advising on the selection of military targets in Germany.

After Germany's unconditional surrender in 1945, the immediate postwar situation once again reversed Kirchheimer and Schmitt's *Lagen* and their roles. In this *fourth phase*, which lasted up until 1961, each had seen their professional situation change fundamentally. Schmitt had lost his job as a prestigious German professor and, for the first time in his life, Kirchheimer had a well-paid position. He was on the victorious side and Schmitt found himself on the side of the defeated. While Schmitt did not feel liberated by the Allies, he was happy that the war was over and that he had survived. He prepared to serve as a defense attorney for German war criminals. As will be described in more detail in Chapters 13 and 14, he saw himself pursued by returning émigrés and people he considered "Jewish-American enemies" seeking to take revenge, enrich themselves, or go after him personally. He considered the trials of German war criminals to be victor's justice perfidiously executed by enemies. From Schmitt's perspective, Kirchheimer was one of the enemies one had to be very cautious of. Kirchheimer, for his part, in his legal opinions for the Nuremberg Trials, took great care to argue within the framework of the international rule of law.

In the years that followed, the basic constellation between them continued to solidify. For the first time in his life, Kirchheimer attained a tenured university position, whereas Schmitt was forced to remain outside the German university system. Nevertheless, he was able to create a large network of contacts with influential legal scholars and young academics that became known as an "invisible college." The next personal encounter between Kirchheimer and Schmitt occurred almost exactly seventeen years after they had last met at Schmitt's Berlin home in November 1932. Kirchheimer's motives for visiting Schmitt at his home in Plettenberg in November 1949 have largely been misrepresented in

20 Hermann Horstkotte, "Universität Bonn will Otto Kirchheimer rehabilitieren," *Bonner Generalanzeiger*, 6 November 2023, 8.

the literature to date. As will be explained in Chapter 14, it was Schmitt who took the initiative to resume their personal contact. He had been arrested in Berlin in 1947 and asked about Kirchheimer's fate during his interrogation by Ossip K. Flechtheim. He also asked Flechtheim to give his best regards to Kirchheimer. Based on Wilhelm Hennis's recollections of his discussions with Kirchheimer,[21] I interpret Kirchheimer's visit to Schmitt's home in Plettenberg two years later as a demonstration to Schmitt that he, who had been forced to leave Germany with Schmitt's applause in 1933, had managed to survive. He also wanted to show Schmitt how the *Lage* had changed after 1945 and the extent to which the political tide had turned. As described in more detail in Chapter 15, Schmitt wrote in a letter to his wife that Kirchheimer had confronted him about his unwillingness to grapple self-critically with his own responsibility for the Nazi regime's policies and told him so during his visit.

After this one and only trip of Kirchheimer's to Schmitt's home in Plettenberg, the two did not resume their relationship as it had been before 1933. Not only their different roles in the years 1933 to 1945 but just as much their differences in dealing with the Nazi past had created a deep rift that could not be patched up with friendly phrases of address in their letters over the following years. The sparse correspondence between the two at this time shows no signs of an intimate personal connection. Their letters contained mostly polite phrases on both sides and occasional critical remarks from Kirchheimer toward Schmitt or his students. They also sent each other their publications from time to time. They briefly met again in person in Cologne in June 1953 but Kirchheimer did not respond to Schmitt's multiple offers by letter to enter into a debate with him again. He contented himself with brief replies, the only exception being his letter of November 1952 in which he expounded on the methodological differences between him and Schmitt and repeated the objections he had raised against Schmitt's conceptual realism twenty years earlier in his article "Remarks on Carl Schmitt's *Legality and Legitimacy.*" In 1932, Schmitt had not responded to Kirchheimer's fundamental and detailed criticism; he did not react to it twenty years later, either.

During the 1950s, Kirchheimer also publicly attacked Schmitt and his followers multiple times in his articles and book reviews. In 1957, he summarized his substantive and methodological objections to Schmitt in a compact form: Schmitt's ever-present negation of the *Rechtsstaat*; the discrepancy between the traditional liberal concept of traditional international law and the rejection of an alien liberalism as part of the domestic constitutional order; the omnipresence of the people's constituent power combined with its incapacity to act as a constituted organ; the indeterminate character of the values underlying concrete decisions; and the lack of any clear-cut criteria for differentiating between violence and *nomos*. Again, Schmitt refrained from publicly reacting to these allegations. On the basis of the letters surviving in the archives, he did not respond to Kirchheimer by letter, either. By the end of the 1950s, Kirchheimer had stopped commenting on Schmitt's work in his publications altogether. Their contact practically dried up. Kirchheimer had obviously lost interest both in debating with Schmitt in person and in debating about him in public. Their exchange of letters was limited to sending offprints, which took the form of small mutual jibes.

21 Wilhelm Hennis in a conversation with the author on 26 September 2009.

Kirchheimer formulated his opinion about dealing with Schmitt in a 1958 letter to Arvid Brødersen, his colleague at the New School for Social Research, who had asked him about his relationship to Schmitt:

> I still think today that nobody should be held criminally or pseudocriminally responsible for their writings or their intellectual production. To a writer, the authority is the reaction of the audience and their own conscience. The question of employment sanctioned and paid for by the state is of course a different matter.[22]

In Kirchheimer's view, the decision not to put Schmitt on political trial in Nuremberg with the intent to punish him for his writing during the Nazi regime was correct. Schmitt could now enjoy all the liberties of life and public expression a *Rechtsstaat* guaranteed its citizens—but Kirchheimer thought he should be barred from the opportunity to continue disseminating his doctrines at a state-funded university as he had successfully participated in the destruction of the *Rechtsstaat* and in the establishment of a fascist terror regime.

5. The godfather of left-Schmittianism?

The narrative that Schmitt and Kirchheimer rekindled a friendly relationship after World War II has developed a life of its own in the literature. This narrative serves two legitimating functions. First, it makes Kirchheimer an apologetic witness to Schmitt's alleged personal sympathies for Jewish intellectuals—before and after the Shoah. Second, it portrays him as a kind of godfather and patron of today's left-Schmittianism. And, depending on one's perspective about Schmitt and his writing, this characterization is used by contemporary left-Schmittians either to enthrone Kirchheimer as their forerunner or, with a critical intention, to turn Kirchheimer's work into the starting point of a fateful dead end for the political left.

I define left-Schmittianism as the transformation of Schmittian concepts or categories into the framework of legal or political theories with emancipatory political intentions. Left-Schmittians insist that Schmitt's work provides crucial contributions to understanding our modern political condition. The label "left-Schmittianism" is used in very different ways. For some, it is a title of honor used for political theories that claim to be of service to their emancipatory cause through a productive reception of Schmitt's writing.[23] To others, this label is tantamount to a stigma because they view Schmitt's theories and concepts as fundamentally incompatible with any emancipatory goal. They are convinced that Schmitt's key concepts such as democracy, parliamentarism, international law, and the political cannot be divorced from his reactionary political intentions.[24]

---

22 Letter from Otto Kirchheimer to Arvid Brødersen dated 2 March 1958. Otto Kirchheimer Papers, Series 2, Box 1, Folder 25.

23 See Mouffe (1999) and (2005), and Kalyvas (2009).

24 See Müller (2003), and Scheuerman (2020).

Kirchheimer did not use the term "left-Schmittian" to describe himself or his work. This points to the term's complex history over the past hundred years or more and to the particular role Kirchheimer plays in this. Its prehistory dates back to the publication of Schmitt's book *Dictatorship* in 1921, which he finished writing during his first professorship at Greifswald University. Only a few months later, the Austro-Marxist Max Adler happily resorted to using some of Schmitt's definitions and analytical distinctions for his legal theoretical foundation of a theory on the dictatorship of the proletariat (see Adler 1922, 193–197). He was followed by Arkadij Gurland, a young socialist from the left wing of the SPD (see Gurland 1930a, 77–80). Both authors were major inspirations to the young Kirchheimer as he developed his own theoretical considerations. Another of his close colleagues at the time and a personal friend of his for the next twenty-five years, Franz L. Neumann, concurred with Schmitt's critique of the Weimar federal system and pluralism in his book *Der Hüter der Verfassung* [The guardian of the constitution], published in 1931 (see Neumann 1931, 81–85). Karl Korsch, who provided crucial intellectual inspiration in the early phase of the Frankfurt School, also praised the book and the potential of Schmitt's critique of parliamentarism. Although he evidently did not share Schmitt's sympathies for a fascist state, he agreed with his "critical analysis of the hitherto dominant bourgeois-liberal" (Korsch 1932, 205) theory of the state. While all of these authors relied on certain concepts and considerations by Schmitt, none of them connected to the Frankfurt School would identify as a left-Schmittian.

To the best of my knowledge, the first time that someone from the left was criticized for using Schmitt to make his own case was in connection with the disputes between communists and socialists toward the end of the Weimar Republic. The allegation was made in the communist magazine *Unsere Zeit* [Our era], and its intent was unequivocally denunciatory. It was published only a few days after the transfer of power to the Nazis. The target of the attack was Otto Kirchheimer. He was accused by an anonymous author under the headline "Mister Carl Schmitt's Key Witness" of left social democratic "uniformity"[25] with Schmitt. The author used Kirchheimer's references to Schmitt as further evidence of the communist narrative that the SPD was partly to blame for the establishment of the new fascist regime in Germany. Thus, Kirchheimer in fact stands at the beginning of the genealogy of alleged left-Schmittianism, although the term itself was not actually used.

"Left-Schmittianism" as an explicit label appeared in Germany in the late 1950s under completely different political circumstances. It was used with positive intentions and meant as a self-ascription. German philosopher Wolfgang Wieland coined it to designate a group of younger West German academics who met with Schmitt on a regular basis and who read his works from liberal and social democratic perspectives (see Lübbe 1988, 428).[26] Schmitt—when he heard about it—liked this label for this group (see van Laak 1993, 238). Ten years later, in a different political situation, the term returned with a new wave of Schmitt reception by a number of authors of the New Left. Political activists and academics such as Mario Tronti in Italy, European theorists of the guerilla movements

25 *Unsere Zeit* (15 February 1933, 244).

26 Besides Hermann Lübbe, Odo Marquard, Martin Kriele, and Ernst-Wolfgang Böckenförde also belonged to this group.

in Latin America such as Joachim Schickel, and Johannes Agnoli in Germany in the late 1960s and early 1970s[27] used Schmitt's critique of parliamentary democracy for their revolutionary purposes. Now self-identification as left-Schmittian took on a proud militant tone.

Another decade later, not much was left of such self-assured declarations in the circles of the academic left. Nevertheless, in 1983, Alfons Söllner, following some observations by Volker Neumann about parallels between Schmitt and Kirchheimer in their Weimar writing,[28] found it "attractive" to analyze some of Kirchheimer's Weimar works "under the label of left-Schmittianism" (Söllner 1983, 222). His reintroduction of the term as a key for interpreting Kirchheimer's critical analysis of the Weimar Constitution was gladly accepted by American political theorist Ellen Kennedy. She used the same attribute for other prominent authors of the Frankfurt School too, namely Franz L. Neumann, Walter Benjamin, Herbert Marcuse, and the early Jürgen Habermas. According to Kennedy's controversial interpretation, their contributions to political theory continued—albeit with different political goals—the anti-liberal substance of Schmitt's thought uncritically (see Kennedy 1987a). The discussion that resulted from her allegations was intense—and has not been consensually resolved to this day.[29] What is striking, however, is that most authors from the academic left in this debate insisted on Kirchheimer's—and their own—distance from Schmitt.

The collapse of the Soviet Union led to a temporary acceleration of processes of globalization and the emergence of a unipolar world—two developments that have been identified as key factors behind another change in the discursive field from the 1990s on. In addition, a number of academic political theorists in the West seemed to tire of the normativism and rationalism of political philosophers such as Jürgen Habermas and John Rawls (see Bernstein 2011, 403–404). The discursive field changed again when the reception of Schmitt's writing spread to a new international level. Schmitt became a kind of "sage of the Left and the Right" (McCormick 1998, 830). On the right, he has become a point of reference for authors of the French Nouvelle Droite such as Alain de Benoist. On the left, Schmitt has been hailed by a new generation of authors in Italy, France, Germany, the UK, and the US as an incisive and stimulating author for their political purposes, too. Among a number of contemporary political theorists, left-Schmittianism has become a positive label again. Just like their predecessors in the 1970s, contemporary self-confessed left-Schmittians claim that Schmitt's concepts and arguments are not necessarily contaminated by his lifelong antisemitic attitude and his preference for authoritarian and fascist regimes. They treat Schmitt as our contemporary with important messages that cannot be found in the work of other theorists of the past.

---

27 See Schickel (1970) and Müller (2003, 169–180).

28 See Neumann (1981) and (1983).

29 The literature on this subject has become legion. See Habermas (1987), Jay (1987), Kennedy (1987b), Preuß (1987), Schäffer (1987), Söllner (1987), Tribe (1987), Perels (1989), Kohlmann (1992), Scheuerman (1994), Scheuerman (1996), Heil (1996), Scheuerman and Caldwell (2000), Richter (2001), Müller (2003), Schale (2006), Bavaj (2007), Landois (2007), Mehring (2007), Kemmerer (2008), Hitschler (2011), Llanque (2011), Turner (2011), Breuer (2012), Mehring (2014b), Neumann (2015), Olson (2016), Zwarg (2017), Buchstein (2021a), Mehring (2021), Klein (2022), Simard (2023), and Klingsporn (2024).

There are numerous examples of this selective and reconstructive strategy of reception in contemporary political theory: Gopal Balakrishnan in his plea for radical democracy and his critique of US imperialism; Giorgio Agamben in his work on the state of exception; Michael Hardt and Antonio Negri on the genuine capacity of the deterritorializing power of the multitude; Chantal Mouffe in her theory of agonistic democracy and her rejection of a unipolar world; Andreas Kalyvas on the relationship between constituent power, sovereign decision, and democracy; Danielo Zolo on humanitarian international law; Luiza Odysseos on spaceless universalism and cosmopolitanism; Horst Bredekamp in his reflections on the eternal constitutive role of political myths; and Jean-Francois Kervégan on the challenges to liberal democracy.[30] Some of them transform and dilute the meaning of some of Schmitt's concepts and theories to such an extent that the resulting syntheses are only partially Schmittian—at least in my opinion. In any case, all of them choose specific concepts, categories, or theorems from Schmitt and incorporate them into their own theoretical framework, sometimes significantly modifying them in the process. And all of them are keen to do their political due diligence, making all the necessary caveats about Schmitt's completely different political intentions, his antisemitism, and his role in Nazi Germany.

Their differences notwithstanding, contemporary left-Schmittians agree that Schmitt provides unique resources for contemporary political theory, diamonds in the rough, as it were, that can be excavated from his work. The following five theoretical contributions in Schmitt's work are highlighted in the works of the authors listed above: Schmitt's antagonistic concept of the political; his theory of the exceptional state and sovereignty; his declaration of an irreconcilable antagonism between democracy and liberalism; his critique of parliamentarism; and his critique of universalism in international law. Kirchheimer already addressed all five of these subjects, albeit with different results.

What exactly is the meaning of left-Schmittianism in Kirchheimer's case? Is it a fitting characterization of his work at all and, if so, for what timeframe and in what respects? Among other things, my book illustrates how problematic it is to separate some of Schmitt's theoretical concepts and impulses from the overarching context of his legal and political thought. Kirchheimer realized after a short period of time that his original intention to make productive use of Schmitt's key concepts in order to fill the gaps in left-wing political thought was a lost cause. But he still made use of Schmitt's work in several other ways that are well worth exploring—not least in reference to the theoretical weaknesses of today's left-Schmittianism.

---

30 See Balakrishnan (2000, 2011), Agamben (2003), Hardt and Negri (2004), Mouffe (1999, 2005, 2007), Odysseos (2007), Zolo (2007), Kalyvas (2009), Bredekamp (2016), and Kervégan (2019).

6. Sources

Most aspects of Schmitt's biography are now fairly well known.[31] This is not the case at all for Kirchheimer; his biography has yet to be written. Consequently, the source base for this book is asymmetrical. With regard to Schmitt, I made use of a number of primary sources published in the past decade, almost all of them only in German. These include a new critical edition of his books *The Concept of the Political* and *Staatsgefüge und Zusammenbruch des zweiten Reiches* [The structure of the state and the collapse of the Second Reich] as well as a few minor works; the second and revised edition of his diary-like *Glossarium*; a collection of his Nazi works; his diaries from the years 1925 to 1934; and various editions of his vast correspondence running to many thousands of letters, including with his wife Duška, his editors, journalists, his colleague Rudolf Smend, as well as philosophers, legal scholars, economists, historians, historians, theologians and writers such as Jacob Taubes, Alexandre Kojève, Hans Blumenberg, Gerhard Ritter, Eric Voegelin, Hermann Heller, Ernst-Wolfgang Böckenförde, Moritz Julius Bonn, Carl Brinkmann, Nicolaus Sombart, Reinhart Koselleck, Dietrich Braun, Armin Mohler, Roman Schnur, and Waldemar Gurian.[32] With regard to Kirchheimer, my research had to rely on unpublished documents to a much greater extent. As a matter of fact, I was fortunate to find far more material than expected, a great deal of it scattered in various archives in Germany and the US. The documents also include the correspondence between Schmitt and Kirchheimer. My interviews with a number of witnesses mentioned below also became invaluable sources for reconstructing the relationship between the two.

Every book in the field of history of political ideas has a history of its own, and this book's history is slightly longer than usual. A longer history does not necessarily make a better book; in any case, the reason I mention this history at all is because some of the sources are interwoven with my own academic biography. My interest in both Schmitt and Kirchheimer was piqued at the beginning of my academic career in the early 1980s. Time and again over the past forty years, I have touched upon certain parts of the subject of this book—be it in connection with my early publications about some unknown manuscripts by Franz L. Neumann (1983 and 1986), my monograph on the history of political science in Berlin after World War II in 1992, my books on the history of public and secret voting (2000) and on democracy and lottery (2009), or my editorial work on collections of the writing of Franz L. Neumann (1989), Arkardij Gurland (1991) Ernst Fraenkel (1999–2011), and Hermann Heller (2023). My research related to the edition of Kirchheimer's collected works published in six volumes between 2017 and 2022 brought to light several hitherto unknown aspects of Kirchheimer's political biography and a number of as yet unknown articles and manuscripts of his. After finishing this edition, I initially wanted to move on to new subjects in my academic work. But I changed my mind in response to the reactions to a talk I gave at the New School for Social Research in New York

31 The best biography is by Reinhard Mehring (see Mehring 2014a). A revised German edition was published in 2022 (see Mehring 2022a).

32 All published exchanges of letters with Schmitt are listed by the Carl-Schmitt-Gesellschaft e.V., accessed 5 March 2024, https://www.carl-schmitt.de/wp-content/uploads/CSC_Briefe-von-und-an.pdf.

in February 2019 titled "The 'ugly Jew' and the 'Man of Darkness'—Otto Kirchheimer and Carl Schmitt." The intense discussion that followed my lecture and the critical questions raised by some members of the audience became the starting point for writing this book.

Over the course of my academic career, I have had the good fortune to meet quite a few people who were happy to tell me about their experiences with Kirchheimer and/or Schmitt—be it Richard Löwenthal on the discussions among the radical left in the Weimar Republic about Schmitt's theory of dictatorship; Susanne Suhr on Otto Kirchheimer's eloquence in the political discussions in Café Dümichen, the meeting place of the socialist intelligentsia in Berlin at the end of the Weimar Republic; Henry W. Ehrmann on the intense atmosphere of Schmitt's seminar in Berlin that he and Kirchheimer attended; Herta Zerna and Dieter Emig on the political activities of Kirchheimer and Arkadij Gurland in the left wing of the Weimar SPD; Ludmilla Müller on Kirchheimer's work in the law firm of Franz L. Neumann and Ernst Fraenkel; Reinhard Bendix on the Weimar debates about reforming criminal law; Peter Gay on social scientists' and legal scholars' "hunger for wholeness" in their theories during the Weimar Republic; Albert O. Hirschman on Kirchheimer's poor living conditions in his Paris exile; Leo Löwenthal on the tense relationship between Kirchheimer and Max Horkheimer at the Institute of Social Research; Ossip K. Flechtheim on his and Kirchheimer's contributions to Franz L. Neumann's book *Behemoth*; Lili Flechtheim-Faktor on the difficult living conditions as a German émigré in the United States; Raul Hilberg on the research about antisemitism at the Institute of Social Research; John H. Herz on his collaboration with Kirchheimer at the Office of Strategic Services (OSS) during the war and while preparing for the Nuremberg Trials; Nicolaus Sombart's vivid recollections of Schmitt's antisemitism; Wilhelm Hennis's lively report on Kirchheimer's motives for visiting Schmitt in 1949; Ernst-Wolfgang Böckenförde's impressive portrayal of Schmitt's charisma as a conversation partner; Helge Pross on Kirchheimer's failed attempts to obtain a professorship in Frankfurt; Jürgen Habermas on Kirchheimer's hospitality at his home in Silver Spring; Arthur J. Vidich's anecdotes about Kirchheimer teaching at the Graduate Faculty of the New School for Social Research; David Kettler about Kirchheimer and Franz L. Neumann at Columbia University; Winfried Steffani on Ernst Fraenkel's personal relationship with both Kirchheimer and Schmitt; Jürgen Fijalkowski on the dismissive way Schmitt reacted to criticism of his work; Karl Dietrich Bracher on Kirchheimer's thoughts on the destruction of the Weimar Republic; Wolfgang Mommsen on his interpretation of Max Weber as a forerunner of the theory of plebiscitarian presidential dictatorship and Schmitt's praise for this controversial reading; Gilbert Ziebura on Schmitt's and Kirchheimer's reflections about Charles de Gaulle's *Coup d'Etat permanent* in France; Kurt Sontheimer and Michael Th. Greven on Kirchheimer's comparative research on party systems; Jürgen Seifert on his attempt to ask Schmitt about his relationship with Kirchheimer; Horst Ehmke on Kirchheimer's work on his book *Political Justice* and his decision to return to Germany which he was unable to realize because of his untimely death; Johannes (Giovanni) Agnoli and Angelo Bolaffi on the interest of the Italian radical left in Schmitt's work in the 1960s and '70s; Claus Offe on Kirchheimer's contribution to Frankfurt School critical theory; Ulrich K. Preuß on the rediscovery of Schmitt's work for the German New Left; Alexander von Brünneck on the rediscovery of Kirchheimer's work for critical legal studies; Rainer

Erd on his visit at Schmitt's home in Plettenberg and the conversation with him about Kirchheimer in 1980; and Jacob Taubes's entertaining stories about Schmitt and the friendly correspondence with his—as he used to call him in his seminars—"enemy *par excellence*." In addition to the publications and the documents available in archives, the memories these individuals shared with me are an invaluable source of information for this book.

In addition, I had the opportunity for numerous conversations with Hanna Kirchheimer-Grossman and Peter Kirchheimer about their father's life and work. I came away from these stimulating conversations with new information I could not possibly have obtained any other way. Poring over the documents and photos in their family archive sparked memories they shared with me and helped fill in gaps concerning details of their father's biography.

I am deeply grateful for the opportunity to have spoken with each and every one of these contemporaries of Otto Kirchheimer and Carl Schmitt, all of them witnesses of a vanished epoch. Not least because some of their recollections exist only in my correspondence, in scattered handwritten notes or in my memory only, I felt it important to share them in this book and I hope my readers also find them newsworthy.

The Weimar Republic

Chapter 2:
The Beginnings in Bonn (1926–1928)

The story of Otto Kirchheimer and Carl Schmitt first began in the fall of 1926, when 21-year-old Otto Kirchheimer arrived at the University of Bonn to continue his law studies. At this point, Schmitt had already established himself as a successful and illustrious teacher of constitutional law. He had come to the University of Bonn from the University of Greifswald as of the summer semester 1922, replacing Rudolf Smend who had accepted an offer from Berlin University.

The six years Schmitt spent in Bonn until his own subsequent move to Berlin in 1928 were a particularly productive phase in his academic career. He had previously published his *Political Romanticism* (1919) and *Dictatorship* (1921) and, during his time in Bonn, published a number of other equally famous monographs in rapid succession, including some that are still seen as his most important works to this day. The first was *Political Theology* (1922), which he wrote in the misty atmosphere of Greifswald on the Baltic coast. This was soon followed by *Roman Catholicism and Political Form* (1923), *The Crisis of Parliamentary Democracy* (1923), and books on the Rhinelands as an object of international politics (1925), on the key question of international law (1926), and on referenda (1927). In 1927, he wrote the first version of his essay "The Concept of the Political," which was later expanded into a book. He also completed his magnum opus, *Constitutional Theory* (1928), during this period.

Schmitt established an extensive network of people he deemed culturally interesting or who were important in the academic and political communities. He was a star at the University of Bonn. Students of all faculties enthusiastically attended his lectures, and his presence in Bonn attracted students from other universities, too. Among them was Kirchheimer, who studied intensively with Schmitt for three semesters and completed his First State Examination in law with him. Schmitt also supervised Kirchheimer's doctoral dissertation, which he defended in 1928. Its subject dovetailed with Schmitt's areas of interests: a comparison of the state theories of Western social democracy and Soviet Bolshevism.

1. Schmitt at the first high point of his academic career

In Bonn, Schmitt was at the first high point of his academic career.[1] Born in the western German town of Plettenberg in 1888, he had completed his doctorate in 1910 and his *habilitation* in public and administrative law, theory of the state, and international law in 1916, during World War I. He was first appointed professor in Greifswald in 1921 and moved to the University of Bonn just a year later. Within a short period of time, he had written his way into the top tier of German legal scholarship; he was highly regarded as an author in literary and artistic circles for his brilliant style and his diverse interests which transcended the law by far. The style of his suggestively worded prose oscillated between cynical frostiness and political agitation. He would switch abruptly from rigorous and heavily footnoted academic writing and analysis to the prophetic tone of metaphysical reflections. The experience of taking in his work with these fascinating transitions is surely part of the secret of his success—to this day.

Schmitt agreed with the political views of the overwhelming majority of the constitutional law professors in Germany at the time, a group of approximately 100 people. Constitutional law during the Weimar period was mainly anti-liberal and anti-democratic, and the majority of scholars were shifting to the extreme right (see Stolleis 1999, 120–122). What usually distinguished Schmitt from his colleagues was not his positions but, rather, his original lines of argument, his pointed hypotheses, and his brilliant use of language. Besides the monographs he published at breathtaking speed, it was in Bonn that Schmitt wrote the first essays in which he cautiously inched his way toward the controversies of the day in constitutional law. The theoretical foundation for his legal opinions was his criticism of legal positivism, liberalism, and parliamentarism laid out in his monographs.

These initial practical opinions interpreting the constitution include essays on Schmitt's reading of Article 25 of the Weimar Constitution on the right of the President of the Reich to dissolve the Reichstag; he granted the President very extensive competence in this matter (see Schmitt 1924b). They also include his interpretation of the dictatorial power of the President of the Reich in accordance with Article 48 of the constitution (see Schmitt 1926c). Here, Schmitt advocated strictly regulating the dictatorial power of the President. The social democrats were also calling for such a law to prevent abuse of Article 48. Both the right-wing conservative President of the Reich Paul von Hindenburg and the Reichswehr (the armed forces) saw such a law as an unwelcome limitation of his presidential power and did everything in their power to ensure that nothing of the sort was adopted. A few years later, Schmitt revised his position on limiting presidential power.[2]

Although Schmitt cannot be declared a supporter of the right-wing parties on the basis of his initial legal opinions, he was unequivocally perceived to be on the political right during his time in Bonn, even if various facets of his positions were difficult to pin down.

---

1 Most of the biographical information in this section is based on Neumann (2015) and Mehring (2014a and 2022a).

2 See Chapter 3, p. 91–96.

He was part of the complex discourse of right-wing Catholic intellectuals vacillating between authoritarianism and anarchism. He published some of his essays in the Catholic magazine *Hochland* and considered the French antisemitic magazine *Action française* essential reading. He quoted the anti-Enlightenment theoreticians of the Catholic counterrevolution Joseph de Maistre, Louis de Bonald, and, above all, Donoso Cortés to support his political views, and had already positioned them to refute the arguments presented by the allegedly shallow authors of the Romantic period in his book *Political Romanticism* (see Schmitt 1919). In the final chapter of *Political Theology*, written in 1922, he drew on these theoreticians of the Catholic counterrevolution and their plea for a dictatorship to develop a methodological distinction between a type of legal thought oriented toward norms and one based on fundamental decisions (see Schmitt 1922, 53–65). In his book on parliamentary democracy, he praised Cortés's vision of a "bloody, definitive, destructive, decisive battle" (Schmitt 1923a, 69) against both the liberal and the revolutionary actors of the Enlightenment. It was abundantly clear to all his readers at the time that Schmitt was waging a war on two fronts: against liberalism and against Marxism.

Schmitt did not commit himself to a particular political party during his Bonn years. In 1923, he took a few days to consider whether to accept the offer from the Catholic Zentrumspartei (Center Party) to run for the Reichstag. In the end, he turned it down because he felt that the party's support for the Weimar Constitution and parliamentarism was too strong (see Mehring 2014a, 144). He held no sympathies for Adolf Hitler's small extreme right-wing party, either. Like many nationalist-minded university professors of his time, his views were most closely aligned with the Deutschnationale Volkspartei (DNVP), which was *völkisch* (of the *Volk*, chauvinistic-nationalistic, antisemitic; see Glossary), right-wing, and hostile to the Weimar Republic. Looking abroad, he was fascinated by Italian fascist Benito Mussolini's dictatorship and traveled to Italy multiple times for this reason.

Despite all his successes, which resonated far beyond Bonn, Schmitt felt himself to be "in a terrible state" and an "outsider"[3] vis-à-vis his Bonn colleagues—as he wrote to Smend, with whom he was still on friendly terms at the time. As a Catholic who disagreed with the church's prevailing dogma, and as a social climber, he was never really comfortable with most of his peers. He maintained a closer relationship only with theologian Erik Peterson, who insisted on the authoritative and dogmatic core of Christianity and advocated for Catholicism to make a fundamentalist about-face. Schmitt also agreed with Peterson about the eschatological role of Judaism as the "delayer," which guaranteed the existence of the Christian Church in the interim.[4] Schmitt felt he was both ignored socially and intellectually superior. His successes notwithstanding, he was extremely sensitive to criticism even in his years in Bonn. When critical comments on his work on parliamentarism had been published in a number of journals, he wrote to Smend that he felt "isolated in my profession" and complained that "4 Jews against one Christian—pounce on me in all the journals."[5] Unlike during the short time he had spent in Greifswald, Schmitt

3 Letter from Carl Schmitt to Rudolf Smend dated 11 September 1928 (Schmitt 2010, 76).

4 See Meier (1994).

5 Letter from Carl Schmitt to Rudolf Smend dated 21 May 1925 (Schmitt 2010, 44).

felt beleaguered by what he suspected to be Jewish old boy networks at the University of Bonn, as evidenced by his diary entries from 1925 to 1928.

The Bonn years marked a watershed moment in Schmitt's private life. He had met burlesque dancer Pauline Carita Dorotić in Düsseldorf in 1912; she had led him to believe that she was descended from Croatian nobility. They married in 1915,[6] and thereafter Schmitt proudly signed his publications "Schmitt-Dorotić." It was only some years later that Schmitt realized he had been duped by an impostor who had fabricated her name, age, and descent. Upon taking his position at the University of Bonn, he dropped her name and began to pursue an annulment of the marriage on the grounds of fraud in the strictest criminal sense. Only an annulment would have enabled Schmitt, who had grown up in a strict Catholic milieu, to marry again in the Catholic Church. The protracted divorce proceedings under civil law were not concluded until the spring of 1924, and Schmitt then continued to seek an annulment from the Catholic Church, but unsuccessfully. Schmitt needed some Serbo-Croatian documents translated for the civil proceedings to prove his wife's criminal intent and approached a young student from Serbia who, like him, had arrived in Bonn in 1922. Schmitt and the 19-year-old student Duška Todorović soon became a couple (see Tielke 2020, 15–18). They married in 1926. From the perspective of the Catholic Church, Schmitt was now living in concubinage and so was excommunicated.

The many students who flocked to study with Schmitt were attracted by his charisma. His classes had already been popular when he taught in Munich and Greifswald. He was generally seen as a captivating teacher and a lenient examiner. In Bonn, he quickly managed to create a distinct school of intellectual thought. Those who met him when they were young wrote time and again in their later letters or their published memoirs that meeting him in person was nothing less than an epiphany.

Schmitt personally selected a group of students to participate in a weekly seminar with talks on current topics, and he invited guest speakers. He also presented initial versions of his own deliberations for lectures and essays and encouraged open critical discussions that helped him improve his hypotheses and writings. In 1940, he wrote in retrospect about his famous article "The Concept of the Political" that it only developed "in my seminars in Bonn in 1925 and 1926" (Schmitt 2019, 123). Looking back 50 years later, Ernst Rudolf Huber appreciated in a celebratory speech how all the issues the participants were interested in were discussed "candidly and passionately in an ongoing conversation with Schmitt and each other" (Huber 1980, 131). Schmitt's seminar was a laboratory not only for his students' theses and dissertations but also for his own works.

A number of well-known and influential jurists were trained in Schmitt's Bonn seminar, including Ernst Forsthoff, Ernst Rudolf Huber, Werner Weber, and Hans Barion, who, like most of the attendees, were on the political right wing. Ernst Friesenhahn, who sympathized with the Catholic Center Party, became Schmitt's assistant in 1926. He was the only one of Schmitt's other students Kirchheimer became friends with. Another participant in the seminar was Waldemar Gurian, who had been born a Russian Jew in St. Petersburg and had converted to Catholicism. Schmitt's second wife Duška, the same age as his students, attended occasionally, the only woman to do so. All the people mentioned

6 On Schmitt's first marriage, see Mehring (2014a, 41–60).

here played a role in the eventful relationship between Schmitt and Otto Kirchheimer over the following forty years.

2. Kirchheimer's early studies and his decision to study with Schmitt

When Otto Kirchheimer became part of Schmitt's circle in Bonn, he stood out as a Jew and a socialist in this heavily Catholic milieu with right-wing leanings. Kirchheimer was born in 1905 as his affluent parents' sixth child, seventeen years after his next youngest brother.[7] Little is known today about his childhood except that as a late-born child, he enjoyed the full attention of his parents and the domestic staff. Yet this happy phase lasted only a few years. Even before he began school in 1911, his mother died at 49, and his purportedly less patient older sisters took on the task of raising him. After elementary school, Otto Kirchheimer attended the *Städtisches Gymnasium* (municipal academic-track high school) in Heilbronn. In April 1918, aged 12, he switched to the academic track of the Pädagogium Neuenheim-Heidelberg, a private school in Heidelberg, because his father had fallen ill.[8] His father died just one year later and left his children a considerable estate. Whereas his older brothers used their shares for their own businesses, Otto Kirchheimer's inheritance was held in trust to finance his education. Although his legal guardian was his uncle Ludwig Rosenthal from Nuremberg, his brothers exercised these rights until Otto came of age in late 1926. In particular, his brother Friedrich (Fritz), who was 17 years his senior and advancing his career in the Heilbronn branch of the Dresdner Bank, considered himself responsible for this. Otto Kirchheimer later explained his increasing alienation from his brothers by pointing out how much he had suffered under their patronization and bullying.

Kirchheimer attended the private school for five years up until the summer of 1923 and began to develop a special interest in politics, literature, and history. After the 1919 revolution, as a young student, he met older students who sympathized with the communists and the leftist socialists. His involvement with Die Kameraden, the German-Jewish branch of the Wandervogel movement, began during this time.[9] Die Kameraden had been established as a nationwide organization in 1919 because many of the other Wandervogel groups discriminated against Jews, some even refusing to accept them into their own ranks. Open to Jewish as well as non-Jewish youths and students, Die Kameraden was strictly anti-Zionist and had several thousand active members in various locations across Germany. Equal rights for all members, coeducation, promoting special communal experiences, and a love of nature were among its principles (see Trefz 1997). Kirchheimer participated in events and hikes organized by Die Kameraden on a regular basis and became an eloquent speaker propounding socialist ideas in the group.

7 The biographical information in this section is based on Anschel (1990), Kirchheimer-Grossman (2010), Buchstein (2017a), and a number of unpublished documents mentioned in the footnotes.

8 At the time, the school year in Germany began after Easter, following a Christian tradition, not after the summer vacation.

9 Originating in the early twentieth century in Germany, Wandervogel was a movement mostly of school and university students who, inspired by Romanticism, sought to liberate themselves from the constraints of modern industrial society on hikes in the great outdoors.

He had to move to a new school in 1923 to be able to complete his *Abitur* (academic-track high school degree). He passed the entry examination of the *Städtische Realgymnasium* in Ettenheim near Lahr (Baden) and spent the 1923/24 academic year there, completing his *Abitur* in March 1924. According to the school's register of grades, Kirchheimer's grades were mixed; he did particularly well in the literary subjects and less so in the natural sciences.[10] After obtaining his brothers' permission to study law, Otto Kirchheimer took up his studies at the university in Münster in the summer semester of 1924.[11] He did not study law as he had promised his brothers, however, but enrolled in the Faculty of Philosophy instead. He had decided on Münster to attend the lectures of philosophy professor Karl Vorländer, a neo-Kantian whose writings on socialism and Marxism he had already read as a high school student. Kirchheimer financed his studies with his inheritance, managed by his brothers. He left them in the dark about his departure from the studies they had agreed on and, on Vorländer's advice, subsequently switched to the field of law.

As a university student, Kirchheimer was politically active in the local Sozialistischer Studentenverband (Socialist Students' Association). At this time, he had already joined the SPD, the German Social Democratic Party.[12] He remained active with Die Kameraden. Eugene Anschel, his closest friend during his student years, reported in his memoirs that when he began his studies at 18, Kirchheimer proudly called himself a Marxist and tried to get his fellow members of Die Kameraden excited about discussing philosophical problems on their long hikes. He also described how Kirchheimer identified with the leftist wing of the SPD in the political discussions and was familiar with various contemporary socialist and communist theories such as those of Max Adler, Rosa Luxemburg, Paul Levi, and Lenin. He had constantly read aloud on train rides, either from the newspaper or from philosophical texts by Plato and other classical philosophers.[13] Kirchheimer only spent one semester in Münster. Besides Vorländer's classes, he also attended ancient historian Friedrich Münzer's lectures. At the time, Vorländer was working on a comprehensive history of the theory of the state spanning the period from the Renaissance to Lenin (see Vorländer 1926) and was already lecturing on the subject—which piqued Kirchheimer's intense interest.

Following the advice of some of his socialist friends, Kirchheimer switched to the University of Cologne, a Catholic reform university, in the fall of 1924. He enrolled in *Staats- und Rechtswissenschaften* (law) for two semesters, but then spent most of his time

---

10 The information on Kirchheimer's career at high school is based on the research findings of Reinhard Mehring (see Mehring 2014b, 39–41).

11 The information on Kirchheimer's studies is based on: Otto Kirchheimer, *Lebenslauf* (27 December 1927). Universität Bonn, Archiv der Juristischen Fakultät, Prüfungsakte Otto Kirchheimer, Promotionen 521/28, No. 500–524.

12 According to his daughter Hanna Kirchheimer-Grossman in a personal conversation on 12 September 2021. The exact date when he joined the party is not known.

13 Anschel also recounted anecdotes from the time they spent in the German-Jewish hiking club in his memoirs: during a hiking trip with Die Kameraden in 1924, Kirchheimer suggested that everyone put the food they had brought with them on a large table and then distribute it following the communist formula "to everyone according to their needs." According to Anschel, this went terribly wrong (see Anschel 1990, 79–80).

studying with sociologist Max Scheler instead of attending law lectures. Scheler was famous for his theories of capitalist culture and his sociology of knowledge and ideology (see Frings 1997). As a member of the Sozialistischer Studentenverband, Kirchheimer quickly made more friends with like-minded students in Cologne. Anschel reported that when he visited Kirchheimer's place in Cologne, he found a portrait of Lenin on the bookshelf. Asked about its political meaning, Kirchheimer had responded that he admired Lenin as a politician impassioned with a strong will but rejected his belief system and the Russian communists' ideology (see Anschel 1990, 83).

Kirchheimer's relationship with his future wife Hilde Rosenfeld also began during his year in Cologne. He had met her by chance on a train while traveling with Anschel. She was a law student at the nearby University of Bonn. It was not so much law that immediately brought them close but, rather, their political discussions. Hilde Rosenfeld held strong sympathies for the KPD, the German Communist Party, and her political preferences oscillated between the SPD and the KPD. Otto Kirchheimer was proud of having won her back for political work in the SPD after spending long nights discussing politics with her. To Kirchheimer, his relationship with Hilde Rosenfeld also meant direct access to the leading figures of the leftist wing of the SPD. She was the daughter of Kurt Rosenfeld, Prussian Minister of Justice from November 1918 to January 1919 and a member of the left wing of the social democratic faction of the Reichstag from 1920 onward. Rosenfeld had a colorful political past and was a celebrity in leftist socialist circles.[14] Along with Paul Levi, he had been Rosa Luxemburg's lawyer and one of her closest confidants for many years. Rosenfeld enjoyed a legendary reputation as the successful defense attorney for the Rote Hilfe, the magazine *Weltbühne*, and prominent authors such as Carl von Ossietzky and Kurt Tucholsky.[15]

On Scheler's advice, Kirchheimer enrolled in Berlin for the 1925/26 winter semester. He spent two terms there; the Rosenfelds helped him find accommodation in the western part of the city. He matriculated in law, enrolling in lectures and seminars taught by public law experts Rudolf Smend and Heinrich Triepel and criminal law expert Eduard Kohlrausch.[16] He also took advantage of what Berlin had to offer, attending lectures and discussions at the Deutsche Hochschule für Politik (DHfP), or German Academy for Politics, which was located opposite Berlin university in the center of the city across from Berlin castle.

Kirchheimer and Smend soon grew closer. Born in 1882, Smend had become a professor in Greifswald back in 1909, during the German Empire, and taught at the University of Berlin from 1922 on. His political sympathies were with the right-wing Deutschnationale Volkspartei (DNVP), which opposed the Weimar Republic, but he kept his distance from the extreme right wing. Despite his conservative political orientation, he quickly began

---

14 On Kurt Rosenfeld's biography, see Ladwig-Winters (2007, 247–249).

15 *Rote Hilfe* was a KPD organization to provide support and care for communists persecuted by the state. The *Weltbühne* was considered the most sharp-witted weekly of the leftist-socialist intelligentsia in the Weimar Republic.

16 Otto Kirchheimer, *Lebenslauf* (27 December 1927). Universität Bonn, Archiv der Juristischen Fakultät, Prüfungsakte Otto Kirchheimer, Promotion 1927/28, No. 500–524.

to appreciate Kirchheimer, a young leftist socialist. [17] During the year Kirchheimer was studying in Berlin, they had a number of discussions, and their political differences did not stand in the way of an amicable relationship.

It would be no exaggeration to describe Smend as Kirchheimer's first academic mentor. Kirchheimer became familiar with Smend's particular anti-positivist approach to legal theory and tried to combine it with the revolutionary Marxist Georg Lukács's critique of legal positivism (see Lukács 1923). At the time, Smend was completing his magnum opus *Verfassung und Verfassungsrecht* [Constitution and constitutional law], published in 1928. This book presented his theory of integration, which he is known for to this day. Smend promoted a kind of early cultural turn in constitutional law, examining the factors binding a state together.

At this time, Smend was a member of the DNVP but he referred to the social democratic legal scholar Hermann Heller multiple times in his theory. The methodological core of the theory of integration is a type of thought that seeks to resolve opposite positions within one another. With this approach, he opposed both Marxism and Carl Schmitt's theory (see Smend 1928).[18] It was Smend who suggested to Kirchheimer that he should follow his interests in the theory of the state with Schmitt in Bonn, and recommended him personally to Schmitt,[19] with whom Smend had an almost friendly relationship at the time.[20]

Kirchheimer switched to the University of Bonn in the fall of 1926. Relocating to Bonn also suited him very well in terms of his private life because Hilde Rosenfeld wanted to complete her studies in Bonn, too, and their relationship had become serious. Kirchheimer followed Smend's advice to read Schmitt's most famous writings in the summer before he moved to Bonn. Schmitt's books, with their diverse topics and literary references, must have seemed like a treasure trove to a young left-wing intellectual like Kirchheimer.

According to political scientist Wilhelm Hennis's report about a conversation he had with Kirchheimer about his decision thirty years later, Kirchheimer was impressed by Schmitt's broad knowledge and his polemic style and he wanted to learn as much as possible to apply to left-wing politics from Schmitt, his theory of dictatorship, and his pointed criticism of parliamentary democracy.[21] On the basis of Schmitt's book *Dictatorship*, in which he had claimed that he intended to update the counterrevolutionary theory of dictatorship to provide a response to the Marxist theory of dictatorship of the proletariat, Kirchheimer viewed Schmitt as a kind of "Lenin of the bourgeoisie."[22] It is

17 Wilhelm Hennis, who had studied with Smend and was a friend of Kirchheimer's from the 1950s, recounted this in a conversation with the author on 26 September 2009.

18 On Smend's legal theory, see Korioth (1990).

19 Wilhelm Hennis, a political scientist and one of Smend's most famous students, in a conversation with the author on 26 September 2009.

20 The volatile relationship between Schmitt and Smend is documented in their correspondence (see Schmitt and Smend 2011).

21 Wilhelm Hennis in a conversation with the author on 26 September 2009.

22 John. H. Herz used this expression in a conversation with the author on 15 November 1985. On Schmitt's and Lenin's structurally similar views of politics, power, state, and dictatorship, see Bolsinger (2001).

difficult, however, to judge whether this second-hand explanation is correct because it may have been a retrospective rationalization.

3. The famous professor and his student

Kirchheimer was only twenty-one when he moved to Bonn in late September 1926. He contacted Schmitt shortly after his arrival with a letter from Smend in hand. Schmitt mentioned his new student's courtesy call on 11 October in his diary: "The student Kirchheimer came and enrolled in the seminar."[23] Kirchheimer studied with Schmitt for three semesters. In the 1926/27 winter semester, Schmitt ran a seminar entitled *Staatstheorien* (Theories of the state), lectured on *Völkerrecht* (International law), and taught *Verwaltungsrechtliche Übungen* (Tutorials in administrative law). Kirchheimer attended all three classes. Schmitt's lectures were crowded but his seminar was generally attended by no more than ten students, with doctoral candidates forming the core. Kirchheimer, the recent arrival from Berlin, was the only one whose political orientation was clearly on the left. He was also the only Jewish student in Schmitt's inner circle.

During those years, Schmitt regularly kept a diary in which he praised Kirchheimer multiple times. For example, he noted on 2 February 1927: "Good seminar [...], [student Heinrich] Oberheid and Kirchheimer speak very well."[24] In the summer semester 1927, Schmitt offered a seminar on *Einheit und Undurchdringlichkeit des Staates* (Unity and unimpeachability of the state) and lectured on *Politik – Allgemeine Staatslehre* (Politics – General theory of the state) and *Deutsches Reichs- und Landesstaatsrecht* (Public law of the German Reich and the *Länder*). In the 1927/28 winter semester, he taught a *Staatsphilosophisches Seminar* (Seminar on the philosophy of the state), and again gave lectures on *Völkerrecht* (International law) and *Allgemeine Staatslehre* (General theory of the state). Kirchheimer took all of these classes.

While his girlfriend Hilde Rosenfeld focused her studies on criminal and civil law and did not participate in Schmitt's seminars, Kirchheimer was directly admitted to Schmitt's inner circle in his first semester in Bonn. He quickly impressed the group with his intelligent and pointed statements and became one of the undisputed "stars in the seminar" (Mehring 2014a, 203). Yet he made just a single friend in the seminar, Ernst Friesenhahn, who had a liberal outlook and sympathized with the Center Party.[25] Friesenhahn became Schmitt's assistant in 1926. The position involved grading the written exams for Schmitt's classes, finding literature for him, and occasionally teaching his classes as a substitute. Beginning in the autumn of 1927, he had another major task, namely supporting Schmitt in correcting the proofs for his magisterial book *Constitutional Theory*. Schmitt worked so feverishly on correcting the proofs and adding text to them in the early months of 1928 that Friesenhahn could not keep up with him. After

23 Carl Schmitt, diary entry of 11 October 1926 (Schmitt 2018, 97).

24 Carl Schmitt, diary entry of 2 February 1927 (Schmitt 2018, 118).

25 For a biographical sketch of Friesenhahn and his relationship to Schmitt, see Meyer (2018).

Kirchheimer had completed his doctoral dissertation, he helped Friesenhahn with the corrections,[26] learning about Schmitt's rapid way of working in the process.

Schmitt was happy to spend time with his students after class. He was thirty-five when he arrived in Bonn and among the younger professors at the university. He liked to take his students out for walks to discuss their work and academic plans and also enjoyed going out with them after his seminar. He invited them to his place for discussions over beer and wine; sometimes they even got drunk together. Schmitt repeatedly made positive comments after those discussions about Kirchheimer in his diary: "nice conversation with Kirchheimer" about *ius belli*,[27] "nice, especially Kirchheimer,"[28] "Kirchheimer was intelligent and nice."[29] Sometimes he invited Kirchheimer to spend some time with himself and his visitors, for example to walk around town with himself and Waldemar Gurian.[30] Or he met Kirchheimer together with other students such as Werner Weber in the library.[31] The two of them had shorter or longer conversations on an almost daily basis, either in Schmitt's office or in the university seminar room, which housed part of the library for the students. The notes in Schmitt's diary evidence that he valued Kirchheimer as a youthful and stimulating discussion partner even though his political views were diametrically opposed to his own. This seemed to make debating with Kirchheimer all the more alluring and interesting. Schmitt also supported him early on. His diary mentions in a note dated March 1927 that he had arranged for an article by Kirchheimer to be printed.[32]

No other doctoral candidate had a presence comparable to Kirchheimer's in Schmitt's diary during his last two years in Bonn. He had become Schmitt's favorite, his "prize student" (Anschel 1990, 85). To Kirchheimer, his relationship with his mentor in Bonn had an even stronger emotional component. Eugene Anschel, who had been friends with Kirchheimer since they were adolescents, reported in his private memoirs that Kirchheimer had a sink-or-swim attitude and did not normally help other students if they were having trouble with their studies, believing they had to manage on their own (see Anschel 1990, 84). He was driven all the more by Schmitt's approval and persuaded Anschel and other friends of his to study with Schmitt, too. Anschel failed his oral examination with Schmitt and was also appalled by his "military-minded" glorification of the Germans and his rejection of the "pragmatic British trade attitude." To Kirchheimer, whose political views were significantly further to the left than Anschel's, this was no reason to reject Schmitt. John H. Herz, another of Kirchheimer's longtime friends, interpreted the emotional component psychologically in retrospect: Schmitt had been a kind of "father substitute" for Kirchheimer (Herz 1989, 12). Herz even used Sigmund Freud's term "patricide"

26 Information from Ossip K. Flechtheim in a conversation with the author on 13 February 1988.

27 Carl Schmitt, diary entry of 2 June 1927 (Schmitt 2018, 143).

28 Carl Schmitt, diary entry of 23 June 1927 (Schmitt 2018, 148).

29 Carl Schmitt, diary entry of 30 June 1927 (Schmitt 2018, 149).

30 Carl Schmitt, diary entry of 12 September 1927 (Schmitt 2018, 162).

31 Carl Schmitt, diary entry of 14 September 1927 (Schmitt 2018, 163).

32 Carl Schmitt, diary entry of 30 March 1927 (Schmitt 2018, 130). Unfortunately, Schmitt's note in his diary is vague. No article written by Kirchheimer in 1927 could be found in any of the journals or magazines Schmitt had close connections to.

for the nothing less than obsessive way in which Kirchheimer conducted his disputes with Schmitt.[33]

Carl Schmitt generally set his doctoral candidates' dissertation topics. Their stimulating conversations during the "ambulatory office hours" when they went walking together inspired Schmitt to assign Kirchheimer a comparison between the theories of the state of socialism and Soviet communism. Kirchheimer consented enthusiastically.[34] He saw it as an opportunity to define his own position in terms of political theory more precisely between communists, social democrats, and leftist socialists; Schmitt in turn hoped Kirchheimer would critique Bolshevism. Kirchheimer began the writing phase of the dissertation in the summer break of 1927. He submitted the work to the Law Faculty six months later, on 27 December 1927.

4. Evaluating Kirchheimer's dissertation

The title of Kirchheimer's dissertation was *The Socialist and Bolshevik Theory of the State*.[35] The original version of the dissertation has been lost to this day; it is nowhere to be seen in Kirchheimer's or Schmitt's estate or in the files of the Bonn faculty or in the library of any other university. Kirchheimer was not required to submit the dissertation as a printed book to the faculty in order to receive his doctorate. He had applied to the faculty for permission to submit 120 copies of an essay which would be published in the *Zeitschrift für Politik* instead of the printed version of the entire dissertation. "The essay represents a summary of the findings of my dissertation," Kirchheimer wrote in the application to the faculty.[36] Schmitt had already consented to this procedure in advance: "The enclosed essay is a condensed summary of the dissertation and is of particular scientific interest."[37] It is worth taking a closer look at it as a starting point to ascertain how Kirchheimer dealt with Schmitt's theory.

Kirchheimer's essay "The Socialist and Bolshevik Theory of the State"[38] impresses readers in the original German version not least because of its lively style and its rhetorical exaggerations, which do not really come to their full effect in the otherwise very good English translation by John H. Paasche. In some places, his wording displays great similarities to Schmitt's language, also and sometimes precisely in passages where he clearly differs from Schmitt in substance. Kirchheimer begins his article by criticizing the "insufficiently political orientation of bourgeois liberalism" (3). He accuses liberalism

33 John H. Herz in a conversation with the author on 15 November 1985.

34 Ossip K. Flechtheim recounted this in a conversation with the author on 13 February 1988.

35 Letter from Otto Kirchheimer to Dean Heinrich Göppert dated 27 December 1927. Universität Bonn, Archiv der Juristischen Fakultät, Prüfungsakte Otto Kirchheimer, Promotion 1927/28, No. 500–524.

36 Letter from Otto Kirchheimer to the Dean of the Faculty of Law, University of Bonn, Heinrich Göppert dated 2 March 1928. Universität Bonn, Archiv der Juristischen Fakultät, Prüfungsakte Otto Kirchheimer, Promotion 1927/28, No. 500–524.

37 Letter from Schmitt to the Dean of the Faculty of Law, University of Bonn dated · March 1928. Universität Bonn, Archiv der Juristischen Fakultät, Prüfungsakte Otto Kirchheimer, Promotion 1927/28, No. 500–524.

38 See Kirchheimer (1928a). The following page numbers refer to this article.

of trusting too naively in constitutionalism and in the notion of the bourgeois *Rechtsstaat* (the bourgeois state under the rule of law) when struggling against the feudal powers in the early nineteenth century. In the meantime, the working class had matured into a relevant political factor. Owing to their "common front against feudal semi-absolutism" (3), bourgeois liberalism and the working class had entered into a closer relationship in the final third of the nineteenth century, which, in Kirchheimer's view, had left its mark on the political identity of Western European socialists that could still be observed. This historic alliance had only come apart when universal and equal suffrage had been established because then the democratic principles were being directed against the social strata supporting liberalism itself. These differences became trenchant in terms of the various conceptual interpretations of democracy. First of all, he claims, democracy meant the political "participation of all individuals" (4) in a very general sense.

For his further conceptual differentiations, Kirchheimer borrows from the terminology of Max Adler, an author on the revolutionary left wing of the Austro-Marxists, which Schmitt had commended. He also borrows Adler's methodology. Adler had given his book *The Marxist Conception of the State* the subtitle *A Contribution to the Differentiation of the Sociological and Juristic Method*. Kirchheimer agrees that all juristic forms were an expression of societal class relations. Adler distinguishes between "political" and "social" democracy.[39] Whereas "merely political democracy" granted all citizens the same political participation rights in principle but was otherwise based on the social heterogeneity of a capitalist class society, he believed only "social democracy" in a classless society was "true democracy." In Adler's opinion, since the character of contemporary bourgeois democracy was based on societal class, it would not be wrong to describe it as a dictatorship of the bourgeoisie. Referring to Marx and Luxemburg as well as Schmitt's book on dictatorship, Adler argues that Marx's formula of the "dictatorship of the proletariat" should be adopted again in the language used by theorists of Austrian and German social democracy.

In Adler's view, Schmitt had provided an "extremely lucid analysis" (Adler 1922, 165) of the problem of dictatorship. Adler praised Schmitt's non-formalist treatment of the problem of dictatorship in particular. Insofar as Schmitt had defined dictatorship as a "concrete exception," he had illuminated the fact that the substance of every dictatorship depends on the existence of the enemy it professed to eliminate.[40] Max Adler's writings and notably his terminological differentiation between the two forms of democracy were embraced by the leftist circles of the Weimar Republic's Young Socialists in the *Jungsozialistische Blätter* and the milieu of the magazine *Klassenkampf–Sozialistische Politik und Wirtschaft*.[41]

Kirchheimer takes up this differentiation, too, but develops his own terms for it: "formal democracy" and "value-based democracy" (5). Following Adler, he considers formal democracy under liberalism to be the condition of general political equal rights, which sees "in the absence of values a value in itself" (5). During a certain phase of the

39 See Adler (1922, 83–94) and Adler (1926).

40 On Adler and Schmitt on dictatorship, see Ananiadas (1999, 121–128).

41 On Adler's theory of democracy and his major influence on the left-socialist theory formation of the day, see Pfabigan (1982), and Bavaj (2005, 201–218).

class struggle, formal democracy was the political form in which the opposing social forces would organize in groups until they had reached a historic decision. In contrast, a democracy of values was based on all citizens recognizing "common values" (4), on a "certain understanding of social homogeneity" (6) going beyond equal rights in the merely political sense. Kirchheimer also follows Adler with respect to his hypothesis, based on Marx's deliberations in his *Eighteenth Brumaire of Louis Bonaparte*, that formal democracy was not stable. It functioned only as long as there was an approximate balance between the social classes struggling against each other and a "tacit agreement" (6) between them resulting from that approximate balance to let the outcome of "each election decide" (6) who would form the government that particular time. Since formal democracy was based on a compromise, all social groups would attempt to secure their visions of social policy by having them included in the constitution.

At this point, Kirchheimer addresses deliberations of his contemporaries and reform socialists Heinrich Cunow and Karl Renner, who advocated for closely circumscribed legal limitations on governmental power to give a bourgeois government as little room as possible to take action opposing the interests of the working class. The risk of this happening was the reason why these socialist theorists had also opposed the *Freirecht*[42] school of legal thought and had advocated strictly binding the judiciary to legal positivism. In connection with the strategy for legal policy proposed by Cunow and Renner, Kirchheimer used the German term *Verrechtlichung* (juridification), which he understood to be the expansion of the legal codification of state administrative action as well as an attempt to "avoid [...] all factual decisions" (7).

The German term *Verrechtlichung* was coined in 1919 by the social democratic legal theorist Hugo Sinzheimer in the context of the *Räteverfassung* (Council Constitution); Kirchheimer expanded it to cover all areas of the law. Only when the juridification of social relationships had become widespread would the "true epoch of the *Rechtsstaat*" (the rule of law) (7) have dawned. Then the value of a decision would no longer lie in the factual reasons given for it but exclusively in the fact that it was a decision based on the law. Kirchheimer critically commented on this development, stating that this kind of state "lives off the law; yet it is no longer law (*Recht*), it is only a legal mechanism, so that those who think they are guiding the affairs of the state actually wield only a legal machinery which claims their attention in the same way a machinist is tied down by the apparatus he serves" (8).

Against the background of this general characterization of contemporary mass democracy under the *Rechtsstaat* in terms of legal policy, Kirchheimer presented two theories of the state, namely socialism and Bolshevism. He did not present them in a systematically organized fashion but meandered between the two theories and various topoi. Nor did Kirchheimer separate his presentation from his critique of the theories but reconstructed them from a critical perspective from the outset. To characterize the Russian and Soviet doctrines and circumstances, Kirchheimer drew on the relevant comments by Marx and Engels on Russia (to the extent these were known at the time),

42 In the early twentieth century, the *Freirecht* school of legal thought (Hermann Kantorowicz, Ernst Fuchs) demanded discretionary power for judges in order to infuse progressive ideas into a reactionary legal system.

statements by Lenin and Stalin, and older Menshevik literature that had been translated. In his presentation of the socialist theory of the state of the Second International, he mostly referenced the writings of Russian social democrat Plekhanov, French socialist Jean Jaurès, and Karl Kautsky, the most important theoretician of German social democracy of his time. Kirchheimer accused the social democrats of paying homage to a naive theory of twofold progress (10) according to which progress in capitalist economic development almost automatically also entailed progress toward humanism in the development of humankind, for which reason political conflicts could be dealt with in a more civilized way. Kirchheimer claimed that this theory fueled the illusion of a peaceful majority of socialist forces in the existing formal democracy, and its logical consequence would be to abandon the concept of dictatorship in the name of the cause of socialism.

According to Kirchheimer, however, Marx had never espoused such a humanist theory. The true kernel of Marx's theory is the doctrine of class struggle. One of Marx's achievements as a political theorist had been his acknowledgment of the existential intensity of class-based enmity. In Russia, it had been Lenin who had effectively rejected ideas like this, replacing them with a theory of relentless class struggle which did not recognize any supra-class morals. Kirchheimer saw parallels in these hypotheses of Lenin's both to Nikolai Berdyaev's Russian Orthodox philosophy of religion with its pointed emphasis of the relentless struggle between Christ and the Antichrist and to socialist theorist Georges Sorel's celebration of political violence and myth.[43] Similarly to Carl Schmitt in his chapter on the irrationalist theories of direct use of force in his work *The Crisis of Parliamentary Democracy* (see Schmitt 1923a, 65–76),[44] Kirchheimer recounted the hypotheses put forward by Sorel and Lenin in a way that revealed his fascination for these two propagandists of relentless political action. He was impressed by the Bolsheviks' ability to evoke the myth of glorifying the world revolution, claiming that it unveils and finally overcomes the enmity between the classes. Soviet mythmaking appeared to be more effective than reformist strategies.

Kirchheimer devoted particular attention to the Bolshevik concept of dictatorship following Schmitt's terminological differentiation between commissarial and sovereign dictatorship in the fourth chapter of his book *Dictatorship* (see Schmitt 1921, 112–131). The Soviet leadership had fully understood the centrality of the "principle of emergency" (14) for their political goals. Kirchheimer classified Lenin's understanding of dictatorship under the label of sovereign dictatorship since it sought to prepare the ground for the establishment of a socialist state of social equality using all means available and in a targeted fashion. Kirchheimer's creativity here in drawing on deliberations he was familiar with from Rudolf Smend's theory of integration is striking. Although Smend did not publish his comprehensive theory of integration until the autumn of 1928 in his book *Verfassung und Verfassungsrecht* [Constitution and constitutional law], preliminary deliberations on the matter can be identified in earlier publications; Kirchheimer had learned about the fundamentals of this theory in the classes he had taken with Smend in Berlin in 1926 (see Smend 1923, 84–86).

43 See Berdjajew (1924) and Sorel (1906).

44 On Schmitt's theory of the myth, see, in more detail, the discussion in the next chapter.

The sovereign dictatorship of the Bolsheviks would change the status of the law within the state by breaking with the liberal concept of the judiciary as a neutral third party ranked above the disputing parties, instead issuing rulings exclusively in line with Bolshevik values, thus attempting to integrate the lower strata of the population into the new state. The status of elections would change in the new state, too; instead of the liberal notion of keeping voting secret, open and unconcealed voting would be reshaped into a factor integrating the state. The everyday practice of the legal system and holding elections were also two key mechanisms of state integration in Smend's constitutional theory.[45]

The Bolshevik theory also changed the doctrine of international law used by the sovereign dictatorship to define its relations to other states. Departing from the irreconcilability of class antagonisms, the Bolshevik power elite did not consider international law to be a law of peace but, rather, a law of ceasefire and, consequently, they were opposed to the Geneva League of Nations as a matter of principle.[46] That had implications for the concept of sovereignty, too. Whereas Kirchheimer believed that notions of sovereignty had been diminished in Western Europe—in political theory by authors such as Harold Laski and in political practice through manifold international contractual relations—"the USSR singled out a definite and well-known locus of sovereignty that is sensational when held against the present-day tendencies of masking and concealment" (20). This locus was the proletarian class, in whom sovereignty was now newly vested. By deeming the international proletariat to be officially granted sovereignty, the Bolshevist theory of the state was the first in political theory to propose the "intentional separation […] of the concepts of state and sovereignty" (20). In its political tendency, this sovereignty was not bound to the borders of nation-states but was universal, just like the proletarian class.

Finally, Kirchheimer raised the question once again whether Soviet Russia was actually to be characterized as a state. He answered his own question in the affirmative: unlike bourgeois democracy—from which the majority of social democrats hoped to be able to begin the peaceful transition toward socialism one day—the political system established by Lenin "restored the integrational character of law and elections" (21). The Soviet Union had succeeded in invigorating the political forces using the political myth of the world revolution. The formal democracies of the West were in a different situation. Although the mere shell of the state still existed there, the state had become something "less than itself, a mere legal mechanism" (21); its citizens' participation in and enthusiasm for the state was only just enough to support the theory of twofold progress (10), which in turn was also an option for exiting the bourgeois *Rechtsstaat*. Such a "state—which no longer is one," Kirchheimer wrote at the end of his essay, "can no longer have an enemy; for it lacks tangible forms of political expression" (21).

The polemical thrust of the essay was obvious: moderate social democrats' basic flaw was to agree to compromises with the bourgeoisie instead of taking up the fervent struggle for socialism. Kirchheimer viewed the class relationship between the capitalist property owners and the working class as irreconcilable enmity. The current-day balance of

45 See Smend (1928, 154–157 and 207–212).

46 On Kirchheimer's discussion of the international law, see Chapter 4.

the class forces was only temporary and precarious and would tip over to one side or the other sooner or later. The socialists were to be under no illusions about a gradual transition to socialism along a sedate administrative path. Its academic status notwithstanding, Schmitt's work was not the only—not even the predominant—influence on Kirchheimer's thought in his essay. It was an eminently political contribution to leftist political theory in which Kirchheimer had mixed elements of the theories by Adler, Smend, and Schmitt in a truly original way.

In addition to his written dissertation, Kirchheimer had to undergo an oral examination with Schmitt. He selected general theory of the state as his major and international law and law of criminal procedure as his two minors for the oral examination. On 14 February 1928, Schmitt noted in his diary that he had conducted Kirchheimer's First State Examination in Law and given him the grade "very good, with distinction"—only two other students of Schmitt's in Bonn received such an exceptionally high grade from him. He read Kirchheimer's dissertation on 19 February 1928 and submitted his report to the Faculty of Law at the University of Bonn the following day. Compared with what is customary today, the report is relatively brief. Its complete text reads as follows:

> There are too many hypotheses and ideas that were not expanded upon in the work. The following are to be mentioned as particularly interesting and scientifically valuable: the hypotheses on the structure of social balance of the modern industrial state and the statement that socialism nowadays encompasses a dual concept of progress (the "theory of twofold progress"). These are complemented by outstanding conceptual deliberations such as the differentiation of utopia and myth, the integrating function of the judiciary, etc. Almost every single one of these hypotheses and opinions—expanded upon and presented soberly and systematically—would have sufficed for a dissertation, whereas now, the reader's overall impression suffers from the overabundance of ideas that are not expanded upon. This is not to say that they are superficial or dilettantish aperçus; rather, this is simply a typical case of youthful productivity. In other words, I would not like to accuse the author of having too many ideas; instead, I would like to emphasize his doubtless very great scientific talent and his independent and valuable discussion of particularly current and important concepts (such as democracy, liberalism, parliamentarism, or socialism) which in my opinion justify assessing the work as excellent.[47]

No secondary report is to be found in the files; at the time, the secondary reviewer would often assent to the first reviewer's assessment simply by commenting "agreed." Kirchheimer defended his dissertation jointly with Werner Weber, another doctoral candidate of Schmitt's, in a two-hour session. After submitting 120 copies of the essay that had been published in the *Zeitschrift für Politik* to the faculty in Bonn, he received his doctoral title on 15 May 1928.[48]

---

47 Dissertation report written by Carl Schmitt, 19 February 1928. Universität Bonn, Archiv der Juristischen Fakultät, Prüfungsakte Otto Kirchheimer, Promotion 1927/28, No. 500–524.

48 In contrast to the correspondence, Schmitt's report, and the article published in the *Zeitschrift für Politik*, the doctoral certificate gives the title of the work as follows "*Zur Staatstheorie* […]" and not "*Zur Staatslehre des Sozialismus und Bolschewismus.*" The grade indicated on the doctoral certificate

One episode at the end of Kirchheimer and Schmitt's time together in Bonn was striking. On the evening of 25 February 1928, after the successful defense, Kirchheimer met Erik Peterson at Schmitt's house for a few glasses of wine to celebrate his exam. This time, Schmitt no longer reacted to the political disagreements between himself and Kirchheimer in a sympathetic and paternal manner. It might be that Kirchheimer expressed his criticism of the mostly rightist and radical rightist students in the Bonn circle around Schmitt more openly now that he had passed his exam or that he attacked Schmitt's political position more directly than before—for Schmitt, in any case, the get-together at his house ended on a sour note. For the first time, he wrote a negative comment about Kirchheimer in his diary: "Kirchheimer lacks any national sentiment, horrendous."[49]

5. Conclusion: Lessons from Bolshevism for Social Democrats

In the same year as Kirchheimer published his article, the SPD received almost 30% of the vote, its best election result since 1919. From the perspective of the SPD leadership, nothing seemed to stand in the way of the reformist dream of twofold progress. These optimistic expectations were vigorously rejected by the communist left, who denounced every social democratic policy success as that of agents of the capitalist system. The criticism of the party leadership's policies from the leftist wing of the SPD—including many Young Socialists as well as highly regarded members of the Reichstag such as Kurt Rosenfeld, Kirchheimer's partner's father—was more complicated. They had less trust in the political neutrality of the institution of the state and its bureaucracy, military, and judiciary. Kirchheimer's critique of the concept of the state held by the reformist party leadership of the SPD was an important contribution to the leftist debates of the day in that, building on Schmitt's theory of sovereignty, he doubted the permanence of the Weimar Republic's model of social balance, thus asserting that its existence was precarious. Instead, he reminded readers of the irreconcilability of the social class basis of the Weimar state.

It is hardly surprising that his first longer essay was quoted a few times in leftist journals such as *Klassenkampf* and *Sozialistische Tribüne* during the Weimar Republic. Yet it was Schmitt in particular who emphasized that the essay was a truly "noteworthy" contribution (Schmitt 1931b, 142) to the theory of the modern state derived from Marxist discussions. He repeated his praise in a number of publications and called it an outstanding analysis of the precarious social balance in the relationship between the bourgeoisie and the working class in modern industrial countries such as Germany.[50] To Schmitt, Kirchheimer's dissertation proved that the Marxists considered the current-day state and its constitution merely a transition period toward a better socialist future. Kirchheimer thus became a key witness for Schmitt's deep conviction that the stance of the left toward

---

is "very good," the best possible grade (Otto Kirchheimer's doctoral certificate; original, owned by Hanna Kirchheimer-Grossman).

49 Carl Schmitt, diary entry of 25 February 1928 (Schmitt 2018, 208).

50 See Schmitt (1929a, 99) and (1930c, 183).

the existing state was one of uncompromising enmity. Schmitt read Kirchheimer's hypotheses as confirming his own expectations of instability, which—in contrast to Kirchheimer—led him to seek authoritarian mechanisms for stabilization. After the Grand Coalition had collapsed in the summer of 1930, Schmitt even sent a copy of Kirchheimer's essay to his colleague Gerhard Anschütz. The leading liberal defender of the Weimar Constitution, however, reacted in a somewhat perplexed way, writing to Schmitt: "As so often with such writings from the camp of the youngest generation, I had the uneasy impression: everything is faltering nowadays, everything. Where will this lead?" [51]

Nevertheless—at the young age of 22, Kirchheimer had already made a name for himself in the Weimar debates on legal theory. In the following decades, his essay became part of the Marxist canon on state theory. More than fifty years later, Jürgen Habermas referred in his seminal *Theory of Communicative Action* to Kirchheimer's term *Verrechtlichung* (juridification) and turned it into a paradigmatic concept of critical theory to illustrate the "colonization of the lifeworld" (see Habermas 1981, 356–373). The piece by the young Kirchheimer has been interpreted by some in the secondary literature as an argument in favor of Bolshevism or at least as evidence of certain sympathies with the development in Soviet Russia.[52]

Yet Kirchheimer emphasizing the power and strength of Bolshevism must not be confused with him supporting it. After all, Kirchheimer saw no realistic political opportunity for Soviet Russian-style communism to take hold in Germany. In order to justify this view, he referenced letters Marx wrote in 1881 to his Russian translator Vera Zasulich about the specific features of the Russian Empire. These letters were first published in 1924 and triggered heated discussions among socialist and communist leftists at the time. Kirchheimer also made no secret of the fact that although he was impressed by the political power of the myth of the class struggle preached by Sorel, he agreed with the French ethnologist Lucien Lévy-Bruhl that the mythical consciousness was a pre-logical irrationalism. Such an irrationalism belonged to the emotional and spiritual world of "primitive peoples"[53] and could not detect any rational form of consciousness reconcilable with Marxism in any way, shape, or form. Kirchheimer was fascinated by Lenin's strong political will. But he rejected any mythical foundation of left-wing politics.

When Kirchheimer emphasized the strength of Bolshevism, this did not mean that he identified with it. However, the Bolsheviks taught the social democrats a lesson about the conditions for the success of left-wing politics. For him, this lesson was about the courage to formulate a socialist program that deliberately placed itself outside the political form of liberal democracy, which was fetishized by contemporary social democracy. Social democracy was to take on the courage for an active, decisive, and militant policy from the Bolsheviks.

51 The letter from Anschütz to Schmitt is quoted in Schmitt and Smend (2011, 85).

52 See Scheuerman (1994, 24–26), Bavaj (2007, 42), and Breuer (2012, 114).

53 Kirchheimer (1928a, 4), see also Lévy-Bruhl (1922, 94–97).

Chapter 3: Democracy in Disagreement (1928-1931)

In 1928, independently of one another, Kirchheimer and Schmitt moved from the small provincial town of Bonn to Berlin, the capital of the Reich at the time. They were both familiar with Berlin from their student days. Schmitt had studied law there from 1907 to 1909 and then Kirchheimer had done the same almost two decades later. Whereas Schmitt had to establish new social circles for himself in Berlin in 1928, Kirchheimer had political friends there as well as the Rosenfelds, his partner's family. When the two arrived in Berlin, Weimar democracy seemed more stable at last. The SPD was the strongest party by far after the May 1928 Reichstag elections and formed a Grand Coalition led by Chancellor Hermann Müller with the Catholic Center Party and the two liberal parties, the left-liberal DDP and the right-liberal Deutsche Volkspartei (DVP) or German People's Party. The coalition had a comfortable majority in the Reichstag and, initially, it was able to find compromises for various major reform projects. Adolf Hitler's NSDAP had remained marginal with support from only 2.5% of the electorate and, at this stage, the communists had lost votes, too. Yet stability was soon to prove elusive.

1. The changing political *Lage*

Only a year later, the political lull of 1928 was a thing of the past. The Grand Coalition went into a tailspin in the second half of 1929. Because they had had to take their parties' wishes into consideration, members of the government had not had much leeway for political compromises from the outset. Now the bourgeois parties, including the Center Party, were moving ever further to the right, making compromises with the SPD even more difficult. In the winter of 1929/30, the economic and financial crisis in Germany rapidly came to a head as a direct consequence of the global economic crisis following the New York stock market crash in October 1929. The previous bourgeois governments had used up all the Reich's financial reserves in 1926/27, leaving behind a barely concealed budget deficit and thus making a fundamental overhaul of the Reich's finances necessary. Nonetheless, the bourgeois parties categorically opposed any tax increases. The economic crisis exacerbated these conflicts, and Social Democratic Finance Minister Hilferding was forced to resign on 20 December 1929 after a new fiscal plan had failed. On the social democratic side, Hilferding was considered a guarantor of a socially just reform policy; Kirchheimer knew him, but not well, through a discussion group around the social democratic theory journal *Die Gesellschaft*. Hilferding's successor was a politician from the right-liberal DVP. The conflicts between the parties in the Grand Coalition reached a crisis in 1930. The key sticking point was financing unemployment insurance, which had only been introduced a few years previously. Influenced by industrial and agricultural interest groups, the DVP refused to allow higher financial burdens on businesses. The SPD called for increasing unemployment insurance contributions in order to raise the necessary funds. Leading politicians of the DVP and the Center Party, which was moving to the right, were already aware that President Paul von Hindenburg and his advisors intended to remove the SPD from the government as soon as the law on the Young Plan, planned to regulate the Reich's debts following the Versailles Peace Treaty was adopted by the Reichstag on 12 March 1930. Then the SPD no longer had any opportunities to find further compromises for funding unemployment insurance. The Grand Coalition collapsed on 27 March 1930.

In this new political *Lage*, the differences between Kirchheimer and Schmitt now became more pronounced. To understand what questions and issues sparked their theoretical conflicts from then on, we need to look at the political events that followed. President of the Reich Hindenburg appointed Heinrich Brüning, a politician from the right wing of the Center Party, as the new Chancellor in early April 1930. Brüning was installed to govern against the Social Democrats and thus without the parliament, if necessary, using emergency decrees as provided for in Article 48 of the Weimar Constitution. This is precisely what he did and he pursued an all-out policy of austerity. When the Reichstag refused to agree to Brüning's budget, as expected, the President of the Reich implemented it nonetheless by means of an emergency decree. According to Article 48, the parliament had the right to overturn emergency decrees with a simple majority. The Reichstag did so, with a considerable majority. In return, the President of the Reich made use of his competence to dissolve the Reichstag, with the support of a legal opinion by Schmitt (see Schmitt 1930d). The President and his advisors thus sabotaged the opportunity to enable the parliament to form a new government, which would continue to be able to act and assemble democratic majorities, in favor of establishing a presidential dictatorship.

On 12 September 1930, during the worsening economic and social crises, the SPD suffered slight losses in the new elections but it still gained the most votes by far. Votes for the NSDAP skyrocketed from 2.6% to 18.3%. The other right-wing political parties that had supported Brüning's policies had to accept losses, some of them massive. Brüning could still have approached the SPD to form a joint government. Hindenburg's informal circle of advisors refused to make any concessions to the Social Democrats and supported Brüning continuing to govern on the basis of Article 48. Although this decision spared the SPD a presumably agonizing internal discussion, it did put the party in a precarious position. The Social Democrats wanted to keep the NSDAP out of political power by all means available. At the same time, they depended on a certain amount of cooperation by Brüning's Center Party in order to keep the SPD-led government of "Red Prussia" in power unimperiled. In light of this constellation, the SPD party leadership agreed with Brüning in several confidential talks in late September 1930 to tolerate his government. In other words, the SPD would not support a no-confidence vote in the Reichstag against the government. In return, the SPD expected Brüning to make certain informal allowances for its political goals. Brüning now governed solely by emergency decree on the basis of Article 48. His government managed to hold on to power for over two years, up until 30 May 1932.

Not surprisingly, the policy of toleration was controversial within the SPD. There was resistance against the party leadership's strategy particularly among the younger members and in the leftist wing. Kirchheimer joined these critics, expressing his opinions in multiple speeches and articles in newspapers and journals. At the opposite end of the political spectrum, Carl Schmitt provided legal support and backup to the cabinet with legal opinions, a number of publications, as well as personal conversations and meetings. Kirchheimer and Schmitt were often in touch and exchanged views during this politically turbulent time. Their differences, which they had only discussed in person up until this point, deepened and now saw the light of day in published articles.

2. Two jurists move to Berlin

Carl Schmitt had accepted his appointment to the vacant general professorship for law at the Berliner Handelshochschule as of the summer semester 1928. This chair had first been held by Hugo Preuß and thus, somewhat ironically, Schmitt became the successor of the father of the Weimar Constitution. The Handelshochschule did not enjoy the same status and reputation as a university because it trained businesspeople and vocational school instructors, not jurists. Schmitt was willing to accept this loss of reputation because he hoped that moving to Berlin would enable him to gain access to the political power centers of the Reich. He was now a direct local competitor of the leading jurists Rudolf Smend, Heinrich Triepel, and Erich Kaufmann at the University of Berlin. With his advanced seminar, which had already been successful in Bonn and which he opened to external participants at the Handelshochschule, he offered an alternative forum to the University of Berlin for discussions on public law and legal theory. Kirchheimer was one of these external participants from the beginning, and while he was serving as a *Referendar* (legal trainee) in Erfurt, he traveled to Berlin multiple times specifically to attend.[1]

Carl Schmitt and his wife moved into an apartment in the Tiergarten district, northwest of the center. After completing *Constitutional Theory*, which immediately became a standard work in academia, he plunged into the world of politics, going beyond academic legal studies. He was not interested in reaching the cultural critical, literary, and Catholic circles as in previous years, but rather the readers of the daily newspapers. He made it clear right in his first publications after moving to Berlin that he was now willing to play a directly political role.

Now interested in economic policy, he gave public lectures and wrote essays on various issues of the day, including, among others, property law, emergency law, and the remuneration of civil servants. He sent reprints of his publications to a large circle of

1 Henry W. Ehrmann in a conversation with the author on 7 June 1988. Ehrmann participated in Schmitt's seminar in Berlin between 1930 and 1932, see Schmitt (2010, 61 and 197).

recipients, targeting potential contacts close to the President of the Reich (see Mehring 2014a, 228–229). He came into contact with Johannes Popitz, who held a top position in the ministry of finance, through his work at the Handelshochschule. Popitz helped him gain access to the political stage (and it was also Popitz who encouraged Schmitt to work for the Nazi regime after Hitler had taken power).[2]

Schmitt attended social events and lectures at the city's leading conservative clubs such as the Herrenhaus and the Deutsche Gesellschaft, and he became acquainted with some of the Reich's conservative elite there. Yet it took him longer than he had hoped to meet highly influential individuals at the center of political decision-making. In contrast, his university teaching in Berlin was highly successful, as it had been in Bonn. His lectures at the Handelshochschule were very popular, drawing a mixed audience of business and vocational teaching students and numerous external guests. As Hans Mayer, later a literary scholar, recalled, "Schmitt was brilliant at putting ideas into words." (Mayer 1988, 148)

Conversely, Kirchheimer's career in Berlin developed more slowly and arduously. After successfully completing his law degree in Bonn with the First State Examination and then his doctorate, Kirchheimer was initially determined to seek a career in politics for the Social Democratic Party (see Herz 1989, 13). Yet he was much more attracted to the world of academia. In light of the difficult career prospects on the academic labor market, he decided to complete the Second State Examination in Law, which would fully qualify him as lawyer, in order to give him more opportunities in the future. He applied for the *Referendariat* (a mandatory post-graduate legal training period) and was appointed *Referendar* on 29 March 1928.[3] This made him a Prussian civil servant with a temporary appointment for the following three years, with a secure, albeit low, income. Hilde Rosenfeld had also successfully completed her law degree in Bonn and had applied for the *Referendariat*. The two married in Berlin on 31 March 1928.[4] Kirchheimer's father-in-law, Social Democratic member of the Reichstag Kurt Rosenfeld, lived in Berlin's affluent Grunewald district in the family's house. Rosenfeld helped his daughter and her husband begin the *Referendariat* by putting them in contact with the relevant agencies in his constituency in Erfurt (which at the time belonged to the province of Saxony, and now to Thuringia).

Berlin remained the city to which Kirchheimer retained the strongest ties, and he sought to keep his contacts to academia there alive. He began his *Referendariat* in Erfurt in April 1928, working first in the *Staatsanwaltschaft* (public prosecutor's office), followed by stints at the *Arbeitsgericht* and finally the *Landgericht* (see List of German Courts).[5] He moved back to Berlin in September 1929 and worked in the *Arbeitsgericht* in the district of

2 See Pyta and Seiberth (1999, 430–432). On Popitz's leading role in the transformation of the Reich Ministry of Finance into a loyal pillar of the Nazi regime, see Middendorf (2022).

3 This information is to be found in a letter from the President of the *Oberlandesgericht* to the Prussian Minister of Justice, 14 October 1929. Bundesarchiv Berlin, R 3001/6322, Ministry of Justice file concerning Dr. Otto Kirchheimer, p. 5.

4 A copy of the marriage certificate can be found in Kirchheimer-Grossman (2010, 60).

5 The individual periods of Kirchheimer's *Referendariat* are documented in the files of the Bundesarchiv (Bundesarchiv Berlin, R 3001/63222, Ministry of Justice file concerning Dr. Otto Kirchheimer, pp. 5, 10). I would like to thank Simone Ladwig-Winters for making me aware of these records.

Spandau for six months. Then he began working in a labor court in Berlin headed by his friend Otto Kahn-Freund. Through Kahn-Freund, he made two other legal contacts from the generation of Young Socialists, Franz L. Neumann and Ernst Fraenkel, who were five and seven years his senior, respectively. They and Kahn-Freund had studied with the social democratic labor law expert Hugo Sinzheimer in Frankfurt. Kirchheimer soon made friends with Neumann, whereas his personal relationship with Fraenkel was not as close. Kirchheimer's legal training during the *Referendariat* also included periods of working with criminal defense lawyers from the leftist socialist political milieu, including in the law firm of Wilhelm Liebknecht, a son of the eponymous founder of the SPD and confidant of Marx's in London and younger brother of Karl Liebknecht, the KPD co-founder who had been murdered. Kirchheimer's final training period was at Berlin's *Kammergericht*. On 2 June 1931, he passed the *Große Juristische Staatsprüfung* (bar exam) and received the title *Volljurist*.[6]

After Kirchheimer had returned from Erfurt to stay in Berlin full time, he immediately began to re-establish his connections to academia. He stayed in touch with Schmitt and Smend in particular, who both received him positively. Smend invited Kirchheimer to his seminar at the University of Berlin as a speaker and assigned his essay on the socialist and Bolshevik theory of the state alongside texts by Schmitt and Leon Trotsky (see Schmitt and Smend 2011, 80). Kirchheimer also continued to see Schmitt on a regular basis and audited his permanent seminar *Contemporary theories of the state* at the *Handelshochschule*. It was Kirchheimer who brought two more auditors to the seminar, Neumann and Fraenkel, piquing Schmitt's interest in their work, too.[7]

Another law professor Kirchheimer became closer to in Berlin was Hermann Heller. The non-Marxist Social Democrat Heller represented a third approach of critical positivism in Weimar legal theory, alongside Schmitt and Smend. Heller had been teaching at the University of Berlin from 1928 on. He propounded a sociological approach in his theory of the state and so was a precursor of political science, which was not established in Germany until after 1945.[8] Both politically and methodologically speaking, Kirchheimer was much closer to him than to Schmitt and Smend, but after Heller accepted a chair in Frankfurt am Main in 1931, they met less frequently.[9]

Kirchheimer's and Schmitt's family constellations developed in parallel over these years. Hanna, the daughter of Hilde and Otto Kirchheimer, was born on 16 December 1930. Schmitt and his wife had a daughter eight months later; Anima was born on 31 August 1931 (see Tielke 2020, 18). Kirchheimer's marriage, but not Schmitt's, broke down after a short time. Hilde Kirchheimer-Rosenfeld turned away from the SPD and directed her attention to the KPD, and their increasingly frequent political disputes contributed considerably to their separation in 1931. They did not file for divorce so as to be able to retain joint custody of Hanna, and Hilde and her daughter moved in with her parents (see Kirchheimer-Grossman 2010, 60). Hilde joined her father's law firm as a lawyer in 1932.

6 Bundesarchiv Berlin, R 3001/6322, Ministry of Justice file concerning Dr. Otto Kirchheimer, p. 14.

7 Henry W. Ehrmann in a conversation with the author on 7 June 1988. On the exchanges between Fraenkel, Neumann, and Schmitt, see Breuer (2012, 111–142).

8 On Heller's approach, see Henkel (2011).

9 Henry W. Ehrmann in a conversation with the author on 7 June 1988.

She worked for the Rote Hilfe, defending the leader of the Central Europe section of the Comintern Georgi Dimitroff and KPD party leader Ernst Thälmann, among others (see Ladwig-Winters 2007, 195–196), and rapidly gained a reputation as a brilliant criminal defense lawyer.

3. Trouble with political justice

Schmitt had devoted a section of *Constitutional Theory* to political justice (see Schmitt 1928b, 176–180). Kirchheimer experienced the issue first-hand a year after completing his doctorate. Even during his *Referendariat* in Erfurt in 1928, he had occasionally written commentaries on legal policy for the local socialist daily press, and he continued to do so in Berlin. His article "50 Jahre Deutsches Reichsgericht" [50 years of the *Deutsche Reichsgericht*] was published in two regional social democratic newspapers in Thuringia on 1 October 1929.[10] Kirchheimer castigated the highest German court in this piece on the occasion of its anniversary; his criticism was as brief as it was vehement.

Kirchheimer wrote that the decisions of the *Reichsgericht* (see List of German Courts) provided "a faithful reflection of the views and notions of Germany's ruling classes" (187). The *Reichsgericht* had never attempted to break away from this worldview or considered its responsibility for social justice. Kirchheimer thought that the positioning of the *Reichsgericht* in Leipzig with respect to the question of the judicial review of laws was particularly hypocritical. During the imperial period, the court had steadfastly refused to review the constitutionality of laws with negative social impacts on the lower classes. In its criminal jurisprudence, it had also actively helped suppress the right of the labor movement to form coalitions and had permitted the unconstitutional Anti-Socialist Laws to stand. Conversely, under the Weimar Constitution, it was now torpedoing social legislative projects, suddenly claiming the right to judicial review of laws, thereby elevating its own status to a "highly dubious guardian of the constitution" (187).[11]

Kirchheimer strongly criticized the decisions of the *Reichsgericht*. As high treason offenses fell within its jurisdiction, it sentenced a large number of socialist activists in the workers' and soldiers' councils to heavy fines and long jail terms. Later, it had meted out disproportionately severe sentences to supporters of the KPD, whereas it had been nothing less than obsequious to supporters of right-wing terrorist groups from the Black Reichswehr and the Organisation Consul.[12] "The enemy of the state from the right," Kirchheimer wrote, "is seen [...] as a decent person by the *Reichsgericht*; after all, he is not an

10 See Kirchheimer (1929a). The following page numbers refer to this text.

11 At an event hosted by the Vereinigung Sozialdemokratischer Juristen (Association of Social Democratic Jurists) in Berlin, Kirchheimer explained the claim to the right to judicial review with the "bourgeoisie's calls for security" (*Vorwärts* of 18 October 1929, Berlin edition, p. 14). I would like to thank Detlef Lehnert for making me aware of this source.

12 Black Reichswehr was the name of the illegal paramilitary units supported by the German Reichswehr in violation of the Versailles Peace Treaty. They included the antisemitic and right-wing extremist secret unit Organisation Consul that attacked and murdered former Reich Finance Minister Matthias Erzberger (1921) and Reich Foreign Minister Walther Rathenau (1922), among others. See Sabrow (1998).

enemy of the bourgeois order" (190). Kirchheimer's criticism of this kind of jurisprudence lacked nothing in terms of critical acuity:

> The technique applied in political trials at the *Reichsgericht* is tantamount to that of Soviet Russia in this matter. Punishment on the basis of active Communist Party membership, medieval-style punishment for printers of newspaper articles, punishment for reciting revolutionary poems are all in the same spirit as the work so successfully supporting the camouflaging of the Black Reichswehr (190).

At the same time, the *Reichsgericht* had succeeded "with an amount of courage and determination admirable from the bourgeois perspective" (189) in summarily declaring unconstitutional the new laws of the *Länder* that sought to put limits on private property in accordance with Article 153 of the Weimar Constitution. In its decisions, the court had protected private property from any and all interventions through lawmaking far more than it had during the German Empire. Kirchheimer summed up the past ten years of *Reichsgericht* decisions laconically: "The 'guardian of the constitution' guards the constitution as it sees fit" (190). To remedy the situation, he called on the social democratic politicians in the *Länder* of the Reich to replace the personnel at the *Reichsgericht* with new judges loyal to the constitution and the republic.

It would not have taken much for this article to put an end to Otto Kirchheimer's career as a jurist. It was only thanks to the social democratic influence in the Prussian Ministry of Justice that he was permitted to continue his *Referendariat*. Two weeks after publication of his piece, the President of the Prussian *Oberlandesgericht* in Naumburg wrote an outraged letter to the Prussian Minister of Justice in which he demanded disciplinary measures against Kirchheimer.[13] He claimed that Kirchheimer's "most highly outlandish, superficial, and one-sided criticism of the highest court" was "in conflict with his duties as a civil servant" and might "undermine the authority [of the court] in his profession." In light of the momentousness of this violation of the political restraint required of a *Referendar*, it was essential to "take measures against the author."

On 22 October 1929, the President of the Prussian *Kammergericht* was asked to prepare a legal opinion, and he presented his four-page analysis of the newspaper article just two days later.[14] Its intention is clear: to protect Kirchheimer. For example, although its author also identified a number of objectionable phrasings and took exception in particular to comparing the jurisprudence of the *Reichsgericht* in political trials with that of Soviet courts, stating that this and other parts of the article proved a "regrettable lack of restraint and factualness," he advised against further disciplinary measures against Otto Kirchheimer. For one thing, he had only been a civil servant for a short time, for which reason he had not yet had the opportunity to adjust to the spirit of the civil service. For

13 The following quotations are taken from the letter from the President of the *Oberlandesgericht* to the Prussian Minister of Justice dated 14 October 1929. Bundesarchiv Berlin, R 3001/6322, Ministry of Justice file concerning Dr. Otto Kirchheimer, p. 3.

14 The following quotations are taken from the legal opinion by the President of the *Kammergericht* to the Prussian Minister of Justice dated 24 October 1929. Bundesarchiv Berlin, R 3001/6322 Ministry of Justice file concerning Dr. Otto Kirchheimer, p. 5.

another, the article had appeared in an insignificant press publication. The strongest argument against disciplinary measures, however, was that they would provide grounds to fear that Kirchheimer would then "attempt to go into his deliberations factually and prove the truth of his assertions, thereby becoming overly absorbed in those views, and might put on airs of a political martyr."

Kirchheimer was not quite out of the woods, though. He was summoned to his supervising judge at the labor court to explain himself on 2 December. In this situation, Kirchheimer decided to distance himself from the polemical wording in his commentary. According to the files, his supervisor urgently recommended that he exercise his "general civic rights" with "greater restraint" in the future.[15] No further disciplinary measures were ordered, and Kirchheimer was permitted to continue his *Referendariat* after this affair. He disregarded the well-meaning advice to exercise political restraint as a civil servant with a temporary appointment and immediately published opinion pieces on legal policy again under his own name. At the same time, as of 1930, he sought out public confrontations with Schmitt concerning his prominent role as a legal advisor and supporter of the presidential dictatorship.

4. Structural changes of parliamentarism

Carl Schmitt's relatively short book *The Crisis of Parliamentary Democracy* is probably the most famous (and most notorious) of his works to this day. Schmitt wrote it in 1923, the Weimar Republic's turbulent and crisis-ridden year. In January, Belgian and French troops occupied the Ruhr. The Rhineland threatened to secede from the Reich. The government of the Reich used military force to remove the socialist-communist governments in Saxony and Thuringia from power. Adolf Hitler's attempted coup in Munich failed. Hyperinflation rattled the country and its economy. The new Chancellor Gustav Stresemann proclaimed the need for a dictatorship and new emergency decrees in accordance with Article 48 of the constitution. Thereupon, the SPD left the multiparty coalition government in protest. The Reichstag was no longer able to form a government, and the country's military brass openly planned the transition to a presidential dictatorship. In 1923, Weimar parliamentarism was mired in its first major crisis. It was not until the end of the year that the republic began to stabilize.

Opposition to parliamentary democracy in the Weimar Republic came from three different political groups: the extreme nationalists such as Hitler's small party; the monarchical circles and political parties such as the right-wing DNVP; and the radical communist left, which preferred a Soviet-like dictatorship of the proletariat. Schmitt wrote his book on parliamentarism in the midst of these months of crisis and the debates about the point and pointlessness of parliamentarism. As a starting point, he used an essay by Smend from 1919, the year in which the constitution for the Weimar parliamentary democracy was drafted (see Schmitt 1923a, 34).

---

15 Letter from the Prussian Minister of Justice to the President of the *Kammergericht* dated 2 December 1929. Bundesarchiv Berlin, R 3001/6322, Ministry of Justice file concerning Dr. Otto Kirchheimer, p. 4.

Smend distinguished two phases of parliamentarism: an initial one in which the parliament had been the institution for finding political truth as independent dignitaries exchanged arguments and a second one, now underway, in which it had become a kind of instance for registering preconceived political tendencies represented by stable party blocks. Smend considered this transformation to be a loss. The "actual creative political dialectic" (see Smend 1919, 64) of the process of political integration had been sacrificed. He attested that the newly established parliamentarism had become a "facade [...] behind which the decisive party discussions took place in secret." (see Smend 1919, 62) The establishment of stable party blocks had made the rationalistic justification of parliamentarism an ideology: "One can hardly say anymore of our parliamentarism that it is still government by talking." (Smend 1919, 62)

Smend believed the structural transformation was caused by the strengthening of the political parties, which had led to them becoming monopolists in a limited pool of potential of electoral candidates. At the end of this article, Smend did not consider Weimar parliamentarism to be dead, but instead focused on the opportunities for the new democratic state which might emerge from the transformation. He called for a "sociological turn" as the basis of a "new constitutional theory" to put the ongoing transformation into perspective (see Smend 1919, 67). This was a methodological demand that Kirchheimer, with his Marxist views, was immediately able to support.

In contrast, Schmitt decided on a different and politically far more radical conclusion based on a similar ideal-typical model of the historical development of parliamentarism.[16] He, too, spoke of a first historical phase of parliamentarism whose "essence [had been] public deliberation of argument and counterargument, public debate and public discussion" (34). This original parliamentarism functioned "without taking democracy into account" (35). Its social basis was a homogeneous social stratum, namely the bourgeoisie. However, as soon as parliamentarism was merged with democracy, this triggered a dual process of transformation. In place of strong individual personalities, it was candidates from democratic mass organizations who were running for office. And in place of a socially homogeneous class, the elected parliament now represented the heterogeneity of society. In Schmitt's opinion, parliament was thus transformed from a place of common exchange of arguments to a place where party-line proclamations were simply read aloud.

Unlike Smend, Schmitt drew two negative conclusions from his descriptive model. First, he inferred that parliamentarism has lost its original ideological essence, its ultimate core, its fundamental principle, and that purely pragmatic reasons were now the only way to justify it. Thus, parliamentarism was missing its intellectual foundation and its legitimacy as a major political institution. Although Schmitt's criticism assumed a discrepancy between the idea and the reality of the modern parliament that had become evident, he did not measure its reality using the idea as the yardstick—as others have often understood and, consequently, criticized his work. He was only concerned with the determination that the great principles of the great institution—public debate and independent political representation—were no longer credible at the time. For this reason, it was relatively unimportant to Schmitt whether or not the same discrepancy between idea

16 See Schmitt (1923a). The following page numbers refer to this text.

and reality had already existed in the nineteenth century. He was concerned solely with the fact that citizens had lost their belief in the reasons legitimizing the great institution (see Hofmann 1995, 96–101). His second negative conclusion was that a parliamentarism in which the social inequalities of a society collided was only able to function temporarily and would certainly have to end in civil war sooner or later. It was logical for Schmitt to call for overcoming parliamentary democracy not only in the agonizing phase of the Weimar Republic from 1930 on, but even during its stable phase in 1928: "A solution outside of these democratic-political methods must be sought." (Schmitt 1928d, 49)

It was in keeping with Schmitt's line of argument that the attempt to establish a parliamentary democracy based on the Weimar Constitution was necessarily doomed to failure from the outset. This failure was not—as Ellen Kennedy has argued—due to any specific "constitutional failure" of the Weimar Constitution (see Kennedy 2004, 154–182); instead, Schmitt thought that it was inherent to any constitution in which parliamentarism and democracy had been merged. To Schmitt, this conclusion was imperative from the moment when he had made a sharp distinction between liberalism and democracy in his small 1923 book at the latest. Liberalism as a "metaphysical system" (35) necessarily included the belief in reasonable discussion, parliamentarism, the balance of political powers, and the *Rechtsstaat*. Democracy was contrary to this, and—as he added in 1928—he sought to "rescue [it] from being concealed by liberal attributes" (see Schmitt 1928d, 47).

Schmitt defined democracy as the total "identity of governed and governing" (26). It was an identity that was never entirely real because the masses were never completely heterogeneous, but always "sociologically and psychologically heterogenous" (25). It was all the more imperative to produce "identifications" (27) through political action. The liberal constitutional state with its parliamentarism and its division of powers aimed to prevent the populace from melding at the emotional level with those governing them in the sense of creating total identity freed from the liberal shackles; a democratic people could express its political will through acclamation with everyone physically assembled in the same place. Schmitt's ideal was a kind of "soccer stadium democracy," in Stephen Holmes's words (Holmes 1993, 93). Schmitt believed that acclamation was the natural and necessary political expression of the life of a people, whereas parliamentary democracy amounted to ignoring the assembled people. The liberal security of the secret ballot destroyed this publicly proclaimed unanimity of a people's political will.[17]

To Schmitt, assenting to a dictator was a genuinely democratic act. In his introduction to the second edition of his book in 1926, he added the following to explain his views on homogeneity: "Every actual democracy rests on the principle that not only are equals equal, but that unequals will not be treated equally. Democracy requires, therefore, first homogeneity and second—if the need arises—elimination or eradication of heterogeneity." (Schmitt 1926b, 9) Initially, it made no difference here whether democratic homogeneity was based on common religious, ethical, and cultural convictions or on racial characteristics or socioeconomic equality. In the final analysis, this meant that class societies or multicultural or multiethnic societies could never be democracies.

17 On Schmitt's fundamental critique of the secret ballot, see Buchstein (2000, 597–600) and (2002).

Schmitt had used these hypotheses to declare that the young parliamentary democracy established only four years previously was stillborn and the Weimar Constitution was an anachronism that was misguided from the outset. He thought the constitution contained two antagonistic principles, one liberal and the other democratic. In other words, precisely at the historic moment when liberalism and democracy had come together in the Weimar Constitution, Schmitt not only separated them again but played them off against each other. He identified the two antagonistic principles with two competing institutions of the republic, the Reichstag (parliamentary vs. liberalism) and the President (plebiscitary vs. democracy).

If parliamentarism was a political project doomed to fail, what were the political alternatives? Schmitt discussed two radical alternatives in his book. One was Marxism and its approach of scientific socialism. According to Schmitt, traditional Marxism still argued within the metaphysical system of liberal rationalism. The "irrationalist theories of the direct use of force" (65) such as George Sorel's theory of the myth were more promising. Sorel stated that within Marxism, radical leftist theorists such as Lenin and Leon Trotsky had learned from the myth of class struggle described by Pierre-Joseph Proudhon and Mikhail Bakunin as well as from counterrevolutionaries such as Donoso Cortés. The theory of the myth had discovered "a new belief in instinct and intuition" (66). The great mythical heroic and warlike enthusiasm sprang out of the depths "of a genuine life instinct" (68), not out of reason or pragmatism.

To Schmitt, the decisive political question arising at that point was, "Who, then, is the vehicle of the great myth today?" (68). It could not be found in the bourgeois ideal of peaceful exchanges of arguments and parliamentary deliberation. Believing in parliamentarism was "cowardly intellectualism" (69). With their appeals to the instinct for class struggle, George Sorel and Russian Bolshevism had taken a more promising direction. Yet they remained half-hearted and in the grip of bourgeois rationalism because they sought to organize the socialist economy using methods of rational planning.

Schmitt also discovered in Bolshevism elements of a second myth: the Russian national myth. Only the proletarian use of force had brought the country back to its deeper traditional cultural roots and made Russia Moscovite again, despite Bolshevism's internationalist propaganda. Schmitt added to these comments a decisive hypothesis that expressed all his political thinking throughout his life: whenever it comes to an open confrontation of the socialist and nationalist myths, such as in Italy 1922, the "irrational power of the national myth" will "always be victorious." (75) At the end of his book, Schmitt quoted extensively from a speech Benito Mussolini gave in Naples in 1922 in which he proclaimed the superiority of the national myth in fascism over that in socialism. To Schmitt, the theory of the myth was "the most powerful symptom of the decline of the relative rationalism of parliamentary thought." (76). He saw the serious disturbances of 1923, the year of crisis, as confirmation of his criticism of an anemic liberal parliamentary democracy.

Not surprisingly, Schmitt's fundamental critique of parliamentary democracy triggered a heated debate immediately upon its first publication, provoking numerous responses from contemporaries, among others the legal scholars Hans Kelsen, Richard Thoma, Rudolf Smend and Moritz J. Bonn, and the sociologist Ferdinand Tönnies. This debate has continually entered new rounds and has not been concluded to this day. I mentioned Smend's considerations on this topic above, not least because they evidence that many of the ideas in Schmitt's work on parliamentarism were by no means as original as the frequent references to this text might suggest. Smend's considerations also substantiate that describing the development of parliamentarism reflecting a theory of decline does not necessarily have to lead to a vigorous rejection of parliamentarism and a glorification of the fascist myth.

Against this background, as Otto Kirchheimer worked on his reflections on parliamentarism in the late 1920s, he saw himself confronted with the task of finding his own position coming from the Marxist side between the alternatives offered by Smend and Schmitt, his two conservative right-wing mentors. Kirchheimer began publishing on parliamentary democracy immediately after completing his dissertation. At this point, Schmitt had published further essays with variations of his hypotheses as well as his book *Constitutional Theory* in which he had laid out his considerations on liberalism, parliamentarism, and democracy more systematically and in more detail (see Schmitt 1928b, 253–378). Up until 1931, Kirchheimer used some of the ideas from Schmitt's theory of parliamentarism in various articles as a key to understanding the current political situation in order to promote the cause of socialism.

His essay "Bedeutungswandel des Parlamentarismus" [The Transformation of the Meaning of Parliamentarism] was published in the October 1928 issue of the journal *Jungsozialistische Blätter*, the theoretical organ of the leftist wing of the Young Socialists in the SPD.[18] The Young Socialists were split into various wings, the majority of which took positions to the left of the party leadership. In this essay, Kirchheimer described the development of modern parliamentary democracy in historical sequence. It is a stage model clearly inspired by Schmitt's theory. Right at the outset, Kirchheimer disallowed the widespread panegyric that it had been only the constitution of the Weimar Republic that had created the democratic form of government and introduced the parliamentary system in Germany. Such language, he wrote, used the terms "parliamentary" and "democratic" together and side by side, inadvertently conveying the impression that they were indivisible and that they had always meant the same thing over the course of history. This, however, was a "theoretical error with far-reaching consequences" (157); not only had Karl Marx and Friedrich Engels treated it with contempt in various writings, it had also resulted in catastrophic mistakes in political practice time and again.

Like Schmitt, Kirchheimer distinguished between democracy and parliamentarism conceptually and historically. In its classical form, parliamentarism was a political institution through which the bourgeoisie exercised its rule over other classes of society, and its details were negotiated exclusively by the bourgeoisie. Classical parliamentarism was distinguished by three components: first, the claim to political power on the part of the bourgeois social strata enjoying property ownership and access to education; second, the belief that what was sensible for the nation could be identified through public parliamentary discussions; and third, adherence to the principle of the *Rechtsstaat*, whereby Kirchheimer also emphasized that the essence of the principle of the *Rechtsstaat* had changed alongside the societal changes from the nineteenth century on. Kirchheimer contrasted classical parliamentarism with an understanding of democracy he ascribed to Marx and

18 See Kirchheimer (1928b). The following page numbers refer to this essay.

Engels: "They considered 'democracy' to be the rule of the entire, the working people, in contrast to the rule of a parliament constituted by census suffrage" (159).

Kirchheimer then outlined how the three components of classical parliamentarism had successively crumbled during the second half of the nineteenth and the first quarter of the twentieth century. The electoral law reforms had afforded all societal strata political access to parliament. Creative political discussions in parliament had been replaced with the representation of class interests, and the parliament had also lost political power to the executive. The principle of the *Rechtsstaat* no longer served the interests of the bourgeoisie alone and was instead caught "between the proletariat and the bourgeoisie" (161). Kirchheimer assigned the *Rechtsstaat* in modern parliamentary democracy the active function of "creating an equilibrium" (162) between the proletariat and the bourgeoisie, thus fighting the social struggles between the classes with legal means and "neutralizing questions of social power by transforming them into problems of finding justice" (162).

Kirchheimer's deliberations in this article could be characterized as a kind of historical semantics of key political concepts founded in materialism. His thoughts again clearly show the influence of Max Adler's writings and seek to borrow from Marx and Engels but his choice of words occasionally displays parallels to Schmitt's writings on parliamentarism, too. Yet, in contrast to Schmitt, who described the transformation of the meaning of parliamentarism as a historical downfall, Kirchheimer welcomed this transformation. In his view, the structural transformation of parliamentarism to an institution of mass democracy, organized by competing political parties, was a thrust toward political and social emancipation. He even considered—again in contrast to Schmitt—that the neutralization of social conflicts by legal means certainly could be successful, at least in principle. Essentially, Kirchheimer contradicted Schmitt in his normative evaluation of the transformation of parliamentarism and was more optimistic than Smend about its potential progressive force.

5. Fascism and socialism as alternatives

Hermann Heller had spent time in Italy in the summer of 1928 to lay the groundwork for a book on fascism. In the spring of 1929, Schmitt also traveled to Italy for eight days. He visited the Senate and various tourist sights in Rome and called on the Italian political theorist Gaetano Mosca at the university. On the fourth day of his stay, he went to Piazza Venezia. It was one of the Duce's major reconstruction projects, including a broad avenue crossing the Forum Romanum to the Colosseum. Schmitt met the Kirchheimers, who were also vacationing in Rome, there. He noted in his diary: "met Kirchheimer and his wife, we chatted at Café Venezia for more than an hour, about socialism, the state, etc."[19] Regardless of all the differences that had emerged, the conversational tone between him and Kirchheimer apparently continued to be unconstrained and friendly.

Schmitt pointedly reiterated the hypotheses presented in his work on parliamentarism in multiple publications. He found his approach to an anti-liberal interpretation

19 Carl Schmitt, diary entry of 14 April 1929 (Schmitt 2018, 283).

of the Weimar Constitution in the fundamental differentiation between liberalism and democracy. While he was working on *Constitutional Theory*, he had written to Smend in October 1927 that the book's essence was "to remove liberalism's death mask."[20] In his lecture "Der bürgerliche Rechtsstaat" [The bourgeois rule of law] a year later, he asserted that the Weimar Constitution was dependent on the Treaty of Versailles and proclaimed the "posthumous" character of Weimar parliamentarism and liberalism. He formulated his creed on constitutional policy: "What matters for the development of the constitution in the near future is to rescue democracy from being concealed by liberal attributes" (Schmitt 1928d, 47). In a book review on Italian fascism, he affirmed the "democratic" character of Mussolini's rule, which was legitimized by acclamation "by plebiscite" (Schmitt 1929b, 110). The fascist state in Italy achieved "political unity of the people," he claimed, and ultimately served the "socialist interests of the workers" (Schmitt 1929b, 113). Readers of this review who were familiar with Schmitt, for example Heller and Schmitt's former Bonn student Waldemar Gurian, considered these statements unequivocally supportive of fascism.[21] Such clear words were not yet to be found in Kirchheimer's publications at that time.

In August 1929, about six months after he and Schmitt had both spent time in Italy, Kirchheimer published a new piece on the issue of acceptance of parliamentary democracy in which he also touched on the fascist option. The article appeared in the journal *Der Klassenkampf* [Class struggle], which was published by followers of Rosa Luxemburg from the leftist wing of the SPD. The article was titled "Verfassungswirklichkeit und politische Zukunft der Arbeiterklasse" [Constitutional reality and the future of the working class].[22] Kirchheimer took the tenth anniversary of the adoption of the Weimar Constitution on 11 August 1919 as an occasion to look back on constitutional policy and to diagnose the current situation. He thought that without much ado, the mass of war-weary soldiers had entrusted the political power they had received following the November Revolution of 1918 to the social democrats forming the majority. When the SPD then sought to begin implementing the social promises made to the working class, "the bourgeoisie had already come back out of the woodwork" (180). Both sides agreed to a compromise and created the constitution for a parliamentary democracy the following year, which, however, lacked a principle "that would have formed the people into a community based on political will in the long term" (180). Here, Kirchheimer meant that the Weimar Constitution had not come to a decision about the question whether the future German republic was to be a capitalist or a socialist democracy. With this assessment, he followed Schmitt's derisive words in *Constitutional Theory* about the Weimar Constitution as a "*dilatorischer Formelkompromiß*" (dilatory formulaic compromise) (Schmitt 1928b, 85).[23]

Kirchheimer then elucidated how the power relations in society had changed over the past ten years. The bourgeoisie's concerns about an expansion of socialism in Western industrial societies had disappeared. Thus, "Europe's bourgeoisie [is now] no longer required to conceal its true face behind a social and democratic mask" (182). After the

---

20 Letter from Carl Schmitt to Rudolf Smend dated 17 October 1927 (Schmitt and Smend 2010, 65).

21 See Heller (1929, 489 and 541–542) and Gurian (1929, 508).

22 See Kirchheimer (1929b). The following page numbers refer to this text.

23 See the Translator's Preface regarding the translation of "*dilatorischer Formelkompromiß*."

constitution had been adopted, the bourgeoisie in Germany had launched a counterattack and begun to dismantle the social advances in the Weimar Constitution step by step. Here, he referred to the elimination of the constitutionally guaranteed eight-hour working day, the ailing education system, and lacking implementation of the Works Council Act mentioned in the constitution. In light of these developments, only "dreaming socialists" (183) could hope that the current-day bourgeois state could be overturned legally.

Considering these general tendencies, Kirchheimer called his SPD's involvement in the parliamentary government of the Grand Coalition its "share of patronage" (183), which could accomplish little against the increased power of the conservative state bureaucracy. Large sections of the bourgeoisie were unwilling to settle even for that. They wanted to go a step further and abolish the present situation of the constitution in favor of a bourgeois dictatorship following the pattern of Mussolini's fascist rule in Italy. That was a clear reference to the option of an authoritarian state as promoted by Schmitt. At the moment, Kirchheimer believed, the bourgeoisie was struggling to ensure that the decision that had not been made in 1919 was not postponed any longer and that it was made one-sidedly in its favor. The SPD leadership was unaware of this danger, instead seeking to continue avoiding such a decision. Yet avoiding it was impossible: "The only choices are forward or backward" (185): in other words, either major steps toward a socialist democracy on the basis of the Weimar Constitution or authoritarian rule in the interest of the bourgeoisie. Kirchheimer knew from his numerous conversations with Schmitt that he and others had strong sympathies for transitioning to an authoritarian state.

Against the background of this diagnosis, Kirchheimer advocated a socialist approach to policy, making it necessary to fight "from below, following a plan" (184) within the party to "replace an old body of functionaries with one in a new spirit" (184). Up until this point, social democratic realpolitik had done more to block than to enable the path toward socialism. The Weimar Constitution continued to be "the book of opportunities" (186). To that end, it was necessary to have the courage to imagine utopias and the strength to follow through with political mobilization. Kirchheimer used words that were as passionate as they were vague to appeal to his readership: "we must again want to learn" (185) and "we must be prepared for the great tomorrow that we can win or irretrievably lose in these years" (186).

Kirchheimer's article is full of verve and polemic power. Both its content and its style fit seamlessly into the general line of the journal *Der Klassenkampf*: presented in the style of a sober analysis of class struggle, with vehement criticism of the SPD party establishment, based on a concept of socialism as a comprehensive cultural movement, and concluding with appeals using pointed and voluntaristic vocabulary in the style of Rosa Luxemburg. Kirchheimer did not see the parliament as having any progressive role. He believed socialist mobilization had to take place outside of parliament. Kirchheimer seemed to have lost the optimism about the possibilities of parliamentary democracy he had expressed a year earlier, and precisely during the brief phase in which the Grand Coalition under Social Democratic Chancellor Hermann Müller was successful. In historical retrospect, however, his warnings proved prophetic.

6. Weimar—and what then?

In his book *Constitutional Theory*, Schmitt had introduced the fundamental distinction between *Verfassung* and *Verfassungsgesetz* (which he defined as "unified constitution in its entirety" and "the individual constitutional law") (Schmitt 1928b, 67). He believed a unified constitution in its entirety was always the expression of a "single instance of decision" by the political unit "in regard to its peculiar form of existence" (Schmitt 1928b, 75). Schmitt analyzed the Weimar Constitution against this theoretical background in his *Constitutional Theory* and concluded that, in many of its individual constitutional laws, it consisted of compromises where the maker of the constitution had attempted "to evade a decision." (Schmitt 1928b, 82) In addition, "the great choice, bourgeois or socialist order, was seemingly settled only through a compromise." (Schmitt 1928b, 83) In the following years, he used this hypothesis to assert the notorious fragility of the Weimar democracy. Schmitt sought potential anchors of stability as counterweights, initially within the framework of the constitution. Ultimately, he found them in the presidential dictatorship based on Article 48.

Competing with Schmitt's analysis of the constitution, Otto Kirchheimer published a short book titled *Weimar—and What Then? An Analysis of a Constitution* in May 1930.[24] Four weeks earlier, the Grand Coalition under the leadership of the SPD had collapsed, and the era of the presidential dictatorship had begun with Chancellor Brüning. The skeptical and radical essay-like book was to make Kirchheimer—who was only 24 at the time—known beyond his previous circles overnight. The literary scholar Hans Mayer reported in his memoirs that it was "eagerly quoted and commented on" in the discussions among the young socialist and communist intellectuals (see Mayer 1988, 128). The book was published in the *Jungsozialistische Schriftenreihe* [Young Socialists' publication series] that was edited by Max Adler in collaboration with the socialist activists Engelbert Graf and Anna Siemsen in the Laubsche Verlagsbuchhandlung publishing house in Berlin. The series had a first print run of 4,000. It published works by its three editors mentioned above, prominent names such as Ernst Toller and Leon Trotsky, and authors from the younger generation of socialists including Franz L. Neumann, Ernst Fraenkel, and Arkadij Gurland.

Kirchheimer introduced his essay in a coolly sober manner, stating he would generally limit himself to presenting "the facts" (33). He described his project as a "socialist analysis of the constitution" (33), drawing a strict distinction between this and the errors of a purely legal and liberal analysis of the constitution. Whereas liberal analysis of the constitution, which often appeared in a democratic guise, gave the false impression of nonexistent societal unity, a socialist analysis of the constitution had to reveal all those contradictions associated with the current-day organization of society and its political form. He began with a longer quote from Rosa Luxemburg's 1899 polemic pamphlet *Sozialreform oder Revolution?* [Social reform or revolution?] which clearly states: "Every legal constitution is the product of a revolution. In the history of classes, revolution is the act of political creation, while legislation is the political expression of the life of a society that has already come into being." (33).

24 See Kirchheimer (1930e). The following page numbers refer to this text.

In the nine sections of the work, Kirchheimer discussed the emergence of the Weimar Republic, the relationship between democracy and dictatorship, electoral law, parliamentarism, fundamental rights, government formation and governance, the *Rechtsstaat* and the status of government officials, the position of President of the Reich, and finally his general characterization of the Weimar Constitution. His analysis culminated in the hypothesis that the Weimar Constitution was fundamentally a "constitution without decision" (71), echoing Schmitt's analysis. In line with Rosa Luxemburg's words quoted above, he made his statements even more pointed by categorically declaring that it was the purpose of all constitutions "to proclaim a specific program of action in whose name the organization of a new social order [was] to proceed" (72).

This sounds very similar to Schmitt in his *Constitutional Theory*. However, Kirchheimer's aspiration that a new constitution had to be creative like this and that it had to change society went far beyond Schmitt's understanding which required a constitution to make an overall decision about the type and form of political unit. Schmitt had demonstrated that the Weimar Constitution certainly had made some fundamental decisions (*Rechtsstaat*, parliamentary democracy), but had left many other controversial issues unresolved. In his critical diagnosis, Kirchheimer focused on a single question: capitalism or socialism? The constitution had become bogged down at this point. He stated that the fact that it had been impossible to come to a clear decision in favor of a socialist society during the course of the revolution was "the basic and irreparable error of this constitution" (72).

Kirchheimer criticized that the concept of "democracy" had lost any and all concrete meaning, thereby repeating his plea for a narrow concept of the term following Max Adler's social democracy. Such a social democracy, however, could not exist in a society divided by class. Moreover, democracy in capitalism entailed "a considerable portion of bourgeois dictatorship" (41). Since Kirchheimer was in contact with Schmitt on a regular basis, he was aware of the latter's friendship with Johannes Popitz, State Secretary in the Reich Ministry of Finance, and knew that the two of them agreed on many political issues.[25] He used this knowledge to take a swipe against Schmitt in his book by quoting an essay by Popitz in which he bitterly complained that the mass of the less well-to-do electorate was plundering the rich.[26] The bourgeoisie now wanted to put an end to this situation.

Kirchheimer considered Article 48 of the Weimar Constitution to be the instrument with which the ruling classes in Germany could manage "to achieve by means of a dictatorship what the will of large segments of the people prevent them from achieving in a legal manner" (41–42). He referred to Schmitt's distinction between provisional and sovereign dictatorship and applied this terminology to two different forms of exercising bourgeois dictatorships: the temporary measures to suppress the leftist opposition in Saxony and Thuringia in 1923 and fascism in Italy, which had been established indefinitely. He thought that it was impossible to accurately predict when the political democracy of the bourgeoisie would suddenly transition into one of the bourgeois forms of dictatorship since the bourgeoisie considered such a regime change to be purely a question

25 The close relationship between Schmitt and Popitz is explored in Kennedy (2004).

26 See Kirchheimer (1930e, 64–65). The following page numbers again refer to this text.

of expedience and of the opportunities for enacting it. As a defender of Brüning's presidential cabinet, Schmitt is portrayed as an ideologist for the authoritarian wing of the bourgeoisie.

Against the background of his hypothesis of a systematic destruction of the order of legality laid down in the constitution, Kirchheimer presented a diagnosis and structural analysis of the republic's political institutions. It included the democratic right to vote which had been introduced at the same time as the republic. Here, Kirchheimer appealed to his readers that even the most perfect electoral law could only support an intense political will but could not replace it. It also included a parliament, the Reichstag that had transformed from an assembly for common discussions to a place for class struggles. At this point, Kirchheimer explicitly rejected calling the list of fundamental rights a "compromise"; he also considered Schmitt's talk of a dilatory formulaic compromise unhelpful because it lacked any conceptual and political clarity. The fundamental rights in the Weimar Constitution "are in their essentials not a compromise but constitute rather a unique linking and acknowledgement of the most varied value systems, which is without precedent in constitutional history" (54).

Kirchheimer concluded his structural analysis of the political institutions with the office of the President of the Reich. The entire section is directed against the "erroneous perception" (68) that the President of the Reich was far removed from the interests of parties and special interests and was thus the only true representative of the nation; a critique directly aimed at Schmitt. Kirchheimer thought that the election of the President of the Reich was also dominated by the political parties. The Schmittian notion that his office was beyond classes was a politically misleading "fiction" (71). Kirchheimer also indirectly countered Smend's monarchy-like concept of the office of the President by stating that a President could not generate an integrating ideal overall will in the absence of the societal and political preconditions it required, namely a classless society. Kirchheimer's short book *Weimar—and What Then?* ended with a negative conclusion concerning the accomplishments claimed by the SPD party leadership after a decade of the Weimar Republic. Its tone was unmistakably ominous about the prospects for such optimism about reforms.

Shortly after its publication, Kirchheimer's book was met with spirited criticism and animated approval alike. It is one of Kirchheimer's most often quoted works to this day.[27] But even the people closer to him responded in very different ways. Franz L. Neumann wrote that Kirchheimer's analysis of the Weimar Constitution was "very [close] to communist trains of thought." (Neumann 1930, 76) He accused Kirchheimer of trivializing the significance of fundamental rights and not going further than denouncing them as a hodgepodge of incompatible value judgments instead of resolutely taking up the jurist's toolbox and attempting to achieve a unifying legal systematization. The general antithesis to Kirchheimer that Neumann's criticism entailed was that the constitution was not to be understood as contradictory, but rather as open; therefore, the labor movement could help push it toward its socialist goals. His prompt riposte to the question Kirchheimer had posed in the title of the book was the imperative: "Erst einmal Weimar!" (First of

27 On the later receptions and debates about this piece, see Schale (2006, 42–46) and Buchstein (2017a, 68–73).

all Weimar!).[28] Hermann Heller's response was negative too. To him, it was part of the cheap criticism by "our aesthetic-heroic revolutionary romantics from the left and the right" who "in extraordinary agreement" lambasted the constitution instead of defending it against all ideologies of violence—"if necessary, arms in hand." (Heller 1930, 376 and 377)

Another critical piece was written by Arkadij Gurland. He had just published a book on the dictatorship of the proletariat in which he also made use of Schmitt's theory. Gurland was an outstanding voice of the left wing of the Young Socialists of the SPD (see Buchstein, Emig, and Zimmermann 1991, 9–22). His review in the left-socialist *Bücherwarte* was the strongest rejection of any author from the left. He started by praising Kirchheimer for pointing out that all constitutional questions were ultimately questions of power. After this introductory remark, he criticized him all the more sharply for limiting his deliberations "to unfortunately more summary statements" rather than specifically illustrating such interconnections from the sociology of law. Gurland identified Schmitt as Kirchheimer's inadequate teacher for such abstractions. Since Kirchheimer had followed Schmitt, his thinking resulted in the dangerous supposition that Weimar parliamentarism had too little potential for the labor movement. Gurland thought that "what was alarming about this piece" was its political finding; he concluded by stating that Kirchheimer's book "was not fitting for an educational library, which the *Jungsozialistische Schriftenreihe* is supposed to be" (Gurland 1930b, 136). Gurland's strong criticism was the beginning of his life-long close friendship with Kirchheimer.

In contrast to the criticism from the left authors, Carl Schmitt's reaction to *Weimar—and What Then?* was much more positive. He sent a copy of Kirchheimer's book to Ferdinand Tönnies, the sociologist and prominent interpreter of Thomas Hobbes,[29] and told him he "understood the sentiment" of Kirchheimer's short book. Tönnies had recently become a member of the SPD and appealed to the public to defend the republic against attacks from the right. Schmitt explained to him that the book "begins with the hypothesis that the Weimar Constitution does not contain any political decision at all" and added: "What should one do, as a teacher of positive constitutional law, when faced with such confusion?"[30] Tönnies reacted similarly to Neumann: his response was that the claim that the constitution did not contain any political decision was "surely untenable."[31]

Schmitt, however, did not distance himself in this way in his publications. On the contrary. He quoted the piece in a positive light as a paradigmatic socialist interpretation of the constitution (see Schmitt 1932d, 182). In 1932, he praised it as a "highly interesting piece" (Schmitt 1932d, 195), referring mostly to Kirchheimer's hypothesis of a constitution without decision. One year later, Rudolf Smend took up this praise when he criticized Schmitt publicly in a ceremonial lecture on 18 January 1933, affirming that

28 Neumann (1930, 74). Neumann repeated his critique two years later (Neumann 1932, 39). On his criticism of Kirchheimer, see Rückert (1993, 446–448).

29 On Tönnies and Schmitt on Hobbes, see Chapter 9, p. 227–228.

30 Letter from Carl Schmitt to Ferdinand Tönnies dated 10 July 1930 (Schmitt and Tönnies 2016, 112).

31 Letter from Ferdinand Tönnies to Carl Schmitt dated 18 July 1930 (Schmitt and Tönnies 2016, 115).

Kirchheimer had performed a "logical execution" (Smend 1933, 319) of Schmitt's decisionism. He continued, directing his criticism toward both Schmitt and Kirchheimer: "It is, however, not the purpose of a constitution to be a 'decision' in the sense of any objective and logical political system of thought, but to bring living people together to form a polity in an orderly way." (Smend 1933, 320)

The reception of Kirchheimer's piece reveals a curious constellation. Public criticism came from his socialist comrades and from Smend, a conservative. Public praise came from Schmitt alone. Yet such praise should not be misunderstood as agreement. To Schmitt, Kirchheimer's piece served as incontrovertible evidence that on the left, social democrats called the constitution into question, too.

7. Property rights and expropriation

If, according to Kirchheimer, the Weimar Constitution was a constitution without decision, how open was it to policies with socialist goals? This question placed the interpretation of property rights and expropriation in the constitution at the center of the struggle for positions regarding the constitution.

The text of the constitution of the Weimar Republic is a synthesis of socialist and capitalist ideas about the economic and social order (see Gusy 1997, 342–352). This synthesis was based on two fundamental decisions: one in favor of a fundamentally private economic system and against a planned state economy, and one against an entirely free play of market forces in an unregulated liberal market. This hypothesis also included Article 153 on property. Paragraph 1 of Article 153 guaranteed the right to private property but left it to the future parliamentary legislator to precisely define the concept of property and the limits of property. Paragraph 2 permitted expropriation for the common good if provided for by law, and only with appropriate compensation. If a law adopted by the Reichstag stipulated this, compensation would not be mandatory. In addition, Article 156 enabled the socialization of businesses. These provisions had been included in the constitution in the founding phase of the Weimar Republic as a compromise for the socialists who could then hope that they would be able to realize some of their economic policy ideas in a manner consistent with the constitution.

The *Reichsgericht* in Leipzig very soon thwarted this calculation by the socialists. Two decisions were key here: first, the court claimed its right to judicial review of laws under Article 153, thus disempowering the parliamentary legislator, and second, the court broadened the scope of that guarantee of private property to such an extent that even monetary losses due to inflation were considered expropriations by the state. Other court rulings expanded property rights and the state's obligations to compensate in cases of expropriation even further. This even applied to the protection of historic buildings, which was guaranteed in Article 150. The courts' decisions had transformed property law, potentially an instrument for reforms inspired by socialism, into a wall protecting existing legal positions regarding property. The text of the constitution and the decisions of the *Reichsgericht* provoked socialist, liberal, and conservative legal scholars to influence the entire political development of the republic by interpreting Article 153 to suit their own ends. Schmitt and Kirchheimer were on the front lines of these debates.[32]

Again, it was Schmitt who tossed the first pitch. In a 1926 legal opinion on the expropriation of princes, he had declared expropriation without compensation to be unconstitutional (see Schmitt 1926d). In *Constitutional Theory*, he explained his position on property rights more systematically. He viewed the Weimar Constitution as a constitution of a state with a bourgeois *Rechtsstaat*. The "fundamental decision" in favor of the bourgeois *Rechtsstaat* included a fundamental sociopolitical decision: "the decision must already have been made to go with the existing social *status quo*, in particular the retention of the bourgeois social order." (Schmitt 1928b, 84) By defining the constitution as a bourgeois *Rechtsstaat*, property law attained an outstanding legal position in the fabric of constitutional norms in the classical liberal sense and became an "institutional guarantee" (Schmitt 1928b, 208).

Since Schmitt defined the Weimar Constitution as having a one-sided bias toward the *Rechtsstaat* and ignoring the social aspects that were also enshrined in other articles, it was only logical that he prioritized the protection of private property in his interpretation of Article 153. Its wording was "contradictory and unclear" (Schmitt 1928b, 210), yet the absolute guarantee of private property was imperative against the background of the alleged primacy of the bourgeois *Rechtsstaat* in the constitution. Schmitt also discussed the possibility of expropriation mentioned in Paragraph 2 of the article against the background of the alleged primacy of the *Rechtsstaat*. The *Rechtsstaat* implied the general character of the legal norm, i.e., the ban on making laws targeted only at individual people or groups of people. This would mean falling back into "absolutism" (Schmitt 1928b, 190) albeit under the premise of democracy. Following this logic, laws that concerned individual cases and directly enabled specific cases of expropriation were a priori unconstitutional.

In 1929, Schmitt stated his views specifically on the question of expropriation in another essay. He criticized the courts' practice of interpreting practically every limitation of property rights as expropriation. He considered this to be *Auflösung* (Schmitt 1929c, 110), a misguided, overly broad application of the concept of expropriation to the extent that it no longer meant anything. However, he insisted that general legal norms never gave anyone a right to compensation. He also deemed undertakings to conduct expropriations through laws adopted by parliament to be examples of inappropriate application of the concept. These attempts were "an abuse of the form of legislation for the purpose of specific acts of expropriation" (Schmitt 1929c, 116). In contrast, he reminded readers that "the protection of private property under Article 153 was fundamentally determined by the legal situation of the year 1919." (Schmitt 1929c, 116) The device Schmitt used for his constitutional theory consisted of considering the revolution of 1918/19 as being a mere change of political institutions, while the continuity of the bourgeois system since the German Empire was retained at the same time. The political thrust of the narrowing of expropriation law arising from this is obvious: Schmitt's purpose was to prevent socialist aspirations to potentially carry out expropriations without compensation.

This legal position was unacceptable to the socialist left, whereas the communists saw their position, namely agitating against the bourgeois capitalist system, borne out. From

32 On the debate between Kirchheimer and Schmitt on expropriation, see Klein (2022).

a socialist perspective, the constitution had stipulated in Article 153 that property did not enjoy absolute protection as a fundamental right, but that its protection was at the disposal of the parliamentary legislator in terms of its substance and scope.[33] To them, Schmitt's position meant erosion of the social basic rights and an amputation of the competences of the legislature. Ernst Fraenkel and Franz L. Neumann, and other younger socialist legal experts followed Hermann Heller in taking critical aim at the factual nullification of expropriation law. The most extensive and thorough works on expropriation in Article 153 were by Otto Kirchheimer who had already complained in *Weimar—and What Then?* that the original purpose of Article 153 had been perverted into its opposite (see Kirchheimer 1930e, 57). For one thing, Kirchheimer was responding to the judicial review conducted by the *Reichsgericht*, which Schmitt had already criticized; Kirchheimer agreed with Schmitt on this point. For another, he formulated a clear position opposing Schmitt's attempt to block potential expropriations without compensation.

Kirchheimer wrote three larger pieces on this subject: two essays and one monograph. In the essays, he attacked the decisions of the *Reichsgericht* with polemic verve.[34] The target group of his main contribution to this subject, an academic book on the limits of expropriation, was exclusively legal experts. The book was published a few months after *Weimar—and What Then?* and can also be read as his answer to the question as to how open the Weimar Constitution was to socialist policies. All these pieces have the same thrust in criticizing ideology. Kirchheimer believed that the legal terms "property" and "expropriation" were not neutral concepts but were embedded in certain traditions. Using them uncritically in the legal context would convey the sociopolitical values of the past along with the terms and concepts themselves. With these three pieces, he aimed to overcome the sociopolitical persistence of these legal concepts.[35]

The weightiest of these publications is the 75-page monograph *The Limits of Expropriation*.[36] Kirchheimer thanked two mentors in the preface of the book: Carl Schmitt "for the research question itself as well as for some of its aspects" and Hermann Heller for the "interest he showed in the work."[37] Schmitt and Heller's positions served as the two points of reference forming the basis for the structure and argument of Kirchheimer's study. In his interpretation of the fundamental rights under the Weimar Constitution, Heller, unlike Schmitt, had explicitly deemed the competence granted the legislature to expropriate in order to create a social democracy to be necessary.[38] However, Kirchheimer also emphasized that "the only critical analysis" (207) of the conceptual expansion of the term "expropriation" was that by Schmitt.

In his line of argument in the first part of the book, Kirchheimer did not follow a purely legal methodology but argued along lines of the sociology of law by discussing the transformation of the meaning of property since John Locke. The function of the article in

---

33 See Neumann (1930, 68–73). On this debate see also Ridder (1977, 174–177).

34 See Kirchheimer (1930b) and (1930g).

35 See Bumke (2002, 189–203), Meinel (2011, 196–200), and Buchstein (2017a, 57–65).

36 See Kirchheimer (1930h). The following page numbers refer to this text. On Kirchheimer's writings on expropriation see also Simard (2023, 52–60).

37 Unfortunately, the preface is omitted in the English translation of the text.

38 See Heller (1924, 310–316) and Heller (1926, 375–409). On Heller's concept of democratic socialism, see Henkel (2012, 454–482) and Buchstein and Jörke (2023).

the constitution concerning property could only be recognized in the social and political context of a particular case. Kirchheimer gave a rough outline of the theory of property in terms of its intellectual history from Locke to Montesquieu to Ferdinand Lassalle, Karl Marx, and Austro-Marxist Karl Renner. The liberal view with its orientation toward private law could no longer be sustained in the age of industrialization since the large number of infrastructure measures at the time made expropriations necessary. The Weimar Constitution had recognized this fact in Article 153 and had developed it qualitatively.

In the second part of his book, Kirchheimer turned to the debate among legal scholars about interpreting Article 153. He conceded that it was indeed possible to speak of an institutional guarantee with respect to property. Referencing deliberations in Schmitt's *Constitutional Theory*, he argued against the author that the significance of such an institutional guarantee was only minor, in contrast to the institutional guarantee regarding the status of civil servants in Germany[39] since the potential substance of property was ultimately always subject to determination by the legislature. Kirchheimer criticized the theories championed by other legal scholars such as Martin Wolff, Gerhard Anschütz, Heinrich Triepel, and Walter Schelcher as attempts to turn the intentions of the legislator of the constitution upside down. These leading legal scholars, he claimed, expanded the area of expropriation in their theories, thus creating "the convenient possibility to characterize every act of intervention as expropriation" (115). Their reinterpretation of Article 153 negated the fundamental essence of legislation in a democratic state and also weakened the legislature by introducing a material right to judicial review. To Kirchheimer, this was driven by openly expressed anti-socialist and anti-parliamentarian resentment and was an expression of altered societal power relations. Since, at the time, the bourgeoisie feared that parliament would enact legislation on property that was contrary to its private interests, legislation on this matter was to be subject to a new authority the bourgeoisie believed was more favorably disposed to it, namely the courts.

Kirchheimer again named Carl Schmitt as a key supporter of his hypotheses, stating that Schmitt's interpretation could be explained against the background of his general framework of analysis for understanding the current state of the constitution as the result of the interplay of the bourgeois state under the *Rechtsstaat* on the one hand and democracy on the other. Kirchheimer thought this type of interplay would become problematic long-term. He asked Schmitt the rhetorical question as to how far the mass democracy of the twentieth century could retain bourgeois *rechtsstaatlich* elements without in the long run suffering severe damage to its basic democratic character (116). Explicitly directed against Schmitt, he wrote:

> The Constitution has, however, directed established rights and their incorporation in the state into the sphere of legislation. Since it is a democratic constitution, this has been done not only, as Schmitt argues, to generally restrict property but to give Legislature a free hand in the initiation of individual acts of expropriation. (115)

39 To this day, the status of civil servants (*Beamte*) in Germany differs from that of civil servants in the US in two ways: for one thing, it requires civil servants to exercise a special duty of loyalty toward the state (for example, they do not have the right to strike), and for another, civil servants enjoy certain privileges (such as a tenured position, and additional health insurance and pensions).

Although a general right to property continued to exist, Kirchheimer thought the legislature had the power to define its "substance and limits": "As long as there is a category of property, then property signifies an absolute right of domination, although this 'absolute' quality is only valid for the sphere of private law and is subordinate to the sovereignty of the state and hence the Legislature" (111). In other words, a socialist majority in the parliament would have considerable leeway when carrying out expropriations without compensation.

Kirchheimer repeated his main objections to Schmitt in another article on property titled *Eigentumsgarantie in Reichsverfassung und Rechtsprechung* [The right to property in the constitution of the Reich and in jurisprudence] in the August 1930 issue of the journal *Die Gesellschaft*. Kirchheimer began with a reference to Karl Renner's book on the functional transformation of the legal institution of private law (see Renner 1929).[40] He added a compilation of Catholic voices critical of property—a clear swipe at Schmitt, who even after his excommunication portrayed himself as a Catholic intellectual. Again, Kirchheimer accused Schmitt of overlooking the "democratic origin" (Kirchheimer 1930g, 343) of the legislation on expropriation and making overly great concessions to liberal theory for this reason.

At their core, the differences in argument between Schmitt and Kirchheimer can be traced back to their assessments of the 1918 revolution relating to constitutional policy (see Klein 2022, 47–49). To Schmitt, the 1918 revolution had ultimately changed nothing about the bourgeois system; the only change was to its political form. Kirchheimer saw the events of 1918 as the beginning of a dual revolution: the transition from the bourgeois capitalist system to a balance of capitalist and socialist elements and the transition from the bourgeois *Rechtsstaat* to a social *Rechtsstaat*. In contrast to Schmitt, Kirchheimer emphasized the revolutionary aspect of the upheavals of the societal structures in the founding of the Weimar Republic as well as the democratic sources of parliamentary lawmaking. The practical political consequences arising from Kirchheimer's reading of the revolution of 1918/19, the opposite of Schmitt's, were just as evident: it was necessary to turn to the political struggle for parliamentary majorities in order to harness Article 153 for socialist ends.

Schmitt was unruffled by Kirchheimer's criticism and responded with friendly and even appreciative words. When Kirchheimer applied to the Rockefeller Foundation for a one-year research stipend in the US in December 1930, Schmitt praised Kirchheimer's book in his letter of recommendation as "one of the best German works on the concept of expropriation."[41] He praised the book in an essay published in 1931 because Kirchheimer had made clear that the idea of socially responsible use of property was actually "directed against property" (Schmitt 1931d, 161). He agreed with Kirchheimer that the institutional guarantee of property rights under purely democratic auspices would become precarious and could even "be accepted by the most extreme communists" (Schmitt 1931d, 162). In that essay, he thanked Kirchheimer and Franz L. Neumann for acquainting him with Austro-Marxist Karl Renner's socialist doctrine of property in his university seminar (see

40 The first edition of Renner's book was published in 1904.

41 Letter of recommendation by Carl Schmitt for the Rockefeller Foundation concerning Otto Kirchheimer's application dated 4 December 1930. Carl Schmitt Papers, RW 265–13422/1-2.

Schmitt 1931d, 168). To Schmitt, Kirchheimer's book was instructive for the simple reason that he saw his view of the vitality of the socialist threat confirmed.

A response along the same political lines, but less friendly, came from Ernst Rudolf Huber, Schmitt's student and Kirchheimer's classmate in Bonn. Huber regularly published articles under the pseudonym Friedrich Schreyer in *Der Ring*, the journal of the Deutsche Herrenklub, in which authors of the Conservative Revolution sounded out the political situation. Huber attacked Kirchheimer for providing the ideology to support "creeping toward socialization through the back door" (Huber 1931b, 163).[42] From Kirchheimer's perspective, Huber's assessment was only partly correct. He would have granted "creeping toward socialism." But not the "through the back door"—he thought Article 153 provided the opportunity for socialism to enter through the front door.

8. Presidential dictatorship

Shortly after the collapse of the Grand Coalition in March 1930, rumor had it in Berlin that Hindenburg's camarilla had long been planning Chancellor Müller's overthrow and had already selected Brüning as his successor. The SPD saw itself once again relegated to the opposition at the Reich level. It was also leaked that Schmitt had been asked informally in advance to write a legal opinion for the new government. Schmitt had personally met Brüning in early 1928 and had noted in his diary how much he looked forward to working with him in Berlin.[43] As things stood, Brüning and his cabinet could not rely on a majority in the Reichstag but depended completely on the support of the President of the Reich. This posed a problem for them inasmuch as Article 85 of the constitution stated expressly that budgetary powers rested with the Reichstag alone. Even before Brüning took office, the wily advisors in Hindenburg's circle had already come up with the following idea: in the event that parliament rejected the government's legislative proposals, they would all be declared essential emergency measures so they could be enacted in the form of emergency decrees applying Article 48 of the Weimar Constitution. This course of action was politically controversial, and it was the biggest topic of constitutional law of the day.[44] In the summer of 1930, Schmitt was at the center of creating constitutional legitimacy for this existential basis of Brüning's presidential dictatorship; a few years earlier, however, he had championed the opposite position on Article 48 (see Kennedy 2011).[45]

Article 48, paragraphs 1 and 2 of the Weimar Constitution reads as follows:

> [1] If public order and safety are substantially disturbed or endangered in the German Reich, the Reich President may take the requisite measures to restore public safety and order, if necessary, with the help of armed forces. [2] To this end, he may temporarily

42 The reference to Huber as the author of this article is to be found in Breuer (2012, 182–183).

43 See Carl Schmitt, diary entry of 4 January 1928 (Schmitt 2018, 193).

44 For an overview of the contemporary debate on the applicability of Article 48 before the Brüning era, see Gusy (1997, 107–109) and Stolleis (1999, 114–116).

45 For a comparison of presidential emergency power in the Weimar Constitution and the constitution of the United States see Kronlund (2022).

> annul, completely or in part, the basic rights laid down in Articles 114, 115, 117, 118, 123, 124, and 153.[46]

The third paragraph stated: "[3] The Reich President must immediately inform the Reichstag of all measures taken" and: "[4] The measures must be revoked by the at the request of the Reichstag." The second paragraph of Article 48 had been one of the most hotly debated provisions during consultations about the constitution. On the one hand, the young republic was to be protected effectively against uprisings. On the other, sufficiently clear limits were to be placed on the state's security agencies. The intended balance between these two goals did not find its way into the text of the constitution because the majority in the constitutional convention favored the political system's ability to act. The law provided for in Article 5, paragraph 5 regulating the competencies of the President in more detail was never enacted because of the resistance by President Hindenburg and the *Reichswehr* (the armed forces). The predominant interpretation of Article 48, paragraph 2 was that in exceptional cases, the President had the competency to suspend the seven fundamental rights mentioned. It followed from this that all the other provisions of the constitution were considered untouchable. The logical consequence of this predominant interpretation was that the constitution did not permit the President and the Chancellor appointed by him to adopt the budget without the consent of the Reichstag.

Kirchheimer's prompt criticism of the new government followed this logic, too. On 4 April 1930, five days after Brüning took office, he reacted to the change of government in a newspaper commentary. Brüning had announced in his government policy statement that he would present a package of measures to the parliament and that it would be his only attempt to solve the current problems in collaboration with the Reichstag. This could be interpreted as a publicly declared threat to establish a presidential cabinet. Kirchheimer's commentary was printed in the socialist *Tribüne* under the title "Artikel 48 – der falsche Weg" [Article 48—the wrong course].[47] He focused on the question whether the future course of lawmaking announced by Brüning was in line with the constitution. In terms of substance, Kirchheimer characterized the measures in economic, financial, and social policy proclaimed in the government policy statement as systematically implementing business associations' programs to one-sidedly shift as many costs as possible to blue-collar workers and the unemployed in the "struggle for internal distribution of the burdens" (202). He vehemently opposed applying Article 48 in order to do this, not least for procedural reasons. He quoted the text of the constitution in his argument. According to the constitution, Article 48 could be applied only in cases in which "public order and safety are substantially disturbed or endangered." Yet it was apparent that such disturbances or endangerments of public order and safety were not substantial. Kirchheimer referred equally to liberal, conservative, and German-nationalist constitutional law professors and described the difference between how former President of the Reich, Social Democrat Friedrich Ebert, had applied Article 48 "temporarily" only for brief and limited periods of time and how Brüning's Presidential Cabinet planned to do so permanently.

---

46 The translation of the Weimar Constitution is taken from Tribe (2020, 195).

47 See Kirchheimer (1930c). The following page numbers refer to this text.

Kirchheimer also referred the newspaper's social democratic readership to an article by the "well-known German constitutional law professor Carl Schmitt" (204) from 1924 on this issue. Schmitt had made the important distinction between a measure that would remain temporary and could be covered by Article 48 and a legislative procedure that was not covered. Schmitt spoke of an "abuse" of Article 48 if it was applied to expand the right of the President of the Reich to put a budget into effect (see Schmitt 1924c, 208–221). To Kirchheimer, it was clear that the actions threatened by Chancellor Brüning and President of the Reich Hindenburg were unequivocally "outside the constitution" (204). He even used Schmitt's theory of dictatorship to declare Brüning's entire government unconstitutional as a matter of principle:

> Schmitt's definition of the nature of dictatorship illuminates in a flash who upholds the constitution and who violates it: 'Dictatorship is like the act of self-defense: never just action, but also reaction. Therefore, implicitly, the enemy will not conform to legal norms that the dictator regards as a binding legal norm' [Schmitt 1921, 118] (204).

The legal basis for Brüning's dictatorship could only be found in the constitution. If Brüning enforced his government program against the Reichstag, then this had nothing to do with Article 48; in fact, his course of action was outside the constitution. Readers should be reminded that there was no constitutional court in the Weimar Republic to resolve such conflicts. For that reason, Kirchheimer conceded that it was virtually impossible to prevent such unconstitutional actions on the part of the Presidential Cabinet through such a channel. He placed his hopes in other courts. He encouraged the financial and revenue courts of Prussia and other German states and also the *Reichsfinanzhof* (see List of German Courts) to rain on the Brüning government's parade. However, the only options were protest and political mobilization against the government dictatorship and its "entrepreneurial ideology" (202). In his article, Kirchheimer referred mainly to Schmitt and his criticism of Brüning's regime of emergency decrees. Not only did he draw on Schmitt's book on dictatorship and the 1924 article he quoted, but also on Schmitt's plea two years later for a limitation of the President's extensive dictatorial power. In light of the danger of a "boundless dictatorship," Schmitt had argued along the lines of Article 48, paragraph 5 for a solution "based on the *Rechtsstaat*," namely adopting a law including "a detailed list of the preconditions" and the "substance of all dictatorial powers." (Schmitt 1926c, 38 and 41)

While Kirchheimer referred to Schmitt as the key witness for the unconstitutionality of Brüning's actions, Schmitt had already long begun to take the opposite position. In the summer of 1930, Schmitt wrote a legal opinion for Brüning about the existential question for the government whether the President had the competence to determine the budget by emergency decree on the basis of Article 48, paragraph 2. Schmitt now stated that he did. He delivered his legal opinion on 28 July 1930, when it was already apparent there would be new elections, and he had made his support for the government's strategy and its austerity policies clear in preliminary talks. Brüning needed a legal opinion supporting his position because he assumed he would not have a majority in the Reichstag after the new elections, either.

At its core, Schmitt's complex argument[48] was based on four steps.[49] First, he disputed the predominant interpretation according to which only the seven fundamental rights listed in Article 48, paragraph 2 could be suspended. This made no sense, he claimed, if other measures were necessary to prevent an emergency. Second, he asserted that practical experience, court decisions, and also the academic literature had long recognized that emergency decrees could be applied in economic and financial matters, too. Third, Schmitt declared the President's decrees to be equivalent to laws adopted by a parliament. Fourth, Schmitt introduced his particular definition of a commissarial dictatorship (see Schmitt 1921, 1–19) in order to grant the President all measures he deemed necessary. Since the commissarial dictatorship was the temporary limited negation of the norm that was to be protected, a dictator who was to preserve the constitution had to have the power to disregard the constitution to this end. Schmitt concluded his deliberations with a political statement of allegiance: "The state of emergency reveals, if I may say so, the core of the state as such." (Schmitt 1931c, 259) The modern state was a state driven by the economy and finance, and it would be an anachronism to desire to turn the development back to the nineteenth century— in order to limit the instruments available to the modern state as a last resort—to those of the traditional state of emergency governed by the military and the police.

Pointing out the fact that the state had transitioned to become a modern *Wirtschaftsstaat* (state committed to promoting economic development) was not controversial under Weimar constitutional law; after all, there were a number of articles in the second part of the constitution regulating this new reality (see Gusy 1997, 342–369). What was controversial was the extensive expansion of the competencies of the President of the Reich contrary to the wording of the constitution. Schmitt's position on granting the President of the Reich competencies derived from Article 48 was the most far-reaching of the Weimar constitutional law scholars. Applying the authority of the dictatorship, which he had supported in his 1930 legal opinion and later in his articles, was not just a continuation of the earlier practice of applying the article of the constitution. Brüning's emergency decrees differed from the previous ones in their scope, their period of validity, and ultimately also in their intent regarding constitutional policy.

Just a few days after Schmitt's legal opinion supporting Brüning, Kirchheimer published an incensed attack on the new emergency decree regime. It was published in the socialist journal *Der Klassenkampf*, titled "Artikel 48 und die Wandlungen des Verfassungsystems" [Article 48 and the transformations of the constitutional system]; Gurland was one of the journal's editors.[50] The title of the article expresses part of Kirchheimer's diagnostic hypothesis: the emergency decrees of Brüning's government had transformed the system of the Weimar Constitution in a move toward an authoritarian state in the interest of the bourgeoisie. The previous system of parliamentarism with its search for compromises between the various social groups had been replaced by an "independent representation of the bourgeoisie alongside their parliamentary parties"

48 Schmitt incorporated parts of his legal opinion (which has not been published in full to this day) in several publications, see Schmitt (1931b), (1931c) and (1932a).

49 For a detailed discussion of Schmitt's position, see Neumann (2015, 174–198).

50 See Kirchheimer (1930d). The following page numbers refer to this text.

(351). What was "fundamentally different" (351) about this form of government was that the "method of giving ground reciprocally [had] finally been abandoned" (351). It was no longer necessary to take the interests of the labor movement into account at all. In this situation, the extensive interpretation of Article 48 had the function of safeguarding the new power relations by constitutional means. Kirchheimer criticized this extensive interpretation using arguments from constitutional law. For one thing, the preconditions for applying the article on dictatorship, namely that public safety was seriously threatened or disturbed, had not been met. And for another, these were not temporary emergency decrees but permanent laws.

Kirchheimer considered what Schmitt had presented as his new ingenious interpretation of Article 48 to be simply unconstitutional. He felt that besides this finding, "one more thing [must be] added" (352): the previous cases in which Article 48 had been applied under Social Democratic President Friedrich Ebert, who had been in office until February 1925, had remained within the realm of tacit or open compromises between the Social Democrats and the bourgeoisie. For the first time, this was no longer the case now. Article 48 was going to be applied not only without regard for the interests of social democracy, but explicitly against the interests of workers. The bourgeoisie was thus undermining the founding document of the Weimar Republic. The republic was built on the foundation of social compromises: "The democracy of compromise has transformed into the democracy of hostile (*feindliche*) military camps" (353). The extensive interpretation enabled the bourgeoisie to revoke the class compromise with the working class without risk and to depart from the parliamentary basis of Weimar democracy. In this transformation of the republic into authoritarian rule by the bourgeoisie, Schmitt had the role of the constitutional law ideologue.

Kirchheimer hoped that the outcome of the new election would make it impossible to form a government without the Social Democrats. But this was not to be. In the elections on 12 September 1930, the SPD suffered slight losses, and the communists gained some votes. There were dramatic changes on the bourgeois side. The right-wing parties that had supported Brüning had to weather serious losses. The biggest winner of the election was Hitler's NSDAP, coming in second at 18.3 percent. It would have been theoretically possible for a majority to form a Grand Coalition in the Reichstag. The Social Democratic Prime Minister of Prussia, Otto Braun, came out in favor of such a "coalition of the reasonable" directly after the election. Yet Brüning invoked Hindenburg's "mission" to make sure the SPD would not be part of a government again and rejected the proposal. He was intent on continuing his policy of austerity on the basis of Article 48.

In this situation, Schmitt's legal opinion provided the legitimation—based on constitutional law and urgently needed by Brüning—for a system of emergency decrees that also abolished the parliament's right to adopt budgets and take out loans. The expanded system of emergency decrees remained highly contested among scholars of Weimar constitutional law. The majority of legal scholars opposed it and renewed demands for a law in which the dictatorial competences of the president were to be clearly regulated—a demand that Schmitt had abandoned by this point. At the Tagung der Deutschen Staatsrechtslehrer (Conference of German Constitutional Lawyers) in Halle in October 1931, the conflict broke out into the open. The majority of attendees voted for a resolution urging the government of the Reich to monitor the situation more closely and ensure that the

President did not continue to abuse Article 48. Schmitt received only two votes from others supporting his opposing position (see Huber 1981, 729–730). He was in the absolute minority with his extensive interpretation and complained in his diary of "the nastiness and malice"[51] of Smend and others who had contradicted him in Halle. The next day, he noted about his stay in Halle: "bought Nazi writings; informational booklets."[52]

By quickly accomplishing what Chancellor Brüning had asked him to do, Schmitt had hoped to be included in his circle of advisors and to enjoy direct access to the center of political power in the Reich from then on (see Neumann 2015, 174–175). Yet, after his initial rapid rise, his contact with the ruling political elite came to an end for the time being. The circle around Brüning did not approach him again, leaving Schmitt to lick his wounds. During the almost two years of Brüning's term as Chancellor, Schmitt noted his personal "infuriation about Brüning"[53] in his diary multiple times after hearing how the latter was maligning him. The failure of Schmitt's first attempt to attain a greater political role did not frustrate his pleas for a presidential dictatorship, however. He now sought new contacts with confidants of politically influential Reichswehr General Kurt von Schleicher (see Pyta and Seiberth 1999, 430–432). When Brüning had been forced to resign in late 1932, Schmitt supported Franz von Papen, who had been selected as his successor in the presidential dictatorship.

Kirchheimer continued to pursue the strategy of argumentation he had taken in the spring of 1930 to play the "old" Schmitt off against the "new" Schmitt. For example, he referred to an essay by Schmitt from 1925 to argue against overly far-reaching competencies of the Reichstag to dissolve (see Kirchheimer 1932c, 399 and 405). Elsewhere, he cited Schmitt's *Constitutional Theory* which mentioned certain limits to changing the constitution (see Kirchheimer 1932d, 411). He attacked the presidential dictatorship with ever sharper words, consistently polemicizing against Schmitt as its proponent. He avoided using the term "fascism" to characterize the political system preferred by Schmitt. Instead, he chose "authoritarian state" as an umbrella term that included all dictatorial alternatives to the political system of the Weimar Republic. Kirchheimer's choice of terms also illustrates that his political language was quite different from the vocabulary of the Communist Party that accused Mussolini's Partito Nazionale Fascista, Hitler's NSDAP, German conservative parties and the SPD alike of being fascists.

9. Who is the guardian of the constitution?

Despite their disagreements about presidential dictatorship, the personal relationship between Kirchheimer and Schmitt was obviously still positive, at least until the summer of 1932. There are 18 entries about Kirchheimer in Schmitt's diaries between November 1930 and November 1932. These include notes about regularly going out to eat after the seminar, going on walks and traveling by *S-Bahn* (commuter rail) with him, as well as about brief visits to Schmitt's house, and Schmitt visiting Kirchheimer and his wife and

51 See Carl Schmitt, diary entry of 29 October 1931 (Schmitt 2010, 141).

52 See Carl Schmitt, diary entry of 29 October 1931 (Schmitt 2010, 141).

53 Carl Schmitt, diary entry of 26 September 1931 (Schmitt 2010, 138).

their baby. The entries from this period are almost identical to those from his time in Bonn. The discussions with Kirchheimer in the seminar were "quite nice" and he went for a walk with him afterwards.[54] Kirchheimer came over to his place in the evening "and drank a bottle of wine again"[55] and they had a three-hour conversation about Soviet foreign policy and the repression by the German police. A number of times, the two of them walked to the railway station together after the seminar while they continued their discussions. When Kirchheimer visited him in March 1931, Schmitt wrote in his diary "I like him" and that he had bought chocolate for Kirchheimer's baby Hanna.[56] In June 1931, Schmitt and his wife Duška came over to Kirchheimer's place to visit little Hanna. On this particular occasion, the adults talked about the chances of having new national elections. Schmitt noted in his diary that Kirchheimer was "smart and sympathetic."[57] He praised his wit and intelligent contributions again in an entry about his seminar session on fundamental rights and the *Rechtsstaat*.[58] All that was soon to change during the dramatic political events of 1932.

To grasp the complexity of the personal relationship between the two men during the growing crises of the republic, we must not forget Schmitt continued to support Kirchheimer. In December 1930, when Kirchheimer applied to the Rockefeller Foundation for a one-year research stipend in the US, Schmitt endorsed his plans as a reviewer. In his letter of recommendation, Schmitt praised the "particular merits of Kirchheimer's way of working and producing", stating he had a "good eye for the sociological and historical circumstances and developments from which he derived both the legal concepts and the theoretical arguments."[59] However, in contrast to the young philosopher Leo Strauss, whose application Schmitt had also supported, Kirchheimer was unsuccessful.

According to Schmitt's diary entry, Kirchheimer was disappointed after learning about the decision and became desperate.[60] He decided to follow two tracks at the same time for his future career. One the one hand, he still tried to obtain a position in academia. On the other hand, he had to make living and so started working as a lawyer. On 2 June 1931, he passed the *Große Juristische Staatsprüfung*, completing his *Referendariat*; he received the grade "sufficient" on the first day of examinations and "good" on the second; his overall grade was "fully satisfactory."[61] After passing his exams, he was unsure about what career path to pursue. His dream job was to be an academic but he considered opportunities at German universities to be hardly realistic at the time. He gained some experience teaching occasionally at the Gewerkschaftsschule (Trade Union School) in Berlin. This had been established by the Räte (council) movement of 1919 and had evolved from a revolutionary educational institution into an institution for

54 Carl Schmitt, diary entry of 6 November 1930 (Schmitt 2010, 53).

55 Carl Schmitt, diary entry of 28 November 1930 (Schmitt 2010, 62).

56 Carl Schmitt, diary entry of 14 March 1931 (Schmitt 2010, 97).

57 Carl Schmitt, diary entry of 13 June 1931 (Schmitt 2010, 116).

58 Carl Schmitt, diary entry of 30 July 1931 (Schmitt 2010, 128).

59 Letter of recommendation by Carl Schmitt for the Rockefeller Foundation concerning Otto Kirchheimer's application dated 4 December 1930. Carl Schmitt Papers, RW 265–13422/1-2.

60 Carl Schmitt, diary entry of 13 March 1931 (Schmitt 2010, 97).

61 An excellent grade at the time. Bundesarchiv Berlin, R 3001, 6322, Ministry of Justice file concerning Dr. Otto Kirchheimer, p. 9.

industry-specific training for employee representatives in preparation for their work as functionaries.[62] Kirchheimer taught courses in labor law and modern European history.[63]

He also applied for a job in the Prussian public service in late 1931 because he desperately needed work, but this was also unsuccessful.[64] He had to earn money somehow. There was nothing left of his inheritance, partly because his brothers had lost money speculating on the stock market. He was now expected to contribute to the family's living expenses. After a visit to his place, Schmitt noted in his diary: "Kirchheimer was depressed because he isn't earning any money."[65] His father-in-law Kurt Rosenfeld, with whom Kirchheimer still had a good relationship despite separating from his daughter in late 1931, not least because of their political differences, advised him to open a law firm. Kirchheimer followed this advice and decided to try his professional luck as one of more than 3,000 lawyers in Berlin. According to his long-time friend Eugene Anschel (see Anschel 1990, 101), his overall personal situation and unclear professional prospects plunged him into a deep personal crisis, which may explain the pause in publications in 1931.

For Schmitt, conversely, 1931 was another golden year of enormous productivity. He wrote essays on international law and the League of Nations.[66] He published articles on a reform of the Reich and the constitution and completed a major commentary on the fundamental rights and duties of citizens according to the Weimar Constitution. He gave several lectures on the competencies of the President of the Reich and wrote pieces defending the system of emergency decrees that Chancellor Brüning was using on an ongoing basis to govern. In addition, in May 1931, he published the book *Der Hüter der Verfassung* [The guardian of the constitution], in which he summarized his criticism of all forms of judicial review and highlighted its political consequences.

The debate about judicial review in Weimar constitutional law had been triggered by a *Reichsgericht* decision in November 1925. That decision asserted that every court in the Reich had the competence to review laws adopted by the parliament with respect to their substantial constitutionality, in other words, to reject them as unconstitutional. The decision divided scholars on Weimar constitutional law along a political front line. Those constitutional law professors who were reserved toward the Weimar Constitution or even rejected it—and they were in the majority—welcomed the broad interpretation of judicial review because they saw it as a fitting check on parliamentarism, which they rejected. The smaller group of jurists who were liberal, leftist, and loyal to the republic rejected the broad interpretation of judicial review just as emphatically. They feared it would bring about a shift of the Weimar class compromise that would disadvantage the working class. Their fears were not unfounded since the *Reichsgericht* decision pertained to problems of inflation that affected not only the working class but especially the middle

62 On the history and development of the Trade Union School in Berlin, see Feidel-Mertz (1972, 70–86) and Olbrich (2001, 185–192).

63 Otto Kirchheimer, Curriculum Vitae (undated, ca. 1939). Emergency Committee in Aid of Displaced German/Foreign Scholars, Public Library, New York. I, A Grantees 1933–46, Box 18, Folder 13 (Kirchheimer, Otto).

64 See Bundesarchiv R 3001, 6322, Ministry of Justice file concerning Dr. Otto Kirchheimer, p. 14.

65 Carl Schmitt, diary entry of 21 November 1931 (Schmitt 2010, 146).

66 On this subject, see Chapter 4.

classes who held their savings in the banks. Since the Weimar judges also belonged to the middle class, their critics called judicial review a method of class justice. The most vocal critics of this judicial review on the side of the Social Democrats included Franz L. Neumann, Ernst Fraenkel, and Otto Kirchheimer.

Only two prominent professors of constitutional law took positions unconnected to this political front line. One was liberal Hans Kelsen, the father of the Austrian Constitutional Court. He argued against granting judicial review to all courts in Germany and in favor of establishing a special constitutional court instead.[67] The other legal theorist deviating from this political front line was Schmitt. In *Der Hüter der Verfassung*, he attacked Kelsen and rejected any form of judicial review, including the establishment of a special constitutional court. Up until that point, Schmitt's position on this question had fluctuated somewhat (see Wendenburg 1984, 175–179). In his 1925 legal opinion on the expropriation of the princes, he had still granted the courts substantial judicial review of laws adopted by the parliament. He was able to protect the German nobility against expropriation with this opinion. In his *Constitutional Theory* of 1928, he had mentioned arguments on both sides concerning judicial review but had himself not taken a consistent position on its desirability. This changed from 1929 on when Schmitt began to voice his view in several publications directed against Kelsen that a constitutional court would be pointless and impossible. He developed his arguments systematically in the book *Der Hüter der Verfassung* and combined them with conclusions about necessary changes to the political system.[68]

Schmitt's key objection to a constitutional court was that settling political issues in the courts would automatically have the problematic consequence of a "politicization of the judiciary" (22). He substantiated this objection with two arguments. The first was methodological. Schmitt believed that the way a constitutional court worked was to apply a general legal norm to another general legal norm. This contradicted the judiciary's characteristic way of working, namely to subsume a matter under a general legal norm. In a case before a constitutional court, "nothing is subsumed; all that happens is that a contradiction is stated, and then a decision is made about which of the norms contradicting each other holds and which one is not to be applied" (43). His second argument was that every real decision by a judge occurs *post eventum*, in other words, "always too late, politically speaking" (33). This was all the more true the more carefully the proceedings were conducted, following judicial procedure and the *Rechtsstaat*.[69]

Both arguments boil down to Schmitt's assertion that, on closer examination, constitutional jurisdiction was not part of the judicial system. Decisions about disputes or doubts pertaining to constitutional law were not matters to be settled by the courts but were always highly political. Taking this assertion as a starting point, Schmitt set out in the following chapters of *Der Hüter der Verfassung* to identify a functional equivalent of the role that Kelsen and others assigned to a constitutional court. He found this equivalent, for Germany, in the role of the President of the Reich. Schmitt preceded this finding

67 On Kelsen's view, see Olechowski (2020, 507–513). On the controversy between Schmitt and Kelsen on this subject, see Vinx (2015) and Olechowski (2020, 507–513).

68 See Schmitt (1931b). The following page numbers refer to this text.

69 On the critical analysis of these two arguments of Schmitt's, see Neumann (2015, 229–232).

with a political diagnosis of "the specific situation of the constitution in the present time" (70), an unequivocal criticism of parties, pluralism, and federalism. The political parties and the "polycracy" of interest groups had caused the parliament to degenerate to nothing more than a stage on which the pluralist state would perform. German federalism made it more difficult to reach uniform political decisions. An "unstable coalition-party-state" (88) had destroyed the "unitary, indivisible unity of the entire German people" (89), which was required by the constitution. In this situation, the composition of a constitutional court would merely reflect the splintering of state unity because of the pluralistic system and would be unable to make decisions with pacifying effects.

Schmitt believed that only a truly independent institution could remedy the situation: a "*pouvoir neutre et intermédiaire*" (132) following Benjamin Constant. This could only be the President of the Reich. For he alone represented neutrality and independence of party politics. The reasons Schmitt gave were the position of the President according to the constitution, i.e., direct election by plebiscite, his long seven-year term of office, his independence from the parliament, and the difficult procedure to remove him from office. He also listed the President's special powers: to represent the Reich in matters of international law, to promulgate laws, to dissolve the Reichstag at any time, and to appeal to the German people directly, bypassing parliament. Of the powers assigned to the office of the President, the most important were ultimately those under Article 48 of the constitution. It authorized the president to declare a state of siege in times of crisis and to rule by emergency decree. Schmitt interpreted the strong constitutional position of the President of the Reich as an "error in terms of legal theory" because it provoked a breach with the organizational principle of the bourgeois state under the *Rechtsstaat*. This breach could have dangerous consequences and tear the constitutional order apart from within. Yet this danger could be averted if the doctrine of *pouvoir neutre* was "developed further" (137). And this was precisely Schmitt's goal.

In his view, the President of the Reich was the only possible true guardian of the Weimar Constitution. Not only would a special constitutional court be entirely superfluous but it would be impossible in the framework of the Weimar system. At the end of the book, Schmitt did not mince words in his explanation of the role he ascribed to the President in the current political situation: only the democratically legitimized President of the Reich could be an effective "counterweight against the pluralism of social and economic power groups" (159). He alone could act as the "guardian and upholder of the constitutional unity and integrity of the German people" (159). He alone had the "authority" (159) to make state politics "capable of taking action" (159) in the midst of all conflicts and to maintain that capability. In this piece, Schmitt elevated the role of the President in constitutional policy by adding a further component: he liberated his political preference for the President of the Reich from the odium of dictatorship and also distanced himself from the monarchist doctrine of a "superior third party."

Kirchheimer had already criticized hypotheses like this previously. It was an "erroneous conception" (Kirchheimer 1930e, 68) to assume that a President of the Reich could liberate himself from all political ties and act completely independently of special interests, and he declared that such an assumption was sociologically uninformed nonsense.

10. Conclusion: The art of quoting each other

One of the special features of Schmitt's theory was that it equated popular sovereignty with the constituent power of the people, thus devaluing all firmly institutionalized elements of democracy. Schmitt consistently played the people, which he asserted was not bound by law, off against all established institutions of democratic decision-making: the people "remains the *Urgrund* (origin) of all political action, the source of all power, which expresses itself in continually new forms, producing from itself these ever renewing forms and organizations" (Schmitt 1928b, 128). He rejected domesticating, as it were, "the people" itself, which would make it a "state body." This concept of the people does not presuppose an ontologizing *völkisch* (of the *Volk*, chauvinistic-nationalistic, antisemitic; see Glossary) mysticism. But it ontologizes something else: the element of non-organization. Kirchheimer did not agree with this anti-institutional core of Schmitt's theory, as can be seen clearly in his defenses of parliamentarism and democracy.

During the four years from 1928 to 1931, the personal relationship between Kirchheimer and Schmitt continued to be good, and they often met and went for walks in Berlin; after a very short time, the constellation of teacher and student was a thing of the past. They now exchanged manuscripts and reprints frequently so that they could even quote from the other's as yet unpublished texts.[70] Kirchheimer, who was only twenty-three when he arrived in Berlin, self-confidently produced a number of publications on various topics. Schmitt began to quote Kirchheimer as early as 1929. Not surprisingly, Kirchheimer quoted from Schmitt's writings much more often. Even in the absence of a precise quantitative analysis of citations in Kirchheimer's works from 1928 to 1931, it is easy to detect that he quoted no other expert on Weimar constitutional law as often as Schmitt. After he had moved to Berlin, Schmitt's political positions became more radical. The more openly he advocated for a presidential dictatorship after 1930, the more often Kirchheimer cited him in order to almost address him directly.

Ellen Kennedy's statement that Kirchheimer had only begun to criticize his former teacher Schmitt in the summer of 1932 (see Kennedy 1986, 399 and 416)[71] is incorrect in light of the many differences between them described above. The assertion by Stephen Turner about Kirchheimer's "dependence on Schmitt," which was hidden "under a layer of dismissive references to Schmitt" (Turner 2011, 120) in his writings of this time, is also inaccurate. A summarizing comparison of their writings after the end of the period when they were both in Bonn through the end of 1931 shows that their differences extended across the entire spectrum of the topics they worked on: the purpose of parliamentarism, the role of political parties in modern democracies, the potential of the Weimar Constitution for stability and development, the function of Article 48 and the presidential dictatorship, property rights and expropriation, and their assessments of Italian fascism. Regardless of these substantive differences, Volker Neumann has rightly

70 For example, in his book on expropriation, Kirchheimer quoted from an unpublished legal opinion of Schmitt's on a German-Polish agreement dated October 1929 on regulating questions relating to property. In *Legality and Legitimacy*, Schmitt quoted from the manuscript of the eponymous essay by Kirchheimer, without indicating page numbers.

71 Following Kennedy, see also Bavaj (2007, 44–49).

pointed out that there are some similarities between their writings at the formal level (see Neumann 1981, 236–239). Both preferred the format of shorter works inspired by topical political events, revealing an intention to intervene politically; this may explain why neither left an oeuvre with a systematically developed theory. Both emphasized style and rhetoric in their works. Both used strong words and prized new terms to bring things to a head. Their texts often resound with bold sentences that assert radical acts of will. And both occasionally argued in an openly agitational manner. Each in their own way, Kirchheimer and Schmitt represented a type of political thinking in which theoretical analysis and political intervention were inseparable.

The pattern of communication was almost always the same in their publications from these years: Schmitt took the first step by formulating a position on a particular question and then Kirchheimer grappled with it in his criticism. Yet he did so in five different ways. One was to take up Schmitt's concepts and theorems and frame them differently in social theory, thus arriving at a different assessment. The best example of this is the hypothesis of the structural change of parliamentarism. To Schmitt, it was proof of the historical demise of parliamentarism; Kirchheimer interpreted it in a positive light, as a new phase of mass democracy. In a way, Kirchheimer exploited Schmitt's outstanding reputation for his own purposes. He borrowed the authority of a constitutional law professor recognized across all political camps to support his own argument as long as it seemed to fit. A second way was to present Schmitt's hypotheses and then formulate them more pointedly in the next step of the argument. The best example is Kirchheimer's *Weimar—and What Then?* In his *Constitutional Theory*, Schmitt referred to "dilatory formulaic compromises" and Kirchheimer to a "constitution without decision." A third way was to present Schmitt himself as a witness against Schmitt. The best example of this was Kirchheimer's criticism of Schmitt's extensive interpretation of Article 48 to justify the presidential dictatorship. Kirchheimer reminded readers of earlier works by Schmitt on the subject in which he had promoted strict regulation of emergency powers, a position Kirchheimer agreed with. The fourth was to "expose" Schmitt, either as a bourgeois ideologist as in the case of property rights and expropriation or as an anti-constitutional supporter of authoritarianism in his interpretation of Article 48. A fifth way, finally, was to go on the offensive and attack Schmitt and his positions as naive nonsense—for instance, in Kirchheimer's critique of Schmitt's panegyric on the nonpartisanship of the President of the Reich.

Schmitt's reactions to this barrage of criticism did not follow a uniform pattern, either. He usually ignored it—at least in public; it is not difficult to imagine that they spoke about these topics in their frequent conversations. When Schmitt felt it incumbent upon himself to respond publicly, he heaped great praise on Kirchheimer. He extolled Kirchheimer's book on the problem of expropriation, and even more his analysis of the constitution in *Weimar—and What Then?* But using a similar tactic to Kirchheimer, who had done so on occasion, he placed Kirchheimer's hypotheses in the context of a completely different theoretical frame of reference. He considered Kirchheimer's book on expropriation particularly instructive, not least because he could use it as proof of the socialist threat. In Schmitt's view, the pointed analysis of the constitution in *Weimar—and What Then?* became evidence of the socialist movement's vitality and determination to fight.

Visualizing the cascade of the fundamental differences described in this chapter, we wonder once again what drove the socialist jurist Kirchheimer—and *cum grano salis* also the leftist trade union attorneys Franz L. Neumann and Ernst Fraenkel—to seek such proximity to Schmitt during these years. It is easy to become lost in speculation when attempting to answer this question. Yet one of the reasons is certainly fascination with Schmitt's personality, which has been widely discussed. One facet of it was that although he reacted to the Young Socialists' criticisms in a friendly manner, he simultaneously gave them the impression that he considered their opposing views to be taken seriously because they were dangerous. They were political enemies, but following *The Concept of the Political*, this did not necessarily mean they had to become personal enemies, too. Another reason was certainly that Kirchheimer and the two other Young Socialists hoped their academic careers could be promoted by Schmitt since the latter had the reputation of being very tolerant in those days.

In my opinion, there were another three even more important reasons. First, Kirchheimer and Schmitt shared the diagnosis that the tensions in the constitution would not be tenable for long but would have to be resolved in one political direction or the other. This diagnosis of the crisis was easier for Kirchheimer to formulate if he was in contact with Schmitt, who was a master of evoking ever new crises. Second, Schmitt was a widely known critic of the Weimar parliamentary democracy. Proximity to him offered the opportunity to observe firsthand, as if in the lion's den, which new lines of argument he was devising to support his positions. Third, Kirchheimer (as well as Neumann and Fraenkel) found Schmitt to be one of the very few German legal scholars—besides Hermann Heller—who were genuinely interested in socialist theory, albeit, in Schmitt's case, as a form of observing the enemy.

When reconstructing the influence of Schmitt's writings on Kirchheimer, we must not forget how much Schmitt benefited from Kirchheimer. Schmitt had no deeper knowledge of the work of Marx and Engels. He was familiar with some writings by Lenin and Trotsky but had only limited knowledge of the debates among the different strands of current-day Marxism. It was in particular through Kirchheimer that he gained insights into Marxist discussions and the radical leftist groups' worlds of ideas to which he would otherwise have had no access.[72] Not only did Kirchheimer convey valuable information from the socialist debating circles but later, in Berlin, he also facilitated Schmitt's personal contact with his father-in-law, socialist lawyer Kurt Rosenfeld, whom Schmitt met a couple of times. He also connected Schmitt to Franz L. Neumann and Ernst Fraenkel. Kirchheimer, Neumann, and Fraenkel as a group had their own significance for the development of Schmitt's legal theory. As he argued with this younger generation of socialist jurists, he was able to readjust and substantiate the positions he considered appropriate for the changing political *Lage* of the republic at the time.

72 See Neumann (1981, 239) and Breuer (2012, 111–140).

Chapter 4: Two Versions of Anti-Imperialism

Right at the beginning of his studies in Bonn in the winter semester of 1926/27, Kirchheimer attended Schmitt's seminars and lectures on international law. From the outset, he learned about Schmitt's deliberations on international law in great depth—the subject was also part of his oral state examination in law, which Schmitt administered. In his doctoral dissertation entitled *The Socialist and Bolshevik Theory of the State*, Kirchheimer took up this thread from his legal training once again by approaching Schmitt's hypotheses on international law in a positive, if somewhat unconventional, way. Yet this positive perspective was soon to change. As with his work on topics of domestic and constitutional policies of the Weimar Republic, Kirchheimer clearly distanced himself from the influence of Schmitt's theories on international law from 1930 on. He finally came down on the side of the position Schmitt had previously attacked so vigorously.

1. Schmitt's early writings on international law

Schmitt's legal interest in international law can be dated precisely to 1923, the year of crisis in the Weimar Republic. Biographical scholarship on Schmitt considers the political events of this year to be the catalyst mobilizing his nationalism and his hatred of the French (see Mehring 2014a, 137). From then on, not only did he intensify his attacks on the Weimar Constitution but also felt compelled to take up the struggle against "Versailles" and "Geneva" at the level of foreign policy, too. Here, Versailles was code for the peace treaty conditions imposed on Germany and Geneva for the League of Nations, which had been established in 1920 and was headquartered there.

What had happened in 1923, the year of crisis in the republic, to cause Schmitt to react so strongly? On 11 January 1923, French and Belgian troops occupied the Ruhr (see Winkler 2001, 434–438). The reason given for this invasion was a pretense. In line with the decisions taken by the Allied Reparations Commission, the German Reich was accused of a culpable breach of its obligations to supply coal and other goods. Although the accusation of Germany being in breach of contract was factually correct, the reason why the German government had failed to fulfill its obligation was economic hardship. Nonetheless, this breach of contract was at least negligent inasmuch as the French government had apparently only been waiting for an occasion to occupy the Ruhr ever since the German Reich had signed the Treaty of Rapallo with the Soviet Union in April 1922. This was partly motivated by France's own security concerns as well as by the drive to underpin the French claim to supremacy in continental Europe. In Germany, the response to the occupation of the Ruhr was the "policy of passive resistance," which was supported by a broad political coalition from the right-wing parties to the Social Democrats. Here, passive resistance meant not following the occupying forces' orders.

This strategy put an enormous financial burden on the Reich until the government officially abandoned it in the autumn of 1923. In response to a series of violent acts of resistance, the occupying forces increased their repressive measures and even sentenced some of those involved in serious acts of sabotage to death. At the same time, the French government supported local efforts in the Ruhr to leave the German Reich. From the perspective of his university in Bonn, in the Rhineland, Schmitt believed that this region was in danger of no longer belonging to the German Reich, as were Alsace and Lorraine.

Schmitt began his struggle against Versailles and Geneva, which he fought with academic prowess, in late 1924 with a brief essay entitled "Die Kernfrage des Völkerbundes" [The core question of the League of Nations], which he turned into a book two years later.[1] He gave a number of talks on questions of international law over the following months. International law became one of Schmitt's key research topics. As of 1925, he was also responsible for the introductory lecture in international law at the University of Bonn as a substitute for a colleague. Kirchheimer attended this lecture of Schmitt's in the winter semester of 1926/27 and again, to prepare for his state examination, in the winter semester of 1927/28. When he began his tenure at the Handelshochschule in Berlin in 1928, Schmitt continued to lecture—albeit irregularly—on fundamentals of international law until 1933. It was not until he took on his new position in the legal hierarchy of the Third Reich in the spring of 1933 that he stopped teaching international law; from then on, he rarely published in the field. Until the setback in his career in 1936, he focused entirely on providing legal support to the Nazi regime as it established itself domestically.

During the Weimar Republic, Schmitt's work on international law focused consistently on three major issues: first, the status of the Rhineland under international law;[2] second, the Geneva League of Nations;[3] and third, the legitimacy of US foreign policy under international law (see Schmitt 1932b). All three subject areas were inextricably linked, particularly after Germany had joined the League of Nations in September 1926 under Foreign Minister Gustav Stresemann. From a domestic policy perspective, this step was the result of negotiations to form a coalition for a new Reich government in 1923. The SPD had been able to negotiate with the bourgeois parties that not only would Germany undertake additional foreign policy activities to solve the question of onerous reparations but it would also apply for membership in the League of Nations. As various German governments had persistently demanded, the Reich immediately became a permanent

1 See Schmitt (1924a) and (1926a).

2 See Schmitt (1925a), (1928a), and (1930a).

3 See Schmitt (1924a), (1925b), (1926a), (1930b), and (1931a).

member of the Council of the League of Nations, its most important organ. While opinion pieces in the right-wing party press furiously opposed Germany's membership of the League of Nations, the Social Democrats celebrated it as a shining hour of international law. Earlier and more systematically than all the other parties, the SPD had called on Germany to join the League of Nations, and its leading politicians also believed it would open up diplomatic opportunities to revise the Treaty of Versailles.

In other words, Schmitt produced his writings on international law in a context of very dynamic events and developments; this also applied to his work on domestic policy. Nevertheless, Schmitt's fundamental position on international law, which he expressed continuously in these various works, can be discerned clearly. This fundamental position resulted from his rejection of theories that, in response to the devastating World War I, promoted an international movement to establish a peaceful world order no longer centered around the sovereignty of the individual nation-state. Members of this movement in Germany included pacifist writers such as Walther Schücking and Hans Wehberg as well as Hans Kelsen's Vienna School of International Law. In his book *Das Problem der Souveränität und die Theorie des Völkerrechts* [The problem of sovereignty and the theory of international law] (see Kelsen 1928), published first in 1920 and in a revised edition in 1928, Kelsen had formulated the position of monism according to which international law and national law were parts of a uniform legal order. The two parts formed a contradiction-free system that could be traced back to a common "basic norm." Unlike the pacifists, Kelsen did not place war outside the realm of the law but considered war to be a reaction, necessary under certain circumstances, of the community of states to grave international injustice (see Olechowski 2020, 513–519). Among Kelsen's best-known students in the 1920s and 1930s were international law scholars Alfred Verdroß and Josef Laurenz Kunz, who also influenced the works of Kirchheimer and Schmitt after 1933.

In his works on international law, Schmitt took a position decisively countering pacifism under international law and Kelsen's monism as well as the relativization of state sovereignty these both entailed. His position developed rigorously from his understanding of the concept of the political and his determination of the state as the status of a nation's political unity. In May 1927, Schmitt had presented the fundamental ideas of his later renowned *The Concept of the Political* as a lecture in Berlin at the Deutsche Hochschule für Politik (see Schmitt 1927a). Schmitt explained some years afterwards that the hypotheses of the lecture had arisen from lively discussions with the students in his advanced seminar in Bonn (Schmitt 1940, 313)—and Kirchheimer had already belonged to that circle for about a year at the time. In the first version of 1927, Schmitt still applied the concept of the political exclusively to relations between states, not to relations within them (see Walter 2018, 286–289).

The prerequisite for politics, Schmitt wrote, was a state's internal unity. Politics developed out of the existential friend-enemy relationship between political entities fighting each other. He believed the political was the most intense and extreme antagonism. Enmity meant the existential negation of the other's existence. The other was the stranger seeking to eliminate one's existence. The struggle between enemies was a struggle for life and death. War was a manifestation of enmity. Schmitt defended the right to war and the willingness to die and to kill for reasons of a political unit's existential self-assertion, thus arguing nationalistically in the German struggle against the Treaty of Versailles and the

League of Nations. He did not say what constituted the characteristics of existence that could mobilize enmity and war. Anything and everything could inflame political mobilization. Although a world without politics was possible in principle, it would be boring without the excitement of existential struggles and would merely be "entertainment," "culture," "art," or "economics" (Schmitt 1932a, 53). Schmitt's political ethics focused on the self-assertion of a collective existence through battle. In this context, political existence did not mean mere survival but, rather, the struggle for one's own collective identity and dignity.

Schmitt's understanding of the concept of the political had four serious consequences for his theory of international relations. First, directed specifically against Kelsen, as long as the political existed, there would also be a pluralism of fully sovereign states in the world. Second, directed against the pacifists, the *ius ad bellum* belonged to the state as an essential political entity. Every state, Schmitt argued, had to have the real possibility of deciding in a specific situation who the enemy was and whether it would start fighting a war with that enemy about whatever issue. Third, only states capable of defining themselves via friend-enemy relationships in their foreign relations retained their right to exist. And, fourth and finally, nations that could not keep up with military technology or feared the effort involved in and the risk of political existence had given up their independence and sought protection under the domination of stronger nations. These consequences for international politics remained unchanged in terms of their substance in all four editions of *The Concept of the Political* from 1927, 1932, 1933, and 1963. In the 1932 edition, Schmitt added a "secondary concept of the political" (see Schmitt 1932a, 30–32) in response to the criticism that he had limited his discussion to foreign policy. In the 1933 edition, he adapted his text linguistically to the Nazi regime's vocabulary, adding antisemitic wording in various places,[4] all of which was gone in the 1963 edition.

Shortly after publication of the 1927 edition, Hermann Heller[5] and others promptly challenged Schmitt about basing his concept of the political on foreign policy and the state-centered perspective. The alterations to the text in subsequent editions were Schmitt's response to this criticism. The secondary literature provides only a few informative references to the intellectual history of the sources of Schmitt's friend-enemy definition in his concept of the political. In his editorial comments on the 1927 version of Schmitt's work, Günter Maschke mentioned reflections on foreign policy in the ancient *Persian Book of Kings* and seventeenth-century Spanish texts as two of Schmitt's sources and inspirations (see Maschke 2005, 221–223). That may well be the case. Kirchheimer's colleague Ernst Fraenkel later alluded to different genealogical clues in his 1941 book *The Dual State*: Joseph A. Schumpeter's famous article on the sociology of imperialism from 1919 and an article by Rudolf Smend, the other professor with whom Kirchheimer had been close during the Weimar Republic, about the aimless quest for power as the central

---

4 Herbert Marcuse was among the first to call attention to Schmitt's linguistic adaptations (see Marcuse 1934, 103). For a detailed comparison of all four editions, see Walter (2018).

5 See Heller (1928, 425) and (1933, 646).

element in imperialist expansionism (Fraenkel 1941, 101–103).[6] Fraenkel suggested that Schmitt had simply given a more abstract theoretical expression to Schumpeter's conception of imperialism. Kirchheimer might have shared details from the discussions in the Bonn seminars with Fraenkel that are not to be found in written sources. Be that as it may, both sources again highlight the close connection between foreign policy and the concept of the political in Schmitt's thinking.

The position laid out in his understanding of the concept of the political was Schmitt's starting point from which he derived his rigorous rejection of all developments aiming to supplant the state from its traditional central position in international law and instead seek the juridification and institutionalization of a global legal order, which was the goal pursued by Kelsen, Schücking, and Wehberg. A global state of any kind had lost all its political character as it no longer had any enemies; universal humanity was not a political category as it did not permit any internal differentiation between friend and enemy. This also included the unconditional right of states to wage war.

Schmitt did not accept the argument that some wars were waged on behalf of humanity and were therefore legitimized by international law. This type of war, he believed, was in reality not a war of humanity but remained a war waged by one or more states against another one. Invoking humanity was nothing but a propaganda slogan. Formulating the essence of his hypotheses in his *Concept of the Political*, Schmitt repeated the jarring aphorism by Pierre-Joseph Proudhon: "whoever invokes humanity wants to cheat." (Schmitt 1932a, 54) Invoking humanity was even dangerous inasmuch as those who were thus declared enemies of humanity were, in the final analysis, denied the quality of being human. Which was why, Schmitt added, such a war was conducted using particularly inhuman means.

Roughly one-sixth of Schmitt's publications during the Weimar Republic were about international law. These works include his "Die Kernfrage des Völkerbundes", which Kirchheimer quoted multiple times in his dissertation. Discussing the occupied Rhineland and the collapse of the "policy of passive resistance," Schmitt accepted the initial situation that a defeated country had become an object of international politics. Yet the issue he declared to be the "core question" was whether the currently existing postwar order could develop and be stabilized in the future as an order of peace and law. The article was written at a time when the German Reich had applied for membership in the League of Nations but was not yet a member. The condition for peaceful development was a functional League of Nations, which Schmitt considered to be one that gave Germany the opportunity to be liberated of the restrictions imposed by the Treaty of Versailles. Agreement upon this matter was a precondition for it truly being a league of nations.

This hypothesis was based on Schmitt's political intention in positing the term since he wanted to develop the legal basis for the League of Nations—*Völkerbund* in

---

6 See also Schumpeter (1919, 13) and Smend (1923, 81). Schumpeter was Schmitt's colleague in Bonn from 1925 to 1928. According to William Scheuerman, archival materials suggest that in 1926, Schumpeter had encouraged Schmitt to complete what later became the famous *Concept of the Political*. The original essay of 1927 was published first in Schumpeter's journal *Archiv für Sozialwissenschaft und Sozialpolitik* (see Scheuerman 2020, 230).

German—from a particular understanding of the term *Bund*. His historical example was the Holy Alliance from 1815 on. A *Bund* was designed to be more than an international office to coordinate intergovernmental matters such as the Universal Postal Union. The legitimacy of a true *Bund* was in a certain shared understanding of its members and a minimum of guarantee (see Schmitt 1926a, 115). The mutual guarantee between the members of a *Bund* arose from the cohesion within the *Bund*, that is, guaranteeing a jointly recognized constitutional standard in solidarity, which Schmitt called homogeneity. He used the example of the Soviet Union to illustrate that such fundamental commonalities were, in his view, the prerequisite for joining a *Bund* (see Schmitt 1926a, 125–126). Prior to Germany joining the League of Nations, he did not take a clear position on whether Germany should do so—this question was still open until 1926—but simply argued vehemently against admitting the Soviet Union because it did not fulfill the requirement of a shared understanding with the capitalist members of the international organization.

Schmitt's deliberations on this requirement of homogeneity at the international level remain remarkably contradictory. He did not clarify whether it referred to each member state's citizens having the same personal characteristics (for example, ethnicity) or the same opinion, or whether it referred to all states in the *Bund* sharing a jointly recognized constitutional standard, in other words, homogeneity of each state's citizens or homogeneity of the various states themselves. He mentioned both understandings but failed to differentiate them properly (see Neumann 2015, 437–439). As the result of his deliberations, Schmitt stated that the League of Nations had not given an answer to the question regarding the characteristics of a *Bund*, at least, no clear answer. Instead, it was showing the states a Janus face. To the victorious Western powers, it appeared to be a construct for a particular purpose and ready for service—to Germany, in contrast, it seemed like a system of harsh and strict rules.

A complex body of treaties was signed in London in late 1925 following the Locarno Conference in October, including various individual treaties between the major European powers as well as ones with Germany. The goal of the body of treaties, negotiated largely between Stresemann and French Foreign Minister Aristide Briand, was to establish a collective security system for Central Europe within the framework of the League of Nations under equal participation of the German Reich. Key to the body of treaties was that France committed to ending its occupation of the Rhineland ahead of schedule and Germany abandoned its claim to a revision of the Western border with Belgium and France. The Locarno Treaties entered into force on 1 September 1926 as Germany joined the League of Nations. In Germany's domestic politics, the treaties sparked fierce controversies and turbulences, prompting the right-wing DNVP ministers to resign from the Reich's government. Schmitt, too, joined those opposing both the Locarno Treaties and Germany's membership in the League of Nations.

A notable feature of Schmitt's writing on international law is his resolute and consistent criticism of the US's capitalist imperialism. Even in his first two articles, "Die Kernfrage des Völkerbundes" and "Die Rheinlande als Objekt internationaler Politik" [The Rhinelands as an object of international politics], he referred to a new "age of imperialism" (Schmitt 1925a, 28) in which economically powerful countries exercised their domination over formally independent states in various indirect ways.

The prime example Schmitt gave was the US, which, he asserted, invoked the old Monroe Doctrine of 1823 to "protect private property" (Schmitt 1924a, 19) and expand its claim to economic domination of all of North and South America. He claimed the Treaty of Versailles and the League of Nations had both been created in the unilateral interests of the US, and he considered the fact that this country had neither signed the Treaty nor joined the League of Nations to be a particularly perfidious form of exercising domination in an indirect imperialist way. As far as Europe was concerned, the US was officially not involved but effectively came on the scene whenever it was in its interests as a "controlling suzerain" (Schmitt 1927b, 243). It corresponds to this view that Schmitt had nothing but scorn and contempt for the Kellogg-Briand Pact, which was signed in Paris in August 1928; its signatories denounced war as a means of solving international disputes and declared renunciation of war to be a means of national politics. Schmitt ridiculed the fact that the Pact was concluded as a permanent treaty from which parties could not withdraw: "A people which exists in the sphere of the political cannot in case of need renounce the right to determine by itself the friend-and-enemy distinction" (Schmitt 1932a, 50). In his writing, he notoriously called the treaty the "Kellogg Pact" after the US Secretary of State in order to express his opinion that it served the economic imperialist interests of the US unilaterally.

Toward the end of the Weimar Republic, the tone in which Schmitt wrote about the US became even harsher. He discussed the Monroe Doctrine once again in a lecture titled "USA und die völkerrechtlichen Formen des modernen Imperialismus" (The US and forms of modern imperialism in international law) which he gave in Königsberg in February 1932. This time, however, he concluded that it "had done its duty" (Schmitt 1932b, 355). The US had gone from being a debtor state to a creditor state and was using the instrument of lending to dictate to other countries. Thus, the US was, at the time, the country of the "most modern" form of imperialism, namely "economic imperialism" (Schmitt 1932b, 349). In Germany's current situation, what mattered most was to see through the "veil of words and concepts" (Schmitt 1932b, 365) of the universalist vocabulary of international law and to defend its own identity using combative means. By the end of the Weimar Republic, Schmitt had become an ardent critic of US capitalist capitalism in the name of German nationalism.

2. Kirchheimer's early writings on international law

Kirchheimer did not attend Schmitt's seminars and lectures on international law entirely unprepared. He had already studied with Heinrich Triepel, an expert in constitutional and international law in Berlin, in 1925 and 1926.[7] It was practically inevitable that his interest in questions of international law would intensify while studying with Schmitt. In foreign (as in domestic) policy, Schmitt and his young leftist student did not agree on anything concerning the political issues of the day. Although Kirchheimer's party, the

7 Otto Kirchheimer, Curriculum Vitae. Archiv der Juristischen Fakultät der Universität Bonn, Promotionen 1927/28. Prüfungsakte Nummer 521–528, Otto Kirchheimer.

SPD, was also among the critics of the Treaty of Versailles, regarding Germany's accession to the League of Nations and the conclusion of international peace treaties, the SPD was one of the most ardent supporters of a policy aiming for reconciliation and endorsed the policies of national liberal Foreign Minister Gustav Stresemann. The party leadership hoped that accession to the League of Nations would mark the beginning of the institutional establishment of a world peace order. The SPD's leading foreign policy experts also hoped that involvement in the League of Nations would finally bring about a revision of the Treaty of Versailles, specifically as far as Germany was concerned, on the basis of a Europe coming together in peace in the medium term. The leftist wing of the party agreed with the party leadership on this matter; its views coincided with those of the communists, if at all, in their criticism of capitalist imperialism.

Kirchheimer's earliest surviving statements on questions of international law are to be found in his dissertation with Schmitt, *The Socialist and Bolshevik Theory of the State.*[8] Kirchheimer described Soviet law as by no means intended to last for eternity. It was meant to be "temporary law to the highest degree" (18). The *clausula rebus sic stantibus* in contract law, i.e., the right to alter contracts if decisive circumstances have changed, would not have to be added to Soviet law ex post. On the contrary, Soviet law was itself nothing but *clausula rebus sic stantibus*. In Kirchheimer's view, Soviet thinking on international law was a particularly clear expression of the Soviet Union's view of the law being conditioned on goals and situations. According to the Soviet theory of the state, it was only the irreconcilability of class antagonisms that had made the Soviet state at all necessary and thus established it. It was this state alone that the legal system was to serve, and the Bolsheviks hoped to launch a successful world revolution in the foreseeable future with the aid of the Soviet state.

Viewed from the perspective of world revolution, traditional international law was, in the eyes of the Soviet state, the most dubious of all bourgeois legal constructions. The Soviet Union recognized nothing more than purely technical common interests among states, as in the international postal system, for example. For this reason, it did not respond to any endeavor to support peace propaganda or to create a closer international legal community such as the Geneva League of Nations. "It considers international law as the rules of truce, not of peace" (18). The Soviet Union saw the international organizations of the day, such as the League of Nations, as an attempt to stabilize the legislation and traditions of a dying age. For the transitional period until the global victory of communism, it would have to find makeshift solutions through precisely worded individual agreements with the various powers.

At this point in his deliberations, Kirchheimer went back to the postulate of homogeneity referring to the League of Nations from Schmitt's *Kernfrage des Völkerbundes*. However, he did not use Schmitt's term *Gleichartigkeit* but the term *Homogenität* instead. In the "absence of any homogeneity of interests and views, however tenuous, that could become the prerequisite of a decision in a legal sense" (18), the Soviet Union was forced to reject any internationally recognized court as well as the majority principle in international practice. When Kirchheimer spoke of a lack of homogeneity, he was referring to a lack of shared interests and views of the members of the League of Nations, not a lack of

8 See Kirchheimer (1928a). The following page numbers refer to this text.

homogeneity of their substantial characteristics. The Soviet Union had become hostile toward the Geneva League of Nations "as a matter of principle" (19) and not only occasionally, as Germany had. Because of its self-understanding as protector and advocate of formal democratic principles, the League of Nations took a stance of equally sharp hostility toward the Soviet state.

While the British referred to the Covenant of the League of Nations to legitimize its intervention in the Polish-Soviet war in 1921, the Soviet government referred to the incompatibility of such an argument with the sovereignty of the Russian working people. To Kirchheimer, this brought about a seemingly paradoxical constellation. Whereas "Europe de-emphasizes the concept of sovereignty, practically and theoretically, Bolshevism strangely engages in ushering in new victories, de facto, for the very concept of sovereignty from which, theoretically, it withholds recognition" (29). As evidence of this hypothesis, Kirchheimer referred to an article published in French by Soviet international law expert Evgeny A. Korovin in the previous year, according to which it was the "*intérêts réels*" (20) of the socialist republic when dealing with the capitalist powers that had made it reasonable to use the concept of sovereignty for tactical reasons.

Kirchheimer believed that international law was further evidence of the extent to which the traditional nation-state was already "on the wane" (20) in Europe. This could be seen, for instance, in the changing justifications for the colonial annexations carried out after the Versailles Peace Treaty had been concluded. Colonial rule was no longer morally defended by the traditional arguments related to the national unity of peoples. Instead, in order to legitimize the desired result, states had to take recourse to legal alternatives such as the construct of the League of Nations as a trustee. These changes in international law highlighted the practical weakness of the concept of nation-state sovereignty for states with formally democratic structures. Formally democratic states were confronted with a theoretical impasse. They aimed at a social equilibrium and did not find a satisfactory answer to the question of who wielded sovereignty, i.e., "who makes the actual decision in a conflict situation" (20). Conversely, Kirchheimer called the Soviet locus of sovereignty "sensational clarity" as opposed to the "present-day tendencies of masking and concealment" (20) in the bourgeois nation-states.

It should be noted that Kirchheimer did not consider the actual purpose of the Soviets insisting on the concept of sovereignty in international law to be an attempt to restore traditional international law, but rather something entirely new: "the intentional separation—performed for the first time—of the concepts of state and sovereignty" (20). The sovereignty of the state that saw itself as the first state of the imminent proletarian world revolution was not bound to traditional state borders. Its sovereignty was potentially universal since the claim of the working class to domination potentially extended to the entire globe. Any member of the working class in any country of the world could occasion the intervention of the Soviet state, either in order to gain its protection or to use it to influence the fate of other countries. A country that claimed or practiced an unlimited right of intervention would forgo the specific characteristic of a state, namely self-limitation at some geographical line. In Kirchheimer's view, this did not mean that the Soviet Union was less than a state. On the contrary, it had restored state integration to a new level. By making use of this power and of the myth of the world revolution, the Soviets had regrouped the international political forces. They had been able to tear open the gap at the very place where the traditional state had stood until the nineteenth century. What still existed in the West was the mere shell of a state.

It is not difficult to see how strongly these early statements of Kirchheimer's on questions of international law intersected with areas of domestic policy. He was concerned with the broader question of state sovereignty in Western capitalist democracies. His brief excursion into the field of international law, contrasting it with Bolshevik doctrine, was primarily to provide evidence for the large extent to which sovereignty was subject to masking and concealment in Western capitalist democracies, although that had not made it disappear entirely. The proximity of 23-year-old Kirchheimer's deliberations to Carl Schmitt's theory is palpable, as is his profound fascination with Bolshevism's strong assertion of sovereignty. Kirchheimer's view of the League of Nations also demonstrates how close his thinking was to Schmitt's. Following Schmitt's postulate, Kirchheimer considered homogeneity of interests and views to be prerequisites for legal decision-making in the realm of international law. Moreover, he believed it logical and plausible that the revolutionary Soviet Union would position itself outside the League of Nations.

What was completely absent in this work of Kirchheimer's and also in all his other early essays was Schmitt's notorious lament about the injustices that Germany allegedly had to suffer in terms of international law under the Treaty of Versailles since it had been concluded. Not a trace of nationalism or yearning for strengthening of the traditional nation-state was to be found in Kirchheimer's work. His early interest in international law was motivated exclusively by the question to what extent developments in international law could provide information about state sovereignty.

3. Kirchheimer's critique of capitalist imperialism

Two years after leaving Bonn, Kirchheimer began to take an interest in other questions of international law and to clearly distance himself from Schmitt's hypotheses on constitutional and international law. This can be seen in his review of Korovin's 1924 book *Das Völkerrecht der Übergangszeit* [International law of the transition period], which was published in German translation in 1929 (see Korovin 1929). As mentioned above, Kirchheimer had taken up a 1925 essay of Korovin's in French in his dissertation. Both of Korovin's works[9] fall within the transition period from the "heroic epoch" of permanent world revolution and "war communism" to the more national reformist and revisionist diplomacy of isolation and the "New Economic Policy" and "socialism in one country." Korovin's theory was not officially replaced in the Soviet Union until 1935, by, among others, Evgeny Pashukanis's *Essays on International Law*, in which he argued that the Soviet Union's accession to the League of Nations in 1934 made sense.

In his book, Korovin had rejected the notion of universally valid international law. On the contrary, he argued that the Soviet Union and the capitalist world as a whole lacked the minimum "Gemeinschaft und Einheit" (community and unity)[10] necessary for

9 On the major importance of this book of Korovin's for the debate of the day on international law, see Flechtheim (1936, 56–78).

10 See Korovin (1929, 24).

any community of international law. In other words, Korovin used a similar argument to Schmitt to counter the possibility of a League of Nations. That argument can also be found in Kirchheimer's article quoted above in which he translated Korovin's term "Gemeinschaft und Einheit" as "Homogenität." Kirchheimer presented Korovin's deliberations in his review,[11] referring to them as a "voluntaristic concept of international law" (324). And he did not hold back his criticism. Kirchheimer believed that Korovin was seriously wrong in two respects. First, he had a completely mistaken idea of the alleged homogeneity of the world of capitalist states. In light of the leading industrialized states' imperialism and their competition with one another, it was absurd to claim that the entire capitalist world was homogeneous in terms of its values and interests. The traditional system of international law was certainly "no such coherent political circle" (326) in stark contrast to Korovin's assumption. The reality of imperialism and international competition was, in fact, evidence of the complete opposite. Second, Korovin had an entirely misguided concept of homogeneity in Kirchheimer's view:

> In reality, the homogeneity of the community of international law does not rest on considerations of principle. [...] Instead, it rests on a vast number of constantly growing necessary technical and economic agreements, and it is these agreements which have the effect of forming a community. (325)

Drawing on Max Huber's classical work *Die soziologischen Grundlagen des Völkerrechts* [The sociological foundations of international law], published in 1910, Kirchheimer opposed Korovin in positing that modern international law had successively "grown beyond [...] states' intensive interest in consistent rules" to the situation at the time, which had "shifted from the quantitative to the qualitative" (325). With this fundamental objection to Korovin's notion of homogeneity, Kirchheimer simultaneously attacked Schmitt's position, which he himself had advanced two years earlier without criticism. Both Korovin and Schmitt assumed a certain degree of homogeneity, which constituted a firm basis, a precondition for finding rules under international law. These few sentences in Kirchheimer's review show that he, conversely, defended a dynamic concept of homogeneity at the international level reminiscent of Rudolf Smend's theory of integration, which addressed domestic policy. The formation of an international community was in itself only the result of experiences of positive international cooperation. The same applied to international law. It developed successively, usually beginning with a web of technical treaties, followed by precedents and mutual recognition of agreed rules. According to Kirchheimer's logic, Schmitt's polemic comment about the Universal Postal Union was practically turned on its head: it was only such positive experiences of cooperation that prepared the ground for more far-reaching international agreements.

Kirchheimer called the voluntarism of Bolshevik theory just as naive and erroneous as the "traditional optimism" (325) found in international law at the time. Yet he also discovered positive elements in Korovin's book. Compared with the thinking along the traditional lines of the existing system of international law, in his opinion, it had the virtue of viewing international law more realistically. Kirchheimer considered two of Korovin's

11 See Kirchheimer (1930a). The following page numbers refer to this text.

suggestions to be important in this context. The first concerned the traditional doctrine of the right of intervention. In light of the ban on interventions demanded by many Latin American states, he considered the contemporary law on interventions to be "one of the sharpest fault lines of international relations" (326). Here, Kirchheimer was referring to US imperialism. Instead of continuing to obscure US intervention policy behind legal terms, international law required objective analysis of the occurrences in Latin America in his view. The second suggestion from Korovin's book had methodological implications. Korovin was one of the first theoreticians of international law to comprehensively establish to what extent a number of additional factors and actors besides states had emerged that de facto had a strong influence on international law. These included international financial markets as well as internationally operating businesses and interest groups. Although they did not always act as recognized subjects of international law, the "economic agreements between interest groups in two states [could] be just as consequential" (326) as traditional diplomatic contacts between states. The roles of these factors and actors had to be "assessed differently in international law in the future, both actively and reactively" (326). Such an expansion of the area covered by international law was necessary in order to expose the "class structure in today's world of states" (326) with sufficient clarity.

Compared with the remarks he had published two years earlier on the Bolshevik doctrine of international law, Kirchheimer's review reads like a complete reversal. Korovin's book contained the same fundamental statements as his 1925 article in French which Kirchheimer had referred to in his dissertation. It was Kirchheimer who had reversed his position. He argued in favor of a "doctrine of international law that seeks to be truly realistic" (327), striving to be universally valid specifically for this reason. To him, the source of this type of international law was not shared values but the interest present in all states—and thus also expected in the Soviet Union in the future—in joint economic and technical agreements.

By 1930, there was no longer any mention in Kirchheimer's work of his fascination for the Bolshevik doctrine's strong claims to sovereignty. His understanding of requirements of homogeneity as a condition for the existence of international law had also changed fundamentally in favor of a dynamic concept of homogeneity. Finally, at least implicitly, he adopted a pluralistic concept of homogeneity for the level of international cooperation by calling attention to the differences in values between competing interests among the capitalist states. Even if Kirchheimer did not formulate it explicitly, his revisions, overall, amount to a plea for an expansion of international law and the League of Nations which even made it imaginable to admit the Soviet Union to the League of Nations. The critical sting of the "doctrine of international law that seeks to be truly realistic" that he called for was aimed not only at the US interventions in Latin America that had been virtually unsanctioned through international law to date, but more generally at the imperialism practiced by capitalist states.

The sociological foundation of Kirchheimer's approach to international law can also be seen in a review of Adolf Grabowsky's book *Politik* [Politics] which Kirchheimer wrote in late 1932 for the social democratic theory journal *Die Gesellschaft* [Society] (see Grabowsky 1932). Grabowsky belonged to the political milieu of young conservatives close to Schmitt and taught foreign policy at the Deutsche Hochschule für Politik (German Academy for Politics). Most of his book is devoted to foreign policy. Kirchheimer sharply criticized Grabowsky's analysis of global politics. Focusing exclusively on states as actors was just as misguided as the assertion it entailed about foreign policy unity in a state. Grabowsky's approach suffered from a crude "overestimation of all intellectual factors" (Kirchheimer 1933a, 513) in foreign policy. This was particularly true of his attempt to interpret modern imperialism. He had limited his analysis "to a purely subjective interpretation of the meaning of the historical events" (Kirchheimer 1933a, 513) and had entirely disregarded economic explanatory factors.

With this criticism, Kirchheimer simultaneously took a position in a debate that the young historian Eckart Kehr had sparked among his fellow historians in 1928 when he presented the findings of his sociohistorical study on German fleet policy directed against England around the turn of the century. In his book *Englandhaß und Weltpolitik* [Hatred of England and world politics], Kehr had reconstructed the armaments and foreign policy decisions of the German Empire as the result of the interest and power relations within society. He spoke incisively of a "primacy of domestic policy" (see Kehr 1928, 500) rather than foreign policy allegedly following its own laws. As one of the editors of the journal which had published Kehr's article, Grabowsky had followed it with a response of his own titled "Der Primat der Außenpolitik" [The primacy of foreign policy] (see Grabowsky 1928). In the methodological debate between those studying social historiography on the one hand and those tracing foreign policy constellations that were forever following their own laws on the other hand, Kirchheimer was a clear advocate of the first group.

He reviewed another book on international law for *Die Gesellschaft* in 1932. This time, it was by an author from his own political camp. Georg Schwarzenberger was a student of the two Social Democrats Carlo Schmid and Gustav Radbruch and belonged to a generation of young Social Democrat jurists specializing in questions of international law (see Steinle 2004). His book *Die Kreuger-Anleihen* [The Kreuger Bonds] (see Schwarzenberger 1932) discussed a topic hotly debated domestically in the early 1930s. Since the mid-1920s, Ivar Kreuger's private Swedish corporation had given loans to several governments in Europe that were under financial pressure. These loans were of particular importance to Germany, enabling it to pay the reparations agreed in the Versailles Peace Treaty. In return, Kreuger's corporation had insisted on a match monopoly, which meant that only matches produced by Kreuger could be sold in Germany. In the course of the Great Depression, the German Reich had become so unstable financially that the government saw no other option but to accept Kreuger's offer of a loan. In January 1930, the Reichstag, with the support of the Grand Coalition, voted to guarantee the monopoly through 1983.[12] Once Kreuger's corporation had received the monopoly, it immediately raised the prices for matches in Germany. Public resentment of the monopoly and the price increases was aired constantly during the Weimar Republic.

Kirchheimer took Schwarzenberger's detailed description and analysis of the contract with Kreuger's corporation as an opportunity to lay out his own fundamental reflections founded in international law on state sovereignty and the questions of whether

12 The Federal Republic of Germany took on this obligation, which meant that up until 1983, no matches made by other companies could be sold in the country.

and how states were bound by contracts at all.[13] After all, Kirchheimer stated—intentionally using Schmitt's wording—that this problem was a "core problem of international law" (375). He criticized the fact that German scholars in international law had largely neglected the legal analysis of relationships between states and foreign corporations. The main question raised by the Kreuger bonds was whether they were contracts valid under international law or contracts subject solely to the law of a single country. Basing his argument on Schwarzenberger's, Kirchheimer maintained that these contracts were in the realm of a single country's law, not international law—a hypothesis with important consequences for fiscal policy.

The polemic thrust of Kirchheimer's interpretation was directed against property claims across different countries. This was evident from his opposition to the "propagandists of an allegedly general principle of international law concerning the unconditional protection of acquired rights of foreigners" (374) vis-à-vis national legislators. He accused the International Law Association (ILA), headquartered in the US at the time, of going even one step further. The goal of its activities was to stipulate "an international standard of law and justice, a kind of capitalist civilizational minimum" (374) as a ticket for membership in the community of international law, thus protecting foreign companies' property claims from being seized by nation-states in the name of an allegedly universal international law.

Although Kirchheimer shared Schwarzenberger's opposition to these ambitions, he arrived at the same position using a different line of argument. Schwarzenberger—"presumably following Korovin" (374)—established a doctrine of two fundamentally different doctrines of international law: one bourgeois and founded on the idea of the rule of law, the other socialist and founded on the idea of international cooperation and social justice. On the basis of this premise, a future socialist government of Germany would not be obligated to continue to recognize Kreuger's claim to a monopoly. Kirchheimer came to the same conclusion, but he did not follow Schwarzenberger in dividing existing international law into two different worlds existing side by side and isolated from each other. Rather, Kirchheimer thought it was not least a pragmatic argument concerning the application of the law that countered Schwarzenberger's premise. Following Schwarzenberger, every government would have to switch back and forth between the two postulated worlds of international law, depending on its political orientation. Kirchheimer believed this could not be implemented in practice. It was much "more reasonable to take international law applying to everyone as a starting point" (375). Employing such a strategy, however, it was all the more important to place precise limits on the domain of international law. "Only those [principles of international law] that all potentially suitable states in the world are able to recognize without endangering their social status" should be considered to be such principles because "it is wise for every jurisprudence, especially international law, to define its own limits" (375). Kirchheimer was convinced that the unlimited right to private property was unequivocally not included in such a catalog of minimum requirements of international law.

Kirchheimer's brief review gave further clarity to the anti-capitalist and anti-imperialist thrust of his thinking on international law. In the process, he simultaneously took

13 See Kirchheimer (1932b). The following page numbers refer to this text.

up the suggestion from Korovin's book to focus international legal analysis on the non-state factors and actors on the international stage and applied it to the specific material of an issue that was hotly debated by the general public. Kirchheimer strongly advocated for uniform international law. Yet its rules and regulations were to be limited to questions that would not encroach on the social order of the individual countries. The implication this statement about international law had for Germany in early 1932 was unmistakable: should there be a socialist-led government in the future, it was to be at liberty to terminate the contracts with Kreuger's corporation. As Kirchheimer saw it, the ensuing legal conflicts would no longer fall in the domain of international law but would lie solely within the Reich's domestic civil law jurisdiction.

4. Conclusion: Left-wing versus right-wing anti-imperialism

The anti-imperialist thrusts of both Schmitt's and Kirchheimer's works are evident. Equally evident are their differences. In the name of militant German nationalism, Schmitt argued against a specific enemy, Anglo-American imperialism, which, he asserted, acted indirectly. He believed he had unmasked universalist international law and the prospect of an institutionally secured world peace order as perfidious claims to power by the Anglo-American enemies. He was of the opinion that the right of every state to start a war at any time, for whatever reason, must never be restricted. Schmitt maintained this basic position after 1933 and even after 1945.

Kirchheimer's advocacy for a juridification of international politics did not emerge during World War II but had its origins in his Weimar writings and his detachment from Schmitt's patterns of argumentation. In his dissertation, Kirchheimer was still under the spell of Schmitt's theses on the one hand and Korovin's Soviet doctrine on the other. Shortly afterwards, however, he opened up to the perspective of international law that aims at long-term cooperation and peace among states. France did not appear in his writings as an "archenemy" but as a friendly country with a great democratic and revolutionary tradition. There was no trace of revanchist thoughts in his work. Nor were there negative comments about the Treaty of Versailles or the Geneva League of Nations. On the contrary. Kirchheimer wanted capitalist imperialism to be fought within the framework and with the means of improved and intelligently restricted international law. He countered the concepts of mutually exclusive legal systems represented by Schmitt and Korovin with a dynamic concept of homogeneity based on positive experiences of cooperation. In doing so, he transposed Rudolf Smend's theory of integration to the international level.

A direct path leads from this basic position to his later criticism of Schmitt's theory of the *Großraum*[14] and on to his involvement in the preparations for the Nuremberg war crimes trials.[15] The diametrically opposed positions taken by Kirchheimer and Schmitt after 1945 concerning the appropriate legal and political handling of German war crimes are contained *in nuce* in their Weimar differences on international law.

14 On Schmitt's theory of the *Großraum*, see Chapters 11 and 12.

15 On the preparations for the Nuremberg Trials, see Chapter 13.

Chapter 5: Escalating Antagonisms (1932)

The year 1932 was ill-fated for the Weimar Republic, marking the beginning of the end. Since 1930, parliamentary democracy had been replaced by a government that had introduced a presidential emergency decree. Almost two years later, Chancellors Franz von Papen and Kurt von Schleicher took over the reins of politicians who had planned to give the republic a different constitution and, consequently, a different form of government in the long term. In the course of 1931, Chancellor Heinrich Brüning had increasingly lost support for his policies in the Reich President's Office. July 1931 saw a serious bank crash. Brüning tried in vain to stabilize the situation by issuing a third emergency decree on 6 October and a fourth one on 8 December 1931. There were 5.66 million unemployed by December with no sign of an economic turnaround. Brüning's policy resulted in mass misery and a rise in right-wing extremism. On 30 May 1932, he was forced to resign, and Hindenburg appointed Franz von Papen as the new Chancellor. Papen had ambitious plans for a constitutional reform that would result in the re-establishment of the monarchy. In the course of the escalating economic and social crisis in the republic, Papen also resigned on 17 November 1932, with Kurt von Schleicher replacing him as Chancellor on 3 December. He attempted to find a compromise to assemble a majority tolerating his policy across the parties, relying on the NSDAP splitting. His plan failed and so he considered establishing a military dictatorship. Since he lacked the President's support for this plan, he finally resigned, too. On 30 January 1933, Hindenburg installed the leader of the Nazi party, Adolf Hitler, as the new Chancellor of a coalition government with the DNVP.

During these turbulent political events, Schmitt and Kirchheimer were among those who took an active part in the constitutional and political discussions about the changes proposed by Papen and Schleicher. In the case of Schmitt, another factor was that he no longer restricted himself to the role of a legal commentator. 1932 was the year he had managed to gain direct contact with the political leaders in the Reich. He rushed to the government's aid as a legal representative, providing informal advice and legal opinions. His restless activities in 1932 earned him the sarcastic title of "crown jurist of the Presidential Regime" (Gerlach 1932, 343) from the leftist magazine *Die Weltbühne*. Ultimately, however, Schmitt's excursions into practical politics failed. In this respect, at least, there was very little difference between him and Kirchheimer.

Kirchheimer continued to fight for a professional existence in 1932. He was admitted to the Berlin bar as a lawyer in January. Like many lawyers at the time, he ran his law firm from his private apartment. His father-in-law Kurt Rosenfeld helped him with the difficult task of acquiring clients as a young professional during the economic crisis and assigned him some criminal law cases. He was also asked to take on some labor law cases in the Berlin law firm of Franz L. Neumann and Ernst Fraenkel, who were on a retainer for the Deutsche Metallarbeiter-Verband (DMV, German Metalworkers' Union) and the Deutsche Baugewerksbund (DBB, German Union of Building Trades). Kirchheimer regularly came to their representative offices on Alte Jakobstraße in the Berlin district of Kreuzberg. Other visitors and participants in the group that met there were Otto Kahn-Freund and Otto and Susanne Suhr. Kahn-Freund worked in a labor court in Berlin; Kirchheimer had spent six months of his *Referendariat* with him in 1929. Otto Suhr had a position at the economic policy department at the headquarters of the Allgemeine freie Angestelltenbund (General Free Employees Association). As early as 1932, he predicted in a newspaper article that the Nazis would take power and that their regime would last twelve years; Kirchheimer, Fraenkel, and other participants in the discussions took him for a pessimist.[1] These political discussions turned into a kind of seminar on constitutional law for the younger generation of Social Democratic lawyers.

Referred by Kahn-Freund, Kirchheimer gave courses at union schools and lectured at events organized by the Republikanische Richterbund (Republican Judges' Association), a small association of lawyers loyal to the republic. In these circles, Kirchheimer enjoyed the reputation of being extremely open to discussion. Looking back, Susanne Suhr described him as follows: "He was a brilliant young intellectual, but ultimately incapable of practical politics."[2] Kirchheimer continued to keep in touch with his comrades of the magazine *Klassenkampf*, who had migrated from the SPD to the new left-wing party SAP, the Socialist Workers' Party of Germany, and with dissidents of the KPD, who, like the young legal scholar Wolfgang Abendroth, found themselves in the KPO, the Communist Party of Germany (Opposition). What the two new small parties SAP, which Kurt Rosenfeld had joined, and KPO had in common was that they wanted to overcome the schism of the labor movement and to unite the SPD and the KPD in their fight against Nazism. Although Kirchheimer remained in the SPD, he sympathized with these unification efforts. Like Neumann and Fraenkel, Kirchheimer began to publish several essays on current constitutional issues in addition to his legal work after a one-year hiatus. He also started to prepare his *habilitation* with Rudolf Smend at the Law Faculty at Berlin University. However, his professional dreams were shattered with the handover of power to Hitler's government.

1 See Fraenkel (1957, 380).

2 Susanne Suhr in an interview with Alfons Söllner, quoted in Erd (1985, 42).

1. Legality and legitimacy

In 1931, Schmitt had contributed an article called "Grundrechte und Grundpflichten" [Fundamental rights and fundamental duties] for a commentary on the Weimar Constitution. He again vehemently rejected the idea of an "integrative function of the judiciary" (see Schmitt 1932d, 192), which Kirchheimer had claimed in his dissertation, taking up Smend's work. Schmitt added:

> The general hypothesis formulated by O. Kirchheimer in my constitutional-theory seminar in the summer of 1931, namely that a state founded on the supremacy of the judicial branch (instead of the supremacy of the legislative and executive branch) is the only one that could have fundamental rights, should at least be mentioned here. (Schmitt 1932d, 192)

Schmitt disagreed with Kirchheimer but was fair-minded enough (and also proud) to quote his former doctoral student in this contribution to his outstanding commentary on the Weimar Constitution.

If we strive to understand exactly what the constitutional thrusts of the articles published by Schmitt and Kirchheimer in 1932 were, we must be aware of their precise dates because events unfolded at breakneck speed during that last year of the Weimar Republic. Kirchheimer continued to attend Schmitt's seminar, and soon they were both debating another subject: legality and legitimacy. After he settled into his new job as a lawyer, Kirchheimer published his first major article in the July 1932 issue of *Die Gesellschaft* [Society]. The title was "Legality and Legitimacy". The genesis of this article is closely linked to his discussions with Schmitt and his supporters. These led to Schmitt publishing an article with the same title shortly afterwards. This in turn prompted Kirchheimer to refute this new contribution by Schmitt point by point in even more detail. This next article was completed in November 1932. Its publication in February 1933 was, however, overshadowed by the new political balance of power after Chancellor Adolf Hitler took office.

The idea of writing an article about the relationship between legality and legitimacy had first arisen in discussions in Carl Schmitt's seminar. Kirchheimer started to work on the subject in November 1931. He may have been prompted to do so by a radio lecture by Schmitt in early November 1931 in which he opposed "formalizing the alternative of legality or legitimacy in the sense of subaltern, formalistic disputes about words" (Schmitt 1931e, 15). Schmitt went ahead with discussing the concepts of legality and illegality in his seminar on constitutional theory in late January 1932 (see Mehring 2014a, 254). He gave a radio lecture titled "Was ist legal?" [What is legal?] in February, but without referring to the opposite concept of legitimacy.[3] Among the group of younger jurists in Schmitt's circle, it was Ernst Forsthoff who also started to reflect on this subject. In a January 1932 letter to Schmitt, Forsthoff used what was to become Schmitt's constitutional policy credo: "To my mind, what matters is not legality, but only legitimacy, the

3 See Schmitt (1932d) and letter from Ernst Forsthoff to Carl Schmitt dated 8 April 1932 (Schmitt and Forsthoff 2007, 41).

political stance toward the fundamental constitution."[4] Schmitt championed this idea of a kind of superlegality of certain elements of the constitution, too. Kirchheimer's article also grappled with this idea, albeit with a sociological approach and a different political thrust. He completed a first draft in early April 1932 and sent it to Schmitt. A few days later, Schmitt forwarded the manuscript to Forsthoff to keep the discussion going. The published version of the article includes additional details and references, which leads to the conclusion that Kirchheimer submitted the final manuscript to the editors of *Die Gesellschaft* in late May 1932.

Schmitt sent Kirchheimer's text to Forsthoff in Freiburg on 14 April 1932.[5] A comparison of this version of the manuscript, including revisions in Kirchheimer's hand, and the published version yields a total of thirty-four modifications in wording, none of which are major substantive changes. The only significant change is in the references to Schmitt's works. Only one of the four references to texts by Carl Schmitt is to be found in the manuscript of the first draft (Footnote 5 in the printed text). Kirchheimer added another reference to Schmitt (Footnote 3) later by hand, and two references to him (Footnotes 15 and 25) were not yet included; Kirchheimer obviously added them later after discussions with Schmitt, shortly before the article went to press.[6]

While writing his article "Legality and Legitimacy,"[7] Kirchheimer was still unaware of the clandestine preparations to remove Chancellor Brüning from power, which were to bring about his resignation on 30 May 1932. Nonetheless, the piece reads like a conclusive record of the changes in the republic made during the Brüning era. The central hypothesis of Kirchheimer's diagnosis of the current constitutional policy was that "the concept of legality is undergoing a structural transformation" (48) in Germany, whereby the changes to the constitution were profound but not formal. The period of the parliamentary democratic legal order of the republic had been replaced by a new order of legitimacy. The "new form of legitimate power" (45) in the state was the *Berufsbeamtentum* (professional civil service) in collaboration with and supported by the Reichswehr (the armed forces) and the judiciary. Kirchheimer supported his transformation hypothesis using texts by three authors whom he identified as major legal scholars promoting this development: Carl Schmitt, Ernst Rudolf Huber, and Otto Koellreutter, who had been openly sympathizing with the NSDAP as early as 1930.

A key characteristic of legal orders that had become rational was that they applied the law equally, irrespective of the person concerned, formally guaranteeing equal treatment to the opponents of the social system prevailing at the time. And in order to guarantee this opportunity in practical terms, legislative and executive powers had to be separated. If this separation were to be suspended by a "government which now fuses legislative and executive authority" (44), as Kirchheimer stated Brüning's regime of emergency decrees was effectively doing, then this would mean no more equal treatment in formal terms.

---

4 See letter from Ernst Forsthoff to Carl Schmitt dated 23 January 1932 (Schmitt and Forsthoff 2007, 40).

5 See the editors' explanations in Schmitt and Forsthoff (2007, 359).

6 I would like to thank Jürgen Tröger for generously making a copy of this manuscript from Ernst Forsthoff's papers available.

7 See Kirchheimer (1932a). The following pages numbers refer to this article.

Such a regime would then have to endeavor to compensate the loss of the indisputable legal basis of its action due to a parliamentary decision by seeking legitimatory authority going beyond that legal basis. Appointing a person to the authority of the office of the President of the Reich served to fulfill this function. In fact, however, Kirchheimer continued, the large number of presidential authorizations of government action on the basis of Article 48 had created a situation whereby the professional civil service could take on the function of the new legitimate power in the republic. The Brüning government's practice of emergency decrees was characterized by vague norms, unclear wording, frequently changing rules, and giving carte blanche to the executive branch. This allowed officials to execute the regulations as they liked and claim that all measures taken were legitimized.

Kirchheimer's key analytical concept for this constellation was "dual legality" (46) (in the German original: *zweistufige Legalität*; "two-stage legality" or "two-tier legality" would be closer translations). The idea for this concept dates back to French legal theoretician Maurice Hauriou, who saw a *superlégalité constitutionelle* inscribed in the French constitution (see Hauriou 1932, 297). Kirchheimer thought that this problem was particularly salient in Germany because a large number of substantive legal provisions in the second part of the constitution essentially had to be understood as an invitation for the executive branch, i.e., the President, to confront the legislature with the claim that it was violating the constitution every time it took a decision that was not to its liking. Yet in contrast to what Carl Schmitt claimed a few months earlier in *Der Hüter der Verfassung* [The guardian of the constitution] (see Schmitt 1931b, 91), the large number of provisions in the second main part made it more difficult to systematically formalize and legally engineer the concept of laws in Germany but did not make it completely impossible. However: "no 'pluralism of conceptions of legality,' as Carl Schmitt describes it, has emerged yet" (47). Even before Brüning's regime of emergency decrees was installed, the bureaucracy in Germany occasionally became the keeper of the seal of this dual legality as a result but it was usually kept in check by functioning parliamentarism.

In his article, Kirchheimer showed how the republic's basis of legitimacy had successively shifted in four areas: the Reich government, the *Länder* governments, the political parties, and the system of labor courts. Expanding the application of Article 48 for emergency decrees with an undefined or unlimited period of validity destroyed any opportunity to review the executive branch using the law as the yardstick. Any criticism of the obvious illegality of a measure decreed under Article 48 or its interpretation by the officials was deflected by reference to the legitimacy of the government and the indisputable validity of its goals and actions. This meant that all legal barriers to government action had disappeared; the government was legitimizing itself. Kirchheimer detected a transformation occurring in parallel in the increasing number of acting governments installed at the *Länder* level. If acting governments were replaced by Reich commissioners as recommended by Ernst Rudolf Huber in the spring of 1932 (see Huber 1932), this would be a further step in the transformation process he had diagnosed.

Kirchheimer saw a similar development in the way political parties were being dealt with. Following the letter of the Weimar Constitution, all political parties had to be treated equally in principle. The Reichstag had rightly always rejected placing individual political parties and groupings under special criminal statutes. The social ideas expressed in a party's goals had no bearing on their status of legality; limitations arose at most from the two *Republikschutzgesetze* (Laws to Protect the Republic) of 1921 and 1930. Their provisions, however, referred to specific punishable offenses committed with the aim of undermining the republic. Kirchheimer saw the fundamental equal political treatment of the political parties, unaffected by this, eroded by Otto Koellreutter's construction of a legal concept of a "revolutionary party." Kirchheimer explained this transformation in the policy of legitimation as he grappled with a contemporary essay by Koellreutter which ultimately argued for an end to all legal limitations on the NSDAP and its combat units as well as a ban on the KPD (see Koellreutter 1932). Kirchheimer countered the construct of the legal concept of a "revolutionary party" (53) by arguing against any positive assessment of the NSDAP, speculating that "the question of the transformation of private property is really what is of concern to Koellreutter" (53) when determining what was revolutionary, and that the latter was in fact concerned with rejecting that transformation.

Kirchheimer also raised fundamental concerns against developing the law in such a way that it differentiated between legitimate and illegitimate parties. The Weimar Constitution, he claimed, did not provide for superlegality of selected elements of its system of norms. Consequently, no additional material criterion besides the concept of legality could exist for legally assessing a political party. Yet this was exactly what was already the case in Germany. Kirchheimer referred to a January 1932 decree by Minister of the Reichswehr Wilhelm Groener permitting members of the NSDAP and its combat units to apply for positions in the Reichswehr, while continuing to ban supporters of the KPD from doing so, as an example of this practice.

In another section of his article, Kirchheimer briefly discussed changes in labor law, largely following the criticism by Otto Kahn-Freund and Ernst Fraenkel of the more recent decisions of the *Reichsarbeitsgericht* (see List of German Courts).[8] For one thing, the *Reichsarbeitsgericht* had presumed the right to limit the legitimacy of parties to collective bargaining in its rulings on not recognizing *Betriebsräte* (works councils) and trade unions. For another, the court essentially limited the freedom to form and join trade unions by claiming sole authority to define whether a labor conflict had a legitimate economic goal or an illegitimate political one.

The result of Kirchheimer's four-part analysis was that "both the origins and the significance of the concept of legality [...] presently appear[ed] to be undergoing a process of decay, emptying it of its original meaning" (46). In retrospect, the parliamentary democratic system of the republic from 1919 to 1930 had proven to be an intermediary stage on the way to rule by the professional civil service in collaboration with the Reichswehr and the judiciary. The officials legitimated themselves as the government, limited the freedom of their enemies through the concept of the legitimate party, and governed labor law with the concepts of the legitimate party to address collective bargaining and the legitimate labor conflict.

With this diagnosis, Kirchheimer had evolved and radicalized deliberations from the previous three years. In his 1929 essay "Verfassungswirklichkeit und politische Zukunft der Arbeiterklasse" [The Constitutional Reality and the Political Future of the Working

8 See Fraenkel (1932) and Kahn-Freund (1932).

Class], he had condemned the tendency of the bureaucracy to take on a life of its own, and in 1930, in *Weimar—and What Then?*, he had criticized the legitimating function of the President of the Reich and the judiciary. He now synthesized these distinct tendencies into a single foundational tendency of transformation with his formula "dual legality." As fundamental as it was by design, he doubted that the contemporary system of rule by the bureaucratic aristocracy could stay in power long term. He considered the current situation to be merely a further "intermediate stage" (45) in terms of constitutional policy.

In contrast to Huber's plea for an authoritarian regime for economic purposes (see Huber 1931), Kirchheimer argued, "the social basis of this system is too weak to permit the bureaucracy to function as a truly independent mediating force" (58) in dealing with large industrial and business companies. The bureaucrats would only be able to remain in power for any length of time if they relied on extremely conservative societal groups from agrarian, small business, and military circles attempting to turn back the current process of capitalist development. Yet a countervailing societal force still existed. He recognized it in the "progressive will of the democratic populace" (59) but without making a prognosis about the outcome of this situation of conflict. In the weeks following the publication of his article, Kirchheimer gave a number of public lectures and talks in Berlin and Leipzig. He also put his hypotheses on legality and legitimacy up for discussion at a meeting of the Association of Social Democratic Jurists in Berlin.[9]

However, the reception of Kirchheimer's work was overshadowed by Schmitt's public appearances. In *Der Hüter der Verfassung*, Schmitt had still supported the separation of powers between the parliament, the government, and the judicial branch, but had considered this to be largely eroded in practice. He then revoked his essential agreement with the principle of the separation of powers in a number of lectures and shorter written contributions in the first half of 1932. The ascendancy of the President of the Reich to a new legislator had brought about a fundamental transformation from a legislative to an administrative state. Only a few days after Papen took office, Schmitt offered his publisher a manuscript compiled from lectures he had held during the previous weeks and borrowing the title of Kirchheimer's article. To Schmitt, the title of this pamphlet *Legality and Legitimacy* signaled the transition to a new political order, the transition from the previous order of legality to one of a higher legitimacy. Whereas to Kirchheimer the problem was the state becoming independent of society, to Schmitt it was exactly the opposite: the root cause of the crisis was a pluralist society encroaching upon the state.

Schmitt's *Legality and Legitimacy* was published a month after Kirchheimer's in mid-August 1932. He added a prefatory statement to the book that "this essay was completed on 10 July 1932" (Schmitt 1932e, 6).[10] This was important because immediately upon its publication, the work was viewed as a legal justification of the coup by the Reich against

---

9 According to an announcement of the event in the main social democratic newspaper, *Vorwärts*, Kirchheimer put his hypotheses on legality and legitimacy up for discussion at a meeting of the Vereinigung Sozialdemokratischer Juristen (Association of Social Democratic Jurists) in Berlin on 2 October 1932. See *Vorwärts* (Berlin edition), 1 October 1932, page 12. I am grateful to Detlef Lehnert for calling this reference to my attention.

10 This statement is not included in the English translation.

the *Land* of Prussia on 20 July.[11] Schmitt's book[12] is a continuation of his earlier one on the crisis of parliamentary democracy. In his new book, he discussed the practical consequences of parliamentarism, which he regarded as a foregone conclusion. Schmitt's *Legality and Legitimacy* has been described as characterized by "rhetorical magnificence accompanying snide *Schadenfreude*" (McCormick 2004, xxiii).

The book is divided into two parts corresponding to its distinction between liberalism and democracy. Liberalism had created the bourgeois state of the *Rechtsstaat* and the parliament had a monopoly on legislation. Majority rule required a legal disposition and the refusal of a governing majority to capitalize politically on its "premium on the possession of power" (35) by excluding the minority from power in the future. This political premium was relatively predictable in calm and normal times. In troubled times, as in Germany at the time, however, it was very unpredictable. Parliamentary democracy had a wide door through which its enemies could gain power and then close the door behind them. Schmitt warned of the possibility of a "legal revolution" (36).

In the second part of the book, Schmitt sought to show that the creators of the Weimar Constitution had designed a political order that was doomed from the start. He played the separate parts of the constitutional system off against each other in "Mephistophelean fashion" (McCormick 2004, xxxiv). The constitution, he asserted, provided for three "extraordinary lawgivers" (39) whose effects, if combined with one another, would lead to a dissolution of the system. These were not lawgivers in the sense of parliaments yet the impacts of their decisions were just as binding as laws adopted by the parliament. First, there were the courts, which could rule in favor of basic rights and were thus open to interpretation to serve any and all ideological purposes. Second, there were referenda, which were in plebiscitary competition with the parliament. And third, there were the extraordinary competencies of the President to decree measures.

The parliamentary system of lawmaking was crushed by these contradictions. Schmitt even went so far as to claim the existence of two constitutions within the Weimar Constitution. He referred to a contradiction between the value neutrality in the first, organizational, part of the constitution and a number of value commitments in the second part. According to Schmitt, "the decision must fall for the principle of the second constitution and its attempt to establish a substantive order" (94). This meant that the parliamentary, democratic, and federal organizational principles of the constitution codified in the first part were at the disposition of the political decision-maker on the implementation of the substantive order. In other words, Schmitt gave the President of the Reich the power to liquidate not only the democratic parliamentarism of the republic but also the federal structure of the Reich. In so doing, he emphatically abandoned the organizational part of the Reich's constitution *in toto*. This was the "sensational outcome" (Muth 1971, 111) of Schmitt's considerations on legality and legitimacy, an outcome that provided a justification for the Reich government's coup against the state of Prussia.

In light of the enormous power to make constitutional changes on the organizational level assigned to the President of the Reich, it was all the more astonishing that Schmitt

11 See the next section of this chapter.

12 Schmitt (1932c). The following page numbers refer to this text.

did not make it clear which articles of the salvageable substantive "core" (92) from the second part of the constitution he considered to be crucial. Since he had regularly mocked the dilatory formulaic compromise in this second part of the constitution in his writings of previous years, he could only have been concerned with a few of these articles. But which ones? His book reveals nothing about this question apart from the vague formulations "to liberate it [the constitution] from self-contradictions and compromise deficiencies" (94) and "recognition of the substantive characteristics and capacities of the German people" (93). This could mean everything or anything—and, in any case, it was the sole responsibility of a President elected by plebiscite to decide. The "substantive values" (93) harbored by conservatives and authoritarians such as Hindenburg and his camarilla were undoubtedly the preservation of the privileges of sociopolitical elites.

Schmitt explicitly referred to Kirchheimer's analysis in his book. He acknowledges he "accept[s]" (9) Kirchheimer's pointed wording about the legitimacy of parliamentary democracy now consisting solely in its legality. However, Schmitt changed the wording of Kirchheimer's finding significantly to support his own argument. The English translations disregard this subtle modification. Kirchheimer had written that the legitimacy of the legislative state was "*allein in ihrer Legalität*" (in its legality alone) (Kirchheimer 1932g, 382).[13] Schmitt misquoted him in the original German version as writing that "*nur noch in ihrer Legalität*" (all that remained [of the legislative state was] its legality) (Schmitt 1932h, 14), thus changing the meaning of Kirchheimer's statement. Schmitt had added a general diagnosis in line with a theory of decline to Kirchheimer's statement about a trend that could still be corrected; there was no such diagnosis in Kirchheimer's text, but it fitted well with Schmitt's own legal theory.[14]

Relating Schmitt's *Legality and Legitimacy* to the dramatic course of events in the summer of 1932 reveals his political intentions. In playing off the presidential system against the legality system of parliamentary democracy and federalism, he was not only motivated by gloomy forecasts about political developments. The tension that runs through the text reflects how anxious its author was about the future. In 1932, the state of Prussia was the last democratic "stronghold of the republic" (Winkler 2001, 413). Schmitt was concerned with dismantling this Prussian bulwark and not primarily with building a wall against the Nazis (Blasius 2001, 29–31). As clear as Schmitt was in his deconstruction of the system of the Weimar Constitution in *Legality and Legitimacy*, he remained vague about the alternative he preferred. In retrospect, Schmitt stated that his piece was part of his genuinely desperate attempt to rescue the Weimar Constitution.[15] However, there are also indications supporting the hypothesis that he was aiming at nothing less than a fundamental alternative to the system of the Weimar Republic at this time, in the name of the "recognition of the substantive characteristics and capacities" (93) of the German *Volk*.

---

13 The wording is identical to the manuscript version of his essay that Kirchheimer had sent Schmitt in advance.

14 On this "additional twist" (Andreas Anter) by Schmitt when citing Kirchheimer, see Neumann (2015, 236–239), Anter (2016, 106), and Buchstein (2017a, 89–91).

15 See Chapters 14 and 15.

Opinions in the secondary literature differ concerning Schmitt's political intentions in this book.[16] The question is whether Schmitt wanted to warn against an outcome—the collapse of the Weimar Republic—or whether he actually encouraged this outcome (see McCormick 2004, xxii). To Kirchheimer, the answer was obvious. Schmitt's book could not be seen as a purely analytical diagnosis of Weimar democracy that lacked a political agenda of its own. His arguments did not conform with a temporary suspension of parliamentary democracy. The logic of his reasoning pointed toward eliminating parliamentary democracy for good because it was becoming obsolete. Viewed from a purely formal point of view, Schmitt still made the case for a commissarial dictatorship. Within the logic of his criticism of the Weimar Constitution, however, it amounted to a sovereign dictatorship. A few weeks after Schmitt's book had been published, Ernst Fraenkel bluntly called it a blueprint for the permanent supersession of parliamentary democracy and the establishment of a permanent fascist dictatorship (see Fraenkel 1932, 507).[17]

Following the publication of Schmitt's *Legality and Legitimacy*, the conversations between Schmitt and Kirchheimer became fierce political arguments. Schmitt, who was known for taking criticism as a personal insult, now changed his personal opinion of Kirchheimer, too. After the two of them went for a walk together in Berlin's Tiergarten park in late August 1932, he no longer called him "nice" or "intelligent," but a "*scheußlicher Kerl*," a "vile fellow."[18] This was the first time since January 1928 that Schmitt commented negatively on Kirchheimer in his diaries.

2. The coup against Prussia

Carl Schmitt's political role in the final year of the Weimar Republic is still the subject of scholarly disputes to this day. In 1957, when assembling some of his essays on constitutional law from the final days of the Weimar Republic, Schmitt himself promoted the interpretation that it had been an "outcry in an emergency," a desperate attempt to safeguard the republic from the onslaught of both the communists and the Nazis.[19] A number of authors sympathizing with Schmitt have followed this self-interpretation. Among independent authors, the debate has focused on answering two questions.[20] First, whether the emergency decrees planned with Schmitt's involvement, had they been successful, would have stabilized the republic or transformed it long-term into a different state order. And second, the extent to which Schmitt (willingly or unwillingly) had promoted the transfer of political power to the Nazis through his activities. The first question can be answered unequivocally but not the second.

16 On the political ambiguity of this piece, see Hofmann (1995, 99–104), McCormick (2004), and Mehring (2014a, 253–258).

17 John McCormick draws similar conclusions in his interpretation that Schmitt's line of argument ended at a "military junta" (McCormick 2004, xxxix).

18 Carl Schmitt, diary entry of 25 August 1932 (Schmitt 2010, 210).

19 See Chapter 15.

20 See Muth (1971), Berthold (1999), Pyta and Seiberth (1999), Seiberth (2002), Blasius (2001, 15–70), and Neumann (2015, 169–302).

In the summer of 1932, the *Land* of Prussia was at the center of the political power struggles. It was by far the largest *Land* in the Reich with two-thirds of the land area and three-fifths of the population and had been governed by coalitions led by the Social Democrats since 1920. After the elections of 24 April 1932, the government under the social democratic Prime Minister Otto Braun no longer had a majority in the Prussian parliament. More than 50 percent of the seats had been won by the NSDAP and the KPD, both of them extremist parties. The Prussian government of Social Democrats, the Center Party, and leftist Liberals remained in office only in an acting capacity. The governing coalition had changed the procedural rules of the Prussian parliament just before the elections. The Braun government could not be voted out by the Nazis after the elections because the new rules required an absolute majority of votes for the office of Prime Minister of Prussia.

Shortly after the elections in Prussia, there was momentum for change in the governing constellation in the Reich. Support for Chancellor Brüning from the office of the President waned because of his policies and his distancing himself from the NSDAP, and he finally had to submit his resignation on 30 May 1932. There was no compelling reason to remove Brüning from power; this can only be interpreted as an attempt by the President's circles to install a right-wing government. General Kurt von Schleicher had already conducted confidential talks with Adolf Hitler in the office of the President and had been able to gain his support for the NSDAP tolerating a different presidential cabinet. Therefore, the presidential cabinet was no longer dependent on toleration by the SPD, as von Hindenburg had demanded of his staff as an ultimatum. In return for his policy of toleration, Hitler was promised, among other things, that the President of the Reich would dissolve the Reichstag once the new government had been appointed, thus bringing about new elections. Hitler's rationale in these confidential agreements was that his party would form a government by legal means after these elections. The new presidential cabinet headed by Chancellor Franz von Papen had moved even further to the right because of its members' political stance. Papen was a member of the rightist wing of the Center Party. His government had virtually no parliamentary support at the time. When he took office, the first and moderate phase of the presidential regime, which the parliament still tolerated, came to an end. Now a second phase began, the openly anti-parliamentarian and authoritarian phase.

The new Chancellor took office on 1 June 1932 and then on 4 June, the President fulfilled one of Hitler's conditions for tolerating the new government, namely dissolving the Reichstag. Schmitt supported the decision to dissolve the parliament on the basis of Article 25 of the Weimar Constitution. The President scheduled new elections for 31 July. Papen sought to transform the Weimar Republic into an authoritarian state; he personally even preferred re-establishing the state in the form of a monarchy. It was for this reason that Prussia came into the sights of the government of the *Reich*. What is commonly known as the Weimar Coalition of the SPD, Center Party, and DDP had lost its governing majority in the Prussian *Landtag* in the elections of late April 1932, remaining in office only in an acting capacity. The same was true of the governments of some other *Länder*. Papen's cabinet had been taking concrete steps to prepare the abolition of the Prussian government, which had been demanded by various German nationalist politicians and publicists since early July. The only thing lacking was a sufficient reason for such a massive intervention in the *Länder* rights, which were guaranteed by the constitution.

The summer of 1932 saw fierce, violent political conflicts with many casualties across the entire Reich. Comparable to civil war, the riots culminated on 17 July 1932 in what is now known as "Altona Bloody Sunday" when the police massively intervened in street fighting between Nazis and communists in Altona, then in the Prussian province of Schleswig-Holstein and now part of Hamburg. The police were completely overwhelmed and responsible for the deaths of a number of uninvolved civilians. The Prussian government was ousted three days later on 20 July 1932; the reason given was that after the catastrophe in Altona, public order and safety in Prussia could only be maintained by the Reich government. Justified in formal legal terms by Article 48 of the Weimar Constitution, this act was a "scarcely concealed coup" (Bracher 1955, 513). A Reich commissioner installed by the Reich government took over political power in place of the Prussian government. Public safety and order were certainly disrupted considerably in the summer of 1932, and not only in Prussia but across almost all of the Reich. The reason why the Reich authorities intervened only against Prussia was that they wanted to put an end to this Social Democratic stronghold. The Braun government, which was caught unawares by the coup, did not even attempt to put up any resistance. Instead, it chose the path of legal appeal before the *Staatsgerichtshof* in Leipzig.

Schmitt was not involved in preparing the coup against Prussia on 20 July 1932. He only found out about it from reading the newspapers. But he had been informed through his close contacts to officials in the Ministry of the Reichswehr that a coup like this would soon take place (see Pyta and Seiberth 1999, 435). He noted in his diary on 20 July: "sad that I wasn't part of it."[21] Back at the university, he immediately spoke with his colleagues, defending the course of action that had been taken, and contacted General von Schleicher's circle. Two days later, the government commissioned Schmitt as one of three attorneys to represent the Reich in the upcoming proceedings before the *Staatsgerichtshof* (see Seiberth 2002, 146–148). As early as 1 August, he published the article "Die Verfassungsgemäßheit der Bestellung eines Reichskommissars für das Land Preußen" [The constitutionality of the appointment of a Commissar of the Reich for the *Land* of Prussia].

Schmitt invoked three arguments in favor of the coup. First, he denied the legality of the provisional Braun government. He called the changes to the procedural rules mentioned above an "event similar to a coup." (Schmitt 1932f, 954) Second, he granted the President unlimited discretion concerning the question of whether public safety and order in Prussia were considerably disrupted or not. Third, the *Land* of Prussia and the government of the Reich had to apply the same political assessment to all parties that were to be combated for being inimical to the Reich. The Prussian government had fought against both the KPD and the NSDAP as enemies of the republic. Not so Papen's Reich government, which at this point in time persecuted only the communists as enemies of the republic. Schmitt granted the right to determine who was an enemy of the republic exclusively to the government of the Reich. Thus, he defined the communists as the "parties truly inimical to the state" and saw the NSDAP as having been deprived of the "legal

21 Carl Schmitt, diary entry of 20 July 1932 (Schmitt 2010, 201).

opportunities for state will formation" (Schmitt 1932f, 258) by the measures taken by the Prussian government. From the perspective of the deposed Braun government, this rationale was sheer mockery. It had tried with all its might to combat the extremists from the right and the left, and its attempt to ban the Nazis' militant mass organizations had been thwarted by a dictatorial government of rightist politicians.

Kirchheimer reacted to this turn of political events with a number of articles and public speeches. The first article "Die staatsrechtlichen Probleme der Reichstagsauflösung" [The constitutional problems of dissolving the Reichstag] appeared in the August issue of *Die Gesellschaft*.[22] He claimed that the last semblance of formal neutrality of the President of the Reich had passed with Papen and his cabinet. This had also changed the situation in Germany in terms of constitutional policy to such an extent that he described it as politically illegitimate. In the new situation, he believed that "every group [had to] itself review under its own responsibility which government actions deserve[d] obedience as required by the constitution" (396). As long as the parliament remained a place of political decision-making, there was a duty to obey. As far as he was concerned, however, it ceased to be a source of legitimacy for the "confessors of democratic socialism" (396) like himself if a government attempted "to annihilate" (397) the institution of the parliament itself on the basis of unconstitutional interpretations of Article 25 of the Weimar Constitution, which governed the President's right to dissolve parliament. Kirchheimer saw the 4 June declaration by the President to dissolve parliament as constituting precisely such a case of illegitimacy.

In his argument, Kirchheimer differentiated between the right to dissolve parliament in a constitutional monarchy and the same right in parliamentary democracies. In British constitutional practice, the rules for dissolution of parliament had evolved in a continuous process. Referring to Schmitt's *Constitutional Theory*, Kirchheimer pointed to various cases that the right to dissolve parliament provided for (400–402).[23] He considered that none of these legally permissible cases existed in the event of the dissolution of the Reichstag, which was based on what Hindenburg and Papen had promised Hitler. It was not permissible under constitutional law, neither in formal terms nor with the reasons given. The fact that a President of the Reich did not want certain parties to be involved in a government coalition or that he sought to "help [certain parties in parliament] attain a better position" (406) was not a reason for dissolving the parliament covered by Article 25 of the constitution.

After the Presidential government's overthrow of the Prussian government, the SPD felt, presumably justifiably, that the chances of successful resistance against this coup were not promising. Instead, trusting that the law would prevail, the party leaders opted for a course of strict legality in a lawsuit against this breach of the constitution before the *Staatsgerichtshof*. The spectacular trial took place from 10 to 17 October 1932. Two of Kirchheimer's mentors—Schmitt versus Heller—were at the center of this decisive court battle about the future of the Weimar Republic with Schmitt representing the Reich government and Hermann Heller representing Braun's former Prussian government.

22 See Kirchheimer (1932c). The following page numbers refer to this text.

23 See Schmitt (1928b, 373–378)

Kirchheimer voiced his opinion in his article "Die Verfassungslehre des Preußen-Konflikts" [The constitutional theory of the Prussia conflict] while the case was being prepared. This was published in the September issue of *Die Gesellschaft*.[24] Kirchheimer called the Prussian coup an obvious breach of the constitution which was simultaneously of eminent political significance. He again embedded his argument in a historical course of events that developed as a result of socioeconomic changes. The successive dissolution of the parliamentary legislative state of the Weimar Republic, he stated, could be divided into three phases retrospectively. In an initial phase from 1919 to 1922, it was based on coalitions of the social forces which were represented by the Social Democratic Party, the Catholic Center Party, and the liberal bourgeois parties. The chancellorship of Gustav Stresemann (1923) had made a first step to reduce the bourgeois attitude to its core economic interests, which was reflected politically in the fact that the purely parliamentarian government was replaced by a balance of the social forces according to their social positions of power. This enabled the state bureaucracy to ascend to the role of an arbitrator in a second phase between 1924 and 1930. Since Brüning was appointed Chancellor in March 1930, the republic had mutated in the third phase to an authoritarian form of government that had suspended important material provisions of the second part of the constitution. Because of the "20 July coup" (423), parts of the first, organizational part of the constitution had also become the focus of the driving social forces, which were progressively eroding the constitution.

Kirchheimer accused the supporters of this "process of shrinking the Weimar Constitution" (410) from the ranks of German constitutional law of having long ceased using the Weimar Constitution as their point of reference. Instead, they were practicing a "science of concrete circumstances" (410), which was beyond the constitution. It is clear from the wording quoted that Kirchheimer was addressing Schmitt directly in this article. He insisted that every constitutional question had to be answered exclusively on the basis of the Weimar Constitution. He also reminded Schmitt of a dictum from his own *Constitutional Theory* according to which there were fundamental institutions of established constitutional law which were immune to parliamentary decisions to change the constitution and "thus also to interventions by the President of the Reich" (411) (see Schmitt 1928b, 77–82). Kirchheimer considered federalism one of these fundamental institutions. The major importance of federalism in constitutional law, he claimed, was evident not least because the Weimar Constitution, unlike the constitution of the former German *Kaiserreich*, had specifically installed the *Staatsgerichtshof* as the decision-making body for disputes between the Reich and the *Länder*.

Again, Kirchheimer directly addressed Schmitt. Of course, in his book *Der Hüter der Verfassung*, Schmitt was right in principle to urge restraint in decisions regarding conflicts between the Reich and the *Länder* (see Schmitt 1931b, 4). In this particular case, however, when the court had been brought in as the legislative body deciding the dispute, those doing so were "fully aware of the highly political nature of such differences" (413). For that reason, it was "impermissible to demand of the court a degree of abstinence which would in reality put the internal organization of the Reich completely at the disposal of the President of the Reich" (413). Kirchheimer attacked Schmitt about a third

24 See Kirchheimer (1932d). The following page numbers refer to this text.

point, too. His decisive objection was that the purpose of the changes to the procedural rules of the Prussian parliament in the run-up to the April elections could not be viewed as preventing a takeover of power. These changes had "only fully brought the parliamentary principle of Article 17 of the Constitution of the Reich to bear" (416). For it was not until after these changes had been introduced that a prime minister elected merely with a relative majority would not be deposed immediately by a vote of no confidence.

Kirchheimer concluded that the events of 20 July "amounted to such a serious case of abuse of discretion that in the face of this, a presumption of subjective good faith on the part of the Reich government could no longer apply" (421). The conflict between Prussia and the Reich showed that the Reich government no longer placed any value at all in sustaining the legal bonds between the German federal states and the national government. Not only the significant social fundamental rights of the constitution, but now also the foundational provisions of the constitution relating to the internal organization of the state were subject to a "systematic process of annihilation by the current Reich government" (421). The "postdemocratic state" (423) thus created was an "authoritarian state" (423). To Kirchheimer, the prospect arising from this diagnosis was not completely pessimistic. At least the new situation had the advantage of being unmistakably clear, and it "forced the working class to adopt new forms of struggle" (423). He cited a famous quote from Alexis de Tocqueville's book *Democracy in America* (1835) to illustrate the political *Lage* after the Prussian coup:

> Le législateur resemble à l'homme qui trace sa route au milieu des mers. Il peut aussi diriger le vaisseau qui le porte, mais il ne saurait en changer la structure, crée les vents, ni empêcher l'Océan de se soulever sous ses pieds.[25]

With this comment, Kirchheimer demonstrated his keen sense of the strategy with which Schmitt was to argue the Reich's case before the *Staatsgerichtshof* one month later. In his argument relating to constitutional law, which he then unfolded meticulously, Kirchheimer adhered closely to the wording of the constitution, referring to relevant constitutional commentaries and providing details of police law which would also have to be taken into account when the *Staatsgerichtshof* heard evidence. He named Smend as a key supporter of the rule-of-law principle according to which the bounds of discretion were subject to review by the *Staatsgerichtshof*, also in the case of individual acts of government (see Smend 1931).

The lawsuit about the legality of the coup against Prussia was brought before the *Staatsgerichtshof* in Leipzig in October 1932. In the words of historian Hagen Schulze, "it was the democratic *Rechtsstaat* and the authoritarian power state that were litigating" (Schulze 1977, 761). Schmitt was the leading attorney for the Reich in what was the most important political case in the Weimar Republic. It was his remarks that gained the most attention from the press at the time (see Blasius 2005, 115).

---

25 "The law-maker resembles a man who plots his route in the middle of the sea. He too can navigate the ship that carries him, but he cannot change its structure, raise the wind, or prevent the ocean from heaving under his feet." (Tocqueville 1835, Volume I, 264).

Schmitt's performance in Leipzig was the apex of his public political impact prior to 1933. In his plea, he followed the lines of argument from his article published on 1 August about the constitutionality of the appointment of a Commissar of the Reich for the *Land* of Prussia and accused the Braun government of violations of duty and lack of loyalty to the Reich.[26] Once again, he attempted to liberate the Nazis from the odium of being enemies of the constitution. He accused the old Prussian government of having been a "partisan government" that had arrogated the right to declare parties illegal for the sole reason that this served to preserve its own power. In contrast, the Papen government of the Reich had "finally" attempted to rule on this question independently of party interests. The Papen government had taken the decision solely on the basis of being "just and objective" and had therefore resolved "to repeal the equation with the Communist Party of a movement with which millions of Germans were not only sympathetic, but which they had voted for; such an equation was insulting to that movement."[27] These words of Schmitt's in Leipzig essentially endorsed the legality of the NSDAP.[28] Hermann Heller and Arnold Brecht, as attorneys for Prussia, countered Schmitt's remarks by pointing out that the government of the Reich was by no means politically independent because its course of action had been founded on prior agreements with the Nazis (see *Preußen contra Reich* 1933, 76–77).

The ruling of the *Staatsgerichtshof* on October 23 was to the satisfaction of the Reich government; the court divided power in Prussia between the Reich and the old government but left the Reich with most of its competencies. It confirmed Papen's position as *Reichskommissar* for the *Land* of Prussia during the state of emergency, reporting only to President Hindenburg. Schmitt was dissatisfied with the ruling even though the Reich had prevailed. In his view, the greatest flaw was that the *Staatsgerichtshof* had exceeded its competencies in parts of the ruling and had thus excessively limited the President's absolute authority of dictatorship (see Neumann 2015, 282–286).

The losers of the case in Leipzig were not only the *Land* of Prussia but also the SPD. After the coup against Prussia, the party leadership had not even attempted to put up resistance and—similarly to the situation after the Kapp Putsch in 1920—call for a general strike to defend the democratic republic. The young socialists in the SPD were outraged by the decision against active resistance, which they viewed as a capitulation in the face of violence. More than 40 years later, Kirchheimer's fellow attorney Ernst Fraenkel spoke about his and his colleagues' deep disappointment when it became clear in July 1932 that the party leadership and the trade unions were not willing to call for active mass protests by the working class to defend the republic.[29] Historian Karl Dietrich Bracher viewed the coup against Prussia in retrospect as the last opportunity for legitimate and promising resistance before the parties and the trade unions were dissolved in 1933 (see Bracher 1955, 523). The ambivalent and apparently mediating ruling of the court ultimately ca-

26 For a summary of Schmitt's oral pleadings, see Schuller (2008).

27 Schmitt's oral pleading in *Preußen contra Reich* (1933, 39).

28 See Blasius (2001, 44–50), Blasius (2005, 75–77, 114–117).

29 Letter from Ernst Fraenkel to Karl-Dietrich Erdmann dated 31 January 1973, Fraenkel (1999, 675–677).

pitulated before the facts on the ground put in place by the Reich government, thereby becoming a "milestone in the demise of the republic." (Stolleis 1999, 121)

3. Constitutional reform?

The new elections to the Reichstag on 31 July 1932 had disastrous results both for supporters of parliamentary democracy and for the previous presidential dictatorship. The NSDAP doubled its seats and replaced the SPD as the largest party. Once again, no parliamentary majority was in sight. Even a coalition of SPD and KPD, which many leftists desired although it was entirely unrealistic, would not have had a majority in parliament. The political basis for the previous regime of the presidential dictatorship had also dwindled.

Although Schmitt was aware of the precarious situation of the presidential dictatorship, he continued to rely solely on this political option in order to stabilize the existing state. There was the danger that the Reichstag, unable to form a government, would be dissolved once more, and that new elections would have to be held yet again. In late August 1932, Schmitt was involved in deliberations by the Papen government to prepare an emergency plan.[30] The constitutionally problematic core of this "September plan" consisted of again dissolving the Reichstag and then postponing new elections indefinitely; this was based on reasoning contravening the wording of Article 25 of the constitution. Schmitt formulated several possible ways to argue for the president to dissolve parliament and had his student Ernst Rudolf Huber work out additional legal details. It was mentioned explicitly in the minutes of the ministers' discussion of the plan that Schmitt would appear in public "certainly in accordance with the wishes of the current cabinet."[31] Hindenburg agreed with the plan, too. Yet, ultimately, it was not implemented because the majority of the Reichstag preempted it by deciding to dissolve itself. After this disaster, Papen made a new attempt to install an authoritarian solution. He summoned Schmitt to the Chancellery in late September and asked him to prepare a draft for a completely new constitution for the Reich (see Mehring 2014a, 260–262). Schmitt noted his great pleasure about this assignment in his diary, as well as his "alarm" in light of the magnitude of the task.[32] The project was called off shortly afterwards when new elections to the Reichstag had been scheduled again, for 6 November 1932.

The election results, however, were surprising. The parties openly supporting Papen's presidential cabinet achieved a slight increase in votes. But the SPD's opposition to Papen had not paid off, either: it lost a few votes, whereas the KPD became stronger yet again. The most notable and surprising outcome of the election was the fact that the NSDAP's apparently inexorable success of the two previous years seemed to have been halted. Hitler's Nazi party had lost more than four percentage points. Nazism had lost its mystique of an invariably burgeoning movement. All of a sudden, the party was faced with financial bankruptcy. It was rife with internal divisions, its relationship to the

30 See Berthold (1999, 32–36) and Blasius (2001, 51–70).

31 Minutes of the meeting of the ministers of 14 September 1932. Printed in *Akten Papen* (1989, 587).

32 Carl Schmitt, diary entry of 25 September 1932 (Schmitt 2010, 219).

Sturmabteilung (SA, the uniformed and armed political storm troops of the NSDAP; see Glossary) was extremely tense, and many observers thought the collapse of the movement was looming. Under these circumstances, it no longer seemed out of the question to return to parliamentary democracy before long.

Yet Papen did not undertake any efforts to this end. Instead, he considered a new emergency plan which Schmitt was to be involved in again. Schmitt spoke openly about these plans with Huber; it is hardly surprising that he did not inform Kirchheimer of them. In his memoirs fifty-six years later, Huber recounted an evening walk with Schmitt and Kirchheimer along Unter den Linden and through the Brandenburg Gate. The three of them had talked about politics and in particular about the strike at Berlin's municipal railroad company which was started by the communist party in collaboration with the Nazi party on 3 November 1932, but neither Huber nor Schmitt mentioned their activities to prepare a new emergency decree: "In Kirchheimer's presence we did not speak of the emergency plan." (Huber 1988, 46)[33] Schmitt concealed from Kirchheimer how much he was involved in the current political events. In late November, Papen invited Schmitt for breakfast and spoke with him about the constitutional side of his emergency decree (see Mehring 2014a, 269). But Papen no longer had the support of the Reichswehr for his project. He resigned and was replaced by General Kurt von Schleicher on 3 December.

The debates about the purpose, the direction, and the potential for revising the constitution, as well as the plans for a permanent emergency regime, were eclipsed by the turbulences of the final days of Franz von Papen's chancellorship. Kirchheimer contributed three articles to the debates about constitutional reform. He also gave a talk to the Sozialistische Studentenverband (Socialist Students' Association) on the topic in Berlin on 15 November 1932.[34] The title of his article "Constitutional Reaction in 1932" in the November issue of *Die Gesellschaft*[35] already made it clear that this was an analysis of reform proposals from the rightist political camp. He accused the office of the President and Papen's government, which was still in power at the time, of presumably no longer considering legality important at all when introducing their preferred constitutional reform. He then took up Schmitt's work *Legality and Legitimacy* again, this time to call attention to the marked change in the reform plans circulating in the President's office and among his advisors. If it had been deliberations within the framework of the Weimar constitutional order after Hindenburg's election in 1925, now it was a "constitutional revolution" (77).

Kirchheimer's criticism zeroed in on two authors from different schools of political thought. First, he again took on Schmitt and his sympathies for a presidential dictatorship to be established through a coup. His prognosis was: "When a later epoch takes inventory of the intellectual content of this one, Carl Schmitt's book *Legality and Legitimacy* will prove itself a work superior to others because it bases itself on the foundations of political theory" (77). He also pointed out that Schmitt exercised restraint in his conclusions rather than presenting clear political solutions. Nevertheless, he strongly rejected as contradictory Schmitt's attempt to underscore the legitimacy of the President of the

33 The walk probably took place on 9 November 1932 (see Schmitt 2010, 232).

34 See the announcement in *Vorwärts* of 13 November 1932.

35 See Kirchheimer (1932e). The following page numbers refer to this text.

Reich by conducting a plebiscite in order to provide an approach of constitutional revolution from his side with the distinction of enhanced democratic legitimacy. He stated that Schmitt's plea for a constitution that was "postdemocratic" (78) at its core rested on a "predemocratic" (79) political anthropology; he alluded to the passive role of the populace in the conduct of plebiscites, which Schmitt had repeatedly described (see Schmitt 1932e, 93–95).

Kirchheimer countered Schmitt's opting for the presidential dictatorship with the political self-understanding of "modern democracy" (79). This form of democracy was possible because in a long and painful process in the course of industrialization, the mass of the population had developed from being solely passive supporters of political affairs to people actively participating in organizations. This sociological fact had to be borne in mind with respect to all reform considerations because he fundamentally questioned the option favored by Schmitt. The authoritarian state only shifted the problem of unifying the political will in a heterogeneous society; it did not solve it. The advocates of a permanent presidential dictatorship would have to be able to answer the question of how to deal with the problem of the "constitutional dynamic" (79) that would necessarily arise from the social basis of politics, which was constantly changing. Despite all its shortcomings, modern democracy "is after all the sole form of government which constitutionally makes possible the cooperation or the alternation of different groups at a time of increasing social and national heterogeneity" (80).

To counter Schmitt's position, Kirchheimer quoted an argument from Smend's theory of integration stating that the social conditions for institutionalizing the personal charisma of a political leadership figure on a permanent basis were no longer given in Germany at the time (see Smend 1928, 142–148). Economic crises, lost battles, or the sudden death of an incumbent were notorious for exposing such a regime to the danger of political instability because of the unceasing social dynamics. He simultaneously made use of Schmitt when he attacked the conservative reform agenda for constitutional reform. Papen's idea to turn Germany into a corporate state would not solve the problem of the unification of the political will; it would merely serve to postpone it. Schmitt's "all is sinecure, nothing lives," which he applied to parliamentary democracy, "applies more adequately to a political system in which all dynamism is suppressed in favor of an illusory static condition" (85–86). He concludes his articles with a quote from Marx in his *Poverty of Philosophy*: "Only in an order of things where classes and class contrasts have ceased to exist will social evolutions cease to be political revolutions."

A month later, Kirchheimer published his second contribution to this debate, his article "Die Verfassungsreform" [The constitutional reform], in the December issue of the monthly *Die Arbeit*, which was published by the Allgemeine Deutsche Gewerkschaftsbund (General German Trade Union Federation).[36] He began by reiterating his criticism of the Papen government's plans for an "authoritarian constitutional reform" (443), placing them in the same context as similar plans in Austria. And again, he targeted Carl Schmitt. This time he did so by aligning himself with Herman Heller's criticism of Schmitt that he was wrong to state that the origin of the problems of the current state order was only,

36 See Kirchheimer (1932f). The following page numbers refer to this text.

or even mostly, to be found in the constitutional norms specific to the Weimar Constitution (see Heller 1932b, 413). Instead, he and Heller insisted that the cause of the current constitutional crisis was an unsound strategy for overcoming the economic and social crisis. They believed Schmitt's notorious criticism of the text of the constitution was misguided because he simply ignored the deeper causes of the crisis. Kirchheimer devoted his attention in this article to proposals for constitutional reform from social democratic circles. He thought they had very limited room for maneuver if the goal was to maintain the major lines of a democratic constitution—sovereignty of the people, the parliament, personal liberties, and basic social rights. All that remained was to change constitutional provisions in an attempt to help political parties and social associations collaborate. Kirchheimer discussed various ideas that had been put forward: to permit votes of no confidence only once a year in the course of budget debates and with a simple majority; to establish a new Chamber of Commerce; to change electoral law to resemble the model practiced in England; and to limit the options for ballot measures and referenda.

Kirchheimer focused in particular on Ernst Fraenkel's proposals in his essay "Verfassungsreform und Sozialdemokratie" [Constitutional reform and social democracy] (see Fraenkel 1932b), which had been published the same month. Fraenkel, who had invited Kirchheimer into the law firm he shared with Franz L. Neumann the same year, became the subject of uncompromising objections from his younger colleague. In his article, Fraenkel had sought to create a new balance between the Reichstag, the Reich government, and the President of the Reich. His three-part proposal included introducing a constructive vote of no confidence, making it more difficult for the President of the Reich to dissolve the Reichstag and enabling him to turn directly to the populace with a plebiscite if the parliament rejected an emergency decree. Kirchheimer weighed the individual components of Fraenkel's proposals one by one. He did not arrive at his ultimately skeptical rejection because he considered some of the individual proposals wrong but, rather, because of a fundamental consideration. Fraenkel's proposals, he claimed, would not make any decisive changes to the "political and social structural relationships" (452) of the republic. If a constitutional order risked at every turn that its current or future organizational positions could be abused in order to destroy democracy itself, then it did not suffer from problems that a constitutional reform could remedy, but from structural problems in the realm of its social basis. The path promoted by Fraenkel was a "futile race" (452) against the proponents of dictatorship.

At the end of his article, Kirchheimer restated his commitment to defending the foundational institutions of the Weimar Constitution. However, he also expressed that he was perplexed and had become more skeptical. If society was to be ripe for democracy again, then that did not require a well-intentioned constitutional policy, but a "rapprochement of the two labor parties" (454), a "new order of the societal relationships themselves" (457), and a "breakthrough to new social forms" (457). Kirchheimer obviously thought that the only choice for the Weimar Republic in this situation was between a presidential dictatorship and a socialist democracy.[37] Since at the time, in 1932, there was no question of unifying the two labor parties, the conclusion of his article remained

37 On 18 December, Kirchheimer also gave a lecture on the subject in the Arbeitskreis Abraham, a discussion club of Jewish Social Democrats, in Berlin (see *Vorwärts* of 17 December 1932).

a helpless appeal. Even if the KPD adopted a milder tone toward the SPD from May 1932 on, the conflicts between the two parties in fact intensified because the KPD repeatedly voted with the Nazis for strategic reasons.

Despite their political differences, Schmitt adopted a stance similar to Kirchheimer's on the issue of constitutional reform. He, too, considered such projects unrealistic. He had argued in a lecture back in 1930 that the call for fundamental constitutional reform meant that the alternative "capitalism or socialism" would have to be decided; that, however, would bring about the catastrophe of a civil war (see Neumann 2015, 301). He took up the subject again in November 1932 in two lectures, the so-called Langnamverein addresses, speaking in front of industrialists from the coal, iron, chemical, and textile industries in the Rhineland and Westphalia, who were organized in the eponymous trade association.[38] He referred to the "egregious mistakes in the construction" of the Weimar Constitution and the fact that, in any case, it was only to be considered as an "emergency construct" and a "provisional solution" from the outset (Schmitt 1932g, 55). Moreover, all the constitution's key elements were "entirely denatured" (Schmitt 1932g, 56). A "strong state in a free economy" (Schmitt 1932g, 60) was needed for the future. In such a system, the state was not to interfere in the market.

Schmitt did not opt for "authoritarian liberalism," as Renato Christi called it (see Christi 1998) but, rather, for authoritarian capitalism. He believed that orderly constitutional reform would not be able to yield the strong state it required, so he inverted the order of the actions to be taken: "First of all, we need a strong state up to its tasks and able to act. Once we have it, we can create new organizations, new institutions, new constitutions." (Schmitt 1932e, 83) In other words, Schmitt certainly did express his support for a new and different constitution—but he felt that because of the existing political blockades, such a transformation would not take the orderly path of changing the constitution under Article 76 of the Weimar Constitution; instead, a new constitution would have to be brought about by strong political action. For those who would still like to see Schmitt as a friend of the Weimar Constitution in this phase, he revealed his "precarious friendship with the constitution" (Roth 2005, 155) at this point. What he had in mind was similar to an idealized image of Italian fascism.

Despite all their differences, Kirchheimer and Schmitt did have one thing in common when it came to constitutional reform: a realistic understanding that such a reform would be impossible under the prevailing conditions. Their views were also similar about the implications of this even if they had opposite strategic goals: a new, more stable order could only be achieved through a successful political struggle.

The new Chancellor Schleicher wanted to continue to govern on the basis of Article 48. In contrast to his predecessor, he did not call on Schmitt to provide the legal legitimation of his actions. The reasons for ousting Schmitt were personal; President Hindenburg also disapproved of Schmitt's draft for Papen's emergency decree (see Mehring 2014a, 269–271). Schmitt's response to this dismissal was depression and "political withdrawal

38 On Schmitt's two Langnamverein addresses, see Hermann Heller's early critique (Heller 1933a) as well as Maus (1976, 152–159), Christi (1998, 193–204 and 212–232), and Scheuerman (2020, 250–253).

symptoms" (see Pyta and Seiberth 1999, 607). In December 1932, his standing had suddenly been diminished, from the "crown jurist of the Presidential Regime," as the journalist Hellmut von Gerlach had called him, to just one among several Weimar constitutional scholars.

4. Conclusion: Defending or destroying the republic

At the end of the Weimar Republic, Kirchheimer and Schmitt were astonishingly similar in their opinions on constitutional reform. Both rejected the idea. Instead, they opted for a political solution to the constitutional crisis, albeit with diametrically opposed goals and strategies. Kirchheimer trusted in the labor movement's parties and unions, Schmitt in the Reichswehr and the civil service. In retrospect, it is clear that neither Kirchheimer's urging for socialism nor Schmitt's opting for the presidential dictatorship were successful prescriptions for safeguarding the Weimar Republic from being handed over to the Nazis. The labor movement was already too weak and divided, and the conservative bourgeois elites had long abandoned their rejection of the Nazis. Schmitt and Kirchheimer viewed themselves as political opponents for whom there was no longer any common ground. Schmitt wanted to prevent the transition to a trade union state or even a socialist republic at all costs; Kirchheimer saw that Schmitt had drifted toward an authoritarian economic liberalism seeking to eliminate key elements of the Weimar Constitution. They each framed the other's arguments on constitutional law and political positions in opposite ways. Schmitt saw Kirchheimer's analyses and calls to fight as further confirmation of both the vitality of the socialism he feared and of his own civil war scenarios. To Kirchheimer, the constitution was a principle of societal organization whose flaws were to be eliminated through political struggle. Although he wanted to defend the existing republic, he pressed for transcending the private capitalist order and for substantially expanding the socialist elements that were also part of the constitution.

In the final year of the Weimar Republic, Kirchheimer, alongside Neumann, Fraenkel, and Kahn-Freund, belonged to the circle of younger Social Democratic legal experts who had followed in the footsteps of Hermann Heller and Hugo Sinzheimer and had specialized wholly or in part on constitutional matters. Despite their differences of opinion—for example, between Fraenkel and Kirchheimer on questions of constitutional reform—, they shared multiple fundamental positions. Their common starting point was pronounced support for the republic, for parliamentary democracy, and for the Weimar Constitution. Whereas Kirchheimer, in his 1930 *Weimar—and What Then?*, still saw the root of the constitution's failure in its lack of decision regarding social policy, he switched to a more optimistic and constructive position in light of the presidential dictatorship, of all things. Although he did not view the constitution as fulfilling all his political desires, he defended it in 1932 as the relevant regulatory framework for the path to its fulfillment; he defended it as a political form open to positive changes. This was not a merely tactical relationship to the Weimar Constitution. Certain flaws notwithstanding, Kirchheimer viewed it as the best possible alternative at the time for introducing the interests of the labor movement into the political process and realizing them. What had not been accomplished in the constituent National Assembly was now to be achieved through political struggle conforming to the constitution. This required political mobilization. Thus, Kirchheimer undertook a marked correction of his original euphoria regarding decision and a cautious convergence with Franz L. Neumann's model of social compromise. In his comments on the crisis events, morale-boosting slogans such as "wait and tough it out" alternated with calls for active mobilization of the labor movement. After the bourgeoisie had rejected the fundamental principles of bourgeois democracy, Kirchheimer viewed the labor movement in the role of guardian of the democratic substance of the republic.

Schmitt had long begun to seek ways to overcome the Weimar Constitution. It is indisputable that he was not a member of the Nazi party in the final phase of the Weimar Republic; incidentally, virtually no German professor of constitutional law had joined it before 1933. But that did not make Schmitt a defender of the Weimar Constitution and its parliamentary democracy by any means. The combination of his interpretations of Article 25 (parliamentary dissolution), Article 48 (presidential powers), and Article 76 (constitutional reform) would have allowed constitutional changes to secure the Weimar Constitution only in principle (see Kennedy 2004, 168). Schmitt himself, however, aimed at a political system that would end parliamentary democracy in favor of an authoritarian regime. In his defense of the coup against Prussia, as already in *Legality and Legitimacy*, his concern was not to erect a defensive wall against the Nazis, as he claimed after 1945 in his interpretation of his behavior at the time, but to smash the last major bulwark of a political system which he regarded with contempt. By justifying the coup against Prussia before the *Staatsgerichtshof* in Leipzig, Schmitt de facto supported the political movement that ruthlessly eliminated the presidential system a few months later in the spring of 1933.

Enlisting support for an authoritarian capitalist order in his two Langnamverein addresses was the "bridge over the Rubicon" (Christi 1998, 179) that allowed Schmitt to join the Nazi party a few months later. Then, in view of these later events, it was quite accurate for Schmitt to praise the coup against Prussia and his own writings on questions of constitutional law as important preparations for the Third Reich.[39] In the spring of 1933, Kirchheimer did not make these connections in his writings. Only in his later exile did he agree with Fraenkel's *Dual State*, where he stated in retrospect that "in view of this speech [the two Langnamverein addresses] it cannot be said that Schmitt's conversion to National Socialism a few weeks later represented any significant inconsistency." (Fraenkel 1941, 61)

---

39 See Chapter 7.

Chapter 6: The Methodological Debate and Weimar's Final Days (1933)

Even before Schmitt's first political fall from grace in late 1932, his political activities and his book on legality and legitimacy prompted Kirchheimer to present his criticism of Schmitt more extensively and systematically in an article written for *Archiv für Sozialwissenschaft und Sozialpolitik* [Archive for social sciences and social policy], the leading sociological journal during the Weimar Republic. This time, Kirchheimer expanded the field of conflict to include fundamental methodological questions in legal studies as they related to the social sciences. During the entire time of the Weimar Republic, the German debate on constitutional law was immersed in methodological questions concerning positivist and anti-positivist approaches.[1] The three major protagonists of the schools of thought decidedly critical of positivism were Carl Schmitt, Rudolf Smend, and Hermann Heller. Their alternative approaches were limited to their common thrust against the legal positivism driven to greater heights by Hans Kelsen. They shared a plea to expand the methods of constitutional law used to gain insights in order to include other scientific disciplines. Smend and Schmitt had the humanities in mind, and Heller also the empirical social sciences.

At the end of the Weimar Republic, Kirchheimer also participated in this methodological debate, focusing his criticism on Schmitt. In contrast to other critical arguments with Schmitt, he linked his methodological criticism to the question about the status of the empirical social sciences for legal and political theory. Kirchheimer became a precursor of the discipline of political science because of his methodological program, which offered an alternative to Schmitt.

1 For an overview, see Gusy (1997, 427–447) and Stolleis (1999, 153–186).

1. Schmitt on his method

Schmitt said comparatively little about his own methodological approach. His texts were intended to speak for themselves with their succinct terms and concepts, the suggestive power of their language, their rich imagery, and either by surprising and overwhelming readers or by horrifying and disgusting them. It was his opinion that theories of constitutional law were developed through specific political debates, with the result that the academic discussion took on the tensions of the political struggle. Referring to specific political situations like this made it impossible to conceive of legal science as methodological and systematic work in order to gain universally valid insights. Every insight was an insight into the present; there was no insight that could be understood intersubjectively by everyone but only statements about specific situations that were always the objects of dispute. Schmitt's terms, concepts, and theorems were designed less to be rationally reconstructed and more to be directly self-evident.[2]

One exception to Schmitt's methodological abstinence was a programmatic section in his 1922 book *Political Theology*. Using the concept of sovereignty as an example, he called his own method the "sociology of legal concepts."[3] His aspiration was to surpass both the sociological explanatory approach put forward by Max Weber and that of historical materialism by Karl Marx and his successors. Schmitt argued against Weber by stating that he was merely seeking "the typical group of persons" (44) for certain ideas and intellectual constructions and then relating them to the "peculiarity of their sociological situations" (44). However, this was the determination of a certain kind of motivation of human action, hence psychology and not sociology of a juristic term. Schmitt criticized the Marxist explanation for making "separate consideration of ideology impossible" (43) since all it saw was "reflexes," "disguises," or "reflections" of economic relations. Marxism worked "with suspicion" (43) toward individuals and their ideas. Paradoxically, it was precisely because of its massive rationalism that historical materialism could easily turn into an irrationalist conception of history "since it conceives all thought as being a function and an emanation of vital processes" (43). To Schmitt, the theory of George Sorel was evidence of such a switch from rationality to irrationality.

Neither Weber nor Marx were sufficiently radical in their thinking. Compared with theirs, his approach "transcend[ed] juridical conceptualization oriented to immediate practical interest" (45). Schmitt was aiming not at the individual representatives and bearers of certain ideas but, rather, at a transpersonal level. The target of his sociology of concepts was not, as its name might suggest, the relationship between concept and social reality. Instead, Schmitt conceived of the sociology of concepts as their "basic, radically systematic structure" (45) related to the "social structure of a certain epoch" (45). Schmitt sought to find analogies between the semantic fields of the individual sciences and the conceptual form they shared. His approach consisted of two steps. First, it aimed to discover the basic and radically systematic structure of a certain type of legal thought. In a second step, this conceptual structure was "compared with the conceptually represented" (45) social structure of a certain historical situation. Schmitt

2 See Neumann (1981, 236–238) and Hofmann (1995, XIII).

3 Schmitt (1922, 42). The following page numbers refer to this text.

thought it made no difference for his approach—in contrast to those of Weber and Marx—whether the idealities produced by conceptualization were a reflection of social reality or whether social reality was the result of a particular kind of thinking. The only presupposition of his approach was "a radical conceptualization," i.e., "a consistent thinking that is pushed into metaphysics and theology" (46). According to Schmitt, "the metaphysical image that a definite epoch forges of the world has the same structure as what the world immediately understands to be appropriate as a form of its political organization" (46).

In Schmitt's view, the metaphysical image in question was—as a rule, but not always—theology. Schmitt explained this idea in *Political Theology* for the concept of sovereignty. When it was developed, the analogy to Christian theology was evident. In the seventeenth century, the absolute monarchy had corresponded to Western Europeans' state of consciousness because the position of the absolute monarch corresponded to the Cartesian concept of God prevailing at the time. Then, in the eighteenth century, the theoreticians of democracy had replaced God with the people, coining the term "sovereignty of the people." The metaphysical system of the legal positivism emerging in the nineteenth century and the liberal theory of parliamentarism consisted of "a political relativism and a scientific orientation that are liberated from miracles and dogmas" that were "based on human understanding and critical doubt" (42). Schmitt was of the opinion that modern scientific thinking was just one of multiple metaphysical systems. The purpose of legal theory had to be to reveal the metaphysical fundamentals of the sets of concepts pertaining to various ideologies. This applied both to socialism with its belief in science and to liberal parliamentarism with its belief in deliberative reason. Against the background of this methodological credo, it is easy to understand why, in his book *The Crisis of Parliamentary Democracy*, published one year after *Political Theology*, Schmitt insisted that an "ultimate core of the institution of modern parliament" and/or "ultimate intellectual foundations" (Schmitt 1923a, 20 and 33) of parliamentarism had to exist as part of a more comprehensive metaphysical system.

Now this begs the question which metaphysical image Schmitt thought that the immediate present constructed from the world and its political organization. His answer in *Political Theology* was brief and inconclusive but a line of thought can still be discerned. The development of constitutional law since the nineteenth century, he asserted, was characterized by two paradigmatic changes. For one thing, the traditional monarchical legitimacy had lost its persuasiveness. For another, all theistic concepts had disappeared from legal thought. Therefore, "legitimacy no longer exists in the traditional sense" (51). It had been replaced by "decisionist thinking" (51) that did not explain its rationale, and thus the political option of "dictatorship" (52).

In just over ten pages in the third chapter of *Political Theology*, Schmitt formulated an ambitious program for a conceptual history that has similarities to the later approaches by both Reinhard Koselleck and Quentin Skinner and was widely received.[4] He contrasted concepts expressing a complex subject matter antithetically with a *Gegenbegriff* (counterconcept); for example, democracy vs. parliamentarism; constitution as decision

4 See Mehring (2006); Müller and Schmieder (2016).

vs. dilatory formulaic compromise; *Rechtsstaat* vs. democracy. In relation to parliamentary democracy: since democracy implied the identity of those governing with those governed (and not their representation of the governed), the representative parliament was automatically and always in contradiction to democracy.

The antithetical contrasts were to express an irreconcilable contradiction for which Schmitt had identified the following well-known formula in *The Concept of the Political*: "All political concepts, images, and terms have a polemical meaning" (Schmitt 1932a, 30). These were bound to a specific situation and focused on a specific conflict. The result was "a friend-enemy grouping" (Schmitt 1932a, 30). Schmitt was of the opinion that concepts distinguished themselves in polemical (i.e., combative) usage through friend-enemy distinctions to such an extent that they firmly established a certain semantic meaning. Accordingly, succinct concepts in Schmitt's combative sense were concepts that were fixed for a certain period of time and with which political actors identified in order to guide their actions. Schmitt's activist concept of a concept was directed against the notion of value-neutral terminology, as Max Weber had asserted for his ideal types. Even in its form, his conceptual thinking was directed against allegedly liberal neutralizations and/or depoliticizations. Not only the concrete substance of the concepts but even the approach itself of his "radical conceptualization" was an expression of Schmitt's combative understanding of politics.

2. The Weimar debate about Schmitt's method

Kirchheimer was not the first to critically examine Schmitt's method and his combative use of concepts. Besides numerous substantive analyses of Schmitt's work, there had also been a few objections concerning methodological aspects. Publicist and writer Hugo Ball was the first to pick up the issue of Schmitt's method. In 1924, he praised Schmitt highly and called him an "ideologue of unusual conviction" (Ball 1924, 263) for whom ideas emerged and entered life where extremes gathered around them: "The extreme is the starting point for his concepts" (Ball 1924, 278). Schmitt's concepts were illuminating because they followed on from extreme, final decisions.

Not all of Schmitt's readers agreed with this positive verdict, however. The first criticism of Schmitt's politics of concepts was formulated by positivist legal theorist Richard Thoma in 1925. In his review of Schmitt's book on the crisis of parliamentary democracy, Thoma accused him of "overemphasizing the literary appearance of things" (Thoma 1925, 80) in his definition of parliamentarism. If one sought to examine the foundations of an institution in intellectual history, he stated, one could not limit oneself to the study of a single coherent ideology that had been used to justify it. Thoma also reminded Schmitt that ideological justifications of an institution might change over time because of new social realities. Political institutions underwent "metamorphoses of purpose and changes in structure" (Thoma 1925, 80). Such changes were simply changes and were not necessarily to be understood as degeneration of a previous ultimate core ideal. There was no reason to stop idealizing parliamentarism as a government by discussion and, instead, to justify it on the basis of purely practical considerations. It was not an ideology but, rather, its usefulness, its vitality, and its adaptability that would make or break a political institution. In his riposte to Thoma, Schmitt insisted that like "every great institution," parliament "presupposes certain characteristic ideas," an "intellectual foundation of a specifically intended institution" (Schmitt 1926b, 2 and 3), and more than practical considerations or justifications regarding its social and technical usefulness.

Anti-positivist Rudolf Smend joined this debate two years later with a brief critique of Schmitt's methodology. In his long-awaited book *Verfassung und Verfassungsrecht* [Constitution and constitutional law], he commented on the debate between Thoma and Schmitt. He agreed with Schmitt where he accused Thoma of thinking technically about constitutions. And he insisted, as did Schmitt, that modern parliamentarism must not be decoupled from the principle of public debate and shifted to clandestine backroom deals. The details of the institutional organization of the relationship to public debate, however, were subject to historical transformations. Smend accused Schmitt of a lack of understanding of the changeability of institutions and their justifications: "In parliamentarism, the original ideology is only a moment of integration [...]—the belief in the exclusive significance of ideology is rationalism or (in C. Schmitt's writings) conceptual realism." (Smend 1928, 153) Smend did not explain the significance of conceptual realism to his readers in more detail, and this would have been superfluous because it was already an established term in the philosophical discussion of the day. Since the beginning of the twentieth century, the term had served in Germany to denote philosophical theories in the tradition of Plato according to which general terms were assigned real existence. In other words, the term was meant from the outset as a delimitation from the various strands of philosophical idealism and their metaphysical presuppositions.

The same year, Georg Lukács also complained from a Marxist perspective about Schmitt's method. He criticized the fact that Schmitt did not transcend the usual methods of *Geistesgeschichte* (intellectual history) in his book *Political Romanticism*. A "social analysis and explanation" (Lukács 1928, 308) of romantic occasionalism was missing. The sociologist Hans Speier, whom Kirchheimer knew from Berlin from the Deutsche Hochschule für Politik (German Academy for Politics), attacked Schmitt with a similar objection. He called his *Concept of the Political* a "witty treatise" but saw the problem precisely in this wittiness. Speier asserted that Schmitt considered the formal distinction between friend and enemy to be an "ontological" one that could not be derived further, and he thought that Schmitt was completely wrong. He was of the opinion that any political theory that deserved to be taken seriously required "sociological considerations" (Speier 1932, 203 and 204) of conflicts and their causes.

The most far-reaching criticism of Schmitt's methodology prior to the essay by Kirchheimer was published by the young philosopher Eric Voegelin in *Zeitschrift für öffentliches Recht* in 1931. It was a detailed analysis of Schmitt's *Constitutional Theory* against the background of Hans Kelsen's legal theory.[5] On the last two pages of the essay, Voegelin addressed Schmitt's "style of thinking" (Voegelin 1931, 106) and "categorical tone" (Voegelin 1931, 108). He stated that Schmitt did not approach the constitutional problems from the perspective of an external observer but deliberately from the internal perspective of a person who was involved. Even if Schmitt operated with a "conceptual apparatus bound by tradition" (Voegelin 1931, 107), all the terms he coined were creative interpretations

5 On Voegelin's various points of criticism, see (Heimes 2004) and Henkel (2005, 44–51).

driven by political intentions. Voegelin did not find fault with this but, rather, with the fact that Schmitt left his readers in the dark about the role he had assumed and that he also confused the two different roles himself. "The standpoints of the politically creative thinker and the external observer are always confused, and the categorical tone arises from this confusion" (Voegelin 1931, 108).

Interpretations emerging from Schmitt's political views were presented as scientifically objective ideal types in such a categorical tone. Voegelin was critical of the way that Schmitt reversed things, as it were: "No facts in reality correspond to any these concepts, they are not fulfilled by any sensory perceptions, yet they themselves are part of political reality as beliefs and as political motives." (Voegelin 1931, 109) Even if Voegelin did not use sharp words, his criticism ultimately amounted to accusing Schmitt's state and constitutional theory of being founded on ideological constructs and therefore being unsuitable for grasping the real structure of Weimar statehood. Schmitt reacted to Voegelin's criticism, and in a friendly manner. Voegelin had sent him the proofs of his article prior to publication. In his response to Voegelin, Schmitt conceded that "for the first time, [I] encountered a criticism that moves me to the greatest personal and factual respect."[6] After reading the essay, he had already noted "very good"[7] in his diary. He did not react to Voegelin's criticism publicly, however.

3. Against conceptual realism

It was in 1932 that Schmitt's fellow legal experts commented critically on his methodology, too. Johannes Heckel found fault with the "tension between theoretical construction and historical reality" (Heckel 1932, 284) that Schmitt constantly created. Richard Grau accused Schmitt of deriving specific legal consequences from the "concepts he created himself" (Grau 1932, 279). Both authors thus also opposed Schmitt's theory of presidential dictatorship.

Kirchheimer's criticism went far beyond the cursory remarks from Schmitt's fellow legal experts. His long 30-page article combined sharp political criticism of Schmitt's work with a fundamental methodological attack on his legal thinking. He joined forces with Nathan Leites, a sociology student at the Berlin University from Saint Petersburg who was aged only 21.[8] Most of their text, however, was authored by Kirchheimer himself. It rose above the multitude of other voices critical of Schmitt during the Weimar Republic by sagaciously placing Schmitt's *Legality and Legitimacy* within his oeuvre. In contrast to previous arguments with Schmitt, Kirchheimer and Leites linked their criticism to the essence of the debate on methodology in legal studies as they related to the empirical social sciences. Not only did they reject the results of Schmitt's book but, above all, his methodology. Thus, they connected the previous leftist criticism of Schmitt about his lack of sociological perspectives with that of conservative authors regarding Schmitt's conceptual realism.

---

6 See letter from Carl Schmitt to Eric Voegelin dated 30 March 1931 (Schmitt and Voegelin 2014, 186).

7 Carl Schmitt, diary entry of 27 March 1931 (Schmitt 2010, 101).

8 On Leites's biography see the memoir essays in Rand Corporation (1988).

The substance of this article concerning methodological criticism has strangely remained largely ignored in the literature to date.[9] This is all the more astounding because a number of authors raised similar objections about his methodology after 1945. No other contribution on Schmitt during the Weimar Republic reached the level of the essay co-authored by Kirchheimer and Leites in terms of criticism of methodology. The authors had completed their manuscript within a very short timeframe. Schmitt's *Legality and Legitimacy* had been published in August 1932 and then Kirchheimer informed Smend in early November that their text had been already accepted for publication and would appear in the January 1933 issue of the journal *Archiv für Sozialwissenschaft und Sozialpolitik*.[10] The *Archiv* had been established by Max Weber, Werner Sombart, and Edgar Jaffé. Now it was edited by Emil Lederer in collaboration with Joseph Schumpeter and Alfred Weber. At the time, it was considered the most prestigious publication in Germany for the social sciences. The journal already had a reputation for discussing Schmitt. An initial version of Schmitt's *Concept of the Political* had been published in the *Archiv* in 1927, as had Leo Strauss's critical comments on Schmitt's concept of politics (see Strauss 1932).

The essay by Kirchheimer and Leites is titled "Remarks on Carl Schmitt's *Legality and Legitimacy*."[11] At the beginning of their article, the two authors directly linked up with the final passages of Eric Voegelin's essay. In *Legality and Legitimacy*, Schmitt was attempting to prove that there was "a contradiction between democracy's underlying justification and specific elements contained in the Weimar Constitution or arising from its application" (64). Once again, the authors identified Schmitt not as a concerned defender of the Weimar Constitution but as its fundamental opponent. This time, however, they shifted the attack on Schmitt to the methodological level. Kirchheimer and Leites reconstructed Schmitt's legal theory as an artifact of methodologically inadequate deliberations and stated that Schmitt "fail[ed] to discriminate sufficiently between providing a justification for a particular system of normative ideals [...] and an examination of specifically political forms" (64). They accused him of ignoring the question about empirical political reality, thus not even considering the possibility that a system of normative ideals "[could] 'function' properly when put into effect" (64).

The authors claimed that Schmitt conflated two different tasks—a logical analysis of normative political ideas and an empirical examination of political forms—and that he implicitly championed the assumption that the contradictory nature of a system of political norms would result in a reality that would not function properly if this system of political norms were applied. Kirchheimer and Leites called this implicit supposition "signs of a strand of conceptual realism" (64) in Schmitt's theory. At this point, they referred to Voegelin's essay; Smend, from whom they apparently had borrowed the term "conceptual realism," was not mentioned at this point although he, too, had placed the

9 See Blau (1980, 457–460), Neumann (1981, 243–245), Scheuerman (1994, 87–89), Scheuerman (2000, 9–11), Schale (2006, 78–81), Breuer (2012, 129–130), and Olson (2016). Mehring (2021, 199–204) is something of an exception, yet this author's defense of Schmitt's position is hardly convincing.

10 See letter from Otto Kirchheimer to Rudolf Smend dated 7 November 1932. Rudolf Smend Papers, Cod. Ms. R. Smend A 441.

11 See Kirchheimer and Leites (1933). The following page numbers refer to this text.

"function" of political institutions for the integration of the state at the center of his legal theory. Kirchheimer and Leites attempted to prove in detail how Schmitt's approach based on conceptual realism took its suggestive power from selectively combining theoretical postulates and empirical examples.

They continued their criticism of Schmitt, which took a methodological approach, by addressing his concept of democracy. The basic error in his *Constitutional Theory*, they wrote, lay in the idea that social homogeneity had to be both the prerequisite of democracy and its outcome. Only by overstating the postulate of equality was Schmitt able to conclude that the modern constitutional state was entirely unable to function in political practice. The two authors argued against Schmitt's one-sided derivation of democracy from the postulate of equality and the conceptual strategy arising from it used to play democracy and freedom off against each other. Following Hans Kelsen, they pointed out that the political norms of equality and freedom had the same origins. Criticizing Schmitt's postulate of homogeneity, they also drew on intellectual history to object that even Rousseau had recognized that special interests always exist in any society. "The total transcendence of all differences in opinion has to be seen as constituting a utopian idea because it would imply the destruction of individuality itself" (66).

Kirchheimer and Leites also attacked Schmitt's concept of liberty. Schmitt's definition of liberty placed a special emphasis on the liberty of the individual. In addition, he distinguished between the liberty of the isolated individual and the liberty of individuals interacting with other individuals. Since Schmitt conceived of the sphere of liberty in terms beyond the scope of the state, he failed to relate individual liberty to the process of democratic will formation. Thus, Schmitt was "incapable of acknowledging the distinction between the rights of citizenship and private rights" (66). Schmitt's concept of liberty obscured "the dual character" of liberty in modern democratic states. Contrary to Schmitt's views, it was this dual concept of liberty that was "the basis for the Weimar Constitution" (67). It also formed the basis for the justifications for all the other modern democratic systems. It followed from this concept of liberty that there would always be a certain amount of heterogeneity and differences of opinion in a society. Total homogeneity would lead to the total destruction of everything individual and ultimately of individuals. In order to protect people from such homogenization, a political order that realized equality and liberty "as fully as possible" (66) was all the more important.

Above all, however, the two authors insisted that Schmitt should have had the courage in *Legality and Legitimacy* to take an open empirical look at real-existing modern democracies. In contrast to Schmitt, Kirchheimer and Leites noted that, indeed, all populations were by necessity heterogenous. They also observed that there appeared to be a trend in all modern societies toward increased heterogeneity. They countered Schmitt's hypothesis that democracy in a heterogeneous society was not only unjustifiable but in fact dysfunctional with empirical findings pointing "to a whole series of phenomena that are difficult to square with his [Schmitt's] thesis" (68). A large part of the article is filled with a comparative view of the political systems in France, Belgium, the United Kingdom, and the US. An "ongoing trend toward heterogeneity" (69) was to be seen in all four countries, without democracy suffering any losses of function. In Belgium, which was extremely heterogeneous in national and social terms, they observed a trend toward a "transformation of political parties into typical integrative parties" (69). Kirchheimer and Leites had borrowed the term "integrative parties" from Berlin scholar of political parties Sigmund Neumann, who had recently described this new type of party with reference to Smend's theory of integration in a study on the German party landscape.[12] Schmitt's description completely ignored such empirical findings. His epitaph to modern democracy was based on an "inadequate inductive basis for the argument and significant empirical evidence to the contrary" (69). Owing to a lack of empirical evidence, Schmitt's proclamation of the end of the Weimar Constitution could not be taken seriously intellectually.

Not only was Schmitt's empirical diagnosis incorrect but he was also deluded by ideology. His yearning for authority blinded him to "new potential solutions" (76) with which modern parliamentary democracies were able to respond to social changes. In the US, a "new prosperity" could already be discerned as an "instrument of social integration" (69). The two authors stated that the US was using a skillful policy of an "instrumental view" (70) to stabilize democracy. Such an instrumental approach had also been adopted at the beginning of the Weimar Republic in the form of the Stinnes-Legien Agreement, an accord concluded by German trade unions and industrialists on 15 November 1918, before the bourgeoisie had withdrawn from it. Kirchheimer and Leites stated that this withdrawal from the agreement was a more significant factor in the current crisis of the republic "than those factors described by Carl Schmitt" (70) in his *Legality and Legitimacy*.

A larger section of the article "Remarks on Carl Schmitt's *Legality and Legitimacy*" consisted of an extensive explanatory analysis of types of legal norms and of the interpretation of fundamental rights and individual articles of the constitution in Schmitt's book. The authors interpreted the elements Schmitt had described as unresolvable contradictions of the construction of the Weimar Constitution as potentially integrative bridging principles which could help lead to more effective social compromises and thus stabilize parliamentary democracy. Schmitt had made the point that introducing material standards in the second section of the constitution altered the organizational core of parliamentary democracy in such a way that parliamentary sovereignty was abrogated in favor of a system based on the primacy of judicial review. Kirchheimer and Leites agreed with the criticism of this development in legal practice, adding that these trends "emerge[d] where the causes Schmitt identifie[d] [were] not present" (72). Schmitt was of the opinion such a structural change occurred if a constitution included special material clauses. Again, the two authors countered this statement with an empirical finding: "The most significant example of a 'jurisdictional state' is the United States" (72), and its constitution included virtually no material clauses.

Kirchheimer and Leites also took up Schmitt's distinction between the constitution and constitutional laws in his *Constitutional Theory* (see Schmitt 1928b, 80–82) where he claimed that some constitutional norms were unalterable. They agreed but deviated from Schmitt's views in terms of what exactly was included in the unalterable elements of the constitution. "If we identify democracy's basis with an ultimate decision in favor of the principles of liberty and equality, [then we would arrive at a] very different assessment of the constitution's unalterable core [than Schmitt]" (75). The universal, equal, secret, and proportional right to vote was untouchable. All norms that "contribute[d] to an unrestrained process of political will-formation," and the "rights to citizenship" (76), were

12 See Neumann (1932, 108–110) and Raulet (2000, 55–58).

unalterable, too. In contrast, all other personal liberties could be the object of changes to the constitution: "All 'private' rights can be amended" (76). Readers of the day understood that this wording declared drastic limitations of private property rights through changes to the constitution permissible without Kirchheimer and Leites having to say this explicitly.

The two authors then analyzed Schmitt's criticism of parliamentary democracy. The advantage of parliamentary democracy, they asserted, was that it was "the only political system [...] that provides an institutional guarantee that even the most decisive transitions of power need not threaten the continuity of the legal order" (82). Parliamentarism deserved democratic legitimation because of this unique feature. Schmitt, conversely, substituted the democratic legitimacy of the parliament with what he believed to be a superior democratic legitimacy, namely direct democracy. Kirchheimer and Leites took into account the empirical fact that the previous liberal justifications of parliamentarism that Schmitt had laid out in *The Crisis of Parliamentary Democracy* had long been "on a decline" (87). At that point, the parliament was justified primarily as a "plebiscitary intermediary" (84). For this reason, there was no longer a fundamental contradiction between democracy and parliamentarism. This change in beliefs concerning what constituted legitimation, which could be observed empirically, had to also include a legal theory about parliament and political parties.

Readers of this essay were left to conclude that Schmitt simply refused to acknowledge this reality. He divided the distinction between legality and legitimacy, which he considered decisive, between two institutions and played them off against each other. To him, legality referred to the underlying justification of parliamentary lawmaking, whereas legitimacy referred to the justification of direct plebiscitary lawmaking. In contrast, Kirchheimer und Leites argued that the institutional difference consisted merely of "different organizational forms of the same type of legitimacy" (86) and that Schmitt followed his incorrect "diagnostic thesis" with the "prognostic thesis" (87) according to which a "caesaristic modification" of the constitution was politically more stable than parliamentary democracy. They declared this prognosis to be a question to be decided empirically—regardless of the normative desirability of such a regime change—and added a number of historic examples where political regimes had stood the test of time despite all the negative prognoses. Kirchheimer and Leites were convinced that such questions could not be answered on the basis of constitutional theory alone:

> We need to take every conceivable extra-constitutional factor into consideration. It seems that only if constitutional theory tackles this task by working in close cooperation with all those disciplines concerned with social experience will it gradually be able to convey general solutions to such problems (88).

They referenced John Dewey's book *The Public and Its Problems* (see Dewey 1927) for the interdisciplinary approach to the social sciences they were promoting.

The continued accusation of conceptual realism in the version spelled out by Kirchheimer and Leites in their essay against Schmitt amounted to the complete destruction of his approach. Schmitt was a conceptual thinker. This meant that not only did he think in certain concepts but he also made the conceptions the subject of his own reflection within their substantial and historical sets of problems.[13] When Schmitt defined terms and concepts, he coined them in his own way to an extent that they became a specific vocabulary, for instance, the concepts of democracy, of *Rechtsstaat*, and of dictatorship in his *Constitutional Theory*. Schmitt saw himself as a participant in a battle for the authority to interpret terms and concepts. Terms and concepts were tools in political struggles and, following Reinhard Mehring, Schmitt considered himself a "military technologist of terms and concepts" (Mehring 2014b).[14] In his ideological struggle, staking out a concept semantically was as important to him as conquering a fortress in war (see Quaritsch 2018, 20). In 1941, Schmitt's student Ernst Rudolf Huber summarized Schmitt's approach to the politics of terms and concepts similarly to Kirchheimer and Leites, the difference being that he considered his summary to be praise:

> The method of this struggle consists in the fact that the device of definition determines the genuine concept of a political institution, and it is precisely thereby that the deterioration of the factual institutions compared to their own essence is made clear. This makes the *Entartung* [degeneration or decline due to biological or cultural factors; see Glossary] of political institutions visible (Huber 1941, 4).

The confrontation of a "genuine" concept with dismal reality was inseparable from Schmitt's methodological approach. The study by Kirchheimer and Leites was corroborated by a report presented by ancient historian Christian Meier during a colloquium in honor of Schmitt in 1988. On the basis of his numerous personal encounters with Schmitt from the 1960s on, Meier observed that Schmitt believed he could "veritably see" concepts and that, to him, they "represented realities" (Meier 1988, 605). Meier also claimed that "it was possible to completely hamstring [Schmitt in discussions] by using terms and concepts in a way contradictory to his." (Meier 1988, 607) If a term or concept that he believed did not fit cropped up in a political debate, "then the entire web of order with which he generally overlaid things fell apart. […] Then he could be quite desperate." (Meier 1988, 607–608) It appears that Kirchheimer had similar experiences in his conversations with Schmitt much earlier than Meier. He concluded that it was easy to attack Schmitt at the methodological level and to point out that terms and concepts such as democracy and liberty had an idiosyncratic meaning in Schmitt's vocabulary.

Referring to Schmitt's way of dealing with political and legal terms as conceptual realism was accurate in the sense that he gained knowledge about reality exclusively by explaining the inner logic of an essential idea inherent to the concept in question. Empirical evidence on functional processes (and their problems) was irrelevant at this level of argument. Schmitt did not confront political institutions with their pragmatic justifications, either; conversely, he understood them as the embodiment of principles free of contradictions. To be precise, his *Constitutional Theory* was a theory of constitutional

13 See Meier (1988) and Kraus (1998).

14 In his inaugural lecture in Cologne in June 1933, Schmitt stated: "Terms and concepts […] are not nominalist labels. […] They are immediate carriers of political energies, and part of their real power is that they are capable of forming convincing juristic terms and concepts. That is why the struggle for them is not an argument about empty words, but a war of terrific reality and presence" (Schmitt 1933l, 198).

terms and concepts.[15] Really existing political institutions, which as a rule fulfill multiple different functions in practice and have various justifications, some of which were in tension, could only fail in the face of the doctrinaire purity of a Schmittian concept. For this reason, Schmitt's method must not be confused with the process of critiquing ideology (see Preuß 1987, 407–409).

Schmitt's conceptual realism placed virtually all of the Weimar Republic's political institutions in an unbridgeable dichotomy of abstract principles. His method of conceptual realism proved so explosive during this time not least because he transferred his polemic conceptual juxtapositions, which he expressed in apodictic formulas, to specific institutions in the Weimar Republic. Anyone who, like Schmitt, traced every important element of the constitution back to a single, pure, and inherent idea destroyed the inner rationality of any constitution. At the same time, this methodological operation opened up the potential for existential political decisions that could not be contained rationally.

The methodological criticism of conceptual realism formulated by Kirchheimer and Leites struck at the heart of Schmitt's entire oeuvre from the era of the Weimar Republic. At the same time, it offered a methodological alternative to the triad of methods following Weber, Marx, and himself that Schmitt had outlined in his *Political Theology*. Kirchheimer and Leites retraced the steps leading back to Weber and Marx. They took on these two authors' guiding principles of situating political terms, concepts, and theories within social history. In contrast to Schmitt, however, their next step was not to seek concealed metaphysical systems, but to argue—now closer to Weber than to Marx—for empirical social sciences to take on an interdisciplinary direction following the American pragmatism of John Dewey.

At this point in the essay, the transition from legal studies to political science—as propagated by Hermann Heller the same year (see Heller 1933b)—was palpable. Kirchheimer did not yet take this step while he was still in Germany. This hesitation was presumably due to the way in which political science, a new scientific discipline at the time in the country, presented itself. In a book review published in the February 1933 issue of *Die Gesellschaft*, he still rejected the idea. The book in question was the first attempt to present a textbook in German for the emerging discipline of the "Science of Politics." Its author Adolf Grabowsky had taught at the Deutsche Hochschule für Politik in Berlin since 1921 and was part of the nationalist conservative group among the faculty. Kirchheimer denied the raison d'être of the new scientific discipline that Grabowsky was promoting. The reason he gave was: "it is common knowledge that the character of the 'political' cannot be determined unambiguously and that quite different opinions exist about this in various countries." (Kirchheimer 1933a, 511) In particular, he criticized the overemphasis of a foreign policy perspective in the description of political systems and the overestimation of ideological factors in the presentation of political processes. It was only in exile, after he had become familiar with other books in the field, that Kirchheimer found a positive relationship to political science—and that had nothing to do with Grabowsky's ideas.

To return to Carl Schmitt: In the following years, the label of conceptual realism that Kirchheimer and Leites had attached to him stuck in three ways. First, through Kirchheimer, who repeated this accusation many times both in his writings in exile and in his

15 See Muth (1971, 141) and Gusy (1997, 439).

correspondence with and about Schmitt after 1945.[16] Second, through authors of the secondary literature, beginning with the entry on Schmitt in the encyclopedia *Meyers Lexikon* during the Nazi period in 1942 stating that he had "worked too much with conceptual templates at the expense of clarifying his worldview" (Meyer 1942, 1176). Conceptual realism became a standard accusation against Schmitt in the 1950s and 1960s.[17] Third, however, Schmitt himself finally adopted this label, too. He never responded to Kirchheimer's fundamental criticism in an article or even a footnote. Yet, after 1945, he accepted the methodological label selected for him, but not the methodological criticism it entailed. Looking back in his diary-like *Glossarium* in March 1948, he praised his own work on *Legality and Legitimacy* as an outstanding academic testimony from the end of the Weimar Republic and explained his supposed masterly achievement as "properly applied conceptual realism as it is part of the science of public law."[18] He noted "my pride in my conceptual realism."[19] Schmitt, the seasoned politician of terms and concepts, had repackaged Kirchheimer's verdict into an honorary title without further ado.

Just like Voegelin had done a year earlier, Kirchheimer had given Schmitt the proofs of his article prior to its publication. He also gave a copy to Smend. Both received their copies in late October 1932. A week later, on 6 November, Kirchheimer met Schmitt at his home to discuss the article over coffee and cookies for a few hours. They sat together the entire morning without reaching an understanding about Kirchheimer's criticisms of Schmitt's book and its political conclusions. Unlike his response to Voegelin, Schmitt's reaction this time was negative and furious. His diary entry about the conversation with Kirchheimer read: "there's no point in talking with him, he simply doesn't want to see a thing." Followed directly by: "*Scheußlich, dieser Jude*" ("Vile, this Jew").[20] It was the first time Schmitt had noted an antisemitic slur in his diary in reference to Kirchheimer. And it was the last time that Kirchheimer was mentioned at all in Schmitt's diary during the Weimar Republic.[21]

---

16 See Chapters 11, 15, and 16.

17 See Schneider (1957, 29–26) and Sontheimer (1962, 78–82).

18 *Glossarium* entry of 2 March 1948 (Schmitt 2015, 81).

19 *Glossarium* entry of 2 March 1948 (Schmitt 2015, 81).

20 Carl Schmitt, diary entry, 6 November 1932 (Schmitt 2010, 231).

21 After this entry in October 1932, Schmitt did not mention Kirchheimer in his diaries through the end of 1934; these have been published. Attempts have been made since 2020 to decrypt parts of Schmitt's extensive handwritten texts from 1939–45, written in difficult-to-decipher Gabelsberger stenographic script, in the research project "Transkription und Hybridedition der Tagebücher Carl Schmitts aus der Zeit des Zweiten Weltkrieges" [Transcription and hybrid edition of Carl Schmitt's diaries during WW II] under the direction of Philip Manow and Florian Meinel, funded by the German Research Foundation (DFG). See https://gitlab.com/arbeitsgruppe-carl-schmitt/tagebuecher, accessed 2 March 2024. At the time of writing, it is impossible to say whether Schmitt mentioned Kirchheimer in his diaries from this period. He did mention Kirchheimer after his visit to Schmitt's home in November 1949 (see Chapter 15).

4. The intense final days of the republic

As regards his professional work, Kirchheimer continued to keep various irons in the fire during these politically turbulent weeks of late 1932. For one thing, he continued his untiring efforts to gain a foothold as a lawyer. He also had a small income from the fees for his essays in *Die Gesellschaft*. And he continued to pursue his unwavering academic ambitions. His goal was to gain a *habilitation* in constitutional law with Smend at Berlin University's Faculty of Law.[22] He submitted an application to this end to the Notgemeinschaft der deutschen Wissenschaft (Emergency Foundation for German Science), the precursor of the German Research Foundation (DFG), in November 1932, aiming to obtain funding for "work on some broad questions of democracy," as he wrote when asking Smend for a reference.[23]

At the same time, he turned to Schmitt for support as a reviewer, informing him that he was interested in researching the legal theory and legal sociology of the American authors Oliver Wendell Holmes, Felix Frankfurter, and Charles Beard.[24] Schmitt supported him despite the conflicts they had had just a few days earlier. Nevertheless, Kirchheimer's application was unsuccessful and he began to consider new ways to finance his academic work. Together with Franz L. Neumann, he had started to take private classes in English conversation to improve his prospects to work abroad.[25] He also kept providing Schmitt with bibliographical references from leftist US writings. He recommended, for example, the 1928 book *American Foreign Policies* by the leftist US political scientist James W. Garner and, on 16 November 1932, the new book *Government by Judiciary* by the American Marxist Louis Boudin.[26]

During the Christmas holidays of 1932, Kirchheimer sat down at his typewriter to write a third piece on the debate about constitutional reform. It was published in the January 1933 issue of *Die Gesellschaft*.[27] After Schleicher had assumed the position of Chancellor on 3 December, Berlin was buzzing with rumors about an imminent reform of the Reich by means of a government coup. Kirchheimer's essay had the same title as one by Fraenkel the previous month in the same journal, *Verfassungsreform und Sozialdemokratie* [Constitutional reform and social democracy]. Kirchheimer rejected all proposals coming from social democratic circles, addressing Fraenkel's proposal in particular detail. This time, his criticism was considerably sharper, both in tone and in substance. He began to come to Fraenkel's defense against Peter Stein, the author who claimed in the

---

22 Memo, Academic Assistance Council (AAC) of 4 March 1934. The AAC file from London is to be found in: Emergency Committee in Aid of Displaced German/Foreign Scholars, Public Library, New York. I, A Grantees 1933–46, Box 18, Folder 13 (Kirchheimer, Otto).

23 See letter from Otto Kirchheimer to Rudolf Smend dated 7 November 1932. Rudolf Smend Papers, Cod. Ms. R. Smend A 441.

24 See letter from Otto Kirchheimer to Carl Schmitt dated 7 November 1932. Carl Schmitt Papers, RW 265–7595.

25 See the account by Neumann's later partner Helge Pross in Erd (1985, 59).

26 See letter and postcard from Otto Kirchheimer to Carl Schmitt dated 24 December 1931 and 16 November 1932. Carl Schmitt Papers, RW 265–7596 and RW 265–7597.

27 See Kirchheimer (1933d). The following page numbers refer to this text.

December issue of the KPD publication *Roter Aufbau* to have detected "theoretical interconnections" between Fraenkel and the "fascist constitutional theorist Carl Schmitt."[28] Kirchheimer called this a "deliberate distortion" (500), based on an easily recognizable communist debunking strategy.

Nevertheless, his analysis of Fraenkel's bundle of proposals lacked the sympathetic tone he had used a few weeks earlier in his article "Die Verfassungsreform" [The constitutional reform].[29] Kirchheimer accused Fraenkel of not taking an appropriate approach to the question, of making a fetish of the value of a constitution, of adapting the constitutional norms to the constitutional reality, and of thus ultimately legalizing rule by the bureaucratic and military apparatus. Fraenkel, he claimed, did not go beyond "constitutional deduction" (499), which was legally tenable but "sociologically irrelevant in the decisive point" (500). He disregarded the fact that the theory of emergency or *Lückentheorie* (gap theory), which was an integral part of constitutional law at the time, "could sociologically speaking certainly represent a usurpation of power by a societal class that would otherwise remain insignificant" (500). Here, Kirchheimer emphasized that it was only possible to fully understand the 20 July coup against the Prussian government by not taking the one-sided view that its initiators sought to shake off the SPD but by appreciating that they also sought to secure the republic against the NSDAP taking over power.

In voicing his opposition to those positions arguing exclusively on the basis of constitutional law, Kirchheimer used a broader Marxist approach and vocabulary incorporating socioeconomic factors and methods. He first quoted from the polemic by Friedrich Engels and Karl Kautsky against "lawyers' socialism" (see Engels and Kautsky 1887), at the time a classic in the eyes of Marxist jurists. Kirchheimer was of the opinion that only a "reordering of the distribution of economic power" (499) could potentially resolve the current tension between the Weimar Constitution and the social power relationships, not a change of the constitution. In this sense, Germany at the time was a case in which the ideological superstructure of the legal order was "hobbling ahead" (499) of the actual social relationships. Kirchheimer argued that when proposals for revising the constitution were discussed, it was essential to review what effects they would trigger in the specific society. In this regard, he was convinced that everything pointed to retaining the constitutional status quo. At the moment, any feasible reform, as well-intentioned as it may be, would be instrumentalized against the labor movement in light of the existing social power relationships. Kirchheimer thus provided quasi-materialistic reasons for a conservative stance toward the constitution. Accordingly, he considered it pointless to deal with the question of a future constitution under democratic socialism.

Chancellor Kurt von Schleicher's fundamental plan to secure his chancellorship was based on the success of his efforts to achieve tolerance of his policies across a majority of the party factions in the Reichstag, in particular the Center Party and the SPD, and to split the NSDAP. Yet his attempt to reach an agreement with Gregor Strasser, the leader of the "leftist wing" of the Nazi party, failed. Schleicher then resorted to the previous year's "September plan" that Schmitt, among others, had prepared for Papen. Again, the core of

28 *Unsere Zeit*, No. 24, December 1932, p. 1144.

29 See Kirchheimer (1932f). See Chapter 5, p. 139–141.

this project was to suspend new elections to the Reichstag indefinitely.[30] Schmitt himself, however, was no longer asked to participate in the preparatory discussions about this new attempt.[31] His proposal in a letter to the Minister of the Interior that the President was to publicly declare that he would not recognize a future no-confidence vote of the Reichstag against the Chancellor did not produce a response.[32] Hindenburg rejected the unconstitutional and more far-reaching proposal to suspend elections with reference to his oath to the constitution; he feared he would be indicted before the *Reichsgericht* in Leipzig for breach of the constitution.

Meanwhile ex-Chancellor Papen had sought and found an agreement with Hitler behind Schleicher's back. Schleicher resigned on 28 January 1933 after Hindenburg had again rejected his alternative proposal to establish a temporary dictatorship. Papen was able to convince Hindenburg to accept Hitler as the Chancellor of an NSDAP/DNVP coalition government. On 30 January 1933, Hindenburg appointed the new government and swore it in. On Hitler's demand, the Reichstag was dissolved again on 1 February. The elections on 5 March already suffered manifold forms of state repression as well as terrorist action by the NSDAP and its combat units. Even though the Weimar Constitution formally remained in force, the Enabling Act of 24 March 1933 ensured that the Nazi regime was safeguarded.

Prior to 1933, Schmitt was in fact not a Nazi. As part of the educated bourgeoisie, he initially felt a good deal of contempt for the party and its troops of thugs, and especially for Hitler himself. The authoritarian transformation of the Weimar Republic that Schmitt desired did not include an important role for Hitler. It should be noted, however, that Schmitt's rejection of social democracy and of a return to a functioning parliamentary legislative state, which he often expressed in venomous words, were far greater than his reservations about Hitler. This was also, and in particular, true of the final days of the Weimar Republic. When it seemed for a short time in January 1933 that Chancellor Schleicher might succeed in organizing a parliamentary majority for his policies including the Social Democratic Party, the Center Party, and the right-wing parties, Schmitt vented about this prospect in his diary without restraint: "Saw the disgusting swamp of parliamentarism and social despotism rise again. Braun and Kaas are triumphing."[33] Two days before Hitler was appointed Chancellor, Schmitt noted: "Fear of the political things to come. Disgust for the social democrats and for what will return, foul liberalism."[34] In his view, returning to democratic parliamentarism was an option to be thwarted under any circumstances.

The only remaining alternative was to involve Hitler and the NSDAP in a new authoritarian government of the Reich, however possible. Schmitt's published writings from before 1933 do not include any explicit comments about Hitler or his party. Ernst Rudolf

---

30 See Huber (1984, 1227–1230), Berthold (1999, 25–31), and Seiberth (2001, 156–160).

31 See Huber (1988, 47–49) and Blasius (2001, 62–66).

32 See Berthold (1999, 38–40) and Pyta and Seiberth (1999, 607–608).

33 Carl Schmitt, diary entry of 22 January 1933 (Schmitt 2010, 254). Ludwig Kaas was the leader of the Center Party; Social Democrat Otto Braun was Prime Minister of Prussia until the coup of 20 July 1932.

34 Carl Schmitt, diary entry of 28 January 1933 (Schmitt 2010, 256).

Huber reported in his memoirs that he had heard only derogatory remarks from Schmitt about Hitler in the second half of 1932 (see Huber 1988, 60). Other published sources concur that Hitler's person did not appear decisive for Schmitt favoring Nazism. His diaries reveal increasingly friendly statements about the Nazi movement overall from 1931 on. He praised the bullying NSDAP walkout from the Reichstag in February 1931 as a "magnificent"[35] move. He considered his fellow legal expert Erwin Jacobi "a fine fellow" because he sympathized with the NSDAP.[36]

In the run-up to the presidential elections, a plebiscite between Hindenburg, the incumbent, and his challengers, Schmitt recorded in his diary: "[I will] vote for Hitler in the first round of voting."[37] When the NSDAP did very well in the Landtag elections in April 1932, he felt downright euphoric for several hours.[38] On the evening of 30 January 1933, when the President of the Reich had appointed the new government under Hitler's leadership, Schmitt wrote in his diary: "Then to Café Kutschera, where I heard that Hitler had become Chancellor of the Reich and Papen Vice Chancellor." And he added: "Excited, glad, delighted."[39] The following day, he wrote: "Angry about stupid, ridiculous Hitler."[40] In those days, there were numerous entries about enjoying good conversations with party members and SA men. Whatever one might think about these and other diary entries, they reveal two things. First, that Schmitt definitely rejected a return to parliamentarism as provided for in the Weimar Constitution. And, second, that he considered overcoming Weimar parliamentarism to be so important that entering into an alliance with the Nazis to this end was acceptable, although he certainly did not favor the option of appointing Adolf Hitler Chancellor.

It is not without irony that Kirchheimer was attacked as a fascist collaborator by the communists at this very time when the SS and SA (see Glossary) had started to terrorize the political opposition. In response to his defense of Fraenkel against the accusations in *Roter Aufbau*, an attack against Kirchheimer appeared in the communist newspaper *Unsere Zeit* [Our era] in mid-February 1933. Under the headline "Mister Carl Schmitt's Key Witness," an anonymous author accused him of left social democratic "uniformity in the political direction"[41] along with Schmitt. The author finished their article with the rhetorical question whether Kirchheimer had plagiarized Schmitt or whether Schmitt had plagiarized the fascist coup plans from Kirchheimer. The author used the references to Schmitt in Kirchheimer's writings as evidence of the communist narrative that the SPD was to blame for the establishment of the authoritarian state in Germany.

35 Carl Schmitt, diary entry of 9 February 1931 (Schmitt 2010, 88).

36 Carl Schmitt, diary entry of 10 March 1931 (Schmitt 2010, 97).

37 Carl Schmitt, diary entry of 27 February 1932 (Schmitt 2010, 181).

38 Carl Schmitt, diary entry of 25 April 1932 (Schmitt 2010, 189). The NSDAP emerged as the strongest party in the elections to the Landtag in four German *Länder*—including Prussia.

39 Carl Schmitt, diary entry of 30 January 1933 (Schmitt 2010, 257). The transcript of the excerpt of his diary for this day, which Schmitt prepared himself and which his first biographer Paul Noack referred to (Noack 1993, 160), had been deliberately falsified by Schmitt in that he had left out the last three words quoted here.

40 Carl Schmitt, diary entry of 31 January 1933 (Schmitt 2010, 257).

41 *Unsere Zeit* (15 February 1933, 244).

The last article Kirchheimer was able to publish while he was still in Germany, before emigrating to Paris, appeared in mid-March 1933, a week before the Enabling Act entered into force. It was the essay "Marxismus, Diktatur und Organisationsform des Proletariats" [Marxism, dictatorship, and the proletariat's form of organization] in the March issue of *Die Gesellschaft*.[42] Some of the wording in the first sentence and the footnotes indicate that Kirchheimer had completed the article a few days before Hitler's cabinet had taken office. This means he could not have known at the time that Hitler would take over the government or what position Schmitt would publicly adopt with respect to this decisive political event.

Once more, he devoted his attention to an important element of Schmitt's work, the theory of dictatorship. In his book *Dictatorship*, Schmitt had examined Marx's concept of dictatorship of the proletariat (Schmitt 1921, xxxix–xlv). Kirchheimer had quoted this book multiple times. In this latest article, however, he conducted a debate entirely internal to Marxism and did not mention Schmitt's name or his book at all, not even where it would have been appropriate with regard to the differentiation between commissarial and sovereign dictatorship. The article addressed readers from the leftist spectrum who were seeking political orientation between reformist social democracy and the communists in the fight against fascism. Kirchheimer's text was mostly exegetical and embedded his arguments in socialist and communist interpretations of classical texts. He first explained the Marxist concept of dictatorship found in the work of Rosa Luxemburg and Paul Levi: dictatorship as the circumstance of actual social rule of one class or group over the others, irrespective of the legal forms within which it evolved. He then discussed the understanding of democracy in the Marxist tradition including Arkadij Gurland's book on proletarian dictatorship. Kirchheimer stated that there were no indications at all in the works of Marx and Engels that democracy as a form of government necessarily had to be the antecedent of the proletarian dictatorship. Of course, the greatest chances of peaceful transformation of the bourgeois state to a proletarian one were to be found wherever there was a democracy that the proletariat had been involved in fighting for.

This, however, was no longer an option due to the emergence of "phenomena commonly summarized under the term fascism" (517). The fascists were recruited mostly from the "lumpenproletariat" (518) which Marx had identified as the social group supporting Bonapartism eighty years previously. Under the current political conditions in Germany, an "independent armed private political army which considered [itself] not primarily a party, but an armed combat troop" (518) had been added to the social groups of capital, the military, the Junkers, and the bureaucracy, with the goal of gaining political power. This type of rule would no longer permit the labor movement any political freedoms at all so as not to lose ground: "Fascism has no choice here. Following the law under which it came to power, it must keep these forces down using the harshest bureaucratic coercive apparatus" (519). Kirchheimer argued for a precise sociological understanding of the concept of fascism, referring approvingly to a distinction made by Franz Borkenau—a communist member of the early Frankfurt School—between "true fascism" (519) as the forcible transition of backward countries to industrial capitalism on the one hand and Nazism as the form of government in a country with fully developed

42 Kirchheimer (1933b). The following page numbers refer to this text.

capitalism on the other hand (see Borkenau 1932). The latter form of fascism blocked the democratic path to socialism for the labor movement in Kirchheimer's view.

In this last publication of Kirchheimer's during the Weimar period, he doubted whether the form of government preceding the rule of the proletariat must necessarily be bourgeois democracy. There were two reasons for him to shatter the expectation that history followed a certain stage model. First, there was the seriousness of the challenge of fascism, which was victorious in various European countries. Second, there were the voluntaristic elements of Kirchheimer's political theory, which he shared with Schmitt. In the current historical situation, maintaining bourgeois democracy's emancipatory potential was becoming a combat mission of the working class. However, the fascist offensive of the bourgeoisie demanded a redefinition of the means of struggle. The defense of constitutional legality was not to be limited to blind trust in the automatic mechanisms of the legal system. Kirchheimer considered this to be the dawning of a constellation similar to that mentioned in the Austrian Social Democratic Party's Linz party platform of 1926 in which "the working class can seize government power only through a civil war forced upon it" (520).

Despite the bellicose language in his article, there was no indication that he was particularly optimistic about the prospect of winning or even starting a civil war in order to defend democracy. His long exegetical analysis of Lenin's concept of the party and his "primitive" (521) understanding of democracy were pointedly critical: its authoritarian orientation was comprehensible against the background of repressive Russian absolutism but in the further course of the Russian Revolution, its hostility to democracy and freedom had had dire consequences. In contrast, he recalled Rosa Luxemburg's criticism of Lenin and the democratic potential of her belief in the spontaneity of the masses but also faulted her for underappreciating that hierarchies took on a life of their own, which was always necessary to a certain degree. Kirchheimer called on his readership to find a reasonable "middle ground" (526) between these two traditions for the ongoing and upcoming political struggles. The vague wording at the end of the article mirrored the extent to which most German leftist intellectuals had lacked orientation when political power was handed over to Hitler's coalition government.

5. Conclusion: Two politically active legal theorists taken by surprise

Nothing in Kirchheimer's writings indicates that he could have expected Schmitt to enthusiastically join the Nazis in 1933. More than fifty-five years later, Henry W. Ehrmann reported in a conversation that Kirchheimer was "perplexed" about this but had also commented laconically that Schmitt had "always been good for a surprise."[43] In other words, in late 1932, he could not yet accuse him of collaborating with the Nazis. What he did accuse him of, however, was that the Weimar Republic had been transformed with Schmitt's support into an authoritarian regime long-term. And this was the kind of transformation that Kirchheimer had wanted to prevent. But his attempts to rescue the parliamentary democracy of the Weimar Republic seemed nothing less than desperate.

43 Henry W. Ehrmann in a conversation with the author on 7 June 1988.

Kirchheimer propagated a dual strategy relying on the defensive on the legal level and the offensive on the social policy level. Hermann Heller pursued a similar dual strategy and was even more direct than Kirchheimer about Schmitt by unceremoniously calling him a fascist in February 1933, even before he had joined the Nazis: “For all intents and purposes, he [Carl Schmitt] acknowledges just a single ‘authoritarian state,’ namely the fascist dictatorship following the pattern of Mussolini.” (Heller 1933a, 647)[44]

It is hardly surprising that in the volatile political situation at the end of 1932, neither Kirchheimer nor Schmitt were successful in convincing the other of their own political positions. Their convictions were anchored too deeply for that to be possible. In particular, their normative theories of democracy showed the high level of their substantial differences at the end of the Weimar Republic. Schmitt repeated the sharp conceptual difference between democracy and *Rechtsstaat* that he had asserted from 1923 on and then took sides for a dictatorship on behalf of democracy. Kirchheimer's understanding of democracy and *Rechtsstaat* had a different conceptual structure. Against Schmitt's deriving of democracy from the postulate of equality, Kirchheimer thought that the norms of equality and freedom were mutually dependent. With this understanding of democracy, he was a forerunner of Jürgen Habermas's theory of the constitutional state in *Between Facts and Norms* with the normative “co-originality” (Habermas 1996, 122) of democracy and the rule of law.

There was no longer any prospect of rapprochement between Kirchheimer and Schmitt on the seemingly more abstract level of methodological questions, either. Kirchheimer failed in his attempt to persuade Schmitt of his methodological criticism. Yet Kirchheimer still agreed to Schmitt's overarching idea of reconstructing the ways in which political concepts were transformed and used by theorists and actors, and how they helped to mobilize actors and construct their goals. Concepts were created in specific historical situations and by specific actors with shifting and antagonistic motives and aims. Kirchheimer's analyses of different stages of parliamentarism and *Rechtsstaat* and of different types of dictatorship in his Weimar writings indicate that he too kept an eye on the ways in which the original meaning of a concept changed over time as a result of historical events. However, he attempted to connect such re-semantizations of political concepts with particular social settings and struggles between groups in society with different socioeconomic interests. Here, he followed the Marxist tradition of historical materialism. In contrast, Schmitt appeared to be an idealist in the sense that he emphasized the active role and power of intellectuals to redefine terms and to create re-semantizations.

Schmitt's antisemitic sentiments against Kirchheimer were no longer distinguishable from his substantial differences with his former student. Of course, Kirchheimer did not know about Schmitt's antisemitic notes in his private diary. Nevertheless, these notes raise the question to what extent he was aware of Schmitt's antisemitism prior to 1933.[45] This question is difficult to answer because there is no original source material of

44 On Heller's astute critique of Schmitt's authoritarianism see Malkopouluo (2023) and Buchstein and Jörke (2023).

45 See Chapter 10 for more details on Schmitt's antisemitism.

Kirchheimer's about it from that period. He had often experienced Schmitt in the classroom, at lectures, and in private conversations. Schmitt was well-known for his outspoken language in personal conversations.

There is, however, an indirect indication of how Kirchheimer may have experienced Schmitt in situations with direct oral communication. Eugene Anschel, who participated with Kirchheimer in some of Schmitt's classes in Bonn in 1927, said it was obvious that Schmitt was an antisemite. He reported that Schmitt had linked the allegedly specific mentality of English and American merchants and shopkeepers with a denigrating characterization of Jews in his lectures on international law (see Anschel 1990, 85). Another piece of evidence supports the likelihood that Kirchheimer had a similar perception of Schmitt during the Weimar Republic. In 1962, during a doctoral defense at Columbia University, an argument erupted between Kirchheimer and the doctoral candidate George D. Schwab about Schmitt's stance toward Jews before 1933.[46] Schwab, who is also Jewish, told the dissertation committee that he was fully convinced that Schmitt's attitude toward Jews was not based on Nazi notions of *Rasse* (see Glossary) but derived from Catholic and Protestant teachings. Schwab reported in his memoirs that Kirchheimer had insisted during the debate in the defense that Schmitt "was already an anti-Semite during the Weimar period" (see Schwab 2021, 175).

It was in keeping with the logic of the development beginning with the coup against Prussia that preventing a supposedly looming civil war—as Schmitt conjured up dramatically in 1932—would be the first step toward conducting a permanent civil war against the purported enemies of the Reich. The leaders of neither the SPD nor the KPD had a clear vision of the fact that the actions of Hitler's new government had been a turning point, in March 1933 at the latest. Most leftists thought they had been driven back only temporarily by a fascist government. They were under the illusion that they had not suffered a permanent loss because the labor movement's actual struggle had not yet taken place.

Kirchheimer did not analyze Nazism as a militant and growing mass movement even once prior to 1933, incidentally in contrast to his fellow Berlin lawyer Fraenkel (see Fraenkel 1930). In the only, and brief, passage about Hitler—in a 1932 review of a book by Italian fascist leader Curzio Malaparte—Kirchheimer depicted him as "un dictateur manqué" (a would-be dictator) (Kirchheimer 1932i, 372) and otherwise praised the strength of the German proletariat as an opponent of Nazism with words full of enthusiasm. What a grotesque error of judgment. This blind spot in Kirchheimer's political analyses is astounding inasmuch as he had emphasized time and again in his Weimar writings how important determined political action was. He shared this political voluntarism with Schmitt. Kirchheimer of all people, who in his dissertation in 1928 had accused the Social Democrats of succumbing to the illusion of believing in twofold progress, now himself had illusions about how prepared the working class was for battle. Just as he had overestimated the defensive capacity of the workers' movement, he underestimated the determination and ruthlessness of Hitler and his ilk—determination and ruthlessness that conversely profoundly impressed Schmitt.

46 On this subject, see Chapter 17, p. 454–456.

Kirchheimer's underestimation of the Nazis was also due to reasons immanent to his theories. He thought the main danger to the parliamentary republic stemmed from a bureaucracy that had taken on a life of its own with a presidential dictatorship—in other words, precisely what Schmitt had declared to be his political ideal prior to 1933. So, ironically, it was presumably partly because Kirchheimer knew Schmitt very well that he lost sight of the danger of a successful Nazi mass movement. Similar to his friend Gurland (see Gurland 1931, 120–124) and many other Marxists of the day, he interpreted Italian fascism as a phenomenon that could prevail only in industrially backward societies. What had distinguished some of Kirchheimer's analytical acuity in the years 1930 to 1932—his view, inspired by Marxism, of the social functions of the state and politics—no longer helped him. He, too, was one of the leftists who after the end of the Brüning era apparently perceived only minor differences between Papen and Schleicher on the one hand and Hitler on the other. Kirchheimer underestimated the residual protective function of bureaucratic state institutions. It was only after he was forced to emigrate that he and many other socialists fully realized the rupture of civilization caused by the Nazi regime.

Schmitt in Nazi Germany and Kirchheimer in Exile

Chapter 7: The Consolidation of the Third Reich (1933-1934)

The lives of Otto Kirchheimer and Carl Schmitt took diametrically opposed paths after the collapse of the Weimar Republic. Schmitt started a successful career as the "crown jurist of the Third Reich" in the capital, Berlin, with a salary that allowed him to move into a villa with domestic staff. In contrast, Kirchheimer was detained briefly before he managed to escape to Paris. The French capital soon became the intellectual headquarters of the exiled resistance against the Nazi dictatorship. Kirchheimer survived his exile in miserable circumstances, with virtually no income and constantly keeping an eye out for a room that was even cheaper than his current accommodation.

In intellectual terms, however, the paths of Schmitt and Kirchheimer crossed again a few times during the consolidation phase of the Nazi Reich. Once Schmitt had decided to support the Nazi *Führer* state (see Glossary), he soon emerged as the most prominent Nazi legal theorist. In newspaper articles, he took a strikingly aggressive position as he insulted Germans such as Kirchheimer who had been forced into exile. Kirchheimer, on the other hand, closely observed Schmitt's activities for the new regime and commented on them. Before he had to leave the country, Kirchheimer had experienced the various ways in which conservative anti-positivists reacted to the new regime. Rudolf Smend, for example, had come to a different conclusion than Schmitt and did not provide his legal expertise to support of the Nazi regime.

The newer research literature on Schmitt provides plenty of material for assessing his role in the Nazi regime, connecting Schmitt's publications and his recently published diaries with various pieces of archival material. This makes it possible to trace the individual stages of Schmitt's collaboration with the regime in detail. He quickly grew into his new role and was adept at translating his prominence into power in the media. This period was an enormously productive phase in his life. He published over sixty pieces of writing between 1933 and 1936 in which he supported the establishment of the new regime. He wrote prolifically in the weeks, months, and years after Hitler came into power: speeches, front-page essays for the Nazi party press, articles for law journals, and a few relatively short books. At times, he even preempted the political developments.

Up until early June 1933, Otto Kirchheimer was able to follow the beginning of Schmitt's activities by reading his newspaper articles and through conversations with his remaining political friends in Berlin. The last time he met Schmitt in person before fleeing Germany was probably in November 1932, when they discussed *Legality and Legitimacy* at Schmitt's home. There is no indication that they met again in Berlin after that. They had no personal contact and no exchange of letters for the next 17 years, and their communication was indirect as they were far apart, both politically and geographically.

1. Kirchheimer's escape from Germany

After 30 January 1933, there were few opportunities for Otto Kirchheimer to participate in the opposition to Nazism within Germany. After the Reichstag fire during the night of 27 to 28 February, the police and the paramilitary wing of the Nazi party, the Sturmabteilung (SA; see Glossary), which had been granted police powers, launched into a first wave of arbitrary arrests and abuses. Kirchheimer had spent the evening of 27 February in the library of the Reichstag and had been one of the last people to leave the building. He feared that he would be considered a suspect for that reason.[1] The law firm of Fraenkel and Neumann recorded reports about the SA torturing the people arrested that night. A number of active leftist politicians fled the country. One of them was Kirchheimer's father-in-law, Kurt Rosenfeld. He was one of the first to be banned from his profession because of "communist activities" and persecuted by the SA, which is why he fled to Prague with a group of political friends (see Ladwig-Winters 2007, 248).

The waves of arrests and abuses assumed ever greater proportions after Georgi Dimitroff and the others allegedly responsible for the Reichstag fire were arrested on 9 March. Franz L. Neumann was among the approximately 50,000 people who were arrested and taken to mostly illegal camps where the SA and the SS abused them and murdered 500 to 600 prisoners. Roughly 65,000 people fled this orgy of violence during the first year of the Nazi regime, leaving the country, either legally or illegally. Then developments unfolded in rapid succession. On 14 March, the government banned the Republikanische Richterbund (Republican Judges' Association). On 7 April 1933, the *Gesetz zur Wiederherstellung des Berufsbeamtentums* (Law for the Restoration of the Public Civil Service) was passed as well as a *Rechtsanwaltsgesetz* (Law on Attorneys) that excluded "non-Aryan" lawyers or those "engaging in communist activities" from the bar.

Kirchheimer's friend Arkadij Gurland had succeeded in escaping to Belgium in April. He emphatically implored Kirchheimer to leave the country as soon as possible, too.[2] Yet Kirchheimer stayed. He was still in Berlin when, on 2 May, the SA henchmen occupied the building of the Deutscher Metallarbeiter-Verband (German Metalworkers' Union) on Alte Jacobstraße, where the law firm of Fraenkel and Neumann was housed, and terrorized its staff.[3] The party leaders of the SPD moved their seat to Prague on 4 May; the official notification banning Neumann from representing clients as a lawyer was issued

1 Information provided by Peter Kirchheimer on 3 May 2023.
2 Ossip K. Flechtheim recounted this in a conversation with the author on 13 February 1988.
3 See what the then secretary Ella Müller recounted in Erd (1985, 55–57).

on 9 May and the book burnings were instigated on 10 May. To Neumann, these were unmistakable signs that it was time for him to leave the country. Fraenkel, on the other hand, decided to make use of an exemption clause in the *Rechtsanwaltsverordnung* (Regulation on Attorneys) that applied to soldiers decorated in World War I, enabling them to continue representing people suffering political persecution.[4]

Kirchheimer did not have that option. Still, he had not yet made plans to emigrate but wanted to wait and see what would happen and go underground for a while in Heilbronn, where his brother Friedrich (Fritz) lived. He was still hoping that Hitler's new coalition government would soon collapse.[5] However, Friedrich Kirchheimer had assumed a leading position with the local branch of Dresdner Bank, and he threw Otto, who was begging for his protection, out of the house, stating that his brother's political troubles were his own fault and that he was unwilling to get dragged into them, and sent him back to Berlin.[6] A few days later, on May 19, Kirchheimer was arrested in Berlin "on the suspicion of political machinations."[7] As chance would have it, he shared a cell in pretrial detention with Paul Kecskemeti, a young sociologist from Hungary who had come to Germany in 1927 and occasionally worked for the US news agency United Press as a correspondent (see Frank 2009, 444). The two had not met before but immediately became friends because of their shared interests in sociological theories.[8] Kecskemeti was freed after the US embassy intervened with the German authorities; he insisted that he would accept the authorities' demand not to publish a newspaper article about his experiences in detention only if his "friend Kirchheimer" was released, too (see Kirchheimer-Grossman 2010, 60–61). As the Gestapo did not find any evidence against Kirchheimer, he was discharged along with Kecskemeti on 22 May.

His three days in jail finally made it urgently clear to Kirchheimer that he should follow Gurland's advice and leave the country as quickly as possible. One of the first things he did after his release from detention was to explore professional opportunities in the US. There were very few employment opportunities abroad for German legal experts like him, and hundreds of refugees who were qualified for academic positions were in a similar situation once they escaped from Germany. A handwritten letter of Kirchheimer's dated 25 May in which he turned in despair to Rudolf Smend read: "I would just like to inform you briefly that I tried to reach Prof. Friedrich in Heidelberg today, but found out

4 On Fraenkel's motives to stay in Germany for as long as possible, see Ladwig-Winters (2009, 106–109).

5 Ossip K. Flechtheim recounted this in a conversation with the author on 13 February 1988.

6 Hanna Kirchheimer-Grossman in a conversation with the author on 11 March 2016. Friedrich Kirchheimer managed to emigrate to Argentina in 1937.

7 The date is to be found in a letter from Staatspolizeileitstelle Berlin to the Geheime Staatspolizei (Geheimes Staatspolizeiamt) dated 1 February 1938. Auswärtiges Amt (German Federal Foreign Office), Politisches Archiv, RZ 214, R 99744 (69. Ausbürgerungsliste, Ausbürgerungsakte betreffend Otto Kirchheimer).

8 Kecskemeti's sociological interests were later also documented in English translations of Karl Mannheim's writings. Kirchheimer's papers in Albany include letters documenting the connection between the two over many years.

to my consternation that he had left [...] just 1/2 day before."[9] Carl Joachim Friedrich had been at Harvard University since 1926. He was responsible for German-American academic relations in the Akademischer Austauschdienst (Academic Exchange Service) he had co-founded[10] and, consequently, also for granting scholarships to German-language early career scholars. Kirchheimer had introduced himself to him, referring to the fact that they both had connections to Carl Schmitt, on the occasion of a lecture by Friedrich at the Deutsche Hochschule für Politik in Berlin in the summer of 1931. He now implored Smend: "I would appreciate it very much if you were so kind as to inform Mr. Friedrich of my failure should you meet him" and added, "as soon as I have more clarity about where I can stay temporarily"[11] and "when I have an address, I will take the liberty of informing you, dear Herr Professor, of it."[12] In other words, he informed his mentor Smend that he was planning to escape from Germany. In early June 1933, he went to see the Porta Nigra in Trier and, posing as a hiker, he fled across the unsecured border to Luxembourg and from there to France.[13] Thus began his long and difficult exile.

2. Schmitt's decision to support the Nazi *Führer* state

After some hesitation, Schmitt, in contrast to Smend, opted for the new *Führer* dictatorship at a time when it was already taking brutal actions against the opposition on the left. In retrospect, Schmitt described the experience of his initial involvement for the regime, namely helping to draft a law amending the constitution, as a "truly fabulously important moment" and he later also found much "joy in [his] work."[14] In the first few weeks of the new government, Schmitt kept a low public profile. In late March, however, he became involved in formulating legislation for the new regime, namely the *Reichsstatthaltergesetz* (Reich Governor's Law) and hoped this work would lead to a personal introduction to Hitler. When Papen had promised Schmitt that he would be invited to a joint consultation on the law with Hitler, he noted in his diary: "Left very excited and exalted."[15] The law gave legal form ex post facto to the liquidation of the federal order by the NSDAP *Gauleiter*. Appointed by Hitler and reporting directly to him, a *Gauleiter* was a Nazi party official who governed a *Gau* (region) and held powers otherwise exercised by the state (see

9 Letter from Otto Kirchheimer to Rudolf Smend dated 25 May 1933. Rudolf Smend Papers, Cod. Ms. R. Smend A 441.

10 Founded in 1924/25 by Friedrich, the sociologist Alfred Weber, and the political scientist Arnold Bergstraesser, Akademischer Austauschdienst. Deutsche Vereinigung für staatswissenschaftlichen Studentenaustausch (Academic Exchange Service. German Association for Exchange of Students in Constitutional Law) was a precursor of the Deutscher Akademischer Austauschdienst (DAAD, German Academic Exchange Service).

11 Letter from Otto Kirchheimer to Rudolf Smend dated 25 May 1933. Rudolf Smend Papers, Cod. Ms. R. Smend A 441.

12 Letter from Otto Kirchheimer to Rudolf Smend dated 25 May 1933. Rudolf Smend Papers, Cod. Ms. R. Smend A 441.

13 Peter Kirchheimer recounted this in a conversation on 3 May 2023.

14 Schmitt in a 1971 conversation with Klaus Figge and Dieter Groh (Hertweck and Kisoudis 2010, 105, 106).

15 Carl Schmitt, diary entry of 4 April 1933 (Schmitt 2010, 278).

Schmitz-Berning 2007, 251, 313). Schmitt authored a legal commentary to this law in the form of a monograph shortly afterwards (see Schmitt 1933g). One outcome of this first specific project was his personal relationship with Hermann Göring, who quickly took a liking to him. Göring was a leading Nazi politician whom Hitler had appointed Reich Minister without Portfolio, Reich Commissioner for Air Transport, and Reich Commissioner for the Prussian Ministry of the Interior; on 11 April 1933, he was also appointed Prime Minister of Prussia. Göring became one of Schmitt's two powerful mentors from Nazi leadership circles.

Schmitt made his first public comment on the changed political environment on 1 April 1933. He published a piece on the *Ermächtigungsgesetz* (Enabling Act) of 24 March in the *Deutsche Juristen-Zeitung*, Germany's top law journal (see Schmitt 1933a). Even the day before the Enabling Act was passed by the Reichstag, Schmitt agreed to prepare a commentary explaining the new legal situation and seeking approval for it.[16] From his previous perspective as an expert in Weimar constitutional law, he would have been compelled to reject the law as unconstitutional because of its far-reaching abolition of fundamental rights guaranteed by the constitution (see Koenen 1995, 235–239). Furthermore, he would have had to reject it because it had come into existence illegally since it had entered into force only on the basis of a previous change to the Reichstag's procedural rules, which were also unconstitutional. Yet Carl Schmitt, who was now politically active, believed that not only was the law acceptable but that it was urgently needed on the path toward the authoritarian state. He emphasized three fundamental special features of the law in his article. First, he stated that the legally disputed procedure of law-making was not a routine matter, but rather a decisive "turning point of relevance in constitutional history" (Schmitt 1933a, 456), Second, he stressed that the government of the Reich had obtained the right to enact not only new laws within the framework of the current constitution but also laws changing the constitution. And third, he highlighted that this right of the government, which the Reichstag had initially granted for four years, was not subject to any substantive limitations at all. Schmitt himself raised the question whether and to what extent the newly appointed ministers in Hitler's cabinet had their own scope for decision-making in relation to the *Führer* and his response was to use wait-and-see wording that gave Hitler free rein:

> The extent to which, besides the political *Führer* rising above any limitations on his power, any change to these components of the current government of the Reich touches on its identity or even abolishes it is a political question which cannot be answered in advance and without regard to the situation. (Schmitt 1933a, 457)

With his commentary on the Enabling Act, Schmitt demonstratively took a stand for the new legitimacy of Nazism. A new state also required a new theory of the state, he claimed: "We should take care not to undermine the legal foundations of the new state using the sophistry of the old party state. Along with the state itself, constitutional law and the theory of constitutional law must be cleansed and renewed" (Schmitt 1933a, 458).

16 Carl Schmitt, diary entry of 22 March 1933 (Schmitt 2010, 272).

On 7 April 1933, Hitler's government enacted the *Gesetz zur Wiederherstellung des Berufsbeamtentums* (Law for the Restoration of the Professional Civil Service) on the basis of the Enabling Act. Contrary to its official name, it actually served to abolish the professional civil service because its purpose was to dismiss all political opponents and individuals who were not of "Aryan" descent from the public service: the prerequisite for employment in the civil service was no longer exclusively professional qualification, but belonging to the *Rasse* (see Glossary) favored by the Nazis. This also pertained to the universities in Germany; the faculties of law lost 36 percent of their professors, for example (see Stolleis 1999, 254–299). At the Law Faculty in Cologne, where Schmitt had been a professor since accepting his appointment in the autumn of 1932, the law impacted Hans Kelsen. In mid-April, members of the faculty sent a subservient letter to the ministry in Berlin requesting to make an exception for Kelsen and to refrain from banning him from his profession because of his merits in World War I. Only one faculty member refused to sign the letter: Carl Schmitt. Instead, on 12 May, he published the article "Das gute Recht der deutschen Revolution" [The undeniable right of the German revolution] in the *Westdeutsche Beobachter* [West German Observer], a Nazi newspaper, in which he used antisemitic words to defend the civil service law against criticism:

> The new provisions concerning public servants, physicians, and lawyers cleanse public life of non-Aryan *fremdgeartete Elemente* [elements foreign/alien to the *Volk*, in an exclusionary and antisemitic sense; *Elemente* was a contemptuous term for opponents; *Volk*: people/nation in a racial sense, of common blood and with a common destiny; see Glossary]. At last, the reorganization of admission to German schools and the establishment of a university student body of German descent secure the *eigenvölkische Art der deutschen Geschlechter* [German houses'[17] uniformity as a *Volk* of their own]. *Kein Fremdgearteter* [No one foreign/alien to the German *Volk*, in an exclusionary and antisemitic sense] should interfere in this great [...] process of growth. Such people would interfere with us, even if they might have good intentions, in a detrimental and dangerous way. We learn once again to differentiate. Above all, we learn to properly differentiate friend and enemy. (Schmitt 1933b, 28)

Previously, Schmitt had noted in his diary about the events in Cologne: "I did not sign the ridiculous submission of the faculty, what a wretched body, to take such a strong stand for a Jew while they cold-bloodedly let a thousand decent Germans starve and go to rack and ruin."[18] Kelsen's Cologne colleagues' submission to the ministry was unsuccessful. In September 1933, Kelsen was sent into early retirement and went into exile in Geneva. Colleagues and former students of Schmitt's discussed his behavior in this matter widely, as they now understood the full extent of his support for the regime's policies.

Otto Kirchheimer had also read Schmitt's defense of the Law for the Restoration of the Public Civil Service in the newspaper a week before he was detained. The Law on Attorneys was adopted at the same time. Of the 3,400 lawyers in Berlin, the government classified over 1,800 as "Jewish" and excluded them from the bar. To Kirchheimer, this

---

17 Houses in the sense of: kinship groups of virtually noble lineage; emotionally charged term evoking mystical blood ties (see Translator's Preface).

18 Carl Schmitt, diary entry of 18 April 1933 (Schmitt 2010, 283).

law meant the end of his livelihood as a lawyer. The same applied to his wife Hilde Kirchheimer-Rosenfeld. After her father had fled the country, she had initially attempted to maintain his law firm. One of its clients was Ernst Torgler, who was charged with the Reichstag fire. She was also threatened for being an attorney for the *Rote Hilfe* and for defending Thälmann and Dimitroff; in mid-April, she fled via Switzerland to Paris with her two-year-old daughter Hanna.[19] Kirchheimer had only been released from detention a few days earlier and had begun to prepare his escape into exile in France, following his wife and daughter, when Schmitt took aim at the émigrés on 31 May 1933 in another article for the *Westdeutsche Beobachter*. In his article "Die deutschen Intellektuellen" [The German intellectuals], Schmitt declared that German intellectuals who had emigrated and were criticizing the Nazi regime from their exiles could not in fact be considered part of the German nation: "They never belonged to the German *Volk* (people/nation in a racial sense, of common blood and with a common destiny; see Glossary). And not to the German spirit, either" (Schmitt 1933c, 32). He proclaimed: "They have been spit out of Germany for all time." (Schmitt 1933c, 32) He welcomed the book burnings which had taken place three weeks earlier, verbally attacked émigrés' critical comments about Germany as treason against the country and the *Volk*, sneered at the "Jewish relativism" of Albert Einstein's theory of relativity, considered revoking émigrés' German citizenship, and threatened further measures directed against them. He praised the laughing SA trooper as the idealized figure of the German man in the new Reich. Kirchheimer must have understood this article as a personal threat directed against him, too.

After the transfer of power to Hitler, people thronged to join the NSDAP. Schmitt waited in line for hours in Cologne and managed to submit his application to join the party and buy a party badge on 27 April,[20] just in time before the party enacted a freeze on new members, which was in place for a number of years. The official date he joined the party was 1 May 1933. With the support of Göring and Hans Frank, a legal expert and party member since 1923,[21] with whom he had made friends in early 1933, Schmitt rapidly obtained a number of influential leadership positions in the regime's legal system. Frank admitted Schmitt into the Akademie für Deutsches Recht (Academy of German Law), which he founded in the summer of 1933, and installed him as Reichsfachgruppenleiter der Hochschullehrer (Reich Director of the Professional Group of University Professors) in the Bund Nationalsozialistischer Deutscher Juristen (Association of National Socialist German Legal Professionals, BNSDJ), which had been founded back in 1928 as the organization of legal scholars who were members of the Nazi party.

In the autumn of 1933, Schmitt returned to Berlin, capital of the Reich, after only one semester in Cologne. Göring appointed him to the prestigious Chair of Constitutional Law at Berlin University. Schmitt moved into a villa at Schillerstraße 2 in Berlin-Steglitz. The same year, he became academic advisor of the Kaiser-Wilhelm-Institut für

19 Hanna Kirchheimer-Grossman in a conversation with the author on 25 April 2023.

20 Carl Schmitt, diary entry of 27 April 1933 (Schmitt 2010, 287).

21 In 1939, Hitler appointed Hans Frank Governor General of Poland, where people soon called him "slaughterer of Poles." Frank was sentenced to death in the Nuremberg war crimes trials and was hanged.

ausländisches Recht und Völkerrecht (Kaiser Wilhelm Institute for Foreign Law and International Law). In May 1934, Frank appointed him lead editor of the *Deutsche Juristen-Zeitung*. A year later, he also took on the role of legal advisor of the University Commission, which was under the personal supervision of the Deputy *Führer* Rudolf Hess and was responsible for assessing *habilitations* and appointments to the chairs of law at all German universities. In early 1936, Schmitt was additionally appointed director of the "Academic Division" of the BNSDJ. Holding so many official positions, Schmitt had advanced to become the linchpin for academic study of law in the Nazi system and remained so for three years. For anyone seeking an academic career in law in Nazi Germany, there was no getting around Schmitt during this period.

In addition, Göring, who had taken control of Prussia in early April 1933, appointed Schmitt a *Preußischer Staatsrat* (Member of the Prussian State Council) on 29 May. The *Preußische Staatsrat* (Prussian State Council), newly established by Göring, had sixty-eight members, including well-known leading Nazis as well as prominent artists and scientists such as actor Gustav Gründgens, physician Ferdinand Sauerbruch, and conductor Wilhelm Furtwängler.[22] Schmitt hoped this function would give him greater and more direct political influence. He figured that the institution of *Preußische Staatsrat* would be the first step toward establishing a *Führerrat* (*Führer's* Council), which would give him the opportunity to advise and assist Hitler himself. The ceremonial inauguration of the *Preußische Staatsrat* took place on 15 September 1933 in the auditorium of the University of Berlin. Schmitt spoke on "Wesen und Gestaltung der kommunalen Selbstverwaltung im Nationalsozialismus" [The nature and organization of home rule under National Socialism] in the presence of Prussian Prime Minister Hermann Göring, *Reichsführer* of the SS Heinrich Himmler, and SA commander Ernst Röhm. Göring subsequently appointed him to the position of rapporteur of a commission tasked with preparing a new municipal constitution.

During the Weimar Republic, Schmitt had already seen home rule as an attack by society on the unity of the state. The *Preußische Gemeindeverfassungsgesetz* (Prussian Municipal Constitution Act), which entered into force on 1 January 1934, followed this line of thinking; the explanations in a circular directive of the ministry were authored by Schmitt (see Blasius 2001, 106) and stated the guiding principle of the new law as follows: "A certain form of home rule corresponds to each form of the state."[23] The principle of the new state was that of unlimited responsibility on the part of the *Führer*. However, this did not imply the abolition of any or all forms of home rule, but rather the establishment of "truly National Socialist home rule."[24] The concept was then explained in detail. The head of a municipality was no longer elected by the citizens but appointed by higher state authorities after conferring with the *Gauleiter* of the NSDAP. There was no longer a representative body with the authority to make decisions, either; instead, merely members of the public volunteering in a consultative role. This arrangement was also in place in the major cities of Prussia. The local party organs and the highest-ranked SA and SS

22 On Schmitt and these three individuals mentioned in their roles as *Preußische Staatsräte*, see Lethen (2018).

23 As cited in Blasius (2001, 107).

24 As cited in Blasius (2001, 107).

leaders were members of these municipal councils as part of their official duties. Kirchheimer, too, was to examine questions of home rule a short time later in his Paris exile, but with an entirely different thrust.

Contrary to Schmitt's hopes, working on the municipal constitution was the only task he was assigned in his new position as a member of the *Preußische Staatsrat*, which was to convene only occasionally in the following years. The Nazi leadership did not develop it to take on the function of a *Führerrat* but, instead, limited it almost exclusively to representative duties. There are no expressions of internal reservations, much less aversion to the Nazi regime, to be found in Schmitt's diaries surviving from this period. Far from it. He even began to feel enthusiasm for Adolf Hitler, whom he had long held in contempt. After Hitler's speech concluding the Leipziger Juristentag, a conference for legal experts, on 3 October 1933, Schmitt wrote in his diary: "Wonderful speech by Hitler about the total state. Much comforted."[25]

In the spring of 1933, Schmitt had consciously decided to help establish the Nazi regime in the areas of propaganda and organization. From the outset, he made it clear both to himself and to his audience that Hitler taking over the government amounted to a fundamental caesura in terms of legitimacy. The boundary of the parliamentary state based on the *Rechtsstaat* had been transcended in favor of a dictatorship legitimated on *völkisch* (of the *Volk*, chauvinistic-nationalistic, antisemitic; see Glossary) grounds. He was also well aware that the Nazi regime was an antisemitic state from the very beginning.

Schmitt was not forced into any of his many and diverse activities at the time. Everything he did in the early years of the regime was of his own free will. He could have taken his older colleague Rudolf Smend as a role model, shifting his professional interests to niche topics and otherwise living a relatively undisturbed life under Nazism as a renowned conservative professor. No German scholar of constitutional law was persecuted after 1933 for being silent. Anyone writing articles supporting the regime wanted to be part of it—in whichever way. When Schmitt opted for the *Führer* state, he made new friends. But his decision also broke up a number of older friendships and severed old connections such as his relationship with Otto Kirchheimer.

3. Exiled in London and Paris

Kirchheimer's life in exile was entirely different from Schmitt's and the latter's successful career. He had fled Germany without any specific professional or financial prospects. After crossing the border to Luxembourg near Trier, he continued on to Paris, where one of his older brothers—a ballet dancer—had been living since the early 1920s. In Paris, he also met his wife Hilde Kirchheimer-Rosenfeld—they had been separated for two years at this point—their daughter Hanna, and his father-in-law Kurt Rosenfeld. The latter was forced to flee with his wife from impending political persecution in early March 1933. After Hilde had fled to her parents with their two-year-old daughter in mid-April, Otto Kirchheimer was the last family member to arrive in the French capital.

---

25 Carl Schmitt, diary entry of 3 October 1933 (Schmitt 2010, 305).

Besides Prague, Paris was the main refuge for political émigrés from Germany. France had been considered the traditional country of asylum in Europe since the nineteenth century, and leftist German intellectuals had viewed Paris as an exciting and livable city since the 1920s.[26] The first wave of emigration to France consisted mainly of scholars, physicians, lawyers, artists, and politicians; Jews and members of the opposition had immediately been banned from these professions in the first few months after Hitler took office. Most had had to leave their homes in panic with only a few belongings. They frequented the small number of émigré cafés in Paris but were unable to gain a foothold in the established Paris community.[27] Kirchheimer spent most of the next four years in Paris. A Francophile, he had often drawn on French legal theorists—for example, Carré de Malberg or Maurice Hauriou—although he never explicitly addressed issues related to France in his works during the Weimar Republic. Moreover, political ideas from the French Revolution had played a key role in arguing his leftist-socialist critique of the Weimar Constitution; time and again, he had juxtaposed the Weimar Constitution, which suffered from compromises, with the shining examples of the French revolutionary constitutions and their democratic vitality (see Schale 2011, 295–301).

Kirchheimer had arrived in Paris "almost penniless."[28] After failing to secure financial support through his connection to Carl Joachim Friedrich, he was more fortunate shortly after arriving in Paris and obtained a stipend from the London School of Economics and Political Science (LSE) for several months. Franz L. Neumann had helped Kirchheimer secure the stipend. Neumann had been a legal advisor to the SPD party leaders, and as late as March 1933, he had written an extensive brief detailing that the special press orders following the Reichstag fire were unlawful (see Neumann 1933). After an SA squad had raided his law firm on 2 May, he left Germany for England by ship. He was acquainted with Harold Laski, a prominent member of the Socialist League, the leftist wing of the British Labour Party, through his party contacts. Laski had been a professor of political science at the LSE since 1926. Neumann had decided to start a completely new career and began working on a doctorate in intellectual history and political theory under Laski. He also advised Laski about how to help persecuted social scientists from Germany at the LSE.

Shortly after arriving in Paris, Kirchheimer visited London from 13 to 23 June 1933.[29] The UK had also become a refuge for scientists driven out of Germany, and a private solidarity fund, the Academic Assistance Council (AAC), provided some financial support.[30] Kirchheimer visited the LSE, which played a major role in the AAC, in order to establish personal contacts.[31] Neumann introduced him to Laski; Kirchheimer had already acknowledged his writings on the theory of pluralism and on democratic socialism in

26 See Badia (1998) and Frank (2000).

27 On the difficult conditions of émigrés from Germany in Paris, see the descriptions by other exiles: Aufricht (1969, 120–125), Fabian and Coulmas (1982), and Sperber (1982, 45–61).

28 Conversation between Hanna Kirchheimer-Grossman and the author, 25 April 2023.

29 Certificate of Registration of the English Aliens Registration Office (original, owned by Hanna Kirchheimer-Grossman).

30 On the emigration of German scholars to the United Kingdom, see Hirschfeld (1985).

31 The Academic Assistance Council (AAC) was established by William Beveridge, then Director of the LSE, in May 1933, to support scholars persecuted by the Nazi regime (see Beveridge 1959). The

his own work during the Weimar period. As a result of this trip, he was granted an AAC research stipend for a project in England on the constitutional theory and legal sociology in the works of the renowned US Supreme Court judge Oliver Wendell Holmes, left-wing Harvard legal theorist Felix Frankfurter, and the Marxist historian of the making of the US Constitution, Charles A. Beard.[32] Kirchheimer spent the period from September to November 1933 as well as February and March 1934 in London as a research fellow of the AAC, making extensive use of the libraries there.[33]

During his initial stay in London, Kirchheimer completed his first academic publication after escaping from Germany. It was a retrospective essay on the history and end of the Weimar Republic, titled "The Growth and the Decay of the Weimar Constitution" (see Kirchheimer 1933c). He also attempted to secure his future living expenses while he was in London and sought to make contacts through his acquaintances among the émigrés in London to help him. Besides Neumann, a few others from the former circle of the Berlin journal *Die Gesellschaft* had found refuge in London, among them Otto Kahn-Freund. Kirchheimer also met up again with Georg Rusche, a fellow student from his time in Münster. Rusche had received funding from the Institut für Sozialforschung (IfS) in Frankfurt to work on a major study on the links between unemployment, on the one hand, and crime and its sanctioning, on the other, and was supposed to work toward completing the study for publication in London on behalf of the institute.

Kirchheimer thanked Smend in a letter from England dated October 1933 for his "recommendation for the Academic Assistance Council"[34] and reported on his work plans: "I have also started collecting materials to work on comparative democratic constitutional law." With respect to England, he noted that "at the moment when we are abandoning democracy once and for all, a whole lot of predemocratic institutions still exist here." It seemed to him "—quasi surrendering intellectual integrity—, generally useless to attempt to pick out the major democratic [...] institutions as still conceivable at all in our period of transition." Developing and formulating a democratic constitutional legal system that would take a less arbitrary approach would, however, "be difficult" in light of "Schmitt's skill in luring [people into rejecting parliamentarism]." It was also ques-

AAC later became the Society for the Protection of Science and Learning (SPSL) and continues to operate as the Council for At-Risk Academics (CARA) to this day.

32 The AAC files indicate Kirchheimer's field as constitutional law; reference is made to the fact that Kirchheimer had sought to obtain his *habilitation* in this area before fleeing Germany. These AAC memos are in the files of the Emergency Committee in Aid of Displaced German/Foreign Scholars, New York Public Library, New York. Otto Kirchheimer, Correspondence, b3.—The AAC also enabled the philosopher Theodor W. Adorno, who later joined the Horkheimer group in New York, to find employment at Oxford University in England in 1934 (see Müller-Doohm 2011, 283).

33 Concerning the dates, see the information provided in Kirchheimer's application for US citizenship. Otto Kirchheimer Papers, Series 2, Box 1, Folder 1.

34 This and the following quotes are from the letter from Otto Kirchheimer to Rudolf Smend dated 16 October 1933. Rudolf Smend Papers, Cod. Ms. R. Smend A 441.—Rudolf Smend declined to be involved in the Nazi regime's academic annihilation of Jewish scholars' contributions in other ways, too. In the summer semester 1933, besides Schmitt's *Legality and Legitimacy*, he also discussed Kirchheimer's eponymous essay in his class on an equal footing (see editor's note 296 in: Schmitt and Smend 2011, 90).

tionable "whether I can find a material basis for such a project." He wanted to "write to Friedrich at Harvard" again concerning this matter.

It was finally his connection to the Institut für Sozialforschung (IfS), which Kirchheimer had established through the LSE while he was in London, that charted the path for his professional future. The IfS, which had been founded in Frankfurt in 1923, was financed by Hermann Weil, one of the world's most eminent grain traders, with funds from a private foundation. Max Horkheimer had been appointed director of the institute in 1931 and had laid out a comprehensive research agenda in the social sciences and humanities titled "Interdisciplinary Materialism." He was the new dominant figure at the institute and remained so into the 1960s, both in organizational and in programmatic matters.[35] The new program he proclaimed after taking office found its strongest expression in the contributions of the institute's own *Zeitschrift für Sozialforschung* (ZfS, Journal for Social Research), which had been established in 1932. The institute's leaders had already decided to begin preparing to emigrate after the Reichstag elections in September 1930, when the number of NSDAP parliamentarians soared from 12 to 107. In light of the tense political situation in Germany, the foundation's endowment, which had been invested in securities, was transferred to the Netherlands as a precaution, and in the summer of 1932, the institute opened a branch in Geneva as "temporary emergency quarters" (Horkheimer).

After Hitler took power, the foundation in Frankfurt was replaced by the *Société Internationale de Recherches Sociales* (SIRES), which was based in Geneva, creating the legal basis for the foundation's endowment to remain outside Germany. Only a few weeks later, these measures proved to be essential for securing the existence of the institute, as its building in Frankfurt was raided by the SA in March 1933 and the IfS in Germany was shuttered. Universities abroad showed their solidarity with the IfS. The École Normale Supérieure (ENS) in Paris offered to make space available for a branch of the exiled institute on rue d'Ulm. Horkheimer also accepted the LSE's offer to make offices at the Institute of Sociology available to the IfS for another branch. The institute was determined to continue the work it had begun in Frankfurt. In early April 1933, Horkheimer wrote from Geneva to philosopher and literary critic Walter Benjamin, who had also fled to Paris, "we will try to continue our research and the journal as before, even more intensively because it appears that we will not be teaching at the university, which was quite time-consuming."[36]

When Horkheimer visited the London branch of the IfS in early 1934, Kirchheimer took the opportunity to meet with him and ask about a position or at least a temporary job at the institute in Paris. Horkheimer's response was positive. Horkheimer then traveled to New York to explore whether the IfS should open another branch there. He decided to relocate the institute. The institute found a new home for its headquarters at Columbia University in New York in the summer of 1934. In New York, the institute was renamed (International) Institute of Social Research (ISR). Paris remained the main location of the institute in Europe until German troops invaded France in May 1940. Paris was of key importance to the ISR because in 1934, the Paris publishing house Librairie Félix Alcan had

35 On Horkheimer's leading role at the institute, see Abromeit (2011).

36 Quoted in Wiggershaus (2010, 38).

agreed, in an act of solidarity, to enable continued publication of the *Zeitschrift für Sozialforschung* as a German-language scholarly journal. The Paris branch was headed by Paul Honigsheim up until 1936 and then by Hans Klaus Brill. The institute in Paris supported a number of scholars who had been forced into exile with larger and smaller sums of money and without applying strict criteria through its *Société Internationale de Recherches Sociales*. The monthly payments, which were actually disbursed on a more or less regular basis, were granted for independent research projects, essays, and reviews for the *Zeitschrift für Sozialforschung* and for specific research assignments for the institute's work on the *Studien über Autorität und Familie* [Studies on authority and family].[37]

As of mid-1934, Kirchheimer received a small stipend and occasional extra payments from the institute in Paris,[38] as did economist Henryk Grossmann, historian Franz Borkenau, and Walter Benjamin, who had also all been forced into exile. The latter had called the Bibliothèque Nationale his "most coveted place to work."[39] As Kirchheimer was affiliated with the Paris branch of the ISR, he was also entitled to a permanent library card. And, like Benjamin and others in the circles of the Paris branch of the ISR, Kirchheimer hoped to obtain a position at the institute—Benjamin finally succeeded in doing so temporarily in the autumn of 1937 (see Jäger 2017, 282). During the years of persecution, the foundation of the ISR supported over 130 scholars who had had to emigrate by providing larger or smaller amounts of money as well as guarantees for their residency status in the countries where they found refuge (see Wheatland 2009, 215–217).

Kirchheimer hoped he would continue to be able to obtain financial support from other foundations for his academic work in exile, too. He applied to the AAC again in autumn 1934, describing his project in his curriculum vitae in much the same way as he had a year earlier to Rudolf Smend:

> All this time I have been collecting material for a greater work on democratical [*sic*] institutions. This work, based on the empirical material as evidenced by the experiences of the democratically governed countries in the last ten years, is intended to discuss the effects of democratical [*sic*] institutions and the possibilities of democratic ideas within the different structures of society.[40]

This time, he did not receive any funding owing to the large number of academics in exile asking for support. Now he had to survive solely on the small amounts he received from the ISR fund as well as occasional fees he received from Gurland for research he conducted for exile news agencies in Paris on the economic situation in Germany.

---

37 The three-volume *Studien über Autorität und Familie* was published in Paris in 1936.

38 The date is provided in Kirchheimer's application for US citizenship. Otto Kirchheimer Papers, Series 2, Box 1, Folder 1.

39 Walter Benjamin in a letter to Theodor W. Adorno dated April 1935, quoted in Kambas (1983, 189).

40 Otto Kirchheimer, Curriculum Vitae (undated; around November 1934). The document is in the files which the London AAC left to the EC in New York. Emergency Committee in Aid of Displaced German/Foreign Scholars, New York Public Library, New York. I, A Grantees, 1933–46, Box 18, Folder 13 (Kirchheimer, Otto).—There is no additional material on Kirchheimer in the archive of the AAC, which is now housed in the Bodleian Library of Oxford University.

Kirchheimer's situation in Paris soon became increasingly difficult. Between 1933 and 1939, France had taken in roughly 65,000 émigrés from Germany, the largest number of any country by far (see Möller 1984, 48). This was despite the fact that the situation for émigrés in France was fundamentally different from that in most other countries where they found refuge. The legal provisions and administrative measures regulating residency and work permits made their social, economic, and cultural integration virtually impossible.[41] The French residency regulations were still based on the laws on aliens from 1849 and 1893 which made it easy to order disfavored individuals to leave the country. Every foreigner had to apply for temporary residence with the prefecture of the relevant province within eight days of arrival. Applicants had to prove they had sufficient funds to support themselves. If they were granted residency, they receive a *carte d'identité*. The prefects were under the direct control of the Ministry of the Interior and could refuse residency, revoke it, or refuse to extend it without giving reasons. Rejected applicants were ordered to leave the country or deported to their countries of origin.

The first émigrés to arrive, including Kirchheimer, still benefited from a generous practice of granting residency that evoked memories of Karl Marx and the poet and essayist Heinrich Heine in the nineteenth century. Unlike most other European countries, France also permitted émigrés to engage in public political activity (provided it did not interfere in French internal affairs) and allowed self-employed businesspeople and artisans, academics, and journalists to work. However, refugees were seldom granted work permits owing to the difficult economic situation. French policy toward accepting refugees from Germany changed gradually in light of their rapidly rising numbers. As early as the second half of 1933, it was virtually impossible for new arrivals to stay in the country with a longer-term perspective.[42] Further restrictions on issuing and renewing *cartes d'identité* were introduced when France experienced a wave of antisemitic and xenophobic actions in the course of a scandal involving financial fraud, and the ruling Radical-Socialist Party was replaced by a government of national unity under the leadership of the conservative Gaston Doumergue in February 1934. The regulations were tightened again in autumn 1934 after the French foreign minister and the Yugoslav king were assassinated in Marseille by Croatian nationalists who had entered France on forged German papers. This event immensely escalated xenophobia in France and, consequently, the French bureaucracy extradited émigrés from multiple countries to their persecutors. Many of those seeking refuge in France therefore traveled on to other countries, mostly to North, Central, and South America, after a time. Of the staff employed by the ISR in Paris, Franz Borkenau left for Panama and Henryk Grossmann for the US.

Eugene Anschel, his old friend from the German-Jewish Wandervogel movement, recounted in his memoirs how Kirchheimer lived in poverty in Paris:

---

41 On these aspects of the situation of German émigrés in France, see Vormeier (1981) and Fabian and Coulmas (1982).

42 For an overview of France's checkered policies with regard to taking in refugees between 1933 and 1940, see Badia (2002).

> He was living a precarious existence. [...] He had a small room in a third-class residence hotel incongruously called 'Le Home', where I stayed with him during my visit. [...] He had friends and acquaintances among the German refugee intellectuals. Without a regular job he spent a good part of his days in the reading room of the Bibliothèque Nationale, doing work for the Institute of Social Research (Anschel 1990, 127).

Moreover, the German passport office in Paris had confiscated his German passport in 1935.[43] This automatically rescinded his German citizenship and made him stateless. All he had was residency papers that could be revoked at any time. He repeatedly moved from one cheap furnished room in downtown Paris to another if it was a little cheaper.[44] Any documents that might provide more information about the specific amounts paid to Kirchheimer by the ISR and the relevant time periods appear to have been lost. The institute's stipend apparently did amount to at least a minimal financial basis.[45] Kirchheimer officially enrolled as a student at the Faculté de droit of the Université de Paris in order to be able to do this work and his work for the ISR at the university libraries.[46] His personal circumstances were complicated. Although they were separated, both parents still felt responsible for their daughter Hanna. After fleeing Germany, she first lived in Paris and was enrolled at a Montessori boarding school in northern Italy in 1935.[47] A number of other German socialists and communists who were persecuted were concerned about their children's safety and enrolled them at this school. It was financed partly by the parents and partly from international solidarity funds. Hanna Kirchheimer-Grossman and her father's letters report that both parents regularly visited their daughter in Italy.[48] Otto Kirchheimer desperately sought a way out of this difficult financial, political, and family situation.

4. Schmitt as an ambitious theorist of the Third Reich

All of Schmitt's writings from 1933 to 1936 on questions of the internal order of the Nazi regime are now finally available in a single volume published in 2021 (see Schmitt 2021).

43 Letter from the German Embassy in Paris to the German Foreign Office in Berlin dated 8 November 1938. Bundesarchiv, Akten des Auswärtigen Amtes. Politisches Archiv, RZ 214, R 99744 (69. Ausbürgerungsliste, Ausbürgerungsakte betreffend Otto Kirchheimer).

44 He first lived in a room on rue Massenet, then with his wife on rue Lombards, then on rue Brancion. His last residence, in 1937, was a room of his own at 7, Square Grangé, rue de la Glacière, Paris III.

45 A letter from Neumann states that he received a monthly salary of just 2,000 French francs in 1937. Letter from Franz L. Neumann to Otto Kirchheimer dated 9 February 1937. Otto Kirchheimer Papers, Series 2, Box 1, Folder 122. – 2,000 French francs in 1937 is equivalent to roughly 450 euros in 2024.

46 Kirchheimer's *carte d'immatriculation* for the *année scolaire* 1936/37 at the Faculté de droit of the Université de Paris is owned by Hanna Kirchheimer-Grossman.

47 Conversation between Hanna Kirchheimer-Grossman and the author, 25 April 2023.

48 In his report to Franz L. Neumann, Kirchheimer wrote: "I had traveled to see my child for 10 days, I found everything to be in excellent order there, and I experienced only 2 car crashes, but nobody was injured." Letter from Otto Kirchheimer to Franz L. Neumann dated 10 March 1937. Otto Kirchheimer Papers, Series 2, Box 1, Folder 122.

Previously, interested readers had to painstakingly track down the source material scattered throughout various publications. This editorial gap in the work of Schmitt, who has often been portrayed as a "classical representative of political thought," obfuscated his impact in the Nazi period for a long time and encouraged apologetic characterizations of his work—particularly in English-speaking countries. The complete corpus of his shorter, longer, and monographic works, numbering over sixty in total, reveals that and how Schmitt changed his linguistic style to that of Nazism over time. Throughout virtually all his publications in the early years of the new regime, Schmitt used key Nazi terms that were by no means ambivalent: On the one hand, he railed against "corrupt parliamentarism," the "*Parteibonzen*" (derogatory term for the bosses of the democratic parties in the Weimar Republic, see Brackmann/Birkenhauer 1988, 41) in the "*degenerierte Weimarer System*" (degenerate Weimar *System*, *System* was a derogatory term for the Weimar Republic), the "*artfremde Geist*" (spirit/intellect [itself derogatory] foreign/alien to the *Volk*) of "*fremdrassige Rabulisten*" and "*fremdrassige Literaten*" (shysters and literati of a foreign/alien *Rasse*, in an exclusionary, antisemitic sense) who, as "demons of *Entartung*" (degeneration or decline due to biological or cultural factors; see Glossary), were "poisoning the brains" of Germans. On the other, he extolled "the voice of German blood," purging of "*nichtarische fremdartige Elemente*" (non-Aryan elements foreign/alien to the *Volk*; *Elemente* was a contemptuous term for opponents), and "annihilation of enemies of the *Volk*", combined with singing the praises of "our SA and SS" and the "national revolution" whose goal was the "*Gleichartigkeit* [see Glossary] of the German *Volk*," which was to be achieved by "eliminating all *Fremdgeartete*" (all those foreign/alien to the *Volk*, in an exclusionary, antisemitic sense). The words quoted here are not "ambivalent," nor does their semantic content amount to merely tactical "concessions to Nazism" (Bendersky 2004, 23), as Joseph Bendersky in his book on Schmitt would have readers believe. This is the language of Nazism plain and simple (see Translator's Preface and Glossary for more detail). Schmitt's choice of words emphatically refutes the proposition often put forward in the secondary literature by Bendersky and other authors that the difference between Schmitt and Nazism was that he had not argued along the lines of biologistic racism. Moreover, new and informative studies on Schmitt's activities are now available that make it possible to accurately reconstruct his impact and his role in the Third Reich. These studies enable scholars to more precisely map Schmitt's position within the field of Nazi ideology production, which was by no means homogeneous.[49]

Among the variety of issues that Schmitt discussed in the early years of the Nazi regime, six in particular piqued Kirchheimer's interest during his exile in Paris.

The first is Schmitt's characterization of the Enabling Act as the decisive "turning point of relevance in constitutional history" (Schmitt 1933a, 456) as already mentioned above. The law had transferred a measure of constituent power to the government of the Reich. In retrospect, Schmitt interpreted the Reichstag election of 5 March 1933, after which the coalition of NSDAP and DNVP was able to continue to govern, not as an optional election but as a clear plebiscite in favor of Hitler, the *Führer*. With this legitimating basis, the Enabling Act had become a "provisional constitutional act of the new Germany"

49 See Blasius (2001) and (2009), Mehring (2014a, 275–348), Neumann (2015, 303–374), and Mehring (2021).

(Schmitt 1933d, 8). Thus, the Weimar Constitution was de facto no longer in force, although officially it was (and remained so until the capitulation in 1945). Schmitt rejected all attempts "even merely to grasp today's constitutional situation with the norms, concepts, or categories of the former *Weimarer System* or its constitution" (Schmitt 1933f, 242; "*Weimarer System*" was a derogatory term for the Weimar Republic). In contrast to his students Ernst Rudolf Huber and Ernst Forsthoff, Schmitt also argued against the idea of a new written constitution for the Third Reich. Demands like this, he asserted, were a "notion of a constitution inimical to National Socialism" (Schmitt 1934c, 27). The new constitution, he stated in his essay "Ein Jahr nationalsozialistischer Verfassungsstaat" [One year of the National Socialist constitutional state] published in early 1934, consisted of Hitler's appointment as Chancellor of the Reich, the Enabling Act, and the laws enacted thereafter. After the NSDAP Reich Party Convention in Nuremberg of September 1935 under the propaganda motto "Party Convention of Freedom," Schmitt added one more component: the Nuremberg Laws with their discriminatory legislation against Jews as the "constitution of freedom." At the same time, he elevated the NSDAP to the role of "*Wächter des völkischen Heiligtums*" (custodian of the *völkisch* sanctuary) and "*Hüter der Verfassung*" [Guardian of the constitution] (Schmitt 1935a, 283).

Second, Schmitt described the transfer of power to Hitler as a political occurrence that was strictly legal. He did not devote a single word to the terrorist and illegal measures used against members of the opposition from 30 January 1933 onward. Instead, he praised the "legality of our own National Socialist state" (Schmitt 1933f, 251) dictatorship; to Schmitt, acknowledging legality had an important function in securing power because, in the machinery of a large state, belief in legality was indispensable in order to keep the complicated apparatus running. What mattered was the unimpeded "mode in which the state apparatus of civil servants and public agencies functioned" (Schmitt 1933d, 8). The constitutional construct of legality ensured the loyalty of the civil servants and the military leadership and helped gain the trust of the bourgeoisie.

A third subject was Schmitt's fundamental programmatic reorientation from purely authoritarian statism to a constitutional construct in which the NSDAP as the only party and Hitler as the *Führer* could be positioned in their appropriate place. In his short monograph *Staat, Bewegung, Volk – Die Dreigliederung der politischen Einheit* [State, movement, *Volk* – the tripartite structure of political unity] of autumn 1933, Schmitt suggested the formula of "unity of the tripartite structure of state, movement, and *Volk*" (Schmitt 1933d, 11). He considered the NSDAP, the only party existing in Germany from 14 July 1933 on after all others had been banned, to be the "political body in which the movement [had found] its special political form" (Schmitt 1933d, 13). At this point in time, when other Nazi professors of constitutional law were conceptualizing a constitution for the Third Reich which demanded that Hitler as *Führer* of the movement would be subordinate to the state and its laws (see Stolleis 1999, 351–353), Schmitt advocated for giving the *Führer* unlimited scope for decisions and actions. Right at the beginning of the monograph, he made it absolutely clear that the will of the *Führer* had precedence over all other institutions and rules; the will of the *Führer* was "the nomos of the German *Volk*" (Schmitt 1933k, 69).

Fourth, Schmitt promoted a fundamental methodological revision of legal thinking. In his programmatic essay "On the Three Types of Juristic Thought" published in spring

1934, he abandoned the decisionism he had previously championed in favor of concrete-order thinking. Decisionism was outdated as a method of legal thinking, he now asserted, since it was a type of personal thinking by an individual or a group of people. In decisionism, the reason why the law applied as it did was a process of the will, more precisely, a decision that was not necessarily derived from existing rules. Schmitt also criticized the concept of normativism based on rules and statutes, by which he meant Hans Kelsen's positivist theory of pure law. The characteristics of normativism were impersonality and objectivity. Rule was to be founded on norms, not individuals. The legal concepts of normativism were general concepts. According to Schmitt, however, because they were abstract, they disregarded the concrete order of life that people experienced as reality. He described legal positivism not as an independent form of legal thinking, but as a hybrid of decisionism and normativism. Schmitt called the third basic type of legal thinking, besides decisionism and positivism, "*gesundes, konkretes Ordnungsdenken*" (concrete-order thinking corresponding to the norm of the NSDAP; see Translator's Preface and Glossary) (Schmitt 1934h, 157). It developed in suprapersonal institutions. A precondition for this was a stable normal situation, a *situation établiée*. A necessary consequence of this was that concrete orders embedded the individual in a *Gemeinschaft* (see Glossary) that was structured hierarchically and served a particular purpose. This *Gemeinschaft* also implied strict rejection of individual rights. The original source of law concerning concrete-order thinking was lived normalcy, which also took place independently of positive norms. General clauses, which had already become more important during the Weimar Republic for some areas of the law, had become the "specific method" (Schmitt 1934g, 91) of this new type of jurisprudential thinking. They were "indeterminate concepts of all kinds, references to extra-legal criteria, and notions such as common decency, good faith, reasonable and unreasonable demands, important reasons, and so on" (Schmitt 1934g, 90).

Fifth, Schmitt published on questions of criminal law and criminal legal procedure. In his *Fünf Leitsätze für die Rechtspraxis* [Five guiding principles for legal practice], which he published in July 1933 and which were also printed separately as recommendations for the courts and public prosecutor's offices, Schmitt had taken up general clauses and called for interpreting the existing laws strictly in line with the principles of Nazism. The only measure to be applied during adjudication was the views of "*bestimmtgeartete Menschen*" (people of a certain *Art*, in an exclusionary sense) (Schmitt 1933h, 55) from the Nazi movement. In his article "Nationalsozialismus und Rechtsstaat" [National Socialism and the *Rechtsstaat*], which was based on a lecture for the BNSDJ, he gave more depth to his deliberations on criminal law. In the introduction, he made a fundamental distinction between a *Rechtsstaat* (state based on the rule of law) and a *Gerechtigkeitsstaat* (state based on a certain idea of what is just). He explained this differentiation to his audience using an example from criminal law. The traditional liberal state based on the rule of law was committed to the principle *nulla poena sine lege* (no punishment without law). A year after the Reichstag fire, Schmitt declared the ban on ex post facto laws to be one of the "formal methods, principles, norms, and institutions" (Schmitt 1934d, 25) to which the liberal state based on the principles of the *Rechtsstaat* was bound. Conversely, the *Gerechtigkeitsstaat* of the Nazis was aiming for the "obvious substantial justice of the cause" (Schmitt 1934d, 25), which found expression in the alternative principle *nullum crime sine poena* (no crime goes unpunished). Schmitt thus openly demanded that it should be possible to prosecute even those deeds using criminal law that were not deemed punishable according to the existing laws. He believed concrete-order thinking should replace liberal normativism in this area of the law too, and that the proper Nazi education of the judiciary should replace courts being bound to the law in adjudicating cases.

Schmitt's activities in this area were not limited to developing legal theories, but also included putting them into practice. In June 1936, his mentor Hans Frank appointed him chair of an *Ausschuss für Strafverfahrensrecht, Gerichtsverfassung und Strafvollzug* (Committee for the Law of Criminal Procedure, the Constitution of the Courts, and the Penal System) newly established within the BNSDJ. Schmitt's activities for this committee were to trigger his demotion within the Nazi hierarchy a few months later. In early 1936, Schmitt had already called the law of criminal procedure a core of constitutional law in a lecture on the tasks of constitutional history. He linked this to the hypothesis that a "bourgeois-legitimizing compromise" (Schmitt 1936a, 410) between the independent judge and the public prosecutor, who was bound by directives, was reflected in the traditional law of criminal procedure. In his new function, he prepared multiple opinions and proposals for legal reforms toward a fundamental reordering of the law of criminal procedure (see Schmitt 1936b),[50] demanding that the major lines of Nazi law of criminal procedure also had to be derived from the overall constitution of *völkisch* life. And in the case of Germany, this meant the *Führerprinzip*: "the antiparliamentarian organizational principle of the Third Reich according to which Hitler ruled not within the framework of a constitution, but as the alleged personification of the will of the *Volk*" (Schmitz-Berning 2007, 245); his authoritative decisions were correct by definition (see Glossary). Schmitt proposed, inter alia, that judges were to deliver verdicts "in the name of the *Führer*" rather than "in the name of the *Volk*." Other suggestions he made aimed at replacing legal remedies with decisions of a political authority to be newly created and appointing an NSDAP ombudsperson for legal proceedings in cases where the party considered itself to be affected by the subject of the proceedings.

Finally, Schmitt declared that the *Führer's* will should have absolute priority; he did so in reaction to what was known as the "Night of the Long Knives" of 30 June 1934. That night, Hitler had adversaries within the party, including SA commander Ernst Röhm, murdered without a trial. In total, approximately eighty-five people were killed in various places within the space of three days. The murder operation, which was illegal under prevailing criminal law, was camouflaged by propaganda claiming that a "Röhm coup" was imminent. On 3 July, the government of the Reich promulgated a law which retroactively declared the murders and further breaches of the law to be legal because they were self-defense of the state. Hitler defended his course of action before the Reichstag on 13 July by stating, among other things: "In this hour, I was responsible for the *Schicksal* [see Glossary] of the German nation and thus I was the highest judge of the German *Volk*."[51] The murderous massacre made an extremely bad impression both in Germany and abroad.

50 For more on Schmitt's work in this commission and his proposals, see Gruchmann (2001, 994–1002).

51 Speech by Adolf Hitler before the German Reichstag on 13 July 1933, as cited in Fest (1973, 644).

Conservative supporters of Hitler were now definitively aware that they, too, could be on one of the *Führer's* next revenge lists. In this situation, Schmitt applied a kind of preemptive defense on his own behalf. As the new editor of the *Deutsche Juristen-Zeitung*, he published an editorial titled "Der Führer schützt das Recht" [The *Führer* is protecting the law] on 1 August 1934. It was an "unequivocal homage to the *Führer's* crimes" (Blasius 2001, 120) cloaked in the terminology of constitutional law.

In a nutshell, Schmitt first offered a justification for the murders based purely on constitutional law: thus, Hitler had protected the unity of the authority of the state from a looming second SA revolution. But then he made his constitutional law constructs more foundational. The role of the *Führer* was not that of a "republican dictator" (Schmitt 1934e, 200) who would resign after ending a crisis; instead, a *Führer* grew organically out of the *Lebensrecht des Volkes* (right of the *Volk* to life). Hitler had identified the enemies of the state who had violated their duties of loyalty toward him. In complete agreement with Hitler's Reichstag speech, Schmitt proclaimed the will of the *Führer* to be a direct source of law. The events of the previous days had shown: "When the *Führer* directly creates law as the highest judge in the moment of danger by virtue of his being the *Führer*, he is protecting the law from the most egregious abuse." (Schmitt 1934e, 200) And he continued: "The true *Führer* is always also a judge. The role of judge flows from the role of *Führer*. Anyone who seeks to separate the two or even have them oppose each other makes the judge either a counter-*Führer* or the tool of a counter-*Führer* and seeks to turn the state upside down with the help of the judicial system" (Schmitt 1934e, 200). Schmitt's conclusion was: "In actual fact, what the *Führer* did was genuine jurisdiction. It is not subject to justice, but, rather, was the highest justice itself" (Schmitt 1934e, 200). Characterizing the *Führer* as the highest judge was an implicit criticism of the retroactive legalization of the murders through the law of 3 July, which he considered a superfluous legacy of liberalism. With his Nazi interpretation, Schmitt surpassed even Nazi practice. This far-reaching position of his was met with rejection in the Nazi state's ministerial bureaucracy, however; the only official to support him was State Secretary of the Prussian Ministry of Justice Roland Freisler (see Gruchmann 2001, 453–460), who later headed the infamous *Volksgerichtshof* (see List of German Courts). Schmitt did not say a word in his editorial about the victims of the murder operation; besides Schleicher, whom he had given legal advice for a longer period of time prior to 1933, they also included other people from the conservative milieu who were close to him, such as Edgar Jung and Ferdinand von Bredow.[52]

Never again in his long life did Carl Schmitt write so much and give as many lectures in so short a time as in the initial years after power was handed over to Hitler. He published short monographs, articles, and legal commentaries downright obsessively in which he accompanied and legitimized the consolidation of the Nazi regime. In producing such legitimation, he was not the only German constitutional law professor to largely

52 The earlier secondary literature occasionally reflects the opinion that some wording in this article might indicate that Schmitt might have demanded that at least the murderers of Schleicher and Bredow were to be punished (see Bendersky 1983, 213–217; Koenen 1995, 612–616). This interpretation has been rejected with nuanced arguments by Mehring (2014a, 320–325) and Neumann (2015, 339–341).

welcome the new *Führer* state as if in a "creative frenzy" (Stolleis 1999, 320). What distinguished Schmitt from most of his colleagues, however, was "his intellect and his ability to formulate, which enabled him to capture the new situation in memorable formulas more rapidly and more effectively than others" (see Neumann 2015, 309).

There is a long-standing debate in the secondary literature on Schmitt about the extent to which he actually identified with all Nazi doctrines in his publicly vaunted dedication to the Nazi state or whether he was actually advocating for a political agenda of his own, an attempt that failed flagrantly due to his naivete about realpolitik—similarly to the philosopher Martin Heidegger in this regard. Representatives of the latter line of interpretation are able to rightly point out that at the end of the Weimar Republic and the beginning of the Nazi era, Schmitt was closer to the group of conservative statists than to the streams of the NSDAP that considered themselves a revolutionary movement beyond statehood. In addition, Schmitt's connections to the conservative Catholic milieu of the Weimar Republic have been underscored in this context.[53] Schmitt was not a "conservative revolutionary" in the sense of resisting Nazism, but rather a conservative who was formatively influenced by the German Empire and who had volunteered to serve the Nazi revolution.

The six facets of his oeuvre mentioned above show Schmitt as an eager Nazi. The personal motives for his activities have been analyzed in the biographical literature on Schmitt multiple times and with different accentuations. Yet Schmitt's personal motives are beside the point here. From a perspective like Kirchheimer's, Schmitt's impact alone was of interest, namely as an ardent and eloquent protagonist of the Nazi *Führer* state.

The fact that Schmitt quickly managed to stir up opposition among other Nazi ideologues is an integral part of his enormous public impact. When he attempted to secure the Nazi regime by means of constitutional law, it did not go down well with long-standing Nazis among his colleagues in the legal profession that he, who had only just joined the party, assumed the role of a better interpreter of Nazism, especially compared to them. For instance, he firmly rejected attempts in constitutional law to differentiate between permissible and impermissible deviations from the Weimar Constitution, which had been modified by the Enabling Act, in laws promulgated by the government of the Reich. Not only did he reject these attempts, but he also considered them to be practically acts of sabotage against Nazism (see Schmitt 1933d, 6–8). In Volker Neumann's apt words: Schmitt "put on airs as the authentic interpreter of Nazism and handed out political grades" (Neumann 2015, 324). Neumann also pointed out that after joining the party, Schmitt used language identifying himself with the Nazis—for example, "we, the National Socialists," "us, the National Socialists," and "our SA and SS"—downright obnoxiously, thereby enraging the party veterans among the Nazis even more. All of this behavior of Schmitt's made long-standing Nazis despise and envy him. Their feelings intensified even more when they saw the eloquent Schmitt, under the protection of Hermann Göring and Hans Frank, overtake them as they wrangled for positions in the Nazi state hierarchy.

53 This aspect is emphasized by Meier (1994) and Koenen (1995).

5. Kirchheimer as a theorist of democratic alternatives

Once Kirchheimer had fled to France, the German journals and magazines that had published his work in the previous five years were no longer in reach for him as author. Unlike 1919, 1933 had been a "turning point for all legal journals" (Stolleis 1999, 299) in Germany. All journals on public law experienced exceptional pressure from the Nazi regime because of their evidently political nature. The journals that continued to exist were placed under the control of the Reichsschrifttumskammer (Reich Chamber of Literature). Within a short period of time, all editors of law and sociology journals who were Jewish or politically disfavored were replaced by supporters of the Nazi regime. Social democratic, communist, and other left-wing journals and newspapers were banned in the Reich.

If he was to continue publishing and not only writing to satisfy his own academic curiosity, Kirchheimer had to find new journals. Some exiled authors, particularly journalists and writers, found opportunities to publish in the German-language exile press, primarily in Paris and Prague. Kirchheimer's father-in-law, Kurt Rosenfeld, founded the Internationale Presseagentur gegen den Nationalsozialismus (International Press Agency against National Socialism, Inpress) in the early summer of 1933. Based in Paris and New York, Inpress was a trilingual news service that supplied international newspapers with reports from and about the German Reich.[54] Kirchheimer's estranged wife Hilde occasionally worked there and had him write and edit news items for Inpress from time to time to supplement his income from the ISR. Arkadij Gurland also helped him find paid work. On occasion, he assisted with data collection for a *Documentation de Statistique Sociale et Economique* in Paris. In addition, he helped Gurland write articles for the business section of the weeklies published by the socialist Max Sievers and disseminated illegally in Germany.

Kirchheimer's first publication after fleeing Germany was his essay "The Growth and the Decay of the Weimar Constitution."[55] The article appeared in the November 1933 issue of the *Contemporary Review*, published in London. The journal was well established in English intellectual circles; its orientation in the 1920s and 1930s was leftist-liberal. Harold Laski, who taught at the London School of Economics and had granted Franz Neumann academic refuge after he had fled Germany, and who introduced Kirchheimer to George P. Gooch, the journal's long-standing editor, occasionally published there. The article provided an overview for the British audience of the entire history of the Weimar Republic.

Kirchheimer reiterated his model of three development phases from 1919 to 1924, 1924 to 1930, and 1930 to the handover of power to Hitler's government. The points he focused on and his assessments of individual political actors remained virtually identical to his Weimar writings. But now he put more emphasis on three aspects: the failures of fundamental political reforms in the early postwar years; the potential of the republic to be stabilized in the middle phase; and the severe impacts of the Great Depression on German domestic policy. Kirchheimer also reiterated his opinion that the Weimar Republic

54 On Inpress, see Schiller et al. (1981, 77–79) and Langkau-Alex (1989, 204–205).

55 Kirchheimer (1933c). The following page numbers refer to this text.

had essentially already come to its end with the Brüning government: "While political liberty was still alive, democracy had gone with Brüning's coming into power" (533). A direct path had led from Brüning's "liberal-minded dictatorship" (533) to the Nazi's erection of a "totalitarian State" (533), which would leave no sphere of human life outside the scope of a centralized powerful government.

In order to explain the rapid political transformation to his British readership, Kirchheimer pointed out Carl Schmitt's preeminent responsibility for providing the legal legitimation for the totalitarian regime: Schmitt had developed a doctrine according to which it was the incontrovertible destiny of every democratic system of government to lose itself in internal struggles between various groups until it was worn down to such an extent that it was replaced by a dictatorship. The political doctrines followed the course of events by constructing a new system of political thought. Kirchheimer asserted that Schmitt was crucial to the new ideological constructions. He summarized Schmitt's theory for his British readers as follows:

> Professor Carl Schmitt, who is the theorist of the Nazi Constitution just as Hugo Preuß was the theorist of the Weimar Constitution, developed a doctrine of the totalitarian state amalgamating the ideas of its being the necessary and the ideal goal of historical evolution (533).

When mentioning Preuß's name, Kirchheimer was alluding to the programmatic ceremonial lecture Schmitt held in January 1930, which Kirchheimer had attended. Preuß had been a left-liberal politician and bourgeois Jewish scholar whom Schmitt had revered as the father of the Weimar Constitution in this lecture (see Schmitt 1930c). Kirchheimer stated that Schmitt's "sympathy with the totalitarian idea was so formal and general in nature that it equally favoured the Bolshevist and Fascist forms of government" (533). He only sided with the Nazis after it was obvious that they had come to power. With these words, Kirchheimer implied that Schmitt might well have sided with the communists if they had come into power. In any case, he described him as an opportunist who would have sided with any totalitarian dictatorship.

Kirchheimer also made a distinction between Schmitt as "nothing but a political theorist" and Schmitt as "a Nazi partisan and official framer of Nazi constitutional laws" (534). He obviously took pleasure in using Schmitt's vocabulary to support leftists' paths of resistance against the Nazi regime, which he supported. In his former role as a political theorist, Schmitt had interpreted the "totalitarian idea" in a way that would "justify even the fiercest enemies of his actual party" (534). In Schmitt's political theory, any form of government that emphasized its own power and advocated for dominance of the state over all other social forces could be considered to be totalitarian. The conclusion Kirchheimer drew from such a broad way of defining the totalitarian state was opposed to Schmitt's. The concept of a totalitarian state "might even be true of a democracy, leaving a reasonable sphere of political freedom to the individual" (534). To Kirchheimer, the fact that it was possible to interpret the idea of the totalitarian state in a diametrically opposed way demonstrated once again that the totalitarian idea did "not represent any substantial political conception at all" (534).

This raises the question as to the genuine contribution of Nazism to German political theory. Kirchheimer thought it was an attempt to base all government institutions on a theory of *Rasse*. The Nazi concept of a *Blutsgemeinschaft* (*Gemeinschaft* founded on *Blut*; see Glossary) was closely connected to the concept of the *Führer*. via a "primitive conception of giving obedience and receiving protection" (534). Kirchheimer called both the theory of the *Führer* and the references to the German *Volk* in Schmitt's Nazi writings an expression of "primitivity of thought" (534). This type of political and legal thought would turn German society into a place with convictions once held by prehistoric tribal societies and of feudal and religious communities of the Middle Ages.

Over the following two years, Kirchheimer wrote three articles about three different subjects: constitutional courts, the problem of sovereignty, and the role of municipalities within the state. All three subjects had also been taken up by Schmitt, either during the Weimar Republic or in his role as legal commentator for the Nazi regime on the six issues mentioned above. In all three cases, Kirchheimer wrote in a kind of internal dialogue with Schmitt, sometimes mentioning his name and sometimes omitting it. In all three cases, he attacked Schmitt's positions and contributed to the analysis and the theory of democratic alternatives to the totalitarian state. Specifically, these articles address the Supreme Court of the United States, the theory of sovereignty, and the role of municipalities within the French state. These subjects themselves contain references to Schmitt's writings.

The first article dealt with the role of the Supreme Court in the United States. Kirchheimer had already mentioned it several times in his Weimar writings. At the time, he had had a negative view of the Supreme Court and had blamed it for policies against the interests of the working class. In 1932, he had even provided Schmitt with references to works by leftist critics of the Supreme Court.[56] Schmitt had rejected proposals to establish a constitutional court in Germany because it would not create a juridification of politics, but rather a politicization of the judiciary. From 1933 on, he believed such proposals for political reform were no longer an issue. Constitutional jurisdiction, he wrote in his commentary on the *Reichsstatthaltergesetz* (Reich Governor's Law) of April 1933, was "no longer of interest" (Schmitt 1933g, 26) because the *Führer* was now the only source of law in the German Reich.

Kirchheimer's new contribution to this debate was his essay "Zur Geschichte des Obersten Gerichtshofes der Vereinigten Staaten von Amerika" [On the history of the Supreme Court of the United States of America]. This was obviously part of a plan he had mentioned to Smend in a letter in October 1933 to collect material for a comprehensive study of "comparative democratic constitutional law."[57] The essay was published in late 1934 in issue 3 of *Zeitschrift für Öffentliches Recht* (ZÖR). It was in fact a publication by a German émigré abroad since ZÖR was the Austrian journal of public law, established in Vienna in 1914 on Hans Kelsen's initiative. In 1934, *International Journal* was added to its title to enable its continued distribution in the German Reich and to keep Kelsen on the editorial board. The ZÖR was the place where other German émigrés including Karl

56 See Chapter 5, p. 158.

57 Letter from Otto Kirchheimer to Rudolf Smend dated 16 October 1933. Rudolf Smend Papers, Cod. Ms. R. Smend A 441.

Loewenstein, Hugo Sinzheimer, and Helmut Plessner were able to publish their work over the next few years.

Kirchheimer wrote the article[58] as a report for German-speaking readers about the controversies in the US legal literature on the Supreme Court. He also attempted to outline his own interpretation of the history and sociology of its impact. And even though he highlighted the limits of the Supreme Court's rulings with respect to property in the US, a distinct shift in emphasis can be discerned compared with his writings on the Supreme Court from the Weimar period.[59] This change in perspective was due not least to his reading of the works of Felix Frankfurter and Charles Beard with whom he shared the fundamental methodological concern that the history of the court had to be embedded in a "sociohistorical account" (117). Kirchheimer thought that the economic structures were particularly important—here, he drew in particular on the American Marxist Louis Boudin's works on legal theory which he had recommended to Carl Schmitt in November 1932. At the same time, the article was a response to the accounts and assessments of developments in public law in the US as represented by Carl Joachim Friedrich in German-speaking countries (see Friedrich 1931).[60]

Against the background of Schmitt's position and form of reasoning during the Weimar Republic, Kirchheimer's essay reads like a completely opposite approach to the subject of constitutional jurisdiction. Right at the outset, he calls for a "sociohistorical presentation" (117) in which the institution of the Supreme Court was located as "an element of all that happens in society" (119). The Court had been established after the founding of the American republic. Referring to the first decision under Chief Justice John Marshall, the famous decision *Marbury vs. Madison* of 1803, in which the Court had for the first time claimed the competence to review the constitutionality of federal laws, Kirchheimer drew the historical parallel to the case in Germany in which Schmitt had represented the Reich against the *Land* of Prussia: "In this highly political situation, Marshall was faced with the same question as, for example, the German *Staatsgerichtshof* in the conflict concerning Prussia in 1932" (122). In Germany, the *Staatsgerichtshof* had failed. But not the Supreme Court. From the beginning, it viewed itself as a powerful political institution and spent considerable energy establishing and defending itself as such. Kirchheimer outlined the history of the Supreme Court in a phase model. In the first phase, the majority of its decisions favored the seigneurial aristocracy of the large Southern plantation owners. Around 1830, the Court took a turn to "competitive capitalism" (124). Kirchheimer emphasized that the Supreme Court had increasingly intervened in the social conflicts of the day, above all in the conflicts of interest between capital and labor. In a number of decisions about the right to unionize, limitations on working hours, and income tax provisions, the Court took clear positions favoring capital.

58 See Kirchheimer (1934a). The following page numbers refer to this text.

59 In his essays on property rights and expropriation, Kirchheimer described the jurisprudence of the Supreme Court as serving exclusively capitalist interests (see Kirchheimer 1930b, 339–340 and Chapter 3 in the present book).

60 On Friedrich's crucial role interpreting the political system of the US for German readers, see Lietzmann (1999).

Kirchheimer attributed a further transformation of the Court's jurisprudence in the early twentieth century to two factors. First, the tough intransigence with which Justices Oliver Wendell Holmes and Louis Brandeis worded their dissenting opinions for many years. Kirchheimer considered these dissenting opinions to be in stark contrast to how things worked in the courts of the Weimar Republic. Whereas the German courts acted as if they were nonpartisan actors, the fact that dissenting opinions were published in the US showed that its Supreme Court did not consider itself the sole guardian of the constitution but, rather, as part of a political process and open to future revisions of opinion. The second decisive factor for the transformation of jurisprudence was the "pressure of mass democratic movements" (128) during the presidencies of Theodore Roosevelt and Woodrow Wilson. It was because of this public pressure that, in contrast to its previous interpretations of the law, the Court had seen itself forced to rule that a larger number of social policy measures were in line with the constitution. Kirchheimer sharply criticized the Court's more recent rulings since they eliminated much of the legal basis for Theodore Roosevelt's government's stabilization measures.

Nonetheless, Kirchheimer concluded his article with cautious optimism. Arguing against Carl Joachim Friedrich, he stated that it would be a mistake to tie the Supreme Court's jurisprudence to certain values long term. This view was also misguided "because it involved a certain overestimation of the opportunities of a court to influence the course of political events" (131). Instead, Kirchheimer trusted that the Supreme Court would let most of the reform laws Franklin D. Roosevelt was planning stand if political pressure were exerted, as had been the case in the Wilson era. There was only one instance where Kirchheimer expected that the Court would not change course: the protection of private property would continue to take the "most outstanding position" (131) in the future.

Although Kirchheimer did not come to an overall conclusion about the Supreme Court's decisions at the end of his article, it is clear that he shared Kelsen's position in the controversy between the latter and Schmitt. For one thing, he emphasized that the Court had essentially become a protective wall against "individual state laws' reign of terror against the freedom of opinion" (130). And for another, he thought that the decisions of the Court that he criticized sharply in substantive terms could in principle be revised by mobilizing mass democracy for a "welfare state" (131) and recruiting new judges. Kirchheimer was remarkably accurate in his assessment of the Supreme Court's future jurisprudence. The "four horsemen," as the group of four conservative justices was called, continued to block Roosevelt's policies up until 1937, when the Court, under public pressure and in a different composition, allowed much of the New Deal reform agenda to stand. From 1939 onward, Harvard professor Felix Frankfurter was one of the new justices. And it was Frankfurter from whom Kirchheimer received a letter concurring with his article in October 1934. Frankfurter praised his knowledgeable and deep insights into the US Supreme Court, but corrected him on one point: "In time I ought to say however that you are a prophet rather than a historian in saying that the Supreme Court has already sustained the Roosevelt legislation. Not yet."[61]

61 Letter from Felix Frankfurter to Otto Kirchheimer dated 12 October 1934. A copy of the letter is in the files that the London AAC left to the EC in New York. Emergency Committee in Aid of Displaced

The second subject Kirchheimer examined had also been of intense interest to Schmitt in his Weimar writings: namely, the problem of sovereignty. His article "Remarques sur la théorie de la souveraineté nationale en Allemagne et en France" [Remarks on the theory of national sovereignty in Germany and France] appeared in French in 1934 and again took a comparative view.[62] It was published in the journal *Archives de Philosophie du droit et de Sociologie juridique* [Archives of legal philosophy of law and legal sociology], which had been founded only four years previously and was published in Paris up until 1939. The journal was edited at the Sorbonne and sought to combine legal, philosophical, and sociological research. The members of the journal's international advisory board included Germans Gerhard Leibholz, Gustav Radbruch, and Hugo Sinzheimer as well as Harold Laski from the LSE. Kirchheimer compared the theories of sovereignty in the French and German legal literature from the late eighteenth century onward from the perspective of intellectual history. Regarding the changes occurring during this long period of time, he was again concerned mostly with elucidating the crucial link between economic and social developments on the one hand and their political and legal implications on the other.

Kirchheimer emphasized the self-confident victory of the French bourgeoisie in the late eighteenth century, citing Emmanuel Joseph Sieyès's theory of *souveraineté nationale*. In the course of the nineteenth century, the bourgeoisie had even succeeded in winning over the rural population who identified with the concept of the nation. Yet, according to Kirchheimer, not long after the victory of democratic sovereignty of the nation, the bourgeoisie in France had begun its constant struggle against this sovereignty and had begun to demand security privileges for its class. He identified this "contradiction between bourgeoisie and nation" (137) in the political theories of scholars ranging from François Guizot to Ernest Renan and noted that its formative power still persisted in contemporary France. Overall, however, he painted a positive picture of the French bourgeoisie in the nineteenth century. But, Kirchheimer alleged, this conflict had been intensified in recent years, and he expressed serious doubts about the stability of the French bourgeoisie's democratic tradition. As evidence of his concern, he referred to Maurice Hauriou's statement that the individualistic tendencies in France were becoming stronger, for which reason a national consciousness drawing clear boundaries to the external world was becoming increasingly important as a factor for integration; a view shared by Schmitt. Kirchheimer stated that such a position retracted the "democratic conception of sovereignty" (140) in favor of the propagandist establishment of a "front against the foreigner" (140). He insisted on defending the democratic conception of sovereignty against such tendencies of no longer defining the French nation with reference to the ideals of the French Revolution but, instead, through antisemitism and xenophobia, yet he refrained from making a prognosis.

Kirchheimer described the development in Germany to his French-language readers by clearly contrasting it to France. He called the history of the German bourgeoisie in the nineteenth century nothing but a disaster. The major theoretical designs of Georg

German/Foreign Scholars, New York Public Library, New York. I, A Grantees, 1933–46, Box 18, Folder 13 (Kirchheimer, Otto).

62 See Kirchheimer (1934c). The following page numbers refer to this text.

Wilhelm Friedrich Hegel, Friedrich Julius Stahl, and even Lorenz von Stein lacked the vigor that would have been necessary to finish off the concept of absolute monarchy. Concerning the legal debates during the Weimar Republic, Kirchheimer went into Hans Kelsen and Carl Schmitt in more detail. Kelsen had no problem abolishing sovereignty. Sovereignty did not exist, he claimed, but, rather, only legal mechanisms of attribution. At the end of these, basic rules were to be found that existed in international law, not in the constitutional law of individual countries. Following Heller—but not Schmitt—Kirchheimer criticized Kelsen for artificially separating the study of law from social reality and asserted that this separation disregarded the personal factor of government. Quoting Soviet legal theoretician Evgeny Pashukanis, he called Kelsen's approach "a kind of mathematics of the cultural sciences" which was, under the current political circumstances, damned to "remain pure theory forever" (147).

His sharpest words, however, were directed at Schmitt. Despite his emphasis on sovereignty, Schmitt's contribution to legal theory had failed because he "never made the effort to postulate a theory of sovereignty in the context of the Weimar Republic" (148). Kirchheimer went on to examine Schmitt's decisionism. Schmitt's move away from decisionism to "concrete-order thinking" had not yet been published, and Kirchheimer had not yet been able to read the new foreword to Schmitt's *Political Theology* of November 1933 in which he had announced this revision. Quoting Schmitt's famous first sentence of *Political Theology* from 1922, that "[s]overeign is he who decides on the exception," Kirchheimer accused Schmitt of using his theory of sovereignty to justify "the victor of the civil war" (148). Schmitt had never attempted to place his theory of sovereignty within the framework of the Weimar Constitution. This failure was "telling" (148), and Kirchheimer saw a somewhat practical reason for it: "Carl Schmitt has always represented the interests of all the powerful social and economic groups that never banked on using democratic reasons to justify their actual power, and he still does so today" (148). These groups could only be satisfied with a theory like Schmitt's that ascribed sovereignty to those actually in power to the exclusion of the traditional established democratic wording of the constitution. The decay of social order in Germany in the course of the economic crisis of the late 1920s lent this theory a semblance of justification.

Ultimately, Kirchheimer considered the changes to the "structure of capitalism" (148) to be causes for the current desire for a strong decision-making authority. Ever since Jean Bodin's day, sovereignty had been seen as the supreme legislative power. A type of legislation limited to a few general laws had been appropriate for the capitalism of the nineteenth century. The capitalism of the twentieth century, however, required the state to intervene in the economic and social realms on a daily basis. This form of regulation could not be achieved by general legislation alone but, rather, increasingly required decisions made on a case-by-case basis. In the fascist Germany of the day and in the writings of Schmitt, Forsthoff, and Koellreutter, this need on the part of the business community was met by means of general clauses. These economic interests were opposed to those of other social groups, in particular those of the "working class" (50). For this reason, their justification for sovereignty could not be a democratic one. The gap in justification was filled by the fascist theory of sovereignty with its return to its transcendent stance—except that the God-given king was now replaced by the concepts of the *Führer*, *Blut*, and *Rasse*. To illustrate his point, Kirchheimer quoted from works by Otto Koellreutter, Hans

Julius Wolf, and Carl Schmitt—and did not refrain from commenting in a footnote that Koellreutter had accused Schmitt's piece *Staat, Bewegung, Volk* of obvious "cynicism" (151).

The third subject of Kirchheimer's research during his exile in Paris was the municipal constitution and the role of municipalities within the state. Schmitt had also discussed this subject during the Weimar Republic and even in a leading role in the legislative process in the early stage of the Nazi regime. Unlike Schmitt, Kirchheimer took a comparative legal perspective on this subject, and he had democratic and socialist intentions. He began to write a paper titled "Die wirtschaftliche Betätigung der französischen Gemeinden und die Rechtsprechung des *Conseil d'État*" [The economic activity of French municipalities and the decisions of the *Conseil d'État*]. This was not published at the time. The essay survived in manuscript form—publication during Kirchheimer's lifetime has not been established to date.[63] References in the text indicate that it was completed in the spring of 1936. The original text, found among Kirchheimer's papers after his death, is the complete 12-page typescript with a few handwritten corrections he made. The fact that Kirchheimer attempted to have this essay published is evidenced by a number of letters. For example, Franz Neumann reported to him from New York in February 1937 that he had forwarded the manuscript to Felix Frankfurter as agreed.[64] In March 1937, Kirchheimer stated again that he, too, would seek an opportunity for its publication.[65]

In his manuscript, Kirchheimer discussed the role of the *Conseil d'État* as a constitutional court for deciding questions of municipal law.[66] Taking up the municipalities' economic activities, he focused on an issue he had already discussed in various contributions during the Weimar Republic: whether and to what extent municipalities should have the right to run businesses. Like many other social democrats, Kirchheimer, too, considered municipal enterprises as a way to stand up to the capitalist private economic system.

As in Germany—Schmitt had sharply criticized this during the Weimar Republic as an attack by society on the sphere of the state—municipalities' economic activity had seen a considerable upswing in France, too. Kirchheimer argued, however, that the legal basis for this development in France had remained unclear. The proponents of municipal socialism referred to a parliamentary law from 1884, whereas the advocates of

63 The text was first published in 2018 in Volume 2 of Kirchheimer's collected works. See Kirchheimer (1936a). The following page numbers refer to this text.

64 Letter from Franz L. Neumann to Otto Kirchheimer dated 9 February 1937. Otto Kirchheimer Papers, Series 2, Box 1, Folder 122.

65 Letter from Otto Kirchheimer to Franz L. Neumann dated 10 March 1937. Otto Kirchheimer Papers, Series 2, Box 1, Folder 122.

66 The French *Conseil d'État* as an institution goes back to Napoleon, who established it in December 1799 following his coup. As a council of the state, it partly also exercised functions of the cabinet, but over the course of the nineteenth century, its actions were limited entirely to the field of legal policy. Unlike the US Supreme Court, it had a dual function in this role. First, it functioned as the supreme administrative court, thereby growing into the role of a constitutional court. It also consulted the government in legislative matters, taking on the role of a justice ministry that reviewed bills presented by other ministries. These two functions of the *Conseil d'État*, which had evolved over time, were defined more precisely by an act of parliament in 1872.

privatizing public enterprises and services invoked laws dating back to 1791. In this article, Kirchheimer first explained how the *Conseil d'État* had sided with the opponents of municipalities' economic activities in its jurisprudence. Taking advantage of the competencies it had attained through the Enabling Act adopted in the course of the French financial crisis of August 1926, the government of French President Raymond Poincaré had amended the 1884 law, thereby expanding the competency of municipalities to undertake economic activities. Referring to various rulings, Kirchheimer described how the *Conseil d'État* had acted toward the new legislation from the 1920s onward and explained that it "openly oppose[d] the will of the legislature" (185). He accused the *Conseil d'État* of "upholding the principles of the individualistic economic order" (186) one-sidedly and also of maintaining a "fundamental claim to control" (188) over municipal economic activities. Kirchheimer argued that it needed to "find its way back to the French tradition of unconditionally applying legally adopted laws and decrees" (190) instead of exercising "a veiled control 'of first principles' by interpreting the text" (190).

In other words, the *Conseil d'État* should finally clear the way for the municipalities' increased economic activity in the areas of public services and municipal housing. This program proposed by Kirchheimer was precisely the opposite of the municipal constitution of the Nazi Reich prepared by Schmitt.

6. Conclusion: Distant reading

The months between February and June 1933 were crucial for both Kirchheimer and Schmitt. After Hitler took office, after the Reichstag fire, after the Enabling Act, after the Reich Governor's Law, after the Law for the Restoration of the Professional Civil Service, and even after the Law on Attorneys that excluded "non-Aryan" lawyers from the bar, Kirchheimer still stayed in Germany, hoping that Hitler's government would soon collapse. Only after his three days in prison did he decide to flee to Paris in early June.

Like Kirchheimer, Schmitt was not sure initially whether Hitler's government would stay in power. For a few weeks, he hesitated to take a clear political position in public. But when the Enabling Act of 24 March 1933 convinced him that the new dictatorship would become stable, he decided to support it. In the secondary literature about Schmitt, scholars are still puzzling over the reasons why he associated himself with the Nazi regime. Reinhard Mehring even prepared a list of forty-two potential explanations in his biography (see Mehring 2014a, 282–284). The hypothesis proposed by some of Schmitt's critics between the 1950s and 1980s that his decision to support the *Führer* state could be traced all the way back to his Weimar writings[67] has not prevailed in the research as a whole. Schmitt in turn claimed after 1945 that he had collaborated because as a professional jurist, he had had to position himself on the foundation of the new legal facts in a positivist manner after the adoption of the Enabling Act: "For me as a jurist, as a positivist, a com-

67 For the German discussion at the time, see most prominently Fijalkowski (1958) and Sontheimer (1962).

pletely new situation began, of course, [with the Enabling Act]."[68] This wording suggests that he had internally maintained his distance from the brutal regime, was forced to collaborate, and had attempted to prevent the worst. His narrative was echoed and intensified in in a number of publications, not only by right-wing German intellectuals such as Helmut Quaritsch and Günter Maschke, but also in the first two major books in English on Schmitt, by George Schwab and Joseph Bendersky.[69] Their apologetic narratives about Schmitt's involvement in the Nazi regime can still be seen in the English-language literature on Schmitt to this day.

A different picture of Schmitt's role in the consolidation phase of the Nazi regime emerges if we attempt to observe it from the perspective of an opponent of the regime who was forced into exile like Otto Kirchheimer. From that distant perspective, it becomes pointless to try to identify the specific personal motives for Schmitt's involvement in Nazi Germany, some of which can perhaps be understood only at an individual psychological level. Kirchheimer disregarded Schmitt's personal motives. He was not interested in whether or not Schmitt may have had second thoughts or his own political plans when he supported the regime, for example, trying to push events in a certain direction. Instead, Kirchheimer focused solely on a sober analysis of Schmitt's actions and functions in the new regime. It was from such a distant perspective that Kirchheimer was the first person to identify Schmitt as *the* "theorist of the Nazi Constitution" in 1933.

We can only speculate as to whether Schmitt had Kirchheimer in mind when he said that the emigrants "have been spit out of Germany for all time" or when he stated that the emigrants "never belonged to the German *Volk*" and "not to the German spirit either" (Schmitt 1933c, 32). Schmitt criticized legal positivism as well as the insistence on the validity of liberal fundamental rights from the perspective of his "*gesundes, konkretes Ordnungsdenken*" (concrete-order thinking corresponding to the norm of the NSDAP; see Translator's Preface and Glossary). It was a form of thinking in categories of suprapersonal collectivities. Schmitt asserted that the normative source of laws was what was known as lived normalcy, which took place independently of positive norms. Consequently, general clauses became the specific method of this new type of jurisprudential thinking. In Schmitt's view, this specific kind of German juridical thinking was inaccessible for legal experts who were not part of the German *Volk*. It did not make sense to start a discussion with them. They were foreign to what the Nazis called the German spirit, and they would remain in this external position forever. They were strangers who could only think in non-German juridical ways. Thus, Schmitt's mode of argumentation in dealing with those who were forced into exile can be characterized as a racism-based critique of ideology.

From his exile, Kirchheimer pursued his criticism of Schmitt in a different form of critique of ideology. He considered changes in the structure of capitalisms to be the main cause for the desire in Germany for an authoritarian political order and accused Schmitt

---

68 Schmitt in a 1971 conversation with Klaus Figge and Dieter Groh (Hertweck and Kisoudis 2010, 91).—This surprising self-description as a positivist, however, is not consistent with the fact that, in a radio interview on 1 February 1933, Schmitt again criticized legal positivism strongly (see Schmitt 1975, 114).

69 See Schwab (1970) and Bendersky (1983).

of using his theory of sovereignty to justify the victor of the civil war. Kirchheimer stated that Schmitt had already represented the interests of all the powerful social and economic groups in society during the Weimar Republic and was now doing so again under Nazi rule. Schmitt's theories of sovereignty and German legal thought as well as his rejection of the *Rechtsstaat*, constitutional courts, and municipal self-governance fulfilled the ideological function of justifying the rule of a small group of power holders in society. In his critique of Schmitt, Kirchheimer practiced a Marxist mode of critique of ideology. At the same time, he turned his academic interest to Western democracies, France and the United States in particular. Some traces of Marxist critique of ideology can be found in these studies, too, but they are overshadowed by detailed descriptions and thoughtful political analyses. In these works, Kirchheimer implemented what he had described as Schmitt's primary shortcoming in his critique authored together with Nathan Leites in late 1932: the empirical analysis of political institutions and political processes. In his articles and manuscripts written in Paris, his approach had begun to shift from pure legal and political theory to the inclusion of empirical political science.

Chapter 8: Confrontations Across Borders (1935-1937)

By early 1935, Kirchheimer and Schmitt had not had any direct personal contact for over two years. As the Nazi regime became established, their perceptions of each other were asymmetrical. While Schmitt was doing well politically, his public statements were often filled with vitriolic words about Marxists, Jews, and émigrés but he did not mention Kirchheimer even once. Kirchheimer's name no longer appeared in Schmitt's diaries from these years nor in his correspondence. There is no indication that Schmitt even took note of Kirchheimer anymore. At most, he may have mentioned Kirchheimer or heard about his fate now and then in private conversations with his Berlin colleague Rudolf Smend. Incidentally, Smend had the courage to discuss and contrast Schmitt's *Legality and Legitimacy* and Kirchheimer's eponymous essay in his seminar at Berlin University in the summer semester of 1933.[1]

Kirchheimer's everyday life in exile in Paris was miserable. It was only with great difficulty and the help of his friends that he managed to survive financially, enabling him to continue his scientific work. He had followed Schmitt's meteoric rise during his last few weeks in Germany and had then kept a close eye on him from his exile in London and Paris. He intensified his confrontation with Schmitt in the summer of 1935, choosing a new and direct tactic that would strike home personally. Using a pseudonym, he wrote a booklet for the resistance in Germany in which he played cat and mouse with Schmitt. The latter was infuriated and assumed that Kirchheimer was the author. He demanded that the German authorities crack down on the printers in Amsterdam and those disseminating the booklet. Nonetheless, German resistance organizations succeeded in circulating several thousand copies. The search for the authors of the booklet, instigated by Schmitt, was unsuccessful and petered out. Paris was still a safe place for Kirchheimer. Nevertheless, expecting that war was imminent, he was determined to try to leave France for the US in late 1936.

That year, Schmitt, in turn, was taken by surprise, experiencing a reality that officially did not exist in his own propagandist publications and speeches: instead of the political

1 See Reinhard Mehring's editorial comment in Schmitt and Smend (2010, 90).

unity of the entire *Volk*, guaranteed by the *Führer* and his party, he experienced firsthand what it was like to get caught up in the machinery of the competing power groups struggling for political influence within the system. He lost his most important party positions in the Nazi legal system's bureaucratic hierarchy. If we seek to understand these events, which Schmitt felt to be humiliating and at times even threatening, then the analytical approach presented by Kirchheimer helps to explain and frame the temporary demise of Nazi crown jurist Schmitt.

1. Kirchheimer camouflaged as Schmitt

Kirchheimer landed his coup against Schmitt in the summer of 1935 when he authored a 32-page pamphlet titled *Staatsgefüge und Recht des dritten Reiches* [State structure and law in the Third Reich] (see Kirchheimer 1935a). This was published under the pseudonym Dr. Hermann Seitz. The booklet was an indictment of the Nazi regime and its legal policies, and Kirchheimer highlighted Schmitt's position as the leading theorist for the legitimation of Nazi legal policy. And, at the same time, he countered the Nazis' invoking of the alleged unity of the German *Volk* by listing their victims and describing the regime's brutal repression measures.

The booklet was distributed illegally in Germany. Kirchheimer did not want his name on the cover for two reasons. For one thing, giving his name might have meant additional danger for his siblings, who were still living in Germany.[2] For another, it would have made it even more difficult to disseminate it in Germany. To boost distribution, the details of the booklet's cover design, color, and typesetting were intended to make it look like part of the series *Der deutsche Staat der Gegenwart* [The German state of today]; even the logo of the publisher, the Hanseatische Verlagsanstalt, seems deceptively genuine. The cover stated that it was number 12 in the series, "edited by Carl Schmitt, Hamburg 1935."[3] The title Kirchheimer chose, *Staatsgefüge und Recht des dritten Reiches*, also alluded to Schmitt, namely to the main title of his epitaph to the Weimar Republic, *Staatsgefüge und Zusammenbruch des zweiten Reiches. Der Sieg des Bürgers über den Soldaten* [The structure of the state and the collapse of the Second Reich. The victory of the citizen over the soldier] (see Schmitt 1934f), which had been published as number 6 in the same series.

Schmitt had begun publishing the series in the summer of 1933. It was one of his favorite political journalistic projects from 1933 to 1936, and he sought to use it to distinguish himself as the leading constitutional scholar of the Nazi regime.[4] Schmitt put considerable effort into the series and was concerned primarily with its political impact. In the militaristic language of the Nazi regime, he saw it as a "shock troop" for a new jurisprudence aimed at legitimizing the regime.[5] The external form of the series was

2 John H. Herz had published his critique of the Nazi regime, which he had written in exile, under a pseudonym for the same reason, see Puglierin (2011, 79).

3 The title of the actual number 12 in the series, which was published in 1935, was *Bericht über die Lage des Studiums des öffentlichen Rechts* [Report about degree programs in public law].

4 On the history of this series and Schmitt's leading role as editor, see Lokatis (1992, 52–59).

5 Letter from Carl Schmitt to Ernst Rudolf Huber dated 28 October 1933, quoted in Mehring (2014a, 308).

also intended to make a political statement: Schmitt prevailed in having the publishing house use the same typeface as in Ernst Jünger's book *Der Arbeiter* [The worker],[6] which it had published in autumn 1932. The series began with a pamphlet by Schmitt with the programmatic title *Staat, Bewegung, Volk – Die Dreigliederung der politischen Einheit* [State, movement, *Volk*—the tripartite structure of political unity] (Schmitt 1933d). Prominent Nazi legal theorists such as Ernst Rudolf Huber, Reinhard Höhn, and Friedrich Schaffstein also published in the series. The booklets had tweny-four to fifty-eight pages and sold for only 1 to 1.50 reichsmarks. Their print run of several thousand copies was relatively high for academic legal works. They were distributed through bookstores and libraries, and large numbers were sent to various institutions and organizations in the Reich and to private law firms.

The camouflaged booklet was written by Kirchheimer in Paris in June and July 1935, and then printed in Amsterdam. The 11th International Congress for Criminal Law and Corrections, which took place from 18 to 24 August in Berlin, was the special occasion for distributing it in Germany and internationally. The decision to stage the congress in Berlin had already been made prior to 1933, and the Nazi government wanted to use this event to take the wind out of its critics' sails and generate a positive image of itself on the international stage—similarly to the much more important Olympic Games the following year. In advance of the congress, the SPD party leaders had sent all congress attendees a position paper from their exile in Prague focusing on the conditions in the Nazi regime's prisons (see Denkschrift 1935). Erwin Bumke, President of the *Reichsgericht* (see List of German Courts), was the chairman of the congress, which was complemented by an extensive program, including visits to German penal institutions and the concentration camp Esterwegen in the Ems moors near Oldenburg, which had been spruced up for this propaganda event (see Müller 1987, 96). Much of Kirchheimer's booklet was written in a decidedly factual, descriptive tone. The booklet is nothing less than a masterpiece of ingenious anti-Nazi propaganda because unsuspecting readers had to read multiple pages before realizing that it was a subversive act and an indictment of the Nazi regime.

Kirchheimer explicitly referred to Schmitt right in the first paragraph of the booklet,[7] immediately after mentioning other prominent Nazis such as Wilhelm Frick, Hans Frank, and Roland Freisler. Schmitt is described as the theorist for their legal policy: "Theoretical clarity concerning how we are to understand the National Socialist version of the rule of law, the so-called 'German *Rechtsstaat* of Adolf Hitler' can be gained in particular from the writings of a member of the state council, Professor Carl Schmitt" (142). This was an allusion to the argument among Nazi jurists in the regime's early phase about the question whether it was still appropriate to call the regime a *Rechtsstaat* (see Pauer-Studer 2014, 61–67). Schmitt had commented on this argument shortly earlier in a lecture at a conference of the Bund Nationalsozialistischer Deutscher Juristen (BNSDJ, Association of National Socialist German Legal Professionals) and had proposed drawing a distinction between a (liberal) *Rechtsstaat* and a (National Socialist) *Gerechtigkeitsstaat* (state based on a certain idea of what is just) (see Schmitt 1934f). In a later, longer version

6 Letter from Benno Ziegler (Hanseatische Verlagsanstalt) to Carl Schmitt dated 28 October 1933. Carl Schmitt Papers, RW 472.

7 See Kirchheimer (1935a). The following page numbers refer to the English translation of the text.

of this lecture, Schmitt had used the formula "German *Rechtsstaat* of Adolf Hitler" (see Schmitt 1935b, 112), which Kirchheimer quoted without indicating the source.

Only at a few points did Kirchheimer note continuities between the development of the law during the Weimar Republic and the Third Reich—for example, the increasing priority of the concept of deterrence over that of betterment as a maxim in criminal jurisprudence and the business community's ambitions for labor law. The dominant overall impression was one of a radical break. Kirchheimer showed that in the way in which the Nazis continued to use the term "*Rechtsstaat*," nothing remained of the liberal pride in the fact that the existing legal order was available to every person irrespective of the individual concerned. The "ethical minimum" (144) of the *Rechtsstaat*, as Kirchheimer wrote, following well-known words carefully chosen by legal theorist Georg Jellinek (see Jellinek 1908, 45), had been dropped in favor of a new law serving exclusively the interests of the groups that had succeeded in seizing power in the state. The new legal theories now put forward by Schmitt and others in Germany were intended to conceal this social fact with the help of an "appropriate timely ideology" (143).

Kirchheimer identified the "transition from competitive to monopoly capitalism" (143) as the underlying cause of this process. In doing so, he foreshadowed an idea that Franz L. Neumann formulated two years later in his famous article "The Change in the Function of Law in Modern Society" (see Neumann 1937, 42). The socioeconomic transformation triggered by the capitalist transition had increasingly caused society's desire for liberal forms of the law to vanish. Security of contract guaranteed by the rule of law was obsolete under monopoly capitalism; corporate monopolies and major banks had increasingly opposed the liberal state under the rule of law because it limited their interests.

By placing the blame for the economic crisis and unemployment after 1929 on the Weimar system of law, the Nazis had deluded the poorer strata of society into believing that a completely new form of the law would improve their lives. Kirchheimer considered the expansion of general clauses to be the foundation of the legal structure of the Nazi state. General clauses made it possible to construe a new version of the concept of judicial independence. If judicial independence had previously meant judges' freedom to come to a verdict while being bound to the law, that commitment to the law no longer existed. One characteristic of the new judicial independence was that laws themselves could be changed without any formal procedures and even repealed with retroactive effect by the *Führer*. Moreover, every judicial interpretation of the law was under the caveat of the general clause "compatibility with the National Socialist worldview," (144) which opened the floodgates to arbitrary decision-making. Kirchheimer quoted verbatim from Schmitt's "Neue Leitsätze für die Rechtspraxis" (see Schmitt 1933h) to illustrate this point.

He described the Nazi system as the order of a new social compromise between multiple social groups that was conducive to the development toward monopoly capitalism at the time. He believed these dominant groups were those with industrial and financial capital, Junkers, the Reichswehr, the state bureaucracy, and the NSDAP. A "system of reciprocal guarantees and obligations" (157) had emerged between them and the Nazi state leadership. Kirchheimer traced the legal structure of the Nazi regime in multiple areas, devoting the most space to an analysis of the developments in criminal law because Nazi lines of reasoning had been established particularly quickly and extensively there. He drew on Nazi literature to provide a detailed overview of the repeal of the ban on ex post facto criminal laws, the expansion of the field of criminal law through broader interpretations, the introduction of *Willensstrafrecht* and *Täterstrafrecht*,[8] the meaning of the formula "*gesundes Volksempfinden*" (assessment of a matter in accordance with the Nazi *Volksgemeinschaft* governed by the will of the *Führer*"; see Glossary), the limitless expansion of what was defined as political crimes, the changes in the Code of Criminal Procedure at the expense of defendants, the abolition of the independence of the judiciary, the introduction of *Sondergerichte* (special courts for political and especially serious crimes, feared for their swift and severe rulings that could not be appealed), the actions of the *Geheime Staatspolizei* (Gestapo), and the "apparently sadistic" (155) toughening of detention conditions for political opponents.

In multiple places, Kirchheimer explicitly referred back to questions circulating in advance of the 11th International Congress for Criminal Law and Corrections in Berlin, concerning the purpose of punishment, the predictability of legislation, the potential curtailment of show trials, and the ban on ex post facto laws. He highlighted capital punishment, which the Nazi regime had been applying forcefully since March 1933, as the most serious violation of general legal sensibilities. For example, it was only on the basis of ex post facto application of increasingly harsh punishments that it had been possible to convict and execute Marinus van der Lubbe, whom the Nazis had accused of setting the Reichstag fire. Kirchheimer listed the names of several people who had suffered the same fate. It was only by drawing on such "murderous constructs" (153) that the regime was able to have its political opponents executed.

The conversion of criminal law into a party political instrument of combat occurred so fast because this was the only way that the groups dominating the state could fight off their opposition and retain power. Yet criminal legislation and the practice of sentencing were subject to the most serious challenges "not only from a liberal or humanitarian viewpoint" (152). Their usefulness was limited even for the ruling elite. This was because their function was purely repressive and they did not positively sustain the system; thus, they did not create political stability in the long run. Kirchheimer also traced the development

8 *Willensstrafrecht* (will-based criminal law) included assumptions about the defendant's will (which was considered reprehensible in most cases) in the assessment of the crime as well as the sentencing decision. Of course, this involved wild speculation and made it possible to punish comparatively minor offenses severely since they had been construed as motivated by the offender's evil will; this was assumed as a matter of principle in the case of communists and Jews. The purpose of *Täterstrafrecht* (offender-based criminal law) was to "identify" a particular type of offender and to punish offenders according to their category. A few examples of offender types were "antisocial types," "notorious criminals," and "fraudulent fellows." According to Nazi doctrine these traits were innate, which meant that resocialization was impossible, leaving punishment "with utmost severity" as the only option. This, too, opened the floodgates for arbitrary decisions by judges and prosecutors. The roots of both theories of criminal law date back to the early nineteenth century, and both cast long shadows on the Federal Republic. For instance, the definitions of "murder" ("out of base motives ..." with no mention of premeditation) still in effect today is based on Nazi legislation, and the wording of the relevant section of the German Criminal Code ("A murderer ... is someone who kills a person ...") relates to the offender, not the crime, unlike in other sections of the code. (See Translator's Preface.)

of constitutional, administrative, and labor law, as well as that of the law of inheritance inasmuch as it concerned bequeathing farms.

Kirchheimer wrote that although Nazi constitutional law declared the unrestrained sovereignty of the *Führer*, the social reality was that various interests, influences, and tendencies collided in the figure of the *Führer*. This social reality also meant that this *Führer* was first and foremost the leader of a civil war party that had managed to take possession of the state apparatus. The rule of the civil war party depended on compromise between various social groups. The regime attempted to accommodate this compromise in various ways. The state bureaucracy was considered to result from the spoils of the civil war and was filled with tens of thousands of members of the NSDAP. However, Hitler in his role as *Führer* of the party could only maintain his devotees occupying prestigious posts and official positions by associating himself with the social groups in power.

Kirchheimer illustrated these interpretations with examples from property law, labor law, and municipal administrative law. In the area of labor law, he demonstrated in more detail how it could be enforced using criminal law if necessary, using the examples of union busting, the powerful position of corporate executives, the mechanisms of determining wages, and the legislation on occupational safety and dismissal. Kirchheimer detected growing discontent with these conditions in industrial companies. He believed he could discern another potential weakness of the regime in the *Reichserbhofgesetz* (Reich Hereditary Farm Law) which was intended to maintain small farms; in fact, its effect was the opposite, and Kirchheimer anticipated this would result in increasing discontent.

As mentioned in Chapter 7, Schmitt had had decisive responsibility for the new Prussian municipal law in 1933. In particular, Kirchheimer emphasized the changes that had taken place regarding home rule after 1933, ascribing Prussian municipal law a pioneering role for the entire Reich. All the top municipal positions had come under the power of the Nazi apparatus as a result of the reorganization following the new law. Moreover, the Nazi state prohibited municipalities from engaging in any activities that could be seen as competing with the private sector. At a time when political debates about taking over the major utilities had erupted, Nazism had handed these companies over to private capital at the expense of the general public.

Kirchheimer concluded his brochure by referring to the otherwise unexposed social instability of the German Reich, not least attempting to show where resistance against the regime was forming. The Reich could retain its current political stability only at the price of a previously unimaginable reign of judicial terror. This analysis of Kirchheimer's was supported by his conviction, which had already become apparent in his writings during the Weimar period, that the social function of Nazism ultimately consisted in preventing the latent class conflict between capital and labor from erupting. For this reason, Kirchheimer saw Nazism as being, among other things, a precarious system of government and, consequently, temporary. He believed the task of forging the law after the end of the Nazi regime would be to put an end to its campaign of annihilation in all areas of the law and to prepare the "groundwork for the legal system of a socialist Germany" (166). At one point in his booklet, Kirchheimer changed his tone and took direct aim at Schmitt and other Nazi jurists: "the jurists of the Third Reich—theoreticians and practitioners alike—will have to take responsibility someday"[9] for their support of the murderous practices in the German Reich.

Leftist émigrés produced and distributed Kirchheimer's booklet as a targeted act of resistance against the Nazis. They had already used false titles to camouflage other publications. For example, the booklet titled *Die Kunst des Selbstrasierens. Neue Wege männlicher Kosmetik* [The art of shaving oneself. New horizons in male cosmetics] was actually the text of the *Prague Manifesto* of the SPD party leaders, who were active from exile, and the *Braunbuch über Reichstagsbrand und Hitlerterror* [Brown book on the Reichstag fire and Hitler's terror] was camouflaged as a booklet published by Reclam publishing house with the title *Goethe: Hermann und Dorothea*. Kirchheimer's forceful critique was distributed mainly through the illegal channels of communist media entrepreneur Willi Münzenberg, who had also fled to Paris.[10] From his exile, Münzenberg worked for the KPD and organized action groups, propaganda materials, conferences, and press services for the anti-Nazi resistance abroad. Kirchheimer's first wife Hilde had already served as a courier transporting anti-fascist propaganda materials to and within Germany (see Ladwig-Winters 2009, 404), which was extremely dangerous, and she was also involved in illegally disseminating Otto Kirchheimer's booklet. Eugene Anschel wrote the following in his memoir about visiting Otto Kirchheimer in his Paris exile: "He also worked on his own, which included the writing of the brochure in which he attacked his old teacher, Carl Schmitt. When I was in Paris, Hilde had already shipped it illegally to Germany" (Anschel 1990, 127).

Carl Schmitt was furious. His distancing *Mitteilung* (note) in the next issue of the *Deutsche Juristen-Zeitung* (DJZ) in the section *Juristische Rundschau* [Legal review] on 15 September 1935 attests to the fact that he felt personally attacked as the addressee of Kirchheimer's piece. Although the *Mitteilung* did not bear Schmitt's name, he had editorial responsibility for these notes to the readership since he had been managing editor of the DJZ since June 1934. The *Mitteilung* stated[11] that the author of the booklet belonged to an "international clique which, according to the *Führer*, is keen on discord in the world."

The complete text of the *Mitteilung* reads as follows:

> Actualizing and securing a true order of peace founded upon the honor of the peoples and the reciprocal recognition of their right to exist is one of the unalterable and inalienable goals of the new Germany. Many attempts are made to reach this goal, and it would be an error based on an outdated view to consider only state actions to be usual or effective means. Above all, what is needed is true intellectual collaboration of

9 My translation (HB). The English version by William Scheuerman and Anke Großkopf does not mention theoreticians and practitioners (such as Schmitt) (page 147). The original German text reads: "*Die Juristen des dritten Reiches – Theoretiker wie Praktiker – werden sich einmal verantworten müssen*" (page 158).

10 It cannot be clarified to what extent Gurland's presumption was accurate that a few passages of Kirchheimer's text had been changed by Münzenberg or his staff; Jürgen Seifert told Volker Neumann about this presumption of Gurland's (see Neumann 2015, 392 and e-mail from Volker Neumann to the author dated 23 February 2017).

11 *Mitteilung*, in: *Deutsche Juristen-Zeitung* of 15 September 1935, vol. 40 (1935), issue 18, columns 1104–1105.

> the peoples based on unlimited reciprocal recognition and not only on an externally organized *coopération intellectuelle*. For this reason, academic work based on responsibility as well as meetings and the exchange of views with scholars from abroad are of particularly high value in order to achieve understanding of National Socialist Germany abroad. Whereas there is no lack of such serious efforts on the German side and encouraging successes can already be ascertained, an international clique which, according to the *Führer*, is keen on discord in the world is attempting time and again to thwart the real understanding between the peoples which is increasingly in the making. They apparently stop at nothing, not even shying away from criminal acts. A particularly drastic example of these attempts to interfere is, in recent days, an inflammatory piece apparently printed in France and Holland which attempts to apply seemingly 'scholarly' objectivity to expose the National Socialist legal structure as the expression and the means of brutal domination on the basis of force and class. It is not worth going into the substance of this piece in more detail; it soon bores the reader as it helplessly turns about in the quandary of presenting both communist/Marxist and liberal/bourgeois/rule-of-law arguments against the National Socialist legal structure. The moral level of this enterprise is just as low as its intellectual level, as this 'exposé' itself resorts to camouflage, which is an outright forgery and a criminal abuse of the generally recognized rights of editors, printers, and publishing houses. In its external design, it presents itself as 'issue 12' of the series *Der deutsche Staat der Gegenwart* (Hanseatische Verlagsanstalt, Hamburg), which is published by the editor of the DJZ, and it bears the promising title *Staatsgefüge und Recht des dritten Reiches*. The author is a completely unknown Hermann Seitz. Even concerning oneself with this instrument of incitement of the people would surely grant it too much honor. But it does seem necessary to guide international attention once again to the criminal methods with which the international front of criminals is trying to thwart the intellectual collaboration of the peoples and thus the actualization of a true peace.[12]

In his *Mitteilung* in the DJZ, Schmitt had been unable to resist taking a swipe at the author of the booklet, namely that it "soon bores the reader." He was certain that the author was Kirchheimer. In a 1958 letter to Arvid Brødersen,[13] then his colleague at the New School for Social Research in New York, Kirchheimer confirmed in retrospect that Schmitt had recognized at once that he was the author behind the pseudonym: "I enclose a perhaps instructive little piece I wrote under a pseudonym in 1935 and that was smuggled into Germany. C.S., who, as he later told me, knew that I was probably the author, responded as can be seen from the blurb."[14] George Schwab, who became a late mouthpiece of Schmitt's, so to speak, claimed Kirchheimer's coup was part of "endeavors to endanger Schmitt's life in Nazi Germany" (Schwab 1990, 81).

---

12 *Mitteilung*, in: *Deutsche Juristen-Zeitung* of 15 September 1935, vol. 40 (1935), issue 18, columns 1104–1105.

13 Brødersen had studied sociology in Berlin in the early 1930s and knew Kirchheimer from this time. He later belonged to the Norwegian resistance against Germany until he succeeded in fleeing to the US, where he was given a chair of sociology at the New School for Social Research.

14 Letter from Otto Kirchheimer to Arvid Brødersen dated 2 March 1958. Otto Kirchheimer Papers, Series 2, Box 1, Folder 25. —The "blurb" is the *Mitteilung* in the DJZ quoted above and "as he later told me" refers to Kirchheimer's visit to Schmitt's home in 1949 (see Chapter 15).

There is no evidence at all to support Schwab's claim. However, the fact that Schmitt called on the Hanseatische Verlagsanstalt to take legal action in the Netherlands for a criminal copyright violation[15] shows how angry he was that his politically ambitious series had been used as cover for criticism of the Nazi regime. But the publishing house decided to respond in a different way, trusting that the affair made for good advertising. In September 1935, it took out an advertisement in the *Börsenblatt des deutschen Buchhandels* [Gazette of the German book trade] with the following heading in large lettering: "Forgery of our series from abroad!"[16] The ad read:

> In agreement with the editor, Staatsrat Prof. Dr. Carl Schmitt, we announce the following to the German book trade: A booklet (communist agitation against the renewal of German law) with the title *Staatsgefüge und Recht im dritten Reich* by Hermann Seitz has been disseminated to a large number of German jurists since mid-August; its appearance completely matches the series we publish [...].

The publishing house also pointed out that a criminal investigation had already been started: "The *Geheime Staatspolizei* (Gestapo) is pursuing the state-police side of the matter." The ad ended with a call for denunciation in which the publishing house made itself a stooge of the Gestapo: "We ask the German book trade to transmit any observations and information which could serve to shed light on the matter and request that any potentially surfacing copies of the piece be handed over to the nearest state police station immediately."

The Amsterdam printing shop evidently produced further editions of the booklet in late 1935 because a typewritten page is pasted into some extant copies explicitly indicating Carl Schmitt as the author of the *Mitteilung* in the DJZ. This added text is quasi a riposte to Schmitt's *Mitteilung* in the DJZ:

> This analysis of the development of fascist law, thousands of copies of which have been sent to Germany, seeks to help German jurists recognize the brutal, cynical reality of the new legal system behind the fog of Nazi phrases. Thus, this piece is a sign of foreign jurists' solidarity with their colleagues in the Third Reich, who have been condemned to silence. It is at the same time a weapon against the barbarous Hitler regime.[17]

The language and style of this additional text make it appear unlikely that it was written by Kirchheimer; it was presumably authored by someone working for Münzenberg's organization. It is no longer possible to ascertain whether the large number of copies of the booklet distributed in Germany were part of Münzenberg's propaganda or whether the

15 Letter from Carl Schmitt to Hanseatische Verlagsanstalt dated 6 September 1935. Carl Schmitt Papers, RW 472.

16 This and the following quotes are from the advertisement of the Hanseatische Verlagsanstalt in the *Börsenblatt für den deutschen Buchhandel* No. 207 of 6 September 1935. The full-page advertisement is reprinted in Lokatis (1992, 58).

17 Quoted in Luthardt (1976, 36), who had received one of these copies from Anne Kirchheimer, Otto Kirchheimer's second wife.

number reaching the German Reich and distributed illegally there was that high.[18] It may have been a few thousand. After all, the activists from the resistance had the bar associations' address lists (so they were presumably lawyers themselves), which they used in order to send the booklets to their members, without indicating the sender's real name. The fact that Benno Ziegler of the Hanseatische Verlagsanstalt, which was based in Hamburg, informed Schmitt in a letter that "the forgery was sent to almost all the lawyers in Hamburg"[19] also supports the theory that a large number, perhaps a few thousand, were distributed across the Reich. A year after the smuggled booklet had circulated, Schmitt complained bitterly in the inaugural lecture he gave at the conference he headed on "Jewry in the Legal Sciences" about activities directed against him from people in exile: "I know from my own experience how strongly hatred drives Jewish émigrés and their allies when they seek to destroy the scientific honor and good name of anyone [who]" (Schmitt 1936c, 484), like himself, supported the cause of the *Führer* and the *Volk*.

2. Sidelining Schmitt

Only about eighteen months after Carl Schmitt wanted the Gestapo to go after the author of the booklet and the Amsterdam printing shop, he himself was confronted with the situation of being the object of harassment by the Nazi regime's apparatus. Schmitt had angered many NSDAP jurists because of how brazenly he had engaged in his pioneering role as theorist of the Reich. The fact that he lectured others about the proper interpretation of Nazi doctrine only shortly after joining the party was perceived as presumptuous by the "old guard." They believed this "late starter" was simply being opportunistic. At the same time, as Hermann Göring and Hans Frank's new favorite jurist, Schmitt was also met with considerable envy since he was able to rise so easily to outstanding positions in the Nazi bureaucracy's hierarchy.

His competitors' envy and the party veterans' political mistrust became factors inflicting a major blow on Schmitt's career in late 1936. The three major protagonists of his takedown were Otto Koellreutter, Nazi legal historian Karl August Eckhardt, and *Reichsführer* of the SS and director of the *Reichssicherheitshauptamt* (Reich Security Main Office) Reinhard Höhn. The latter had been a colleague of Schmitt's at Berlin University from the winter semester 1935/36 on; Smend had been forced to give up his Berlin chair and switch to Göttingen to make the position available to Höhn. Incidentally, Koellreutter, Höhn, and Eckhardt were all imperiled themselves at some point during the Nazi period. All three felt threatened as a result of their political and professional ambitions, and they all lost their political positions at various points in time.[20] The machinations of Schmitt's colleagues to sideline him have been analyzed in detail multiple times in the

18 In the mid-1970s, Wolfgang Luthardt had asked several of Kirchheimer's contemporaries who knew him well about this matter but was unable to obtain any reliable information; see Luthardt (1976, 35–37).

19 Letter from Hanseatische Verlagsanstalt to Carl Schmitt dated 22 August 1935. Reprinted in Schmitt (2013, 77).

20 See Rüthers (1990, 89–92) and Neumann (2015, 398–406).

biographical literature on Schmitt,[21] so I will simply mention the most important events here. A key role was played by the SS and its security apparatus, which in the early years of the regime considered themselves to be custodians of the pure doctrine of the Nazi worldview, so to speak.

Schmitt's career setback was triggered from a completely different direction, namely from exile. The source was Waldemar Gurian, a publicist who had converted from Judaism to Catholicism and had had a close and friendly relationship with Schmitt and his wife in Bonn back in the 1920s. Gurian had had to flee with his family to avoid arrest in July 1934. He was an outspoken opponent of Nazism and after 1933 a sharp critic of the German Catholic Church as well as of Carl Schmitt.[22] He was the editor of the Catholic émigré journal *Deutsche Briefe* [German letters] in Switzerland from 1934 to 1938. Although it only had around 200 subscribers and its print run was not much larger, it was used as source material by several Swiss newspapers, and it was also monitored closely by the German security agencies (see Hürten 1972, 96–127). Gurian sharply attacked Schmitt, whose work he followed at every turn from 1934 to 1936, in multiple anonymous contributions to the *Deutsche Briefe*. In his articles, Gurian proved to be extremely well informed about Schmitt's various changes of course in legal theory. And it was Gurian who in one of his first articles in the *Deutsche Briefe* in 1934 had coined the term "crown jurist of the Third Reich,"[23] soon to be used polemically by Schmitt's opponents and competitors in the Reich. In particular, the passages of Gurian's articles showing where Schmitt had touched up parts of his books for new editions and describing his connections to Jewish friends and political Catholicism were potentially explosive for Schmitt. Gurian asserted that the latter was not a staunch Nazi but an opportunist worthy of contempt who was merely seeking to exploit the regime for his own purposes. He ridiculed the NSDAP for not seeing through Schmitt's opportunism.

Gurian's writings were grist for the mill of Schmitt's opponents in the Nazi system. In the late summer of 1936, Höhn, who had published a booklet in Schmitt's series the previous year, had begun keeping a file on Schmitt in the *Zentralabteilung* (Central Division) II/2 Department 22 of the SS-*Sicherheitsdienst* (SD, the intelligence service of the Third Reich), of which he was director. Schmitt's file ran to almost 300 pages. It included reports summarizing Schmitt's biography as well as lists of his works, which were presumably prepared by Höhn's staff at Berlin University. It also contained reports by informers, some of whom were Schmitt's assistants at the university. As is always the case with such intelligence files, some of the information collected is vague or even incorrect. This is also true with respect to Otto Kirchheimer, who was mentioned twice in the file as an incriminating factor from Schmitt's Weimar days. In one place he was incorrectly referred to as a socialist Jew who had praised Schmitt's *Legality and Legitimacy* in a review.[24] And in another document Kirchheimer was even mentioned as a "most fa-

21 See Bendersky (1983, 219–242), Koenen (1995, 651–764), Blasius (2001, 170–180), Mehring (2014a, 336–348), and Mehring (2022, 384–386).

22 On Gurian's biography, see Hürten (1972).

23 *Deutsche Briefe*, 26 October 1934.

24 Bundesarchiv Berlin-Lichterfelde, Akten des Sicherheitsdienstes des Reichsführers SS. Aktennummer R. 58/854, 76. The essay, co-authored with Nathan Leites (Leites is not mentioned in the SD

vored student"[25] of Schmitt's. Some of the articles written by Gurian accusing Schmitt of political opportunism also found their way into the file. However, the SD files and the accusations and assessments assembled there indicate that the purpose was not to persecute Schmitt politically as an alleged opponent of the regime, but only to reduce the large number of functions he had relating to constitutional law, which made him too powerful, and to diminish his public reputation.[26]

Two attacks on Schmitt were published on 3 and 10 December 1936 in *Das schwarze Korps* [The black corps], the SS weekly with a circulation of just under 340,000, on the basis of material collected by the SD (see Koenen 1995, 726–733). He was accused of political hypocrisy: he had not joined the NSDAP until 1933, and then only to further his career. There were references to Schmitt's connections to political Catholicism prior to 1933. Moreover, the antisemitism he expressed lacked credibility because he had had some Jewish friends before 1933 and some of his students had been Jews. There were quotations in the second article from Schmitt's earlier writings in which he had rejected ideology based on *Rasse* as thinking of the Romantic period. The material used against Schmitt in the two articles was a combination of accusations made by the professors he worked with and Gurian's writings from his exile in Switzerland.

Schmitt had become ensnared in the machinery of the Nazi system. His mentor Hermann Göring intervened with the SS, countering the attacks and shielding him. But Göring could not prevent Schmitt's demotion. Hans Frank, in contrast, dropped Schmitt once he had realized that he could not accomplish anything against the will of the SS. A few days after the accusations, which generated considerable publicity, Schmitt had to cede his leadership position as *Reichsgruppenwart* of the *Reichsgruppe Hochschullehrer* in the NSRB, falsely citing "health reasons."[27] He also lost editorship of the *Deutsche Juristen-Zeitung* against his will; the journal was simply discontinued and replaced with a new journal. He did not lose his membership in the *Akademie für Deutsches Recht*, which was politically less important, or his position as *Preußischer Staatsrat*. However, his nemesis Reinhard Höhn was unable to realize his intention of stripping him of his prestigious chair at Berlin University. Höhn and his comrade in arms Eckhardt were also to become entangled in the intrigues within the apparatus a year later because of disrespectful statements about the Nazis back in the 1920s and contact with Jews. To Schmitt, demotion after four years of intense activities in the service of the Third Reich was a major rupture in his biography. Once a celebrated and feared figure in the Nazi legal hierarchy, he had become "almost an outcast" (Rüthers 1990, 106) virtually overnight. From then on, he again focused more on working as a university professor and traveling to give lectures. He also avoided writing about any topic in constitutional law related to the internal power struggles within the regime.

---

file), was obviously confused with Kirchheimer's previous article with the same title as Schmitt's as well as with a positive review.

25 Bundesarchiv Berlin-Lichterfelde, Akten des Sicherheitsdienstes des Reichsführers SS. Aktennummer R. 58/854, 149.

26 See Koenen (1995, 716–723) and Neumann (2015, 408–414).

27 Mitteilung des NS-Rechtswahrerbundes 1936, quoted in Rüthers (1990, 105).

It is impossible to determine now, in retrospect, how relevant it was for these events that Gurian and Kirchheimer knew each other. Gurian and Kirchheimer had met through Schmitt in Bonn in 1927; the three had occasionally spent evenings together over beer and wine.[28] John H. Herz reported that Kirchheimer had met Gurian when he traveled to Switzerland after 1933 to see his family, especially his daughter, at boarding school in northern Italy. According to Herz, they had even planned to publish together about Schmitt.[29] However, there is no evidence in the archives to confirm this or any speculation about the extent to which Kirchheimer provided Gurian with additional information or material about Schmitt. From his Swiss exile, Gurian reported with satisfaction on what was happening in Berlin to his small readership of the *Deutsche Hefte*, using Friedrich Schiller's words: "The Moor has done his work, the Moor may go."[30] He commented on Schmitt's plummet in the hierarchy using the following words: "Carl Schmitt has become a superfluous assistant today, and he has been dropped."[31] On Christmas Eve, his headline referring to Schmitt was: "On the way to emigration or to the concentration camp?"[32] Yet Schmitt was neither threatened with the fate of a member of the opposition in a concentration camp nor was emigration an option he was seriously considering. Gurian was apparently referring to Göring's personal interventions on behalf of his protégé when he wrote six months later that Schmitt "has to thank only certain personal relationships for the fact that he is not completely done for."[33] Essentially, in 1936, Schmitt was confronted with a facet of the brutal reality of Nazi Germany that simply did not exist in his countless lectures and writings in which he exalted the unity of *Führer*, *Volk*, and party.

3. Kirchheimer's political activities in Paris and his arrival in New York

In the Paris resistance circles, the booklet written by Kirchheimer and circulated illegally under Carl Schmitt's name had earned Kirchheimer his reputation as an expert in questions of Nazi criminal law. This expertise was the basis for him to be able to leave Paris for the US and take up his first position there, as he had long sought to do.

Despite all his efforts, he was unable to significantly improve his situation in Paris. The everyday life he had to deal with in his French exile frayed his nerves and sapped his strength. A brief improvement in the German refugees' legal situation which many émigrés remembered as "the summer of our hopes" (Sperber 1982, 112) came about in 1936 after the *Front populaire* had won the election. Shortly after Léon Blum's Popular Front government had taken office, France was one of the first countries to ratify the Convention on Status of Refugees. Temporary passports were introduced as a result, and Otto Kirchheimer was issued one. However, this did not include a work permit. Things were

28 See diary entries, 7 April and 12 September 1927 (Schmitt 2018, 133 and 162).

29 John H. Herz at the Symposium on the Work of Otto Kirchheimer at the Freie Universität Berlin in West Berlin in 1985 in response to a question asked by Wilhelm Hennis.

30 *Deutsche Briefe* of 11 December 1936.

31 *Deutsche Briefe* of 18 December 1936.

32 *Deutsche Briefe* of 24 December 1936.

33 *Deutsche Briefe* of 9 July 1937

made even more difficult for émigrés in academic professions because their degrees were not recognized in France. Taking stock in retrospect, and considering her own personal experience, Ruth Fabian, a lawyer and leftist Social Democrat who had known Kirchheimer well during their time in Berlin and who had also fled to Paris in 1934, said that "until the end of the war, [there had been] no willingness on the French side to integrate people and no opportunity on the side of the refugees to integrate" (Fabian 1981, 202). As a rule, France's academic establishment did not welcome German émigrés, either, and remained inaccessible.

In his memoirs of these years, sociologist Paul Honigsheim recounted that most German academics had had to reckon with a cold reception. There had been only a few French scholars who had spoken up for their colleagues from German émigré circles (see Honigsheim 1960, 313–314). Among these exceptions were the École Normale Supérieure and the Sorbonne, which provided support to the ISR. German émigrés also reported similar supportive behavior by some scholars in England and Switzerland. In his memoirs, Eugene Anschel sketched the following image of the situation in which Kirchheimer and his wife found themselves in the apartment they shared, despite having separated, in Paris in 1936: "Separated from Hilde [...] and their child [...] he was a lonely figure" (Anschel 1990, 127).

In contrast, Kirchheimer's wife was practically bubbling over with political activism in Paris. Anschel reported that he saw Hilde "quite often. She was deeply involved in communist party affairs [...]. When I saw her in Paris, she had turned her attention to the Spanish Civil War" (Anschel 1990, 128). Hilde was involved in various positions of the communist exile in France and in Willi (Wilhelm) Münzenberg's extensive communist organizational network. There were 4,000 to 5,000 German émigrés in Paris who were members or sympathizers of the Communist Party, far more than any other opposition groups; most social democratic émigrés were in Prague, while many liberals and conservatives had left Germany for Switzerland.

Hilde Rosenfeld-Kirchheimer carried out research and collected material for the *Braunbuch* [Brown book] published by Münzenberg in July 1933; this reconstructed the progression of events of the Reichstag fire and documented political persecution and terror in Germany over almost 400 pages. It also documented her father's work as a lawyer for a KPD member of parliament whom Göring had falsely accused of setting the fire (see Braunbuch 1933, 86). She and her father were also involved in preparing the high-profile "countertrial" to the trial against Dimitroff and the others accused of arson before the *Reichsgericht* in Leipzig, which was held before an "International Legal Commission" in London in September 1933 and gained considerable publicity. At this time, Otto Kirchheimer was in London thanks to his stipend, and he presumably also witnessed the public taking of evidence before the commission (see Ladwig-Winters 2007, 248). In Paris, Hilde Rosenfeld-Kirchheimer also coordinated support services for the *Rote Hilfe* and the Workers' International Relief, organized public demonstrations against the Nazi regime, and was a delegate of the German Women's Commission, a subgroup of the World Committee Against War and Fascism. In late 1935, she went to Moscow for several months. When she returned in 1936, she joined the KPD (see Ladwig-

Winters 2007, 195)[34] and worked for the *Association Juridique*,[35] an international jurists' association organized by the communists.

Otto Kirchheimer never considered switching to the KPD; he remained affiliated with the SPD. For the SPD, Paris was the most important center in exile besides Prague, the seat of Sopade—as the party executive committee working there called itself. Over 3,000 party activists had found refuge in the French capital. The Paris group of exiled Social Democrats sought to collaborate closely with the SAP[36] and took a stance against the party executive committee's Prague Manifesto of January 1934 (see Matthias and Link 1968, 231–233). But the group kept getting smaller and more divided. Like several other active Social Democrats from the Weimar Republic who had fled to Paris, Kirchheimer and Arkadij Gurland, who had headed the Paris group for a time, withdrew from active party work and focused primarily on personal political contacts from 1934 on.[37] Kirchheimer did, however, participate in efforts beginning in autumn 1935 with the goal of bringing together the various oppositional circles among the exile groups in Paris.

On 26 September 1935, a conference called Freedom Committee Meeting, chaired by writer Heinrich Mann, was held at the Hotel Lutetia in Paris. This was the first time that all the political streams among the German émigrés in Paris, from the liberals to the communists, gathered together. Following the new policy decreed by Moscow, the KPD committed to democracy, free elections, and convening of a national assembly following the overthrow of the Nazi regime. A second conference took place at the Hotel Lutetia on 2 February 1936. The surviving list of attendees indicates that the group of "socialists" was represented by Otto Kirchheimer as well as Kurt Rosenfeld. Another "socialist" guest was Max Horkheimer, the head of the Institute of Social Research (ISR) in New York. The other groups mentioned were the "bourgeois-democratic group," the "Catholic group," and "communists"' (see Langkau-Alex 2005a, 330). Horkheimer was at the Paris branch of the ISR from December to mid-February and met not only with Walter Benjamin but also with Kirchheimer on this occasion. The assembly at the Lutetia Hotel decided to establish a *Volksfrontausschuss* (People's Front Committee) headed by Heinrich Mann, a program committee, and a joint press publication. Otto Kirchheimer also took part in a follow-up closed-door meeting on 3 February 1936, where a small group of the SPD who were exiled in France discussed how best to press ahead with establishing this *Deutsche Volksfront* (German People's Front), which had been initiated in collaboration with the bourgeois and communist groups (see Langkau-Alex 2005b, 5).[38] After various negotiations, these efforts failed in 1937 to unify those in political exile in France. It is also documented that Kirchheimer took part in the asylum law conference *Conférence internationale pour le droit*

34 Vivid descriptions of the German communists' diverse activities in exile in Paris can be found in the memoirs of Arthur Koestler (Koestler 1954) and Manès Sperber (Sperber 1982).

35 This is evident from a letter from Franz L. Neumann to Otto Kirchheimer dated 27 April 1937. Otto Kirchheimer Papers, Series 2, Box 1, Folder 122.

36 *Sozialistische Arbeiterpartei* (Socialist Workers' Party); a small party that was founded in 1931 in order to unite the Social Democratic Party and the Communist Party.

37 On the rapid demise of the Paris group of the exiled SPD and the role of Gurland, see Langkau-Alex (2005a, 128–134 and 138).

38 No documents have survived, however, that would reveal the extent to which Kirchheimer continued to support these efforts of his party later.

d'asile (see Schale 2006, 94),[39] which was held on 20 and 21 June 1936, as well as in the preparations for this in the German émigré circles. The most important decisions made at this international conference of leading representatives of the host countries included the draft of a statute on refugees and the establishment of an International Bureau for Asylum Law and Political Refugees.[40]

Kirchheimer's initial efforts to relocate to the US are documented from early 1936 on. They were actively supported by his father-in-law Kurt Rosenfeld. In January 1936, after a personal conversation with John Whyte, the Assistant Secretary of the Emergency Committee in Aid of Displaced German Scholars (EC) in New York, Rosenfeld wrote a letter to the EC recommending Kirchheimer for a stipend in the future.[41] However, the EC responded immediately that Kirchheimer did "not fall within our group"[42] because he was not a refugee from Germany living in the US. So Kirchheimer had no other option but to remain in Paris.

In May 1936, the Popular Front coalition of socialists and communists under Léon Blum had won the election in France. This brought about considerable economic upheaval because unprecedented capital flight abroad set in within a matter of days. The Blum government shied away from reacting by imposing an export ban on currency and gold and instead devalued the franc by 30 percent. This caused prices to increase considerably, and the ensuing waves of strikes exacerbated the crisis. Blum declared in spring 1937 that his financial and economic policies had failed, and he resigned that summer. These events impacted the émigrés not only in terms of asylum law but also with respect to their living expenses. The *Société Internationale de Recherches Sociales* paid its staff and the stipend holders of the ISR in francs from monies from French foundations, so the economic turbulences beginning in the summer of 1936 meant that their economic situation deteriorated dramatically. For example, Walter Benjamin's income dropped below the subsistence level.[43]

In this situation, Otto Kirchheimer spoke with his friend Eugene Anschel and decided in the late summer of 1936 to make new attempts to relocate to the US. Anschel reported the following about Kirchheimer's motives in his memoirs: "Otto had decided to come to the United States because of the bleak future he faced in Paris. He could not

---

39 On the impact of this conference see Schiller et al. (1981, 48) and Langkau-Alex (2005b, 261–266).

40 The outcomes of this conference contributed to Sir Neill Malcolm, the League of Nations High Commissioner for Refugees Coming from Germany, joining the call for international regulation of the status of refugees. A League of Nations Convention on this issue was ratified by the French Popular Front government in late 1936. Yet the League of Nations did not take any further measures, and the efforts toward international coordination of the protection of refugees failed at the Évian Conference in July 1938. It was not until 1951 that the United Nations succeeded in adopting an international Convention relating to the Status of Refugees based on the Paris conference, see Vormeier (2002).

41 Letter from Kurt Rosenfeld to John Whyte (EC) dated 27 January 1936. Emergency Committee in Aid of Displaced Foreign Scholars, New York Public Library, New York. I, A Grantees, 1933–46, Box 18, Folder 13 (Kirchheimer, Otto).

42 Letter from John Whyte (EC) to Kurt Rosenfeld dated 30 January 1936. Emergency Committee in Aid of Displaced Foreign Scholars, New York Public Library, New York. I, A Grantees, 1933–46, Box 18, Folder 13 (Kirchheimer, Otto).

43 See the information about Walter Benjamin's financial situation in Fuldt (1979, 265).

find an academic position, and even if he had found one, he would not have obtained a work permit" (Anschel 1990, 157). Kirchheimer counted on his connections to the Institute of Social Research in New York for support, as well as on his relationships with family and friends in the US. His most important contact at the institute in New York was Franz L. Neumann, who had succeeded in obtaining a position there in October 1936. Kirchheimer again renewed his private connections with both Kurt Rosenfeld and Eugene Anschel, who had been living with his relatives in New York since February 1937.

They were all able to benefit from the circumstance that it was still relatively easy for Germans to obtain a visa to resettle in the US prior to 1938 because of the American immigration quota system (see Appelius 2003, 22–28, 69–73).[44] Max Horkheimer agreed to Kirchheimer's request, and it was arranged that the Institute would make an employment contract with Kirchheimer for a limited period of time and would provide a sworn affidavit, as it had done for Neumann the previous year; these documents were sufficient to receive an immigration visa to the US outside of the quota system. In addition, the institute would cover the costs of passage by ship as ticket prices were exorbitant. It is not clear from the correspondence with the institute whether the idea was for Hilde Rosenfeld-Kirchheimer and their daughter Hanna to move to the US, too; this impression arises at least from the initial letters concerning this matter between Kirchheimer and Neumann, who had taken on the role of in-house lawyer at the institute in New York.

The details of how Kirchheimer immigrated to the US can be reconstructed well on the basis of the surviving brisk correspondence between Paris and New York. The procedures necessary were set in motion in New York in late 1936. Neumann was able to report to Paris in early February 1937 that Friedrich Pollock, the managing director of the institute, had informed him "that all the official documents for your immigration have already been delivered to attorney [Willy] Haas."[45] One week later, Horkheimer specified the institute's offer to Kirchheimer in an official letter to the American authorities for the affidavit:

> Our Dr. Neumann who has returned from Europe has reported to us that you are prepared to join our staff in New York in the near future. We are glad to hear of your decision, and we hope that we shall soon be able to welcome you here. Dr. Neumann has already told you that we are not in a position to employ you on a full-time basis, but that we shall consider your appointment to full-time work after the university summer vacation. We confirm, therefore, that we shall employ you as Research Assistant for at

44 As of 1921, immigration to the US was subject to a quota system. The numbers of immigrants were curtailed in the 1924 National Origins Act. From then on, 153,879 foreigners per year were permitted to immigrate to the US. The German quota was relatively large, at 51,227 immigrants, but was reduced to only 25,957 after the Great Depression began, see Später (2017, 390–392). Mass emigration from Germany set in only after the anti-Jewish pogroms in November 1938, and the US introduced waiting lists. As a result, the waiting time for refugees from Germany averaged two years as early as 1939.

45 Letter from Franz L. Neumann to Otto Kirchheimer dated 9 February 1937. Otto Kirchheimer Papers, Series 2, Box 1, Folder 122.

> least one year on a part-time basis with a monthly salary of $ 100.-, starting from the moment of your arrival.[46]

In March 1937, however, there was mention of problems with issuing the visa. According to Kurt Rosenfeld, the institute's contract with Kirchheimer indicating a salary of 100 dollars per month was not sufficient for obtaining a visa for the US.[47] On 25 March, Kirchheimer reported that he had spoken personally with Pollock that day; Pollock was in Paris and had promised that he would "intervene with the consul himself in early April so that we can hope that things will be all right. I have the necessary papers from the police, even the German ones."[48] A month later, Neumann wrote to Kirchheimer that he had spoken with Pollock and that Kirchheimer would "presumably not have any difficulties with the consulate."[49] Neumann recommended that Kirchheimer "arrange that you can arrive here around 1 September, at the latest 1 October." Kirchheimer was restless and wanted to leave Paris immediately but Neumann informed him that there was absolutely no purpose in him being in New York from June to August because none of the people he would be interested in would be in town during the hot and humid summer. In June, he put him off again. Kirchheimer was able to make his way to London to the branch of the institute there only in mid-October, and then, on 5 November 1937, embarked by ship from Le Havre to New York on the SS Washington with a ticket paid for by the institute. On 11 November, Kirchheimer's 32nd birthday, the ship arrived at New York harbor.

Anschel and Neumann picked him up at the harbor. Anschel penned a vivid report about Kirchheimer's arrival in New York:

> Now on his arrival in America, he came down carrying nonchalantly around his neck a large camera, an incongruous sight for anybody who knew his unfamiliarity with and remoteness from any mechanical device. Both Franz Neumann and I wondered what, in the world, had induced him to walk around with that thing. Naively, Otto explained the reason for it to the customs official who asked him whether it was his own. No, he said, not at all, because he did not know how to use the camera. An acquaintance in Paris had asked him to take it along so that he, the acquaintance, could sell it when he would come to America later on. Of course, that made the camera contraband and the official promptly confiscated it. Otto was most unhappy on the way to the furnished room that we had rented for him. He thought his friend in Paris might believe he, Otto, had sold the camera and pocketed the money. Upset as he was, he left his winter coat in the taxi, the first such coat that he possessed since he had left Germany in a hurry and bought especially for the harsh New York winter. The next morning, Franz Neumann and I were able to convince a soft-hearted official at the Customs House that Otto, in the confusion

46 Letter from Max Horkheimer to Otto Kirchheimer dated 16 February 1937. Max Horkheimer Papers, Letters VI,11, page 119.

47 Letter from Otto Kirchheimer to Franz L. Neumann dated 10 March 1937. Otto Kirchheimer Papers, Series 2, Box 1, Folder 122.

48 Letter from Otto Kirchheimer to Franz L. Neumann dated 25 March 1937. Otto Kirchheimer Papers, Series 2, Box 1, Folder 122.

49 Letter from Franz L. Neumann to Otto Kirchheimer dated 27 April 1937. Otto Kirchheimer Papers, Series 2, Box 1, Folder 122.

of the arrival and due to his difficulty in understanding English had misspoken. We got the camera back, but the coat was gone (Anschel 1990, 156).

Kirchheimer first found accommodation with the Rosenfelds and then with Anschel, who were neighbors in the borough of Queens, where many émigrés lived. Hilde Rosenfeld-Kirchheimer had decided to remain in Paris for the time being with her new partner, pediatrician and KPD politician Rudolf Neumann. Their daughter Hanna stayed at the boarding school in northern Italy, but she was to be brought to the US as soon as possible. In the spring, Hilde Rosenfeld-Kirchheimer was interned for a month as an enemy alien in a camp in Rieucros near the Spanish border. Their daughter Hanna arrived in New York on the SS Washington on 1 November 1939, traveling with friends of the Rosenfeld family. Hilde Rosenfeld-Kirchheimer succeeded in escaping to the US after her release in early April 1940. She lived with her daughter and her parents in Queens for several months. Since the US authorities had banned her partner from residing in the US because of his activities as director of the International Brigades' medical services in the Spanish Civil War, she moved to Mexico with him in early 1941. A larger colony of communist émigrés from Europe had already established itself there.[50]

4. Conclusion: In waiting positions

This and the previous chapter have shown that Andreas Kalyvas's claim that there were "huge differences between Schmitt's political and legal theory and the politics of the NSDAP" (Kalyvas 2009, 443) is correct with respect to the Weimar Republic but does not hold true at all after Hitler came to power. Kirchheimer rightfully characterized Schmitt as the legal theorist *par excellence* of the regime in its early phase. Two more years passed before the first direct confrontation between Kirchheimer and Schmitt in 1935, even though they were not in direct personal contact. Its circumstances, however, were completely different from the previous one in November 1932. Back then, they had been able to have an animated conversation over coffee and pastries about their substantive differences; now, the Nazi regime Schmitt supported had forced Kirchheimer to flee abroad and to communicate indirectly. The result was communicative asymmetry, with Schmitt, on the one hand, as a strident representative of the Nazis in power, finding wide circulation, shouting to the émigrés that they would soon be stripped of their German citizenship, and turning his attention away from Kirchheimer—and Kirchheimer, on the other hand, with many copies of his anonymous booklet parodying Schmitt making the rounds in Germany illegally.

Kirchheimer presented Schmitt as the theorist of the Reich without going into Schmitt's personal motivations for his dedicated work for the regime. Schmitt's op-

50 This information is based on conversations with Hanna Kirchheimer-Grossman on 26 April 2017 and on 12 September 2021 as well as Kießling (1980, 194–196), Ladwig-Winters (2009, 404), and Barth (2010, 946). From 1941 to 1946, their daughter Hanna moved back and forth between her parents in Mexico City and New York; she had "the fragmented childhood characteristic of refugees [...] made even more so by the fact of separated and eventually divorced parents" (Kirchheimer-Grossman 2010, 63).

ponents within the Nazi system and several other émigrés chose a different tactic for dealing with Schmitt. They described his Nazi writings as the work of an unscrupulous political opportunist who during the Weimar Republic had made fun of the idea of *Rasse*, ridiculed the Nazi movement, and socialized with Jewish friends in his personal life. The reasons for the attacks against Schmitt, which resulted in him losing his most important leadership roles in late 1936, were not that he had internally distanced himself from Nazism, much less taken a stance of resistance. Nor were there any signs of him retracting Nazi statements in his writings. The truth was closer to the opposite. Schmitt proclaimed his loyal worldview far more than necessary. The party veterans considered Schmitt to be a competitor who threatened to rise even higher—into the ministerial ranks. They feared he would advance further, which is why they attacked him head-on: "It was not Schmitt who distanced himself from Nazism in 1936/37. It was the other way around; an important and powerful segment of the Nazi authorities had turned away from him" (Rüthers 1990, 107). Ultimately, Schmitt fared much the same as the renowned philosopher Giovanni Gentile in Italy, who had held the position of an official interpreter of Mussolini's state doctrine during the establishment phase of fascism and was useful for the regime's reputation on the international stage until he was pushed aside by the old guard of the fascist movement. The only indications that would support stylizing Schmitt's role after his demotion as a voluntary "inner emigration" are those circulated by Schmitt himself after 1945.

How is the setback in Schmitt's party career in late 1936 to be explained? Waldemar Gurian's writings about him were not the decisive factor. An illegal publication with a small number of hectographed copies and minuscule circulation, the *Deutsche Hefte* were too insignificant to be influential in the Reich. The files of the *SS-Sicherheitsdienst* (intelligence service of the Third Reich) show that Gurian's accusations were used at most as additional material against Schmitt. Analyzing the files of the Security Service as to the actual goal of the activities of Reinhard Höhn and the SS brings us closer to an explanation. This goal is quite bluntly characterized as "sidelining"[51] in one of the files. The goal of "ousting"[52] Schmitt from Nazi leadership positions is mentioned in two other documents in the files. In other words, Schmitt's fall was not a case of a supposed opponent of the regime being persecuted, but a successful attempt to limit his leading role in the institutions of the party and constitutional law.[53]

Yet this explanation is not exhaustive, either. After all, the above-mentioned jealousies alone—of Höhn and other party veterans and of Koellreutter and others who had quickly and opportunistically joined the NSDAP in March 1933—are not sufficient to explain the events. They must be placed within the structure of the system required for the various personal motivations to be able to prevail. The best way to identify a key to such a structural explanation is, of all things, by using Otto Kirchheimer's descriptions of the

51 Bundesarchiv Berlin-Lichterfelde, Akten des Sicherheitsdienstes des Reichsführers SS. Aktennummer R. 58/854, 257 (Aktenvermerk SS-Sturmbannführer J. Lehfeldt). On my interpretation of the files, see also Neumann (2015, 412–414).

52 Bundesarchiv Berlin-Lichterfelde, Akten des Sicherheitsdienstes des Reichsführers SS. Aktennummer R. 58/854, pages 263 and 280). See also Koenen (1995, 660) on his ousting.

53 On the goal of sidelining, see also Blasius (2001, 170–180) and Neumann (2015, 412–414).

social basis of the Nazi system of rule. In contrast to Schmitt, who admired Nazism for overcoming the pluralism of the Weimar Republic and creating a tripartite structure of unity of the German state, Kirchheimer claimed in his analysis of the Nazi regime that no such unity existed.

Starting points for such an analysis are to be found as early as 1935 in his illegally disseminated booklet *State Structure and the Law in the Third Reich*, which was decorated with Schmitt's name. Kirchheimer had called the Nazi regime a "system of reciprocal guarantees and obligations" (Kirchheimer 1935a, 157), thereby describing the relationships of the party, the army, industrial and financial capital, the Junkers, and the state bureaucracy to the *Führer*. In 1941, he further sharpened this analysis in his famous article "Changes in the Structure of Political Compromise." The Nazis had emerged from a civil war as victors and had succeeded in liquidating the democratic parties and the mass organizations of the workers' movement. Yet, contrary to the official ideology, the new state was not a homogeneous entity, but was instead based on a "compromise, as in every other society which has a high degree of social stratification" (Kirchheimer 1941a, 287). Kirchheimer mentioned five major social groups that struggled for influence with and against each other in the Nazi system and had to find new compromises time and again: "monopolies, army, industry, and agriculture as well as the diversified layers of party bureaucracy" (Kirchheimer 1941a, 287). Only in the case of strong conflicts between the competing partners in compromise did the authoritarian *Führer* come into play as the "ultimate arbiter."

Now, Kirchheimer claimed that the relationship of the party bureaucracy to the other four social groups was "subject to sudden shifts" (Kirchheimer 1941a, 287), and the party hierarchy below the level of the *Führer* was also liable to constant regroupings. Under the scrutiny of such an analytical approach, the reasons for the activities of the Security Service of the *Reichsführer* of the SS can be identified not so much in Carl Schmitt's person and more in the complex internal situation of the Nazi system in 1936. As the director of the NS-*Rechtswahrerbund* (see Glossary), Schmitt had positioned himself against the Reich Ministry of Justice directed by party member Franz Gürtner in the question of reforming the law of criminal procedure.

At the same time, Höhn was planning to replace Gürtner with Schmitt's mentor Hans Frank. Schmitt's career came to an abrupt end in this tangle of rivalries between various factions of the NSDAP. The concern in the Reich Ministry of the Interior was that, as the new Minister of Justice, Frank would bring along his loyal assistant Schmitt as state secretary and that he, Schmitt, would put the existing structure of compromises between the two ministries at risk with his characteristic activism.[54] To comprehend the explosiveness of this potential personnel decision, we need to understand that the Nazi laws on *Rasse* were not the responsibility of the Ministry of Justice but of the Ministry of the Interior, which did not want to cede control of these to Frank and Schmitt: "The SS leadership construed the 'case of Schmitt' in order to keep Frank in check. The purpose was to harm *Reichsrechtführer* Frank by disparaging and neutralizing his most important assistant" (Blasius 2001, 173). In other words, Schmitt, the preacher of tripartite state unity,

54 On the details of these events, see Blasius (2001, 170–180).

and his ambitions had become caught in the clutches of the polycratic power structure that Kirchheimer had analyzed objectively.[55]

Both Kirchheimer and Schmitt found themselves in situations that had changed yet again at the beginning of 1937. Schmitt remained in a waiting position at first, seeking a new orientation for his career in the Reich. Meanwhile, Kirchheimer hoped nervously that he would be able to relocate to the US since he feared that Germany would soon start a war. By the end of 1937, decisive changes had occurred in both of their lives which also had consequences for the main areas of their theoretical work. Kirchheimer had succeeded in moving to the US, and Schmitt had found his way back into the top ranks of Nazi jurists by throwing himself into a different topic. Yet the subject areas they both began working on in the following years again touched on each other in remarkable ways even though they still had no direct contact at the time.

55 Incidentally, the hypothesis of polycracy following Kirchheimer is supported by the fact that all of Schmitt's opponents from the ambit of the SD—Höhn, Koellreutter, and Eckhardt—were also disempowered over the course of the following four years, see Gross (2000, 121–122).

Chapter 9: From Leviathan to Behemoth (1938-1942)

The year 1937 had seen both Schmitt and Kirchheimer begin new phases of their oeuvres. Suddenly degraded to the level of an average Nazi jurist, Schmitt sought out new and more anodyne subject areas in which he could distinguish himself in Nazi Germany. Besides international law—which will be discussed in Chapters 11 and 12—he focused on intellectual history and revised his previous interpretation of Thomas Hobbes's *Leviathan* in the process. In his 1938 book on Hobbes, Schmitt now saw him as unintentionally foreshadowing the liberal notion of the rule of law. He accused Hobbes of being misguided in drawing on Leviathan, the sea serpent from the Old Testament, as a myth because it could not prevail over the other biblical beast, the Behemoth, which symbolized rebellion and state collapse.

Kirchheimer, too, broadened the scope of his work. He was forced to do so simply because it was his only means of continuing to receive financing through Horkheimer's Institute of Social Research (ISR) in New York with the vague prospect of obtaining a position in the American academic system. Kirchheimer's studies in criminology account for the largest segment of this broader field of work. In his 1939 book *Punishment and Social Structure*, he briefly mentioned Hobbes in the context of intellectual history, and, like Schmitt, as a precursor of liberal thought. Kirchheimer linked this reference to the empirical question as to the remaining legal protection of the individual under the Nazi regime. He also expanded the liberal idea of protection by reformulating it as the question of social protection of individuals and the obligations of state economic and social policy that this entailed.

In this chapter, the mythical images of Leviathan and Behemoth in Hobbes's theory of the state symbolize four aspects in the works of Schmitt and Kirchheimer between 1937 and 1943: first, the changing role of Hobbes's *Leviathan* as the successful founder of the theory of the authoritarian state in Schmitt's work; second, the Behemoth as a symbol for Kirchheimer's empirical studies on the practice of criminal law under the Nazi regime; third, Schmitt's critique of the *Leviathan* in his search for a more suitable mythical concept to describe a political order engendering and maintaining unity; fourth, Kirchheimer's theoretical conclusions from his empirical analyses of the Nazi system that became part of Franz L. Neumann's seminal book *Behemoth* in 1942.

To Schmitt, Behemoth was a symbol of the pluralistic powers of a society destroying the uniform state. To Neumann and Kirchheimer, it symbolized the anarchical struggle for power playing out in the Nazi regime under the ideological mantle of state unity. Finally, it was Kirchheimer more than Schmitt who considered the return of Leviathan, steeped in liberalism, to be a desirable option.

1. Kirchheimer's early studies in criminology

The monstrous character of any dictatorial regime is reflected in its criminal policy and the way it treats prisoners. Otto Kirchheimer's studies on criminology included an investigation of this. Compared to other subjects, criminology played a lesser role in his Paris exile initially. He then expanded his studies in criminal law into empirical research on crimes and criminals. Kirchheimer published his first article on criminology in the September 1936 issue of the *Revue de science criminelle et de droit pénale comparé (RSC)*. Focusing on reform policy, the journal was not established until the beginning of 1936 and was co-published by the Institut de criminologie (Institute of Criminology) and the Institut de droit comparé (Institute of Comparative Law) of the University of Paris. Kirchheimer wrote this first article, "Remarques sur la statistique criminelle de la France d'après-guerre" (see Kirchheimer 1936b),[1] in French.

In his article, Kirchheimer developed the hypothesis, inspired by Marxism, that there was a clear causal link between the special features of a society's social order and both the crime and the policy toward crime in that society. He used methods of empirical social research to prove his hypothesis. His argument against traditional legal scholarship was that the most recent experiences in the area of policy toward crime showed that the impact of laws and penal reforms on actual crime were overestimated beyond all measure. Kirchheimer also rejected the view championed by a "renowned German criminologist" (101), the Nazi Edmund Mezger, in the 1934 first edition of his textbook *Kriminalpolitik auf kriminologischer Grundlage* [Policy toward crime on a criminological foundation] that hereditary predispositions and individual weaknesses of will were the only factors determining criminal behavior (see Mezger 1934).[2] He positioned his "sociological hypothesis" (83) as an explanatory model to counter both points of view.

Kirchheimer used an abundance of figures and tables based on official French crime statistics data as well as criminological literature about France, the UK, the US, Poland, Switzerland, Austria, and Germany to support his arguments. He found confirmation of his "sociological hypothesis" for France concerning the frequency of the widespread criminal offenses of vagrancy and theft, which reflected "the changes in the economic conditions almost without fail" (85). Comparisons of the fluctuations in the frequency

1 The following page numbers refer to this article.

2 Kirchheimer misspelled Mezger's name as "Metzger" throughout the article, a very telling mistake in light of the fact that he had held the chair of Nazi criminal law in Munich from 1932 on (*Metzger* is the German word for butcher). On Mezger's work for the Nazi regime, see Thulfaut (2000).

of the prosecuted vagrancy offenses with the unemployment rates and the development of real wages—he prepared complex calculations specifically for this purpose—showed that they clearly correlated with one another. The reason he identified to explain why theft was the most listed crime in French statistics was that offenses committed by the lower social classes in France were apparently prosecuted more vigorously than fraud and forgery offenses committed by members of the upper classes.

What demands for improving legal policy did Kirchheimer believe could be deduced from these findings? He answered this question with an analysis of the *correctionnalisation* which had entered into force in France in March 1932. It was a legal reform replacing jury courts that also involved legal laypersons, with criminal courts consisting of three professional judges. He noted similar shifts in responsibility in other European countries, too, such as the Emminger amendments of 1924 in Germany. In France, the administration of justice was also reorganized. The latter resulted in significant declines both in acquittals and in convictions acknowledging attenuating circumstances. In other words, both cases amounted to more stringent punishments issued by professional jurists. Lay judges generally adjudicated cases more sympathetically.

Kirchheimer concluded that France needed far-reaching social changes. Considering the ongoing economic crisis, he thought that French society was in danger of "destroying the function of criminal law to protect its own society in the long run" (116). He described two ways to create alternatives to this rampant loss of security. One was the path taken at the time in several authoritarian and fascist regimes such as that in Germany. It consisted of continuing to "destroy the procedural guarantees" (116) of the defendants, combined with hoping for deterrence. Kirchheimer thought this path had little prospect for success, simply in view of current crime statistics. Conversely, he considered minimizing the social disparities characterizing French criminal law at the time to be the "only effective way" (116) to improve the situation. This included granting better rights to defendants under French procedural law. But, above all, it also included an economic and social policy reducing the social disparities in society. Only in this way "can one hope [to rescue] the liberal procedural system, one of the most noble features of the past epoch" (116) from the attacks of the representatives of a repressive and authoritarian policy.

After relocating to New York, Kirchheimer systematically continued to pursue the criminological research he had begun in Paris; its aim was to create security for individuals and society. Besides a few reviews of new books in criminology, his next major contribution to the field was his article "Recent Trends in German Treatment of Juvenile Deliquency"[3] in which he combined his knowledge of criminal law in Nazi Germany and his newly acquired competence in empirical criminology. He presumably wrote part of the article—or rather a draft in German or French—while he was still in Paris, with the aim of continuing and fleshing out "Remarques sur la statistique criminelle de la France d'après-guerre"; in any case. this impression arises from the statistical material, ending with the figures for 1936. In contrast, the legal sources he quoted date up until May 1938, which supports the assumption that he did most of the legal work in the first few months after arriving in New York. The article was published in the September issue of the *Journal of Criminal Law and Criminology*, the leading journal of criminology in the US at the time.

3 See Kirchheimer (1938a). The following page numbers refer to this text.

It was the first article that Kirchheimer was able to place in a US journal after relocating there.

Kirchheimer relied primarily on German sources for this article. He first described the legal situation, including the reformed juvenile criminal law which had entered into force in 1923, during the Weimar Republic, and which was the first legislation to focus on education and resocialization in dealing with juvenile offenders. The proponents of the official Nazi doctrine were up in arms about it and called for severe and uncompromising punishment of juvenile offenders. Nonetheless, "less dogmatic and less reactionary" (117) recommendations for changing juvenile criminal law prevailed after 1933. Kirchheimer stated that the draconian proposals for amending juvenile criminal law that were disseminated as propaganda after 1933 were mostly ignored in practice. Using official data, he refuted the rapid decline of juvenile crime which Hitler's government claimed as a success. In all the years examined, the lower figures included in the official statistics were the result of amnesties for recently initiated criminal proceedings. Consequently, according to Kirchheimer's analysis, there was no real improved protection from crime. He then meticulously evaluated the official data on punishment of juvenile offenders between 1931 and 1933 and demonstrated that although there were fewer acquittals and more prison sentences for juveniles, it was striking that imposing suspended sentences continued to be common practice.

Another subject Kirchheimer addressed was the situation of juvenile offenders behind bars. He described German jails as generally overcrowded and having dire hygienic conditions. The Nazi principles concerning execution of prison sentences were aimed at deterrence; consequently, the situation in jail was brutal. Yet Kirchheimer discovered an exception in handling juvenile offenders. The 1937 imprisonment guidelines applying to this group still referred to a responsibility for educating juvenile delinquents and leading them back to society. Kirchheimer identified the "spirit of these regulations" (123) in the provisions on sports and recreational activities for detained juveniles. These programs relied on suppressing any formation of spontaneous or stable groups. The juvenile detainees were forced into a hierarchical system of "strictly individual rewards and favors" (124). The only way for them to advance personally was through athletic achievements and proper behavior toward people ranked more highly. Vocational training in jail was not geared toward the real labor market. The training in the crafts and trades provided there was "a romantic gesture" (124) and served to keep the detained youths busy all the time rather than to prepare them for future work. As a result, juvenile offenders had few opportunities on the labor market after their release, which amounted to a new source of conflict for the regime.

Overall, Kirchheimer's findings were ambivalent. The approach to handling juvenile offenders was "caught between the official reactionary slogans on the one hand and the honest desire of officials to save these prisons from some of the worst possible consequences of the new policy, on the other hand" (120). In 1938, Kirchheimer apparently assumed that juvenile crime in Germany would not disappear, neither through repression nor through indoctrination, but would continue to fester as a problem of stability for the Nazi regime. Therefore, he again emphasized—this time from a criminological perspective—that the Nazi regime was prone to internal conflicts.

These two articles of Kirchheimer's already display key approaches, subjects, and hypotheses of his monograph *Punishment and Social Structure* which was published in 1939. They include as the "sociological hypothesis" his explanation of criminal law, the administration of criminal justice, and the practices of punishment, which was based on theories of society. They also include a critical analysis of competing approaches for explaining crime, an analysis of comprehensive data on social statistics and crime, and the comparative perspective on other epochs and countries. Finally, they include embedding current tendencies in criminal law in a critical analysis of capitalism as well as an argument for reforms in society and legal policy in order to generate security for individuals on the path to creating security for society.

2. Thomas Hobbes and the authoritarian state in Schmitt's Weimar works

In contrast to Kirchheimer, Schmitt turned further away from empirical questions and devoted himself to speculative areas of intellectual history. His 1938 book about Thomas Hobbes's mythical image of the Leviathan was the first major fruit of this pursuit. Schmitt had expressed his thoughts on Hobbes several times after World War I and returned to this subject in 1937 after a hiatus of multiple years.[4] Although he did not undertake extensive and original research on Hobbes during the Weimar Republic, he contributed brief and pointed interpretations to the renewed debate about him in Germany. Viewing his pre-1933 images of Hobbes is worthwhile not least because it makes the contrast marked by his 1938 book even more distinct.

Hobbes's theory of the state had increasingly engaged political theory and social philosophy in Germany from the second half of the German Empire onward. The pioneer of this new reception of Hobbes was sociologist Ferdinand Tönnies, the "founder of Hobbes studies in Germany" (Mastnak 2015, 968). Schmitt had been in contact with Tönnies in 1930 about Kirchheimer's *Weimar—and What Then?*[5] Tönnies had edited a number of Hobbes's works that were not known in German-speaking countries at the time and then, in his book *Thomas Hobbes. Leben und Werk* [Thomas Hobbes. Life and work] (see Tönnies 1925), first published in 1896, Tönnies presented Hobbes as a political thinker of the modern period. His book was of fundamental importance to the German discussion about Hobbes. Revised editions of the book, published in 1912 and 1925, laid the ground for a reception of Hobbes during the Weimar Republic that was to take different directions during a very short period of time. Suddenly, representatives of ideological streams across the board from socialist and radical democratic to liberal to right-wing authoritarian positions viewed Hobbes as compatible with their own political ideas.

The new line of interpretation leading to liberalism was launched by Tönnies and underscored by Leo Strauss and others, with different emphases, from 1930 on (see Strauss 1930). In the social democratic milieu that Kirchheimer considered his political home, it was Hermann Heller who claimed Hobbes from a leftist perspective for his theory of

---

4 On Schmitt's changing interpretations of the works of Thomas Hobbes, see Rumpf (1972), Voigt (2009), and McCormick (1994) and (2016).

5 See Chapter 3, p. 85.

a strong democratic state capable of enforcing social reform (see Henkel 2011, 362–365). Heller referred to Hobbes for his own definition of the state as a unit unifying a heterogeneous society. He also drew on Hobbes in his definition of democracy as a form of rule in which "the people as a unity rule over the people as a multiplicity. And through which the people as a unity become the sovereign person" (Heller 1927, 108). Of the Marxist social philosophers of the Frankfurt School, Max Horkheimer and Franz Borkenau supported Hobbes's theory of the state with an ideological critical perspective of the democratic interpretation or continuation of its "explosive historical dialectics."[6]

Schmitt took a different position and changed his view a number of times. In his 1919 book *Political Romanticism*, he had called both Hobbes and Descartes prototypical representatives of an "abstract rationalism" and "mechanistic world view" (Schmitt 1919, 80). Two years later, in *Dictatorship*, he presented Hobbes as the theorist who had argued with absolute clarity that there was no law outside or overriding the state. The logical consequence of the original role of the state to create laws was that "the state cannot do any wrong" (Schmitt 1921, 16).[7] Only the sovereign had the power to define what was useful and what was detrimental to the state. Since people's actions were motivated by their understanding of good and evil, or of gain and loss, the sovereign "also [had to] have the decisive power over the opinion of the people" (17). For Schmitt, it was the logical result of Hobbes' reasoning that "no private conscience exists in a state" (17) and that not only did all citizens have to obey the laws of the state but this also had to be their highest moral obligation. Following this line of interpretation, Hobbes's state as a political unit was "by constitution, essentially a dictatorship" (17). Here, Schmitt deviated blatantly from Tönnies's interpretation of Hobbes—which had become predominant—as a precursor, perhaps even the first representative of liberalism. But Schmitt did agree with Tönnies on another point: Tönnies was correct in observing that, compared to other works by Hobbes such as *De Cive*, "*Leviathan* is a political treatise more than one about natural law" (234). Seventeen years later, Schmitt was to focus on the distinction between the degree of persuasion through argumentation and the actual impact of Hobbes's theory of the state in his book on Hobbes.

In his *Political Theology*, first published in 1922, Schmitt again presented Hobbes as a fundamentally important figure and as a paradigmatic thinker in terms of his methodology. He praised Hobbes as the "classical representative" of the "decisionist type" (Schmitt 1922, 33) in legal thinking, competing with juridic normativism. Schmitt considered himself an adherent of the decisionist type of legal thought, too. The decisionism he traced back to Hobbes was Schmitt's own creation. It referred to the doctrine of the intrinsic value of legal decisions, disregarding the criterion of substantive correctness. Both in *Dictatorship* and in *Political Theology*, Schmitt also linked decisionism to the decision about a state of emergency. Decisionism was a personal type of thinking inasmuch as an individual or a group of individuals came to the decision about a state of emergency. He explained the personalism in Hobbes's theory of the state by a transfer of the habitual ways of thinking along monotheistic lines prevalent in Christian Europe in the seventeenth and eighteenth centuries—according to which it was just a single god who

6 Horkheimer (1930a, 229), see also Borkenau (1934).

7 The following page numbers refer to his book *Dictatorship* (Schmitt 1921).

ruled the world—to the sphere of political thought. This was also why Hobbes, despite his nominalism, his strict natural scientific approach, and "his reduction of the individual to the atom" (Schmitt 1922, 47) postulated an individual person as the decisive final authority. Another consequence of this form of transfer was that Hobbes "heightened [his state, Leviathan ...] into an immense person and thus point-blank straight into mythology" (Schmitt 1922, 47). Here, Schmitt can be understood as claiming that it was only this addition of mythology that unleashed the entire force of Hobbes's theory in its readers' minds.

Schmitt continued his hymns of praise in the original 1927 lecture "The Concept of the Political." In the eponymous book published five years later, he praised Hobbes as "truly a powerful and systematic political thinker" who had drawn the simple consequences of political thought "without confusion and more clearly than anyone else" (Schmitt 1932a, 65, 67). At the same time, he clarified his position in various places where he viewed himself as a legitimate heir to Hobbes's political thinking: his basic anthropological assumptions, his conception of the natural condition between states, his decisionism, his conceptual power, and his clear and systematic way of thinking (see Schmitt 1932a, 59, 65, and 67). He explicitly came to Hobbes's defense countering Tönnies's interpretation according to which Hobbes merely represented a type of thinking specific to his time based on free competition in early-stage capitalist society[8]—a line of interpretation expanded upon by Max Horkheimer and Franz Borkenau at the Frankfurt Institut für Sozialforschung (IfS).[9] In *Constitutional Theory*, Schmitt added a number of points to the list of Hobbes's positives: the grounds given for founding the state on human reason alone, systematic absolutism, the rejection of mixed constitutions, the clear contrast to the traditional idea of the bourgeois *Rechtsstaat*, and his concept of representation (see Schmitt 1928b, 101, 182, 237, and 247).

In 1932, both Kirchheimer and Strauss had approached Schmitt to request his support for their applications to the Rockefeller Foundation for a research stipend. Strauss had shown Schmitt the as yet unpublished manuscript of his essay "Notes on Carl Schmitt, The Concept of the Political." Following a discussion with Strauss, Schmitt agreed to support him and also arranged for the text to be published in a respected journal.[10] Strauss's essay addressed the key hypothesis of his book on Hobbes, which was published in exile in 1936, according to which Hobbes was the actual 'founder of liberalism" (Strauss 1932, 91). Strauss concluded from this interpretation, which was similar to that advanced by Tönnies and Horkheimer, that Schmitt remained "trapped in the view that he [was] attacking" (Strauss 1932, 104) despite his pointed criticism of liberalism. Strauss urged him in no uncertain terms to be more consistent and to depart from the horizon of liberal bourgeois thought.[11] That was a criticism that Schmitt took

8 See Schmitt (1932a, 65).

9 See Horkheimer (1930a, 216–221) and Borkenau (1934, 439–482).

10 See Chapter 3, p. 97.

11 To Strauss, overstepping the liberal horizon implied returning to ancient philosophy, see Meier (1996).

seriously; he sought to preclude it by hastily making some revisions to the 1932 book version of *The Concept of the Political*.[12]

Moreover, an idea is presented again that Schmitt had merely hinted at briefly in a footnote earlier (see Schmitt 1927a, 72) both in his eponymous lecture in 1927 and in the first version of the text, which was published as an essay, and that is not to be found when he had previously mentioned Hobbes. Schmitt introduced the idea in the thematic context of wars between states. When a people was no longer willing or able to organize itself successfully as an independent state, then foreign enemies took over political rule. The new ruler then decided who the enemy was by virtue of the "eternal relation of protection and obedience" (Schmitt 1932a, 52). Schmitt expanded this idea to the legitimacy of state rule, referring to Hobbes and his statement about the "mutual relation between Protection and Obedience"[13] in Hobbes's *Leviathan*. According to Hobbes's line of reasoning, the relationship between ruler and ruled could be summed up in the clear and simple formula: "No form of order, no reasonable legitimacy or legality can exist without protection and obedience. The *protego ergo obligo* is the *cogito ergo sum* of the state." (Schmitt 1932a, 52) Hobbes, Schmitt claimed, had painfully experienced the truth of his own statement firsthand during the terrible times of the English Civil War.

3. Schmitt's second thoughts about *Leviathan*

After 1933, Schmitt's admiration for Hobbes, which he had expressed without reservation, yielded to a significantly more critical assessment. Schmitt began to distance himself from Hobbes not only by making numerous stylistic revisions in the new edition of *The Concept of the Political*, published in June 1933, in line with the ruling Nazi system (deleting references to Jewish authors; replacing wording to reflect Nazi terminology; using fewer words of foreign origin; changing the font on the book cover), but also reclassifying Hobbes within intellectual history. The "powerful and systematic political thinker" from the previous year's edition (see Schmitt 1932a, 65) was demoted to "powerful and systematic thinker" (Schmitt 1933i, 47) and had to forgo the attribute "political," which was eminently important to Schmitt.[14] Schmitt also removed the reference to Tönnies, who had publicly supported the Social Democrats at the end of the Weimar Republic as an act of protest against the Nazis. Yet those passages in which Schmitt reflected on the link between protection of citizens and their duty to be obedient remained unaltered (see Schmitt 1933i, 35).

In his preface to the new edition of *Political Theology*, written in November 1933, Schmitt suddenly made a radical shift with respect to where he positioned himself in methodological terms, resolutely turning away from Hobbes and the decisionism he had ascribed to him. Instead of distinguishing between the two approaches of normativism

12 The revisions are described in detail by Walter (2018, 297–301).

13 Thomas Hobbes in his English translation of *Leviathan* of 1651, quoted in Schmitt (1932a, 52).

14 Heinrich Meier has shown that this change was also a reaction of Schmitt's to Strauss's criticism, see Meier (1995) and (1996). Schmitt gave the attribute "political" back to Hobbes in 1938 (see below).

and decisionism, Schmitt now differentiated between three paradigmatic approaches, adding a third one, "institutional type," (Schmitt 1934i, 2) i.e., concrete-order thinking. After vehemently defending decisionism for years, he now began to criticize it. Since it focused on the moment, it notoriously ran the risk "of missing the stable content inherent in every great political movement" (Schmitt 1934i, 3)[15]—by which he meant the Nazi movement led by Hitler. After Schmitt had distanced himself from Hobbes in this passage, not a single reference to Thomas Hobbes and his work was to be found in his new texts written between 1933 and 1936.

This was all to change after Schmitt was sidelined in December 1936. In the autumn of 1937, on the occasion of the 300th anniversary of René Descartes's *Discours de la méthode*, Schmitt gave the lecture "The State as a Mechanism in Hobbes and Descartes" in Berlin (see Schmitt 1937c). Two further lectures followed in January and April 1938 in which he presented the new findings of his research in intellectual history. The book based on these lectures, *The Leviathan in the State Theory of Thomas Hobbes—Meaning and Failure of a Political Symbol*, held special meaning for its author, as revealed by the date in the introduction: 11 July 1938 was Carl Schmitt's 50th birthday. The publication was his birthday present to himself, so to speak. He again addressed almost all of the points he had made about Hobbes prior to 1933. However, he revised a number of interpretations, coming to a more critical overall assessment of Hobbes's accomplishments in constitutional theory. In an April 1938 letter to Rudolf Smend, Schmitt revealed his claim to originality in the question posed in his book: "Why has nobody ever reflected on what Leviathan actually means as a symbol and a political myth?"[16]

Yet Schmitt's book is also a response to the discussions about Hobbes during the Nazi period.[17] This debate was brief, and it was triggered by a controversy in France. In 1935, Joseph Vialatoux, a Catholic proponent of natural law, had declared Hobbes to be the mastermind of the totalitarian state of Bolshevism and Nazism. Countering him, René Capitant, a French jurist who had been friends with Schmitt since the late 1920s, drew attention to the elements of the rule of law in Hobbes's political theory. Schmitt reacted directly to this debate in his Berlin lecture in 1937.

In a prominent speech, Nazi legal scholar Paul Ritterbusch had followed the individualistic interpretation, arguing that Hobbes's natural condition could not be transcended in a purely rationalist way; that would also require a strong ideological motivation (see Mastnak 2015, 982). In another contemporary publication, German sociologist Helmut Schelsky declared Hobbes compatible with the activist thinking of Nazism. Schelsky rejected the interpretation of Hobbes's theory as mechanistic rationalism. Instead, he applauded him, placing him in the same rank as Nietzsche as a "thinker of political action" (see Schelsky 1938, 193)—meaning highly intense political activity—who aimed to create political unity by means of the resolute will of a leader and to maintain it by instructing

15 It is one of the ironies in the contemporary debates about Schmitt that Strauss complained to a friend from his London exile in October 1934 that Schmitt had turned away from decisionism and embraced concrete-order thinking "because of arguments in my review, which, of course, he does not quote." (Quoted in Meier 1995, 138).

16 Letter from Carl Schmitt to Rudolf Smend dated 14 April 1938 (Schmitt and Smend 2011, 96).

17 See Jänicke (1969), Rottleutner (1983), and Voigt (2009).

the population. Schmitt did not directly contradict Schelsky's activist characterization, but he made clear in his book that he thought the ways to influence people through education or instruction were limited (Schmitt 1938a, 36).[18] He took Schelsky's interpretation as an opportunity to formulate the guiding questions of his book: Was the myth of Leviathan created by Hobbes "a faithful restoration of the original unity of life" (11), and had this mythical image "battling the Judeo-Christian destruction of the natural unity" (11) proven successful or not? Schmitt's book had a special position in the long series of interpretations of Hobbes because of how it "handled its subject [Hobbes], bringing it up to the present, with its author identifying with it." (Jänicke 1969, 402). In the sections in which he reconstructs Hobbes's theory of the state, Schmitt's book was not even original—it was innovative at best with respect to some of his previous hypotheses regarding Hobbes. Again, Schmitt used Hobbes primarily as a source of prompts to expand on his own views.

If we take Kirchheimer's external perspective on the polycratic power structure of Nazism as the starting point for reading Schmitt's book and link that perspective to Schmitt's personal experiences of defeat in the schemes and struggles of the competing groups and factions of the Nazi regime, then six topoi emerge; the subject of Schmitt's increasingly caustic antisemitism will be addressed separately in the next chapter of the present book.

First, from the above-mentioned perspective, it is striking that Schmitt—similarly to Schelsky and, before him, Strauss—presented Hobbes less as an abstract and almost timeless systems thinker, instead emphasizing his polemical side, a reflection of his time, more strongly than before. In contrast to his works on natural law, *Leviathan*—as Schmitt had postulated a year earlier in his lecture on Descartes—was a "preponderantly political treatise" (94), in other words, in Schmitt's language, a text that had particular opponents in mind. Hobbes's purpose with his theory of the sovereign state had been to crush the "medieval pluralism" (71) of the feudal lords, the clergy, and the estates. Schmitt called the feudal estates and the clergy "indirect powers" (73) competing for power and provoking an open or latent civil war. They were unable to offer security or protection to the individual. Hobbes's theory of the state, Schmitt stated, was formulated specifically to counter these "old opponents," who could be named. Politically speaking, Schmitt's criticism was directed first and foremost against the Roman Catholic as well as the Presbyterian clergy in England, which were unduly claiming power—and here, his historicizing view matched that of Tönnies (see Tönnies 1925, 256–266).

Second, it is striking how clearly Schmitt again emphasized the relationship between protection and obedience. The state was responsible for the protection and the security of those subject to the state. If protection ceased, then the state also ceased to exist, and any duty to be obedient ceased to apply. Schmitt now even declared the relationship between protection and obedience to be the "cardinal point" (113) of Hobbes's theory of the state, overshadowing everything else. And he later added a sentence following Ferdinand Tönnies, namely that that relationship "permits a very good reconciliation with the concepts and ideas of the bourgeois constitutional state." (72).

18 The following page numbers in the text refer to Schmitt's Leviathan book.

This—third—leads to the observation that Schmitt no longer equated Hobbes's construct of the state with a dictatorship as he had in 1921; much less did Schmitt go so far as to declare once more that there was no such thing as a private conscience in a state. Instead, he underscored—now even explicitly referring to Tönnies—the elements of the *Rechtsstaat* in Hobbes's theory. Schmitt also tacitly followed Tönnies's interpretation (see Tönnies 1925, 263, 266) in that he focused on the passages about miracles in Chapters 26, 37, and 42 of *Leviathan* for his political interpretation of the book. Here, Hobbes explained that a citizen always had the liberty to believe or not to believe those actions declared to be miracles; after all, because of the realities of human existence, thoughts were free. Schmitt stated that when it came to believing in miracles, Hobbes committed a "non-eradicable, individualistic proviso" (56) with his "distinction between outer and inner" (53), between acting or speaking in public and thinking or believing in private.[19] Schmitt considered Hobbes's granting of internal freedom of belief to be a "rupture of the otherwise so complete, so overpowering unity" (56) of his theory of the state. In terms of intellectual history, Schmitt interpreted this crack as a starting point for what later became liberalism. Schmitt now followed Tönnies's liberal interpretation of the concept of law, too, emphasizing the ban on *ex post facto* laws. He wrote that the well-known formula *nullum crimen sine lege* could be traced back to Hobbes, "including its linguistic character" (73). In addition, he even declared Hobbes to be the actual progenitor of modern legal positivism, placing him on the same level as John Locke. In other words, Schmitt interpreted Hobbes as the founder of everything he had rejected in a large number of lectures and articles over the previous five years as entirely incompatible with the spirit of Nazism.

Fourth, Schmitt combined the liberal reading of *Leviathan* with his well-known hypotheses inimical to pluralism. The "wonderful armarture" of modern state organization required a "uniformity of will and uniformity of spirit" (74). This need ran counter to the spirit of the liberal constitutional state that Hobbes had unleashed. In the process, Schmitt developed his own understanding of Hobbes's *potesta indirecta*, bringing it up to date. The old opponents of the state, the indirect powers, namely the clergy and the estates, had succeeded in reappearing in the twentieth century in their modern forms as political parties, social interest groups, churches, and trade unions, taking advantage of and enjoying the protection of the guarantees of the *Rechtsstaat*. It is also evident that Schmitt meant the Weimar Republic and not Nazism here from his comment that the indirect powers had seized control over the legislative process via the parliament, for which reason they believed "they had placed the Leviathan in harness" (73). The unity of the state had thus been destroyed from the inside out. To describe the condition of anarchy allegedly arising from this process, Schmitt turned to the mythical image of Leviathan's biblical adversary, the Behemoth, at a different point in the book. The Behemoth symbolized the "revolution" (21), civil war, and anarchy. It was the symbol of the non-state. In Hobbes's words, "one of the monsters, the Leviathan 'state' continuously holds down

19 In my opinion, the function of Hobbes's separation of internal and external was to enable the subject to obey a supposedly or de facto non-Christian sovereign without having to abandon their Christian faith for this reason. This function was the decisive motive for the inaccessibility of the internal in Hobbes's thinking, not an honorable and respectful reservation vis-à-vis the internal. Nor was it resignation in the face of the factual impossibility of entering the internal.

the other monster, the Behemoth 'revolutionary people'" (21)—incidentally, a juxtaposition that Schmitt could have taken from Tönnies's interpretation (see Tönnies 1925, 61). There are no indications—or even merely suggestions—that Schmitt, in mentioning the reappearance of the indirect powers, was alluding to the polycratic structure of the Nazi regime's rule with the competitive relationships between the party, the executive branch, the *Wehrmacht*, and other groups.

Fifth, Schmitt articulated the main topic of the book in its subtitle *Meaning and Failure of a Political Symbol*. This centered around the relevance and role of the myth in Hobbes's theory of the state. Schmitt claimed that the arguments of the traditional theories of legitimation were based on the myth of the monarchs' divine origins. Hobbes had radically swept aside this myth, declaring that all rule was man-made. Nonetheless, he had not wanted to abandon the mythical element entirely, which Schmitt believed was evident from the efforts the author had taken with the artistic design of the frontispiece. Schmitt had highlighted the enormous power of myths time and again in his Weimar writings from 1923 onward. He thought that the myths of the Bolshevik, fascist, and nationalist movements were vastly superior to the rationalism of parliamentary democracy in the political struggle (Schmitt 1923a, 79–83). He also began his 1940 anthology *Positionen und Begriffe* [Positions and concepts] with the political theory of the myth and concluded it with a 1939 essay on the myth of the Reich. Moreover, Schmitt placed his book on Hobbes within this mythopolitical arc—whereby he still situated him at the rationalist opposite pole of political thinking up until 1933.

Sixth, Schmitt took the question he raised at the beginning of *Leviathan* in multiple directions, namely what Hobbes had been trying to achieve for his otherwise very rationalist theory by adopting the mythical image of Leviathan. On the one hand, Schmitt presented an extensive genealogy of the biblical sea monster based on the history of the myth from pre-Jewish times to the Book of Job to Hobbes's day. The purpose of this genealogy was for Schmitt to be able to state that in Hobbes's lifetime, the myth of Leviathan as a biblical horror of evil spirits had lost its power and had been transformed in English usage into a humorous label for powerful and rich individuals, or large buildings and ships. In addition, he answered the question by analyzing the frontispiece of the first edition of *Leviathan* as well as the text, coming to the conclusion that Hobbes understood Leviathan in four different ways: as a great human being, a large animal, a machine, and a mortal god. As a result, the goal he had pursued by selecting this symbol had been watered down beyond recognition. Schmitt also hypothesized that a disastrous "*Mißgeschick*" (mishap)"[20] had occurred when Hobbes had selected the mythical image of Leviathan. For one thing, Hobbes should have selected the symbol of land (Behemoth) instead of the symbol of the sea (Leviathan). For another, the metaphor of the machine

20 The translation of the German term "*Mißgeschick*." is missing in George Schwab's English translation (page 79), see in comparison the 1938 German edition (page 119). Incidentally, a *Mißgeschick* happened to Schmitt himself when he analyzed the image in that he erroneously believed that he was in possession of the first edition of *Leviathan* of 1651. In fact, he owned an illegal reprint from the seventeenth century with a somewhat blurry frontispiece. Schmitt therefore felt compelled to use a copy of the book from the eighteenth century with a modified frontispiece "because it was more recognizable" (26) for his interpretation, see Bredekamp (2012, 46–47).

inherent to the symbol of Leviathan had contributed to Hobbes's notion of indivisible political unity being "destroyed from within" (85). Finally, Schmitt emphasized that neither England nor the UK had ever become a state to the same degree as the major powers on the continent. The English people had withdrawn from the continental type of state unity, drawing on its political instinct as a maritime and trading power. To Schmitt, the tragedy of Hobbes's theory of the state lay in the fact that it was felt to be an anomaly contrary to nature in his home country of England. On the European continent, the image of his Leviathan "ran aground" (81), too, because the sea serpent was not an adequate depiction of how the typically territorial shaping of power by military land powers had unfolded. Political myths, Schmitt asserted, had an "arbitrary historical force" (26), and Leviathan was a mythical name that could not be "cited with impunity" (53). Its image was so strong that it had its own impact independent of the person seeking to use it for their own purposes.

What follows from all this? Had Schmitt, with his liberal reading and his renewed emphasis on the relationship between protection and obedience, become a covert critic of the Nazi regime because of his negative experiences with the regime the year before, choosing his wording prudently? This does not appear to be the case because, at the exact same time, Schmitt continued to attack the guarantees of the *Rechtsstaat* in other publications, calling them typically Western, liberal thinking. A better way to understand the goal of his book is to read it against the background of the question Schmitt had grappled with incessantly since the Weimar Republic: the question of whether it was possible to realize a unified political order. Did the state achieve its unity by ending the civil war? Or could such unity be accomplished only by a party that had prevailed in the civil war? In other words, could state unity be founded upon neutralizing conflicts, or was an ideology encompassing the entire population necessary?

Schmitt tended toward the second alternative, referring in his book on Hobbes to the quest for a theory of the state achieving a "restoration of the vital energy and political unity" (81)—whereby "restoration" had to imply that this unity had already existed in the past and that "vital energy" spoke to a vitalistic dimension going beyond purely cognitive elements. This quest was taking place in a new epoch that Schmitt called the "technical age" (82). This was characterized by machines and technical procedures involving all of human life. The image of Leviathan no longer made a monstrous and terrifying impression on the mindset of a "total technology" (82).

On closer examination, Schmitt's central hypothesis about the failure of the symbol of Leviathan because of the separation of internal and external confessions of faith says virtually nothing about Hobbes, but a lot about Schmitt's own political thinking. In 1938, what were the theoretical alternatives for the foundation of a political order, given that the mythological grounding of the state in the symbol of Leviathan had failed? What other ways were there to establish political totality by means of mythologization? Returning to the bourgeois *Rechtsstaat* was not an option for Schmitt. But what then? Apparently, Schmitt did not want to simply relinquish the idea of the state according to Hobbes, instead seeking alternative ways to underpin the state mythologically. This interpretation is supported by the fact that at the end of his book, he heaped great praise on Hobbes despite his mythological blunder as "the great teacher in the struggle against indirect powers" (86) whose teachings would come into full effect only in the fourth century after he had created his work.

Why so much praise for Hobbes if his concept of the state belonged in a museum? What alternatives could there be? To Hobbes, the power of the sovereign representative was "so much more than the sum of all the participating particular wills" (33). Yet his myth of Leviathan had ultimately failed because he "did not unequivocally conjure up a definite and a clear enemy" (85). Only by taking recourse to "*das Elementare*" (the elemental)[21]—in this context, associated with a metaphysical and vitalist philosophy of life and, for Schmitt, again, the unequivocal differentiation between friend and enemy—could such a myth unfold its full "mythical force" (49). Schmitt granted Hobbes that his theory of the state had an "activist character" (85) even though systematic thoughtfulness prevailed—but only to object that not every case of philosophical activism was the same as political thinking. Schmitt was of the opinion that a state could not function in the absence of an activating myth. The image of Leviathan had to be replaced by a myth offering better ideological integrative power. Yet that new myth, he believed, was no longer necessarily coupled to the concept of the state.

Just as he was finishing his work on the book, Schmitt wrote a brief text on Thomas Hobbes's birthday. The wording of this greeting to the German Hobbes Society makes it clear that, to Schmitt, this was more than simply a gesture on that occasion. At that point, he suddenly believed once more that Hobbes was a "great political thinker"; the "undiminished force of his polemics" and the destruction of "opaquely evading all 'indirect powers'" could be properly understood only in Schmitt's day.[22] What Schmitt actually meant was revealed in his essay "Völkerrechtliche Totalität und völkische Totalität" [Totality of international law and *völkisch* totality], which was published the same month as the book. In the article, he showed his admiration for Hobbes: in his criticism of the papacy, the latter had recognized the international dimensions of influence through *potestas indirecta*. Hobbes became Schmitt's key witness for the right of any state to define its foreign policy enemy itself and to start a war against that enemy (see Schmitt 1938c, 621).[23] In the following years, Schmitt was to name Hobbes as a reference who systematically rejected the idea of universalist international law. Several months after his book on Hobbes was published, Schmitt had taken a further step in his deliberations. After the German *Wehrmacht* had invaded the rump Czech lands in the spring of 1939, he replaced the concept of the state with the concept of the Reich in his publications.[24] Schmitt's line of thinking in his book on Leviathan led, in direct consequence, to the "Reich" in a dual function, namely as an organizational formation of order and as an inspiring myth.

The reception of Hobbes under Nazism did not set in until later, unlike the renaissance of Hegel, and its extent was relatively modest—in retrospect, Schmitt's book can be considered its apex. Above all, however, this reception failed. And not only in Schmitt's

---

21 The term "*das Elementare*" is also missing in George Schwab's English translation (page 49), see the 1938 German edition of the book, page 76. In this context, *das Elementare* is associated with a metaphysical and vitalist philosophy of life.

22 All quotes are from Schmitt (1938b, 495).

23 See, already referring to Hobbes, Schmitt (1937b, 51 and 68).

24 See Schmitt (1939a) and Chapter 12 of this book.

book, but also in the contemporary reactions to his book in Nazi Germany. Otto Koellreutter penned an essay a few weeks later in which he viewed Schmitt's book as evidence that Nazism could not be explained from the perspective of Hobbes's theory (see Koellreutter 1938). Even if Ernst Forsthoff, Schmitt's former student, published a more positive review in 1941 (see Forsthoff 1941),[25] Koellreutter's verdict had put an end to Hobbes being well received under Nazism (see Rottleuthner 1983, 252). Schmitt himself did not argue on the basis of Hobbes's theory of the state again until 1945. Instead, he attempted to legitimize Nazi Germany using the concept and myth of the Reich and his theory of the *Großraum*.[26]

After 1945, Schmitt misrepresented his book on Leviathan as a sign of his internal resistance. He even placed it on the same level as Ernst Jünger's *On the Marble Cliffs*[27] (see Schmitt 1950a, 22). In 1947, to prove this claim, he even misquoted from his 1938 book, insinuating his internal resistance to the regime.[28] Helmut Schelsky, Schmitt's younger competitor in interpreting Hobbes at the time, made a similar point in retrospect. Schelsky was one of the most influential sociologists in Germany after 1945 and became the conservative antagonist of Theodor W. Adorno and other members of the Frankfurt School. In 1979, he attested that Schmitt had attempted to "*unterjubeln* [smuggle] the reason of the *Rechtsstaat* into [the actions of] the dictatorial sovereign" (Schelsky 1979, 150). In other words, Schelsky claimed that Schmitt had turned into a defender of the bourgeois *Rechtsstaat* in 1938, but without using this term. Various authors later subscribed to this absurd assessment.[29] Considering the book *Leviathan* as a hidden sign of internal resistance is as far-fetched as the parallel to Jünger is wrong. Rudolf Smend expressed his distance to the book in his letter thanking Schmitt for sending it to him with the following words about the concluding passages: "More restrained in the future—all one can do is limit oneself to showing alternatives, and perhaps we disagree on that matter."[30] They most certainly did.

25 Referring to the Prussian state of the eighteenth and nineteenth centuries, however, he contradicted Schmitt's central hypothesis that a state that permitted freedom of conscience was damned to perish in the further course of history.

26 See Chapters 11 and 12.

27 In his 1939 novel *On the Marble Cliffs*, Jünger described the life of outsiders in a totalitarian dictatorship. The first-person narrator lives a secluded life with his family on the marble cliffs of a great lake and devotes himself to botanical science. Their idyllic life is threatened by the erosion of values and traditions by the regime. The novel was understood as a parable of Nazism, written by an author in what was frequently called "inner emigration."

28 For details of how Schmitt misquoted himself, see Salzborn (2009, 158–159). On Schmitt's writings about Hobbes in his late work see Mehring (2023).

29 The main reception of the book in the English-speaking world by George Schwab (1996), Joseph Bendersky (1983, 245–247), and Tracy B. Strong (2008) followed in the footsteps of Schmitt's and Schelsky's interpretations.

30 Letter from Carl Schmitt to Rudolf Smend dated 10 July 1938 (Schmitt and Smend 2011, 99).

4. Kirchheimer's Behemoth in *Punishment and Social Structure*

The book *Punishment and Social Structure* by Georg Rusche and Otto Kirchheimer, published in New York in 1939, has a back story of almost ten years.[31] In 1930, Georg Rusche had begun working on the book in Frankfurt at the Institut für Sozialforschung (IfS) and had continued his efforts up until 1934 after emigrating to London. After he had relocated to Palestine and the institute had no longer received word from him, Horkheimer desperately sought someone who was competent to complete the half-finished manuscript which had been sent from Europe to New York. Otto Kirchheimer seemed suitable to the institute's leadership because of his expertise in criminal law in Nazi Germany and his brief study on French crime statistics. Franz L. Neumann first contacted Kirchheimer on this matter on Horkheimer's behalf in early February 1937; Neumann was in New York, Kirchheimer in Paris. It had also been Neumann who had suggested Kirchheimer for the project at the institute since he was aware of his knowledge and his financial needs as well as his interest in establishing closer ties with the institute in order to enter the US. Kirchheimer gratefully accepted this new project. His extensive correspondence with Neumann in the following months was equally about substantive matters relating to his work on the book and the paperwork he required to enter the US (see Buchstein 2019a, 12–44).

When Kirchheimer arrived in New York in November 1937, he had the manuscript with him; he had spent the previous nine months working on it and it was finalized in another five months with the aid of Moses Finkelstein, a young historian at City College in New York.[32] Finley, as he was then known, later reported that Kirchheimer wrote the new chapters in German and that he, Finley, translated them into English. In late 1937, Kirchheimer also contacted Nathan Leites, who had also managed to escape Germany and enter the US. Kirchheimer asked for his opinion and comments and subsequently made further changes.[33] The manuscript was finalized for printing in late 1938 after multiple rounds of editing. Rusche was not involved in completing the book, which gave rise to conflicts between him and the institute after its publication. When the book was published by Columbia University Press in early 1939, it became an unexpected success for Horkheimer's institute, which was still in the process of establishing itself in the US. Multiple newspapers and journals reviewed the book, almost without exception in a positive light, which brought the recently renamed Institute of Social Research "a small degree of fanfare." (Wheatland 2009, 143). Today it is without dispute a classic of critical

---

31 On the convoluted genesis of the book, see the preface to the French reprint by Lévy and Zander (1994).

32 Finkelstein later became a renowned classical scholar under the name Moses I. Finley. On the impact of his collaboration with Kirchheimer for his later work, see Perry (2014).

33 Leites called attention to a few minor errors and, importantly, suggested that Kirchheimer should either be more explicit in his implied criticisms of Max Weber where he mentioned him briefly or use more neutral wording. See letters from Nathan Leites to Otto Kirchheimer (no dates). Otto Kirchheimer Papers, Series 2, Box 1, Folder 101.

criminology—not least because of Michel Foucault's praise for the book in his *Discipline and Punish*.[34]

Kirchheimer wrote six book chapters; Rusche had prepared the other seven, but Kirchheimer reworked them, in part extensively.[35] Despite all the effort put into revising the book, the resulting work did not really come together. Since the English translation was edited carefully, it is not owing to the style that there are differences between the individual chapters; instead, it is the form of reasoning, the nature of the substantive argument, which makes these differences stand out all the more clearly. All of Rusche's historical chapters focus on demonstrating the validity of his economic theory of punishment from the Early Middle Ages to the late nineteenth century, which he had already outlined in an article in the *Zeitschrift für Sozialforschung*. In contrast, Kirchheimer stresses the relevance of the political conditions for criminal law and penal policy in all the chapters he authored. There were also methodological differences. The parts based on Rusche's research primarily use economic labor market arguments. Those written by Kirchheimer also refer to the relevance of economic factors but the specific analyses are mostly from the perspectives of law and legal sociology. To put it bluntly, it is difficult to avoid the impression that two books were forced between the covers of one.

The book draws a historical line from the Middle Ages to the year it was completed, 1938. Comparing and contrasting it with Schmitt's humanities-based approach in *Leviathan*, three aspects of Kirchheimer's parts of *Punishment and Social Structure* stand out.

First, Kirchheimer's methodological approach employs a materialist analysis. In the introduction to the first chapter, he formulates the questions posed in the book as follows: "Why are certain methods of punishment adopted or rejected in a given social situation? To what extent is the development of penal methods determined by the basic social relations?" (3). Kirchheimer intended to prepare a materialist sociology of corrections, which aimed to remove the ideological veil and the legal pretense from the institution of punishment. He distinguished between a negative and a positive determinant in the enforcement of different types of punishment. The positive determinant was a given: the level of a society's economic development. That alone, however, was not sufficient to understand the historical transformations of forms of punishment because it would require incorporating the specific aims of punishment as a negative determinant, too. Kirchheimer's point, opposing Rusche's purely economic approach, did not become clear until the final paragraph of his introduction where he wrote that "social consciousness" (7) was acquiring an even broader field of action in the development of methods of punishment. How broad this field of action was and where its limits lay then became the key questions Kirchheimer attempted to answer. The focus of the chapters he wrote alone was on the negative determinants of enforcing different types of punishment, including, in the ninth chapter, the question as to the limits of modern prison reforms.

In this chapter, he describes the improvement in the lower classes' living standards and its impact on crime policy. He details how criminology had gained status as a new

34 On the reviews at the time and the book's later enormous reception, see Buchstein (2019a, 45–51) and Klingsporn (2024).

35 See Kirchheimer and Rusche (1939). The following page numbers refer to this book.

social science in the second half of the nineteenth century, aiming to curb criminal behavior. He makes the "political basis" (143) responsible for changes in criminal law and corrections, and in this context, he argues in the footnotes with Schmitt's student Ernst Rudolf Huber's criticism of the reforms in criminal law during the Weimar Republic. Using analyses of crime statistics, he then examines the development of incarceration in the period of relative prosperity up to the beginning of World War I in various European countries. His description of the outcomes of resocialization measures has a skeptical tone. To achieve deterrence, no reform program had been willing to relinquish the principle according to which the standard of living of those incarcerated had to be reduced. Thus, all reform programs ultimately remained caught in the contradiction between deterrence and rehabilitation, which in turn was an expression "of the antagonistic tendencies in society itself" (159). Kirchheimer saw that the development after World War I was also decoupled from the purely economic basic data and, in light of this, he refers primarily to some successes in humanizing corrections during the Weimar Republic. In Chapter 10, he analyzes the role of fines as a parallel to incarceration in more recent penal practice. He diagnoses extensive "commercialization of the penal system" (175) because of the prevalence of fines. His empirical finding was that the imposition of fines had increased along with the prosperity of a society. Thus, even though fines were politically feasible, their application had ultimately reached its limits in the material conditions of the lower strata of the population.

A second set of issues in clear contrast to Schmitt's writings was the changes in the policies for meting out punishment in countries with fascist governments, which Kirchheimer analyzes in Chapter 11. He continues some of his previous deliberations on criminal law in the Third Reich, adding more recent literature from Italy and examples from the authoritarian regime in Poland. He describes the theory and practice of criminal law in the early years following the fascists' takeover of power in Italy as being partly liberal initially because they provided defendants with a certain amount of predictability about the outcome of lawsuits against them. In contrast to Italy, the German regime had in 1933 immediately set in motion a radical shift away from a penal policy that was predictable to defendants. Only very recently had the authorities in Italy begun "to imitate the Germans" (181). The fascist doctrine now in force had replaced the "separation of law from morality, an axiom in the period of competitive capitalism" with "a moral conviction derived immediately from the 'racial conscience' (*Volksgewissen*)"[36] (179).

Against this background, Kirchheimer lays out in detail the Nazi theory of criminal law, the changes in criminal law and criminal procedural law made between 1933 and 1938, and examples from the practice of adjudication and corrections in the Nazi state. Analyzing official crime statistics, he demonstrates that the number of acquittals had plummeted, the average length of jail sentences had increased by one-third, the share

36 One might argue about this translation by Kirchheimer. To a certain extent, *Volk* and *Rasse* certainly were interchangeable. Another option would be "conscience of the *Volk*," adding a definition of *Volk* as "people/nation in a racial sense, of common blood and with a common destiny." However, this might have been too unwieldy at this point in the publication (see Translator's Preface and Glossary).

of sentences of imprisonment with hard labor had increased, and the conditions of detention had continued to deteriorate. He highlights the return to capital punishment as particularly striking and illuminates how fines were used as a means of systematic dispossession.

Kirchheimer's analysis shows that all these increases in the severity of penalties had originated in the time before Hitler came to power; the programs of the proponents of penal reform had already reached their limits during the Great Depression since sufficient funding for humane corrections was no longer made available. The various ways in which criminal policy had been toughened after 1933 were caused by the economic crisis as a result of which broad segments of the population had lost social stability. The Nazi legal theories, combining elements of a biologistic doctrine of *Rasse* and predestination with the principles of retributive justice typical of the classical German doctrine of criminal law, merely gave them a new ideological justification. Kirchheimer referred to the changes in Nazi criminal procedural law, for which Schmitt, among others, was responsible, and the weakening of the function of the defense. He also mentioned Schmitt's criticism of the "abstractions" of liberal criminal law and his concrete-order thinking, which he had presented in his 1934 booklet *On the Three Types of Juristic Thought*, as the most influential legitimation in terms of legal theory for the changes that had been made (see 250). Concrete-order thinking following Schmitt robbed the defendants of any remaining certainty that their criminal proceedings would have a fair outcome.

Kirchheimer argues in the book for a criminal policy that guarantees the safety of the individual and social stability in society and for the duty of the state to create these two things. This is the context into which Kirchheimer's assessment of empirical findings regarding the link between crime rates and penal policy should be placed. He scrutinizes the effect of sentencing policies on the crime rates in four countries: England, France, Italy, and Germany. On the basis of his analyses of the official statistics from 1900 to 1932, he found that harsher sentencing policies by no means lowered crime rates. On the contrary. He observed that in England, a policy of reducing sentences in favor of probation and fines correlated with a marked drop in the general crime rate. In addition, he counters the ideological self-descriptions of fascist penal policy with empirical evidence. Contrary to its propaganda, there were two limits to the fascist states' attempts to lower the crime rate by means of a more severe penal policy. One limit was due to the economic situation at the time. Times of economic crises lead to an increase in the number of crimes. The second limit was the general "rationalization which modern industrial society requires" (206). Such needs for rationalization blocked the full development of the ideological penal program in fascist states because it would mean wasting human resources. Kirchheimer argues that the realization of the fascist penal program was fully effective only in the (considerably expanded) area of political offenses. Overall, however, the societal need for rationalization ran counter to the universal expansion of a repressive system of corrections.

Now, leftist and left-liberal reform programs were also subject to limits because of the rationalization that modern industrial society required. All attempts to lower the crime rate solely by means of a penal policy relying on lighter sentencing and resocialization were unrealistic. Their success still depended mostly on the societal system: "The crime rate can really be influenced only if society is in a position to offer its members a certain measure of security and to guarantee a reasonable standard of living" (207). Kirchheimer ends his book on a pessimistic note. Although progress in the social sciences had made the problem of corrections easier to understand and solve than ever before in human history, it seemed to him in 1939 that a fundamental improvement in corrections policy was "further away today than ever before because of its functional dependence on the given social order" (207).

The third point that is striking in his book is that Kirchheimer also discusses Hobbes. In Chapter 5, he gives a brief outline of the intellectual history of theories of punishment in the epoch of the Enlightenment. He praises the liberal heritage and makes Thomas Hobbes and his *Leviathan* the classical reference, even more important than the classical authors Beccaria and Montesquieu. For it was Hobbes who had already accomplished the separation of law from ethics as well as "a strict legal formulation of the idea of criminal culpability by placing it in close relationship with a legally defined fact" (73). Although Hobbes still passed moral judgment on human actions, its sphere of application was now limited, and moral judgments remained clearly distinguished from criminal sanctions. This separation was made easier by the principle of nonretroactivity in criminal law formulated by Hobbes in Chapter 27 of *Leviathan*. Kirchheimer extensively quotes passages where Hobbes cites the principle of the rule *nulla poena sine lege* (see 74). Thus, he declares Hobbes to be the precursor of Beccaria and the Prussian Civil Code of 1794.

It cannot be discerned clearly where Kirchheimer got the idea to give Hobbes such a prominent role in the book. He must have changed his assessment of Hobbes at some point, for in an essay completed in late 1932 he had criticized Hobbes's theory for allowing the sovereign to override individual liberties (Kirchheimer 1933d, 76). Considering Hobbes instead to be a genuinely liberal thinker along the lines of Tönnies's reception was part of Horkheimer's interpretation, as mentioned above. Another possible source is Franz L. Neumann, a close ally of Kirchheimer at the institute. Neumann had devoted an entire chapter to Hobbes in his doctoral dissertation *The Rule of Law. Political Theory and the Legal System in Modern Society*, which he had completed under Harold Laski in 1935. He opposed Schmitt's early interpretation of Hobbes in his 1921 book about dictatorship, instead advancing the hypothesis that rudiments of a genuine natural law were to be found in Hobbes's theory of the state which ultimately amounted to a limitation of state sovereignty and the recognition of liberal rights (see Neumann 1935, 100–113). If Kirchheimer had no opportunity to read Neumann's entire dissertation, he was in any case familiar with his 1937 article "The Change in the Function of Law in Modern Society" in the *Zeitschrift für Sozialforschung* in which Neumann summarized his interpretation of Hobbes (see Neumann 1937, 24–25).[37] One can only speculate whether Kirchheimer may also have been aware in his New York exile of Schmitt's new twist in his writings on Hobbes.

---

37 See also Neumann (1940, 346–347), where Neumann once again explicitly mentioned the "impossibility of retroactive legislation, especially in penal law" (Neumann 1940, 361).

5. Controversies over Nazism at the Institute of Social Research

At the same time as Carl Schmitt, in Berlin, began to interpret Nazi rule using a mythopolitical concept of the Reich, Otto Kirchheimer was involved in the internal debates at Horkheimer's Institute of Social Research (ISR) about the proper socio-theoretical categorization of Nazism in New York. The academic success of *Punishment and Social Structure* notwithstanding, Kirchheimer's funding as an employee of the institute was an ongoing issue, notoriously warranting his and the institute leadership's attention.

When it was foreseeable that the book would be completed by the end of 1938, he applied to the Social Science Research Council (SSRC) in New York for funding for a subsequent project, with Horkheimer's support. The proposal followed on from the empirical analyses in the three final book chapters. Horkheimer asserted that the project would fill an important research gap: "it has never been made clear whether there is a relation between a social stratification in a given society and the procedural treatment and the types of punishment."[38] The five-page exposé was titled *Criminal Law and Social Structure*. The study was planned to start at the beginning of the twentieth century. Although it was to focus on a comparison of seven European countries, Horkheimer assured the SSRC that "the corresponding American developments [would] always be kept in mind."[39] It was to investigate under what conditions which crimes and which groups of people were or were not criminally prosecuted. The second part of the project was to examine the various types of punishment once again. The SSRC's rejection arrived in the spring of 1939. Kirchheimer had no other option but to hope for alternative funding through Horkheimer's institute and participated intensely in the studies on Nazism which had just been incorporated into the work program of the institute from 1938 onward.

For the staff members of Max Horkheimer's institute who had fled Germany, the establishment and consolidation of the Nazi regime marked "the traumatic gravitational center of their entire academic and political orientation" (Dubiel and Söllner 1981, 8). Yet in the initial years after 1933, the institute's core team hardly worked on Nazism at all. One of the very few exceptions is an article by Herbert Marcuse on the struggle against liberalism in Nazi political thought.[40] The Weimar democracy, the welfare state, and, above all, the collapse of the Weimar Republic were not examined during the institute's first ten years, either. As late as early 1938, a few months after Kirchheimer's arrival in New York, this thematic gap was still apparent in the programmatic *Memorandum* in which Horkheimer outlined the ISR's research: the institute, he wrote, was concerned with developing a "comprehensive theory of society" (Horkheimer 1938a, 143). There was no mention of specific analyses of fascism or other political systems.

38 Letter from Max Horkheimer to the Social Science Research Council dated 17 December 1938. Max Horkheimer Papers, Na 1, VI 11, 370.

39 Max Horkheimer and Otto Kirchheimer: *Criminal Law and Social Structure. Research Project of Dr. Kirchheimer.* Max Horkheimer Papers, Na 1, IX 59, p. 2–7. —The project was planned to take eighteen months; the amount of funding applied for was $ 4,000.

40 Herbert Marcuse briefly discussed Schmitt in this article, but mainly with reference to his criticism of liberalism during the Weimar Republic (see Marcuse 1934, 4, 11, 21–23).

Almost two decades after the rise of fascism in Italy and more than ten years after the first major electoral successes of Hitler's party in Germany, the subject of Nazism was finally given priority on the institute's research agenda. After Nazi Germany had started the war in Europe, Horkheimer decided that besides antisemitism, the ISR should also address questions concerning the Nazi system's economic and legal order, class structure, culture, and state organization. He hoped that this thematic turn would augment the increasingly limited funding for the institute's staff with monies from US foundations. The institute's move toward research on Nazism kept the staff busy until the end of World War II, producing a series of publications with which the institute quickly made a name for itself in US academia—as Horkheimer had hoped. At the same time, however, profound differences between staff members in terms of political theory came to the fore, resulting in the dissolution of the institute. Kirchheimer was involved in these controversies from the outset and did not mince words when it came to criticizing positions held by Horkheimer and his inner circle that he considered to be misguided.

Although Kirchheimer knew that his employment at the ISR was precarious, he was certainly confident about the substance of his work there. He had grounds for optimism not least because he had been asked to do extensive groundwork for a programmatic essay envisioned by Horkheimer, which was published under the title "Die Juden und Europa" (The Jews and Europe). Horkheimer began work on the essay after the November 1938 pogroms in Germany; it was his first dealing explicitly with the subject of fascism. For almost a year, Horkheimer toiled away at the manuscript but did not agree to its publication in the institute's journal until after Germany's attack on Poland in September 1939—and only after considerable hesitation and several rounds of revision. The essay, which is the source of the famous dictum "Whoever is not willing to talk about capitalism should also keep quiet about fascism" (Horkheimer 1939a, 78), is a mixture of political observations, theoretical deliberations, and historical-philosophical reflections. The text is permeated with the fear that the fascist model of governance and a murderous form of antisemitism would sooner or later spread worldwide. Horkheimer advocated for a functionalist interpretation of antisemitism, which—in contrast to his views in the chapter on antisemitism in *Dialectic of Enlightenment*, his book co-authored with Adorno—assumed a primacy of economic factors for explaining the hatred of Jews.[41] According to this stance, the Jews were the circulation agents par excellence. They owed their emancipation solely to the fact that, as financial market actors, they were among the pioneers of capitalism, and, as lenders, they were indispensable in the sphere of circulation. As free circulation in the financial markets disappeared, the Jews were now "being run over" (Horkheimer 1939a, 89), thereby becoming superfluous.

Horkheimer also championed the same functionalist view in his theory of politics and his legal theory. To him, the sphere of circulation was simultaneously the actual social foundation of bourgeois democracy and the universality of the law. As the sphere of circulation became less important, democracy and ties to the universality of the law became obsolete, thus clearing the way for dictatorship and the abolition of the rule of law.

41 On the transformation of Horkheimer's functionalist interpretation of antisemitism and his transition to an interpretation rooted in the history of civilization, inspired by Adorno, from 1941 on, see König (2016, 220–244).

At Horkheimer's request, Kirchheimer had compiled a twelve-page manuscript providing an overview of empirical findings from various sectors of the economy in Italy and Germany, the two fascist states at the time, in the spring or summer of 1939 (see Kirchheimer 1939). With a variety of quantitative data, Kirchheimer documented a strengthening of private capitalism through reprivatizations under both regimes, the processes of concentration in various sectors at the expense of small and medium-size businesses, the reduction of wages and intensification of labor, the enduring importance of large-scale landholdings in agriculture, the failure of the expansion of the public administration, and increasing bureaucratization. Even if Horkheimer's published essay did not directly include any individual passages from Kirchheimer, let alone his tables of statistics, a precise reading clearly reveals that Kirchheimer's findings were incorporated into Horkheimer's statements about the role of business monopolies, increasing concentration in various sectors, and the expansion of the government apparatus (see Horkheimer 1939a, 79, 81, and 89).

When Kirchheimer's text and Horkheimer's use of it are compared, two things are striking. The first is the skill with which Horkheimer integrated into his general line of argument the loss of importance of the sphere of circulation in the two fascist states, which Kirchheimer had evidenced with copious data. The second is the degree to which he had massaged Kirchheimer's findings; the latter could probably hardly recognize his own work anymore. Horkheimer had made Kirchheimer's findings more pointed, to the extent that they expressed the complete disappearance of the sphere of circulation under fascism—which was not backed up by Kirchheimer's figures at all.

Kirchheimer's first longer article within the framework of the ISR's analyses of Nazism was published in the summer of 1940, in Issue 3 of *Studies in Philosophy and Social Science*, the first English-language issue of the previous *Zeitschrift für Sozialforschung*. The title of the article was "Criminal Law in National-Socialist Germany" (see Kirchheimer 1940b).[42] It is one of Kirchheimer's works most often quoted in the secondary literature on the Frankfurt School (see Buchstein 2019a, 51–59). Written in the complex political situation of the year 1939, the study stands up to later analyses of the development of criminal law in the Third Reich that were based on richer source material.[43]

Kirchheimer distinguished three phases in the Third Reich's theory of criminal law. In the first phase, an authoritarian theory of criminal law prevailed; this was expressed through weakening the status of the defense lawyer and imposing more severe sentences. The theoretical reasoning for the authoritarian theory of criminal law was based on the theory of free will according to which the objective characteristics of a crime were less important for sentencing than its subjective aspects, i.e., the alleged will of the perpetrator. In the second phase, with the proclamation of the racist state, the theory of criminal law was dominated by the phenomenological school of legal theory propagated at the University of Kiel. According to this theory, essentiality and intuition took the place of logical deduction as methods of legal assessment. It was not a particular crime or a particular will that made a person a criminal but the inner essence of their personality. Examination of the criminal's inner essence, their general predispositions, their pre-

42 The following page numbers refer to this text.

43 See the more recent works by Hartl (2000) and Pauer-Studer (2014).

vious convictions, and their character replaced scrutiny of the objective features of the crime. Kirchheimer knew the leading representatives of the Kiel School, including Ernst Rudolf Huber, Friedrich Schaffstein, and Georg Dahm, personally from the period during the Weimar Republic when he had had academic ambitions of his own. He began his remarks on the Kiel School by pointing out that their recourse to phenomenology "in its theoretical foundation" (445) was based on Carl Schmitt's attack on general conceptions, on normativism and positivism, and on his argument for orientation toward a Nazi understanding of the concrete order of life. It was on this basis that intuition and essence had been introduced as the best methods of discovering the criminal agent. In his article, Kirchheimer presented a highly sophisticated analysis that demonstrated how closely he had kept up with the current debates in legal theory as well as with legal practice in Germany. The materials he worked with were laws, legal articles in German academic journals, and collections of rulings handed down by German courts. It is worth examining four aspects of his article in more detail.

First, Kirchheimer describes the massive expansion of the areas in which criminal law was then applied to encompass areas of society previously not subject to it; the expansion went so far as to include foreigners' acts on foreign soil. Second, he sketches the recourse to *gesundes Volksempfinden* (assessment of a matter in accordance with the Nazi *Volksgemeinschaft* governed by the will of the *Führer*"; see Glossary) as the legal standard, which had advanced into the wording of individual laws and which required judges to use the purported will of the *Führer* as orientation. Third, Kirchheimer also notes "remarkable restraint" (449) in the legal practice of the *Reichsgericht*—after severely criticizing it in his Weimar days—in invoking precisely this *gesunde Volksempfinden*. The remaining "normativists" (446)—as Kirchheimer called these judges, using a term from Schmitt—at that court were following the letter of the law more closely. Kirchheimer saw this as a legal policy conflict between the "conservative" (448) normativists on one side and the "extremists" (448) on the other within the Nazi legal community. The extremists oriented their work toward Schmitt's methodological postulate of concrete-order thinking. Conversely, an "evident tendency to maintain rationality in the realm of criminal law" (450) was to be seen in the decisions of the *Reichsgericht*. Kirchheimer considered this to be a type of rationality following Max Weber: "This rationality requires that the statute is preserved as a main focus for the decisions of the individual cases" (450). However, this rationality was suspended whenever the matter in question concerned central ideological goals of the regime, the Nazi *Rassepolitik* (policy on *Rasse*) in particular. Fourth, at the institutional level, Kirchheimer describes the further suspension of judicial self-administration beginning in the second half of the 1930s. The dismissal and forced resignation of judges, several organizational changes, for example the establishment of *Sondergerichte* (special courts for political and especially serious crimes, feared for their swift and severe rulings that could not be appealed), amendments to the Code of Criminal Procedure, and above all the new wartime decrees "mark the last stage in the transformation of the judge from an independent agent of society into a technical organ of the administration" (462). Kirchheimer considered these more recent developments to be a third phase in the development of Nazi criminal law. In this phase, the "fight between normativism and the concrete conception of life" (463) in legal theory ever since Germany had attacked Poland had been settled. The new and "recent campaign for ruthless extermination" (463) of the population in the conquered areas was legitimized using the argument that such practices were the authentic expression of the concrete conception of life as a German.

Yet even this dense and grim description of the situation of criminal law in Germany did not prevent Kirchheimer from seeking potential rifts within the Nazi penal system, which appeared compact from the outside. He found them less in the ideological space of the controversies within legal theory, which were continuing to fester under the surface, but rather at the level of "departmentalization" (462) of judicial policy and the societal fault lines underlying it. The increased efficiency of the state and industrial machinery was paid for by the "complete subordination of man in his productive relationships to the disciplinary and penal machinery built up by the special services and by private combinations invested with the garments of public authority" (462). It was at this point that the inroads of the state on the daily life of the average appeared to be most striking and that "the exclusive predominance of strict power relationships [would] most likely create frictions" (462). Kirchheimer thought it difficult to see how the official goal of improving public morality could be achieved by a state founded on an oppressive political organization supervising and directing all spheres of life.

While Kirchheimer was working on his first article for *Studies in Philosophy and Social Science*, the conflicts at the ISR about the proper socio-theoretical categorization of Nazism were intensifying.[44] Perhaps it was Horkheimer's creative overinterpretation of Kirchheimer's commentary on his article on the Jews and Europe in 1939 that prompted Kirchheimer to criticize him more openly from then on. The first instance of this was in May 1940 regarding Horkheimer's manuscript "The Authoritarian State," in which Horkheimer followed Friedrich Pollock's hypotheses on the transition from monopoly capitalism to state capitalism, even down to the wording of individual passages (Horkheimer 1940, 106). He subsumed both the Nazi and the Soviet systems under the rubric of state capitalism. To Horkheimer, the authoritarian state was a societal form *sui generis*, with socialist and capitalist components. The difference between the authoritarian state and capitalism, he claimed, was in the control of the economy through the political system and the suppression of market competition; the difference between this and socialism was the long-term oppression of the masses and their economic exploitation.

Horkheimer was particularly critical of labor movement organizations. In his opinion, bourgeois institutional structures had prevailed in proletarian mass organizations as a result of the successful corporative integration of the working class into the state apparatus in the 1920s. Even during the Weimar democracy, oligarchical tendencies and the organizations of capital and labor had converged, and fascism was now able to benefit from this. Considering such a comprehensive mechanism for integration and oppression, Horkheimer saw no option for the labor movement to act as organized political opposition; instead, he expected a long period of rule for the authoritarian state. He pointed to the intellectual power of "critical theory" as well as to spontaneity and "human will" leaping out from human history and into the realm of freedom and solidarity as the only opposing forces and the last remaining glimmer of hope (see Horkheimer 1940, 107). His deliberations had consequences that touched the core of previous collaborations at

---

44 On this famous debate, see Wiggershaus (1995, 280–302).

the ISR. If, in the spirit of Walter Benjamin's hypotheses on the philosophy of history, the only purpose for a better world was to escape from the continuum of a history of decline, then the empirically observable societal tendencies and the specific actions of political actors would lose their significance for theory production. For this reason, Jürgen Habermas saw Horkheimer's essay as the beginning of his departure from the collective work at the ISR (see Habermas 1986, 167–169).

Horkheimer's manuscript had been circulating internally among his colleagues since the spring of 1940. Kirchheimer also contributed an internal memorandum with the opposite hypothesis to the internal debates at the institute, directly attacking Horkheimer's deliberations at several points. He referred to Horkheimer's "undisguised optimism with respect to human nature,"[45] despite expressing pessimistic expectations of the future, and criticized Horkheimer's reference merely to a will to freedom as the source of political changes. Above all, however, Kirchheimer objected to Horkheimer's systematic rejection of political mass organizations. These organizations were indispensable, he claimed, in the struggle to bring down the authoritarian state. If such an overthrow were to be brought about, then certain organizational "hardenings" which Horkheimer had criticized had to be tolerated. Finally, Kirchheimer developed an antithesis to Horkheimer's criticism of the bureaucratic administration of society, asserting that it was important to differentiate between bureaucracy as an instrument of domination and bureaucracy as an instrument for providing services. The latter, he claimed, would be essential in a free society of the future with a "socialist subsistence economy," so it was all the more important to contemplate the political forms of organization of such a society.

The discussion with Horkheimer about "The Authoritarian State" and also about his later essay "The End of Reason" proceeded by letter after Horkheimer had moved to his new domicile on the West Coast in the summer of 1942. This written discussion again illustrates his and Kirchheimer's opposing perspectives on fascism's mechanisms of exercising power. Although in a letter dated June 1942 Kirchheimer initially agreed with Horkheimer's criticism of the naive social democratic concept of progress in its struggle against fascism, his subsequent disagreement was sparked by Horkheimer's hypothesis that fascism shook the foundations of bourgeois anthropology because it taught individuals to fear things worse than death. According to Kirchheimer, the fear of death had historically been suspended time and again under different regimes but Nazism did not belong "in this category" because "the focus here seems to be on self-preservation of the individual." Thus, Nazism remained entirely within the scope of bourgeois anthropology. Kirchheimer assumed "conversely" that in fascism "nothing has remained of the individual besides self-preservation and fear of death, which are zealously fostered and cultivated by the 'pseudogroups' as *aréna dominationis* (the arena of domination)."[46]

---

45 This and the following quotations can be found in Kirchheimer (1940a).

46 All quotations in the letter from Otto Kirchheimer to Max Horkheimer dated 24 June 1942. Max Horkheimer Papers, Letters VI, 11 and 328. In his response, Horkheimer conceded that he took this empirical objection "very seriously." Nonetheless, he wanted to continue following up on the suspicion that designating self-preservation under fascism "as mere residuals of the bourgeois character do not do justice to the matter" and, in this context, referred to his current work on "fundamental philosophical questions" within the framework of the "project on dialectics" he was planning (let-

In 1941, Kirchheimer was given the opportunity to present his own view of Nazism more extensively in two essays intended to appear in the institute's journal, "The Legal Order of National Socialism" and "Changes in the Structure of Political Compromise." Yet this did not proceed without difficulties. Kirchheimer worked on these essays while the debate at the ISR about the appropriate way to understand Nazism had already erupted. Initially, the institute's leadership had planned to include the article on the legal order in the special issue of the institute's journal *Studies in Philosophy and Social Sciences* on Nazism but the text Kirchheimer submitted seemed too politically controversial to Horkheimer, so it was decided on short notice to publish the essay on the structure of compromise first. The essay on the legal order of Nazism was published in the following issue of the journal after being reworked multiple times (see Kirchheimer 1941d). The difficulties Kirchheimer faced before his essay on the structure of political compromise in the Nazi Reich, the first published on this subject, could find a place in the institute's journal can be reconstructed from various sources.

Kirchheimer had submitted a preliminary version of the text to the institute's director in the spring of 1941.[47] This version was criticized by Horkheimer and Adorno, and Kirchheimer was asked to fly to the West Coast so that they could all discuss the text. In May 1941, Kirchheimer spent several days in Santa Monica for "extensive discussions" about the text with Horkheimer, Adorno, and Herbert Marcuse.[48] Kirchheimer was asked to rework it to make his deliberations fit better into the theoretical approach championed by the institute's leadership. In July, Adorno reported to Horkheimer from New York that he was processing the essays for the special issue as intensively as possible and that he had "discussed Kirchheimer's with him in detail."[49] Pollock was assigned responsibility for Kirchheimer's essay on behalf of the institute's leadership. It is evident from a letter from Horkheimer to Kirchheimer that the fundamental differences between the latter and the institute's leadership in terms of political theory had not been resolved even after his trip to the West Coast: when Horkheimer asked how much progress Kirchheimer had made in revising the article, he called it "Klassenkompromisse im Staatskapitalismus" [Class compromises in state capitalism].[50] In other words, he wanted to read Kirchheimer's contribution from the perspective of the hypothesis of state capitalism he himself championed.

Kirchheimer stood his ground, however. On 11 July, he explained to Horkheimer that after various changes he thought the essay was now "finished,"[51] and Pollock would certainly inform him about this in more detail. The fact that neither Horkheimer nor Adorno was satisfied with Kirchheimer's contribution is clear in letters in which they considered

ter from Max Horkheimer to Otto Kirchheimer dated 6 July 1942, Max Horkheimer Papers, Letters VI, 11 and 320).

47 The first version is not to be found in the relevant archives in Frankfurt am Main and Albany.

48 Otto Kirchheimer in retrospect in his letter to Max Horkheimer dated 22 January 1943. Max Horkheimer Papers, Letters VI, 11 and 318.

49 Letter from Theodor W. Adorno to Max Horkheimer dated 2 July 1941 (Horkheimer 1996, 96).

50 Letter from Max Horkheimer to Otto Kirchheimer dated 27 June 1941. Max Horkheimer Papers, Letters VI, 11 and 349.

51 Letter from Otto Kirchheimer to Max Horkheimer dated 11 July 1941. Max Horkheimer Papers, Letters VI, 11 and 348.

removing the essay from the issue entirely.[52] However, this was not an option because of a lack of thematically suitable alternatives. For this reason, Horkheimer changed his mind several days later and asked Adorno to attempt to "adapt Kirchheimer's article linguistically in such a way that it might appear in this issue."[53] Adorno soon informed him that he had "already tinkered with it in New York."[54] Horkheimer was obviously worried that the first English-language issue of the institute's journal with a theoretical focus might appear to be too Marxist for an American readership. Consequently, he asked Adorno in late August to reconsider the article once again: "Please take another close look at Kirchheimer's essay from a tactical perspective, especially the beginning and the end of it, [...] and please ask [Leo] Löwenthal, if necessary, to induce Kirchheimer to make changes or cuts."[55] One week later, Adorno reported to Horkheimer, "I got along well even with wild Kirchheimer when I talked through his article with him."[56] Another week later, Adorno added reassuringly: "I have studied Kirchheimer's article again most precisely. I pushed through quite a few minor changes—all of them with a view to tactical censorship. I think we no longer need to worry about the issue."[57]

Which parts of the text and which passages from Kirchheimer's original manuscript fell victim to Adorno and Löwenthal's "tactical censorship" and which additions and improvements were made to the text in the editorial process can unfortunately no longer be ascertained because the source material is incomplete. Regardless, in the version of the essay finally published, Kirchheimer made no mention of Pollock's theory of state capitalism. Instead, he continued the analyses of the Nazi regime he had already submitted from 1933 on, elaborating in particular on the political structure of compromise in the Third Reich.

In his essay "Changes in the Structure of Political Compromise,"[58] Kirchheimer defined political compromise as the agreement between social groups about the system of government and the highest maxims of government action. He presented a three-phase model of the development of political compromise in the political systems on the European continent. In the first phase, liberalism, each social compromise is negotiated between members of parliament. In the second phase, mass democracy, social agreements are made between the major voluntary associations. In the third phase, fascism, the "heads of the compulsory estates" distribute state "power and booty" (264). Therefore, even if economic factors determine the underlying conditions of the social structure of

52 Letter from Max Horkheimer to Leo Löwenthal dated 21 June 1941 (Horkheimer 1996, 78).

53 Letter from Max Horkheimer to Theodor W. Adorno dated 13 August 1941 (Adorno and Horkheimer 2004, 188).

54 Letter from Theodor W. Adorno to Max Horkheimer dated 17 August 1941 (Adorno and Horkheimer 2004, 192).

55 Letter from Max Horkheimer to Theodor W. Adorno dated 28 August 1941 (Adorno and Horkheimer 2004, 208).

56 Letter from Theodor W. Adorno to Max Horkheimer dated 4 September 1941 (Adorno and Horkheimer 2004, 221).

57 Letter from Theodor W. Adorno to Max Horkheimer dated 13 September 1941 (Adorno and Horkheimer 2004, 228).

58 See Kirchheimer (1941a). The following page numbers refer to this text.

compromise, in the final analysis, it is political processes of negotiation in which specific political decisions are made.

Kirchheimer was of the opinion that fascism was a type of government whose "outstanding characteristics" seemed "at first glance" to be "greater independence and power" (273). In fact, fascism was a form of "domination by institutionalized monopolies" (274), which Kirchheimer attempted to explain in more detail using the example of Nazism in Germany. He characterized fascism as a stage in the development of capitalism in which individuals had lost their independence and the ruling groups alone were recognized as partners in forming political compromise. According to this model, the German economic system consisted of various monopolies in which competing elements merely had the status of oases. Such monopolies were to be found in three forms: the private monopolies in industry and agriculture, "the monopoly for industrial and agricultural labor," and "a public monopoly under joint state and party control" (274). The three monopolies formed the "backbone of a new system of guarantees" (274); the traditional organizations of the labor movement were excluded entirely in this constellation, for which reason the capitalists had lost the need for "the pressure potentialities of credit control" (274).

The fascist regime actively promoted the formation of further monopolies and cartels. Kirchheimer took up observations here that had already featured in his initial working papers for Horkheimer and provided structural economic data as evidence of this process. Against this background, he described in a further step the results of direct processes of negotiation between the top representatives of industrial and agricultural sector groups, the state bureaucracy, the Nazi party, and the military in making and applying laws. It was characteristic of the structure of compromise under Nazism, he claimed, that Hitler acted in his role as the "ultimate arbiter" (287) in the struggle among these five powerful groups only if they did not reach a decision among themselves. The leadership could arbitrate these conflicts so successfully "only because the unfolding program of [international] expansion [had] given the various groups the possibility of extending their activities . . . and of satisfying their desires without too much need of getting in each other's way" (287).

Kirchheimer emphasized the continuities of the Weimar Republic and Nazism in terms of their social bases. Both systems were founded on negotiating compromises about power and influence between the heads of the industrial and agricultural monopolies, the state bureaucracy, and the military. In one decisive aspect, however, the Weimar democracy and the Nazi regime were fundamentally different: the participation of the collective representation of labor in reaching compromise had been replaced with the Nazi party. With this description Kirchheimer openly contradicted Pollock's and Horkheimer's hypothesis of state capitalism. In contrast to Pollock and Horkheimer, he also underlined the fragility of the Nazi regime. In his view, fascism had established itself not as a stable political order in Germany, but as a system of constant rivalries between power groups that depended on expansion through military action and successful imperialist policies for its continued existence. If the tide of war were to turn against Germany, the system would be threatened by political instability.

Giving the example of competitive capitalism in the UK, Kirchheimer explained how economic and political institutional factors interacted. In this phase in the UK, money as the "all-embracing medium" (264) had left its mark on political institutions and set the transformation toward mass democracy in motion. Kirchheimer explained the new constellation of conflict emerging in this way using the example of private control over central banks for state credit policy. Liberating central banks from political oversight was itself a political decision made in the interest of financial capital in order to be prepared to withstand potential leftist majorities in parliaments. The independence of central banks from government policy made it possible to keep governments and parliaments under control in modern mass democracies, thus limiting the latitude available to political alternatives. Kirchheimer described the policy-setting role of central banks in monopoly capitalism in more detail using the example of France in the 1920s and during the *Front populaire* after 1936, which he had experienced firsthand during his exile in Paris. He described the same pattern using multiple examples: as soon as the French electorate had shown left-leaning tendencies and the governments had initiated social reforms, the central bank had used its veto power and taken rigid countermeasures, making governments fall and all reforms vanish.

6. Conclusion: A message across the Atlantic

Kirchheimer's deliberations on the structure of political compromise in modern mass democracies and their institutional restrictions remained without positive resonance at the ISR; Adorno even considered cutting them from the article entirely.[59] Nevertheless, his analysis of Nazism took on great importance for formulating the position opposing Pollock's hypothesis of state capitalism at the institute. For these deliberations were also the basis of Franz L. Neumann's hypothesis of polycracy, according to which the Nazi system was characterized by notorious struggles between four pillars of power: monopoly capital, the state bureaucracy, the party, and the *Wehrmacht*.

The publication of Neumann's 1942 book *Behemoth: The Structure and Practice of National Socialism*, a comprehensive description and analysis of the Nazi system running to over five hundred pages, marked the culmination of polemics in the controversies at the ISR. Neumann countered Pollock's programmatic concept of "state capitalism" with that of "authoritarian monopoly capitalism" (Neumann 1944a, 473). The book, several chapters of which Kirchheimer had contributed to with his expertise in criminal and private law, became the standard work of research on the Nazi regime in the United States and the United Kingdom shortly after its publication. An updated version published in 1944 helped staff members of the US occupation authorities understand the Nazi regime following the liberation of Germany from Nazism.

In the two years between the first and second editions of *Behemoth*, Carl Schmitt took up his mythological topic once again in a brief article titled "Behemoth, Leviathan und Greif" (Behemoth, Leviathan, and Griffin) in *Deutsche Kolonialzeitung*, the monthly magazine of the *Deutscher Kolonialbund* (German Colonial Association). In a casual, grandfatherly tone, Schmitt told his readers about the sea monster Behemoth being replaced by Great Britain's maritime global supremacy, which he now called Leviathan. Then he

59 Letter from Theodor W. Adorno to Max Horkheimer dated 17 August 1941 (Adorno and Horkheimer 2004, 192).

guided his readers' attention to the current situation: Germany had established itself as the leading industrial power in Europe since the end of the nineteenth century, and it was at the cutting edge of weapons technology with its *Luftwaffe* (Air Force). The new element decisive for winning wars was no longer the sea but the air. Thus, the two mythical beasts Behemoth and Leviathan were joined by "a large bird," the "griffin" (Schmitt 1943, 526). The new nomos of the Earth could "only emerge through battle," and it could be that following air, fire as the fourth element would either bring the history of mankind to an end or have it begin anew.

Schmitt demonstrated not only in such speculations but also in his book on Hobbes that he did not consider political theory to be an activity resting on the premise of people exchanging arguments intersubjectively in order to learn from each other. He did not argue, he insinuated. This was in stark contrast to Hobbes's rigid and strictly scientistic approach. In light of these methodological differences, it is difficult to understand why Schmitt held Hobbes in such great veneration. The philosopher Wolfgang Kersting boiled down Schmitt's relationship to Hobbes to the image of "picking flowers in a quarry" (Kersting 2009, 99). To Schmitt, Hobbes's oeuvre was like an old quarry he climbed down into, seeking wild and beautiful flowers. He plucked the blooms he was interested in, removing them from the settings of their arguments, and adorned himself with them.

The fragrance of some of the flowers Schmitt picked wafted further than intended. In his introduction to *Behemoth*, Neumann explained its unusual title, first discussing the monsters Leviathan and Behemoth against the background of Jewish eschatology. Then he mentioned Thomas Hobbes's books *Leviathan* and *Behemoth* as his references in intellectual history and explained why he had decided on *Behemoth* as the title of the book:

> Since we believe National Socialism is—or tending to become—a non-state, a chaos, a rule of lawlessness and anarchy, which has 'swallowed' the rights and dignity of man, and is out to transform the world into a chaos by the supremacy of gigantic land masses, we find it apt to call the National Socialist system *The Behemoth*. (Neumann 1944a, xii).

Of course, the rich secondary literature on Neumann's book later picked up on it paralleling Schmitt's choice of title for his *Leviathan*.[60] Duncan Kelly even called Neumann's book a "continuation of his dialogue with Schmitt" (Kelly 2002, 491). Kirchheimer was involved in choosing the title for Neumann's book. In a 1988 conversation, Ossip K. Flechtheim, who had collaborated on the book as well as Kirchheimer and Gurland, speculated in retrospect that Neumann and Kirchheimer had jointly hatched the idea for the title as they worked hand in hand, day in and day out, in their offices on Morningside Park in uptown Manhattan. They had wanted to "get back at Schmitt" and send him a "message across the Atlantic."[61] Whether or not this anecdote is accurate—it is a fitting allegory for the change in how the Leviathan was politically coded as it was transformed into the Behemoth.

---

60 See Rottleuthner (1983, 247–251), Scheuerman (1994, 123–156), Kelly (2002), and Bredekamp (2016, 63–68).

61 Ossip K. Flechtheim in a conversation with the author on 13 February 1988.

Chapter 10: Practicing Antisemitism and Analyzing Antisemitism

There can be no doubt that Carl Schmitt was antisemitic.[1] During the Weimar Republic, he only articulated his stance in his publications indirectly,[2] but he began to do so overtly when the Nazis took over the government in the spring of 1933. In the following five years, Schmitt became one of the constitutional scholars most vocally supporting the Third Reich's policy toward Jews. Whereas Schmitt's antisemitic actions took place before the eyes of the national and international public, it has only recently become generally known that Otto Kirchheimer also addressed the situation of Jews in Europe during the Nazi era. A research paper on antisemitism and the Catholic Church that he wrote in 1943 at the Institute of Social Research was finally published in 2018. He was also involved in the institute's empirical research on the prevalence of antisemitism among blue-collar workers in the US. Schmitt's antisemitism and Kirchheimer's analyses of antisemitism touched on each other multiple times.

The word "antisemitism" was coined in Germany in the latter third of the nineteenth century. From there, it spread to other languages in a matter of years. In contrast to the hatred of Jews in antiquity and Christianity's religious hatred of Jews, antisemitism was—in the antisemites' own understanding—directed against apparent or actual powerful positions of Jews in society, which they called "rule by Jews." It was a reaction to a "Jewish question" that was posed anew, and differently, because of the emancipation legislation. As a political catchword, antisemitism was post-emancipatory. At the same time, German antisemitism was always also a nationalist and anti-liberal movement. Among the various antisemitic streams in the Weimar Republic, Adolf Hitler claimed in *Mein Kampf* that the Nazi movement was not advocating for emotional antisemitism but for a decidedly "scientific antisemitism" that was based on a biologistic theory of *Rassen* (see Glossary). He also called it "*Antisemitismus der Vernunft*"[3] (antisemitism based on rea-

1 Numerous definitions of antisemitism exist, and there has been an extensive debate around them for decades. This is not the place to go into this debate or to make a decision for or against any particular definition.

2 See Gross (2000, 137–261) and Egner (2013, 351–355).

3 See Brumlik (2020, 67). The quote from Hitler is attested in Jäckel and Kuhn (1980, 89).

son). After the NSDAP took over the government, however, the term antisemitism became officially discredited. The Reich Propaganda Ministry admonished people to avoid the word "antisemitic" and instead use "*antijüdisch*" (anti-Jewish).[4] The *Handbuch der Judenfrage* [Manual on the Jewish question], first published in 1907, went through numerous editions, some with extensive revisions, during the Nazi period. Unlike the 48th edition from 1943, the 49th edition from 1944 gave the following directive on linguistic policy: The "incorrect catchword antisemitism" was to be replaced by the word "*Antijudaismus*" (anti-Judaism, Fritsch 1944, 18).

It is useful to look at the history of the word to better comprehend the decades-long debate over whether Carl Schmitt should be deemed antisemitic. Not surprisingly, Schmitt himself rejected this label, indignantly correcting any mention of it in the secondary literature about him by handwriting "critical of Jews" (see Hofmann 1995, XII) in the margin. Another way in which Schmitt tried to downplay his antisemitism is by using the term "anti-Judaism," also found in some of the literature. Those who prefer this term to describe Schmitt use a three-step argument. [5] First, they rigidly differentiate between a biologistic-racist form of hostility toward Jews ("antisemitism") and a rejection, based on the Bible and theology, of Judaism as a non-Christian religious community ("anti-Judaism"). In a second step, they give Schmitt credit for his Catholic background and the allegedly strong Catholic motives in his thinking evident in his work so that he can be allocated to the rubric of the tradition of anti-Judaism, which has less of a negative ring to it. In a third step, they either claim (incorrectly) that Schmitt never made biologistic-racist comments about Jews, or they dismiss[6] these comments as "lip service"[7] only. They consider the assumption that he never supported biologistic-racist thinking to be even more plausible because he did not hold the natural sciences in high esteem.

All this notwithstanding, the claim that biologistic-racist wording cannot to be found in Schmitt's writings is simply wrong, as I will show in more detail below. In the context of this chapter, the second step seems to be important only in terms of whether there had ever been a clear-cut categorial discontinuity between traditional religiously founded anti-Judaism and a Nazi "Aryan" form of antisemitism. Historian Rainer Walz has demonstrated that reducing the dichotomy to that between non-racist anti-Judaism and racist antisemitism does not do justice to the historical complexity of hostility toward Jews. A strict differentiation between premodern anti-Judaism and modern anti-

4 Directive for the press by the German government as of 22 August 1935, quoted in Nipperdey and Rürup (1972, 151).

5 Many arguments have been advanced countering the Catholic interpretation of Schmitt's oeuvre; I will not take this debate up again here. See Gross (2000) and (2016), Mehring (2014b, 20–29), and Neumann (2015, 374–391).

6 Although they do in fact differ in various details, all those who absolutely refuse to call Schmitt antisemitic and instead consider him to be part of the Catholic tradition of anti-Judaism apply this same three-step approach. Among the phalanx of apologists for Schmitt, see Schwab (1970), Bendersky (1983), Koenen (1995, 313–318), Tielke (2007), Quaritsch (2018, 85–88), and also the editors of Schmitt's diaries, Tielke and Giesler (2020, 51–52). Somewhat more nuanced: Strong (1996, xiv–xxiii).

7 Schwab (1970, 101) and Bendersky (1983, 381). Bendersky has slightly softened his position after reading Schmitt's diaries (see Bendersky 2016, 119).

semitism is historically inaccurate, both for the premodern period and for the nineteenth and twentieth centuries (see Walz 1995).[8] Moreover, ultimately, there must have been a particular reason why the struggle against those people whom the Nazis called "non-Aryans" in their theories on *Rasse* was primarily a struggle against Jews. As traditional Christian hostility toward Jews was a widespread belief, it served to amplify the propagandistic success of these pseudo-scientific theories. Racist hostility toward Jews cannot be separated entirely from older Christian anti-Judaism (see Katz 1989). The Nazis used all forms of racism and xenophobia in their agitation and were able to build on the religious resentment pervasive in the two largest Churches in Germany, the Catholic and the Protestant Church. Therefore, even if Schmitt's hostility toward Jews had a different religious source than Nazi antisemitism, it made it easy for him to align himself with the Nazi position (see Meier 1998, 152–156).

1. Schmitt's view of Kirchheimer: The "vile Jew"

In the early 1990s, two very old friends of Otto Kirchheimer's—Eugene Anschel, with whom he was friends from his high school days, and John H. Herz, who became one of his closest friends when they worked together at the OSS—looked back on their personal experiences as students of Schmitt's; Anschel in Bonn in 1927 and Herz in Berlin in 1928. Both referred to Schmitt's *Concept of the Political*. Antisemitism played an important role in both their memoirs. When Schmitt defined the enemy as "the stranger," "the other," as one who is "in an especially intense way, existentially something different and alien," or as an adversary who "must be repulsed or fought in order to preserve one's own form of existence" (see Schmitt 1932a, 27), his readers at the time understood these definitions of the enemy very specifically, their linguistic vagueness notwithstanding. "Nobody faced with such enemy definitions," Herz said after listening to Schmitt in person a couple of times in Berlin, "could escape a hidden, code-word type of reference" (Herz 1992, 308). Anschel, too, wrote that no one reading or listening to Schmitt's words at that time in Bonn could avoid associations with the antisemitic propaganda during the Weimar Republic that portrayed Jews as "alien," as "the other," who, despite all efforts to integrate, would always be outsiders hostile to, and endangering, the German-"Aryan" way of life. The same is true of sociologist Werner Sombart contrasting *Helden* (heroes) and *Händler* (merchants), which Schmitt regularly referenced in his seminars. According to Anschel, Schmitt not only stylized the Germans into heroes, as opposed to British merchants, but simultaneously linked the allegedly specific mentality of merchants and shopkeepers with a denigrating characterization of Jews (see Anschel 1990, 85).

Along with Eugene Anschel and John H. Herz, Kirchheimer had also registered the antisemitic tone Schmitt took in his Bonn days. In 2021, George Schwab reported in his memoir about his own failed dissertation at Columbia University in 1961[9] that Kirchheimer had claimed that Schmitt "was already an anti-Semite during the Weimar period" (see Schwab 2021, 175). Schmitt's diaries from 1925 to 1934, which were published

8 Walz uses the term "genealogical racism" as a common category.

9 This incident will be discussed in Chapter 17.

in 2010 and 2018, provide additional evidence of his antisemitic attitude even before his involvement in the Nazi regime. The diaries are filled with numerous notes about "a vile *Ostjude*;"[10] "vile Jewish women on the Kurfürstendamm;"[11] "creepy: Jews and socialists;"[12] referring to Berlin: "a shame and a disgrace that I am in this Jewtown, insulted and defiled by Jews;"[13] the "vile noise of the Jew-children and rabbis"[14] in the neighborhood; meeting a "likable man, hopefully not a Jew;"[15] "Jews being chosen and depraved;"[16] "disgust about being poisoned by Jews;"[17] and a "dream: the attack of *Juda*, I am murdered".[18] From mid-1930 to early 1933, his statements about Jews became increasingly frequent, disparaging, and aggressive, and, as of spring 1933, his diaries are literally riddled with antisemitic passages. Now, he also used the words "*ordinär*" (vulgar), "*gierig*" (greedy), "*elend*" (squalid), and "*betrügerische Assimilanten*" (fraudulent assimilationists). Readers have the impression that Hitler's takeover of the government opened another outlet for Schmitt to express his antisemitism.

This increase in Schmitt's antisemitic statements can be observed in relation to Otto Kirchheimer, too. He had invited him over for dinner one evening in November 1931. Schmitt wrote about his wife's behavior when Kirchheimer was at their home: "Duška foolishly talks about the Jews." And about the later part of the evening, when Schmitt and Kirchheimer had gone to a tavern to continue their conversation: "had a nice chat, I like him."[19] In the following months, Schmitt repeated multiple times how much he liked Kirchheimer. That changed when the political controversies between the two became increasingly intense in the summer of 1932.[20] It was in August 1932, when Schmitt had taken a prominent stance in relation to the coup of the Reich against Prussia and Kirchheimer had severely criticized him for this in an article, that Schmitt first used the word *scheußlich* (vile) to describe Kirchheimer in his diary: "Went for a walk through the Tiergarten [park] with Kirchheimer, vile fellow."[21] Previously, he had mostly used the word to describe other Jews, but not Kirchheimer. After Schmitt received the manuscript of Kirchheimer's fundamental critique of his work *Legality and Legitimacy*, they met in person to discuss the matter but did not reach a consensus. Exasperated, Schmitt noted in

---

10 "*Ein scheußlicher Ostjude.*" Carl Schmitt, diary entry of 10 January 1930 (Schmitt 2010, 4). *Ostjude* was a derogatory term for an Eastern European Jew.

11 "*Scheußliche Jüdinnen auf dem Kurfürstendamm.*" Diary entry of 30 July 1931 (Schmitt 2010, 128). Kurfürstendamm was the entertainment street in the west of Berlin.

12 "*Unheimlich, Juden und Sozialisten.*" Diary entry of 4 August 1931 (Schmitt 2010, 130).

13 "*Scham und Schande, daß ich in dieser Judenstadt bin, von Juden beleidigt und geschändet.*" Diary entry of 24 December 1931 (Schmitt 2010, 160).

14 "*Scheußlicher Lärm der Judenkinder und des Rabbis.*" Diary entry of 21 August 1932 (Schmitt 2010, 210).

15 "*Sympathischer Mann, hoffentlich aber kein Jude.*" Diary entry of 15 November 1932 (Schmitt 2010, 235).

16 "*Die Auserwähltheit und Verworfenheit der Juden.*" Diary entry of 14 January 1933 (Schmitt 2010, 252).

17 "*Ekel vor der Vergiftung durch Juden.*" Diary entry of 23 January 1933 (Schmitt 2010, 255).

18 "*Traum: der Anschlag von Juda, ich werde ermordet.*" Diary entry of 12 September 1927 (Schmitt 2018, 162). In German, *Juda* is a negative term for the totality of Jews.

19 Carl Schmitt, diary entry of 21 November 1931 (Schmitt 2010, 146).

20 For more on this, see Chapter 6.

21 "*Mit Kirchheimer durch den Tiergarten, scheußlicher Kerl.*" Diary entry of 25 August 1932 (Schmitt 2010, 210).

his diary: "It is pointless to speak with him, he simply does not want to see it. Vile, this Jew."[22]

2. Schmitt as an antisemitic Nazi propagandist

Schmitt agreed with the first rapid steps of Nazi policy toward Jews even before he publicly professed his allegiance to the Nazi regime and began to support it with his work on constitutional law. He noted in his diary that he was positively impressed by the boycott of Jewish stores declared by the NSDAP on 28 March 1933 under the pretense of world Jewry allegedly having declared war[23]—not without expressing his concern about some Jews who were his personal friends.

Just a short time later, Schmitt also declared his agreement with Nazi policy toward Jews publicly.[24] He played an active part in the new regime firing Jews and social democrats from universities and simultaneously made sure that students of his would be appointed to the now vacant chairs.[25] The antisemitism he propagated in numerous lectures and written works intensified in the years 1933 to 1938. As a young man, Nicolaus Sombart, an eccentric Berlin intellectual whose father, economist and sociologist Werner Sombart, had been friends with Schmitt since the 1920s, knew Schmitt well from many personal conversations on long walks in Berlin. He concluded in retrospect that in these public statements, Schmitt clearly and frankly spoke his mind and could at last "reveal his deepest, innermost conviction in an unguarded way" (Sombart 1997, 279).

A closer look at Schmitt's choice of words and way of arguing in his public statements during these years reveals, as I see it, five stages of systematic escalation.[26]

The first stage can be called somewhat muted antisemitism. It begins with Schmitt's article "Das gute Recht der deutschen Revolution" [The undeniable right of the German revolution], which he published on 12 May 1933 in the NSDAP rag *Westdeutscher Beobachter* [West German observer], twelve days after joining the party. The article made clear that combating Jewry and Judaism were at the center of the "national revolution" that Schmitt supported with his work on constitutional law. The specific purpose of the article was to justify the *Gesetz zur Wiederherstellung des Berufsbeamtentums* (Law for the Restoration of the Public Civil Service), which had been enacted three weeks earlier. All the subjects Schmitt addressed were linked in one way or another to the "Jewish question" through the polemic concept of *Gleichartigkeit* (see Glossary), which Schmitt considered to be the

22 "*Es hat keinen Zweck mit ihm zu sprechen, er will einfach nichts sehen. Scheußlich, dieser Jude.*" Diary entry of 6 November 1932 (Schmitt 2010, 231).

23 On the day of the organized boycott, which Schmitt experienced in downtown Jena, he wrote: "Met Koellreutter, had a nice conversation, very much in agreement about the Jews." "Boycott of the Jews, it did impress me." Diary entries of 29 March 1933 (Schmitt 2010, 275).

24 On antisemitism in Schmitt's activities and publications during the Nazi period, see Rüthers (1990, 96–103), Gross (2000, 42–136), Blasius (2001, 157–169), Mehring (2014a, 328–348), Neumann (2015, 174–399), and Mehring (2022, 364–387).

25 For example, Ernst Forsthoff was appointed to Hermann Heller's former chair in Frankfurt am Main.

26 For the following, see Buchstein (2021c).

key to the "entire body of laws at hand."[27] The new provisions on civil servants, attorneys, and physicians "cleanse[d] public life of non-Aryan *fremdgeartete Elemente* (elements foreign/alien to the *Volk* in an exclusionary and antisemitic sense; *Elemente* was a contemptuous term for opponents; *Volk*: people/nation in a racial sense, of common blood and with a common destiny; see Glossary and Translator's Preface), thereby securing the "*eigenvölkische Art der deutschen Geschlechter*" (German houses'[28] uniformity as a *Volk* of their own). "*Kein Fremdgearteter* [No one foreign/alien to the German *Volk*] should interfere in this "great [...] process of growth [...] even if they might have good intentions." On the new law, he wrote: "A *Volk* awakes to awareness of its own *Art* [the state of being characterized by *Blut* and *Rasse*; see Glossary] and recurs to itself and its peers."

A new revised edition of Schmitt's *Concept of the Political* was published at the same time as this article. In that work, he had emphasized that the ability to distinguish between friend and enemy was the decisive prerequisite for every political unit, so an invisible assimilated enemy represented the greatest challenge to his concept of the political. By referring back to his concept of the political, Schmitt supported the Nazis' intention to create ethnic *Gleichartigkeit* by proposing a radical and incisive definition of the enemy. The differentiation between *Gleichartigen* and *Fremdgearteten* (those belonging to the German *Volk*; those foreign/alien to the German *Volk*, in an exclusionary and antisemitic sense) became the epitome of his political thought, as the following maxim from his May 1933 article makes clear: "We learn once again to differentiate. Above all, we learn to differentiate friend and enemy properly" (Schmitt 1933b, 28).

Two things in this article stand out. First, that Schmitt replaced, or concretized, the neutral term *Homogenität* (homogeneity), which he had used during the Weimar Republic, with the term *Gleichartigkeit* here. In so doing, he used *Gleichartigkeit* as a biologistic metaphor. Yet he did not define the substance of this *Gleichartigkeit* more precisely, applying positive criteria. Instead, although Schmitt called it something "deeper," he differentiated it from other words only in a negative sense. Second, Schmitt did not mention Jews explicitly even once in the entire text. Instead, he chose paraphrases and allusions. For example, he criticized the *Rechtsstaat* for making it possible for "usurers" (Schmitt 1933b, 29) to have driven thousands of German peasants into misery. This was one of the most commonly used stereotypes of Jews in German antisemitic propaganda. Schmitt also insinuated that legal positivism was a Jewish way of understanding the law. The legal positivists defended legal equality. Schmitt ironically illustrated this claim with the following words: "baptized was baptized" (Schmitt 1933b, 29). Thus, he insinuated that unbaptized people did not deserve to be integrated as equals into the German legal community. He also criticized the formalism of legal positivism. He wrote: "Thousands of *Fremde* (people foreign/alien to the *Volk*, in an exclusionary and antisemitic sense) were permitted to change their names which would have identified them and received the permission of the authorities to deceive harmless Germans by using names that give the impression of trustworthiness" (Schmitt 1933b, 29). In other words, if Jews were permitted to change their names, then "harmless Germans" would not realize they were in fact dealing with

27 This and the following quotes are from Schmitt (1933b, 28–31).

28 Houses in the sense of kinship groups of virtually noble lineage; emotionally charged term evoking mystical blood ties (see Translator's Preface).

Jews, whom he considered untrustworthy by definition. Here, Schmitt connected antiliberal and antisemitic polemics, as was typical under Nazism. Even if he did not mention Jews explicitly in the article cited —just as in his other legal commentaries from the first months of his renewed involvement in constitutional law[29]—he could safely assume that these whispered intimations made abundantly clear to his readers what he was talking about: namely, that Jews were the beneficiaries of equality before the law and the liberal state under the rule of law, which he vilified, at the expense of non-Jewish Germans.

In the 1933 edition of *The Concept of the Political*, Schmitt even took a further semantic step by connecting the political concept with racist and biologistic ones. A newly added passage reads:

> For this reason, only those involved can identify cases of extreme conflict; in particular, each of them can only decide themselves whether the characteristic of the *Fremde* [foreigners/aliens, in an exclusionary and antisemitic sense] of being different implies the negation of their own kind of existence in the concrete case of conflict at hand, for which reason it must be repelled or fought against in order to rescue their own way of life commensurate with their own way of being (Schmitt 1933i, 8).[30]

In the revised edition, *Fremde* becomes identical to *Feind*. Schmitt complemented the original term *Feind* (enemy) with that of the *Fremde* (foreigner/alien, in an exclusionary and antisemitic sense) in that he still used "enemy" but added the meaning of *Fremde* to it (see Schmitt 1933i, 6 and 8). Thus, he transformed the dichotomy of friend and enemy into the dichotomy of *artgleich* (belonging to the *Volk*, in an exclusionary and antisemitic sense) and *artfremd* (foreign/alien to the *Volk*). Anyone who was not of the same *Art* was therefore automatically an enemy.

The second stage was antisemitism expressed as a form of biologistic racism. It started with Schmitt's lecture at the Deutsche Juristentag, a conference in Leipzig in October 1933. Addressing the audience of 12,000 legal experts as "Dear *Volksgenossen* [*Volks*comrades; see Translator's Preface regarding *Volk* as a prefix]," Schmitt opened the conference on the morning of 3 October and Hitler gave the closing speech in the evening. Schmitt noted in his personal diary that day: "Wonderful speech by Hitler about the total state. I feel very comforted."[31] In his own speech, Schmitt replaced the term he had previously used, *Gleichartigkeit*, with "*Artgleichheit*" (Schmitt 1933f, 67) (belonging to the *Volk*). Although both terms are comprised of the same components *Art* (the state of being characterized by *Blut* and *Rasse*, see Glossary) and *gleich* (equal, similar, alike), there is a subtle but important difference in meaning: whereas *Gleichartigkeit* may or may not be used in a biologistic sense, *Artgleichheit* is unequivocally biologistic (see Glossary). Schmitt's use of the term in October 1933 was certainly biologistic.[32] For one thing, he declared *Artgleichheit* to be the "key concept" of the new Nazi state "indispensable for

---

29 See Schmitt (1933a), (1933g), and (1933h).

30 For a systematic comparison of the different versions of this passage in the book, see Walter (2018, 284–286).

31 Carl Schmitt, diary entry of 3 October 1933 (Schmitt 2010, 305).

32 With this interpretation, I contradict those interpretations that draw a fine line between a biologistic form of antisemitism promoted by the Nazis on the one hand and Schmitt on the other.

reasons of legal systematics." For another, he linked *Artgleichheit* with the concept of the *Führer* (Schmitt had not done so previously with the concept of *Gleichartigkeit*): "*Führen* [literally: leading; see Glossary on the concept of the *Führer*] is not commanding; *Führen* is not a dictatorship; *Führen* is something that rests on the *Artgleichheit* [state of being of the same *Art*] of the *Führer* and *Gefolgschaft* [subordinates, see Glossary]" (Schmitt 1933f, 67).

Although he added that the concept of *Artgleichheit* was "basically clear" and had already "prevailed," Schmitt again avoided giving a more precise definition of this biologistic notion he had elevated to become his key concept. Instead, he gave specific examples with the intent of triggering particular associations among his audience. For instance, he emphasized the legitimacy of the Nazi revolution; Nazi legality had to be "*wesens- und artgetreu* [faithful to its own essence and *Art*, the state of being characterized by *Blut* and *Rasse*; see Glossary]" (Schmitt 1933f, 68). And Schmitt added: "We need not permit a *Volk* with a different opinion or mindset to lecture us about this" (Schmitt 1933f, 68). Thus, Schmitt constructed a connection between *Art* (see Glossary) and *Denken* (thinking). The *Artfremde* (individuals alien to the *Volk*) could not think in a German way, even if they wanted to. They could not leave behind the biological character with which they were endowed by adopting German cultural practices. The thrust of these words of Schmitt's is unequivocal: it was Jewish jurists who remained *fremd* to the German manner of legal thinking.

At the end of his talk in Leipzig, Schmitt offered his audience another subject for antisemitic association: the will of the *Führer* was directly applicable law and "we" Nazis should not permit ourselves to be misled by sophistic antitheses about politics and law. Following the will of the *Führer*, Schmitt stated, "is, as Heraclitus said, also a *nomos*" (Schmitt 1933f, 68). The concept of the *nomos* was to replace the concept of the liberal rule of law. Schmitt used the ancient Greek word *nomos*[33] in the sense of a particular spatial order, a particular social unit, and a particular "*Ortung und Ordnung* [spatial location and order]" (Schmitt 1933f, 68). Schmitt's turn to the new term *nomos* evoked antisemitic associations, too: the concept of the *nomos* liberated allegedly German legal thinking on the one hand from allegedly Jewish legal thinking on the other, with the former rooted in the will of the *Führer* and the latter characterized by thinking in terms of laws and by its purported strictness, fetish for rules, and mercilessness.

The third stage was explicit outspoken antisemitism. It was only now that Schmitt explicitly mentioned the Jews he was fighting by name. He did so for the first time in his essay "Nationalsozialistisches Rechtsdenken" [Nazi legal thinking] in the spring of 1934, in which he presented a summary of German legal history from a Nazi perspective. In this view, the legal system in Germany during the Middle Ages was characterized by the type of the wise, experienced, and just man of "*gesund* concrete-order thinking" (concrete-order thinking corresponding to the norm of the NSDAP; see Translator's Preface

---

See Bendersky (1983, 227–236), Koenen (1995, 313–318), Strong (1996, xiv–xvii), Tielke (2007), and Tielke and Giesler (2020).

33 The ancient Greek word *nomos* has two meanings: *nómos* (emphasis on the first syllable) in the legal sense of "law" and *nomós* (emphasis on the second syllable) in the spatial sense of "district." Schmitt connected these two semantic aspects in his use of the word (see Chapter 12).

and Glossary) (Schmitt 1934h, 157). This harmonious legal concord had been disrupted from the fifteenth century on, when a type of legal thinking oriented toward Roman law began to prevail. With its abstract rules and difficult, delicate differentiations, it became a kind of book learning removed from the problems of practical life. Yet the "normativist type of thinking" did not achieve a breakthrough in Germany until the nineteenth century. Schmitt traced the cause of this development back to the "influx of the Jewish guest-*Volk*" (Schmitt 1934h, 159). He mentioned two reasons why Jews had played a decisive role in this fallacious development of legal thinking in Germany. First, "because of the *Eigenart* (nature, or: its own *Art*, the state of being characterized by *Blut* and *Rasse*, see Glossary) of the Jewish people, which for millennia had been living not as a state and on one soil but only in the law and the norm, and was, in other words, most literally 'existentially normativistic'" (Schmitt 1934h, 159). Second, not only was normativism part of Jewish nature but it was also in their practical interest. "A *Fremder* [a person foreign/alien to the *Volk*]," Schmitt claimed, viewed the right of a *Volk* hosting him "solely from the aspect of legal certainty" (Schmitt 1934h, 159). And, he continued, that was not even surprising, given that "the Jewish guest-*Volk* does not belong to the reality of the *Volk*, the reality in which it lives" (Schmitt 1934h, 159). "*Der Jude*" (see Glossary) simply wanted the law to be "the schedule in order to know when and where he can embark or disembark" (Schmitt 1934h, 159). Although another reason was not mentioned explicitly by Schmitt, it is a subtext at various points of the article: he onviously felt that Jews were the people of the book and of constant interpreting and disputing and could therefore outdo their fellow (non-Jewish) German jurists in such debates.

In the fourth stage, Schmitt expressed his antisemitism not only explicitly but also militantly. It began with Schmitt's commentary on the infamous Nuremberg Laws passed on 15 September 1935 at the Reichsparteitag der Freiheit (Reich Party Rally of Freedom) held in Nuremberg. At the party rally, Hitler proclaimed the *Reichsbürgergesetz* (Reich Citizenship Law), the *Reichsflaggengesetz* (Reich Flag Law), and the *Gesetz zum Schutze des deutschen Blutes und der deutschen Ehre* (Law for the Protection of German Blood and German Honor). One new aspect of these laws was that they were no longer about "non-Aryans" but explicitly about "Jews." According to the Nuremberg Laws (see Reichsgesetzblatt I 1935, 1145–1147), marriages "between Jews and citizens of German or *artverwandten Blutes* [blood related to the German *Art*, the state of being characterized by *Blut* and *Rasse*, see Glossary]" were prohibited, and Jews were banned from hiring "Aryans" as domestic staff or hoisting the Reich flag. Violations of these laws were subject to severe punishment. Only two weeks after the laws were proclaimed, Schmitt commented on them for the *Deutsche Juristen-Zeitung*. He praised the new laws, calling them "the constitution of freedom, the core of our German law today" (Schmitt 1935a, 282). Countering the oppression of Germans as "slaves to interest and wages," countering "Germany's enemies and parasites," countering the "demon of *Entartung* [degeneration or decline due to biological or cultural factors; see Glossary]," countering *geistige Fremdherrschaft* (rule over the minds/spirit/intellect of the *Volk* by foreigners/aliens, in an exclusionary and antisemitic sense)," the Nazi revolution had finally made the concepts of the constitution "German again."[34] In other words, Schmitt considered the Nurem-

---

34 All quotations in this sentence are from Schmitt (1935a, 282–284).

berg Laws to be Germany's new constitution. For decades, German lawyers and legal scholars had been entangled in the "conceptual web of un-German systems" (Schmitt 1935a, 283) and had surrendered to liberalism. This epoch of German legal history had now been brought to an end once and for all. If that were to prove untrue, Schmitt uttered an open threat to Jews:

> The Führer has mentioned that in the event that the current regulation of the situation of the Jews does not achieve the desired result, there would be the possibility of a fresh review, and he presented the prospect that then, a law would assign the solution to this question to the party. This is a serious warning (Schmitt 1935a, 283).

As combative as Schmitt sounded here, he was barely capable of saying anything at all about the characteristics distinguishing "Aryan" and "Jewish" by using expressions such as the "voice of German blood" (Schmitt 1935a, 282) or "German substance" (Schmitt 1935a, 283). That did not change when he spoke about applying the Nuremberg Race Laws in practice at the annual conference of the German branch of the International Law Association in October 1935, giving reasons for the ban on non-Jewish Germans marrying Jews abroad (see Schmitt 1935c, 377–384).

Following the Nuremberg Laws, Schmitt and others provided the academic foundations for the antisemitic measures in the Reich. During a trip to Italy in spring 1936, Schmitt called "the problem of *Rasse*" (Schmitt 1936e, 334; see Glossary) the greatest difference between Nazi and fascist legal science and praised the "biological well-being" of the German *Volk* in the form of its "purity of blood" (Schmitt 1936f, 346), which was the objective of the Nuremberg Laws. In his lectures at the university in Berlin, he also made strongly antisemitic remarks in the presence of his students (see Mehring 2014a, 342). Schmitt's militant antisemitism in his publications and his role in organizing legal scholarship culminated in the autumn of 1936. On 3 and 4 October 1936, a conference of the former Bund Nationalsozialistischer Deutscher Juristen (Association of National Socialist German Legal Professionals, BNSDJ), which had been renamed Reichsgruppe Hochschullehrer des Nationalsozialistischen Rechtswahrerbundes (NSRB; see Glossary), took place under his direction.[35] Preparations for the conference had already begun in early 1936, that is, even before the point in time when Schmitt realized that the SS-*Sicherheitsdienst* (intelligence service of the Third Reich, SD) was gathering information about him.[36]

Schmitt formulated three guiding principles in his keynote speech. He started with a "statement by the Führer in his book *Mein Kampf*" (Schmitt 1936c, 482). Schmitt reminded his audience of Hitler's remarks about Jewish "dialectics"[37] in the book and chose another quote from the book as the first guiding principle for his talk: "By fending off the

---

35 For more details about this conference, see Hofmann (1988).

36 The point in time is relevant inasmuch as some authors cling to the legend that Schmitt had organized the conference merely as a means to protect himself against the SD's harassment; in this vein, see Bendersky (1983, 234–237), Koenen (1995, 709–720), and Tielke (2007).

37 Schmitt (1936c, 482). See Hitler (1925, 225).

Jew, I fight for the work of the Lord" (Hitler 1925, 231).[38] In his second guiding principle, Schmitt stated that legislation on *Rasse* had been successfully completed. But now, what remained was the following task: "tireless education to keep the German *Volk* engaged in recognizing the Jewish danger" (Schmitt 1936c, 483). In his third guiding principle, he reminded his audience of the "deadly poison" (Schmitt 1936c, 483) with which Jewry and Judaism had for decades permeated the German state and science unhindered. Schmitt demanded that his audience and the German legal community as a whole should take a significantly more combative stance toward Jewry and Judaism: "A merely emotional antisemitism and general rejection of some particularly obtrusive and unpleasant Jewish phenomena are not enough; what is needed is security based on knowledge"[39] (Schmitt 1936c, 482). Adolf Hitler, Schmitt told his audience, had already found such security based on knowledge as a young man in Vienna before the war—Schmitt may have alluded to Hitler's wording "*Antisemitismus der Vernunft*" (antisemitism based on reason), mentioned above.[40] Such security based on knowledge was also necessary according to Schmitt because the Germans had proven to be particularly susceptible to "Jewish infection" (Schmitt 1936c, 484) in the past. It was only thanks to the scientific study of *Rasse* that the difference between the Jews and all other *Völker* (peoples/nations in a racial sense, of common blood and with a common destiny; see Glossary) had finally been identified clearly.

In his closing remarks, Schmitt mentioned a number of specific measures that had to be taken in order to make the knowledge gained from the scientific study of *Rasse* directly applicable in the practice of law. Schmitt demanded that a bibliography be prepared with precise information about which author in the field of law "[was] a Jew and which [was] not a Jew" (Schmitt 1936d, 486). Building on this work, the next step would be a "cleansing of the libraries"; thus, students were to be "protected from confusion" (Schmitt 1936d, 487). Third, he thought the problem of citations was of decisive importance. Following a conference like the one that had just taken place, it was "no longer possible to quote a Jewish author like any other author" (Schmitt 1936d, 487). It was nothing less than irresponsible to quote a Jewish author as an authority in an area of the law. Jews per se had no authority in German legal thinking. If a Jewish author was to be quoted nonetheless, then the word "Jewish" had to be added. Then and there, Schmitt helped his audience become accustomed to this new practice of attaching a literary yellow star, speaking of the "Jew Kelsen" and calling Friedrich Julius Stahl, the ideological leader of the Protestant Prussian high conservatives, "Stahl-Jolson"[41] to indicate that he had been born into a Jewish family and had only later converted to Christianity. Mentioning the word "Jewish," he claimed,

---

38 "*Indem ich mich des Juden erwehre, kämpfe ich für das Werk des Herrn*" (Hitler 1925, 231), see Schmitt (1936a, 483).

39 "*Erkenntnismäßige Sicherheit*" (Schmitt 1936c, 482).

40 For a discussion of Hitler's wording in the context of Schmitt's antisemitism, see Gross (2000, 54–55 and 125–126).

41 Schmitt attempted multiple times in various works to expose how the most influential political philosophers of Prussian arch-conservatism had been deceived by Jews, see (1933c, 34), (1935a, 282), (1936g, 262), and (1938a, 69–75). However, it seems he had doubts about exposing Stahl as Jewish and began to research Stahl's genealogy. In a letter to Ernst Jünger dated May 1935, he reported that he wanted to spend a whole day in the archive in Wolfenbüttel to "collar *der Jude* [...]

would bring about a "holy exorcism" (Schmitt 1936d, 488). Fourthly, Schmitt took on the question of future research into legal history. He suggested setting several dissertation subjects examining what he considered to be the detrimental Jewish influence on legal thinking in Germany in the past.

At the end of his closing remarks, Schmitt returned to the fundamental principles once again. "*Der Jude*" (see Glossary) had "a parasitic, a tactical, and a merchantlike relationship" (Schmitt 1936d, 489) to German intellectual labor; with his soul of a merchant, he was to be found wherever he could make a profit (Schmitt 1936d, 489). One of the "diabolical traits" of *der Jude* that Schmitt mentioned was that others could not easily detect him. Jews could even be found among nationalists and conservatives. He explained this phenomenon of ideological diversity, which he described as a problem, with the "overall Jewish behavior of changing masks, reflecting diabolical ulterior motives" (Schmitt 1936d, 490). Schmitt thus linked up with the biblical image of the Antichrist as well as Hitler's social Darwinist notion of Jewish virtuosity in mimicry. The global power of "*der Jude*" did not tolerate *völkisch* (of the *Volk*, chauvinistic-nationalistic, antisemitic; see Glossary) productivity. Otherwise, his own kind of existence would be proven wrong. With reference to Hitler and his warning of "Jewish dialectics" in *Mein Kampf*, Schmitt declared "*der Jude*" to be the "mortal enemy of any true productivity of another *Volk*" (Schmitt 1936d, 490, 491).

Jewish legal scholars who had been driven into exile keenly observed and vehemently commented on the conference Schmitt had organized. The most extensive reaction was from Hugo Sinzheimer in his exile in the Netherlands. During the Weimar Republic, Sinzheimer had been the editor of the journal *Die Justiz*, which had published some of Kirchheimer's work. *Die Justiz* was published by the Republikanische Richterbund (Republican Judges' Association), the small organization of the minority of jurists loyal to the republic. Kirchheimer and Sinzheimer had met at its events and knew each other well from Social Democratic Party circles.[42] Two years after Schmitt's conference, Sinzheimer's book *Jüdische Klassiker der deutschen Rechtswissenschaft* [Classic Jewish authors of German legal science] was published in Amsterdam. In twelve individual portraits, he described the contributions of Jewish jurists to legal thought in Germany during the nineteenth century and the first third of the twentieth century. Sinzheimer directly attacked Schmitt and his talk of the diabolical Jewish masquerades in the final chapter, employing a dual strategy. First, he showed that the image of the masquerade fit best of all to the German Romantic Adam Müller, whom Schmitt had derided in his 1919 book *Political Romanticism*. Second, Sinzheimer accused Schmitt of staging such masquerades himself by supporting the Nazi regime; after all, he had celebrated Hugo Preuß, who was Jewish, as a great and fearless legal expert in 1930, among others. "A mask has fallen"—it was with these words that Sinzheimer summarized his remarks

---

Joel Jolson again." Letter from Carl Schmitt to Ernst Jünger dated 24 May 1935 (Schmitt and Jünger 1999, 48).

42 Hugo Sinzheimer survived the German occupation of the Netherlands in hiding. He died of exhaustion a few days after liberation in 1945. Kirchheimer later wrote a touching commemorative article (Kirchheimer 1968a).

about Schmitt—"but the mask that has fallen is not a Jewish mask" (Sinzheimer 1938, 247).

In the fifth stage, Schmitt cloaked his antisemitism in pseudoscientific terms. The manifest expression of this phase is Schmitt's 1938 book *The Leviathan in the State Theory of Thomas Hobbes*. An initial attempt is to be found in his 1937 lecture in Berlin "The State as a Mechanism in Hobbes and Descartes" in which he believed he could prove the subversive influence of Jewish thinking using the example of Jean Bodin, the French jurist and political theorist of the concept of sovereignty. Bodin had been familiar with kabbalistic writing, and consequently identified Leviathan as being an "all-demanding Moloch or an all-trampling Golem" (Schmitt 1937c, 95). Schmitt stated that it was imperative to conduct "detailed historical research" (Schmitt 1937c, 95) on the kabbalistic utilization of the myth of Leviathan.

Schmitt claimed to fill this alleged gap in the book he published a year later. In his foreword to *The Leviathan*,[43] he emphasized that he would treat the subject "with scientific objectivity" (5). Schmitt's central hypothesis about the failure of the symbol of Leviathan and the separation of the internal and the external has already been described in detail in the previous chapter. In the present chapter, I would like to add how Schmitt argued that it had even been possible that the crack integral to Hobbes's theory of the state had resulted in the entire theory coming apart. Schmitt stated that "a liberal Jew noticed the barely visible crack" (57) in the theoretical justification of the sovereign state shortly after *Leviathan* had been published. Jewish philosopher Baruch de Spinoza had immediately taken the seed planted by Hobbes in separating the internal and the external to its extreme outcome "until the opposite was reached" (57) and the "Leviathan's vitality was sapped from within and life began to drain out of him" (57). As a Jew, Spinoza had approached Hobbes's theory of the state from the outside and had purposefully misrepresented it. Whereas Hobbes had primarily been concerned with public peace, Spinoza had made the opposite, namely individual freedom of thought, his form-giving principle. Spinoza had thus undertaken "*eine kleine, umschaltende Gedankenbewegung aus der jüdischen Existenz heraus* [a small switch in his thinking arising from his Jewish existence]" (Schmitt 1938d, 88–89).

Although Schmitt felt that freemasons, illuminates, Pietists, Rosicrucians, and all kinds of other sectarians were also among the advocates of such inner reservations, it was "above all" the "restless spirit of *der Jude*" that knew best how to "exploit the situation" (60) until the relationship of public and private was turned upside down. Schmitt mentioned Moses Mendelssohn, the German-Jewish philosopher of the Enlightenment, by name; endowed with the "unerring instinct" for undermining state power, he served to "paralyze" the German *Volk* and to "emancipate" (60) his own Jewish *Volk*. Schmitt characterized the role of Friedrich Julius Stahl, who championed the "monarchical principle," as particularly perfidious here, again using the name "Stahl-Jolson." Stahl had been the "boldest" (70) representative of the "Jewish front" (70). He had outright infiltrated the Prussian state and the Protestant Church in order to "confuse ideologically and paralyze spiritually" (70) the inner core of Prussian statehood. Schmitt discerned in Stahl's plea for

43 See Schmitt (1938a). The following page numbers refer to this book.

a moderately constitutionally bound Prussian monarchy what he considered the typical Jewish "deceitful manner to mask his motivation" (70).

Schmitt used this to construct an antisemitic stab-in-the-back myth of his own: the constitutional monarchy à la Stahl, Schmitt claimed, had ultimately resulted in the Prussian military state having to collapse under the severe test of World War I in October 1918. Schmitt argued that Stahl had contributed to "continuing the metaphor, done his part to *verschneiden* a Leviathan full of vitality."[44] The German word "verschneiden" has four semantic fields: (1) to cut incorrectly, (2) to cut back, as in pruning a hedge, and (3) to castrate. Moreover, *verschneiden* is close to the German word *beschneiden* (to circumcise), and (4) it could be construed as "to circumcise incorrectly"; it conjures up associations of a castration performed by or on a Jew. A Jew is circumcised—and to Schmitt, that also symbolized his intellectual castration. In any case, Schmitt intended a negative connotation here.

A second aspect of Schmitt's work on *Leviathan*, which he considered to be strictly scientific, refers to the origins and varying interpretations of the biblical myth of Behemoth and Leviathan. Schmitt closely linked this discussion with the history of the reception of *Leviathan* he described. He made a rough distinction between two lines of argument that had been developing since the Middle Ages: one Christian, to which he devoted only a few lines, and one Jewish, which he explored in more depth. For it was the "traditional Jewish interpretation" (82) of the myth of Leviathan that he considered to be responsible for the failure of Hobbes's symbol. Schmitt believed this interpretation was the specific expression of the "unique, totally abnormal condition and attitude" (8) of the Jewish *Volk* toward other *Völker*. The myth according to Schmitt said that Leviathan and Behemoth symbolized the unbelieving peoples battling each other. "The Jews," he continued, "stand by and watch how the people of the world kill one another" (9). To them, the practices of killing and performing religious slaughter on each other were lawful and kosher. And that was why they ate and lived on the meat of the peoples that had been killed. In another version, God spent a few hours each day playing with Leviathan until the latter was weary. Schmitt believed that Leviathan and Behemoth were "Jewish battle myths of the grandest style" (9). He claimed that seen with Jewish eyes, they were images of fertility and the heathen life force that deserved to be destroyed. Against this background, Schmitt thought, it was all the less astounding that Hobbes's Leviathan had been slain by Jews.

Schmitt did not mention any of the kabbalistic sources he used in his portrayal of the alleged Jewish battle myth. Historian Raphael Gross reviewed the sources Schmitt vaguely alluded to and determined that Schmitt did not read them himself, instead taking them second-hand from notorious anti-Jewish screeds.[45] This is supported by the fact that most of Schmitt's references are not to be found in the original sources; some, for example about kabbalist Isaac Abarbanel, mentioned only one side of his work; and others

44 Laced with innuendo, this wording, found on page 110 of the German edition, is missing in the translation by George Schwab (page 70), who simply wrote "castrating" and omitted "continuing the metaphor."

45 For example, Schmitt made reference to the 1700 diatribe *Entdecktes Judenthum* [Judaism unmasked] by Andreas Eisenmenger, see Schmitt (1938a, 8). See also Gross (2000, 273–275).

were entirely incorrect.[46] Italian philosopher Giorgio Agamben even spoke of "an anti-Semitic falsification" by Schmitt because he jumbled kabbalistic and Talmudic traditions and fabricated ideas of his own.[47] The mythological underpinning of Schmitt's critique of Hobbes is an antisemitic phantasm. One can only agree with Paul Bookbinder's characterization of Schmitt's incredible conspiracy theory as "reminiscent of *The Protocols of the Elders of Zion*" (Bookbinder 1991, 104).

At the end of his book, Schmitt explicitly made clear that he agreed with Hobbes about strict "scientism" (85) in theoretical work. The fact that Schmitt was unwilling to retract a single word on this matter even four years later, when the murderous terror against the Jews in Europe had long reached a new dimension, is very clear from his statement in 1942 about the "complete lack of relation of the Jewish spirit to the German *Volk*" (Schmitt 1942c, 205). His book about Leviathan is not a "cryptic, esoteric [...] and enigmatic" (Maschke 1995a, 227) work; he wrote it with the intention of "burnishing [his antisemitism] academically" (Neumann 2015, 389) in order to "give his antisemitism a scientific form" (Mehring 2014a, 356).

3. Kirchheimer's research on antisemitism at the Institute of Social Research

Up until 1940, Kirchheimer hardly dealt with Nazi antisemitism and the German Reich's policies toward Jews at all, at least in his scientific work on the Nazi regime. He addressed the subject only twice, and briefly, in his 1935 *Staatsgefüge und Recht des dritten Reiches* [State structure and law in the Third Reich], the brochure disseminated illegally in Germany. He stated that Nazi "legislation in the area of race has come nowhere close to realizing all the points sketched out in the party program" (Kirchheimer 1935a, 159)—whereby it must be taken into consideration that this statement was made prior to the Nazis' Party Rally in Nuremberg in September of that year. The second passage illustrates that Kirchheimer considered the development that began with the Nuremberg Race Laws to be within the realm of the possible: "Especially in the case of the 'Jewish Question,' the development of so much of the German legal system remains in a state of flux" (Kirchheimer 1935a, 160). Following the enactment of the Race Laws, Kirchheimer attended a multi-day conference of the International League for Human Rights in Paris where the situation of Jews in Germany was one of the three main subjects discussed (see Langkau-Alex 2005b, 221). The focus of his academic work continued to lie elsewhere, however. In *Punishment and Social Structure*, he mentioned the persecution of Jews only in the chapter on the Middle Ages (see Kirchheimer and Rusche 1939, 20–22). The subject was eventually no longer relegated to the background when he addressed it for the first time in his 1941 essay "The Legal Order of National Socialism." Kirchheimer saw the abolition of the liberal separation of law and morality as the core of the German Reich's legal system. The Nazi moral code "substitutes racial homogeneity for equality" (Kirchheimer 1941d, 456), thereby abandoning the

---

46 See Meier (1994, 236–240), and Gross (2000, 271–278).

47 See Agamben (2014, 41) and (2015, 58). Paul Bookbinder lists additional mistakes of Schmitt's concerning Jewish history, the Jewish sources, Spinoza, and Stahl (see Bookbinder 1991).

notion of shared human existence. He called the legislation concerning Jews beginning with the Nuremberg Laws "thoroughgoing extirpation of the Jews" (Kirchheimer 1941d, 462). He went into more detail only with respect to marital law and the new ways to divorce a Jewish spouse.

Kirchheimer addressed the subject of antisemitism in his academic work in more depth after moving to New York as he sought new professional opportunities at the Institute of Social Research (ISR). This new focus of his work was completely unrelated to Carl Schmitt and the escalation of his aggressive antisemitic diatribes. Kirchheimer was involved in the work on antisemitism at the ISR from the outset and contributed to multiple empirical research projects between 1941 and 1943. To gain a better understanding of his contribution to antisemitism research at the ISR, it is useful to take a brief look at the context of his work at the institute at the time.[48] The first specific plans for a major study on antisemitism at the ISR were made in 1939, when antisemitism was already escalating in the United States and had become the subject of a broader public debate in books such as Sinclair Lewis's *It Can't Happen Here* (see Lewis 1935). The findings of the multi-year research at the ISR finally led to the publication of the famous series *Studies in Prejudice* beginning in 1949. Max Horkheimer's programmatic essay "The Jews and Europe" of 1939 can be seen as the starting point of the institute's work on antisemitism. The article was permeated with the fear that the fascist model of governance and a murderous form of antisemitism would spread worldwide. It ended with a note stating that it had been completed on the first day of the war in September 1939. Horkheimer developed the hypothesis that the Nazi regime had stabilized the monopolistic capitalism of the Weimar Republic, which had fallen into an existential crisis in the form of a historically new post-liberal order.

At Horkheimer's request, Kirchheimer had contributed a short memorandum on an earlier version of this article in the late summer of 1939. His comments include an overview of empirical findings from various areas of the economy in Italy and Germany. Presenting large amounts of data, Kirchheimer documented a strengthening of private capitalism through "re-privatizations" (Kirchheimer 1939, 203) under both regimes, the processes of concentration in various sectors at the expense of small and medium-size businesses, a reduction of wages and an intensification of labor, the continuous importance of large-scale land holdings in agriculture, the failure of the expansion of public administration, and increasing bureaucratization. Even if Horkheimer's essay "The Jews and Europe" did not directly include any individual passages or tables of figures from Kirchheimer, it is evident that Kirchheimer's findings were incorporated into Horkheimer's statements about the role of business monopolies, advancing concentration in certain sectors, and the expansion of the government apparatus (see Horkheimer 1939a, 79, 84, and 90). Horkheimer still advocated a functionalist interpretation of antisemitism, which assumed the primacy of economic factors.[49] He made the decreasing

---

48 On antisemitism research at the exiled Institute of Social Research, see Ziege (2009) and Rensman (2017).

49 On the functionalist interpretation of antisemitism championed by Horkheimer in 1939 and his transition to an interpretation based on the history of civilization (inspired by Adorno) from 1941 on, see König (2016, 220–244).

importance of the sphere of circulation, which Kirchheimer had illustrated with data, fit into his own interpretation, and refashioned it into its complete disappearance. According to Horkheimer, Jews were the circulation agents par excellence. They owed their emancipation to the fact that they served as pioneers of capitalism and were indispensable as lenders in the sphere of circulation. To Horkheimer, the sphere of circulation was simultaneously the foundation of bourgeois democracy and the universality of the law. As circulation disappeared, he believed, Jews were "being run over" and had become superfluous (see Horkheimer 1939a, 89). At the same time, democracy and the universality of the law had become obsolete.

In 1941, the project outline "Research Project on Antisemitism," which had been prepared with Kirchheimer's extensive involvement in 1939, was published in the ISR's renamed journal *Studies in Philosophy and Social Science*. Under Horkheimer's leadership, the institute presented an interdisciplinary and thematic research program on antisemitism that was to provide the basis for intellectual combat. The goal of the project analysis was "to show that antisemitism [was] one of the dangers inherent in all more recent culture" (ISR 1941, 124). The project was to demonstrate in different ways that antisemitism was widespread, had deep historical roots, and was also to be found where one might hardly expect it, for instance, in the work of humanistic Enlightenment philosophers. The institute's hypotheses on the "Foundations of National Socialist Antisemitism" were formulated in Section VI of the project outline, which attributed Jews' weakened role in society to "the change in the functions of money" (ISR 1941, 140).[50]

The empirical evidence presented to demonstrate the new importance of the sphere of circulation in capitalism came from a short paper Kirchheimer had prepared for the project outline, "Funktionswandel und Konzentrationstendenzen im Bankgewerbe" [Functional transformation and concentration tendencies in the banking industry] (see Kirchheimer 1940c). Kirchheimer summarized how Germany's banking sector had changed since 1933, focusing in particular on the increase of political influence in decisions about loans and the Reich's heavy indebtedness to the banks. Overall, he found a shift of profits from private bankers to major banks, and he considered the Nazi "Aryanization" of Jewish property part of the process of growing concentration. Horkheimer used Kirchheimer's figures about changes in the private banks and major banks' total deposits as evidence supporting his hypotheses on economic transformation (ISR 1941, 141). However, he made the findings more pointed by using them to explain the complete liquidation of the sphere of circulation. Horkheimer stated that it was the decline of the power of financial capital and the replacement of the market by a planned economy overseen by government bureaucracy that made the anti-Jewish policies of the Third Reich possible in the first place.

The ISR's first attempts to gain financing from the Rockefeller and Carnegie Foundations for empirical and historical research on antisemitism failed. It was not until early

50 On the functionalist features of the theory of antisemitism in this project outline, see König (2016, 235–240).

1943 that Franz Neumann, working together with Arkadij Gurland and Kirchheimer,[51] succeeded in obtaining funding from the American Jewish Committee (AJC) for two parts of the institute's original research project.[52] It was initially scheduled for one year, from April 1943 to March 1944, and was to be headed by Friedrich Pollock. The researchers in New York were to work on the economic and social causes of antisemitism, and those on the West Coast were to conduct present-day psychological studies under Horkheimer's leadership (see Horkheimer 1943b). The New York group was directed by Friedrich Pollock and Leo Löwenthal, although most of its work was conducted by Arkadij Gurland and sociologist Paul W. Massing with Kirchheimer involved part-time (see Wiggershaus 1995, 362). Kirchheimer's close friend at the institute Herbert Marcuse also contributed to the work of the group and commented extensively on the original research agenda.[53]

In order to be able to present the AJC with initial findings that could be used politically, the group of researchers in New York focused on the political function of antisemitism in European history and also conducted a survey of German émigrés about their experiences and the reactions of the German population to the Nazi regime's antisemitic policies. The findings from the first year of research were presented to the AJC at a conference in New York in the spring of 1944 and delivered in the form of a hectographed research report in August 1944, which was never published. The 150-page report was augmented with several essays and exposés as well as a list of its twenty-one authors, including Kirchheimer (see ISR 1944, 142–144). The AJC leadership was sufficiently convinced of the results that in the fall of 1944, it approved funding to continue and expand the project and to establish a scientific department directed by Horkheimer. The publication of the five volumes of *Studies in Prejudice* in the United States between 1949 and 1951 was a "sensational scientific success" for the ISR and established its legendary reputation as a pioneer of empirical research on prejudice (Ziege 2009, 252).

Two research papers by Kirchheimer stem from this Frankfurt School context, both titled "The Policy of the Catholic Church Toward the Jews." The first is a twenty-one-page typescript, which remains incomplete and presents a number of hypotheses. The second, running to thirty-two typed pages, is a more in-depth version of the same work, with more detailed references. Neither carries a date, yet we can narrow down the time of writing with reasonable certainty based on other sources. Horkheimer had decided in the beginning of June 1943 to give Kirchheimer a grant to write the research paper.[54] Kirchheimer was funded for special projects within the antisemitism project from mid-

51 Kirchheimer's early involvement in the organizational preparations for the antisemitism project is evident from a letter from Kirchheimer to Max Horkheimer dated 16 July 1943. Max Horkheimer Papers, Letters VI,11, 310.

52 On the various attempts to secure funding for the project and Horkheimer's skepticism about the American Jewish Committee, see Wiggershaus (1995, 350–359). The AJC was established in 1906 as an advocacy group for Jews promoting integration into American society (on the AJC and the ISR, see Ziege 2009, 61–72).

53 See letter from Leo Löwenthal to Herbert Marcuse dated 29 June 1943, quoted in Jansen (2000, 101–114).

54 Letter from Max Horkheimer to Friedrich Pollock dated 9 June 1943 (Horkheimer 1996, 456).

June to mid-August 1943.[55] Considering his other activities and the work flow at the institute, the second, more detailed, manuscript can be dated to the autumn of 1943.[56] He certainly worked on the paper in November 1943 since the agenda for a dinner meeting of the Institute of Social Research and the AJC at the Harmonie Club in New York City on 8 December 1943 announced Kirchheimer as the third speaker (alongside Friedrich Pollock and Arkadij Gurland). Kirchheimer's talk was titled "The Catholic Church and Her Jewish Policy."[57]

4. Kirchheimer's *Policy of the Catholic Church Toward the Jews*

Despite Horkheimer's skepticism about whether Kirchheimer could "achieve anything really valuable in a relatively short time,"[58] he produced a research paper of thirty-two typewritten pages titled "The Policy of the Catholic Church Toward the Jews."[59] He conspicuously touched on Schmitt's work at a number of points, starting the research paper with a statement reminiscent of Carl Schmitt's book *Roman Catholicism and Political Form* (see Schmitt 1923b). Schmitt had written in this early book that the Catholic Church had to be viewed primarily as a religious institution with "astounding [...] elasticity." The Catholic Church was a "*omplexion oppositorum*" comprising a practically infinite number of contradictions in an integrative way. That, Schmitt thought, also applied to the theological diversity within the Church. Kirchheimer did not mention this early work of Schmitt's explicitly, but he, too, described the Catholic Church as eminently flexible and adaptable, adding that this was also true of its position toward the "Jewish question."

Kirchheimer developed his deliberations[60] by distinguishing between Catholic doctrine and "the different, sometimes even contradictory, motivations which have guided Catholic attitudes towards the Jews at different periods and in different regions" (498). One constitutive element of Catholic doctrine, Kirchheimer stated, was its "anti-Judaism," which he strictly differentiated from antisemitism. Anti-Judaism was not an accidental element in the doctrine of the Church. It was based on the firm belief that Jews were guilty of not having accepted Jesus as the Son of God and the Redeemer. Instead, they had put Jesus to death. Kirchheimer described various Christian interpretations and evaluations of Judaism, beginning with the New Testament through the Middle

55 This is evident from the project budget for the AJC. See *Re: Antisemitism Project, Preliminary Budget*. Max Horkheimer Papers, 665, 19.

56 In a letter to Max Horkheimer dated 20 September 1943, Kirchheimer writes that "antisemitism" and his other commitments "are literally [eating] up my time." Max Horkheimer Papers, Letters VI, 11, 307.

57 See *Agenda for the Joint Meeting of the Institute of Social Research and the AJC on December 8, 1943*. Arkadij Gurland Papers, Documents and Letters, Na 5/131.

58 Letter from Max Horkheimer to Friedrich Pollock dated 9 June 1943 (Horkheimer 1996, 456).

59 Kirchheimer's research paper was discovered in connection with the edition of his *Gesammelte Schriften*. It was not referenced before, except for in a brief vague footnote by Felix Weil in *The Chicago Jewish Forum*, a minor newsletter that existed for a few years after the war. See Weil (1951, 11). The research paper was first published together with an introduction by the author in the journal *Antisemitism Studies* in 2018.

60 See Kirchheimer (1943). The following page numbers refer to this text.

Ages to the late eighteenth century, as well as the Church's practical policies toward Jews. In this historical reconstruction, Kirchheimer stressed the elements of Catholic theology that protected Jews because, in principle, they could be evangelized. He placed particular emphasis on the question debated among Christians of how and when Judgment Day would come: would Jews return to the flock automatically on Judgment Day, or would it be postponed and everything remain uncertain until their blindness had ended? Kirchheimer quoted the 1935 book *Le mystère des Juifs et des Gentils dans l'Église* by Erik Peterson on this debate in order to make his point.[61]

Schmitt had introduced Peterson and Kirchheimer when they all lived in Bonn. The three had concluded the evening of Kirchheimer's doctoral graduation in February 1928 over wine in Schmitt's apartment.[62] In 1924, Schmitt had called Peterson, who had moved to Bonn as a theology professor, his closest friend; their friendship was to break up later, not least because of Peterson's criticism of Schmitt's portrayal of Jews in *Leviathan* (see Schmitz and Lepper 2021, 277–285). In Bonn, Peterson extolled Schmitt's *Roman Catholicism and Political Form* and promoted a theological approach departing from historicism, instead championing a new dogma according to which only the literal text of the Bible was accepted as valid. One of the main subjects discussed by Peterson and Schmitt was the "Jewish question," more precisely, Peterson's hypothesis that the Church existed only under the precondition that the Jews, as the people chosen by God, had not come to believe in Jesus (see Mehring 2017, 311–336). For this reason, he gave Jewry the questionable eschatological role of a "delayer"—*katéchon*—guaranteeing the continued existence of the Church before the return of Christ; this was a motif that Schmitt took up, too, and spelled out in more detail in his later works after 1945. Peterson thus developed his concept of the Church separately from the "Jewish question," placing his hopes for eschatological salvation in converting Jews to Christianity. In his reference to Peterson's book on St. Paul's passages about the Jews in his *Epistle to the Romans* in the New Testament, Kirchheimer noted that it "takes the text in its literal sense" (499). And criticizing Peterson sarcastically (as Schmitt did later), he commented that the idea that the final fulfillment of religious expectation depended on the conversion of all creatures was not particularly original and could already be found in some Jewish traditions.

Kirchheimer dated the decisive watershed in Catholic policy toward Jews, which was caused by the looming loss of much of the Church's societal power during capitalist modernization, to the early nineteenth century. More than before, the Church had to opportunistically seek potential political allies in order to maintain its powerful position. Kirchheimer selected the examples of the *Kulturkampf* in the German Reich and, in more detail, the Dreyfus Affair in France to examine Catholic tactics concerning the "Jewish question." In these conflicts, the Catholic Church seized the opportunity to assert its public rejection of capitalist modernity by employing antisemitic propaganda more influentially. Catholic theologians condemned "the evils of finance-capitalism personified in the Jews" (509). Using the example of the Weimar Republic, Kirchheimer explains that

61 Peterson, who had relocated to Rome in 1933, had had his most important theological works translated into French for this book.

62 Carl Schmitt, diary entry of 25 February 1928 (Schmitt 2010, 208).

the Catholic Church was more tolerant toward Jews in contexts where Catholicism was also a minority faith.

In the final section of his study, "Catholic Policy and Totalitarianism," Kirchheimer argued that the senior leadership of the Catholic Church was not only prepared to cooperate with those in political power at a given time to maintain the Church as an institution but was even "willing to sacrifice the Catholic parties" (515), as illustrated by the examples of fascism in Italy and Nazism in Germany. During the Weimar Republic, the Catholic Church declared that working together with the NSDAP was out of the question because of its anti-clerical propaganda. Yet immediately after Hitler took power, not even his militant antisemitism could move the leadership of the Catholic Church to refuse to collaborate with the new German government. In the concluding passages, Kirchheimer states that "after 1939 antisemitic policies rapidly spread all over Europe" (515). This simultaneously showed a shift in the Catholic Church's position toward the "Jewish question" resulting from concern about the Vatican's position of power in fascist Italy. Now, the Vatican did emphasize that it nonetheless remained the fundamental task of the Catholic Church to overcome Judaism, but that there were also "certain definite limits" to this goal (516). Considering its universal mission, the Church had to insist that Jews would still be permitted to convert to Catholicism. From this followed the political thrust countering a "race doctrine, which would tend to nullify the Church's freedom to enlist" (516). Kirchheimer's view was that the Catholic Church was still not adopting a friendly stance toward Jews—official Church publications continued to describe them "both as members of the financial oligarchy and as revolutionaries" (517)—but from 1939 on, the Catholic Church saw itself subject to a "struggle with totalitarianism," as did Jews (518).

It is striking that Kirchheimer did not draw the obvious political conclusion that new allies in the struggle against antisemitism could be won in this changed constellation. This was presumably linked to the fact that Pope Pius XII, who had taken office in 1939, was friendly toward Germany, and Jewish circles in the US were aware of this. The new pope raised more fears than hopes for the future of European Jewry (see Feldman 2000, 125–127). It is easier still to understand why Kirchheimer refrained from relying on Catholicism as an ally against antisemitism if we also examine the very first version of the text, which has survived only in part.[63] In addition to the deliberations focusing on Europe in the more detailed version, this text includes several pages on American Catholicism and its role in antisemitism, which had increased sharply since the early 1930s. In these six pages, he reminded readers of the marginalized role of Catholics in American history and also emphasized the special significance of immigration from Ireland for changes in North American Catholicism. In the early twentieth century, the social competition in the lower strata of society between Irish immigrants and Jewish immigrants from Eastern Europe and Russia led to the emergence of "anti-Jewish bias," with contempt for Jewish competitors in particular. This had formed an echo chamber for rapidly increasing antisemitism among Catholics in the American industrial labor force in the preceding years. In the political realm, it had been stoked above all by Catholic "radio

63 This quotation and the following ones are taken from the first version of the research paper. Institute of Social Research, "The Policy of the Catholic Church Toward the Jews" (typescript, no date). Otto Kirchheimer Papers, Series 4, Box 2, Folder 15.

priest" Father Charles Coughlin's aggressive antisemitic propaganda during the 1930s.[64] Kirchheimer linked the "meteoric rise" of Coughlinism in the United States to similar contemporary "popular Catholic antisemitic movements" in Poland and Austria. While antisemitic mass movements led by Catholics were not a new phenomenon, their existence in the United States at the time, however, was new and could prepare the ground for a "native American fascism." The AJC's major concern and reason for financing the study was combating antisemitism, and it placed little trust in other representatives of the Catholic Church in America when it came to that goal. They would do nothing against rampant Coughlinism for the simple reason that it promised to attract new adherents to Catholicism, which had an "inferiority complex" in North America. Consequently, Kirchheimer's pessimistic expectation concerning the potential of the Catholic Church's resistance against the Nazi policies toward Jews was colored not least by his experiences with the success of antisemitic propaganda in the US.

5. Kirchheimer's contribution to the Frankfurt School's research

Kirchheimer had already left the Horkheimer Institute when the final report on antisemitism was submitted to the AJC in August 1944. Nonetheless, his work left its mark on the institute's future research on antisemitism, but less with respect to the philosophical parts of the Frankfurt School in exile. Kirchheimer's deliberations did not directly influence Horkheimer and Adorno in the chapter "Elements of Antisemitism"—also written in the summer of 1943—in their *Dialectic of Enlightenment*. Yet certain parallels can be seen in terms of their differentiation between religious and *völkisch* (of the *Volk*, chauvinistic-nationalistic, antisemitic; see Glossary) forms of antisemitism and their view of Christianity as an institution (Horkheimer and Adorno 1944, 206–208). Conversely, there are no points of contact with their interpretation of Christian antisemitism following Freud's image of patricide. In their reading of antisemitism, Horkheimer and Adorno followed a different path from Kirchheimer, one that presented itself in the empirical studies in the early phase of the antisemitism project. Surveys of German-language immigrants to the United States in 1943—Kirchheimer was also involved in their analysis (see Wiggershaus 1995, 362)—revealed that devout Catholics and conservatives helped Jews far more than Protestants. Horkheimer later used this finding to support his hypothesis that conservatives were often better guardians of critical thinking than liberals. The question of Catholicism played no role in the best-known volume of *Studies in Prejudice*, the book *The Authoritarian Personality*, co-authored by Adorno (see Adorno et al. 1951).

Kirchheimer's empirical observation that antisemitism was widespread among the American Catholic industrial labor force had a greater impact. Gurland and Neumann succeeded in securing funding from the Jewish Labor Committee (JLC) for the ISR to conduct a major study on antisemitism in the US labor force beginning in the summer

64 On Coughlinism and its political success, see Baldwin (2001).

of 1944.[65] The lengthy interim report, *Antisemitism among American Labor*, completed a year later on the basis of 566 interviews, stated that antisemitism was widespread among industrial workers: almost thirty-one percent were classified as "actively hostile to Jews," 38.5 percent as rejecting them, but without assenting to consistent discrimination, and 30.5 percent as "friendly to Jews." The share of respondents who approved of the persecution of Jews in Germany was significantly higher among Catholic workers than among those of other faiths, whereby the highest share was found among Catholics who no longer attended church.[66] The leaders of the institute considered results of this kind so unfavorable for the American labor force that they decided not to publish the study (see Jay 1973, 224–227). Thus, Kirchheimer's thoughts on antisemitism in "The Policy of the Catholic Church Toward the Jews" had little in common with the concept of a "philosophical prehistory of Antisemitism" (Horkheimer and Adorno 1944, xvii) as elaborated in *Dialectic of Enlightenment*. Instead, there were more similarities to the functionalist interpretation Horkheimer had previously championed, but without his apodictic claim regarding an entirely new societal epoch for Jewry.

Clearer parallels can be found, however, to Hannah Arendt's approach in the first part of her 1951 book *The Origins of Totalitarianism*, in which she localized the phenomenon of antisemitism in terms of time and space and placed it within the political histories of Germany and of Europe more generally. Like Kirchheimer, Arendt championed the position that changes in social and religious antisemitism beginning in the last third of the nineteenth century were tied to the history of nation-building and modern imperialism. In the revised German edition of *The Origins of Totalitarianism*, published in 1955, four years after the first American edition, Arendt calls this the transition from "*Antisemitismus als gesellschaftliche Idiosynkrasie*" (antisemitism as a social idiosyncrasy) to "*Antisemitismus als politische Bewegung*" (antisemitism as a political movement, see Arendt 1955, 72 and 77). The terms she used in the English edition were less trenchant: "aristocratic antisemitism" with "mild discrimination without further political significance" and "antisemitic movement" (see Arendt 1951, 32 and 39).[67] In her political history of antisemitism, Jews do not appear as victims, always damned to be passive, but as a social minority actively and sometimes skillfully engaged in political life. Kirchheimer would concur with this view, as indicated by some of his writing in the 1960s.

At the ISR, Kirchheimer's work was continued, in a sense, by his colleague Paul W. Massing in the first volume of *Studies in Prejudice*, entitled *Rehearsal for Destruction: A Study of Political Antisemitism in Imperial Germany*, published in 1949. This book examined the background of the Nazi regime's political antisemitism in the German Empire. With respect to Catholicism, Massing, too, underlined how political antisemitism took root in

---

65 The final report of this study was not published. On the main project findings, see Worrell (2008) und Ziege (2009). The JLC was founded in New York in 1934 as an umbrella organization for the Jewish trade unions.

66 See Institute of Social Research, *Antisemitism among American Labor, 1944–45*, New York, 1455 pages (Max Horkheimer Papers, IX–146, 1–23, 755–760).

67 There is no evidence that Kirchheimer and Arendt met in person while they were both in Paris or in the first years of their exile in New York. The sparse correspondence between the two in the 1950s and 1960s, however, indicates that Arendt valued Kirchheimer's book *Political Justice* and that he agreed with her in the debate about Eichmann in Jerusalem (see Chapter 16).

the Catholic Zentrumspartei (Center Party) in the course of the *Kulturkampf* (see Massing 1949, 219–221). After leaving the ISR in late 1943, Kirchheimer continued to be interested in writing on antisemitism, at least in shorter pieces. Here, the parallel to the activist component in Arendt's position becomes even more evident. In a book review for *The Washington Post* published after the Eichmann Trial of 1961, Kirchheimer called Germany's murder of Jews the "greatest crime of modern history," carried out "in cold blood and with mathematical precision," and reminded readers of heroic Jewish resistance in the camps. Even if their struggles were unsuccessful, their activism made them, "if often only for a fleeting moment, heroes with the sense of a mission and certain of their place in the history of their people" (Kirchheimer 1962a, 6). A few weeks later, he reported to the American public in the same newspaper that there was a group of people in Germany for whom the memory "of their country's shame forms a key to their frame of mind" (Kirchheimer 1962b, 9). Although this group was still relatively small, it was growing and consisted mostly of young people who wanted to face up to the German past.

Shortly before his death in 1965, Kirchheimer again took up the subject of antisemitism and the Catholic Church by writing a review of sociologist Gordon Zahn's book *German Catholics and Hitler's Wars,* which was intended for publication in *The Washington Post* but did not appear in the end. We don't know why not. In his manuscript, Kirchheimer praised the book for its detailed description of the German Catholic bishops' strong support for the German army during the war. "Given the trends of our time," he concluded, Zahn had written "a tragically important book" because it "demonstrate[d] how little guidance the individual may expect in hours of gravest moral doubt from any established institution."[68]

A few weeks before his death, Kirchheimer's review of the book *The Catholic Church and Nazi Germany* by historian Guenter Lewy appeared in *Dissent*, then one of the leading journals in the American leftist intellectual milieu. Kirchheimer saw Lewy's findings on contemporary history, which relied on countless sources, as confirming his own hypotheses about the sociology of power. In Lewy's opinion, the Catholic Church in Germany had cooperated with the Nazi regime after 1933 purely out of power interests, similarly to the industrial and military leaders, and had declared its loyalty to the bitter end in return for the state's financial services. In contrast to Lewy's interpretation, however, Kirchheimer insisted that Catholic hatred of Jews was not the same thing as Nazi antisemitism. Kirchheimer used the contrasting example of France to answer the question of why the Catholic Church in Germany and Italy did not muster more energy against the murder of Jews. In France, the state and the Church were separate, which gave centrifugal forces within the Church greater weight, thus creating greater latitude for individual believers' decisions of conscience. After 1945, the historical opportunity to sever the unholy alliance of Church and state in Germany was wasted: "now the alliance is once again profitable in terms of money, educational privileges, and Church influence in state policy" (Kirchheimer 1965a, 92). In this review, Kirchheimer considered the play *Der Stellvertreter* [The deputy] by Rolf Hochhuth, which premiered in Berlin in 1963, to be the first public questioning of the false image of the Catholic Church as an institution that had resisted

68 Otto Kirchheimer, "Book Review of Gordon Zahn, German Catholics and Hitler's Wars." Unpublished Manuscript, three pages. Otto Kirchheimer Papers, Series 4, Box 3, Folder 22, 2.

the Nazi regime. This image had been widely cultivated by the Catholic Church after 1945 in West Germany, and Kirchheimer criticized it as a "product of [the] imagination" mixed with "elements of misrepresentation" (Kirchheimer 1965a, 94). In personal conversations with German friends, he expressed his optimism about a growing commitment by the younger generation in Germany to face up to the past.[69]

6. Conclusion: The modernity of Catholic antisemitism

While Schmitt expressed and even celebrated antisemitism in his speeches and writing praising the Nazi regime, Otto Kirchheimer's family had to experience its antisemitism firsthand. Most of his closer relatives had been able to flee Europe during the war, but not all of them escaped.[70] Two members of his closer family circle did not make it and were murdered in the Holocaust, his brother-in-law Adolf Rosenthal in Theresienstadt in 1942 and his brother-in-law Ludwig Rosenthal in Izbica in 1943. Among the wider group of family members (around 160 persons in total), twenty-eight of them are recorded as killed in concentration camps. Seven family members are known to have spent greater or lesser amounts of time in a concentration camp but survived. Almost eighty family members managed to emigrate from Germany between 1933 and 1940. Among the émigrés were Otto Kirchheimer's brothers Max, Leo, and Fritz, who all went to Argentina at different times (his other siblings had died before 1933). A few members of the family survived in hiding in Germany, the fate of other family members is unknown to this day.

Carl Schmitt was preoccupied with Jews and Jewishness throughout his life. Raphael Gross took the biographical finding of Schmitt's views about Jews as the starting point for his interpretation of his work, establishing the significance of antisemitism for Schmitt's criticism of parliamentarism, pluralism, the liberal concept of the law, legal positivism, and universal international law even during the Weimar Republic.[71] It is tempting to apply the horizon of interpretation introduced by Gross and to analyze further topics and motives in Schmitt's thinking from the perspective of his antisemitism. Yet I will limit my reflections here to a brief comparison of Kirchheimer and Schmitt.

There are multiple points where the antisemitism analyzed by Kirchheimer and the antisemitism practiced by Schmitt touched on each other. A first and particularly striking instance is Kirchheimer's focus on Catholicism. Because he followed Catholic theological debates as a student in Bonn, Kirchheimer seemed the most suitable among the exiled members of the Frankfurt School to address the subject of Catholicism. Another point is how Kirchheimer, drawing on Schmitt's *Roman Catholicism and Political Form*, emphasized the institutional and ideological elasticity of the Catholic Church. Third, Kirchheimer entered the theological debates about Christianity and Judaism by using Erik Peterson's work, which he was familiar with from working with Schmitt. The fourth point is how he evaluated the role of the Catholic Church as it dealt with the Nazi regime. Kirchheimer

69 Wilhelm Hennis in a conversation with the author on 26 September 2009.

70 The following information is based on Kirchheimer-Grossman (2010) and several conversations with Hanna Kirchheimer-Crossman and Rebecca Kirchheimer in September 2021.

71 See Gross (2000; 2016).

was skeptical as to the role of the Catholic Church in Germany as a potential force of resistance against Nazi policies. In turn, Schmitt as a Catholic had attempted to explain in retrospect that one reason for his support for the Nazi regime was that "Hitler had accomplished the Concordat [between the Holy See and the German Reich] without any resistance at all."[72]

Besides these points where their works touched on each other, the contrast in methodological terms could hardly be greater between Schmitt with his constructions in intellectual history on the one hand and Kirchheimer with his empirical analyses and his assessments critical of ideology on the other. Whereas Schmitt, in his book *Leviathan*, drew on vague second-hand sources on the kabbalah written by antisemites to construe a genealogy of the triumph of an international Jewry operating while concealed by masks, Kirchheimer undertook sober sociological contextualizations of the attitudes inimical to Jews.

Since the nineteenth century, modern antisemitism has reacted to the development of capitalist modernity, i.e., to industrialization, urbanization, the loss of binding worldviews, and the objectification of human relationships. To Kirchheimer, Catholic antisemitism was a modern phenomenon despite its initially religious sources which had to be understood as the expression of negative experiences of modernization. If we apply Kirchheimer's analytical approach to Schmitt, the latter emerges as one among many others socialized in the German Empire who would always reject this modernity.[73] Schmitt stood out from this large group only inasmuch as he used particularly copious words and quotations to articulate his antisemitic resentment.

72 Schmitt in a conversation with Ernst-Wolfgang Böckenförde (Böckenförde 1988, 336).

73 On Schmitt's view that Jews and their threatening "*maskenhafte*" (mask-like) nature embodied the ambiguity of modernity, see Balke (2016, 648–650).

Chapter 11: Preparing Germany for New Wars (1936-1939)

Schmitt's removal from a position of power to influence the domestic policy of the German Reich relieved him from time-consuming commitments, and he took the opportunity to focus on specific questions of shaping the supranational order instead. He managed to enter the political stage again, through the "back door of international law" (Koenen 1995, 783), so to speak, and became a star once more, this time with some international recognition, too.

1937 was the year in which political sentiment in Hitler's Germany reflected disillusionment. President of the Reichsbank Hjalmar Schacht had resigned from his position as Minister of Economics (later he was Minister without Portfolio), and the public's nationalist enthusiasm was on the wane. Hitler overcame his personal crisis during these events with a new wave of activities. He reacted to the economic problems that were becoming apparent and the looming change in political mood by installing a foreign policy aimed at aggressive expansion (see Fest 1973, 738–741). From the second half of 1937 on, the Reich was reorganized more systematically than ever before to support the regime's violent intentions. This included accelerating the buildup of armed SS units and increasing the number of concentration camps. The Red Cross was instructed to prepare for mobilization. At the same time, the *Hitlerjugend* (Hitler Youth) was directed to cover the staffing gaps in the armaments industry that had been quantified. In early November 1937, the press received directives not to report publicly about the preparations for "total war" initiated in all the Nazi party units. These preparations culminated in the secret conference in Berlin on 5 November 1937 where Hitler laid out his plans for a violent expansion of the Reich in eastern Europe to the military top brass in a talk lasting no less than four hours.

The year 1937 was initially a year of personal crisis for Schmitt, too. After his inglorious demotion in the hierarchy of the Nazi regime, he had a mental breakdown in the summer of 1937 (see Mehring 2014a, 358). He felt hindered in his urge to be constantly active, and he suffered from having to watch how the Nazi *Führer* state's discriminatory logic for winning the favor of the Nazi leadership now benefited his rivals, including Otto Koellreutter, Werner Best, and Reinhard Höhn. Just like his *Führer* Hitler, Schmitt decided to go on the offense and push his way out of the crisis. The opportunity soon arose: he was asked at short notice to substitute for Koellreutter, who had a scheduling conflict, and give a lecture on the conventions of war in international law at the Akademie für Deutsches Recht (Academy for German Law) in the autumn of 1937. He immediately accepted the opportunity to fill the gap and turn the tide in his favor. As early as late September, he successfully returned to the stage of academic prominence in the Nazi state, this time in the role of theorist of international law, not with a subject related to domestic policy. The organizers who approached Schmitt knew that he had already studied questions of international law on various occasions in the past, so it was by no means unreasonable to ask him to speak. At the same time, it indicated that Schmitt had not become a pariah of the system. The carefully prepared lecture he gave on 29 October 1937 at the 4th Annual Conference of the Akademie für Deutsches Recht in the Arbeitsgruppe Völkerrecht (Working Group on International Law) was entitled "Die Wendung zum diskriminierenden Kriegsbegriff," later published in English as *The Turn to the Discriminating Concept of War*."

1. Schmitt's "specifically National Socialist insights"

Pinpointing Carl Schmitt's role in the context of Nazi Germany's thinking on international law requires a brief outline of the general development of this thinking between 1933 and 1945.[1] After the handover of power to Hitler's government, people were somewhat unsure initially how international law would retain its political and academic function under the new regime. At first, Germany's de facto withdrawal from the League of Nations on 14 October 1933, which was highlighted propagandistically by broad agreement in the referendum on 12 November 1933, and the gradual relief from the obligations of the Treaty of Versailles paved the way for international law along traditional lines. At least as long as Hitler's government was rebelling against the Treaty of Versailles with legal arguments from a defensive position, it was politically inopportune to negate international law in general. The strategy was to insist on nation-states' equal right to self-determination, and from this basic position, to raise territorial demands as well as the end of all limitations under the Treaty of Versailles in the name of restoring state sovereignty. With this strategic reasoning, Nazi international law thus continued to be in line with universalist internationalism.

In the second half of the 1930s, international law scholars holding Nazi views began to develop their own concept of international law, which was potentially more aggressive. The argumentative core of the new offensive concept consisted in replacing the state with the *Volk* and later the *Reich* as the legal entity of international law to provide the background for denying the validity of a universal internationalist basis of international law. "International law" in terms of "law of the states" became the "law of the *Völker*." The more that the concept of "*Rasse*" was emphasized in the further development of Nazi international law, the more the scope of rules and regulations under international law was lim-

1 On the development of Nazi international law, see Fischer (1974), and Koskenniemi (2001, 179–265), and for an incisive overview Stolleis (1999, 381–400).

ited; they were to apply only to "*Völker* related by *Rasse*." In Nazi thinking on international law, these steps opened up the way for a new ordering of the world along *Rasse*-based perspectives. Only in 1944 and 1945, when Germany's military defeat was foreseeable and merely a question of a few months, were there initial attempts to guide the arguments back to the traditional lines. In preparation for the victorious powers establishing a post-war order, Nazi authors described Germany as a victim of aggressive Anglo-American world imperialism, whereby the Allies, behind the mask of spreading peace throughout the world, were attempting to brand Germany as the aggressor and to rob it of the right to a state of its own. Carl Schmitt went along with all these steps until the early 1940s, albeit to a varying extent. He had even powered ahead of his colleagues at some stages of formulating Nazi international law doctrines and their aggressive turn. However, he also emphasized points of his own, thereby departing from the Nazi doctrine's crude *völkisch* (of the *Volk*, chauvinistic-nationalistic, antisemitic; see Glossary) racism.

It was also Schmitt who, in the summer of 1934, presented a combination of arguments along traditional lines and an initial programmatic formulation of an international law unique to Nazism in a much-noted lecture that was later published as a standalone brochure titled *Nationalsozialismus und Völkerrecht* [Nazism and international law].[2] He claimed that the situation of Germany and Europe under international law "first had to be brought to a state one could call normal" (391). Such normalcy required an end to the German Reich's "defenselessness and deprivation of rights" (391) under the alleged dictate of the Treaty of Versailles. Schmitt used natural rights to justify his position: there were "fundamental rights of *Völker* and of states" (393). These inalienable and enduring fundamental rights included the basic "right to one's own existence" (393) and the rights to self-determination, self-defense, and the means for self-defense derived from this. The German *Volk* would have to have this right immediately and without further limitations, he asserted, because it had "put itself in order internally and under consideration of its own nature" (393) under the Nazi government.

By insisting on "normalcy," Schmitt was in line with a type of universalist and natural law thinking that constituted the conservative and defensive side of this piece. At the same time, there are statements just above according to which there were "as many types of fundamental rights as there are types of human community" (392), which thus contradicted his proclaimed natural law universalism. The hypothesis formulated at the beginning of the brochure that it was a "specifically National Socialist insight" (391) not to derive the law of intergovernmental relations from universal and abstract thinking that followed rules but solely from "a concrete order of states and *Völker* of a certain *Art* and recognized in their concrete *Eigenart* (nature, or: their own *Art*, the state of being characterized by *Blut* and *Rasse*, see Glossary)" (391) fits with this statement. In using such formulations, Schmitt expanded the "thinking in legal orders newly awakened by the National Socialist movement" (392) he had already postulated previously for domestic affairs to the realm of international law, thus simultaneously opening the door to basically unlimited latitude of interpretation of what could or must be considered appropriate in a specific case for a state's foreign policy. He justified Germany's withdrawal from the

2 See Schmitt (1934a). The following page numbers refer to this text.

League of Nations six months earlier as a logical step to regain German honor. Nonetheless, he continued to polemicize not only against the international treaties on which the League of Nations was founded, but also against its internal characteristics, asserting that the League of Nations was nothing but a perfidious system to institutionally safeguard the liberal democracies under Anglo-American supremacy. Both Germany's and Japan's withdrawal was therefore only logical. The Soviet Union's accession to the League of Nations in September 1934 exposed once and for all, Schmitt stated, that the League of Nations was not a league and not a real community; all that remained was an "old-style opportunistic alliance" (405) of the former victorious powers of the war against Germany.

In this context, Schmitt drew on Moscow jurist Evgeny A. Korovin's book *International Law of the Transition Period*, which Otto Kirchheimer had critically reviewed four years previously for the journal *Die Gesellschaft*. Kirchheimer's criticism had been sparked by Korovin's hypothesis that there was an independent Bolshevik legal sphere besides the international legal sphere of the capitalist countries. Kirchheimer countered that Korovin had erroneously underestimated both the differences between capitalist states and the opportunities for the two allegedly unconnected legal spheres to come to an understanding. Kirchheimer's criticism ended in a plea to expand international law and to strengthen the League of Nations, including the Soviet Union.[3] Immediately after the Soviet Union had joined the League of Nations in 1934, international legal theory in the Soviet Union was revised to reflect the position that had been linked with Korovin's name up until this point. From 1934 on, the Soviet legal theorist Evgeny Pashukanis set the tone as the new Soviet doctrine of peaceful coexistence with the capitalist powers in the interest of collective security and preventing war was advanced, and Korovin soon felt compelled to toe the line (see Flechtheim 1936).

Schmitt's comments on Korovin's *International Law of the Transition Period* followed a line diametrically opposed to Kirchheimer's criticism. Not surprisingly, he did not mention Kirchheimer's review at all, although he had read it. Although Schmitt pointed out that changes might be made to the Soviet Union's position on international law after its accession to the League of Nations, he was convinced that the outdated position advanced by Korovin was the authentically Soviet one. There was "no community of international law at all" (399) between the liberal capitalist and the Bolshevik world. There could be no peace between them, at best a temporary ceasefire. Whereas Kirchheimer had critically examined Korovin's theory and had then concluded that a universal system of international law was needed even more at the time, Schmitt's reception strategy consisted of stating that Korovin's hypotheses were further evidence of unbridgeable pluralism in international law, thus using them to justify the foreign policy of the Nazi regime's early years.

2. Challenging the discriminating concept of war

The science of international law in Germany, which was loyal to the regime, soon had to reorient itself once again in parallel to the Nuremberg Laws of 1935 and the Reich's

3 See Chapter 4.

military buildup. When the regime successively began to expand, by committing acts of violence and threatening to do so, Nazi doctrines now foregrounded questions of the international law of war, minority rights, and resettlement policy in place of seemingly defensive demands for a revision of the Treaty of Versailles. Up until 1938, Schmitt published eight longer works on international law in which he defended the German Reich's expansive foreign policy and simultaneously criticized the League of Nations, the US, the UK, and the Soviet Union. In all these articles, he presented Germany as a country surrounded by begrudging enemies.

Schmitt's above-mentioned lecture, "The Turn to the Discriminating Concept of War," which he gave in Berlin on 29 October 1937 at the 4th Annual Conference of the Akademie für Deutsches Recht, must be understood in the context of a tacit reorientation of Nazi science of international law. In this lecture, Schmitt succeeded in making his colleagues forget his temporary demotion and placing himself right back in the vanguard of the Nazi legal community with his clear and distinct words. His lecture was met with his colleagues' great acclaim and approval. It marked the beginning of a new stage in Schmitt's career after he had fallen out of favor with the regime for some time. Such a career boost would doubtless have been impossible without the intensified expansive foreign policy dynamics of the Third Reich. Within a year, Schmitt advanced to become one of the leading Nazi international law scholars, outshining the established proponents of international law in Germany with his pointed hypotheses and formulations and enjoying recognition for his work in the highest government circles.

In the weeks and months following his brilliant presentation, Schmitt revised it and rounded it out and, in late April 1938, it was circulated as a standalone brochure in the academy's publication series. Reich Minister Hans Frank had asked Werner Weber, the publisher of the series, to urge the publishing house to bring the brochure to press faster.[4] At the time of publication, the expansive dynamics of the German Reich had reached a new level of intensity. In a speech at the Reichstag on 20 February 1938, Hitler had vowed to protect German minorities outside the territory of the Reich. Shortly thereafter, on 12 March 1938, Germany invaded Austria. Subsequently, the Reich turned its attention toward Czechoslovakia and further expansion. This direct connection is not obvious at first glance in Schmitt's text. Of all his publications on international law, this brochure is the most technical in its legal reasoning. He selected four contemporary and prominent international contributions to the debate on the theory of international law and aligned his argument closely to theirs. At the same time, he also took up all the topics and concepts of international law he was to address in the coming years until, during, and following World War II. In this text, Schmitt developed the key hypotheses—for the first time within a larger context of argumentation—that constituted the substance of his late work *The Nomos of the Earth*, published in 1950.

Schmitt's *Turn to the Discriminating Concept of War*[5] was a critical review of four international authors' works on international law. At the very beginning, he made two things clear. First, that to him, the history of international law was and had always been a history of the concept of war and that the development of the entirety of international law

4 On the circumstances of publication, see Koenen (1995, 784–786).

5 See Schmitt (1937b). The following page numbers refer to this text.

was reflected in the concept of war. And, secondly, that the global political landscape was one in which "old orders are unraveling just as no new ones come to replace them" (31), in other words, that new armed conflicts were imminent. This was a reference to the current global political context, and among the examples Schmitt mentioned in his lecture were Italy's invasion of Ethiopia two years earlier, the Spanish Civil War, which had begun in 1936, and Japan's invasion of China in the summer of 1937. To Schmitt, these events marked a crisis of international law, which was entering a new phase. The core of Schmitt's argument relating to the theory of international law lay in stressing that two diametrically opposed concepts of international law were on a collision course: a universal world legal order increasingly secured through institutions on one side and a renationalization of the theory of international law on the other—in his shorthand, the opposition between a "universalistic" and a "politically pluralistic worldview" (67, note 168). This dichotomy was "not about new norms," but rather "about new orders—orders whose concrete character" very concrete powers "struggle with" (34).

Against this background, Schmitt did not tire of emphasizing the "practical meaning" (37) of his deliberations. He first analyzed the works of Georges Scelle, professor of international law at the Faculté de Droit de l'Université de Paris, and Hersch Lauterpacht, who taught international law at the London School of Economics. Schmitt presented the two authors as prototypes of new French and Anglo-Saxon thinking on international law, respectively, and the latter as a "native of the Polish region of Galicia" (39)—code for him being Jewish. With this opening, he immediately attempted to undermine their claims of being systematic and universalist by assigning them to independent and fixed national legal cultures. The second half of the article was devoted to works by the two US international lawyers, John Fischer Williams and Arnold D. McNair, on special problems of intervention and neutrality in international law.

Schmitt noted positively that Scelle's two-volume *Précis de droits des gens* from 1932 and 1934 advanced the universalist and individualist positions within international law to their logical conclusions. Individualism, Schmitt claimed, appeared in naming the individual as a legal entity in international law; universalism was expressed in the global military right to intervention. To Scelle, the state consisted solely of individual people, and relations between states were no different from relations between people. In this regard, Schmitt's statement that this approach radically dethroned the state and elevated the individual to the only direct subject of international law was entirely true. Scelle concluded from this that the Geneva League of Nations would have had to intervene against the treatment of Jews in Germany in 1933, which Schmitt mentioned as a particular example of how absurd and far-fetched Scelle's deliberations were (see 44). Scelle followed his assertion of a right to resistance against domestic activities in contravention of international law with a call for an international instance to which individuals could appeal in the event of such violations. Schmitt alleged that this made war into an "intervention" in the interest of protecting individual rights and transformed the classic war between states into a civil war. To Schmitt, this type of system of international law was a mirror image of liberal constitutionalism magnified to universal internationality and the attempt to transform the entire planet into a global state under a single rule of law. Schmitt concluded approvingly that Scelle's linking of the polar opposites of liberal individualism on the one hand and universalism under international law on the other led "to a new systematics of international law with logical consistency" (46) which fed into the hope for an upcoming "trans-state, universal, ecumenical order" (43)—only to add sarcastically that Scelle's lovely view was "obscured today through dictatorships and states that are not liberal democracies" (48).

Lauterpacht came to similar conclusions as Scelle concerning the binding nature of international law and its enforcement, Schmitt stated. He discovered the same tendency in Williams's commentary on Article 16 of the Covenant of the League of Nations on dealing with members of the League of Nations that violated this Covenant. The final article that Schmitt commented on by McNair on collective security, written in 1936, was linked most closely to the title and subject of Schmitt's brochure. McNair explained that the experiences from the gruesome World War from 1914 to 1918 had brought about a fundamental change in how wars were evaluated from the perspective of international law. The firm belief now prevailed that armed conflicts between states could no longer be justified under any circumstances. This conviction was reflected in the Covenant of the League of Nations and in international treaties such as the Kellogg-Briand Pact, which Germany, too, had ratified. Schmitt considered this position to be a ploy to expand international law. He explicitly agreed with McNair's view that the distinction between just and unjust wars, attributed to seventeenth century Dutch legal theorist Hugo Grotius, had disappeared from international law over the course of the nineteenth century. The concept of war that had emerged from this, and which Schmitt considered nondiscriminating, had had its "justice, honor, and worth" (71) in the fact that the enemy was "neither a pirate nor a gangster" (71) but rather a state and a "subject of international law." International law had placed limitations only on *ius in bello*, the conduct of war, but not on *ius ad bellum*, the right to go to war. Schmitt asserted that President Wilson's declaration for the US to join the war against Germany and the Covenant of the League of Nations had ushered in the beginning of the end of this civilizing concept of war.

Several months before his lecture, Schmitt had published a brief article entitled "Totaler Feind, totaler Krieg, totaler Staat" [Total enemy, total war, total state]. Total war, he explained, derived its meaning from the total enemy. Schmitt associated total war with English naval warfare, which, he asserted, was the only form of war that was completely ruthless towards noncombatants, and that this distinguished it from traditional land warfare on the continent (see Schmitt 1937a, 484). Schmitt took up this distinction in *The Turn to the Discriminating Concept of War*, differentiating between two concepts of war: first, "justified war with a compensation for the loss of life as its goal" and, second, a "war of annihilation fueled by a universalist ideology and led against a 'total enemy'" (67, note 168). The clincher at the end of Schmitt's lecture was that he closely linked total war with the doctrine of just war. The doctrine of just war made war a kind of executive measure or purge on the just side. The unjust side was declared to be illegal and immoral resistance led by "vermin, troublemakers, pirates, and gangsters" (67, note 168), and the government of the unjust side was ruled to be war criminals. The discriminating concept of war would lead to an "intensification of war and enmity" and to a policy of preventive military buildup to "fortify for the case of war" (72).

Schmitt's line of argument did not state clearly why military action conducted in the name of protecting human rights automatically has to become a kind of war of annihilation. There is no convincing theoretical explanation in Schmitt's work why universalist theories of international law bring about the dissolution of any and all boundaries of warfare and goals of war. On the contrary, sanctions against those declared to be lawbreakers explicitly do not aim to annihilate them. Using strict logic following legal doctrine, one can derive from all universalist approaches that—and where—the boundaries of warfare lie in the interest of protecting the individual rights of the other side's combatants. Moreover, Schmitt did not explain conclusively why the enemy in the traditional conflict between states should be excluded from appearing to be a criminal monster worthy of being annihilated. Be that as it may, this is not the place to discuss the persuasiveness of Schmitt's hypotheses more extensively.[6]

Schmitt did not tire of specifically emphasizing the most elementary practical significance with respect to a "possible coming war" (63). This applies in particular to the question of neutrality in such a case. Great Britain, France, and the United States, he claimed, had direct interests of their own in a discriminating concept of war and thus for taking a stand against an aggressive country at the international level. What mattered to Schmitt politically in his brochure was the hypothesis that at the time, international law was in a situation in which two competing concepts of war and international law coexisted: on one side, the concept of war in the Covenant of the League of Nations and, on the other, the traditional, nondiscriminating concept of war which Schmitt claimed for Germany and other nations seeking to be considered independent. The political core of his deliberations, which he formulated in decidedly academic language, consisted of reversing the distinction between a war of aggression prohibited under the Kellogg-Briand Pact of 1928 and a justified war in self-defense. Schmitt was convinced that this distinction was nothing but a perfidious means used by the Western powers to curtail Germany's options for action. Ultimately, this was how the *völkisch* foundation of international law developed its aggressive firepower: after all, following this logic, to a *Volk* claiming to fight resolutely for its national interests, a war of aggression was a legitimate war, too.

Schmitt sent his brochure to the highest government circles of the Nazi regime and the military top brass as he had done many times before in the preceding years. Reich Minister of Foreign Affairs Joachim von Ribbentrop wrote Schmitt a personal letter thanking him for the brochure.[7] When he was the German ambassador in London, Ribbentrop had called on Churchill to grant the German Reich the right to expand eastward to Ukraine and Byelorussia; an unconditional acolyte of Hitler's, he had become minister only in February 1938. Schmitt had succeeded once more in gaining the ear of those holding political responsibility in the regime.

3. Echoes in Geneva and New York

At least a brief echo to Schmitt's sensation in the field of international law was to be heard from Otto Kirchheimer in his exile in New York. It was in an omnibus review in Volume 3 of the *Zeitschrift für Sozialforschung* published by Horkheimer's institute (see Kirchheimer

6 For a critical discussion of Schmitt's arguments, see Habermas (2001, 165–203), Cohen (2004), Teschke (2011b), Benhabib (2012), Neumann (2015, 449–451), Teschke (2016), and Blasius (2021).

7 See editor's note (Günter Maschke) in Schmitt (2005, 592).

1938b). Starting with the arrangement of the texts reviewed, Kirchheimer's review was designed to contrast Schmitt's works on international law with two fundamental contemporary alternatives. Kirchheimer presented *Die Völkerrechtslehre des Nationalsozialismus* [National Socialism's doctrine of international law] by Eduard Bristler, which had been published in Switzerland, as a first alternative (see Bristler 1938).

Bristler was the pseudonym of John H. Herz, who had fled Nazi Germany in 1935. Herz had obtained his doctorate under Hans Kelsen in Cologne in 1931 with a dissertation on the identity of the state. After he had been dismissed from the *Referendariat* (a mandatory post-graduate legal training period) in 1933 because he was Jewish, he worked in a law firm for two years. In 1935, he emigrated to Geneva, where Kelsen had arranged for him to prepare his study of the Nazi doctrines on international law at the Institut des hautes études internationales (IUHEI). Herz's study was the first of its kind. It was published by the Swiss publishing house Europa-Verlag in 1938 and was banned immediately in Germany and Austria.[8] Georges Scelle, whom Schmitt had attacked, contributed a brief foreword in which he emphasized the fundamental difference Herz had identified between a universalist and a biologistic-racist concept of international law, thereby indirectly alluding to Schmitt, too (see Scelle 1938). During his exile in Paris, Kirchheimer had occasionally attended Scelle's lectures at the Faculté de Droit de l'Université de Paris but they did not know each other well. When Kirchheimer wrote the review, he and Herz had not yet met; they did so only later at the Office of Strategic Services (OSS), the forerunner of the CIA, where they became close personal friends. Unlike Kirchheimer, Herz's background was the liberal political milieu of the Weimar Republic. His book became an important additional source of information for Kirchheimer in his understanding of Schmitt's work on international law.

The very first quotation in Herz's book (see Bristler 1938)[9] was from Carl Schmitt's programmatic text *Nationalsozialismus und Völkerrecht*. No less than six works by Schmitt were quoted in the book. Herz considered his analysis to be an intrinsic critique of Nazi theories of international law. For this reason, he had sought less to construe them as being a monolithic, fully developed racist theory of international law but had instead hammered out their internal contradictions. He saw a "basic contradiction" running through them in all their "vagueness" and "obtuse ambiguity" (171). One the one hand, they still propagated the idea of a supranational order worthy of recognition; on the other, they went to great lengths to claim the superiority of a single *Rasse*.

In contrast to the German conservative author Adolf Grabowsky, for example, whom Kirchheimer criticized in 1932 for this reason,[10] Herz did not base his analysis on the notion of the primacy of foreign policy; instead, he explained that perverting rational international law into racist international law could only be seen as a function of striving to become a global power due to domestic policy concerns. His analysis ended with works from late 1937 and, of course, he could not be familiar with Schmitt's later career in international law. So the nothing less than prophetic succinctness with which he addressed a problem that Schmitt was to solve just one year later by including the buzzword

8 See Herz (1984, 111–113) and Puglierin (2011, 79).

9 The following page numbers refer to this book by John H. Herz.

10 See Chapter 6, p. 187.

Großraum in his theory of international law, again proving himself to be a legal pioneer, is all the more remarkable. Herz predicted that the German Reich's imperialist tendency toward expansion would sooner or later be brought in line with the ideologically motivated struggle against allegedly Jewish Bolshevism. The "goal of Germany expanding its power in the east and perhaps the southeast of Europe will be considered identical to the formerly proclaimed goal of the Aryan dominating the inferior races and peoples" (192).

In his review, Kirchheimer praised Herz's book as a comprehensive overview as complete as it was outstanding. He agreed with and highlighted how Herz had hammered out the political functionality present in the development of the Nazi doctrines on international law. Nazi international law was international situational law. What could directly serve the regime's foreign policy interests of the day, and nothing more, was to be recognized as international law. In the early phase of the Nazi regime, the lines of argument in international law had come from natural law, as Herz had stated, also with regard to Schmitt (see 83–85). This phase was followed by *völkisch* and racist doctrines that were to give the regime's current foreign policy goals a better legal foundation. Herz had quoted works by Schmitt as evidence of this phase, too (see 110–114). In this second phase, Schmitt no longer had a leading role in the field, but his earlier concepts such as "homogeneity" and the "friend-enemy distinction" provided important terms supporting and easing the transition to a decidedly racist doctrine of international law (see 118–120). In Herz's view, Schmitt was "hesitant to make his theory of race the criterion of his concept of homogeneity" (118) and the basis of "*Gleichartigkeit*" (see Glossary), which Schmitt considered essential, remained "unclear" (204) to the reader.

Kirchheimer in his review[11] followed Herz's strategy of immanent critique, agreeing that the line of argument in Nazi international law was contradictory. On the one hand, its protagonists were attempting to avoid completely negating international law and recognized an international community of laws, albeit a limited one. On the other hand, the *Rasse*-based theory of international law considered itself forced, for political reasons, to state that international law on the basis of bilateral contracts was the only appropriate form of international law at the time, which as a consequence would lead to a "dissolution" (200) of any international law. We can only speculate why Kirchheimer did not praise other aspects of Herz's book that he must have also appreciated. These would include the "legal-sociological" (Bristler 1938, 194) perspective, which Herz took programmatically, and the rejection of the doctrine of the primacy of foreign policy, which was inspired by power politics. Kirchheimer did not go into the parallels between Nazi and Bolshevik theories of international law, which Herz had elaborated at the end of the book, either; parallels to his own thinking are to be found here, too. Nor did Kirchheimer discuss Herz's detailed analysis of Schmitt in his book.

The second book Kirchheimer reviewed was the latest edition of *Völkerrecht* (International Law) by Alfred Verdroß, which was published in 1937. The author was the founder of the Vienna School of International Law, which was characterized by a Catholic and natural rights-based approach, and his works were already received widely in Germany during the Weimar Republic. Verdroß had often been the target of Carl Schmitt's attacks; Schmitt criticized his approach but above all his alleged willingness to give ground

11 See Kirchheimer (1938b). The following page numbers refer to this text.

to the victorious powers of 1918. In his book, Herz had classified Verdroß within the camp of Nazi literature because he had borrowed from "semi-*völkisch*-racist" (Bristler 1938, 136) works, notwithstanding his proximity to Kelsen's international law monism. Kirchheimer acknowledged in his brief comments on Verdroß's book that he considered his fundamental approach based on a monistic doctrine of international law to be a "decisive societal advance" (201) despite all the difficulties in its structure. Kirchheimer was apparently familiar with older works by Verdroß, for he accused him of not having decided whether to take the side of the Christian corporative state or of Nazism when he was working on the book—in other words, prior to the German Reich's invasion of Austria, which was celebrated as the "*Anschluss*."[12] This was the only way to explain that the "emphatically submitted avowal of natural law oscillated between *völkisch* and Christian traits" (201). Otherwise, he did not consider the book to be particularly original or forward-looking.

Finally, Kirchheimer reviewed Schmitt's book on the turn to the discriminating concept of war. This section is also the most polemical part of his review, ending with derisive comments about Schmitt's lack of knowledge. Kirchheimer correctly and concisely recounted the basic thrust of Schmitt's "position"[13] countering those of Scelle and Lauterpacht. Kirchheimer presumably knew the latter personally from his brief time working with Harold Laski at the London School of Economics in 1933/4. Next came a staccato of critical comments on and objections to Schmitt's work. "Following the pattern of all his other writings," Kirchheimer claimed, Schmitt began by showing "the political relevance of the writings he rejected," only to "give the political interests of the Nazis the dignity of a scientific theory using his usual apparatus of specious phrases based on conceptual realism" as the next step. Kirchheimer noted the "obvious contradiction" in Schmitt's latest attempt between the *völkisch* ideological position Schmitt championed and the need of the regime to maintain a consistent legal position. Since he had to "take the position of total and just war" because of his *völkisch* basic ideological assumptions, the position he espoused in his new work concerning international law demanded precisely the opposite, namely containment of a universal right to wage a war of execution against attacking states (thereby raising a war of execution, an intervention against a single state led by the central government to enforce imperial law in German empires, to the international level).

Kirchheimer had nothing but sarcasm for Schmitt's justification of "old-style" limited wars between states. "As we have heard," such a war would allegedly be contained practically automatically. Kirchheimer quoted Schmitt's statement in the brochure about limited wars between states according to which "its right, its honor, and its dignity" lay in the fact "that the enemy is not a pirate or a gangster, but a 'state' and a 'subject of international law.'" Kirchheimer's intention in quoting this passage was obvious to his readers at the time: the German Reich had just wiped the Republic of Austria off the map, a former "subject of international law," and annexed parts of the state of Czechoslovakia in

12 *Anschluß* was the Nazi German term for the incorporation of the Republic of Austria into the German Reich.

13 This and subsequent quotations are from Kirchheimer (1938b, 201–202).

the name of a higher-ranking *völkisch* international law. He concluded his review by ridiculing Schmitt: even an average French legal expert could only be astounded by Schmitt's statement that the French state was a mirror image of the legislative state. He also scoffed at Schmitt considering Harold Laski to be a philosopher of the Second International. Thus, in Kirchheimer's eyes, Schmitt had become a theorist who had made a dangerous fool of himself.

4. Conclusion: Germany attacking Poland

Schmitt's lecture in October 1937 marked the beginning of his aggressive and successful efforts to prevail as one of the leading international law theorists in the German Reich. From the mid-1930s on, when Hitler's rule had stabilized, international law provided excellent opportunities for building a career inasmuch as it enabled legal experts working in the field to do two things. First, it permitted them to avoid the risks of national constitutional law, which had increasingly become both devoid of substance and politically dangerous. Second, they could specifically prove how useful they were to the regime by doing legal work to legitimize the expansive goals of Nazism that were gradually coming to the fore. International law increasingly gained practical relevance to support the interests of the Nazi regime. Schmitt was not the only legal theorist in the Third Reich to shift the focus of his interests like this. The rise of international law compensated for the decreasing relevance of national constitutional law. Once again, Schmitt proved to be resolute and remarkably original when it came to his role as a forward thinker on the law of Nazi expansion policy up until the early war years. Viewed in the context of his oeuvre, not all of what he presented in his lecture of autumn 1937 was really new. His positions were not a strategic maneuver to avoid further attention from the SS, but a "logical extension" (Koskenniemi 2016, 594) of positions he had already taken during the Weimar Republic and in his earlier works on international law after 1933.

Due to the political circumstances, the issues of *Zeitschrift für Sozialforschung*, which was still printed in Paris, appeared almost a year after its official publication date. The readers perusing the 1938 issue of the journal which included Kirchheimer's review a year after its official publication date will have been struck by the contrast between Schmitt's claim that wartime enemies would not be treated like pirates and gangsters on the one hand and the brutal actions and atrocities of the German troops and authorities after the attack on Poland on 1 September 1939 on the other. Kirchheimer had accused Schmitt of conceptual realism as early as 1932/33 in the essay he had authored with Nathan Leites;[14] with respect to international law, he may have felt confirmed in this objection by a similar comment about Schmitt's "obsession with being original" (Bristler 1938, 78) and his method in Herz's book. Herz's accusation that some of Schmitt's work showed that he was completely unaware of the facts of the matter was not new, either.[15] What was new

14 See Chapter 6, p. 151–157.

15 "It arises from the peculiar way of thinking practiced by this theoretician, who is surely very sharp-witted in many ways, who is always striving to think 'concretely,' to derive his concepts from the 'realities,' but who simultaneously thinks constructively and immediately makes approaching what is

in Kirchheimer's review was his clear political prognosis that Germany would start a war in Europe very soon and that Schmitt's role in it was to produce the appropriate legitimation in international law. As Kirchheimer's review makes clear, he had stopped considering Schmitt's oeuvre to be a source of intelligent hypotheses or stimulation for developing further ideas on international law. He now read Schmitt as a legal theorist who was aligning his work entirely with the Nazi regime's situational political needs in foreign policy and at the price of accepting theoretical contradictions. Kirchheimer followed Herz in the strategy of immanent critique and he expanded it with the critique of ideology.

'concrete' the basis of constructions which, once they are generalized, must again lead away from the concrete." (Bristler 1933, 120).

Chapter 12: From *Großraum* Theory to the Escalation of World War II (1939–1942)

Unrestricted *ius ad bellum*, which Carl Schmitt had championed, was also a key foreign policy doctrine of the Nazi state. Preparations for war, intensified from the summer of 1937 on, laid the groundwork for the first practical military operations initiated only a few months later. On 12 March 1938, Germany invaded Austria. The next step was directed against Czechoslovakia. In late September 1938, the Munich Conference sanctioned Hitler's policy of violence toward Sudetenland and agreed to German troops invading parts of Czechoslovakia. This was followed by the invasion of rump Czechoslovakia in March 1939 and the establishment of the Protectorate of Bohemia and Moravia in violation of the Munich Agreement, which had been signed six months earlier.

This was the atmosphere in which a high-level conference on the occasion of the 25th anniversary of the founding of the Institut für Politik und Internationales Recht (Institute for Politics and International Law) in Kiel took place from 31 March to 3 April 1939; on 1 April, Schmitt gave a lecture there titled "Der Reichsbegriff im Völkerrecht" [The concept of the Reich in international law]. The conference provided a "model for the German professors who could make themselves useful to the war effort by providing concepts and catch phrases for educated opinion" (Balakrishnan 2000, 234).[1] This was also true of Schmitt. He explained to his audience that he had recently taken a major step forward in his own understanding of international law. After he had given his lecture on the turn to the discriminating concept of war in 1937, he had been asked what exactly he was able to offer in place of the existing order of states since he "at the time, neither wanted to remain with the old concepts, nor subject [himself] to the concepts of the Western democracies" (Schmitt 1939a, 110). Now, two years later, Schmitt said, "I can give the answer to that question. The new concept of the order of a new international law is our concept of Reich, which proceeds from a *völkisch Großraum* order upheld by a nation." (Schmitt 1939a, 111) Schmitt also mentioned the crucial historic event that prompted replacing the state with the Reich as the key power of a *Großraum* (literally "large space," inherently linked

1 On the context of this lecture, see Blasius (2021).

to geopolitics and Nazi Germany's expansionist policies; see Glossary): "the action of the *Führer* has lent the concept of our Reich political reality, historical truth, and a great future in international law." (Schmitt 1939a, 111) Schmitt's audience doubtless understood what Schmitt meant by "action of the *Führer*"[2]: the German invasion of the rump Czech lands two weeks earlier.

Schmitt's lecture caused a sensation reaching beyond academic circles. It seemed so significant for understanding the current situation that all the major German newspapers ran lengthy reports that included quotes. Two leading British dailies, *The Times* and the tabloid *Daily Mail*, reported on it at length four days later and characterized Schmitt as the theorist behind Hitler's expansionist policy (see Bendersky 1983, 257–259). Outside of Germany, the lecture was generally noted as a quasi-official signal of Germany being on the threshold of a new imperialist era (see Stolleis 1999, 390), which had in fact already begun as far as Austria and Czechoslovakia were concerned. In the years that followed, Schmitt wrote a number of books and articles on international law and politics. Even from New York, Kirchheimer was keen to follow the writings of his former doctoral advisor and political adversary during the Weimar Republic on these subjects. Both Schmitt and Kirchheimer retained their sensitivity to the specific contexts of their writing on international law and politics. Below, I will also embed these works in their particular political contexts.

1. Early critical theory's disregard of international politics

There is no greater contrast to Schmitt's preoccupation with international politics than the subjects at the center of the Frankfurt School's early critical theory. In view of the global political turbulence in the 1920s and 1930s, it is astonishing that none of the works by this first generation of the Frankfurt School were notable contributions on international politics.[3] Occasionally, the *Zeitschrift für Sozialforschung* ran reviews of books on international politics, but these were not written by members of the Institute of Social Research (ISR). Neither the inner circle including Max Horkheimer, Theodor W. Adorno, Friedrich Pollock, and Leo Löwenthal nor the outer circle with Erich Fromm, Franz Borkenau, and Walter Benjamin had much to say about questions of international politics. Whenever institute members did in fact deal with other countries, it was in order to analyze the internal social problems there.[4]

This thematic gap is all the more striking as proponents of the early Frankfurt School placed the Hegel-Marxist concept of critical theory, which they pursued with aplomb, between what they saw as rigid social democratic progressive thinking and the revolutionary Marxism of the Soviet Union, which they also criticized as dogmatic. During the Weimar Republic, however, both the Social Democrats and the Communists in Germany

2 German original: "*Die Tat des Führers.*"

3 On the following, see Buchstein (2022, 112–117).

4 This is the case with Friedrich Pollock's analyses of the Soviet planned economy, Karl August Wittvogel's studies on agricultural economy in China, and Felix Weil's work on the American New Deal.

had clearly taken internationalist positions on questions of foreign policy. The SPD and its foreign policy theorists espoused a type of international politics that we now call the liberal paradigm: an internationalism that relied on a policy of reconciliation with neighboring countries and that sought active participation by Germany in the League of Nations. In his Weimar writings, Kirchheimer, too, had supported this approach.[5] In contrast, not only the Stalinized KPD but also the Trotskyists who remained outside the party (often in literal agreement with the right and with the right-wing extremists) criticized the alleged "disgraceful peace" of the Versailles Treaty and propagated the proletarian world revolution as their version of internationalism. Marx and Hegel cannot have been role models for the silence at the ISR on questions of international politics—there are far too many theoretical considerations and original individual observations about the international *Lage* in their works.

Was it perhaps a certain Frankfurt provinciality that was responsible for this silence, regardless of the sophisticated and worldly lifestyle of some in Horkheimer's group? That could be assumed for the early years of the institute, but certainly no longer for the period from the mid-1930s, when the émigrés who had been forced to flee to various countries looked anxiously at the German Reich, and certainly not after Germany's attack on Poland and the military events that escalated into a world war. The many letters that have come down to us from this period show that most of Horkheimer's group definitely expected a new war in Europe from 1935 onwards. Horkheimer's pessimistic credo from the essay "The Jews and Europe," written in the first few days of September 1939, speaks volumes about the mood at the institute: "Nothing can be hoped for from the alliance between the great powers." (Horkheimer 1939a, 93) At the same time, this statement by Horkheimer documents his great helplessness in view of the international developments at the time. However, helplessness is not a disgrace. It is better to admit that you do not know the right thing to do than to hasten to come up with some new and strange interpretation or theory. At least in this respect, it was wise of the core group of the Frankfurt School to remain silent in public on questions of international politics.

It was not until after Germany had started the war that Horkheimer and his group began to include international politics in their research agenda. In October 1939, Horkheimer reported to his correspondent Franz Bischofswerder, who lived in Seattle and was researching the situation of Jewish refugees, about the institute's work: "This year we have weekly meetings on foreign policy issues." And he added: "The reason for this event is the realization that we, the members of the Institute, have far too little precise knowledge of foreign policy to make more than just amateurish statements about the meaning of current events."[6] Horkheimer's firm expectation in the summer of 1941 that the German military would defeat the Soviet Union in the shortest possible time and Friedrich Pollock's prognosis of an imminent war between the Soviet Union and the US, which he expressed at the same time, are evidence of how difficult it was to assess the current events of the war from the other side of the Atlantic.[7]

---

5 See Chapter 4.

6 Letter from Max Horkheimer to Franz Bischofswerder dated 2 October 1939 (Horkheimer 1995, 651).

7 See letter from Max Horkheimer to Leo Löwenthal dated 26 June 1941 and letter from Franz L. Neumann to Max Horkheimer dated 23 July 1941 (Horkheimer 1996, 81, 107).

The only exceptions to this public silence of the Frankfurter School in exile can be found in some of Otto Kirchheimer's writings during his time at the ISR and in 1942 with the publication of the book *Behemoth* by Franz L. Neumann. Almost a fifth of this book is devoted to the foreign policy of the Third Reich. Kirchheimer collaborated with Neumann on the chapters on Nazi *Lebensraum* ideology. Carl Schmitt's theory of international law and his *Großraum* theory received a prominent place in the book.[8] Kirchheimer continued to follow Schmitt's writing on international law and politics inasmuch as he had access to it in his exile in New York.

2. Schmitt's *Großraum* theory

Following the extraordinary public response to his lecture at the Kiel conference on 1 April 1939, Schmitt wasted no time finalizing the text for publication, and the printed version was already available three weeks later. It immediately caused even more of a stir because Hitler proclaimed the idea of a "Germanic Monroe Doctrine," which Schmitt had laid out in his lecture, a matter of days after its publication in a programmatic foreign policy speech in the Reichstag in Berlin on 28 April.

Hitler reacted in his talk, which of course was broadcast by the German radio stations, to an urgent appeal by US President Roosevelt to him and Mussolini on 14 April 1939 to give a ten-year guarantee of nonaggression to a list of 31 countries in Europe and the Middle East. Following his tried and tested pattern, Hitler's speech was full of avowals of peace and the harmlessness of his intentions. At the same time, he demonstrated all his sarcasm, suggestive logic, and persuasive power, which prompted some contemporary historians to call it the presumably most brilliant speech he had ever given (see Fest 1973, 795). One rhetorical highlight was the idea of a German Monroe Doctrine, which had been formulated by Schmitt. Hitler provocatively asked Roosevelt how he would respond if a German chancellor asked him to change his policy in Latin America:

> In this case, Mr. Roosevelt will certainly invoke the Monroe Doctrine and reject such a demand as interference in the internal affairs of the American continent. It is exactly the same doctrine that we Germans are now advancing for Europe, but in any case, for the area and the interests of the Greater German Reich.[9]

Yet Hitler's speech conveyed not only rhetorical effects, but also an important political decision. Hitler terminated the Anglo-German Naval Agreement with immediate effect as well as the Reich's contract with Poland, and he eliminated the pledge included in such contracts and agreements to resolve all disputes peaceably. From then on, recourse to a German Monroe Doctrine was part of the fixed repertoire of propaganda in Nazi foreign policy, for example in Ribbentrop's meeting with Sumner Welles, the US representative, in March 1940. Constant references to this doctrine served to pursue a practical political

8 See below in this chapter.

9 Hitler's speech is quoted in Gruchmann (1962, 11).

goal: keeping the US out of Nazi wars of conquest in Europe. Hitler held fast to the notion of a German Monroe Doctrine right up until his final days.

There is some disagreement in the historical scholarship whether Hitler actually adopted the formula of a German Monroe Doctrine directly from Schmitt.[10] The latter had examined the Monroe Doctrine in other, earlier publications—some of them already during the Weimar Republic—and had mentioned it time and again after 1933 in his criticism of Anglo-Saxon imperialism. It is most probable that Hitler heard about the idea of a German Monroe Doctrine, a term coined by Schmitt, from high-ranking Nazi legal experts and that he considered it his own creation from then onward. After Hitler's speech, Schmitt was telephoned by his patron Hans Frank, who pointed out how highly the *Führer* valued the originality of his own thoughts and deliberations on the Monroe Doctrine in his speech of 28 April (see Bendersky 1983, 258)—a blunt instruction to Schmitt not to claim in public that he was the one who had originally come up with the idea.

The fact that Hitler and Schmitt's speeches were in harmony contributed considerably to Schmitt regaining his reputation in the Nazi system and to his renown abroad. In December 1940, the Swiss *Die Weltwoche* wrote that Schmitt with his *Großraum* theory was for contemporary Europe what Rousseau had been for the French Revolution (see Neumann 2015, 457). The extended version of Schmitt's lecture appeared in April 1939 as a short book titled *Völkerrechtliche Großraumordnung mit Interventionsverbot für raumfremde Mächte* (The *Großraum* Order of International Law with a Ban on Intervention from Spatially Foreign Powers).[11] Three additional and slightly revised editions were published before the attack of Germany on the Soviet Union in 1941, but none thereafter. Schmitt explicitly addressed the US. The doctrine named after US President Monroe had been proclaimed in December 1823 and was directed against violent interventions by European states in all of North and South America. It was a reaction to the alleged threat against the US and the Americas by the Russian expansion south of Alaska. In addition, the US feared that the counterrevolutionary Holy Alliance on the European continent, under the leadership of France, would recolonize the new republics that had been established following the Latin American wars of liberation against Spain. The ban on interventions claimed by the US in the Monroe Doctrine was thus of a primarily defensive nature (see Gruchmann 1962, 146–162).

Schmitt interpreted the doctrine differently. Although he had criticized it in his writings from the 1920s as the epitome of American imperialism, he now interpreted it in a positive light, viewing it as the historical precursor of a new type of principle of order in international law, "the precedent for a *Großraum* principle" (83). The basic idea of this doctrine, he stated, was also transferable to other friend-enemy constellations, to other historical situations, and to other geographical *Räume*; specifically for Germany, this meant to the central and eastern European *Raum*. Schmitt's intent in establishing the German counter-doctrine to the American Monroe Doctrine was to be able to claim the core of its

10 For the different views, see Winkler (2001, 37) who states that Schmitt did have had a direct influence and Maschke (1995b, 348) who sees only indirect connections.

11 See Schmitt (1939a).

legitimacy for the German Reich, too. He believed he had found a principle in international law supporting the German Reich's policy of expansion which the US would not be able to contradict because it claimed the same for itself.

It is difficult to render the Nazi German definitions and connotations of terms such as *raumfremde Mächte* (powers foreign/alien to the *Raum*) or *Großraum* accurately in English. To Schmitt, the concept of *Raum* was not identical to space, its literal translation. It was also related to *nomos* and *Ort* (place). Schmitt translated the Greek term *nomos* as *Lebensgesetz* (the law of life), thus adding an existential element to it. He republished his book on *Großraum* three times during the Nazi period, constantly adapting it to the events of the war.[12]

His theory of the *Großraum* order of international law had multiple components.[13] Schmitt did not select *völkisch* or biologistic terms but rather purely technical language for his definitions of *Großraum*. A *Großraum* was the result of an "economic-industrial-organizational development" (119) of expansion and was based on a "*Großraum* economy" (79) typical of the modern energy industry. Schmitt considered the origin of the *Großraum* to be in the industrial, organizational, economic, and technical fields, referring to a *Leistungsraum* (the *Raum* required for a country's desired economic performance, thereby providing an economic justification for expansionism) (79). The size of the *Großraum* went far beyond the borders of a traditional nation-state. He contrasted the concept of the *Großraum* order with imperialism, which he characterized as Anglo-Saxon. Imperialism, Schmitt claimed, was based on extending invisible domination to include apparently independent regions. It ruled as an indirect power with indirect methods. In a *Großraum*, in contrast, political power was exercised as direct and publicly visible domination.

The transition to *Großraum* theory had serious consequences for Schmitt's theory of the state, which he had championed since the 1920s. Schmitt had vehemently advocated for the theory of the sovereign state in all his works since the beginning of the Weimar Republic, but now departed completely from the concept of the state as the primary category of political order. The new concept in its place was the Reich. His hypothesis was that the Reich would counter the state, previously the key concept of international law, because it was practicable in international law and, being up-to-date, superior. After the "action of the *Führer*" (111)—the term used by Schmitt in his lecture in Kiel to describe the invasion of German troops in rump Czechoslovakia—he thought that the traditional concept of the state was no longer appropriate to the "political reality and the historical truth" (111) of the political world of the day. According to Schmitt, traditional international law was founded on the postulate of the legal equality of all independent and sovereign

---

12 Timothy Nunan's 2011 English translation is based on the fourth edition of the book from 1941. There are only minor substantial differences between the 1939 and 1941 editions. Overall, it is a very good translation. However, Nunan chose more literal translations of some Nazi German terms that do not always properly convey their ideology and connotations, e.g. "on the basis of nation" (102) for *volkhaft* and "species-determining" (124) for *artbestimmend*; see Translator's Preface and Glossary.

13 See Schmitt (1941e). The following page numbers refer to this text, On the following, see also Gruchmann (1962, 51–65), Maschke (1995b, 343–364), Elden (2010), Benhabib (2012), Neumann (2015, 457–473), and Minca and Rowan (2016).

states. He believed the actually existing real hierarchy of subjects of international law was no longer captured by the traditional vocabulary of international law.

Schmitt gave two reasons why state-related international law was to be laid to rest. First, the technical progress of the previous decades had meant that it was no longer the case that all the states of the world would be able to pass the test of being able to establish and sustain a sovereign state apparatus. Here, he was addressing the capabilities to wage war. Only a few states were in a position to wage a modern war of material on the basis of their own industrial, technical, and organizational power. The idea in international law that territorially small states were sovereign was nothing less than absurd in light of recent developments, for example, in aviation. The attributes of traditional statehood necessary for sustaining such states—internal organization and the capability to defend themselves militarily—could now only be achieved by a few political entities, the *Reiche*. Schmitt's second argument referred to the reality of global politics of the day. The truly fundamental variables shaping the coexistence of *Völker* (peoples/nations in a racial sense, of common blood and with a common destiny; see Glossary) were a few major powers exercising political influence beyond the territories of their states in various ways. In other words, in Schmitt's works, "Reich" referred not only to the German Reich; he also used the term to denote other major powers such as the US or the Soviet Union. The size of the *Großraum* resulted solely from the reach within which the political concept of the Reich could prevail within the *Großraum*. To be precise, the Reich alone determined its reach. The new global *Großraum* order Schmitt had in mind thus consisted of multiple *Reiche*, each of which exercised direct domination over other states that had previously been independent. For continental Europe, Schmitt considered Germany, not France, as having the historical role of hegemonic central power. The German *Großraum* order was legitimized through the ability of the Reich's *Volk* to dominate the *Völker* of the neighboring states within the *Großraum*.

It is striking that when Schmitt defined *Großraum*, he avoided the *völkisch* term *Lebensraum* (*Raum* necessary for autarchic economy of an increasing population, *Raum* assigned to a people by racial destiny, see Glossary) used by Hitler in his speeches and writing as well as by other ideologues of the Nazi system. Schmitt's conceptual parallel term was the "*Leistungsraum*" (Schmitt 1941d, 319) (the *Raum* required for a country's desired economic performance, thereby providing an economic justification for expansionism), which was held together by a strong political will. Consequently, some scholars emphasize the marked differences between his *Großraum* theory and the allegedly official Nazi ideology when interpreting this piece by Schmitt.[14] In fact, however, the differences between them were not so great. First, because Schmitt adopted the "new concrete concept of *Raum*" (123) from the German biologist Viktor von Weizsäcker according to whom any spatial order is produced by activism and movement. In Schmitt's view, a strong political will was the human equivalent to the movement of other biological species. A second reason not to overinterpret the differences between Schmitt's and the dominant Nazi ideology is that he himself also used the term "*Lebensräume*" (Schmitt 1941d, 278).

The similarities between *Großraum* theory and Nazi ideology are even more striking if we move away from Schmitt's idealized conceptualization of the *Großraum* and in-

14 See Maschke (1995b, 358–364), and Neumann (2015, 487–489).

stead focus on the German Reich's practical policies which Schmitt explicitly listed and praised as examples of German *Großraum* policy in the various editions of his text up to the summer of 1941. In these cases, the political idea behind the German Reich was not any supposed respect for the individual *Völker* in the *Großraum* but the cultural genocide of Czechs, Poles, and other Slavic peoples as well as the ghettoization and murder of Jews. In later editions of the book, Schmitt explicitly welcomed the deportation of Poles and Jews in the General Government, which had been newly established by the Reich (see 100). Schmitt's statements about "*der Jude*," which are scattered throughout his writing, expressed his antisemitism downright obsessively. The relationship of a *Volk* to its *Raum* and its soil had to remain incomprehensible to "the spirit of *der Jude*" (121). Moreover, the "*artfremd*" (99) (foreign/alien to the *Volk*, in an exclusionary and antisemitic sense) Jews did not even want to understand it; this mindset would allegedly enable them to promote their intellectual and abstract theories of international law even more successfully and subversively (see 121–122). Schmitt also took care to preface Harold Laski's name with the words "*der Jude*" (see Glossary) (108).[15] The former opposed any kind of "assimilation" or "absorption" and the "idea of [...] melting pots," (99) using racially charged language in this context multiple times. He accused the Western powers of seeking to oppress the German Reich using an abstract and individualistic theory of international law.

The connection Schmitt thereby construed particularly between the US and allegedly Jewish thinking gave him the opportunity, as Timothy Nunan stated, "to repackage his enemy, the Jew, in a discourse of Great Powers and *Großräume*" (Nunan 2011, 16). Thus, Schmitt's image of the US and the Monroe Doctrine served as an "ersatz for a deeper-seated fear of Jews" (Nunan 2011, 16). Yet the core of Schmitt's theoretical construction of the *Großraum* did not require the *völkisch* or biologistic arguments he himself had made, quoted above. This distinguished his approach from the other two Nazi *Großraum* ideologies. The *Lebensraum* ideology as advanced by Reinhard Höhn, Werner Best, and Werner Daitz defined the *Großraum* exclusively based on a biological essence, namely ties of common blood. For this reason, both Höhn and Best attacked Schmitt for not fully representing the pure doctrine of Nazism.[16] Schmitt's concept also deviated from the concepts of the *Großraum* put forward by Karl Haushofer's geopolitical school, which delineated the borders of a *Großraum* on the basis of geographical characteristics such as oceans, rivers, or mountain ranges.[17] To Schmitt, neither blood nor soil defined a *Großraum*, but only the political will of a *Reichsvolk*. In other words, the boundaries of a *Großraum* were determined in practical terms by the Reich alone, in the case of Germany, for example, by the "action of the *Führer*" mentioned above. At the same time, the vague category of political will indicated the point where, because of its elasticity, Schmitt's *Großraum* theory could easily be squared with the fundamental views of the inequality of human beings inherent to the Nazis' theories of *Rasse*. What Schmitt borrowed from *völkisch* terminology was irrelevant for his theoretical construct; however, it enabled links from his own to the other two major Nazi *Großraum* ideologies.

---

15 Laski had helped Kirchheimer and Franz L. Neumann come to the LSE after they had to leave Germany in haste; see Chapter 7, p. 178.

16 See Herbert (1996, 271–298), and Blasius (2021, 470–472).

17 On the two other schools, see Gruchmann (1962, 20–24) and Maschke (1995b, 358–363).

3. Schmitt and the further escalation of the war

Schmitt published the first edition of his *Großraum Order of International Law* four months before Foreign Ministers Ribbentrop and Molotov signed the non-aggression pact between Nazi Germany and the Soviet Union on 23 August 1939. His concept of German regional hegemony was comparatively moderate and did not yet envisage Germany directly dominating the remaining eastern European states. When the Hitler-Stalin Pact and its secret protocol regarding partitioning Poland were concluded, the *Großraum* that Germany had set its sights on expanded considerably. On 1 September 1939, the German armed forces attacked Poland and established the General Government of Poland after a brief period of warfare. In April 1940, Germany attacked Denmark and Norway, and the German western offensive began on 10 May 1940. The Netherlands and Belgium capitulated, and following the rapid victory over France, a ceasefire agreement in the West was finalized on 22 June 1940.

Schmitt adapted his *Großraum* theory to the new *Raum* conditions as rapidly as the Germans advanced into neighboring countries. The *Großraum* theory with a regional hegemon, which he had presented in Kiel in the spring of 1939, was transformed into the concept of a giant, German-dominated Fortress Europe extending from the Atlantic to central eastern Europe. Schmitt advanced his *Großraum* theory in the next two years in multiple publications and numerous lectures. In 1940, at the time of the German invasion of Denmark, which was followed by the invasion of Norway, the Netherlands, Belgium, Luxembourg, and France, he embarked on a lecture tour, speaking in Bremen, Kiel, Rostock, Halle, Naumburg, Cologne, and Berlin. His *Großraum* theory was compatible with this expansion of German power in Europe.

Schmitt made other additions to the text indicating a transformation of the meaning of his *Großraum* order. His preface to the fourth (and final) edition, dated 28 July 1941, one week after the invasion of the Soviet Union, called it a "document" that was not to "take up a foot race with the events themselves." The new events required "their own treatment" (Schmitt 1941a, 77). The new additions he made in this edition show that the invasion of the Soviet Union, Operation Barbarossa, and the immense eastward expansion it entailed went far beyond the limits of Schmitt's concept of the *Großraum*. It is obvious from other published works, too, that in 1941, Schmitt wanted to limit the war to a struggle with Great Britain and that he hoped that neither the Soviet Union nor the United States would get involved. In his opinion, the *Großraum* world powers were the United States, Germany, Japan, and the Soviet Union.

The limits of the territorial framework of his understanding of the *Großraum* were shattered not only by the German attack on the Soviet Union on 22 June 1941. The war in the Far East was developing in a similar way. Schmitt had expressed admiration for the Japanese military success against the colonial European outposts in Asia and the Pacific. He interpreted it as additional confirmation of his *Großraum* theory (Balakrishnan 2000, 239). However, the subsequent Japanese air raid on Pearl Harbor on 7 December 1941 meant that the war had expanded beyond the territorial dimensions of his envisaged *Großraum* and had become a world war between the Axis powers and the Allies. Schmitt was intensely distressed about this global expansion of the war. From then on, he no longer attempted to adapt his concept of the *Großraum* to the new territorial constellations. He considered the German invasion of the Soviet Union a disastrous overstepping of the boundaries of any *Großraum*. When the fourth and final edition of Schmitt's Großraum *Order of International Law* was published, the German army had just invaded the Soviet Union. Schmitt's concluding comments in the preface, dated 28 July 1941, expressed his concerns about being overwhelmed by the turn of the war: "May the reader understand when I give this writing the following motto: 'We resemble navigators on an unbroken voyage, and every book can be nothing more than a logbook.'" (Schmitt 1941a, 77)

Nonetheless, Schmitt continued to examine questions of the *Großraum* economy, not least in an academic advisory function. From 1941 on, he belonged to the scientific advisory body of the Gesellschaft für europäische Wirtschaftsplanung und Großraumwirtschaft (DeWG, Society for European Economic Planning and *Großraum* Economy), which had been newly established in September 1939, at the same time as the invasion of Poland (Teschke 2016, 396). Headed by Werner Daitz, the director of the Nazi party's Division of Foreign Trade, the DeWG was tasked with plundering the territories conquered in the expansion of the Reich. In the remaining war years, however, Schmitt's academic interests turned more to the US than to Europe. In place of an American *Großraum*, he now spoke of the "Western hemisphere" as a part of the world dominated by the US. An "American century for our planet" (Schmitt 1943, 447) was looming, with worldwide American pan-interventionism. In his writing and lectures from 1942 to 1944, he commented on the role of the US as a world power, which was becoming even clearer as the war progressed, with a mixture of aggressive hostility and admiration for successful power politics.[18]

4. Kirchheimer on Schmitt's apologia for the Nazi wars

US government agencies as well as a few scholars of law and political science in the country keenly observed the development of German ideologies used to legitimize waging war. American international law scholar Josef Laurenz Kunz, who had immigrated from Austria in 1932, made the following observation on Schmitt's latest U-turn in an article titled "Germany's Lebensraum" in the *American Journal of International Law* in the autumn of 1940: "Carl Schmitt, professor of law, has, of course, never been a jurist, but a politician" (Kunz 1940, 170). Those driven into exile after 1933, who had been meticulously observing Schmitt's activities from then on, paid far less attention to his shift to *Großraum* theory than to his previous contributions to establishing the Nazi state in Germany and to his antisemitic diatribes. Kirchheimer was among the few who took note of these shifts in Schmitt's political thinking at an early stage. But he did not initially react directly to Schmitt's latest political theory ploy because he was too busy with works commissioned by the Institute of Social Research in the years 1939 to 1942. Yet even in these works, it is possible to find indications that he continued to pay attention to Schmitt's writing—provided he could get his hands on them at all in New York shortly after publication.

18 See Chapter 13.

The first direct indication of this is to be found in the article "Changes in the Structure of Political Compromise," published in 1941 (see Kirchheimer 1941a).[19] Kirchheimer presented his interpretation of what can be called the methodological primacy of domestic policy. How could the relative stability of the Nazi regime be explained even though the social groups within it were wrestling for power and influence? The current compromise structure of Nazism had brought an old question to the fore: How could the interests of the five main partners to the compromise—the monopolies, the army, industry, agriculture, and the Nazi party—be brought to a common denominator? Kirchheimer stated that it had become apparent that the *Führer* had been able to establish political authority acting as an ultimate arbiter in all cases where the monopoly groups involved were not able to reach a compromise by themselves. He thought that the party under Hitler's leadership was able to resolve intergroup differences "with relative ease" (287). According to Kirchheimer, Hitler's decisions were possible "with a minimum of resistance only because of the unfolding program of expansion" (287). German expansion by means of war had given the various groups the opportunity to extend their activities and to satisfy their desires and interests with little need to get in each other's way. The existence of the authoritarian regime in Germany was intimately connected to the execution of Germany's imperialist program. At this particular point at the end of his article, Kirchheimer made a subtle reference to Carl Schmitt: The form of domination which "the large-space (*Grossraum*) imperialism of Germany" (287) represented was not compatible with the fiction of sovereignty limited to the domestic realm. He quoted the following from an article by Ernst Rudolf Huber, his fellow student in Bonn with Schmitt: "The developing large-space order might, contrary to earlier imperialism, constitute a system of direct and open domination" (288). A closer look at Kirchheimer's source shows that Huber's sentence was a practically verbatim quote of Schmitt (see Huber 1941, 14).

Two more, albeit smaller, direct indications are to be found in two book reviews Kirchheimer wrote in 1941 and 1942. In an omnibus review of several books published in the US on the Nazi regime's economic policy, Kirchheimer devoted his attention to the Reich's foreign economic relations. On the basis of the import and export figures for 1929 and 1937, it could be stated that the balance of trade "has not been basically altered by the conquests made up to the spring of 1941" (Kirchheimer 1941b, 363). The parallels to Schmitt's *Großraum* theory, which also argued along economic policy lines, are striking. A "Nazified Europe, the sphere of domination which ends at the Channel ports and at the confines of the Russian Empire," had developed into an integrated economic area. Commenting on the tables in the economic policy analyses of the Brookings Institution, Kirchheimer wrote, "it looks as if Hitler's advisers had studied the same figures" (Kirchheimer 1941b, 364).

Kirchheimer published a review of Leopold Schwarzschild's book *World in Trance* in the 4 December 1942 issue of the *Aufbau* in which Schmitt was also discussed, at least indirectly. The *Aufbau*, a weekly founded in New York in 1934, was considered the most important source of information for Jewish and other German-speaking refugees in the US; it was also a place where they could publish. With its readership of up to 300,000 in forty-five countries around the world, it was the leading periodical for German-speaking

19 The following page numbers refer to this text.

Jewry. Its authors best known today include Hannah Arendt, Gershom Scholem, Max Brod, and Thomas Mann.[20] Institute of Social Research members Max Horkheimer, Theodor W. Adorno, Felix Weil, and Paul Massing occasionally wrote for the *Aufbau*, too. Leopold Schwarzschild was one of the best-known German-speaking émigrés in New York at the time. He was from Frankfurt am Main and had been one of the publishers of the left-liberal weekly *Das Tage-Buch* up until 1933. After fleeing Germany, he founded *Das Neue Tage-Buch* in Paris, which became the most important cultural-political magazine of the German-speaking émigré community. In 1936, Schwarzschild was involved in the efforts of the Lutetia Circle in Paris to establish an anti-fascist people's front; this was where Kirchheimer met him personally. After fleeing to the US in the summer of 1940, Schwarzschild tried to make a living as a writer and journalist in New York. His book *World in Trance: From Marseille to Pearl Harbor* is a kind of sequel to his 1934 *Ende der Illusionen* [End of illusions] in which he lamented the many misconceptions of German politics during the Weimar Republic at the time (see Schwarzschild 1934).[21]

Kirchheimer began his review[22] with the expectation that "the defeat of fascism […] is approaching" (331). He praised Schwarzschild's book as the best-written, most gripping argument with the most logical structure that had ever been penned criticizing both German imperialism and the aimlessness and leniency of Allied policy. It included an exemplary portrayal of the "continuing parallelism between Germany's systematic hegemonic ambition and Allied weakness" (331). Yet he did not hold back his fundamental criticism that Schwarzschild's work suffered from two misconceptions. First, his strict separation of domestic and foreign policy was incorrect. The "constant interaction" between foreign and domestic policy had never been more manifest than in the period between the Paris Peace Conference of 1919 and the Spanish Civil War with the accompanying appeasement efforts. Methodologically speaking, it was important to take account of the "class constellations" in society and the "social and economic forces supporting them" (332) when analyzing these interactions. Second, Kirchheimer believed that Schwarzschild had misinterpreted the role of France as a guarantor of peace in Europe following the war from 1914 to 1918. France's military hegemony in a pacified Europe would have necessitated "France's willingness to use most of its national product for policing functions" (332). Yet it had become apparent that "none of France's social groups wanted to shoulder the sacrifices that would require" (332). Kirchheimer wrote that it was Europe's undoing that at no point since World War I had there been governments of the same political orientation in France and Germany that could have achieved reconciliation between the two countries.

Kirchheimer's analysis of the Third Reich's structure of compromise found its theoretical culmination in his hypothesis that the existing compromise could not draw on its own resources to generate a stable equilibrium. It was sustainable only because decisions were made by the dictatorial head of the regime. And these were successful only if and only as long as every sacrifice by a group could be balanced out by other benefits. Nazi Germany was thus virtually programmed toward an expansive foreign policy that was not at all concerned with existing international law. The decision-making power of the

20 On the history and the authors of the *Aufbau*, see Kotowski (2011).

21 On his political diagnoses of the present while in exile, see Papcke (1993, 13–37).

22 See Kirchheimer (1942). The following page numbers refer to this text.

political leaders in Germany 'rests on its ability to compensate every group sacrifice with advantages which, however, can ultimately be got only in the international field, that is to say, through imperialist policy" (Kirchheimer 1941a, 289).

In order to understand the causes of this imperialism, Kirchheimer insisted on reconstructing the constant interactions between domestic and foreign policy on the methodological level. This included an empirical analysis of the class constellations in society and the particular social and economic forces supporting imperialism in foreign policy.

5. Kirchheimer and Neumann's *Behemoth* on the concept of *Großraum*

An even greater, albeit indirect, indication of Kirchheimer's sustained interest in Schmitt's works on international law is to be found in Franz L. Neumann's 1942 book *Behemoth: The Structure and Practice of National Socialism*, which ran to more than 500 pages. *Behemoth* was the first comprehensive description and analysis of the Nazi system. Soon after its publication, it became a standard work on the Nazi system in the English-speaking world. An updated version was published in 1944. The book became a kind of handbook for the American military forces after the victory against Nazi Germany.[23]

This book was written while Neumann worked at the ISR. Both Gurland with his knowledge of economics and Kirchheimer with his legal and economic expertise had contributed to it in large measure. As Ossip K. Flechtheim, Neumann's research assistant at the ISR at the time, recalled, his brief acknowledgments in the introduction of the book (see Neumann 1944a, 18) were a "colossal understatement"[24] of the actual scope of Kirchheimer's contribution. According to Flechtheim, Kirchheimer was also involved in the analyses of the economic order in *Behemoth*, and Neumann and Kirchheimer discussed Carl Schmitt's more recent works multiple times. In Neumann's book, many passages on the new theories of international law and on various original sources from the German Reich were taken from the book by Eduard Bristler (John H. Herz) that Kirchheimer had previously reviewed.[25] Whereas Neumann's passages on the constitutional law of the Third Reich largely echoed his work *The Rule of Law. Political Theory and the Legal System in Modern Society* (see Neumann 1935), which he had completed when working with Harold Laski in London, his writing on how Nazi international law provided legal legitimacy to the various stages of Nazi foreign policy was new; this further supports Flechtheim's statement that Kirchheimer was heavily involved in preparing it. To a large extent, the passages in the book that deal with Schmitt can be attributed to Kirchheimer as the co-author.

The extensive chapter in *Behemoth*[26] on the Nazi theory of the Greater German Reich read like a black book on Carl Schmitt. It opened with excerpts from Hitler's Reichstag speech of 28 April 1939 in which he had hurled the German Monroe Doctrine, which

23 See Chapter 13.

24 Ossip K. Flechtheim in a conversation with the author on 13 February 1988.

25 See Chapter 11, p. 289–291.

26 See Neumann (1944a). The following page numbers refer to this text.

Schmitt had invented, at the international community. Commenting on the German Reich, Neumann stated: "The ideology of expansion is not complete with tradition, geopolitics, and pro-natalism. A new international law is needed too; more correctly perhaps, a new one at each stage in international relations" (150). He called Carl Schmitt "the leading voice in the National Socialist revisionist chorus" (152).

The first stage of Nazi foreign policy was about deterring foreign interventions and breaking the fetters of the Versailles Treaty. In his 1934 piece "Nationalsozialismus und Völkerrecht," Schmitt had introduced natural law as the leitmotif of his argument, a concept that the Nazi legal theorists had categorically ruled out for the realm of domestic law. Schmitt invented a catalog of natural rights that had to be granted to every state, such as the self-determination of states, equal treatment of states, and the right of states to life and defense. Neumann rejected such an argument for alleged natural rights of states as "purely arbitrary [...] reasoning" (153).

As in Kirchheimer's previous writing, the discussion of Schmitt in *Behemoth* examined his deliberations about the League of Nations and the Soviet Union in more detail. Schmitt and other international law theorists of the German Reich had "duped the civilized world successfully" (153). They and the Nazi propaganda machine knew how to place their hypotheses in renowned international law journals abroad. Their argumentative "trick of excluding Soviet Russia from the international community" (153) was useful in this regard, too. Nazi international law doctrine was borrowed from the domestic policy doctrine according to which a democracy could function only if a certain amount of homogeneity existed within its borders. Neumann claimed that Schmitt analogously championed the hypothesis that "membership in the international community requires homogeneity, a number of common features and beliefs" (153). Although it was possible to make such assumptions of homogeneity plausible at the domestic level, a void at the international level remained in Schmitt's theory: "Just what the elements of this international homogeneity are is never made clear" (153). Yet one thing was crystal clear to Nazi theory, namely its assertion that the Soviet Union shared none of the characteristics of the civilized world and thus was outside the boundaries of international law from the outset. At this point in *Behemoth*, Neumann again made the connection between Schmitt's theoretical deliberations and the policies of the German Reich. Schmitt's "excommunication of Soviet Russia" (153) in his 1934 essay "Sowjet-Union und Genfer Völkerbund" found its counterpart in Hitler's speech at the 1936 NSDAP party convention, which further delegitimized the aspirations of the League of Nations to global legal influence.

Behemoth then critically examined the Nazi regime's supposedly official theory of war. Again, it was Carl Schmitt whose role was considered particularly prominent in this regard. Neumann first reminded readers of the renaissance of the theory of just war in the more recent British and American literature on international law. He drew on the deliberations of former US Attorney General Robert H. Jackson, who had emphasized again in April 1941 that the neutral states had to provide military support to nations that had been attacked by other countries. Neumann argued that a theory of just war had to be compatible with and acceptable to German legal philosophy. Nonetheless, he asserted, Schmitt was attacking the theory of just war, and he was using the oldest and most strongly rationalist arguments of all. "The same Carl Schmitt who invented 'thinking in concrete words' to replace abstract, rationalistic thought has devoted many articles to combating the new theory of war and neutrality. He denies the distinction between just and unjust wars, and that neutrality can be 'halved'" (154). Either, Schmitt asserted, war was merely a policing measure implemented by any supranational authority or it remained a legal institution, and in this case, if a neutral state took a side, this made it a party to the war on one side or the other. Schmitt's attack on the theory of just war and discriminatory neutrality was characterized in *Behemoth* as "nothing more than part of the preparation for the new World War" (156).

Neumann also traced Schmitt's most recent turn in international law theory and placed it in the context of the new expansive stage of Nazi foreign policy: "With the coming of the present war, however, a completely new pattern of international law has been developed: The Germanic Monroe Doctrine" (156). Referring to Schmitt's 1939 brochure *The* Großraum *Order of International Law with a Ban on Intervention from Spatially Foreign Powers* and his 1940 essay "Raum und Großraum im Völkerrecht," Neumann first laid out the broad lines of Schmitt's *Großraum* theory, placing the greatest emphasis on the economic, technological, and organizational aspects of Schmitt's definition of a uniform *Großraum*. Because of the cartelization, monopolization, electrification, and rationalization of German industry, the creation of a far reaching *Großraum* had become imperative. Neumann stated that Schmitt did not view the integrating function of modern technology within the framework of a world with a territorial division of labor. Instead, he considered technology to be an integrating factor for far-reaching territorial expansion of the Reich, which was to be big enough for the products of the large companies: "The intrinsic connection between a monopolistic economy and territorial conquest stands fully revealed" (157).

This interpretation of Schmitt's *Großraum* theory not only saw it as a direct contribution to legitimizing Germany's attacks on neighboring countries but also placed it within his theory of organized/monopoly capitalism. An important issue in Neumann's explanation of Schmitt's theory in *Behemoth* was its antisemitic aspects. After replacing the state with *Raum* as the primary basis of the international order, Schmitt condemned traditional international law as a "creation of Jews" (157) and "a cloak for British imperialism" which, he claimed, was the system of Anglo-Saxon world imperialism lurking behind a facade of general norms of international law. Jewish universalism and abstract thinking were replaced by "thinking in 'concrete orders' and the most concrete of all orders existing is the *Grossdeutsche Reich*" (157). Works by international law scholars from the German Reich no longer referred to a single uniform international law, but to many international laws; their number equaled that of the *Reiche*, that is, *Großräume* with the capability to prevail. Thus, the Greater German Reich claimed the right to create its own international law for its own territory, with a strict ban on other states intervening.

The consequences arising from Germany abandoning universalist standards of international law were described in *Behemoth* using the examples of minority policy and Germany's actions against the population in the areas it had conquered. Schmitt and his followers fundamentally rejected calling the legal relations between the rivaling *Reiche* international law, limiting the term exclusively to the law of the *Volksgruppen* (minorities belonging to non-German *Völker*) living within the *Großräume*: "The ideological aim is clearly to give the German solution of the problem of racial minorities the sanctity of international law" (160). The political consequence was the abandonment of traditional minority protection under international law in favor of the *Volksgruppenrechte* (rights of minorities belonging to non-German *Völker*).In *Behemoth*, Neumann compared and contrasted in detail the differences between a system of minority protection intended to guarantee the legal and political equality of all citizens in a state on the one hand and Nazi *Volksgruppenrecht* (law pertaining to the rights of minorities belonging to non-German *Völker*) on the other. The latter dispensed with all of the individual guarantees for citizens from minority groups. Schmitt's construct of *Volksgruppenrecht* within the *Großraum* made the motherland the arbiter of disputes between the state and the minority living within it. In place of the international community intervening on the basis of rational norms and procedures, Nazi theory propagated arbitrary intervention by the motherland. Or, in other words "racial imperialism" (163). Neumann then went on to describe the racial imperialist measures taken by the German Reich in the conquered areas of Czechoslovakia, Poland, Belgium, Norway, the Netherlands, and northern France in detail. *Volksgruppenrecht* in the areas dominated by Germany meant that the German minority was granted the status of a dominating majority, whereas all that remained for much of the population was the powerlessness of a minority.

The significance of Schmitt discovering the US Monroe Doctrine to legitimize the political interests of the German Reich was highlighted explicitly in *Behemoth*. Neumann acknowledged that the postulates of a German Monroe Doctrine "seem convincing at first sight" (158). More careful examination, however, would reveal that claiming the Monroe Doctrine for Nazi ambitions was entirely misguided. The doctrine had been created to repel European interests in conquests in the Americas. In the early twentieth century, it had briefly been repurposed as the ideological basis for US imperialism under Theodore Roosevelt. Since the 1920s, however, the Monroe Doctrine had begun to lose its interventionist and imperialist sting. During Franklin D. Roosevelt's presidency (from 1933 to 1945), rudimentary forms of pan-American solidarity had developed that had nothing in common with Nazi ambitions of domination. Still, as a result of the collapse of the League of Nations and the numerous violations of international law following this, Schmitt's criticism of existing international law had the advantage "of appearing to be realistic" (158) in the eyes of many of his readers.

Neumann took a staunch position against this purported advantage. Giving up the universalism of international law because of its practical failings to date would be akin to abolishing civil rights because they helped legitimize and obscure class exploitation or democracy since it concealed boss control. Considering corrupt administration of justice, reasonable people would not demand reverting to a war of all against all but would fight for an honest system. "Likewise, when we have shown that international law has been misused for imperialistic aims, our task has begun, not ended. We must fight against imperialism" (159). Neumann affirmed that Schmitt's theory did have a rational core inasmuch as it articulated changes in the international capitalist system: "The decline of the state in domestic as well as international law is not mere ideology; it expresses a major trend" (160). Schmitt's *Großraum* theory was, of course, a particularly unattractive response to the challenges of the contemporary capitalist transformation of the economy. It was not convincing simply because creating a European *Großraum* would be unable to liberate Germany from the pressure of the world economy long-term. The protective function was illusory because even if continental Europe were dominated by Germany, it would remain dependent on foreign trade, especially for raw materials.

It was correctly emphasized in *Behemoth* that to Schmitt, a *Großraum* emerged mostly from economic and military events. It is striking that Neumann's explanations and analyses of Schmitt's *Großraum* theory did not mention differences in emphasis compared with the purely *völkisch* and racist *Lebensraum* theories of Werner Best and Richard Höhn, legal experts of the SS and masterminds of extermination. This has been criticized in some of the secondary literature. Schmitt's confidant George Schwab assumed in an interview forty years later that Neumann took revenge on Schmitt, so to speak, in the historical context of the height of the World War II: Neumann had "sought to settle a score"[27] by presenting the theory in an intentionally simplified way. Volker Neumann believed it was more plausible that Franz Neumann, working intensely on *Behemoth*, did not discern where works by Schmitt and those by Best and Höhn differed; he also had an issue with the fact that although Franz Neumann's other criticisms of Schmitt were justified, he had not paid attention to the major "difference between deprivation of rights and murder" (Neumann 2015, 491). This criticism also applies to Kirchheimer as the likely co-author of the sections in the book on Schmitt. Volker Neumann certainly has a point. This becomes clearly evident if we recall once again the racist passages in Schmitt's book in which he had approved of the deportations in the occupied territories, freely vented his antisemitism, and welcomed the Germans' cultural genocides of Czechs, Poles, and other Slavic minorities as examples of the "new order based on national groups" (Schmitt 1939a, 100).

6. Schmitt lying in wait again

From a political and propagandistic perspective with the intent to prevent powers from outside continental Europe from intervening in central Europe, Schmitt's *Großraum* theory was coherent. Viewed in terms of international law, however, it was hardly suitable for interpreting the war against the United Kingdom and even less so after Germany and Italy declared war on the US on 11 December 1941.

Even before the US entered the war, Schmitt had attempted to grasp the new and foreseeable situation by inventing new categories of international law. In early February 1941, he gave a talk with the title "Staatliche Souveränität und freies Meer" [State sovereignty and the open sea] (Schmitt 1941b) in which he presented an initial attempt to further develop a planetary theory. In this text, Schmitt dated the beginning of statehood to the second half of the sixteenth century. The opposite of the concept of the cohesive territory of the state surrounded by boundaries, which had developed on the European continent, was the open sea, which was not traversed by any state borders at all. In Schmitt's thinking on intellectual history, England now stepped onto his stage as the great adversary. With this new version of his politics of ideas, Schmitt was now able to declare an enemy; Hitler had already done so in the run-up to the German invasion of Poland. Hitler had long wished to cooperate with the United Kingdom. It was only in

27 Schwab in an interview with Rainer Erd, quoted in Erd (1985, 51).

1939 that he finally abandoned his overtures, which the island empire had rejected, and concluded that he would always face this country as an adversary in his efforts to conquer a *Großreich* in the east, for which reason he would have to conquer it first (see Fest 1973, 793). This reorientation became the basis of the German alliance with the Soviet Union.

Schmitt construed more abstract political concepts in his deliberations, but they were no less intensely programmed toward confrontation. England had opted against the state and for the sea because of its geographical situation as an island. Its imperialist methods of domination were indirect. What was new in this text of Schmitt's was the accentuation or idealization of this alleged dichotomy, making it a dichotomy of land and sea covering the whole world and all of history. Schmitt drew far-reaching conclusions from this dichotomy he had asserted. Land and sea, with their different ideas of *Raum*, corresponded to two "entirely different orders of international law" (Schmitt 1941b, 405). On land, the state was vested with providing order, progress, and humanity, and it was the only subject of international law as well as of constrained laws of war. International law relating to the sea, in contrast, engendered a completely different concept of war, for naval warfare was not a war of combatants. Naval warfare was based on a total concept of the enemy which considered all citizens of the enemy state and even everyone who sought to do business with them to be enemies, too. According to Schmitt's theory, two concepts of war and the enemy that were so contradictory could not be reconciled in terms of international law.

As the US entered the war and by the time the German army was defeated in Stalingrad in the winter of 1942 at the latest, the Nazis, too, realized that the war would be lost. It is striking that from that time on, Schmitt largely refrained from publishing new works on international law. Scholars disagree whether this lapse into silence was due solely to the fact that Hitler's Germany was going to lose the war it had started or whether it also indicated that Schmitt was creating a certain distance to the war Germany was waging in the east, which was growing ever more brutal as a war of extermination and genocide. Schmitt increasingly withdrew from offensively legitimizing the foreign policy of the Third Reich. If he said anything at all about international law at this time, then it was mostly about its history. During the final war years, his studies on the history of international law, which he had begun in 1940 and most of which he did not publish until after 1945, became his main occupation besides teaching and administering examinations at the university in Berlin. Yet a transformation can be observed in the few works he did publish after 1941. Whether Schmitt's reasoning ever strictly followed international law is debatable. In any case, it certainly no longer did after the advance of the German military had been stopped by the snow during the war against the Soviet Union in the winter of 1941. In the last three years of the war, Schmitt selected literary narratives as the form for presenting his new conceptual constructs and speculations.

The high point of this literary production was his brief book *Land and Sea: A World-Historical Meditation*. Initially a series of newspaper articles from March 1942 on, it was published as a standalone work in the autumn of that year.[28] The book was presented

28 The postwar second edition of the book in German from 1954 contains substantial alterations to the first edition of 1942. The differences relate mainly to Schmitt's characterization and depiction of Jews and of the British. See Chapter 15.

as a narration told to his 12-year-old daughter Anima. Schmitt interpreted the war as an elemental event, the elements being land, water, air, and fire. The general idea of the book was based on his literary article "Das Meer gegen das Land" [The sea against the land], which was first published in March 1941 for the weekly newspaper *Das Reich* (see Schmitt 1941c). Founded by Minister of Propaganda Joseph Goebbels in 1940, *Das Reich* was his favorite propaganda vehicle and, with its circulation of 800,000 copies, it soon became the single most important journal read by the Nazified German political and intellectual establishment (see Herf 2008, 21). Schmitt also delivered early versions of the book as lectures in occupied Paris in October 1941.

Schmitt claimed in *Land and Sea* that human beings had a privileged relationship to the earth because they were "land-dwellers" (Schmitt 1942a, 5). The central argument of the book involves fundamental tensions between land-based and sea-based cultures. The mythic repetition of the irresistible conflict between terrestrial and maritime worlds was constructed as a backdrop for the battle of the day: Germany versus "Anglo-America," as it was known in Nazi German. It was the battle between the order and stability of solid earth against its disorderly, chaotic, threatening, and anarchic counterpart, the sea. In Schmitt's view, the United Kingdom was the archetypal maritime country, and the United States represented the principle of the sea, too, even though it spanned an entire continent. In the 1942 version of the book, Schmitt's concept of the human being excluded the British people because they had decided against being land-dwellers and instead lived on the water. Schmitt also excluded Jews, claiming they were landless wanderers who lived in tents (see Schmitt 1942a, 10).

What was new in this book was Schmitt's idea of *Raumrevolutionen* (*Raum* revolutions, i.e., major changes in *Raum* orders), which was characterized by the various elements. He asserted that the first *Raumrevolution*, with the transition from land to sea, had occurred during the age when European powers had discovered America and colonized the world. England had evolved as the leading power in this epoch and had revolutionized international law to meet its maritime interests. Schmitt provided a kind of counter-history to Marx's theory of original accumulation by stating that British colonial policies had provided the resources for its industrialization. The industrial revolution of maritime modernization had become a worldwide battle between Great Britain and similarly highly industrialized Germany from the late nineteenth century on. In the meantime, with the invention of electric technology and electrodynamics, a new, second *Raumrevolution* and a new distribution of global *Räume* had begun (see Schmitt 1942a, 89–91).

Schmitt concluded his book with a prognosis: in his day, the battle between the elements would be decided by domination of the air. At least in his newspaper articles on *Land and Sea* of February and March 1941, Schmitt seemed to hope for a new *Raum* order in which Germany might surpass British sea power with the aerial supremacy of the German *Luftwaffe*. In light of the persistent bombing of German cities by the Allies throughout the summer of 1941 and into 1942, however, Schmitt revised his view of the distribution of airpower in the book *Land and Sea*. It did not end with the *Raumrevolution* from the element of water to the element of air, which would have been logical, but with one into the element of fire. In Germany, the aerial war had brought about a collision of the elements: flames were raining down onto the land from the sky, and enemies were invading from the sea.

7. Kirchheimer's career problems

While Schmitt used the secure position of his professorship in Berlin to provide his audience in Germany with legitimacy based on international law for the expansion of the Reich and later with morale-boosting slogans, Kirchheimer had to contend with an existential lack of professional and financial security in his US exile. He was also in trouble at his job at the ISR. He did not have an easy time with the core of the group, Max Horkheimer, Friedrich Pollock, and Theodor W. Adorno. For one thing, relations were strained because of their conflicts about the theory of state capitalism advocated by Pollock and Horkheimer.[29] For another, discord at the personal level dimmed the prospects for continuing his employment there. Besides Franz L. Neumann, his best friends at the institute were Arkadij Gurland and Herbert Marcuse. Yet Kirchheimer had to justify the subjects he was interested in working on to the heads of the institute. He explained to Pollock that he was primarily at home in the field of "political science" (a field that was just emerging), not in economic statistics.[30] Although Horkheimer had informed Neumann in the summer of 1939 "that I would be glad to support every step that could lead to him [Kirchheimer] remaining here,"[31] Kirchheimer was notified by the institute shortly thereafter that he would soon no longer be able to work there.

Kirchheimer's precarious professional situation briefly improved in late 1939 when the Emergency Committee in Aid of Displaced German/Foreign Scholars (EC) agreed, after negotiations that had dragged on for months, to cover 1,200 dollars for the year 1940, i.e., fifty percent of Kirchheimer's salary at the institute. The reason given in the funding proposal was the importance of Kirchheimer's analyses on Nazi Germany.[32] The remainder of his salary was paid by the Oberlaender Trust in Philadelphia and the institute's own funds.[33] One year later—again after lengthy negotiations—the EC approved a one-off extension of his funding for 1941; this time, however, the Oberlaender Trust withdrew from the funding because of a change in its guidelines, and Kirchheimer was left with nothing because the EC made matching funding a condition for its payment. It was mostly Franz L. Neumann who wrote multiple letters to potential funders and colleagues at US universities to ask for support for Kirchheimer.[34] In the end, Felix Weil,

29 See Chapter 9, p. 251–253.

30 Letter from Otto Kirchheimer to Friedrich Pollock dated 6 August 1939. Otto Kirchheimer Papers, Series 2. Box 1, Folder 127.

31 Letter from Max Horkheimer to Franz L. Neumann dated 10 August 1939. Max Horkheimer Papers, Letters VI, 11, page 216.

32 Letter from Betty Drury (EC) to Friedrich Pollock dated 1 December 1939. Emergency Committee in Aid of Displaced German/Foreign Scholars, New York Public Library, New York. I, A Grantees, 1933–46, Box 18, Folder 13 (Kirchheimer, Otto). —The reviewers of the proposal for Kirchheimer were Thorsten Sellin and Carl J. Friedrich. On the EC's funding policy, see Krohn (1987, 28–37).

33 The Oberlaender Trust funded many diverse cultural activities for immigrants from 1931 to 1953 and had supported the ISR before then on other occasions (see Gramm 1956, 65–66).

34 Neumann turned to David Riesman and Nathaniel Cantor (both at the University of Buffalo) as well as Carl J. Friedrich. He wrote in his letter to Friedrich: "I have known Kirchheimer since 1927 and always found him a true friend and excellent scholar. In fact, I brought him over from France to the United States and got him the job at the Institute." Letter from Franz L. Neumann to Carl J. Friedrich

the philanthropist supporting the institute, agreed to make up the difference.[35] The EC's published *Report* for the year 1941 listed Kirchheimer alongside Ossip K. Flechtheim and Erich Hula as one of fifteen scholars in the category "Law" receiving funding.[36] Even if this meant that Kirchheimer again had part-time financing at the ISR, it changed nothing about the basic constellation: since he had no alternatives, he was ultimately at the institute's disposal for a comparatively low salary, without a binding employment contract, and with no real prospects—Horkheimer's occasional assurances of his appreciation for him notwithstanding.

This precarious professional situation was compounded by the fact that it endangered Kirchheimer's residency in the US, which was still dependent on employment.[37] He had lost his German citizenship on 6 December 1938. The Gestapo had taken the initiative to revoke his German citizenship in early 1938 during "rectification" of German citizenship. An initial investigative report about him by the Gestapo dated 1 February 1938 stated[38] that his "current residency [was] unknown" and that he was "presumably [living] in Paris." Hence, Kirchheimer's "behavior had thoroughly violated his duty as a citizen." The report on the police investigations concluded by recommending that the Gestapo initiate the process of withdrawing Kirchheimer's citizenship. This was followed on 16 July 1938 by the Gestapo's "recommendation for revocation of citizenship"[39] to Himmler, the *Reichsführer* of the SS, "to revoke the German citizenship of *der Jude* Otto Kirchheimer" (see Glossary: "*der Jude*") and to expand this legal act to include his wife Hilde and their daughter Hanna. On 2 August, the *Reichsführer* of the SS agreed. Following consultation with the German embassy in Paris in early November, the request was granted on 17 November 1938.[40] The result was that Otto Kirchheimer and his daughter Hanna were stateless. Shortly after he lost his German citizenship, the University of Bonn revoked his doctoral

---

dated 13 February 1941. Emergency Committee in Aid of Displaced German/Foreign Scholars, New York Public Library, New York. I, A Grantees, 1933–46, Box 18, Folder 13 (Kirchheimer, Otto).

35 Letter from Friedrich Pollock to the EC dated 5 March 1941. Emergency Committee in Aid of Displaced German/Foreign Scholars, New York Public Library, New York. I, A Grantees, 1933–46, Box 18, Folder 13 (Kirchheimer, Otto). — A handwritten note from a conversation with Kirchheimer at the EC in February 1941 characterized him as a man of "awkward manner," as "keen," "good humored," and "a brilliant mind."

36 Overall, the *Report* of the EC from 1941 listed the names of 235 scholars supported in the US (see Report 1941, Kirchheimer is mentioned on page 11).

37 Hanna Kirchheimer-Grossman in a conversation with the author on 25 April 2023.

38 Letter from the Geheime Staatspolizei, Staatspolizeileitstelle Berlin, to Geheime Staatspolizeiamt in Berlin dated 1 February 1938. Bundesarchiv, Akten des Auswärtigen Amtes. Politisches Archiv, RZ 214, R 99744 (69. Ausbürgerungsliste, Ausbürgerungsakte betreffend Otto Kirchheimer).

39 Letter from the Geheime Staatspolizei to the *Reichsführer*-SS dated 16 July 1938. Bundesarchiv, Akten des Auswärtigen Amtes. Politisches Archiv, RZ 214, R 99744 (69. Ausbürgerungsliste, Ausbürgerungsakte betreffend Otto Kirchheimer).

40 File Memorandum of the Reichsministeriums des Innern, dated 17 November 1938. Bundesarchiv, Akten des Auswärtigen Amtes. Politisches Archiv, RZ 214, R 99744 (69. Ausbürgerungsliste, Ausbürgerungsakte betreffend Otto Kirchheimer).

degree of 1928; the university did not give it back until November 2023 after a newspaper reported about his case, including interviews with the author of this book.[41]

US Department of Justice files show that Kirchheimer's residency status in the US became precarious when the war began in Europe. In the summer of 1940, a mail carrier from Canaan, Vermont, considered him suspicious and reported him to the FBI "owing to the Fifth Column activities" because he had a German first name and rarely spent time at the vacation house he had rented. The FBI special agent in charge started an investigation in New York; evidently, nothing incriminating was found and so the investigation was closed.[42]

In the meantime, Kirchheimer had found a new partner, Anne Rosenthal.[43] In mid-May 1941, he was divorced from Hilde Kirchheimer-Rosenfeld, who lived in Mexico in the communist community of exiles, among them writers Anna Seghers and Alexander Abusch. Two months later, Kirchheimer married Anne in Chicago, and they moved to Queens, New York. He selected this location intentionally because he did not want to live too close to the Manhattan intellectual circles.[44] After Germany and Italy had declared war on the US on 11 December 1941, Kirchheimer had the status of an "enemy alien." This entailed certain requirements, one of them being that he had to apply for official permission to travel within the US. These requirements were lifted only when he obtained US citizenship on 16 November 1943. Kirchheimer desperately searched for new employment opportunities. He had unsuccessfully applied to a college in Tennessee in September 1939, and Horkheimer had written a glowing letter of recommendation.[45] A year later, he applied to a college in Chicago, also without success.[46]

By this time, Horkheimer, who had moved to the West Coast in the meantime, had very little confidence left in the critics of his theory of state capitalism who had remained at the institute in New York[47] and enhanced the visibility of the ISR in Los Angeles as an institution with workshops and small conferences. He was in contact with Kirchheimer mainly by letter while the latter sought new ways to secure his financing through the

---

41 Hermann Horstkotte, "Zu Landesverrätern erklärt," *Bonner Generalanzeiger*, 31 July 2023, p. 8 and Hermann Horstkotte, "Universität Bonn will Otto Kirchheimer rehabilitieren," *Bonner Generalanzeiger*, 6 November 2023, p. 8.

42 U.S. Department of Justice, Federal Bureau of Investigation, files on Subject Otto Kirchheimer, (unnumbered).

43 Anne Rosenthal was born in Würzburg in 1915. She had studied modern dance in Berlin before fleeing Germany. She succeeded in entering the US in September 1937. After marrying Otto Kirchheimer, she found work as a physiotherapist.

44 Hanna Kirchheimer-Grossman in a conversation with the author on 25 April 2023.

45 Letter of recommendation from Max Horkheimer for Otto Kirchheimer dated 23 September 1939. Max Horkheimer Papers, Letters VI, 11, page 361.

46 Letter of recommendation from Max Horkheimer for Otto Kirchheimer dated 7 October 1940. Max Horkheimer Papers, Letters VI, 11, page 353.

47 Horkheimer wrote Leo Löwenthal in late November 1941, when work on the lecture series at Columbia University was still in full swing: "However, if we continue working with Neumann, Grossmann, Kirchheimer, Gurland, then neither your presence nor even that of Marcuse or of Teddy [Adorno] will have any effect on the rapid disintegration." Letter from Max Horkheimer to Leo Löwenthal dated 29 November 1941 (Horkheimer 1996, 225).

ISR.[48] Kirchheimer received a research stipend from the New School for Social Research in New York from March to July 1942. The Graduate Faculty (GF) of the New School had been established in 1933 for academic refugees from Europe.[49] The core group of the GF was mostly veterans of practical politics from previously democratic European countries, and their research had less of a theoretical focus. Besides the monetary aspect, the stipend at the New School was a particularly great distinction for Kirchheimer, seeing as the New School received about 5,000 requests of this kind per year, but could only grant less than fifty of them (see Später 2017, 384). His project with the vague title "Contemporary Legal Trends" was formally placed with Max Ascoli, Dean of the Graduate Faculty. Ascoli was a liberal political philosopher who had fled to the US after being released from detention in fascist Italy in 1931. However, Kirchheimer and Ascoli did not get along well personally and never really collaborated at the New School.[50] In a letter to Horkheimer, Kirchheimer believed that he could still be happy "that you have not given up the idea of seeing me in Los Angeles."[51] He missed the old debates at the institute and would be happy to move to the West Coast, but in order to do so, he would need a steady income higher than the 125 dollars he was receiving from the institute at the time. Horkheimer's response was noncommittal. He did not make a specific offer but said he would continue to support Kirchheimer's applications to other institutions.[52]

8. On the verge of Germany's liberation

With the support of Franz L. Neumann, Kirchheimer finally succeeded in securing official financing through the institute for 1942. The Carnegie Foundation enabled him to collaborate with Neumann and Gurland at the ISR on a report for the US Senate in 1942/43. The book *The Fate of Small Business in Nazi Germany* resulted from this project (see Gurland, Kirchheimer, and Neumann 1943). This commission to prepare the report was part of US efforts to gain a better understanding of the current situation in the German Reich. Claude Pepper, Member of the Special Committee to Study Problems of American Small Business in the Senate and a leading voice of the left wing of the Democratic Party, underlined the connection to the war effort in the first sentence of his preface: "One of the

48 Letter from Otto Kirchheimer to Max Horkheimer dated 24 June 1942. Max Horkheimer Papers, Letters VI, 11, page 348.

49 On the history of the Graduate Faculty of the New School for Social Research, see Friedlander (2019); on the tense relationship between Horkheimer's institute and the New School, see Krohn (1987, 213–226).

50 As Ascoli, too, stated when queried by an FBI special agent tasked with investigating Kirchheimer's loyalty to the US, he had nothing to do with the substance of Kirchheimer's project; in fact, Kirchheimer worked on his projects entirely independently. Report: Results of Investigation of 26 May 1950 (page 10). U.S. Department of Justice, Federal Bureau of Investigation, files on Subject Otto Kirchheimer, (unnumbered). Kirchheimer's friend Franz L. Neumann described Max Ascoli in a private letter as a "puffed-up idiot of unmatched vanity" (quoted in Friedlander 2019, 146).

51 Letter from Otto Kirchheimer to Max Horkheimer dated 16 July 1941. Max Horkheimer Papers, Letters VI, 11, page 324.

52 Letter from Max Horkheimer to Otto Kirchheimer dated 16 August 1942. Max Horkheimer Papers, Letters VI, 11, page 323.

secrets of fighting an effective war is knowing the enemy, how he operates, and what are his aims."[53] In the study, the methodological toolkits of economic statistics, evaluation of press materials from the German Reich, and legal analysis were applied to document the transformation experienced by the social class of small businesses in Germany over the past twenty years.

Kirchheimer's contributions to the 150-page study consisted of analyses of the position of small businesses in terms of business law during the Weimar Republic, Nazi propaganda for small businesses before taking over the government in 1933, and the Nazi regime's economic policy goals and measures taken in relation to the war effort. The economic sections of the study provided an abundance of figures from various sectors describing the decline of small businesses. In particular, the reorientation of economic policy in Germany from 1936 on to support preparations for war increasingly relied on industrial mass production and rationalized large-scale technologies. This process of decline was accelerated by the fact that the state refrained from directly interfering in the economy and handed responsibility for the politically mandated economic measures to the major trade associations, which were organized as cartels. The period of economic policy studied by Kirchheimer ended just after the devastating defeat of the 6th Army in Stalingrad in February 1943 and included some measures with which the German business community intended to intensify its war efforts and thus also its support for the monopolies. The study concluded with the—in Kirchheimer's view, positive—statement that "in the winter of 1942–43, Nazi Germany underwent her severest domestic crisis since 1939" (Gurland, Kirchheimer, and Neumann 1943, 395).

Kirchheimer was again employed temporarily by the institute in the summer of 1943. Working with Neumann and Gurland, he had succeeded in obtaining funding from the American Jewish Committee for multiple studies on antisemitism in Europe.[54] As part of the project, Kirchheimer prepared two texts on the policy of the Catholic Church toward the Jews. He determined that because of Nazi Germany's military expansion, "after 1939 antisemitic policies rapidly spread all over Europe" and that the Catholic Church had made every effort "to refute the charge of being a tool of 'International Jewry'" (Kirchheimer 1943, 515). Focusing primarily on intellectual history, the texts contained no further statements on international policy or on the war against Germany. While he was working on this antisemitism project in early 1943, Kirchheimer still held fast to the vague idea of moving to the West Coast in the foreseeable future, to Horkheimer and the others. Neumann had asked him whether he was interested in working at the Office of Strategic Services (OSS), which was in the process of being established, in Washington. Kirchheimer immediately agreed to this but was not all too optimistic in his correspondence with Horkheimer.[55] In his response, Horkheimer repeated once again that he regretted

53 Pepper in Gurland, Kirchheimer, and Neumann (1943, 333).

54 See Chapter 10.

55 "As my child is going to stay with her mother in Mexico from next summer on for a whole year, my wife and I are considering different projects and one of them is still to move to Los Angeles [...]. Neumann just now is trying to fix me up in Washington, but I am somewhat less optimistic than he." Letter from Otto Kirchheimer to Max Horkheimer dated 22 January 1943. Max Horkheimer Papers, Letters VI, 11, page 318.

that in his function at the institute, he could do nothing for him.[56] Kirchheimer urgently needed a new way to support himself and his family in his exile in the US.

In early 1943, Kirchheimer applied to become a US citizen. He had succeeded in obtaining a position as Visiting Lecturer at Wellesley College in Massachusetts for the spring term (January through May) in 1943. He taught courses on social change and social theory for the program in sociology.[57] He was still seeking other employment opportunities at universities. His difficulties in gaining a toehold in the US university system were likely largely due to the fact that he was not as well integrated into the academic network as some other German émigrés.[58] In retrospect, one of his American colleagues also saw the difficulties in finding a position at a university as the result of certain problems of acculturation and what he observed as a "psychological failure to adapt to his American surroundings."[59] Kirchheimer's written English remained "teutonic," and he spoke with a strong German accent. But then, to his great surprise and thanks to Neumann's backing, he was commissioned by the Office of Strategic Services (OSS) of the US Government in the summer of 1943 to prepare a paper on German criminal and constitutional law, a project that lasted a few months.[60] When he finally obtained US citizenship in November 1943, at least his residency in the US was no longer precarious, although he still had no job security.

At the same time, Kirchheimer grappled once more with Schmitt's theory of sovereignty in his essay "In Quest of Sovereignty." The back story of the essay began in 1942. It, too, reflected Kirchheimer's complicated relationship to the ISR.[61] Parallel to his deliberations on the theory of state capitalism, Horkheimer had begun in 1939 to seek a concept to encompass the analyses of society at the ISR that took an even more general approach. His idea, which Adorno supported vehemently, was that a theory of "rackets" could accomplish this task (see Horkheimer 1939b).[62] Horkheimer had borrowed the term "racket" from 1930s American sociology, where it was used to characterize Mafia-like trade union structures and corrupt business relationships. In the summer of 1942, Horkheimer invited Kirchheimer to contribute to this project by writing about rackets in the labor movement; an offer Kirchheimer could not turn down. He sent his manuscript to Horkheimer in January 1943, who commented on it three weeks later, praising it, but proposing changes he linked to Carl Schmitt in particular:

56 "As things are, I can only hope that somehow you will arrive here anyway—if in the meantime you have not been appointed chief of some section of the State Department, or in some other powerful agency." Letter from Max Horkheimer to Otto Kirchheimer dated 8 February 1943. Max Horkheimer Papers, Letters VI, 11, page 317.

57 Otto Kirchheimer, Curriculum Vitae (1952), Otto Kirchheimer Papers, Series 2. Box 1, Folder 1.

58 Hanna Kirchheimer-Grossman in a conversation with the author on 1 April 2016.

59 This was the view of H. Stuart Hughes, his colleague at the OSS (as cited in Müller 2010, 389).

60 Letter from Max Horkheimer to Otto Kirchheimer dated 16 July 1943. Max Horkheimer Papers, Letters VI, 11, page 310. For more about Kirchheimer's work at the OSS, see the next chapter.

61 On the details of the complex backstory of this essay, see Buchstein (2020a, 95–100).

62 On Horkheimer's unsuccessful attempts to construct a "theory of rackets," see Fuchshuber (2019, 223–304).

> Should not, in a study bearing the title 'In Quest of Sovereignty,' the concept of sovereignty be more principally discussed? Of course, you touch it in various points, among others when you speak of Schmitt's decisionism. Your own concept, however, is so realistic (fortunately) that it differs strongly from the formalistic conceptions usually harbored under that title. A very short confrontation of both could add to the weight of your theory.[63]

However, the racket project never saw the light of day, buried by Horkheimer without a word. The theory that Horkheimer had begun to develop with so much ambition remained an unfinished torso of critical theory. Kirchheimer subsequently published a revised version of his article in the American *Journal of Politics*.

In his article "In Quest of Sovereignty,"[64] Kirchheimer observed a "peculiar mixture of shrewd analysis and ethical utopias" (161) in the modern theory of pluralism. As for the frequently posed question "what about sovereignty?", he reflected on what would happen if the group conflicts within a society had become so massive that they could no longer be settled. Referring to Schmitt's books *Dictatorship* (1921) and *Political Theology* (1922), he attacked his deliberations on sovereignty theory from the days of the Weimar Republic. "[A]s early as 1922, [Schmitt had] given up the hope of finding a permanent subject of sovereignty" (191) balancing the interests and volitions of different groups and factions. Schmitt attributing sovereignty to the individuals or groups that proved capable of exercising political domination under extraordinary circumstances was not convincing. For even in Schmitt's own words, this idea was too strongly "structurally akin to the theological concept of miracle" (196).

Kirchheimer argued in a strictly logical way to refute Schmitt's theory of sovereignty. It was true both of emergencies and of miracles that they had to be recognized as such, and that recognition required criteria. Emergencies and miracles were exceptions from rules that applied under normal circumstances. Their contours were determined by these rules of normalcy, which they confirmed by being exceptions. The lack of binding rules for recognizing an "emergency in permanence" (190) became the genuine symbol for the lack of a reasonable system of coordination which had historically and traditionally been accorded the attribute of sovereignty. All theories of emergency such as Schmitt's attested to the fact that society had reached a stage where the balance between the various forces within society had become unstable. In such a situation, "the most powerful groups [... were] compelled to resort to building a machine of violence" (191) with which they would eliminate their enemies politically, as fascism was doing in Germany. In this article, Kirchheimer used the term "rackets" to describe the close ties between the ruling fascist elites.

Despite its focus on domestic politics, the article concludes with an outlook on the international dimensions of Nazi domination. Since there was fierce competition in the field of international relations, the first country to complete a policy to coordinate the interests of the major trade associations would receive a "differential rent" (192). By using

63 Letter from Max Horkheimer to Otto Kirchheimer dated 8 February 1943. Max Horkheimer Papers, Letters VI, 11, page 316.

64 See Kirchheimer (1944e). The following page numbers refer to this text.

the term "differential rent," Kirchheimer took up a category from contemporary Marxist economic theory on ground rent, which meant a productivity advantage that made a positive difference to the average profit because of natural production conditions (for example, soil quality). Nazi Germany had gained this type of advantage on the European continent. Its trade policy of imperialist coordination made the Nazi government a "silent partner" (192) to private international business agreements. The Germans had even succeeded in inserting clauses into private contracts that seriously weakened the war potential of the country to which the other party to the contract belonged. Kirchheimer considered the contracts between the German conglomerate IG Farben and the American corporation Sterling Products, which had been concluded in 1937, to be a prime example of this. At this point, he alluded to a brief essay by Schmitt "Über die zwei großen ‚Dualismen' des heutigen Rechtssystems" [On the two great "dualisms" of contemporary legal systems] from 1939. Schmitt had explained that the two traditional dualisms in the law—public and private law as well as international and domestic law—had to be abandoned in legal thinking because they were now historically obsolete. The historical "development toward ever more tightly organized statehood" (Schmitt 1939b, 262) put strong states in the position of prevailing with their interests internationally in the area of private law, too. Schmitt used the term "*Gemeinrecht*" for this fusion in legal thinking (Schmitt 1939b, 263). Besides international trade agreements, he considered the responsibility of German courts for marriages between foreigners as well as the entire field of "*Rasse* law" to be legitimate objects of German "*Gemeinrecht*" (see Schmitt 1939b, 268–269).

Kirchheimer in his article "In Quest of Sovereignty" saw in the foreseeable German defeat in the war "the differential rent which ruling groups in Germany were able to secure through being the firstcomers in the field as bound to disappear" (193). For this reason, the foundations of the compromise between the rivaling social power groups in the Nazi state would again become unstable. The only thing still holding the regime together was the power groups' common interest in keeping the "practitioners of violence" in power; the regime had depended on foreign expansion from the outset. In Kirchheimer's view, the Nazi system was "substantially shaken" (193).

9. Conclusion: Waiting for the end of the war

With the publication of *Land and Sea* in 1942, Schmitt finally abandoned the idea of the Reich as a power creating and maintaining the order of a *Großraum*, moving from the academic discipline of international law to the domain of literary narration and speculation on the history of mentalities. Reinhard Mehring called this change in form of presentation the "literary staging of an exit" (Mehring 2014a, 392). In the final war years, Schmitt published more historical narratives of international politics. In April 1942, his article speculating on the decline of the US was published in Goebbels's prestigious journal *Das Reich*. The text was written in the usual tone of Nazi war propaganda against the US. The war, it claimed, was far from being decided in favor of the Allies. US involvement in it had proven to be obviously "not decisive for the outcome of the war" (Schmitt 1942b, 431). The German Reich had already won the continental land war in Europe; now it was about the victory of the Axis in naval warfare. US domestic policy was splintered, and the country had become a "magnified and coarsened reflection of old Europe" (Schmitt 1942b, 434). President Roosevelt was not an active and major force influencing policy but, rather, an "accelerator against his will" (Schmitt 1942b, 436) of the decline of the US.

Schmitt contributed another essay with the title "Die letzte globale Linie" [The final global line] in an anthology that appeared in the autumn of 1944. He again rejected the alleged aspiration of the US to assume the air of moral judge of the whole world and aim for boundless pan-interventionism. But now his criticism of the US seemed remarkably defensive; several months after the devastating defeat of the 6th Army in Stalingrad in February 1943, it lacked the cocky jargon typical of the Nazis' confidence that they would be victorious. The current world war, he asserted, was the conflict of two competing geopolitical *Raum* orders of planet Earth. On one side, the aspirations of universal planetary control and world domination. And on the other—under Germany's leadership—a "different *nomos* of the Earth whose basic idea is to divide the Earth into several *Großräume* pervaded by their historical, economic, and cultural substance." (Schmitt 1944, 447). For the period following the end of the war, he prophesied that the world would always remain too large for the US, regardless of how the war might end, and that it would be big enough for multiple *Großräume* "in which freedom-loving people [were] capable of preserving and defending their historical, economic, and spiritual substance and character" (Schmitt 1944, 448). Schmitt continued to be very productive in the last three years of the war, but he published very little; his manuscripts from 1942 to 1945 were the basis for his first books after 1949. He shifted the focus of his work to the history of international law. As he redirected his attention to new subjects, he altered his terminology again. From 1942 on, he successively abandoned the term "Reich," which he had not elevated to prominence in international law until 1939, replacing it with "*nomos*."

Schmitt's published writing on questions of international law from the final years of the war are ambivalent. Up to and including *Land and Sea* and later works through 1944, they were clearly written as works of German Nazi propaganda supporting attacks on the neighboring countries. But they were also conceived as a position distant from Hitler's geopolitics, if not a critique of certain strands of them. Even though Schmitt's writing was clearly antisemitic, he did not define his concept of *Großraum* in biologistic terms like the concept of German *Lebensraum*, but rather on the basis of its infrastructural and economic functions. In addition, he was at least implicitly critical of the German attack on the Soviet Union. In the fourth edition of his Großraum *Order of International Law*, which he dated to the end of July 1941, more than a month into Operation Barbarossa, Schmitt retained the passage in the book on the validity of the German-Russian border and the Soviet-German friendship treaty of September 1939, the successor agreement to the Molotov-Ribbentrop Pact.

Schmitt apparently thought that the invasion of the Soviet Union was a violation of international law. His view had previously been that the German invasions of Austria, Czechoslovakia, Poland, Belgium, France, and Denmark were legitimate in terms of international law, but he did not believe the same of the attack on the Soviet Union; this position was not consistent, given that the German government had also broken multi- and bilateral treaties in the previous invasions. Be that as it may. Schmitt had praised his *Führer* for his wars on numerous occasions in printed articles. But he did not do so in 1941 even as it seemed for a few months that the German army might be successful in its blitzkrieg heading for Moscow. These new nuances notwithstanding, the overall continuity in Schmitt's writing on international law between 1923 and 1943 is remarkable. All the major subjects—rejection of the Geneva League of Nations, insistence on unrestricted *ius ad bellum*, the nationalist definition of a *Volk*, his interpretation of the Monroe Doctrine, critique of a universal international law—can already be found in his Weimar works. In developing his theories of Reich, of *Großraum*, and of *nomos*, he did not revise his old theoretical building blocks but simply added some new ones.

Schmitt aspired in his writing on international law to criticize US hegemonic foreign policy. As William Rasch has argued, "Schmitt, the nationalist, might also be Schmitt, the international multiculturalist, who offers those, who 'obstinately' wish to resist the 'West' a theoretical foothold" (Rasch 2000, 1683). Following the US military reactions to the attacks on the World Trade Center and the Pentagon, a number of authors from the left have referred to Schmitt as an inspiring source for their critique of universal international law as a perfidious strategy of domination by Western imperialists.[65] Contemporary theories of imperial rule (see Hard and Negri 2004), critics of colonialism (see Kalyvas 2018), and postcolonial critics of international law (see Blanco and Valle 2014) also refer positively to Schmitt's theory. Kirchheimer, in contrast, already opted during the Weimar Republic for an extension of the juridification of international politics. Kirchheimer's writing contradicted Schmitt's negative view about international institutions and his cynicism about international law. It also contained a normative argument against the Schmittian critique which can be found in the current debate, too: every objection raised against the one-sided or selective application of universalistic standards in international law must already presuppose these same standards. Thus, Schmitt's hermeneutics of suspicion about universalistic international law—aptly expressed with the famous aphorism by Pierre-Joseph Proudhon: "whoever invokes humanity wants to cheat" (Schmitt 1932a, 54)—smuggled moral-normative commitments into his purportedly "realist" diagnosis of international politics.[66] Other authors argue that Schmitt's *Großraum* theory must be understood as "one of the earliest and most well-founded theories of globalization" (Manow 2022, 18). Such an assertion absolutely overstates the originality (as well as the plausibility) of Schmitt's theory and ignores liberal theories of international trade relations as well as the Marxist tradition of theories of capitalist imperialism. Even though Kirchheimer was only marginally involved in the Marxist debates about imperialism, there are some elements of continuity in his work about international politics between 1939 and 1943, too.

Among the Germans who were driven out of their country into exile and who followed Schmitt's activities and writing, Kirchheimer was one of the few who paid close attention to Schmitt's latest shift to *Großraum* theory in 1939. Although he was busy with his commissioned works at the ISR, Kirchheimer continued to keep abreast of Schmitt's latest writing—provided he could put his hands on it in New York. The sections in Neumann's *Behemoth* that Kirchheimer was involved in writing pointed out the key significance of Schmitt's contribution for making use of the Monroe Doctrine in order to legitimize the

65 See Zolo (2007) and Balakrishnan (2011).

66 See Teschke (2011a) and Benhabib (2012).

political interests of the German Reich. In his interpretation of Schmitt's *Großraum* theory, Kirchheimer emphasized the economic, technical, and organizational aspects of the concept. He stated that Schmitt's theory did have a rational core inasmuch as it objectively articulated changes both in the German and in the international capitalist system. Due to the high level of cartelization, monopolization, electrification, and rationalization of German industry, the economy of the *Großraum* had become imperative for the ruling classes. The decline of the state in domestic as well as international law was not merely an ideology; Schmitt's theory was in concordance with the major trends in German society.

In his attempts to understand the causes of German imperialism, Kirchheimer focused on the interactions between domestic and foreign politics. This included an empirical analysis of class constellations in society and the particular social and economic forces supporting German imperialism. Kirchheimer based his analysis of the Third Reich on his general theory of compromises between social groups, which he had already used in his Weimar writing. With respect to the system of Nazi Germany, he argued that the existing compromise that had excluded the working class could not generate a stable equilibrium between the ruling social groups. The compromise was only provisionally stable and was based on decisions made by the *Führer* as the dictatorial head of the regime. Hitler, however, could be successful in the long run only if and only as long as he was able to sacrifice the major group interests. He had to balance their competing interests by providing benefits and additional resources. Concerning economics, fierce competition at the international level was characteristic of monopoly capitalism. The first country to complete a policy to coordinate the interests of the major trade associations would receive a productivity advantage; Kirchheimer used the Marxist term "differential rent" as a label for this advantage. According to his line of argument, Nazi Germany was thus virtually programmed for the expansive foreign policy of *Großraum*. The position of the political leadership in Germany rested on its ability to compensate every group sacrifice. These benefits, however, could ultimately be gained only through imperialist politics which ignored international law. According to Kirchheimer, Schmitt's *Großraum* theory was the most important ideological soundtrack to Nazi Germany's warring imperialism.

Kirchheimer in his own writing and Neumann, too, in *Behemoth* disregarded the differences between Schmitt's *Großraum* theory and purely *völkisch* and racist *Lebensraum* theories of Nazi authors such as Werner Best or Richard Höhn.[67] Since Schmitt had welcomed the Germans' cultural genocides of Czechs, Poles, and other Slavic minorities, approved of the deportations in the occupied territories, and agitated against Jews on issues of international law, Kirchheimer and Neumann probably overlooked the "difference between murder and deprivation of rights" (Volker Neumann) in Schmitt's writing. Schmitt's works in the final two years of the war did not cross the Atlantic during that time, so Kirchheimer was unable to read them. Therefore, he could not write about the nuanced differences between Schmitt's writing from that period and official Nazi propaganda. By that time, however, both Kirchheimer and Neumann had already been recruited by the Office of Strategic Services (OSS) and contributed their expertise to achieving victory against Nazi Germany.

67 On Best and Höhn in contrast to Schmitt, see Blasius (2021, 272–274).

Chapter 13: On the Road to the Nuremberg Trials (1943–1945)

With the aid of Franz L. Neumann, who had been working for the Office of Strategic Services (OSS) since March 1943, Kirchheimer had obtained a part-time position as a research analyst with the OSS in July 1943.[1] His position became full-time in early 1944. He left New York and moved to Silver Spring, near Washington, DC, with his family. Kirchheimer was to work for US agencies for the next twelve years: for the OSS during the war and subsequently for the State Department after Germany surrendered in May 1945. His tasks initially included analyses of Germany as the war enemy, then planning for the period after the Allied victory, and finally studies on the political situation in Europe in the early postwar years. The Kirchheimer family lived in a small apartment in a development in Washington, DC, built specially for government employees, and his daughter Hanna from his first marriage spent some time there, too. For the first time in his life, at thirty-seven, Kirchheimer was earning enough to be able to support himself and his family and no longer depended on support from friends and the Rosenfeld family.

On the opposite coast of the Atlantic, Carl Schmitt wrote a lot but published little in the final two years of the war. The war affected him personally in August 1943 when he found out that his Berlin residence had been destroyed in his absence by an aerial mine. He rented a villa in upscale Berlin-Schlachtensee and moved to the southwest of the city. He still taught at Berlin University and continued lecturing as actively as before at various other universities in Germany and in German-occupied countries.

After the defeat of the 6th German Army in Stalingrad in the winter of 1942/43, it became clear to any dispassionate observer that the German Reich would be defeated in the war. New questions emerged such as what should happen to Germany after the defeat and how should the war crimes committed by the Germans be dealt with.

As an employee of the OSS, Kirchheimer was soon confronted with eminent practical questions about the planning for postwar Germany. Like all the other émigrés, he, too, had been thinking since the beginning of his own exile about the period after the end of Nazism. The question of how to deal with Germans who, like Carl Schmitt, had supported

1 FBI Report by Special Agent Patrick M. Rice on Otto Kirchheimer of 21 June 1950. FBI, US Department of Justice, Federal Bureau of Investigation, File on Subject Otto Kirchheimer (121–13351-5).

the Nazi reign of terror or had even committed war crimes was naturally of deep concern to him as well as to all the others who had escaped Germany.

1. Schmitt's wait-and-see stance

After Stalingrad, Schmitt stopped making public political statements. Although he continued to publish in the regime's propaganda magazines on occasion, he otherwise took a wait-and-see stance. His neighbor and longtime friend Johannes Popitz, who had brought him to the NSDAP in 1933, was involved in planning the assassination attempt on Hitler on 20 July 1944. Schmitt was not invited into these circles. Popitz was arrested and executed after the failed attempt on the *Führer*, which convinced Schmitt to continue to wait and see. Wolfgang Abendroth, a socialist Kirchheimer knew from his Weimar days, reported in his memoirs that Schmitt had cautiously implied to his students in seminars that the war was lost and that "it was necessary to cautiously reorient oneself" (Abendroth 1978, 212) for the postwar period. Schmitt was well aware of the extent of German crimes against the Jews in Europe. In a letter to Schmitt, one of his friends, right-wing author Ernst Jünger, had drawn a comparison between the current situation and the "extraordinary stubbornness of the Jews during the siege of Jerusalem"[2] during the Roman Empire. Schmitt responded with a quote by philosopher Bruno Bauer, a contemporary of Marx: "After all, God created the Jews, too. And if we beat them all to death, we will all be in their place."[3]

In the course of the Reich's cultural-policy offensive in other European countries that had begun in 1942, Schmitt traveled to speak at fourteen universities and academies during the summer of 1944, in locations as distant as the front lines permitted, including Madrid, Salamanca, Lisbon, Budapest, and Bucharest.[4] One of his lecture tours included a reception hosted by Romanian dictator Ion Antonescu, who was chiefly responsible for the murder of hundreds of thousands of Jews, Sinti, and Roma. Although Schmitt was well aware that the Nazis' reign of terror in large parts of Europe in 1943/44 was lawless, the title he selected for his lecture was none other than "Die Lage der europäischen Rechtswissenschaft" [The situation of European legal scholarship]. The lecture was published in 1950 and 1958, in two altered versions; it can no longer be determined precisely what Schmitt actually said on his lecture tours on the role of legal science and its future renaissance. In any case, when the lecture was published in 1950, he had removed any references to the notion of *Rasse* (see Schmitt 1950b).[5] Another lecture he gave on tour particularly at various universities in fascist Spain was titled "Donoso Cortés in gesamteuropäischer Interpretation" [A pan-European interpretation of Donoso Cortés]. In this, he countered the interpretation in Marxist philosophy of history with the radically anti-socialist and anti-humanist position of the religious concept of the Spanish counterrev-

2 Letter from Ernst Jünger to Carl Schmitt dated 10 February 1945 (Schmitt and Jünger 1999, 188).

3 Letter from Carl Schmitt to Ernst Jünger dated 25 February 1945 (Schmitt and Jünger 1999, 190).

4 See Tilitzki (1998) and Quaritsch (2000).

5 See Chapter 15 for more details about this lecture.

olutionary from the mid-nineteenth century. This text, too, was published in 1950, and in a revised version (see Schmitt 1950c).

Besides his lecture tours, Schmitt was primarily occupied with teaching at the university and working on a new book he intended to be about the political order after what he thought would be the last European war. His aim was to rewrite international history in geopolitical terms as an ongoing series of *Raumrevolutionen*, thereby including Hitler's conquests in a transhistorical continuum of land appropriations (see Teschke 2011a, 187). His as yet unpublished diaries mention that he began working on this in January 1943. Because more and more of Berlin, including libraries, was destroyed by the bombings, Schmitt could not finish the book. He completed and revised the manuscript and published it in 1950 under the title *The Nomos of the Earth* (see Schmitt 1950d). Most of the classes he taught in the last two years of the war addressed topics in international law, and he also lectured to soldiers in classes steeped in ideology. He held some of his seminars privately at his home. Schmitt traveled to Hamburg to give another lecture as late as January 1945 and still wrote dissertation reports in February. He continued lecturing until university teaching was stopped after the destructive bombing of Berlin on 2 February 1945. A few days later, at age 56, he was conscripted into the *Volkssturm* (a militia of poorly equipped civilian boys and men drafted by the Nazi regime in a last-ditch attempt to defend the fatherland). Schmitt was tasked to defend the Teltower Kanal, a canal in the southern part of the city, against the Red Army—a futile endeavor given the overall military situation.

In the meantime, the Allies were advancing ever faster, the Reich government's morale-boosting slogans notwithstanding. On 25 April 1945, Berlin was encircled by Soviet and Polish troops. The first Soviet soldiers appeared at Schmitt's house in Berlin-Schlachtensee. His Serbian wife spoke to them in Russian. Nothing happened to her, but some of the other women in the house were raped by Soviet soldiers. Schmitt and his family took cover in a bomb shelter for the next few days, hoping for the American army to arrive soon. On 30 April, Schmitt's *Führer* committed suicide. Coincidentally, on the same day, Schmitt was arrested and interrogated by Soviet soldiers. He took this occasion to offer the Soviet authorities his support as an advisor, but—to his surprise, as he later said—they did not take him up on this (see Wieland 1987, 101). He was let go after a few hours of questioning. He later reported that communist poet Johannes R. Becher had arranged for his release (see van Laak 1993, 31).

Two days later, Berlin capitulated, and the German Reich signed the unconditional surrender on 8 May. Besides the problems of everyday survival faced by those released from the concentration camps as well as the German population whose homes had been bombed, two political questions were of paramount importance once the war had ended: How to deal with past German crimes and how to organize the future administration of the territory of the former German Reich. Not surprisingly, Kirchheimer and Schmitt gave diametrically opposed answers to these questions.

2. Bringing German war criminals to justice

In the pamphlet camouflaged as an issue of a series edited by Schmitt, Otto Kirchheimer had written from his Paris exile in 1935: "The jurists of the Third Reich—theoreticians and practitioners alike—will have to take responsibility someday" (Kirchheimer 1935a, 147).[6] Once the Allies had prevailed over Nazi Germany, he had the opportunity at the OSS to play an active part in seeking out and punishing Nazi criminals.

Two years before beginning to work for the OSS, Kirchheimer had expressed his views on the question of dealing with war crimes and war criminals in a book review of *War and Crime* by criminologist Herman Mannheim. The author had a teaching position in London after fleeing Germany in 1935. He had previously taught criminology as an *Außerordentlicher Professor* at Berlin University from 1924 to 1933. After emigrating to England, he had become one of the internationally leading criminologists. Kirchheimer knew Mannheim from his time in Berlin but did not make allowances for such old acquaintances in his criticism of this author's criminological approach. At the end of his review,[7] Kirchheimer briefly addressed the subject of war crimes. He essentially agreed with Mannheim's plea to treat states like companies in the future and to hold them responsible in a similar way: "That states like corporations should be made responsible for their actions is an excellent program for the future" (428). He also agreed with Mannheim's demand "that an individual citizen who resists the army of an aggressor state should enjoy the protection of international law" (428). But he pointed out that it was one thing to make justified demands, like this one, and another to put them into practice. The current situation was characterized by the "non-existence of an international order" (428). And then he added an objection that sounded more fundamental:

> It is, moreover, a methodologically questionable procedure, and, as past experiences have shown, a politically unprofitable enterprise to establish, on the basis of the dual fiction, (a) of the existence of an international order and (b) of the identification of the people with the ruling group, that a "legal" war guilt attaches to the people of just one warring country. Even the establishment of an enlightened treatment tribunal cannot reconcile us to such procedure (428).

Although Kirchheimer advocated an international order regulated by international law, in 1941 he did not support the establishment of an international court dealing with war criminals. On the normative level, he agreed with Mannheim that war criminals should be prosecuted; on the practical level, however, he was skeptical. His skepticism was based on a view that was similar to Schmitt's analysis of the international order as an anarchic system that lacked any fundamental consensus necessary for success on the international stage. Two years later, Kirchheimer overcame his practical concerns and became part of the legal team that prepared the Nuremberg Trials. The reasons for this change in opinion are not documented. It is possible that Kirchheimer only became aware of the full extent

6 My revised translation, see p. 207 (note 9).

7 See Kirchheimer (1941c). The following page numbers refer to this text.

of German war crimes two years later; however, he may also have considered the chances that such a court would succeed in gaining international legitimacy to be better.

The Office of Strategic Services (OSS) was the first independent US agency established exclusively for the purpose of collecting and processing information. Its founding by the appointment of a Coordinator of Information (COI) in June 1941 and its institutional establishment through Presidential Executive Order 9182 of 13 June 1942 were part of the US war machinery.[8] Considered a supporting agency, the OSS was placed under the direct command of the Joint Chiefs of Staff (JCS) and the Combined Chiefs of Staff (CCS). The directive JCS 155/4/D of 23 December 1942 defined the two central functions of the OSS as follows: "the planning, development, coordination and execution of the military program for psychological warfare" and "the compilation of such political, psychological, sociological, and economic information as may be required for military operations."[9]

Kirchheimer was assigned to the Research & Analysis (R&A) Branch tasked with coordinating the various sources of information and preparing individual studies on strategic, political, geographical, and economic subjects. Its role within the OSS was that of a "final clearinghouse" (Söllner 1986a, 25). There were four more branches besides R&A, including the Secret Intelligence Branch and X-2, which was responsible for counterespionage. Kirchheimer worked in R&A's Central European Section (CES), headed by Eugene Anderson. Its 40 staff members included Neumann, Herbert Marcuse, John H. Herz, Hajo Holborn, H. Stuart Hughes, Arkadij Gurland, Felix Gilbert, Hans Meyerhoff, and Carl Schorske, among others. Within the CES, Kirchheimer was assigned to the unit responsible for Germany and Austria. Since the OSS was subject to constant internal restructuring during the war, it was difficult for R&A to obtain sufficiently informative external and internal intelligence reports for its analyses. These difficulties were exacerbated by the fact that its staff were redeployed to other positions inside and outside the OSS bureaucracy—for example, for the expansion of R&A with new units in London, Chongqing, and Cairo, or as intelligence units in the states liberated from the Axis powers. By the end of the war, the R&A staff had prepared 3,000 research reports, many of which were called memorandums, and as many geographical maps (see Smith 1973, 371).

The working conditions were the result of improvisation, as was the entire structure of the OSS. R&A had a staff of 2,000. Kirchheimer and his colleagues worked at large tables set up in tightly spaced rows at the Uline Arena (later to be renamed Washington Coliseum). Kirchheimer was seated next to John H. Herz, whose work on Schmitt's theory of international law he had reviewed positively several years earlier, and the two became close friends (see Herz 1984, 136–138). One of the difficulties of the everyday operations was the complex way in which the OSS was embedded in the US war machinery. R&A received material from all the government agencies relevant to the war as well as from other OSS units and sites through an internal agency administering information, the Central Information Division (CID). R&A had no say whether the desired information was made available; other agencies, and even other OSS units, held it back at

8 For the history of the OSS, see Smith (1973), Katz (1989), Marquardt-Bigman (1995), and Müller (2010).

9 Quoted in Troy (1981, 431–433).

times. The reports completed by the R&A staff were reviewed by the Project Committee, an oversight body internal to the branch, as to whether they met the standard of neutrality. That was the reason why the Project Committee, headed by the deputy director of the R&A Branch, consisted of members of all the divisions. Their expertise rivaled that of the analysts who had authored the reports. Some R&A reports were handed over to other research teams for revision before being delivered to the government departments and agencies that had commissioned them or offered to other agencies. The reports had to maintain "strict objectivity" (Müller 2010, 50) both in substance and in writing style and had to be easily comprehensible to outsiders. Because of the growing external demand for these reports and the multiple internal rounds of revision, the analysts were seldom able to pursue projects of their own.

By the time Kirchheimer joined the OSS, everyone there had realized that questions of military administration and occupation policy would become more important because of the foreseeable defeat of the Axis powers. Staff turnover notwithstanding, a core group in the Central Europe Section (CES) can be discerned: Anderson met weekly with Neumann, Kirchheimer, Marcuse, Gilbert, Holborn, and Harold Deutsch, the director of the Political Subdivision of the Europe-Africa Division. Regardless of the different disciplines and political orientations of those working at the CES, Neumann was considered "by far the most significant personality among its members" (Katz 1989, 34) not only because of his personality but also because of his book *Behemoth*, published in 1942. The book's structural analysis of Nazism, which had originally been inspired by Kirchheimer, was used in almost all R&A reports, at times verbatim, after Neumann joined the OSS.[10] In the summer of 1943, the Department of War gave R&A the major task of compiling the most important background information and preparing practical advice for a future military government in Germany on the basis of its competence, which was acknowledged by other OSS branches. Over the next two years, the branch prepared a number of comprehensive *Handbooks* with background information and around eighty *Civil Affairs Guides* with recommendations for the officers to be deployed on the ground (see Marquardt-Bingman 1995, 120–122).

Kirchheimer was hired in connection with the new major project for Neumann's group. He had a privileged position in that group inasmuch as he had already worked closely and well with Neumann at the Institute of Social Research. However, Neumann's dominant position within the CES could not hide the fact that there were some substantive and personal differences within R&A. In retrospect, it is remarkable that seasoned US historians from Ivy League universities, German historian Friedrich Meinecke's students Gilbert and Holborn, who had both emigrated, and the critical theorists who had come from the Institute of Social Research were able to develop such a strong esprit de corps. For example, Carl Schorske, from 1944 on Acting Chief of the Europe-Africa Political Subdivision of the Central European Section, later reflected: "The Central European Section remains its brilliant but incoherent, Teutonic, and maladministered self."[11] There was a latent conflict within R&A between the older, more established historians

10 See Erd (1985, 153–157), Marquardt-Bigman (1995, 132–136), and Kettler and Wheatland (2019, 334–347).

11 Quoted in Katz (1989, 173).

born around 1900, such as William Langer, Eugene Anderson, and Walter Dorn, and the ambitious scholars including Carl Schorske, Leonard Krieger, and H. Stuart Hughes, who were an average of fifteen years younger. Although the German émigrés at R&A, like Kirchheimer, belonged to the generation of the older group, they were not established scholars, either.

In the summer of 1943, Kirchheimer was given his first larger assignment: investigating how German war criminals might be punished. By 1942 at the latest, the Allies had agreed that those responsible for the crimes under the Nazi regime were to be brought to justice before an international tribunal, although they had different ideas of how this was to be accomplished (see Heydecker and Leeb 2020, 103–130). The Western powers had established the United Nations Commission for the Investigation of War Crimes in late 1942, but, initially, it was unclear whether the war criminals were to be court-martialed and shot—which would be the responsibility of the military on the ground—rather than prosecuted in court cases requiring considerable time and effort. On 30 October 1943, after lengthy negotiations, the Moscow Tripartite Conference of the Allies issued the "Statement of Atrocities" announcing the punishment of the main war criminals through regulated procedures on the basis of decisions made jointly by the Allies. Various questions remained: How was the circle of the main Axis war criminals to be defined? What procedures were to be applied when prosecuting crimes? From February 1944 on, the United Nations War Crimes Commission, based in London, began to answer these questions. Kirchheimer and John H. Herz presented Memorandum R&A 1482 entitled "The 'Statement of Atrocities' of the Moscow Tripartite Conference"[12] six weeks after the Moscow Declaration as a set of guidelines for the Americans involved in the London Commission for interpreting the statement and taking action.

To better understand the thrust of this report, readers should be aware that the British had been reluctant to permit the Nazi leadership to enjoy formal legal proceedings. In the early years of the war, Churchill thought they should simply be declared outlaws whom every member of the Allied forces should be permitted to shoot on the spot. It was not until 1943 that he changed his mind about this. US Finance Minister Henry Morgenthau had favored summarily shooting those responsible and rejected proceedings he considered unnecessary. Although the Soviet Union had also supported the idea of prosecuting Nazi war criminals in court, Stalin considered that to mean brief show trials followed by mass executions by firing squad as already practiced in the areas the Soviet Union had reconquered from Germany. In addition, US President Roosevelt had shown only mild interest in the problem of German war criminals (see Heydecker and Leeb 2020, 115–119).

Against this background, it is clear why Kirchheimer and Herz first laid out what had led up to the Moscow Declaration and called attention to the fact that it was "the first common announcement of intentions" (452) by all three major Allied powers. They emphasized that there was no longer any reason to fear that disagreements between the three powers after the end of the war might endanger the practical implementation of what they had agreed on jointly. "In many respects it seems to constitute a victory for

---

12 See Kirchheimer and Herz (1943). The following page numbers refer to this report.

those who, led by Russia, as one of the main direct victims of German atrocities, had insisted upon uncompromising prosecution and concrete procedures" (453). Kirchheimer and Herz then explained the individual provisions of the Moscow Declaration: the immediate extradition of all war criminals after the cessation of hostilities with Germany; the opportunity to prosecute other Axis powers and German satellite states; the ban on other countries granting asylum to criminals fleeing justice. The reason they gave for demanding extradition of the war criminals even in the ceasefire agreements was to prevent a repetition of what had happened in Germany after World War I: the "delay and open sabotaging of the Allied demands by Germany, together with the sham trials of Leipzig" (455).[13]

Kirchheimer and Herz then pointed out that in the Moscow Declaration, "for the first time a definite procedure is outlined for those persons whose acts are 'localized' in one of the Allied countries" (454). The perpetrators were to be handed over to the countries in which they had committed their crimes for prosecution before the courts according to those countries' laws. As Kirchheimer and Herz interpreted the declaration, those courts would operate on the basis of the laws applying there, but "it does not seem excluded that they may also apply certain rules of international law" (454). They mentioned the Hague Convention of Land Warfare in this context. Their report was the first in the OSS to prepare the argument for judging the German war criminals according to the rules of international law. The two authors considered the Moscow Declaration to be an effective "means of deterrence" (455). Announcing and broadly disseminating later punishment was "probably" the strongest effort that could be made "to deter Germany or Germans from continuing atrocities" (455). In particular, the "rather broad definition of responsibility" (455), which held all those accountable "who commit, or consent to, atrocities" (455) might induce many Germans to dissociate themselves from the demands or orders of their superiors. As morale in Germany regarding the war could be expected to worsen in light of withdrawal and defeat, the effect of the threat of punishment would increase.

As Roosevelt and Churchill were meeting in Quebec in mid-September 1944 for one of their last war conferences, the US Department of War had the CES under Neumann develop the first comprehensive strategy for punishing German war criminals. The ideas were synthesized by Neumann in the report *Problems Concerning the Treatment of War Criminals*,[14] dated 25 September. Kirchheimer was involved in writing this, and the authors circulated various drafts among their colleagues at R&A before it was finalized. From the outset, the report assumed that the prosecution and punishment of German war criminals was a matter of international law. Then it listed the offenses to be considered war crimes, including shooting hostages, abusing prisoners of war, plundering the civilian population, and atrocities against whole groups in pursuance of a general plan of "annihilation" (for example, massacres of Jews). Participation in such crimes had "directly or indirectly involved [...] a large number of persons" (458). The Allies' previous declarations had not included sufficient principles according to which the responsibility of individuals was to be established. The report sought to fill this gap by presenting a proposal for the

13 This referred to the acquittals in the trials against war criminals—which were part of the Treaty of Versailles—that took place from May to July 1921 before the German *Reichsgericht*.

14 See Neumann (1944b). The following quotations refer to this text.

American position in the negotiations with the Soviets and the British about the future war crime trials.

The wording of the Moscow Declaration ("have been responsible for or have taken a consenting part in") was again interpreted as a broad concept of responsibility. This laid the foundation for being able to indict not only Hitler and his close associates. Then, the question as to how to react to potential defendants' excuses such as superior orders or ignorance was examined in depth. Since the established principles of Anglo-American jurisprudence could not be applied to the system under Nazi rule, the report turned to the legal theory of Nazi Germany for guidance. The line of argument that Neumann and Kirchheimer developed in reaction to this problem can be considered a tactical variant of Marxist immanent critique because they refer to none other than the Nazi *Führerprinzip* (see Glossary). According to this principle, on which all organizations in Nazi Germany were built, the *Führer's* authoritative decision was always correct. Every superior, in his or her role as *Unterführer* (sub-*Führer*; see Glossary) was "responsible for whatever happens within the functional and territorial sphere of his jurisdiction (without legal limitations)" (458). Thus, according to Nazi legal theory, superiors bore the criminal responsibility for all crimes committed under their leadership. The only admissible excuse, according to the report, was if it could be proven that the incriminated person had done "all in his power to prevent the act" (459). Another criterion for assigning responsibility mentioned was whether a person had joined a criminal organization such as the Nazi party or the SS voluntarily. If a person had done so, "he must be assumed to have had full knowledge of the practices and functions of the organization and can therefore not avoid to share his responsibility" (460).

As to the question of who was to conduct such proceedings, Neumann and Kirchheimer argued that Allied military courts should first pursue the most important cases according to international law. This legal opinion also, and importantly, stated that the authors believed that the large number of smaller cases were to be adjudicated by German courts after the first major proceedings had been concluded. "Punishment of Nazi crimes by German courts would go far to prove to the German people and the whole world that Germany repudiates the crimes of its former leaders" (462). Attached to the report was a provisional list of Germans to be treated as war criminals. This appendix, however, is no longer to be found in the archives today (see Laudani 2013, 457).

OSS Director General William J. Donovan immediately sent the report to John J. McCloy, Assistant Secretary of War, with the enthusiastic note "that this was the story" on the war crimes question (see Smith 1973, 58). Donovan's personal views on how best to try Nazi war criminals were strongly influenced by this legal opinion written by Neumann and Kirchheimer. Both had included their personal opinions about the strategic advantages for postwar democratic reconstruction of having the Nazi criminals convicted by German courts for violations of German law. Another reason for assigning the following cases to German courts was the problem of capacity. In a letter to Donovan, Neumann mentioned that the Allies would be able to handle at most 5,000 cases, but that there would be significantly more than that.[15] These figures were based on estimates prepared

15 Letter from Franz L. Neumann to William J. Donovan dated 4 May 1945. Quoted in Slater (2007, 317).

by Kirchheimer. Neumann and Kirchheimer's report served as guidance for the American position in the discussions with the British and the Soviets in preparing trials against war criminals.

From October 1944 on, once the report had been finalized, the CES was tasked with specifically preparing the planned trials of war criminals. Neumann and Kirchheimer's group assembled documents in a form usable by the courts and was involved through late April 1945 in compiling an *Arrest Categories Handbook* that included an authorized list of persons to be arrested immediately. The list was limited to the names of war criminals with major responsibility for operations and did not include intellectuals helping to lay the groundwork for the regime, such as Schmitt, as they seemed less important at the time. Neumann and Kirchheimer also continued to be involved with tactical aspects of the trials. OSS Director Donovan was originally envisaged to serve as the second American prosecutor besides Supreme Court Justice Robert H. Jackson. Neumann was considered to be his "right-hand man" (Müller 2010, 53) whose team at the CES was doing the lion's share of US preparations for the International Military Tribunal (IMT). Neumann coordinated all aspects of preparing the trials and had direct access to Donovan.

A few days after Hitler's suicide, Jackson was named US Chief of Counsel for the prosecution of Nazi war criminals. Donovan was merely subordinate to him, and thus also the group around Neumann and Kirchheimer. It was still in May 1945 that Neumann, as the newly appointed Chief of the War Crimes Unit of the OSS in Europe, reorganized the CES, assigning more than twenty people, with Kirchheimer among them, to an American War Crimes Unit (see Intelmann 1996, 51). In August 1945, Neumann traveled to Europe with a small group of his staff, including John H. Herz, to help prepare the proceedings on site. Kirchheimer stayed in Washington with the others and worked on compiling intelligence materials that were sent to Germany. In the meantime, Jackson had pulled off the feat of bringing the four victorious powers together and getting them to adopt a resolution about an International Military Tribunal and its staff and procedures. One problem relevant to Jackson's tactics for bringing charges was the expectation that some of the defendants would put forward the excuse that they had merely executed the *Führer*'s commands, for which reason they were innocent as charged.

Kirchheimer and Herz addressed this problem in their Memorandum R&A 3110 of 18 July 1945, titled *Leadership Principle and Criminal Responsibility.*[16] The report was produced while the victorious powers were holding a conference in London (26 June to 8 August) and was sent to Neumann, who was a member of the American delegation negotiating the preliminaries of the indictment. It went into more detail about the idea first developed by Neumann and Kirchheimer in September 1944, namely to assign responsibility on the basis of the Nazi *Führerprinzip*. Adopting that strategy, the line of argument in the report was ingenious in that it, too, "allow[ed] the Nazis to do the work for them" (Katz 1989, 52) and to use their own logic against them. First, as in *Behemoth*, it characterized the general structure of the Nazi regime as a system in which, theoretically, all power and authority were concentrated in the hands of Hitler as the *Führer* but, in practice, *Unterführers* in various areas exercised a high degree of unfettered power. They were no tools

16 See Kirchheimer and Herz (1945). The following page numbers refer to this report. Regarding the term "leadership principle," see Translator's Preface.

of Hitler's without a will of their own, but rather people actively contributing to the Nazi system. As such, they were responsible for the policies within their areas of competence. "The more such policies involved a political aspect, the freer they were from any form of legal restraint" (464). In other words, direct orders in writing were often lacking, for which reason it would be very difficult to prosecute this level of leadership following the standard American legal doctrines.

According to Kirchheimer and Herz, the Nazi theory of *Führer* and *Führerprinzip* would prove useful for litigation at this point: "By drawing an analogy to the 'leadership' theory of responsibility[17] as developed by the Nazis themselves, a theory of incrimination in connection with war crimes might be developed which could be applied to fit the special circumstances arising under the Nazi hierarchy" (465). Another advantage of this strategy would be that it would be much more comprehensible to an incriminated member of the Nazi party. The authors then laid out important contributions of Nazi legal theory, the general principles and special features of the *Führerprinzip*, *Unterführerschaft* (the concept of the role of the *Unterführer* level), and *Führungswirtschaft* (the economic system under the *Führerprinzip*).

The second principle of the Nazi state mentioned was that the individual agencies and organizations were largely exempt from legal limitations. For that reason, a *Führer* or *Unterführer* was not dependent on direct instructions; instead, he or she formulated certain principles and guidelines for his subordinates, to whom he or she also delegated their implementation. "One of the reasons why the Nazi system has relied more on the execution of implied policies than on outspoken orders lies in the very illegality or immorality of a great many of its policies" (467). Kirchheimer and Herz mentioned the policy of exterminating Jews, including technical aspects such as the system of deportation and the erection of gas chambers, as a succinct example of this. The authors of Nazi legal theory they cited included Schmitt's political mentor Hans Frank as well as Otto Koellreutter, Werner Best, and Kirchheimer's fellow student in Bonn, Ernst Rudolf Huber, among others. They did not mention or quote Carl Schmitt himself. This may seem surprising, but only at first glance. Although Schmitt had referred to the *Führerprinzip* nothing less than emphatically and multiple times in his Nazi writing,[18] it was right not to mention him here inasmuch as he had in fact not written any elaborated text on interpreting the *Führerprinzip* in administrative law.

Kirchheimer and Herz developed "a new concept of responsibility for actions committed under the Nazi program" (470) on the basis of their analyses. Criminal responsibility for the annihilation of the Jews was to be assigned to all the *Führers* and *Unterführers* who, below the uppermost level of leadership, were responsible for functional and regional implementation of the Nazi's policies toward Jews. They were all aware of the general political guideline to eliminate all Jews from European life once and for all. Whether or not they knew about every detail of its practical execution in every individual case "appears immaterial" (470).

Overall, the report is a preemptive rebuttal of the most common excuses later made by the defendants in public and in the trials, namely that they were simply carrying out

17 The authors surely meant the *Führerprinzip* here; see Glossary.

18 See Schmitt (1933d, 103–105), (1933k, 63–68), and (1936f, 343–345).

superior orders and implementing the current law. First, the report explained on the basis of Nazi legal theory that the unconditional application of the *Führerprinzip* precisely did not mean that tasks were strictly delegated top-down and, second, it pointed out that it was important to avoid the pitfalls of the hyped-up hypothesis of all Germans' collective guilt; those responsible for the crimes later often used this hypothesis as an excuse, applying a false generalization.

Neumann and also Donovan took up Kirchheimer's proposal in Nuremberg (see Slater 2007, 400–402). Donovan used it to develop the idea of cross-examining the defendants, bringing them to pronounce themselves and others guilty through their own words. In contrast, Jackson, who lacked any recent experience of cross-examination in criminal trials, preferred to rely on documentary evidence. Jackson prevailed in the ensuing conflict, and Donovan withdrew from the trials. Neumann and his staff remained on Jackson's team for the time being. They now worked on the trial briefs with Telford Taylor and Benjamin Kaplan from the legal department of the Office of the Secretary of War. Kaplan regularly visited or contacted the group of OSS staff. They and Taylor, too, aimed to cast a relatively wide net and to investigate and indict as many potential defendants as possible. But they could not convince either the British or Jackson to agree to this goal.

3. Defending a German war criminal

As soon as Carl Schmitt was released from Soviet interrogation on 30 April 1945, he continued to read and write in his office at his house. He stayed home for the next weeks without even attempting to make a trip to the center of Berlin or to the university. His former university assistants visited him and delivered books while he began work again on his *Nomos of the Earth*. He also penned an exposé in English about his work for the Nazis in which he compared Hitler deceiving him personally to his first wife cheating on him (see Mehring 2014a, 408). His diary of the first days and weeks after the war is filled with antisemitic comments and his fear of "Jewish revenge."[19] In early June, the new Berlin Magistrate, installed by the occupying forces, ended all employment contracts with universities and pensions received through them. As a result, all former professors had to reapply for employment with the university. The same month, Schmitt reported back to the newly appointed rector of Berlin's university, Eduard Spranger, to take up his duties. He was incensed about the university's questionnaire in which Spranger asked for information about his activities under the Nazi regime and refused to fill it in. It was clear to him that he could not expect to return to the university for the time being.

Financially speaking, however, Schmitt did not live in desperate want, let alone go hungry, in the months following the war, which were extremely difficult for most Germans. In July 1945, Friedrich Flick, the biggest entrepreneur in the Third Reich, whose weapons factories had made extensive use of slave labor from concentration camps, had

---

19 I owe this information to Reinhard Mehring's knowledge about Schmitt's diary between March and September 1945, which has not yet been transcribed in full and is still unpublished. Conversation with Reinhard Mehring on 17 December 2022.

read in *Stars & Stripes*, the magazine for American troops, that leading German industrialists were to be put on trial before an international court (see Quaritsch 1994, 125–133). The description of the group of perpetrators to be tried led him to expect, justifiably, that he would be a defendant sooner or later. Facilitated by a lawyer friend, Schmitt was commissioned to prepare a legal opinion in support of Flick in advance. Flick and his lawyers could not predict the charges against him. They hired Schmitt in case these included participation in and preparation of a war of aggression.

As early as late August 1945, Schmitt delivered his extensive opinion *The International Crime of the War on Aggression and the Principle "Nullum crimen, nulla poena sine lege."*[20] The fact that he was able to complete such a comprehensive text going into the details of international law so quickly can only be explained by him doing preliminary work before the end of the war. Schmitt focused exclusively on the potential charge of participation in a war of aggression, reacting to up-to-date information available prior to the Nuremberg Trials. The statute for the International Military Tribunal of 8 August 1945 had stated that besides war crimes and crimes against humanity, the planning, preparation, initiation, or waging of a war of aggression was a crime against peace, which was to be punished.

Refusing to accept that this last crime was punishable was one of Schmitt's central goals in his legal opinion for Flick, and he explicitly excluded the other two types of war crimes: first, violations against *ius in bello* as codified in the Hague Land Warfare Convention, i.e., violations of the rules of warfare by armed forces, and, second, "atrocities in a specific sense, planned killings and inhuman atrocities whose victims were defenseless humans" (127). Such cruelties were not military actions, he claimed. And he added, "the rawness and bestiality of these crimes transcends normal human comprehension" (128). Such deeds went beyond the scope of the usual measure of international law and criminal law. They proscribed the perpetrator and made him "an outlaw" (128). It is unclear what was to be concluded from these sentences of Schmitt's. Did he mean that it should be possible to punish those committing such atrocities regardless of the existing legal situation, or that they could not be legally prosecuted at all because their atrocities went beyond the scope of the law?

Schmitt countered the hypothesis of the war of aggression contravening international law with his own hypothesis, which he had propounded from the 1920s until the end of the war, namely that according to existing international law, every sovereign state had a *ius ad bellum*; thus, a war of aggression could not be a crime. He discussed the provisions of the Treaty of Versailles, the Geneva Protocol of 1925, and the wording of the Kellogg-Briand Pact in detail. Although the pact condemned unjust war, i.e., war conducted in a manner countering the provisions of the pact, it provided for no other sanction than moral condemnation. Schmitt's interim finding was that international law as of the year 1939 did not include a punishable ban on a war of aggression; thus, Germany's attack on Poland had not been unlawful according to his argument.

If now, after the end of the war, this was viewed differently in international law, then, Schmitt believed, the next question was to what extent was invoking a new state of affairs under international law compatible with the ban on ex post facto laws under the rule of law. Schmitt's ability to transform his views on the rule-of-law principle *nulla poena*

20 See Schmitt (1945). The following page numbers refer to this text.

sine lege in this text is astounding. As a proponent of Nazi justice, he had vehemently rejected this liberal principle and replaced it with the formula *nullum crimen sine poena*, regardless of the specific legal situation. In this legal opinion, he presented himself as an eloquent defender of this principle and called it a "maxim of natural law and morality" (196). He discussed the ban on ex post facto laws in the legal traditions of continental Europe, England, and the US in a detailed and knowledgeable manner. His conclusion was that this principle was undisputed in all three legal traditions. So, even if wars of aggression were considered a crime today, the actors responsible for the German wars of aggression could not be punished for them ex post facto.

In the final part of his legal opinion, Schmitt discussed the question to what extent international law could even apply to principals of and accessories to the (alleged) crime of a war of aggression. Schmitt disputed that individual citizens or institutions of any state could be prosecuted under international law. The only subjects of international law were the individual states. Anyone who had a different view of this matter would have to be able to precisely define the circle of those responsible for a particular war. Drawing a parallel to piracy, he claimed it would be absurd to consider everyone found on board such a ship a perpetrator; that would amount to collective penal custody of the entire population of a country. Wars had to be prepared politically and militarily, for which reason it made more sense to hold the government or the parliament liable. Schmitt supplemented his argument with a structural description of the Nazi regime that was new is his writing. It was "part of the essence [...] of the regime that many power groupings fight amongst one another behind the closed façade of the unconditional unity of the regime" (180). Access to the *Führer*, the sole ruler, was decisive for anyone seeking to have influence. The much-touted *Führerprinzip* had become an opaque "antechamber principle. It was here, in the proximity of the *Führer*, that the actual plot in a criminal sense and the actual conspiracy came into being" (180). Prosecution of the members of Hitler's innermost circle alone should be permissible, but not, in contrast, an "economically active ordinary businessman" (186) such as Flick.

The finding of the legal opinion was that Flick was to be absolved from the accusation of participating in a war of aggression on three counts: first, because this offense did not exist in 1939; second, because it should not be applied ex post facto; and, third, because Flick as an individual had not had the right to resist the Nazi war machinery. Incidentally, the steel magnate's concern about being indicted in Nuremberg proved to be justified. Yet it was not until two years later, in April 1947, that an American military tribunal pressed charges in the Subsequent Nuremberg Trials, Case #5. The court charged him and five of the Flick corporation's leading employees with using tens of thousands of forcibly recruited workers from the occupied territories, prisoners of war, and concentration camp inmates in the corporation's industrial facilities and mines and with taking possession of foreign industrial enterprises. During the trial, they presented themselves as victims of the Nazi system. Flick's attorneys did not present the legal opinion prepared by Carl Schmitt to the court because it pertained to a charge that had not been brought and was therefore not needed. Flick was sentenced to seven years imprisonment on 22 December 1947 for use of slave labor, deportation for labor, plunder of property in areas under German occupation, and participation in crimes committed by the SS. He was released in August 1950 and was able to devote himself to his business again.

When Schmitt wrote the legal opinion in the summer of 1945, he had hoped it would help place him at the pinnacle of the legal discussion in Germany once again (see Quaritsch 1994, 142–144). That summer, American prosecutor Robert H. Jackson had actually intended to indict Flick in the first Nuremberg Trial for participating in preparing a war of aggression. I have not found any source material confirming that Neumann and Kirchheimer were involved in this project of Jackson's. Yet it is not implausible in light of the special attention given to Flick in Neumann's *Behemoth*.[21] Jackson, however, was unable to convince the representatives of the Soviet Union, Great Britain, and France of his idea (see Taylor 1993, 77–82). If Jackson had prevailed, then it is highly probable that Schmitt's legal opinion would have been submitted on Flick's behalf; Schmitt might even have served as another defense attorney. He had sought to play a part on this stage where his two former mentors, Hermann Göring and Hans Frank, had also been prosecuted. In the words of Helmut Quaritsch, he may well have viewed arguing the case in Nuremberg against criminalizing the war of aggression in international law as the "high point of his life." (Quaritsch 1994, 144).[22] In reality, his argument remained unnoticed and was published only posthumously in 1994.[23]

Schmitt's legal opinion provokes a number of critical comments, two of which I would like to address briefly.[24] The first concernes his hypothesis that only states are subjects of international law, but not individual citizens such as ordinary businessmen. To him, it followed from this dualistic concept that individual citizens could not commit crimes under international law, either. The flip side of his dualistic position was that citizens had the duty to obey their states in all matters of foreign policy. They had no duty of disobedience or resistance, not even in the event that the government took criminal action. Citizens had the right *not* to resist their governments and could not be prosecuted for supporting their government in a war of aggression. Schmitt again adapted his position to the changed political conditions. For his argument of 1945 contradicted the line of argument in his 1925 article "Die Rheinlande als Objekt internationaler Politik" [The Rhinelands as an object of international politics]. Then, Schmitt had written about the obligation of every citizen to resist rulers who lacked "publicity" and did not rule "in full openness" (Schmitt 1925a, 33). In the political situation of the day, this was meant as a call to resist the French and British occupying forces in the Rhineland after the end of World War I. Comparing Schmitt's legal opinion with this older text, Timothy Nunan commented "that it is notable that Schmitt makes no attempt to subject Hitler's regime

21 Friedrich Flick, the owner of "the biggest German combine," an "outstanding 'finance' capitalist," and a "close friend of Göring" (Neumann 1944a, 614).

22 Another reason why it would have been unlikely for Schmitt to be able to appear as an attorney on the stage of the Nuremberg Trials is that the American military government only accredited lawyers who could prove they had kept a distance from Nazism. Three years later, Schmitt noted in his *Glossarium*: "I would have gladly died had my August 1945 exposition on the criminalization of the war of aggression been able to be published then or during the Nuremberg Trials." *Glossarium* entry of 20 June 1948 (Schmitt 2015, 126).

23 Schmitt included some systematic sections in part four of *The Nomos of the Earth*, see Schmitt (1950d).

24 For more critical questions, see Salter (1999) and Nunan (2011, 17–22).

to the same kind of analysis" (Nunan 2011, 19) as the one he had conducted twenty years earlier.

My second comment is directed at Schmitt's dichtomization of the German war against Poland and the Soviet Union, namely into a war of aggression and a war of atrocities, creating the impression that the two were separate acts. He did not write a single word about the fact that the Germans had killed Poland's political elite in order to destroy the subject of international law called the Polish state, for example. The German attacks in eastern Europe were not wars in which atrocities happened accidentally alongside warfare; rather the atrocities were, from the beginning, an integral component of the warfare conducted by the *Einsatzgruppen*.[25] It was a specific kind of war from the outset: committing atrocities had become the means of warfare and extermination its end.

4. Preparing for the trials

In June 1945, three weeks before Schmitt was commissioned by Flick with preparing the legal opinion, Herbert Marcuse and Otto Kirchheimer were tasked with summarizing the Nazis' plans for dominating Germany and Europe in two reports for the Nuremberg prosecution. The reports were completed by mid-August. In the meantime, the first estimates of the number of Jews murdered (5.7 million victims) had been presented to the prosecution in another report prepared by the group (see Müller 2010, 56). Marcuse's report described how the Nazis had taken over power and prepared for the war, while Kirchheimer dealt with the domestic crimes of the regime.

Kirchheimer's report R&A 3114.2 was titled *Domestic Crimes.*[26] The fundamental problem Kirchheimer attempted to solve in the first part of his report was the objection expected from the Nazis' defense lawyers that the crimes the defendants were accused of were in fact authorized according to the laws of the Third Reich. In the second main part of the report, Kirchheimer provided an overview of the mechanisms of the Nazi regime's organization of terrorism to the extent that they could be deduced from the files secured by the American authorities at the time. Kirchheimer proposed "the principle 'selective retroactivity'" (523) to solve the problem to be expected during the trials. According to this principle, all the laws, amnesties, and policy measures that protected Nazis from the consequences of their crimes were to be specifically rescinded. Before proposing this, he clarified the question of whether the Nazi regime was constitutional. Kirchheimer followed the hypothesis that after 1933, Schmitt, too, had repeatedly proclaimed a revolutionary break with the order of the Weimar Constitution. A similar break had occurred with the defeat of Nazi Germany.

In his deliberations on selective retroactive revision of a defunct regime's legislation, Kirchheimer referred to precedents from various countries, examining in detail how the US had dealt with the Southern states and providing more recent examples from France,

25 The *Einsatzgruppen* were special units under *Reichsführer* of the SS Heinrich Himmler that carried out mass murders during the war to implement the Nazi genocide policy in the German-occupied territories.

26 See Kirchheimer (1945). The following page numbers refer to this report.

Denmark, and Italy as well as from international law. His list of retroactive rescissions of Nazi laws included what was commonly called "racial legislation" as well as the laws to suppress political opposition. Only if these laws lost their validity retroactively would it be possible to prosecute the members of the official repressive agencies such as the *Volksgerichtshof* or the *Militärgerichte* (see List of German Courts). Viewed in retrospect, it is remarkable that Kirchheimer's proposal of selective retroactive rescission corresponded to the ideas of some members of the German resistance against Hitler, even though the two sides were not aware of each others' positions. In 1943, the Kreisauer Kreis resistance group had also called for a retroactive penal provision for "Nazi acts of desecration of the law" (see van Roon 1967, 553–559).

Kirchheimer discussed Carl Schmitt in particular at one point in his report. It was in connection with whether the murders committed by the Nazi regime between 1933 and 1945 might have been covered by the law. He explained that the Nazi government had "indeed attempted in only one instance to justify specifically a series of political murders" (528). Kirchheimer was referring to the *Gesetz über Maßnahmen der Staatsnotwehr* [Law on measures of state self-defense] issued by the *Führer* soon after the purge of July 1934 and which Schmitt had celebrated a few days later in the editorial titled "Der Führer schützt das Recht" [The *Führer* is protecting the law] (see Schmitt 1934e, 199).[27] Kirchheimer quoted Schmitt and commented that such a position was justifiable only "from the viewpoint of the National Socialist doctrine" (529). His fundamental objection was that Hitler, the perpetrator, had made himself the judge in his own case and that the law could therefore not be given any recognition. But then, his line of argument went in a different direction. The regime would never have adopted this law if the majority of the German people had been willing to accept at the time that state authorities could murder their political opponents without a trial. Issuing such a law was, in fact, to be considered as "confirmation of the thesis that the substantive rules of criminal law, including those pertaining to murder, were never revoked under the Nazi regime" (528). Kirchheimer concluded from this that all the other murders during the Nazi regime were automatically not permitted under Nazi law, either, and could consequently be prosecuted.

Whether and to what extent the recommendations put forward by Neumann's group and specifically by Kirchheimer had any effect in the following months can no longer be determined today. In any case, they had only a minor influence on how the trial against the tweny-four individuals identified as major war criminals, which began on 18 October 1945 and took almost a year, was conducted (see Perels 2002). The further details of taking evidence during the 218 days of the trial with more than 5,000 evidentiary documents and films were no longer the responsibility of the OSS team but of an American prosecution team that had grown to almost 2,000 members in Nuremberg and London within a few weeks. Despite Donovan's support, Neumann was unable to convince Jackson of the idea of prosecuting more people involved in the economic aspects of Nazi aggression; this idea had mainly been developed on the basis of preparatory work done by himself, Herbert Marcuse, and Kirchheimer (see Slater 2007, 384–387). The materials Neumann had prepared for Donovan on this matter included Kirchheimer's report titled *Domestic Crimes* (see Slater 2007, 388). Jackson revoked the responsibility of Neumann's group for

27 See Chapter 7, p. 188.

the economic case and gave it to other individuals who were less critical of German big industry. From late summer on, Neumann's group was increasingly marginalized and mainly used as consultants for preparing evidence. For instance, they tracked down film evidence of the destruction of the Warsaw Ghetto later shown in the trial itself.

Jackson repeatedly stated that Neumann's group with its staff in Nuremberg and Washington had done "excellent work of laying the foundation"[28] for the case. Nonetheless, the conflicts became irreconcilable as time went on. In early December 1945, the émigrés in the OSS, including Neumann and Herz, returned to the US from Europe. Before then, conflicts between the American legal experts recruited by Jackson, who as civilians were specialists in stock company law or family law, and the small group of political émigrés from Germany, had intensified, even becoming personal. John H. Herz wrote in his memoirs: "Having young, mostly Jewish whippersnappers from Central Europe peer over their shoulders and into their papers hurt the egos of the American officers, who considered themselves superior" (Herz 1984, 140). When it came to formulating the closing arguments in the case, the group around Neumann and Kirchheimer was no longer involved. All the R&A staff members who were interviewed subsequently complained bitterly in retrospect that their *Guides* had little practical effect.[29] They were often not distributed on the ground or ended up in the occupying officers' wastebaskets. In this sense, the preparations for the Nuremberg Trials that Kirchheimer was involved in were something of an exception, at least in the initial phase.

R&A's recommendations had no impact at all when it came to denazification. Neumann's group followed the theory of domination approach in Neumann's *Behemoth* according to which the mass of the population was powerless vis-à-vis the four organizational pillars of the Nazi regime. Neumann's staff estimated the number of people whose classification as offenders was to be prioritized at about 220,000.[30] They also identified the names of 1,800 business leaders who were considered to be active Nazis and who were to be taken into custody pending further investigations. Contrary to these recommendations, the American military authorities on the ground were out for numbers. As a result, 1.5 million Germans were registered using a comprehensive questionnaire as early as mid-1946.[31]

If the OSS had prevailed with its recommendations, the American denazification measures would presumably have been more successful; their failure was widely deplored. Instead, the responsible authorities were overburdened with bureaucracy over the following years. This necessitated multiple waves of amnesties, which in turn mainly benefited those considered more serious offenders after many members of the Nazi party with much less important positions had already been sanctioned. Because he worked for the OSS/State Department, Kirchheimer was forbidden to make public

28 Letter from Robert Jackson to Franz L. Neumann dated 14 August 1945. Quoted in Slater (2007, 259).

29 See Erd (1985, 151–182), and John H. Herz in a conversion with the author on 15 November 1985.

30 See Söllner (1986a, 153–155) and Katz (1989, 45–49).

31 The questionnaire distributed in the American zone totaled 131 questions. The prominent right-wing German author and screenwriter Ernst von Salomon took this questionnaire as the point of departure for his derisive autobiography, which became one of the most widely read books in post-war Germany.

comments on these matters. After leaving the OSS, John H. Herz called denazification policy a "fiasco" (Herz 1948, 569). Kirchheimer most probably shared that view, since one of his closest friends in Germany, Richard Schmid, a Social Democrat and senior official in the justice administration in Stuttgart, also published a radical critique that was based on internal information from military government sources (see Schmid 1948). Kirchheimer, who had visited him in Stuttgart shortly before the article was written, was probably the main source for this.

5. Conclusion: Scenes of an indirect dialogue

The immediate postwar situation formed an ironic reversal of Kirchheimer and Schmitt's roles. Schmitt, who from 1933 onward had supported the Nazis' merciless criminal judiciary system and their terror against members of the opposition, now took on the role of a defense attorney and prepared himself mentally for this role in Nuremberg. Kirchheimer, who had once been incarcerated by the Nazis, now joined the team of prosecutors against German war criminals. Because of his expertise in analyzing the Nazi regime, he was able to make major contributions to the legal justification and strategy of the Nuremberg Trials.

In some parts of the lives of Kirchheimer and Schmitt, the year 1945 marked new similarities; in others, their roles had switched. Both were prepared for the defeat of the German Reich from 1943 on. Both were aware of the German crimes against the Jews in Europe, albeit not to their full extent. The differences between the two of them are also obvious. Schmitt took a personal wait-and-see stance and made no connections with German resistance groups; Kirchheimer became an active member of the group later named "The Frankfurt School goes to War." In his public appearances, Schmitt showed complete loyalty to the regime until its final day; Kirchheimer intensified his observation of German domestic politics. Schmitt turned his academic interest toward the history of political ideas; Kirchheimer wrote for the day-to-day administrative purposes of American government and military agencies. Schmitt had lost his prestigious job as a professor, Kirchheimer had a well-paid tenured position for the first time in his life. And, finally, Kirchheimer was suddenly on the side of the victors and Schmitt on the side of the defeated.

Kirchheimer could not have been aware of Schmitt's legal opinion for Flick; it was not until later that the document had circulated among Flick's lawyers during the trial in 1947 and among selected legal experts from the same Nazi milieu. It was not made known outside these almost conspiratorial circles (see Quaritsch 1994, 137–141). Kirchheimer had a very good idea how Schmitt's mind worked and so it is not at all astounding that his ideas about the best prosecution strategy, which he had previously committed to paper at the OSS, read like responses to some of the arguments in Schmitt's legal opinion. Key to his legal opinions for the OSS is his discussion of selective retroactive rescission which countered the "rediscovery" of the liberal *Rechtsstaat* in Schmitt's legal opinion for Flick. Kirchheimer's considerations of how to react to potential defendants' excuses such as superior orders or ignorance can also be read like direct responses to Schmitt. Kirchheimer's strategy was to turn to the legal theory of Nazi Germany and, in particular, to the

Führerprinzip for guidance as the basis for his purely immanent argumentation. From today's perspective, Kirchheimer anticipated what was to be the defendants' main line of defense in the Nuremberg Trials and the trial against Adolf Eichmann, one of the major organizers of the Holocaust, as well as the prosecution's weak point: invoking orders from superiors to shift responsibility away from themselves. What may be more important in the context of the relationship between Kirchheimer and Schmitt is that if one takes Kirchheimer's considerations for the evaluation of Schmitt's activities as a yardstick, then his deeds for the regime were reprehensible but Schmitt was still not a war criminal who had to be prosecuted.

Kirchheimer's personal motive for participating in preparing the Nuremberg Trial was probably the same as John H. Herz's. In his memoirs, Herz wrote that it was not about "satisfaction, a satisfying sense of revenge." What mattered to him was "that the world, and especially the Germans, should get a clear picture of what had happened" (Herz 1984, 142). Even though Kirchheimer in his US exile in August 1945 was fully aware of the war crimes and mass murders committed by the Germans, he stood by his assessment that the majority of Germans had been opposed to the Nazis. Hitler's government had not had a majority in 1933 and had only been able to establish itself in power because of its "system of terror" (Kirchheimer 1945b, 523). This view of Kirchheimer's does just as little justice to the crimes of the *Wehrmacht* as to the fact that the vast majority of Germans did indeed support the regime and its crimes.

Kirchheimer expected that with Germany's military defeat, the majority's rejection of the Nazi system had already become stronger than before. If, he thought at the time, it were possible to successfully eliminate the Nazis and the functional elites that had chosen to conspire with them, then little would stand in the way of reestablishing democracy in Germany. Kirchheimer did not mention Schmitt's name in this context. But it logically follows from Kirchheimer's considerations that Schmitt was among those who had to be categorized as someone banned from regaining a position in the functional elite of a future German democracy. In 1945, Kirchheimer had high expectations with respect to the future of German democracy. He pushed to revitalize democratic parties and organizations. Carl Schorske and H. Stuart Hughes, Kirchheimer's American R&A colleagues at the time, reported independently of each other in retrospect that Kirchheimer—like Neumann—supported democratic socialism at the time and placed their hopes in a rapid revitalization of the trade unions and social democracy as forces of reform (see Erd 1985, 185–199). Viewed from today, his optimism seems naive since it fails to take account of the high intensity of ideological indoctrination and the complicity of the majority of Germans with the Nazi crimes.

Although Kirchheimer failed in his political ambitions, his idea of a political compromise of four ruling groups, which Neumann had taken up in *Behemoth*, curiously enough has a bureaucratic legacy to this day. The Subsequent Nuremberg Trials needed to have an immense number of files compiled and sorted for the proceedings. Just over 35,000 pieces of evidence were ordered in four series. The four series correspond exactly with the structure and the names of the ruling groups of the four-headed *Behemoth* and have served as the organizational principle of the archives for these trials ever since.[32]

32 Raul Hilberg in a conversation with the author on 2 December 2000. See also Hilberg (2002, 82) and Wildt (2023, 68).

Postwar Democracies

Chapter 14: Dealing with the Future—and the Past (1946-1948)

Otto Kirchheimer was included in the American plans for postwar Germany in addition to his work preparing the war crimes trials. He had great hopes for rapid denazification and swift reestablishment of a democratic and socialist Germany. Meanwhile, Carl Schmitt continued to live in Berlin-Schlachtensee after his interrogation by the Soviet military. He started to earn a living as a private legal consultant for the time being, hoping soon to return to a position as professor again. Both saw their hopes dashed within a short period of time.

1. Denazifying and governing occupied Germany

At the OSS, Kirchheimer was involved in plans for abolishing the Nazi laws as well as recommendations for suitable ones to replace them, including laws on the organization of the courts and other judicial institutions, the areas of criminal law and criminal procedural law, as well as public administration in general and the public service. His recommendations are documented in a number of *Civil Affairs Guides* and contributions to *Handbooks* which he prepared between early 1944 and August 1945. Some of his contributions to these official documents, which were classified top secret, are worth looking into since they touch on Schmitt and his work in various places.

The recommendations in the *Civil Affairs Guides* were based on longer *Handbooks* on the same subjects. As reference manuals to provide a factual basis, the *Handbooks* were primarily intended to present information about Nazi Germany in order to enable the US military government to make future decisions appropriate to the situation at hand. They covered geography and society; government and administration; financial, economic, commercial, and social policies; agriculture and mineral resources; communication systems, licensing of publications, and questions of censorship and freedom of expression; transportation systems; public services and public safety; education and culture; and churches and museums. These *Guides* had an impact on American policy in Europe. They were produced in Washington, DC, and after an internal approval process transmitted overseas to the American Civil Affairs Training School in Shrivenham, England. They were to be limited to a maximum of twenty pages since they were designed for practical use in occupied Germany. In contrast to the *Civil Affairs Guides*, the *Handbooks* did not include any recommendations for politics or planning.

Already in early 1944, Kirchheimer was assigned to prepare a *Handbook* to serve as a basis for the recommendations on legal affairs in Germany in the *Civil Affairs Guides*. He was given only four months to complete this task. His *Handbook of Legal Affairs*[1] presented the judicial system in Nazi Germany, including the Reich Ministry of Justice, the bar association, criminal law, the system of administrative courts, civil law, commercial law, and the war-related measures in Germany through 1943. Produced in a rush, the *Handbook* endeavored to convey the absolutely sober tone and objectivity expected of R&A. Kirchheimer discussed factual matters with his co-workers in his division. The *Handbook* was copyedited by native English speakers. Nonetheless, Kirchheimer's analytical hand is apparent in the *Handbook*—not least because of direct references to some of his earlier works and a few brief passages taken from them (but not indicated as such).

There was particular emphasis on the extensive chapters on Nazi criminal law and the judicial administration—Kirchheimer had already conducted research and published multiple times on both subjects over the previous years. He now drew attention to long-term historical trends dating back to the mid-nineteenth century. This gave his readers the impression that Nazi legal policy was not only opposed to that of the Weimar Republic but also to that of the German Empire, which Kirchheimer described as a "liberal age" (319). It also rejected the notion that German history had somewhat inevitably resulted in Nazism.[2] He characterized Prussia's historical distinctiveness not by using the concept of militarism but that of the *Rechtsstaat*. Kirchheimer thus set a clear counterpoint to Carl Schmitt's historical interpretation in his 1934 book *Staatsgefüge und Zusammenbruch des zweiten Reiches* [The structure of the state and the collapse of the Second Reich],[3] yet without mentioning Schmitt's name.

Kirchheimer reflected on the political views of the members of the judicial system in a relatively detailed section, linking up seamlessly with his writing from the 1920s. Politically challenged by the defeat in World War I and the November Revolution and financially ruined by inflation, "the judge felt humiliated by all these developments and instinctively sought to take revenge on those whom he held responsible for the evil days which had befallen him—the war profiteers, the foreigners, the radical workers. In following this course, he became a nationalist as well as a champion of the cause of the newly disinherited of the nation—the independent middle class" (259). Even though Kirchheimer strongly criticized the reactionary and authoritarian character of the German judicial system, the prosecutors and judges appeared to have the function of "a brake" (260) in the system of Nazi legal practice. Their attempts to hold fast to

1 See Kirchheimer (1944d). The following page numbers refer to this text.

2 In so doing, Kirchheimer, along with all of R&A, opposed the idea of a German-Prussian *Sonderweg* (special or unique path) propounded at the time by Robert Gilbert Vansittart and Hans Morgenthau, among others.

3 See Schmitt (1934f). See Chapter 7 for further details.

proper court proceedings at least lessened the totalitarian state's claim of unfettered domination.

Kirchheimer emphasized the existing "frictions" (261) between the justice system and the party. Although the early politicizing of the judicial system by the Nazi regime underscored the "all-embracing character" (261) of Nazism, it did not succeed in completely amalgamating the judicial system and the party, at least not until 1942/43. For this reason, Hitler's 1942 speech in the Reichstag on legal policy and the decision of the German Reichstag of 26 April 1942, which Kirchheimer interpreted as attempts to toughen legal policy, were mentioned multiple times in the *Handbook*. Kirchheimer characterized the national conservative legal experts who had offered their services to Nazism—besides Franz Gürtner, he singled out Erwin Bumke and Franz Schlegelberger—as representatives of an authoritarian but bourgeois era. He contrasted them with the type of the explicitly Nazi legal expert, mentioning Hans Frank and Roland Freisler by name. Kirchheimer's typology did not mention Carl Schmitt, neither directly nor indirectly.

One of the specific questions studied in advance by the occupation authorities in their planning was which Nazi German laws would have to be repealed and which ones could remain in force for the time being to allow everyday life to be reasonably governed. The question as to the personnel who was to develop the new laws and regulations in detail seemed at least equally important. In January 1944, Major General John H. Hilldring, who directed the newly established Civil Affairs Division within the US Department of War from spring 1943 on, tasked R&A with preparing recommendations for legal policy. Franz Neumann was the formal director of the project but had internally handed it over to Kirchheimer and Herz, who was to work under Kirchheimer's direction.[4] Since the recommendations had to be harmonized internally with other branches of the OSS, competition between those branches made for friction. Nonetheless, the *Civil Affairs Guide: The Abrogation of Nazi Laws in the Early Period of Military Government*, for which Kirchheimer was responsible, was completed rapidly by March 1944.[5] In a letter to Crane Brinton, head of the R&A Branch in London, from late March 1944, Neumann spoke in retrospect of "innumerable difficulties" between R&A and the other branches involved. He also complained of resistance from the Editorial Committee in which conservative American legal experts—he meant Ernst J. Cohn—rigidly defended property rights, overlooking "the need for political and social transformation"[6] in Germany.

Kirchheimer's *Guide* listed the Allies' legal policy measures that were to be taken by the American military directly after occupying German territories. Kirchheimer quoted the ideals of the United Nations and the Moscow Declaration and applied the goals of their occupation policy for Italy ("that Fascism and all its evil influence and configuration shall be completely destroyed and that the Italian people shall be given every opportunity

4 Interoffice Memorandum dated 18 January 1944. Record Group 226, Entry 44, Box 2, Folder: Status of Reports. Records of the Office for Strategic Services, National Archives at Maryland.

5 Record Group 226, Entry 44, Box 2, Folder: Status of Reports. Records of the Office for Strategic Services, National Archives at Maryland.

6 Letter from Franz Neumann to Crane Brinton dated 30 March 1944. Entry 146: Miscellaneous Washington Files, Box 83, Record Group 226, Records of the Office for Strategic Services, National Archives at Maryland.

to establish governmental and other institutions based on democratic principles"[7]) to Germany. He linked these considerations regarding policies for freedom and democracy to aspects of security policy. Preventing "chaotic conditions" (229) was top priority.

The *Guide* begins with a brief analysis of the problem, followed by an extensive list of recommendations. Two groups of laws had to be abrogated immediately. The first was laws that contradicted the principle of all citizens being equal before the law. Kirchheimer placed all laws and other legislative rules that were expressions of racial discrimination in this category. The second was laws that restricted fundamental personal liberties as well as the liberties of certain social groups. This included immediate release of and care for all concentration camp inmates. However, the former concentration camp sites were to be kept available: "the facilities of existing concentration camps will be needed for the detention of the large numbers of active Nazis and similar categories of persons to be detained by MG for reasons of security" (245). Kirchheimer recommended the prompt release of political prisoners as well as a general amnesty with immediate effect. He also advocated for revoking the validity of any acts concerning Nazi symbols such as swastikas or uniforms without delay. Likewise, the enforcement of Nazi eugenic legislation "would seem undesirable even during [the] transitional period since it has been one of the mainstays of Nazi ideology and propaganda" (232). The 1934 Act for the Ordering of National Labor with its provisions on *Führer* and *Gefolgschaft* (subordinates, see Glossary) in enterprises had to be repealed at once.

Kirchheimer laid out in detail which particular laws were to be abolished immediately and which other measures were to be taken in judicial policy. Among others, these included the entire complex of "Aryan Legislation" (234), in particular, the Nuremberg Laws of 1935. The laws regulating expatriation were also to be declared invalid. Individuals whose German citizenship had been revoked—such as Kirchheimer himself—and who had taken on the citizenship of another country were not to be reinstated automatically as German citizens; instead, they would need to apply for German citizenship individually after returning to Germany. One argument against automatic restoration of German citizenship was that it might lead to unnecessary harm to former Germans who had chosen to reside abroad.

As for politics, Kirchheimer placed his hopes in a swift handover of political responsibility to democratic Germans. He apparently assumed that the majority of Germans had not agreed with the Nazi regime but had been oppressed by a "bellicose and rapacious minority" (238). The military government "has no desire to perpetuate its rule" (238) for which reason it was to commit to a program of "speedy restoration of liberties" (238). Restoration of democracy in Germany would be rapid "when the German people have gained some experience once more in the full development of social and cultural patterns of their own" (238). That was why the military authorities were to grant the Germans as many political freedoms and rights as possible. Following these fundamental considerations, Kirchheimer presented a detailed list of measures for restoring the freedoms of speech, assembly, and the press, as well as establishing political parties and organizations. The Nazi party and comparable successor organizations were to be disbanded and banned. The *Guide* ended with a list of sixty-five laws to be rescinded immediately. This

7 Kirchheimer (1944a, 230). The following page numbers in the text refer to this *Guide*.

included two laws as formulated in 1922 and 1925, both of which codified privileges for police officers of the *Länder* and members of the *Wehrmacht*.

Another group of measures where "immediate action seems imperative concerns the denazification of the judiciary" (232). At this point in his *Guide*, Kirchheimer used an argument that was founded in legal history and based on his own experiences during the Weimar Republic. It might seem at first glance that the fate of the German judiciary was not of immediate concern. For instance, it could be argued that the elimination of the *Sondergerichte* (see List of German Courts) for prosecuting political opponents had abolished the most objectionable parts of the judicial system and its most undesirable elements. Yet this was a misperception because it overlooked the historical role the judiciary had played in German politics after 1918. Kirchheimer listed multiple examples from the Weimar Republic and summarized: "Thus the judiciary constituted one of the chief benefactors of groups thriving upon aggressive nationalist policies" (232). The few members of the judicial apparatus who had remained politically neutral had been weeded out in the repeated and vigorous purges beginning in 1942 at the latest. Kirchheimer concluded: "it seems necessary that all the 12–13,000 judges and public prosecutors be suspended from office until each of them has been thoroughly investigated" (233).

Consequently, Kirchheimer concluded that "the activities of the courts must be suspended for a certain period" (233). Since civil jurisdiction had practically ground to a halt in Nazi Germany since 1943, such a hiatus would hardly amount to a change for the general public. Kirchheimer recommended that all judges and prosecutors be suspended for two months. The personnel records of all those suspended were to be reviewed, and new staff was to be recruited. No new appointments to positions of judge or prosecutor were to be made without a public hearing. The military government was also to consider for these positions those who appeared specially qualified without having the required professional qualifications, even if this might provoke resistance among the legal profession. All positions were to be filled only on a temporary basis. Overall, Kirchheimer estimated that altogether 9,000 to 10,000 positions would have to be filled with new staff. He thought this could be accomplished in just two months by only twenty officials processing seven cases per day each. Then, as now, this recommendation is astonishing for how unrealistic it was and how much it underestimated the necessary administrative manpower enforcing it.

In July 1944, Kirchheimer completed another *Guide* with the title *General Principles of Administration and Civil Service in Germany*.[8] Neumann was originally supposed to prepare this but he decided to entrust Kirchheimer with the task. This *Guide* also examined the question of dealing with existing staff after Germany's liberation from the Nazi dictatorship. Kirchheimer used the example of *Preußische Landräte* (Prussian head officials at the local level), 80 percent of whom had obtained their positions after 1933, to illustrate his estimate that approximately 7,000 higher officials in Germany would have to be terminated in the early phase of occupation. The numbers of staunch Nazis were also very high in the middle and lower ranks, which was why they were to be prohibited from moving into these newly available positions.

---

8 See Kirchheimer (1944b). The following page numbers in the text refer to this *Guide*.

For this reason, it would be necessary to consider the question of whether replacements could be found "outside the present ranks of the German Civil Service" (308). One source would be the civil servants who had been dismissed or demoted by the Nazis. Their immediate rehiring would be useful "both for psychological and for service reasons" (308). In addition, new civil servants were to be recruited from opposition groups even if they did not have the training required for their new positions. "Their wide range of political and social experience and the trust which the population places in them would amply compensate for their lack of formal training" (308). Members of resistance groups such as trade unionists, intellectuals, or the church opposition "should be given preference over members of the Civil Service in responsible nontechnical jobs" (315). According to Kirchheimer, the question of democratizing Germany essentially depended on the types of people serving in the civil service. The traditional narrow esprit de corps of the German bureaucracy had to be broken: "One of the ways in which its structure might be democratized may be the granting of the right for most classes of civil servants to organize on trade-union lines" (309). From today's perspective, Kirchheimer's optimism about the civil service after 1945 is striking. In retrospect, his proposals to perhaps rehire administrative personnel from pre-Nazi times as well as amateurs appear to underestimate the totality of the collapse of non-Nazified civil society in Germany.

Kirchheimer added to his general deliberations an extensive and detailed list of all the German agencies that would have to be abolished immediately. A second list included those that would have to be reorganized. He also recommended establishing a new health and welfare agency that would have to be organized at the national level, departing from the German tradition of federalism. He paid particular attention to the question of the civil servants who had been dismissed, believing that "no pension should be granted" (316) them. And he added: "Whatever the hardships which may be imposed on some of them, it would seem unjustified to grant former officials' preferential treatment in the form of pensions" (316). This recommendation of Kirchheimer's directly affected Carl Schmitt, among others, because university professors in Germany have the status of civil servants. Kirchheimer did not mention any names in this section of the *Guide*.

A *Guide* titled *Administration of German Criminal Justice under Military Government*[9] of July 1944 also reflected the latent trilemma between rapidly handing over governmental responsibility to the Germans, the Allies' security-policy considerations, and the revocation of individual Nazi legislative acts. Once more, Kirchheimer raised the question of an amnesty for individuals imprisoned by the Nazi regime. He considered it a particularly urgent problem and again highlighted the positive psychological effects of a speedy release of all political prisoners.

In this *Guide*, Kirchheimer explicitly addressed a fundamental reform of German criminal law. His recommendation here was that the entire system of German criminal law was "a problem whose solution may be left to the German people and is not a matter of concern for Military Government" (319). He derived this principle from international law. At the same time, it had "to satisfy the requirements of military occupation and to fulfill the purposes and policies of the United Nations" (319). Kirchheimer was full of praise for the criminal law of the Weimar Republic because it was to some extent "influenced by

9 See Kirchheimer (1944c). The following page numbers in the text refer to this *Guide*.

progressive reforms" (319). This development had been discontinued abruptly under the Nazi regime, and Kirchheimer listed in detail the changes in substantive and procedural criminal law made after 1933. His analysis was almost identical, in parts practically verbatim, to the articles he had written prior to his time at the OSS. He proposed "limited abrogation" (327) of all those penal provisions that violated human rights or posed a threat to the military security of the military government. He listed a number of laws, statutes, and decrees—more than fifty in total—in this *Guide* that were to be rescinded immediately by proclamation of the military government. These were to be replaced provisionally by laws and regulations existing before Franz von Papen's government had taken power in May 1932. With this recommendation, Kirchheimer expressed a position he had already championed at the end of the Weimar Republic.

Since the political situation of an occupied Germany after the war was still unclear when he wrote his *Guide*, Kirchheimer devised two scenarios. The first assumed short-term occupation of Germany. Defining political crime and the machinery with which it had been implemented during the Nazi regime would prove to be a particularly difficult problem in this case. Kirchheimer advocated a broad concept of political crime that included hunger and destitution as well as violations of labor laws, for example in the case of forced labor, as motives for taking action. If occupation was brief, the military government would not have sufficient time and personnel to investigate all cases, so all criminals sentenced to less than eighteen months' imprisonment should be granted amnesty, as well as all political prisoners. In the second scenario, with a longer period of occupation, the military government was to establish an agency responsible for reviewing longer sentences.

Finally, Kirchheimer devoted an entire section to the prosecution of Nazi offenders against German citizens. Even if it was likely that the top Nazi criminals would be tried as war criminals in Allied courts, plenty of other perpetrators would remain unpunished. Kirchheimer recommended that these cases were to be tried only in German courts. In cases of flagrant offenses, he assumed that a sufficient number of German prosecutors would be available to indict the perpetrators. It would be more difficult in cases where perpetrators invoked the doctrine of superior orders, or because the statute of limitations would have run in favor of the defendant, or because a Nazi amnesty may have intervened. Solving these problems was not the responsibility of the military government. It would be such a drastic intervention in the German system of criminal law "that the Military Government should not take such a step in the absence of a specific directive [...] from the political authorities" (343). On the organizational level, Kirchheimer proposed a division of responsibilities between German criminal courts and military government courts. The decision of the latter would be final in the event of conflicts about competences between the two courts.

Kirchheimer's recommendations were not undisputed within the OSS. In an internal review of his *Civil Affairs Guide: The Abrogation of Nazi Laws in the Early Period of Military Government* of March 1944, Magdalena Schoch, an émigré like Kirchheimer, criticized his recommendations for not going far enough as they did not include some of the obviously discriminatory laws such as the *Reichstagsbrandverordnung* (Reichstag Fire Decree) and the *Reichserbhofgesetz* (Reich Hereditary Farm Law).[10] Whereas Kirchheimer was against automatically reinstating German citizenship to those who had been stripped of it and had called for them to have to apply for it in Germany, Schoch considered this an unnecessary burden on displaced individuals, many of whom were stateless. Instead, she demanded an agency be established to settle these cases.[11] Schoch's criticism was put even more succinctly words in another considerably more polemic report by H. Bowen Smith of the Foreign Economic Administration (FEA): he considered the strategy of the *Guide* to entrust the Germans and not the Allies with certain matters as an unnecessary threat to the political mission of the US. If Kirchheimer saw the danger posed by a "bellicose and rapacious minority" of Nazis in Germany after liberation, Smith wrote, why was he advocating a rapid handover of governmental responsibility to the Germans? Smith also rejected banning judges and prosecutors for a predetermined period of two months. Instead, the courts should be reopened only very slowly as the Allies saw fit. Overall, he considered the *Guide* a "failure."[12]

Kirchheimer's reaction to this criticism was a kind of controlled offensive. In a report he signed jointly with Herz, he responded positively only to the criticism about dealing with individuals stripped of their citizenship. He had not devoted enough space to this matter because of the brevity of the *Guide*. Otherwise, he reiterated his position: which of the many Nazi statutes in the various areas of the law would also have to be rescinded or changed was a lesser problem than purging the staff and was to be decided later by the (anti-fascist) Germans themselves.[13] Presumably after speaking with Kirchheimer and Herz, Neumann also raised "grave objections"[14] to the FEA's proposal to include the *Reichstagsbrandverordnung* (Reichstag Fire Decree) and the *Ermächtigungsgesetz* (Enabling Act) of 24 March 1933 in the list of laws to be repealed immediately. That would automatically reinstate the second part of the Weimar Constitution and thereby implicitly also its first part. The basic rights that would then apply again—in particular, the constitutional provisions protecting property—would unnecessarily hinder the military government. Moreover, they would promise personal liberties that would apply only to a limited extent during the period of occupation. The two positions differed fundamentally in their assessment of the Germans' political activities: although both sides were convinced that German society was deeply permeated by Nazism, Neumann, Kirchheimer, and Herz

10 The *Reichstagsbrandverordnung* of 28 February 1933 abolished some of Germans' civil liberties; the *Reichserbhofgesetz* of 29 September 1933 excluded "non-Aryans" from farming.

11 See letter from Magdalena M. Schoch to David M. Levitan dated 7 April 1944. Entry 44: Civil Affairs Guides, Box 3, Record Group 226, Records of the Office for Strategic Services, National Archives at Maryland. Levitan was the Chief of the Property Control Division of the FEA and Chairman of the Draft Committee.

12 Memorandum from H. Bowen Smith to David M. Levitan dated 11 April 1944, page 3. Entry 44: Civil Affairs Guides, Box 3, Record Group 226, Records of the Office for Strategic Services, National Archives at Maryland.

13 Otto Kirchheimer and John H. Herz: Memorandum to Sherman Kent dated 9 May 1944. Entry 146: Miscellaneous Washington Files, Box 83, Record Group 226, Records of the Office for Strategic Services, National Archives at Maryland.

14 Franz L. Neumann: Critique on FEA Guide: Elimination of Fundamental Nazi Political Laws in Germany of 10 May 1944. Entry 146: Miscellaneous Washington Files, Box 83, Record Group 226, Records of the Office for Strategic Services, National Archives at Maryland.

thought that the majority of democratic Germans, along with anti-fascist committees with close ties to trade unions, would seek to establish a new constitution with a democratic socialist orientation. This was why the further process of denazification was to be carried out by the Germans, not the Allies. Schoch and the FEA considered this expectation too optimistic and thought that the measures proposed by R&A were wrong.

Nowadays, it is virtually impossible to determine which specific role Kirchheimer's list of Nazi laws to be rescinded played when the Special Legal Unit prepared its own list of laws in the spring of 1945. Both lists mention many of the same laws. A number of repeals recommended by Kirchheimer are also found in the laws enacted by the Allied Control Council (ACC) (see Hochstein and Schale 2021, 53–56). By contrast, the Allied denazification measures corresponded with Kirchheimer's proposals only to a limited extent. He had called for all judges and prosecutors to be dismissed for a limited period of time; the Allies limited this to "members of the Nazi party [...] and all other persons directly involved in the penal methods of the Hitler regime."[15] The proposed two-month closure of all courts did not come about; the first *Amtsgerichte* (see List of German Courts) opened as early as May 1945. The most clear-cut difference between Kirchheimer's proposals and the early Allied Control Council laws is to be found in the handling of the Nazi laws on eugenics. Whereas Kirchheimer emphasized their background in *Rasse* biology and demanded they be repealed immediately, neither the Western Allies' military government nor, for a long time, the Federal Republic abrogated these laws.[16]

Awareness of the institutional positions of the CES and R&A within the OSS should prevent current-day readers of these documents from overestimating their real influence on practical US policy. The political impact of the studies and reports remained dependent on the ability both of the individual branch and of the OSS overall to prevail over other competing military and intelligence units. But even then, it was uncertain whether the analyses and recommendations resulted in the political consequences the staff members desired. Looking back on their work at the OSS, Kirchheimer's friends and co-workers from the years 1944 to 1946, such as John H. Herz, Carl Schorske, and H. Stuart Hughes, all expressed their frustration about the purpose of their work at the time. The intelligence provided by the OSS was only one factor among others and was usually ignored on the ground in the complex institutional arrangement of countless information sources and the at times chaotic decision-making processes.[17]

2. Schmitt's imprisonments and his return to Plettenberg

Carl Schmitt spent the first postwar months in southwest Berlin, which had suffered little destruction and was in the American sector. To his great surprise, he was arrested by the American Public Safety Division on 26 September 1945 and taken to the Interrogation Center in Berlin-Wannsee.

15 Allied Control Council Law No. 4 of 30 October 1945.

16 The decisions on sterilizations issued by the *Erbgesundheitsgerichte* were not repealed until 1998.

17 See Rainer Erd's interviews with Herz, Schorske, and Hughes; Erd (1985, 153–165).

Schmitt was not detained as part of the automatic arrest of all Germans the Allies considered to be a security threat because of their membership in a Nazi organization, as he falsely reported later (see Schmitt 1958a, 75) but at the behest of Karl Loewenstein, a former acquaintance from his Weimar days (see Bendersky 2007). Loewenstein was a German lawyer and political scientist with a *habilitation* who had had to flee the Reich in 1933. He was familiar with Schmitt's writing and had received and further developed Schmitt's *Constitutional Theory* in his own deliberations on the limits of constitutional amendments (see Lang 2007, 130–158). From August 1945 on, he was legal adviser to the American military government and the Allied Control Council in Berlin on matters of denazification (see Loewenstein 2023, 223–258). He had Schmitt's library and some of his Nazi writing seized to secure them as evidence in a potential war crimes trial.[18]

Kirchheimer and Loewenstein knew each other personally from the Weimar period and had stayed in contact in US exile since then. However, there is no evidence that Kirchheimer was involved in Loewenstein's actions against Schmitt. After Schmitt was arrested, Loewenstein prepared an extensive memorandum titled *Observations on [the] Personality and Work of Professor Carl Schmitt*. He made it available to the groups working in Nuremberg to prepare the planned Subsequent Nuremberg Trials, with the goal of putting Schmitt on trial. Whether the group around Neumann and Kirchheimer also received a copy of the memorandum can no longer be ascertained today.

In the memo, Loewenstein called Schmitt "one of the most eminent political writers of our time" and "one of those rare scholars who combine learning with imagination." The crime Schmitt was accused of was that in his writing after 1933 he had become a "defender of totalitarianism" who "has done more to support the regime than most other people." In particular, Loewenstein accused him of defending "the aggrandizement of Germany at the expense of weaker powers." Schmitt is described less as a Nazi acting out of a strong sense of conviction but rather as a ruthless opportunist, including, and in particular, with respect to his relationship to Jews: "Suddenly he became an enthusiastic antisemite."[19] Loewenstein stated that Schmitt's arrest would be viewed as an act of justice by all Germans with a sense of responsibility and that his speedy release would have to be considered an attack against public opinion abroad and against emergent democracy in Germany. Loewenstein's memorandum had no negative consequences for Schmitt. It is no longer possible to determine the reason why the American authorities rejected Loewenstein's recommendation. After multiple interrogations and on the basis of favorable reports by colleagues who were his friends, Schmitt was released from the civilian detention camp in October 1946 after a total of twelve months.

Schmitt felt that his detention was deeply unjust and humiliating. Even though he was officially banned from writing, he succeeded in penning a number of brief texts during that time. These were published several years later in the booklet *Ex captivitate salus* [Salvation coming from captivity] and provide illustrative insights into how Schmitt per-

18 Karl Loewenstein: Library of Carl Schmitt. Memorandum dated 10 October 1945. Karl Loewenstein Papers. Box 28, Folder 1.

19 All quotes: Karl Loewenstein: *Observations on [the] Personality and Work of Professor Carl Schmitt*. Memorandum dated 14 November 1945. Karl Loewenstein Papers. Box 28, Folder 2.

ceived himself and the world in the postwar period.[20] Around six months later, on 17 March 1947, he was arrested again. This time, he was transferred to Nuremberg to be interrogated by Robert M. W. Kempner, assistant US chief counsel during the International Military Tribunal at Nuremberg; the first war crimes trial had begun there the previous November. The circumstances of this second arrest are largely unclear to this day. Perhaps Loewenstein's memorandum did play a role after all. But it is not even apparent from the surviving sources whether Schmitt was taken to Nuremberg as a potential defendant or a potential witness for the prosecution (see Quaritsch 2000, 16–27).

The second arrest was initiated by Robert M. W. Kempner, who had been dismissed from the Prussian civil service in 1933 by Schmitt's mentor Göring because of "political unreliability in concurrence with continued Jewishness." Like Loewenstein, Kempner was familiar with Schmitt and his writing from that time. Ossip K. Flechtheim, who had collaborated with Kirchheimer at the Institute of Social Research, worked in the position of US army lieutenant colonel in the Berlin division of the American War Crimes Council for a year beginning in the summer of 1946. His task was to prepare the Subsequent Nuremberg Trials, in particular the Judges' Trial and the Wilhelmstrasse Trial (see Keßler 2002, 171). As a member of Kempner's staff, Flechtheim sent a letter to Schmitt ordering him to present himself for an interrogation with him. Schmitt had rejected Flechtheim as a doctoral candidate in Cologne in 1933 as a "*Fremdrassiger*" (a member of a foreign/alien *Rasse*, in an exclusionary and antisemitic sense).[21] When Flechtheim interrogated him in Berlin, Schmitt invoked his role model Thomas Hobbes, "who would have made himself available to any regime as well."[22]

It was also Flechtheim who had had Schmitt transferred from Berlin to Nuremberg (see Wollenberg 1991, 106). Schmitt was listed in the directory of the Nuremberg detainees as a "leading Nazi propagandist in the field of International Law & Nazi theories" (see Tielke 2020, 36). As far as the historical facts can be reconstructed today, Schmitt took the opportunity arising from his interrogation by Flechtheim on 27 March 1947 to initiate contact again with Kirchheimer.[23] Schmitt must have been aware that Flechtheim knew Kirchheimer because he asked him about Kirchheimer's fate. He also asked Flechtheim to convey his best regards to Kirchheimer, which Flechtheim did when they met a few months later in Frankfurt.

We can only speculate about Schmitt's motives for this initiative. Perhaps he was hoping that Kirchheimer would help him, perhaps he was afraid that Kirchheimer could harm him, or perhaps he was just curious. After Schmitt had found a place to stay in his hometown Plettenberg in the Sauerland region after his first release from detention in 1947, he still regarded his situation as precarious and suspected "victors' justice" everywhere. He was unsure about his role in Kempner's strategy for the trials, whether he was

20 See Chapter 15, p. 364–366.

21 Schmitt later denied this version of how he had rejected Flechtheim (see Tommissen 1990, 144–147; Bendersky 2007, 23). Schmitt's version contradicts Flechtheim's written description of the events (see Keßler 2007, 77 and Keßler 2011, 150) and his statements in conversations with Jörg Wollenberg (see Wollenberg 1991, 105–107) and with the author on 13 February 1988.

22 Quoted in Wieland (1987, 108).

23 The following information is based on statements by Ossip K. Flechtheim and Lili Flechtheim-Faktor in a conversation with the author on 13 February 1988.

a possible defendant or whether he was to serve as an expert witness for the prosecution. In this uncertain situation, he may have hoped for some help from Kirchheimer. Another possible explanation for his attempt to contact him is related to his professional plans. Schmitt considered returning émigrés in particular to be responsible for him being banned from returning to a chair at a German university. At this point in time, he had not yet given up hope that he would be appointed professor again after his release, like many other professors who had had successful careers during the Nazi period. From right after the end of the war, he kept abreast of the policies of filling positions at universities in the Western zones, casting his net widely. He may have feared that Kirchheimer could harm his career prospects in academia. A third possible explanation is that Schmitt was merely wondering about Kirchheimer, without any particular ulterior motive. All three possible explanations are merely speculation.

In 1947, Kirchheimer was well aware "that C.S. was sent to a camp,"[24] as documented in a letter he sent Arvid Brødersen, his colleague at the New School for Social Research in New York, ten years later, thus confirming Flechtheim's report indirectly. Kirchheimer made his first trip to Germany after the war in the summer of 1947.[25] During this trip, he was in contact with Karl Loewenstein, his colleague at the State Department, as well as with Flechtheim. He also met Rudolf Smend and his wife in Göttingen again and the two social democratic legal experts Carlo Schmid and Richard Schmid in Frankfurt. Even though there is no documentation of the subjects of their discussions, it is plausible that they talked about the prominent case of Carl Schmitt when they discussed current affairs. There is no indication that Kirchheimer contacted Schmitt at this time.

Schmitt arrived in Nuremberg six months after the Allies' major trial against the top war criminals had ended. These proceedings were followed by twelve more trials addressing specific issues that were also held in Nuremberg. Schmitt's interrogations by Kempner in Nuremberg must be viewed in this context. They focused on the question whether and to what extent Schmitt's writing on international law constituted the offense of participation in preparing a war of aggression within the meaning of Article II of Allied Control Council Law No. 10 of December 1945 (see Quaritsch 2000, 16–26). Kempner interrogated Schmitt three times during the two months of his pretrial detention and gave him the opportunity to respond to multiple charges and questions in writing.[26] The German-language *Aufbau*, a newspaper of Jewish émigrés from Germany that Kirchheimer read regularly in New York, reported about Schmitt's detention in Nuremberg on 2 May 1947. The article with the byline "c.m." called Schmitt a "charlatan" who had justified everything "that had been committed by the state in Germany against the law and humanity" during the Nazi period. The author speculated that Schmitt seemed to be "slippery as an eel in escaping his deserved punishment" and was deeply satisfied that Schmitt was allegedly to be indicted in Nuremberg.[27]

24 Letter from Otto Kirchheimer to Arvid Brødersen dated 2 March 1958. Otto Kirchheimer Papers, Series 2, Box 1, Folder 25.

25 More on this trip below.

26 All documents are collected in Schmitt (2000). Only some of them have been translated into English (Schmitt 1987).

27 All quotes c.m. (1947).

In the letters Schmitt wrote during his detention in Nuremberg, he called himself a "hostage." It is worth noting that in the private setting of his numerous letters to his wife, he showed no indication that he felt his involvement after 1933 was problematic, much less that he had any sense of guilt—in keeping with his stance in public. He again saw himself put in a situation more dire than that of Europe's Jews during the Nazi period. "What the Nazis did was beastly. But what is happening to me (and to thousands of honest Germans) is fiendish."[28] In other letters, he complained that vindictive émigrés were destroying his files, thereby sabotaging his release. Yet in contrast to Schmitt's expectations and the *Aufbau* journalist's hopes, once Kempner had interrogated Schmitt and read the written statements he had prepared in response to questions, he no longer saw a reason to detain him in Nuremberg. Schmitt was released on 21 May 1947 under the condition that he serve as a voluntary witness to the prosecution.

After his release, Schmitt went to his birthplace Plettenberg in the Sauerland region in the American zone, where he lived in his family's house with his wife Duška and their daughter and their housekeeper, who had all moved there from Berlin. Schmitt's fear of being treated and sentenced with the same harshness and mercilessness that the Third Reich had exercised against its opponents had proven unfounded. He refused to undergo denazification proceedings and lost his professorship once and for all. His dismissal had been declared "with immediate effect" in a letter from Berlin University's Rector Johannes Stroux to Schmitt dated 29 December 1945 as he had been a member of the Nazi party up until the fall of the regime.[29]

Schmitt named his house in Plettenberg San Casciano after the place where Machiavelli had retreated after the Medici had banned him from the Florentine Republic. His drive and his productivity as a writer remained unwavering. A number of manuscripts were almost ready for publication, and he worked on several new ones. From August 1947 on, he also regularly entered notes in a diary he had titled *Glossarium*. He entered into increasingly extensive correspondence with some of his previous acquaintances. After losing his chair, he was not isolated and did not fall into poverty. Affluent friends and students founded an association for him named Academia Moralis. Its members supported him with a fixed salary and ensured that he was commissioned with well-paid legal opinions. Nevertheless, he briefly considered emigrating to Spain or Argentina in 1947/48.

In his interrogations by Robert M. W. Kempner, Schmitt had claimed that he sought to retreat from public life into the "security of silence"[30]; it is not clear from the source material whether these are Schmitt's or Kempner's words.[31] Some authors sympathetic to Schmitt made his supposed "path of silence and privateness"[32] into an established topos.

---

28 Letter from Carl Schmitt to Duška Schmitt dated 1 May 1946 (Schmitt and Schmitt 2020, 200).

29 Letter from University President Johannes Stroux to Carl Schmitt dated 29 December 1945. Humboldt-Universität zu Berlin, Universitätsarchiv, Akten der Juristischen Fakultät, Nr. 159a, Prof. Dr. Carl Schmitt, Blatt 73.

30 Quoted in Wieland (1987, 105).

31 Quoted in Quaritsch (2000, 39–42).

32 The journalist Karl Korn in an article in the *Frankfurter Allgemeine Zeitung* in 1950, quoted in van Laak (1993, 127).

Yet it was a very telling silence; to be precise, it was simply posturing. The German historian Dirk van Laak differentiates various facets of the meanings of this "silence" from Schmitt's own perspective (see van Laak 1993, 126–133). As van Laak writes, silence, as an immediate reaction after 1945, meant the absence of intense activity, a pause in a confusing time. It also had the meaning of refusing to provide information, thereby fundamentally refusing to acknowledge the right of any worldly authorities to pass judgment on misconduct; divine instances alone could do this. And finally, silence was a synonym for an elitist understanding of the relationship of the general public and an intellectual elite, an understanding that expressed itself in the concern that the stupid masses would either ignore or misunderstand him. Therefore, withdrawing into such "silence" was a way to shirk his individual responsibility for his past deeds. To Schmitt, silence was therefore not a passive status; instead, it was his way of "intentionally taking a position" (van Laak 1993, 128). Silence meant refraining from making statements.

Yet, rather than stopping writing, on the contrary, Schmitt's "silence" produced a number of new publications. It simply meant being selective in what he said, and that was determined exclusively by his own estimation. Thus, it functioned as a protective cloak, safeguarding him from arguments in which opposing views were presented, and simultaneously opening up the space for establishing a small circle of people in the know surrounding themselves with secrets and legends.

Schmitt's first postwar works were published in Spain and Portugal since the Allies had banned him from publishing in Germany. That ban was lifted after the founding of the Federal Republic of Germany on 23 May 1949 and so his first articles published in Germany appeared in newspapers and magazines later that year. The constitution of the new West German state, the Basic Law, guaranteed freedom of expression, and Schmitt was among the first to profit from this constitutional right. If we look at his work alongside what Kirchheimer was doing and writing about at that time, we see striking parallels in the selection of two subjects Schmitt addressed in particular detail up to 1949: international law and looking back at the Nazi past.

The most important subject for Schmitt was his theory of international law. He worked intensively on revising and completing his monograph *The Nomos of the Earth* that he had begun writing during the final years of the war. When Kempner interrogated him, Schmitt insisted on a rigorous dividing line between his works on international law and Hitler's foreign policy. He had been an "intellectual adventurer" (Schmitt 1987, 103) and had been grossly misunderstood. Now, he claimed, he had to suffer as a result of that: "I take the risk. I have always accepted the consequences of my actions. I have never tried to avoid paying my bills." He claimed his works on international law had always been nothing but a "serious scholarly [...] diagnosis" (Schmitt 1987, 104) with no connection to actual practice. He rejected Kempner's accusation as unfounded that he had supported the theoretical underpinnings of German *Großraum* policy.

His first postwar publication in Germany appeared—anonymously—in the summer of 1949. The Catholic magazine *Die neue Ordnung* [The new order] published his essay "Francisco de Vitoria und die Geschichte seines Ruhms" [Francisco de Vitoria and the history of his renown] (see Schmitt 1949b), an excerpt of *The Nomos of the Earth*. Expert readers immediately identified Schmitt as the author and vehemently criticized the fact that a long-standing propagandist of the Nazi regime was again granted space in magazines (see van Laak 1993, 37–39). Between 1948 and 1950, Schmitt also wrote a *Repetitorium zum Völkerrecht*, i.e., a review course in international law for law students preparing for their examinations (see Schmitt 1948). The four-part course materials were published by a publishing house specializing in educational materials without indicating the author's name. Schmitt had taken on this project to improve his financial situation. Long-standing associates such as Werner Weber praised it highly in their book reviews—"originality and eminent skill" and "strong individuality" (Weber 1949, 819)—and did not reveal his name, either. The extensive course materials were partly based on his lectures during the Nazi period but were cleansed of Nazi vocabulary. Schmitt added chapters on the law of occupation and on postwar developments in international law.[33] He raised the question multiple times whether occupied Germany continued to be a state with full sovereign rights after unconditional capitulation and answered it in the affirmative—in contrast to Hans Kelsen (see Schmitt 1948, 726–728, 733–736, and 742–743).

Schmitt used unemotional language in his didactically structured educational text. Only in his discussions of the future expansion of international criminal courts did he switch to a sarcastic tone. Continuing his line of argument in his legal opinion supporting Flick, Schmitt accused the Nuremberg Trials—both those of the International Military Tribunal and the American Military Tribunal—of having introduced a historically unique system of criminal justice that was new to international law. Regarding the rulings of the Nuremberg courts, he emphasized that multiple defendants had been acquitted of conspiracy for a war of aggression (see Schmitt 1948a, 769–772); again, there was no mention in the book of the massacres and mass murders committed by the Germans. The fact that it was international courts handing down decisions, Schmitt asserted, had transformed the citizens' right to resist into "a duty to resist in the name of international moral judgments, and it amounted to dethroning the state" (Schmitt 1948a, 7[illegible]6). In contrast, *ius publicum Europaeum* was based on the inviolable right of all peoples to self-determination. Since the majority of Germans rejected the international courts, Schmitt's characterization of *ius publicum Europaeum* ultimately amounted to a general amnesty for all war crimes.

Schmitt's second major topic was dealing with the past, including his own. The term *Vergangenheitsbewältigung*—the German effort to come to terms with, or literally "overcome" the past—has become established in German. In September 1947, Schmitt noted the following in his *Glossarium*, his postwar diaries, about his own role: "Back then, in the years 1933–36, I abandoned less of myself and the dignity of my thoughts than Plato abandoned of himself and his thoughts because of his trips to Sicily."[34] Shortly thereafter, he wrote in his diary: "Annihilating feeling of betrayal." "Defeated? Oh no, annihilated and trampled."[35] He saw himself as a victim betrayed by Hitler on a personal level. The "true victims," he wrote in the spring of 1949, in Germany were the members of the Nazi party with a membership number higher than two million, for they were "the victims of the

33 On Schmitt's concept of belligerent occupation see Butha (2005) and Cohen (2007).

34 *Glossarium* entry of 26 September 1947 (Schmitt 2015, 14).

35 *Glossarium* entry of 8 October 1947 (Schmitt 2015, 22).

Nazis as well as the persecutors of the Nazis."[36] Schmitt himself fulfilled his own criterion for being a victim of the Nazis since his party membership number was 2,098,860.

In his legal opinion for Flick, Schmitt had already narrowly defined the circle of perpetrators responsible for the Nazi system. He only qualified as perpetrators those individuals who had direct personal "access to the peak" (Schmitt 1945, 180). As executors, all the rest were mere participants without responsibility or culpability. Not a word is to be found about potential differences in the degrees of perpetration. Nor did he consider indirect access to the "peak," for example through publications or as a consultant to a consultant who in turn had direct access to the "peak." To Schmitt, actions based on economic interests were also no reason for someone to be considered a perpetrator in terms of criminal law.

In his responses to Kempner's questions during his detention in Nuremberg, Schmitt explained his understanding of access to the power holder in more depth. He emphasized the fundamental abnormality of state organization and the legislative process under the "Hitler regime" (Schmitt 1987, 118). It was a new form of totalitarian dictatorship, and he had no experience in dealing with it. Because of the "most extreme concentration of all power in one hand," he believed the question of access "became the most important internal political problem of the German *Reich*" (Schmitt 1987, 118). In his written statement to Kempner, he outlined a structural representation of the Nazi regime. After Hitler and his inner circle had de facto disempowered the various ministries, a "new 'super-ministerial' structure" (Schmitt 1987, 119) with a highly personal character of power had developed. Its personnel were recruited from "three 'pillars' of the regime: the party, the military, and the state" (Schmitt 1987, 119). Hitler's orders and commands were implemented through these three control centers. Schmitt considered himself to have been outside the exclusive "essential circle of 'loyalists'" (Schmitt 1987, 117) that constituted the core of the regime.

It would have been logical for Schmitt to flesh out his description of the structure of the Nazi system, which he had sketched out in only a few sentences, at this point. For one reason, because he painted a picture of a harmonious inner circle of the regime even though he had experienced firsthand that rivalries and intrigues had been commonplace within the ruling elites. And also because he had conjured up the image of a structural triad to describe the Nazi regime 14 years earlier. In the fall of 1933, he had offered the formula of the "unity of the tripartite structure of state, movement, and *Volk*" (Schmitt 1933d, 11). He considered the Nazi party to be the political body in which the movement had found its specific political form. Both of Schmitt's triadic structural models are harmonious but they differ in that he replaced the *Volk* with the Reichswehr (the armed forces) in 1947. The model Schmitt had merely sketched out comes close to Kirchheimer's structural model of the Nazi regime, although what had mattered most to Kirchheimer was the conflict dynamics between the main actors. Schmitt's new description fulfilled the function of absolving himself of any and all political responsibility. In his interrogation by Kempner, Schmitt even went so far as to claim that the *Bund Nationalsozialistischer Deutscher Juristen* (BNSDJ) or Association of National Socialist German Legal Professionals in which

36 *Glossarium* entry of 5 May 1949 (Schmitt 2015, 181).

he had taken over a leading position in 1933, had "extracted, so to speak, from my mouth" (Schmitt 1987, 106) and against his will his writing in support of the regime.

During his twelve months in the internment camp in 1945/46, Schmitt was able to smuggle letters and short essays to his wife. As mentioned above, he published these diaries and essays several years later, in 1950 in a book titled *Ex captivitate salus.*[37] His prison writings, however, are quite different from the Italian communist Antonio Gramsci's voluminous *Prison Notebooks*, not only because of their brevity. Their original title was *Salus ex captivitate*,[38] which made the reference to a well-known Bible passage in John 4:22 even more explicit. In the commonly used Latin version, the wording in this passage is "*salus ex Iudaeis est*"[39]—Schmitt replaced the Jews bringing salvation with his detention (by the returning Jews, in his view) doing so instead. Nonetheless, the title might have prompted some of his readers at the time to take notice or even expect some self-criticism. They would have found nothing of the sort. Schmitt did not make a single statement about his complicity in the crimes of the Nazi regime, brusquely rejecting such an idea: "Whoever wishes to confess, go and present yourself to the priest" (61).

In place of a "confession", Schmitt presented an attempt to justify his behavior that was virtually unparalleled in its hypocrisy and self-righteousness. The previously enterprising Nazi legal expert styled himself as a "contemplative person" (14) with keen observational skills and inclined to use "sharp formulations" (14) but lacking any practical ambitions. He commented on his role in the Nazi system using only a few general words. As a professor of constitutional law, he had had to be a "loyal citizen" (21) like everyone else as a matter of principle; he believed that, for this reason, "the mutual relationship that results from protection and obedience" (54) had applied to him until the end. He vehemently rejected the idea that there might be a duty to sabotage, much less resist a state, even if it is a terrorist one. At one point in his text, Schmitt added a methodological hypothesis to this argument reminiscent of his works on Hobbes: "The material" from which the scholar of public law assembled his concepts and upon which he relied "binds him to political situations" (47).[40] But Schmitt refrained from explaining why defending a government and not criticizing it should automatically follow from this reference to the political situation at hand.

Ex captivitate salus lacks any empathy for the victims of the regime. Schmitt cynically attested that those who—like Kirchheimer—were driven to emigrate enjoyed the divine benevolence of a "guardian angel" (18). He replaced his lack of compassion with all the more self-pity, speaking of his "persecutors" who had compiled "black lists" and "card files" (73). He claimed they had transformed the means and methods of the justice system into a total deprivation of rights, indeed "into means and methods of annihilation" (48). Schmitt transformed the perspective he had taken as one "of the vanquished" (29)—as he explained, referring to Tocqueville—into the privileged epistemic position of an objective standpoint. The hypothesis he left unsaid was that emigration had not entailed any increase in knowledge because the pain of the German defeat could not be sensed from

37 Schmitt (1950a). The following page numbers refer to this text.

38 Letter from Carl Schmitt to Armin Mohler dated 29 August 1948 (Schmitt and Mohler 1995, 47).

39 On this Vulgata quote and the theological debates within Christianity, see Gross (2000, 347–349).

40 See Chapter 9.

a distance. Schmitt even went so far as to consider himself to be the only remaining law professor on the entire planet who had recognized and experienced the allegedly occurring global civil war in all its deep-rooted causes and manifestations; he viewed himself as "the last knowing representative *of ius publicum Europaeum*" (60).

Ex captivitate salus is Schmitt's most personal book. Yet it remains a mystery to the reader what exactly constituted the "*salus*" he acquired in the "wisdom of the cell" (62). It may have been a personal turn of Schmitt's to the Christian faith, or his steadfast subjective conviction that as a person who had been defeated, he had an edge in terms of political knowledge. He saw his imprisonment as providing a space for solitary illumination. By contrast, the book's strategic intention is easier to recognize. It is the passionate attempt to distance his writings from the Nazi context and to reinvent himself as a member of the German public.

Schmitt was right to anticipate that there would be a sympathetic audience for an unapologetic statement by a former Nazi member. His unwillingness to criticize himself was not merely a psychological matter. It was a reflected political decision. As of 1948, he sent more than 100 copies of his *Ex captivitate salus* manuscript to a wide circle of recipients. In a cover letter to legal scholar Luis Cabral de Moncada in fascist Portugal, Schmitt called it "my intellectual cry for help by a legal scholar in today's global constellation."[41] Neither the manuscript nor the printed book is to be found in Otto Kirchheimer's estate. He was not on the list of persons Schmitt had sent the manuscript to. But, as a matter of fact, Kirchheimer must have had an opportunity to read it prior to its publication in 1949. He probably received it from a former colleague who was on Schmitt's mailing list, probably Werner Weber in Göttingen.[42]

3. Post-Holocaust antisemitism

After 1945, Schmitt also worked intensively on two other subject areas that had close connections to Kirchheimer: the role of the (few) émigrés returning to Germany and Judaism. Both of these merged into a single topic in his hands. When Kempner had asked him repeatedly about his attitude toward the Jews and about the Holocaust, he had answered tersely, "It was a great misfortune and, indeed, from the very beginning" (Schmitt 1987, 53). When Kempner had responded that his allegedly purely scientific writing had ended in the murder of millions of people, Schmitt responded just as tersely that Christianity had also ended in the murder of millions of people.

None of Schmitt's works intended for speedy publication in the early postwar years include open statements on Judaism; such statements are only to be found in his conversations, correspondence, and diaries. There is no difference between these and his public antisemitic tirades during the Nazi period. Even obdurate apologists of Schmitt were shaken by his postwar notebooks, his *Glossarium*, when this was first published in 1991. The corrected and revised second edition of 2015 includes even more drastic remarks

41 Letter from Carl Schmitt to Luis Cabral de Moncada dated 25 June 1948 (Schmitt and de Moncada 1997, 34).

42 See Chapter 15.

by Schmitt. There is a consensus among scholars that Schmitt wrote his *Glossarium* for posthumous publication.

In September 1947, several weeks after his release from detention in Nuremberg, he noted: "Jews will always be Jews. Whereas the communist can better himself and change. That has nothing to do with the Nordic *Rasse*, etc. The assimilated Jew in particular is the true enemy."[43] Schmitt felt that when German society began opening up to Jews and enabling their social integration in the nineteenth century, it had been swamped by foreigners. It was the "tragedy" of Jewish assimilationism that Jews were unable to "be part of the great step from Goethe to Hölderlin." Instead, the Jews developed an "infuriatingly unsuspecting feeling of superiority," never even beginning to grasp "the step from *Begriff* [concept] to *Gestalt* [totality] and what that meant in the view of the German spirit."[44] Schmitt called non-Jewish authors such as Thomas Mann "traitors to the spirit"[45] and Karl Jaspers and Gustav Radbruch, who attempted to spark a critical and public debate on the Germans' responsibility for the mass murders, "liberated Germany's pin-ups from the humanities."[46]

Schmitt drew a sharp and unbridgeable dividing line between "us Germans" and "the Jews" in his *Glossarium*, and he did so multiple times. When Robert M. W. Kempner wrote a piece in the magazine *Der Monat* about Nazi ideologue Alfred Rosenberg's diary, which had just been found, Schmitt made the following comment: "What business of the Jews is that?"[47] He called the Nuremberg Trials the "enormous clattering fiasco of the justice system" and a Jewish-Bolshevik undertaking. He felt it was a "scandal" that in the trials, "Russian professional revolutionaries" and "Jewish émigrés" "were imposed on us as the yardstick of normal German behavior."[48] He interpreted the trials as a "call for revenge"[49] as in the Old Testament. The image of the triumphant and vindictive Jew flares up once again in statements like these.

Schmitt viewed the Western zones and the newly emerging Federal Republic of Germany as constructs of Jewish revenge. The West Germans, he believed, were being "sacrificed to the émigrés,"[50] for "it is the returned émigré who makes the decisions in the Western half."[51] These émigrés were "unpredictable," "mostly partially deranged in a moral sense," and even demanded "300 percent exploitation of rights,"[52] in other words, three times as much compensation as what they had lost. Formally speaking, the Americans were the new lords but, in fact, it was the Jews who were in power: "To this day, for 5 years, I have never spoken with an American [...], but only with German Jews, with Mr. Löwenstein [sic!], Flechtheim."[53] He mockingly called the American occupying

43 *Glossarium* entry of 25 September 1947 (Schmitt 2015, 14).
44 *Glossarium* entry of 23 May 1948 (Schmitt 2015, 116).
45 *Glossarium* entry of 26 May 1949 (Schmitt 2015, 198).
46 *Glossarium* entry of 12 July 1949 (Schmitt 2015, 194).
47 *Glossarium* entry of 23 July 1949 (Schmitt 2015, 197).
48 *Glossarium* entry of 12 July 1949 (Schmitt 2015, 194).
49 *Glossarium* entry of 1 May 1948 (Schmitt 2015, 107).
50 *Glossarium* entry of 14 March 1948 (Schmitt 2015, 86).
51 *Glossarium* entry of 13 August 1949 (Schmitt 2015, 199).
52 *Glossarium* entry of 4 July 1949 (Schmitt 2015, 191).
53 *Glossarium* entry of 17 August 1949 (Schmitt 2015, 200).

force a "peculiar lord of the world" that was infiltrated by Jews: a "newfangled Yankee with its age-old Jews."[54] The Jewish émigrés active in Germany, Schmitt wrote, refused to understand how the Germans of 1945 were living and suffering; instead they only wanted "to be right all along with their positions from 20 years ago." Above all, however, they sought to "snag their compensation."[55] The image of the moneygrubbing Jews is also to be found in Schmitt's notes on financial compensation for victims of the Nazis: "In the meantime, we Germans are suffering because of compensation on open account."[56]

There is no mention of the Holocaust in Schmitt's *Glossarium* except for a disgraceful note from August 1949 in which he first speaks of twelve million dead Jews only to reject this figure in the same sentence as a "horrific episode."[57] By arbitrarily doubling the number of murdered Jews known at the time, Schmitt apparently sought to express that the precise number of victims did not matter. He interpreted the Nuremberg Trials as a perfidious Jewish annihilation strategy: "'crimes against humanity' is merely the most general of all general clauses to annihilate the enemy."[58]

At multiple times in his *Glossarium*, Schmitt equated the annihilation of European Jewry by the Nazi regime with his own professional situation after 1945: "Genocides, what a touching term; I experienced an example of it firsthand (translator's note: in German: "*am eigenen Leibe*," literally: on my own body): extermination of the Prussian-German civil service in the year 1945."[59] He even called his own fate even harsher than that of the Jews in the Nazi period as he noted in August 1949, using the superlative: "But when it comes to me, people are committing the most shameless *Ideocidium*,"[60] in other words, the most shameless murdering of his political ideas. The statements quoted above show, in the apt words of Raphael Gross, a "new strategy of political antisemitism after the Holocaust" (Gross 2000, 352), namely styling oneself as the actual victim.

4. Kirchheimer's struggle with the FBI

Schmitt's antisemitic assumption that a few Jewish émigrés from Germany had been able to control the policy of the American administration appears all the more absurd in light of the fact that these émigrés had been in the spotlight of the FBI. Kirchheimer had to struggle for almost ten years with a secret service that had put him under suspicion and he had reasons to believe that he might soon lose his job.

Once the Allies had defeated the Axis powers, the OSS no longer had a raison d'être. President Harry Truman, newly in office, issued Executive Order 9621 on 20 September 1945, dissolving the OSS as of 1 October. William Donovan's attempts to secure the OSS's continued existence in peacetime were unsuccessful. The individual departments were integrated into other government agencies. The Research and Analysis Branch with its

54 *Glossarium* entry of 17 August 1949 (Schmitt 2015, 200).
55 *Glossarium* entry of 14 March 1948 (Schmitt 2015, 86).
56 *Glossarium* entry of 23 May 1948 (Schmitt 2015, 116).
57 *Glossarium* entry of 23 August 1949 (Schmitt 2015, 202).
58 *Glossarium* entry of 6 May 1948 (Schmitt 2015, 110).
59 *Glossarium* entry of 21 August 1949 (Schmitt 2015, 201).
60 *Glossarium* entry of 21 August 1949 (Schmitt 2015, 201).

1,655 positions became part of the State Department, and a majority of its staff members—among them Kirchheimer—were offered the opportunity to work in a division to be established there.[61] Some of his close colleagues and friends at the OSS, such as Franz L. Neumann, Herbert Marcuse, John H. Herz, and Ossip K. Flechtheim, used this situation to obtain positions in academia.

Kirchheimer struggled with making that decision. Once his son Peter was born in 1946, he felt he had a duty to hold a job with sufficient pay to support his family and to set aside his academic aspirations if necessary. He reluctantly accepted an offer in the research division of the State Department as of 30 September 1945.[62] His position at the State Department was officially designated as Research Analyst in the Department of State, Division of Research for Europe.[63] He worked in the Germany section, first under Neumann and later, from 1948 on, as section head. In 1950, he was promoted to the leadership position of Chief of the Central European Branch, Division of Research for Western Europe. He remained in this job until 1955.

One of Kirchheimer's duties was to travel to Germany to gather information on the ground about the political situation and to summarize it in reports for the State Department. He used these official trips for private visits, too. There was no one left of his family in Germany—they had either succeeded in fleeing to Argentina or had been murdered in the Holocaust—, but he still had various acquaintances and political friends whom he visited on these occasions. On his first trip from March to July 1947, his first stop was in France,[64] where he met up again with old friends from his days in exile, among them Charles and Leonore David as well as Fritz Meyer, an antiquarian bookseller and friend since the 1920s; they had kept some of Kirchheimer's books after he had fled Paris in 1937 and returned them to him after the war.[65] From Paris he traveled to Berlin with Gurland[66] where they were to prepare a study on the political party landscape in Germany for the State Department as consultants to the Berlin Office of Military Government, United States (OMGUS).[67] The report formed the basis for the OMGUS report *Bureaucratization Trends in Postwar German Society*, which Kirchheimer wrote jointly with Gurland and sociologist Hans H. Gerth (see Kirchheimer, Gerth, and Gurland 1947).

One of the first people he visited in Germany was his former mentor Rudolf Smend, whom he met in Göttingen in July 1947. Unlike Schmitt, Smend had tried to continue

61 The institutional setup was somewhat more complicated: in 1945, the unit was first called the Interim Research and Intelligence Service (IRIS), from 1946 on the Office of Research and Intelligence (ORI), which was then incorporated in the Office of Intelligence Coordination and Liaison (OCL) and was renamed the Office of Intelligence Research (OIR) in 1947; see Müller (2010, 59–61).

62 Office Memorandum of US Government dated 7 December 1948, FBI, US Department of Justice, Federal Bureau of Investigation, File on Subject Otto Kirchheimer (unnumbered).

63 Otto Kirchheimer, Curriculum Vitae (1965). Private collection of Hanna Kirchheimer-Grossman (Arlington).

64 FBI, Report by Special Agent Patrick M. Rice on Otto Kirchheimer of 21 June 1950, page 20. FBI, US Department of Justice, Federal Bureau of Investigation, File on Subject Otto Kirchheimer (121–13351–5).

65 Peter Kirchheimer in a conversation with the author on 21 September 2019.

66 See OMGUS Order of 4 June 1947. Otto Kirchheimer Papers, Series 1, Box 1, Folder 1.

67 Otto Kirchheimer, Curriculum Vitae (1965). Private collection of Hanna Kirchheimer-Grossman (Arlington).

supporting Kirchheimer even after the handover of power to Hitler's government. And, in contrast to Schmitt, Smend had been no propagandist of the regime. In Göttingen, he had limited his work to matters of church law and maintained contacts with Christian resistance circles. In the spring of 1947, Kirchheimer had asked Pastor Martin Niemöller to convey some of his newer publications to Smend.[68] From then on, the two resumed close and regular correspondence. Kirchheimer often sent packages from the US to Germany and periodically visited Smend and his wife Gisela. In unpublished notes, Kirchheimer's long-time friend Eugene Anschel reported on the special emotional ties to Smend: "In Smend, Otto saw not only a scholar, but also an upstanding human being. After the war, he regularly sent Smend care packages; by contrast, he intentionally did not send any to Schmitt."[69]

Smend proposed Kirchheimer as his successor in Göttingen. Kirchheimer was pleased but skeptical about his chances. He "doubt[ed] [...] that everyone familiar with my lectures and publications has a very positive view of my utility on the German university scene."[70] The ministry appointed Schmitt's former student Werner Weber to the chair instead of Kirchheimer, much to Smend's resentment. [71] It was hardly surprising that Kirchheimer and Smend immediately took up the topic of Schmitt. Three years after their first postwar meeting, Kirchheimer wrote Smend that he agreed that it was necessary to introduce students to the work of Hermann Heller, which had fallen into oblivion. In the same letter, he emphasized the necessity they had both acknowledged of "combating Schmitt's conceptual framework and way of thinking"[72] in German constitutional law.

From 1947 on, Kirchheimer traveled to Germany almost every year. In the summer of 1948, he spent several weeks in the Western zones, including a number of days in Heilbronn and Heidelberg in his old southern German home as well as in Frankfurt am Main. From October 1949 to January 1950, he again flew to Frankfurt for a longer period, this time as a consultant to the US High Commission for Germany (HICOG). His visits in 1951 and 1952 were not as long. He worked at the American Embassy in Bonn from February through May 1953.

Besides his work on legal opinions, reports, and various other internal papers for the State Department, Kirchheimer had begun in 1946 to publish smaller contributions in academic journals. He initially published his ideas in book reviews he wrote for *Political Science Quarterly*, *American Political Science Review*, and the *Yale Law Journal* on the subjects of France (see Kirchheimer 1946), opinion polls (see Kirchheimer and Price 1949), and the legal system of the Soviet Union (see Kirchheimer 1947 and 1949b). It was only in 1950 that he began publishing longer essays in academic journals again. In an interview fifty years later, Kirchheimer's wife talked about how he felt about his work at the State Department:

68 Letter from Rudolf Smend to Otto Kirchheimer dated 10 June 1947. Rudolf Smend Papers. Staats- und Universitätsbibliothek Göttingen, Nachlass Rudolf Smend, Cod. Ms. R. Smend A 441.

69 Handwritten note by Eugene Anschel from 1985 to John H. Herz. John H. Herz Papers, Folder 58.

70 Letter from Otto Kirchheimer to Rudolf Smend dated 4 May 1948. Rudolf Smend Papers. Staats- und Universitätsbibliothek Göttingen, Nachlass Rudolf Smend, Cod. Ms. R. Smend A 441.

71 For more about this attempt to have Kirchheimer appointed, see Günther (2004, 161–163).

72 Letter from Otto Kirchheimer to Rudolf Smend dated 9 June 1951. Rudolf Smend Papers. Staats- und Universitätsbibliothek Göttingen, Nachlass Rudolf Smend, Cod. Ms. R. Smend A 441.

"he got a well-paid job—but as soon as he had a chance, he left."[73] His daughter used even stronger words: "He strongly disagreed with American foreign policy. And he hated this daily job and only did it for economic reasons."[74] Kirchheimer had various reasons for disliking his work at the State Department.

For one thing, he was annoyed by the bureaucratic procedures in his everyday work at the State Department and by compiling and painstakingly revising analyses and recommendations, which he and many of his colleagues increasingly felt to be pointless, only to see them either shelved or not read at all by the upper echelons.[75] As he wrote in a letter to Rudolf Smend, he felt "marginalized; sometimes you can prevent something stupid, but that's all."[76] John H. Herz reported that Kirchheimer could not bring himself to vote for the Democrats in the 1948 election because he rejected the "bourgeois-capitalist system"[77] governed by Truman. When Republican Dwight D. Eisenhower took office in the White House and John Foster Dulles at the State Department in 1953, Kirchheimer felt even more strongly opposed to his job than before.

A second motive for his desire to change jobs was being able to work freely and without being censored as an academic. He wanted to work on subjects he himself was most interested in; for this reason, he had begun collecting material on political justice. In addition, he did not want to have to obtain permission from his superiors at the State Department for his academic publications (see Herz 1989, 13).

The third reason for his dissatisfaction, and the one burdening him personally most heavily, was his recent harassment by the Federal Bureau of Investigation (FBI). Kirchheimer had already experienced this in 1940. As US Senator Joseph McCarthy went after real or alleged communists, the intelligence services again focused their attention on Kirchheimer in 1948.[78] The immediate reason was his first wife, whose name after remarrying was Hilde Neumann. After their divorce, Kirchheimer avoided direct contact with her to the extent possible. He had received custody of their daughter Hanna and met Hilde only during her brief visits from Mexico to see her daughter and to reach an understanding with him about her schooling.[79] In April 1947, Hilde Neumann returned to Germany. She went to Berlin and rose to become one of the key figures alongside Hilde Benjamin[80] in establishing the justice system in the Soviet zone of occupation and the German Democratic Republic.

From then on, Kirchheimer took the utmost care not to be associated with his ex-wife. Except for questions of custody, he maintained no direct contact with her,

---

73 Anne Kirchheimer in a conversation with political scientist Frank Schale on 6 October 2002 (Schale's personal notes).

74 Hanna Kirchheimer-Grossman in a conversation with the author on 10 September 2021.

75 A number of Kirchheimer's colleagues said the same, see Erd (1985, 151–182) and Söllner (1986b, 30–33).

76 Letter from Otto Kirchheimer to Rudolf Smend dated 13 February 1949. Rudolf Smend Papers. Staats- und Universitätsbibliothek Göttingen, Nachlass Rudolf Smend, Cod. Ms. R. Smend A 441

77 Herz quoted in Söllner (1986b, 45).

78 On the role of the CIA and the FBI in the McCarthy years, see Jeffreys-Jones (1989, 137–170).

79 Anne Kirchheimer in a conversation with Frank Schale on 6 October 2002 (Schale's notes).

80 Hilde Benjamin was Walter Benjamin's brother Georg's wife who was murdered in concentration camp Mauthausen in 1942.

merely following her career by reading East German newspapers available at the State Department and by talking with friends. For example, Gerhard Kramer, senior prosecutor in Hamburg, told him in 1947 that he had recently met Hilde Neumann in the Eastern sector of Berlin and that she had been "extremely unhappy" about Kirchheimer's work for the US government; she called the US "a pre-fascist country."[81] Kirchheimer was piqued by such communications. He informed Kramer that he would be "most unwilling to see my name mentioned in any way—in private or in public—in any relationship or connection to that of Mrs. Neumann."[82]

An intelligence source in France had told the FBI in October 1948 that Kirchheimer had some links to communists in East Germany and had also had contact with communists during his exile in Paris.[83] As a result, investigations were initiated that wore on for a total of almost five years, with interruptions. His colleagues, supervisors, landlords, neighbors, and even the people delivering milk and newspapers to his door were questioned by FBI agents; multiple supervisors and colleagues in the State Department were asked for written reports, and undercover FBI staff attempted to obtain more information.

Kirchheimer himself was summoned to multiple interviews. These investigations weighed very heavily on him. He safeguarded his old SPD membership card as "anti-totalitarian evidence,"[84] akin to a political insurance policy. In the course of the investigations, the FBI confronted him about his contact with Daniel Guérin, a French historian of the French Revolution who sympathized with the anarcho-syndicalists, and had fabricated connections of Kirchheimer's to Trotskyites in the US.[85] It is all the more remarkable how vehemently some of Kirchheimer's colleagues—including Herbert Marcuse, Franz L. Neumann, and Fred Sanderson—emphasized his loyalty to American democracy when they were questioned.

Kirchheimer's FBI file became more voluminous over the years, and the investigations began to focus their attention on Neumann, too. Kirchheimer even registered as a Republican for the 1952 elections "for tactical reasons."[86] The investigations were ultimately discontinued in March 1953 with no incriminating findings. In the end, the FBI was apparently convinced that he was not a communist but a social democrat who did not support the demand to nationalize important industrial companies in the US. On 19 March 1953, J. Edgar Hoover himself signed a note terminating the FBI's investigations into Kirchheimer. Nonetheless, the suspicion of disloyalty dogged him as long as he worked for the State Department. Rumors were circulated that Kirchheimer was se-

81 Letter from Gerhard Kramer to Otto Kirchheimer dated 13 September 1947. Otto Kirchheimer Papers, Series 2, Box 1, Folder 92.

82 Letter from Otto Kirchheimer to Gerhard Kramer dated 8 May 1950. Otto Kirchheimer Papers, Series 2, Box 1, Folder 92.

83 Report: Results of Investigation as of 26 May 1950 (page 10). US Department of Justice, Federal Bureau of Investigation, File on Subject Otto Kirchheimer (unnumbered).

84 Hanna Kirchheimer-Grossman in a conversation with the author on 25 April 2023.

85 FBI Report by Special Agent Patrick M. Rice on Otto Kirchheimer of 21 June 1950, page 4. US Department of Justice, Federal Bureau of Investigation, File on Subject Otto Kirchheimer (121–13351-5).

86 Hanna Kirchheimer-Grossman in a conversation with the author on 15 April 2019.

cretly still in touch with his wife in East Germany—"they didn't trust him."[87] In 1955, he was again subjected to a "loyalty investigation" whose outcome was also in his favor.[88]

5. Kirchheimer's dashed hopes for a socialist democratic Germany

After the OSS had become part of the State Department, Kirchheimer was assigned fixed subject areas. An overview prepared by Neumann listed them as follows: constitutional problems, law, Catholics, trade unions, churches, and the French zone, which included his old home in southwestern Germany (see Söllner 1986b, 12). Kirchheimer traveled to Europe regularly to collect material for the reports on the French zone for the State Department. An important part of his trips was conversations with people on the ground, enabling him to reconnect with many of his old contacts and get to know the younger political actors in Germany at the same time. Even though he found uniforms abhorrent, he needed a US Army uniform to travel in the military zones (see Stiefel and Mecklenburg 1991, 133).

Kirchheimer prepared his initial studies on Germany for the State Department as the breakup of the anti-Hitler coalition was becoming increasingly visible. Analysts had already warned of this before the end of the war because they feared it would put democratic reconstruction at risk. Ever since the Potsdam Conference, it had been virtually impossible to mediate the conflicts. The diverging interpretations of the "eventual reconstruction of German political life on a democratic basis" stipulated in the Potsdam Agreement resulted in rapidly intensifying arguments about reparations and tensions concerning economic and/or political unity. The analysts who had previously worked for the OSS reacted to this constellation with "anti-anticommunism" (Hughes). Although they had no illusions whatsoever about the communists' actions, which they characterized either as authoritarian or as totalitarian, they attempted to prevent the mutual security interests and political decisions in the East and the West from becoming a quasi-perpetual system conflict in order to retain some room for maneuver when negotiating new policies between the two adversaries that were now forming blocs. They failed miserably, however

In August 1946, Kirchheimer presented the report *The General Character of a Future German Government*.[89] He welcomed the American demand to retain German unity as well as the necessity to nationalize businesses, which was to be seen as implicit agreement with the land reform in the Soviet zone of occupation that was beginning. In his report, he discussed four general guidelines for a future political system for Germany: federalism, electoral law, parliamentarism, and basic rights. In light of the Basic Law of the Federal Republic of Germany, it is breathtaking to see how many structural elements of the new constitution were already formulated in Kirchheimer's report.

87 Anne Kirchheimer in a conversation with Frank Schale on 6 October 2002 (Schale's notes).

88 Report on Otto Kirchheimer via Army Courier of 20 January 1955. US Department of Justice, Federal Bureau of Investigation, File on Subject Otto Kirchheimer (100–400640-2.1).

89 See Kirchheimer (1946a). The following page numbers refer to this text.

Overall, Kirchheimer's considerations aimed to achieve a stable democracy. This included, among other things, clear conditions for peace that were to give a certain measure of self-confidence back to the Germans. At the same time, a certain skepticism about granting sovereignty rights too quickly was evident in passages indicating that social democracy alone "fully guarantees liberal-democratic policies in the Western zones" (341), whereas both the Communist Party of Germany (KPD) and the bourgeois parties—with the exception of the Christian Democratic Union (CDU) in Greater Hesse and Berlin—were ultimately "authoritarian parties" (341). The reasons behind making exceptions of the Berlin and Frankfurt local CDU parties were probably a combination of the role of former members of the resistance and the fact that majorities there supported the concept of Christian socialism. Kirchheimer pushed the US authorities to help form political coalitions of Social Democrats and the "left or progressive wing within the middle-class parties" (341), stating that "so far only one major party (the Social Democratic) has emerged which fully guarantees liberal-democratic policies" (341).

Here, we need to look at Kirchheimer's precise understanding of "liberal-democratic policies" in his report. He listed freedom of opinion and of assembly and the right to protection against arbitrary arrests and punishment, but not property rights. Unlike Schmitt with his fundamental criticism of federalism in his Weimar writings, Kirchheimer did support the idea of a centralized government, albeit with a federalist component similar to the solution found in the Basic Law. Kirchheimer was critical of the political system of the US, writing "it has been demonstrated in other federally organized countries like the US that under modern conditions major state functions increasingly devolve upon central government" (342). In a letter written from Washington to Herz and Gilbert in Wiesbaden, Kirchheimer claimed their preliminary reports "confirmed our thesis [...] that decentralization as provided by the Potsdam Declaration, really amounts to centralization on a regional level."[90]

Another contrast to Schmitt's Weimar writing as well as to the political system of the US was Kirchheimer's advocacy for a parliamentary system and against a presidential one. A future president was to be restricted to purely representative functions with no real power to influence politics. Responding to the anticipated objection that a parliamentary system would be unable to defend itself against totalitarian parties, he wrote, "the danger of power being captured by a totalitarian or otherwise undemocratic party or group has traditionally derived from too much concentration of power in the executive, rather than in the legislative branch" (347). Nonetheless, Kirchheimer opposed both an overly comprehensive right of the parliament to self-dissolution and a vote of no confidence that was too easy to accomplish. He wanted to counter these by means of a mandatory waiting period between the proposal of self-dissolution and the vote on it and by requiring higher majorities for certain parliamentary decisions. Strangely enough, he did not go into the "constructive vote of no confidence" proposed by Ernst Fraenkel at the end of the Weimar Republic.

Kirchheimer also argued against the single-member constituencies electoral system of the US and UK. Proportional representation had not been the reason for the collapse

90 Letter from Otto Kirchheimer to Felix Gilbert and John H. Herz. Entry 81, Box 3, Record Group 226, Records of the Office for Strategic Services, National Archives at Maryland.

of the Weimar Republic and a single-member system would work out undemocratically by overrepresenting conservative middle-class parties. Kirchheimer's proposals already went in the direction of the system later used in the Federal Republic: a moderate electoral threshold and having single candidates elected by relative majorities in small election districts while utilizing remaining minority votes for additional minority representation. To Kirchheimer, such questions of electoral law were less important than the actual problem: "the internal structure of German political parties" (348).

In the report, the basic rights were initially defined as negative rights protecting against arbitrary acts by the executive. Kirchheimer took up the debate conducted during the Weimar Republic about binding all laws to the basic individual rights in the constitution. This time, however, he came up with a different answer than twenty years earlier. He argued for immunizing basic rights (personal freedom and integrity, freedom of expression, and freedom of assembly, among others) against the legislature. This is where his trust in the future parliamentary legislature had its limits. The greater significance of the basic rights compared to the Weimar Republic led him to support a constitutional court.

Another point is striking in the context of basic rights. The legislature was not to be bound equally by all basic rights because "the condition under which economic reconstruction in Germany must proceed will not permit absolute protection of private property" (349). This relativization of the alleged basic right to property echoed Kirchheimer's 1930 studies on Article 153 of the Weimar Constitution when he had crossed swords with Schmitt about the question of expropriation without compensation.[91] The nationalization of private property, Kirchheimer asserted, was necessary for pragmatic reasons of rapid reconstruction, too. Moreover, large sections of the population would support such measures. Absolute protection of private property rights would serve only "reactionary forces as the remnants of the Junker class" (349). It is obvious that Kirchheimer wanted to leave the door open for a future democratic socialist Germany after the next elections.

A comparison of the text with Kirchheimer's works written for the OSS before 1945 reveals what the report does not mention: the liberal and emancipatory forces of democratic reconstruction he had previously invoked combined with the recommendation to give political responsibility back to the Germans as soon as possible. Neither is even mentioned here. Nonetheless, it is noteworthy how strongly Kirchheimer's deliberations and recommendations anticipated the decisions taken two or three years later by the Parliamentary Council about the Basic Law which had emerged from the three Western zones. Kirchheimer had had personal contact with several Social Democrats who were later members of the Parliamentary Council during his visits to Germany from 1947 on. Besides Ludwig Bergstraesser, Fritz Eberhard, and Otto Suhr, he had close friendly ties in particular to Carlo Schmid, the chief negotiator of the Social Democratic members of the Parliamentary Council. However, the extent to which Kirchheimer's ideas found their way into the consultations for the Basic Law through these channels cannot be determined today. According to John H. Herz, it was Kirchheimer's close relationship with

91 See Chapter 3, p. 86–91.

Carlo Schmid that helped create the mixture of centralism and federalism in the German Basic Law.[92]

The report *Current Political Tendencies in Germany Bearing on Its Future Governmental Structure*[93] from the late summer of 1946 reflects Kirchheimer's skeptical interpretation of the first year of occupation, based on his view that denazification policy was entirely misguided. The Americans' early efforts had at least accomplished the goal of removing Nazis from public administration, the considerable shortcomings in mechanically reviewing the questionnaires notwithstanding. Conversely, transferring the authority for denazification to

> German boards (purge tribunals) [... has] met with such hostility and resistance on the part of what amounts to a coalition of Nazi sympathizers and conservatives, that frequently they have been prevented from functioning altogether. Intimidation of tribunal members through threats of economic boycott, social ostracism, and personal violence has been reflected in extremely lenient sentences imposed even upon active Nazis, many of whom are permitted to rescue the administrative posts which they were forced to vacate under US procedures (355).

Such failures to purge the administrative bureaucracy had the effect of "re-Nazifying" the German public administration. Regarding the British occupation zone, Kirchheimer took up the concern mentioned in older OSS reports that the Western Allies might abandon denazification in light of the adversities involved in reconstruction and the looming confrontation between the Western democracies and the Soviet Union. He criticized the fact that the British had prioritized functioning public administration over systematic denazification and had used the bureaucracy as the technical guarantor of reconstruction, thereby continuing to apply the civil service law of 1937. He thought this was the "opposite of the desired effect" (354) measured against the goals of denazification.

The main subject of the report, however, was the elections and the party system emerging in the Western zones of occupation. Because of "considerable political apathy" (355), notably among the young generation, older actors were dominating that process, apparently aiming to reestablish the party system of the Weimar Republic. Kirchheimer paid special attention to the ethnic Germans expelled from Eastern Europe who accounted for a considerable fraction of voters in Bavaria. Their anti-Bolshevism would "constitute an influence toward rightist radicalism," which would strengthen a "growing trend toward a revived Nazism" (352). His assessments were prompted by the election to the constitutional assembly in Bavaria in June 1946 in which the CSU received a surprising 58.3 percent of the vote and the SPD only 28.8 percent. Kirchheimer saw continuities with the Weimar Republic here, using the last regular elections to the Bavarian *Landtag* in 1928 and 1932 as benchmarks. He assumed that the voters supporting the conservative and Christian parties at the time had cast their votes for the newly founded CSU, whereas the Social Democrats received almost the same share of votes. But where

92 John H. Herz in his response to Wilhelm Hennis at the Kirchheimer symposium in Berlin in November 1985.

93 See Kirchheimer (1946b). The following page numbers refer to this text.

were the NSDAP voters—6.3 percent in 1928 and 32.5 percent by 1932? Kirchheimer presumed that the overwhelming majority of them had voted for the CSU, prompting him to formulate the hypothesis that it would not be the SPD but rather the CDU/CSU that would play the key political role in postwar Germany.

Another piece of Kirchheimer's research from the first two postwar years was the length of a short book. In late 1947, he completed a monograph titled *A Constitution for the Fourth Republic* (see Kirchheimer 1947). It did not bear his name. In it, he described extensively the circumstances of the French constitutional process and assessed the constitution of the Fourth Republic in detail. Kirchheimer explained the substance of the constitution as the result of a political compromise between the strong social groups in the country, just as he had already explained the Weimar constitution. Parts of his analysis of the provisions of the French constitution read like a discussion of Schmitt's critique of the Weimar Constitution (see Schale 2011). He strictly rejected Charles de Gaulle's attempt to establish a presidential system, instead welcoming the stronger position of the legislative branch.

The book is also instructive in that Kirchheimer devoted almost three times as much space to the first—rejected—draft of the constitution as to the one that was ultimately adopted. The reason for this odd imbalance was likely Kirchheimer's sympathy for the parliamentary system of government, which the political left emphasized even more strongly there. He criticized that in the course of the consultations about the constitution, the new draft presented by the French Christian Democrats, in which the president was granted an at least symbolically more significant role, had prevailed. In Kirchheimer's view, future constitutional conflicts between the president and the parliament were virtually inevitable, and he doubted that the new constitutional order would be stable.

Overall, the reports from 1946 and his book on France document the beginning of the political disappointment Kirchheimer felt, despite their factual tone. He thought that the Western policy regarding occupation and the German resistance against denazification—Kirchheimer even spoke of sabotage—benefited conservative parties and the restoration of prewar capitalism and reactionary politics. At the same time, the democratic left in the Eastern zone was degraded to meaningless "figureheads" of the Socialist Unity Party of Germany (SED), either through pressure or "'natural' attraction" (Kirchheimer 1946b, 353). As early as the late summer of 1946, little was left of Kirchheimer's hopes for a socialist and democratic new order in Germany, despite his attempts to leave the door open for such a development in his recommendations. Nor did he have great hopes for the socialists succeeding in France. He was even more disappointed that the judicial system in Germany had no real new beginning.[94] Kirchheimer's colleagues at the State Department were aware of his increasing political disappointment. Looking back, his longstanding supervisor Eugene N. Anderson said in an interview about the years 1945 und 1946, "I think that Neumann and Kirchheimer expected too much too fast. If I have any criticism of the [...] émigré scholars, then that they expected results too eagerly, too rapidly. Things don't happen that quickly. I have Kirchheimer in mind in particular."[95]

94 On the transition of the judicial system in Germany between 1943 and 1948, see Lahusen (2022).

95 Quoted in Söllner (1986b, 3-).

Kirchheimer intensified his efforts to leave the State Department for a job in academia. But it was exceedingly difficult for him to take this route. From the mid-1940s on, he had applied unsuccessfully for multiple teaching positions at various colleges and universities in the northeast US; but his only teaching experience had been a job for one semester as a visiting lecturer at Wellesley College in Massachusetts, where he had taught two courses in social change and social theory.[96]

Out of the blue, he saw an opportunity to obtain a professorship in 1948 and was confronted with the serious possibility of returning to Germany at least temporarily. After 1945, a number of former students of Carl Schmitt's had become professors in West Germany. Those who had been forced to emigrate after 1933 had a more difficult time of it (see Stolleis 2012, 40–42). In April 1948, the Hessian Minister of Culture and Education, Erwin Stein (CDU), wrote a letter to Kirchheimer, asking him to "take on a position as visiting professor of public law" at the University of Frankfurt in the winter semester of 1948/49.[97] The position was to begin as a visiting professorship and later be made into a regular chair at the Faculty of Law. Kirchheimer responded and stated his interest. He had become the candidate preferred by the Frankfurt Faculty of Law and supported by Vice Dean Gerhard Schiedermair, with whom he was acquainted from his years as a student with Schmitt in Bonn. Yet instead of receiving an invitation from Frankfurt University and before he could review and discuss this matter sufficiently with the university, his family, and his American employer, the offer dissolved into thin air. Hermann L. Brill was appointed in his place. Kirchheimer learned of Brill's appointment from the newspaper.

Although he held Brill, a courageous socialist resistance fighter against the Nazi regime, in high regard both personally and politically, he felt grossly misled by the course of action taken in Hesse. Outraged, he wrote in July 1948 to Ernst Friesenhahn,[98] who served as Dean of the Faculty of Law in Bonn, that he had heard nothing more about the offer from the university since Stein's letter except for various "newspaper polemics," for which reason he had to assume that he had merely served as a "politically strategic" means in the dispute between the university, the government of Hesse, and the Hessian parliament. The "Brill case" caused quite a stir in the newspapers and the West German university environment in the summer of 1948. The debate centered around academic autonomy and the role returning émigrés were to play in restaffing the universities in the Western zones (see Klingsporn and Wilke 2019, 10–13). Frustrated, Kirchheimer remained at the State Department.

---

96 Otto Kirchheimer, Curriculum Vitae (1952). Otto Kirchheimer Papers, Series 2. Box 1, Folder 1.

97 Letter from the Hessian Ministry of Culture and Education to Otto Kirchheimer dated 6 April 1948. Otto Kirchheimer Papers, Series 2, Box 1, Folder 79.

98 Letter from Otto Kirchheimer to Ernst Friesenhahn dated 13 July 1948. Otto Kirchheimer Papers, Series 2, Box 1, Folder 61.

6. Conclusion: Different disillusions

Schmitt had to bury his hopes of being reappointed professor in Berlin or at any other German university as early as late 1945. As a former prominent Nazi propagandist who rejected any and all self-criticism, he was cast into the role of a pariah by official post-Nazi Germany. At the same time, he cultivated this role as an outsider who had not given in to what he called Jewish-American rule over Germany and started a lively correspondence with friends, political companions, and old colleagues that ultimately led to an "invisible college" in the 1950s. Kirchheimer too was active in creating a communicative network with former friends and political allies. Although Schmitt and Kirchheimer had no direct contact until November 1949, they did hear about each other indirectly from other people, starting with Schmitt's greetings to Kirchheimer via Flechtheim in 1947. The only person who corresponded with both Kirchheimer and Schmitt during this time was Rudolf Smend.

Kirchheimer began to visit Germany in 1947. Besides a number of old friends from the Social Democratic Party, he soon got in touch with his former mentor Rudolf Smend and stayed in contact with him. He did not, however, contact Schmitt even though it would have been easy for him to visit him in the American zone. Kirchheimer spent most of his professional energy between 1946 and 1948 on memoranda, reports, and short books for daily administrative use by the State Department. After the loss of his prestigious professorship, Schmitt finished his book on international law and wrote a few essays. The establishment of the Federal Republic of Germany in 1949 guaranteed freedom of expression and lifted the Allies' ban on former Nazis publishing, and Schmitt's articles appeared almost immediately thereafter.

Nevertheless, the years between 1946 and 1948 again mark a number of parallels between Kirchheimer and Schmitt. Today, some of their writing from those years read like an indirect dialogue under the condition of personal absence. Both had realized the extent of the full defeat of Nazi Germany in 1945. Whereas Schmitt was imprisoned twice, Kirchheimer became part of the US State Department, planning denazification and the rebuilding of a democratic Germany. In Kirchheimer's eyes, Germany's future could be managed in a positive way only if the Germans were to develop an honest attitude to the war crimes and domestic crimes committed during the Nazi era. Schmitt's denial was typical of many Germans of his time but he took the denial to its extremes. His writing and private notes lacked any word of empathy for the victims of the regime. He replaced his lack of compassion with self-pity, speaking of his "persecutors" (Schmitt 1950a, 63). He stated that the US State Department had transformed the means and methods of the justice system into means and methods of annihilation against people like him. To Schmitt, it was the expression of a policy of collective guilt against Germany. As a matter of fact, however, the US government and administration in occupied Germany never pursued such a policy. Kirchheimer in particular argued strongly against the assumption of a collective guilt of all Germans in his legal opinions.

Schmitt's antisemitism remained as intense as ever, albeit now coupled with tearfulness and paranoia. In his view, the returning émigrés were either traitors and opportunists or Jews who wanted to take revenge, enrich themselves, or go after him directly. Owing to his position at the OSS and State Department, Kirchheimer was aware of the full extent of German war crimes at an early stage. After being only partly successful with his suggestions for the Nuremberg Trials, he began to devote his energy to proposals for the denazification and the governance of occupied Germany. His main concern was the denazification of the German judicial system and civil service. Following the logic of his recommendations, anyone who had supported the regime as much as Carl Schmitt—of course, without mentioning his name in this context—was to be suspended from any job in the judicial or academic system in a new democratic Germany for the rest of his life. Whereas Schmitt saw such measures as Jewish revenge, Kirchheimer recommended them as a necessary element for building democracy in Germany and as a protection against an authoritarian backlash. It is not without irony that Kirchheimer was confronted with suspicions and ten years of political observation by the FBI at a time when Schmitt characterized the small group of émigrés as the real political power holders in the US administration in occupied Germany.

Whereas Schmitt strongly felt he was on the side of the defeated, Kirchheimer did not see himself in the glorious position on the victorious side. It took him less than a year to realize that prospects for a socialist and democratic new order in Germany were diminishing. Instead, the American military administration safeguarded capitalist private property, and large numbers of former Nazis were given the opportunity to return to their positions in the judicial system and civil service. Kirchheimer had no sympathy for the East German Socialist Unity Party, and his pessimism grew after the communists destroyed the liberal basic rights and established their dictatorship. He became increasingly disappointed politically and experienced his daily work in the State Department as frustrating. A comment of his in a 1961 essay on the relationship of expertise and politics can also be read as a bitter stocktaking of his work at the State Department:

> How are we to evaluate the costs of modern military, paramilitary, diplomatic, and intelligence agencies, where it is questionable whether conclusions can be drawn as to the relation between input and output? [...] There is no proof that this or that form of political propaganda or of military preparation has brought the desired success. [...] And it is exactly in those areas where the relationship between input and output cannot be reliably determined that new projects abound and bureaucratic proliferation flourishes (Kirchheimer 1962c, 376).

Evaluating the success of Kirchheimer's work at the State Department depends on the yardstick used. If it is the creation of a democratic and socialist Germany, he obviously failed. Neither the US administration nor a decisive majority of voters in West Germany wanted to take crucial steps in that political direction. But if we orient our evaluation toward the normative and institutional elements of the West German constitution, put into effect two and a half years after Kirchheimer's recommendations, it is amazing to see how many of his proposals can be found in the Federal Republic's Basic Law that are still valid to this day.

Chapter 15: Renewed Contact and Controversy (1949-1956)

The year 1949 saw the establishment of two German states, the Federal Republic of Germany (FRG) in the West and the German Democratic Republic (GDR) in the East. Each state claimed to be the only one representing German statehood and both constitutions provided for their counterpart's accession. Initially, this changed little in the lives of Kirchheimer and Schmitt. The latter had given up hope of being able to return to a German university. He now focused on expanding an "invisible college" (van Laak 1993, 209). From the 1950s on, he carefully selected and brought young academics into this circle. Among them were men—and all of them were men—who were to become important in the intellectual history of the Federal Republic of Germany, among them historians, philosophers, and legal scholars including Reinhart Koselleck, Odo Marquard, Hermann Lübbe, and Ernst-Wolfgang Böckenförde.[1] In 1950, Schmitt also began to go on the offensive with multiple publications. He optimistically banked on receiving the civil service pension he was entitled to in accordance with Article 131 of the Basic Law when he turned sixty-five in 1953. With the substantial support provided by Academia Moralis, a special account set up on his behalf by a group of entrepreneurs, he was already living an unconstrained and independent life as a private scholar. In the meantime, Kirchheimer struggled increasingly with his work at the State Department. Despite the fiasco in Frankfurt, he did not want to abandon his desire to be a professor in Germany entirely. His wife Anne opposed this aspiration of his and his cautious attempts to make it a reality; she "didn't want Peter to be raised in Germany. [...] My son should be an American boy."[2]

During his third visit to Germany in the autumn of 1949, Kirchheimer attended the reestablishment of the Vereinigung der Deutschen Staatsrechtslehrer (German Association of Professors of Public Law) in Heidelberg on 21 October 1949; he and Karl Loewenstein were present as guests from abroad. There he met Rudolf Smend and Ernst Friesen-

1 The work and the eminent influence of this "invisible college" are described in detail in van Laak (1993, 179–240).

2 Anne Kirchheimer in a conversation with Frank Schale, 6 October 2002 (personal communication between the author and Frank Schale).

hahn, his friend from Schmitt's circle in Bonn; the two had corresponded closely in the following years, exchanging views about Schmitt multiple times. Kirchheimer also met Social Democratic legal scholars Carlo Schmid, Adolf Arndt, Hermann Brill, and Martin Drath again on this occasion as well. Carl Schmitt had not been invited to Heidelberg for the formal reason that he was not a professor; the same applied to Ernst-Rudolf Huber. Ernst Forsthoff, the other particularly prominent student of Schmitt's, declined to attend. Political reasons were behind Schmitt's exclusion, which Smend had previously explained in a newspaper article: "all too eminent standard bearers of the Third Reich" (Smend 1949, 17) were to be barred from the association. Along with his two Nazi rivals Otto Koellreutter and Reinhard Höhn, Schmitt was explicitly denied membership on a permanent basis in 1950 (see Stolleis 2012, 85–88). Schmitt was outraged both by Smend's article and by his former student and assistant Friesenhahn. Before the Heidelberg conference, the latter had given his inaugural lecture as rector of the University of Bonn on the subject of public law scholars and loyalty to the constitution, with Kirchheimer in attendance. Friesenhahn stated that Article 5, Paragraph 3 of the Basic Law, according to which the freedom of teaching did not release any person from allegiance to the constitution, had been included in the constitution "because of the activities of certain public law scholars prior to 1933" (Friesenhahn 1950, 9). Schmitt rightly felt that Friesenhahn meant him, among others. Contemporaries said that his exclusion from the German Association of Professors of Public Law rankled him more than any other right up until the end of his life (see van Laak 1993, 36–38).

A month after the Heidelberg conference, Kirchheimer visited Schmitt unannounced at his home in Plettenberg. After 17 years, almost to the day, they first saw each other again in person on 27 November 1949. Three and a half years later, in June 1953, they met once more in Cologne. As already mentioned in the Introduction to this book, much has been written in the secondary literature about Kirchheimer's visit to Plettenberg.[3] Various authors have claimed that Kirchheimer visited Schmitt at his home in Plettenberg not only once but several times on a regular basis. This assertion has developed a life on its own and can be found in many scholarly contributions on Schmitt and Kirchheimer as evidence of their close personal ties and their renewed friendship. A closer look at the archival sources, however, reveals a different image not only of their two meetings and their relationship in the postwar years in general but also of the circumstances and political context of those meetings.

After the visit in 1949, they began corresponding again. Only twelve letters have survived in the archives, ten from Kirchheimer to Schmitt, and two from Schmitt to Kirchheimer.[4] Schmitt kept all the letters he received; in terms of a typology of literary estates, he was the "paradoxical case of someone who was chaotic but who never threw anything away" (Mehring 2014a, 526). Kirchheimer rivaled him in terms of the chaotic aspect, but he threw away a lot. Moreover, he sometimes used the backs of letters and envelopes for

3 See, among others, van Laak (1993, 135), Quaritsch (1995, 72), Wiggershaus (1995, 470), Mehring (2014a, 432), Bendersky (2016, 137), and Tielke (2019, 377).

4 Schmitt's letters are dated 6 August 1958 and 12 August 1961. Kirchheimer's letters are dated 4 May 1952, 8 September 1952, 27 November 1952, 28 January 1953, (probably) February 1953, 28 March 1953, 1 July 1953, 25 July 1958, and 4 July 1961.

taking notes. So, whereas it is safe to assume that all of Kirchheimer's communications to Schmitt have survived, most of the letters from Schmitt and greetings Schmitt sent along with his books must now be considered lost. Since Kirchheimer apparently thought that most of his correspondence with Schmitt was not worth keeping and discarded it once he had read it, some of the subjects and events mentioned in Schmitt's letters must be inferred from Kirchheimer's replies. It can be deduced from Kirchheimer's letters to Schmitt that at least five letters he received from Schmitt between spring 1952 and February or March 1953 have been lost.[5] There is little hope that these letters and additional notes Schmitt may have attached to offprints he sent to Kirchheimer may yet be found because Kirchheimer's estate has been thoroughly reviewed and catalogued at the German and Jewish Intellectual Émigré Collections at the University of Albany. Adding the twelve surviving letters and the five (at least) lost ones from 1952/53, we arrive at a total of less than twenty letters from the postwar years. The first documented written contact was from 22 November 1949, when Schmitt sent a copy of one of his essays to Kirchheimer.[6] The first surviving letter is dated 4 May 1952 (written by Kirchheimer as a response to a lost letter from Schmitt), the last one is dated 12 August 1961 (written by Schmitt).

Now, Schmitt was a prolific writer of letters. His papers include between 15,000 and 20,000 letters he received from the early twentieth century until his death in 1985,[7] and he presumably wrote roughly as many letters himself. Compared with the published correspondence between Schmitt and other legal scholars of Kirchheimer's generation such as Ernst Forsthoff and Ernst Rudolf Huber, which run to hundreds of letters each,[8] the number of letters he exchanged with Kirchheimer after the war is small.

Viewed together with material from other archives and information from interviews with people involved at the time, their correspondence sheds light on the facts about their relationship after 1945, most importantly indicating that Kirchheimer visited Schmitt in Plettenberg only once; that they met again only once, namely in Cologne in 1953; and that their correspondence was sparse.[9]

1. Amnesty as amnesia

According to Schmitt's *Glossarium*, one of the first books he read after his release from detention in Nuremberg was, in the autumn of 1947, *The Managerial Revolution* by American

---

5 The lost letters were from early 1952, June 1952, autumn 1952, late 1952, and February or March 1953.

6 List by Carl Schmitt about mailing complimentary copies. Carl Schmitt Papers, Versandliste, RW 265–19600. Neither the copy of the essay nor a note or letter which Schmitt may have attached to it have survived in Kirchheimer's estate.

7 I owe this information to Gerd Giesler, e-mail dated 20 December 2022.

8 The number of letters between Schmitt and Kirchheimer is similar to the exchange between Schmitt and Smend in the same period of time (twenty-one letters). Schmitt and Forsthoff exchanged 148 letters in the period between 1952 and 1965 (the year of Kirchheimer's death).

9 With respect to the visit, my research confirms the information George Schwab provided to Ellen Kennedy, namely that Kirchheimer visited Schmitt in Plettenberg only once (see Kennedy 1987a, 392).

popular political theorist James Burnham (Burnham 1941). Among other things, Burnham gave a functionalist rationale for the existence of political opposition. He claimed opposition was useful in democracies because it offered citizens a way to legally vent their anger and also give the government food for thought. Schmitt called this view "optimism about opposition, optimism that is full of contradictions." He was only able to understand this view within the interpretive framework of his geopolitical ideas about *Raum*, which he had developed from the late 1930s on: "That is still a maritime empire's notions about *Raum* and about opposition, an empire whose social group in power can divide the riches of the Earth among itself, with a free ocean."[10] This was based on Schmitt's theory that Great Britain was not a traditional state in the continental sense, but a maritime empire (see Schmitt 1956, 59–65). In addition, he presumably thought that Great Britain, as a predatory maritime country, could afford to allow an opposition to exist. With this dictum, Schmitt directly linked up with his writing on parliamentarism from the 1920s in which he had criticized the idea of political equal opportunity of government and opposition as "metaphysics" of notions of balance (see Schmitt 1923a, 41). Conversely: a political system with a real opposition, he claimed, had to remain alien to Germany's continental existence.

A political system had been established in the Federal Republic of Germany that was based on securing personal liberties and thus on the institutional guarantee of political opposition. As Carlo Schmid, one of the fathers of the constitution, wrote in his memoirs thirty years later, "some [council members] likely also had Carl Schmitt's theory of the state" in mind as an invisible text during the deliberations of the *Parlamentarischer Rat* (Parliamentary Council) on the future Basic Law (Schmid 1979, 335). However, Schmitt's *Constitutional Theory* remained one of his texts that was not to develop formative power (see Lietzmann 1988). During consultations about the future competencies of a constitutional court, Adolf Süsterhenn of the CDU practically implored the gathering with the words: "We are not afraid of the danger of the domination of political decision-making by the judiciary as conjured up by the namesake, spelled with a double "t," of our esteemed colleague Carlo Schmid."[11]

Throughout his life, Schmitt made no secret of the fact that he considered the Federal Republic of Germany unworthy of recognition. In his *Glossarium*, he continued to complain of the "demise of Germany" and the "destruction of identity" of the German *Volk* and called the Federal Republic a "pseudo-sovereign state" that suffered the "existence of a worm in rotten wood."[12] He wrote in his *Glossarium* on 21 September 1949: "In medieval theory, the just war meant that the victor had the right to enslave the subjects of his opponent and to seize his land; today, with more highly organized forms of mass domination it means above all: determining the constitution and regime of the defeated." (Schmitt 2015, 205) He gave an example of what he meant by this in a letter to Ernst Forsthoff after the latter had failed to receive an appointment as a professor in Frankfurt: "What was

---

10 *Glossarium* entry of 16 March 1947 (Schmitt 2015, 32).

11 Speech by Adolf Süsterhenn during the deliberations of the Parliamentary Council on 8 September 1948. Minutes of the meetings of the Parliamentary Council, second meeting, 8 September 1948, 25.

12 *Glossarium* entry of 1 March 1954 (Schmitt 2015, 309).

visited upon you in Frankfurt troubles me as being a case in point of our actual constitution, the *geistigen Morgenthauplan* [the Morgenthau Plan of the mind] under which we must live and find our way in West Germany today. After all, the injustice inflicted on you goes far beyond an individual appointment."[13] The two yea-sayers in 1933 were naysayers in 1949/1950. Unlike most right-wing and conservative public law scholars in Germany, who eventually made their peace with the new state order, Schmitt continued to reject it right up until the end of his life. In his view, it was not even a state, but a pseudo-sovereign entity that would necessarily result in the destruction of German identity and, consequently, the demise of Germany. Even after the Federal Republic had overcome the economic crisis of the 1960s and the new state had lasted longer than the Weimar Republic and the Third Reich put together, Schmitt's rejection and contempt remained unchanged.

In his published works, however, he struck a more moderate tone. During and after the consultations of the Parliamentary Council in 1948/49 about the new constitution of the future Federal Republic, Schmitt published the article *Gegenwartsfragen der Verfassung* [Constitutional questions of our time] and a six-part series of articles *Das Grundgesetz der Bundesrepublik Deutschland* [The Basic Law of the Federal Republic of Germany], both under a pseudonym (see Schmitt 1949c and 1949d). The magazine *Eisenbahnerzeitung* (for teaching and educating railroad staff) where Schmitt's articles appeared had a lay readership to whom Schmitt was to explain the new German constitution, for a large fee. Against Hans Kelsen, he posited that the Reich continued to exist after Germany's unconditional surrender in1945—a view that quickly became the prevailing opinion in West German public law.

Schmitt emphasized that because of its genesis, the Basic Law was not a normal democratic constitution but, rather, a provisional solution—yet he did not bring up his strong concept of willful and revolutionary constitution-making from his *Constitutional Theory* of 1928 to argue against it. He gave a factual presentation of the institutional structure of the Federal Republic according to the constitution—but he did not put his usual hostile stance toward federalism and pluralism in writing here, either. He singled out five innovations in the constitution: first, the great importance of liberal fundamental rights and the protection of fundamental rights; second, Article 21 with its potential for the militant defense of liberal democracy; third, the constructive vote of no confidence in Article 67 which made the Weimar practice of negative votes of no confidence impossible; fourth, Article 139 on the status of occupation, whereby Schmitt considered the denazification measures mentioned there to be a considerable limitation of West German sovereignty; fifth, he devoted considerable space to the competencies of the *Bundesverfassungsgericht* (Federal Constitutional Court) at several points and explicitly called it "the

13 Letter from Carl Schmitt to Ernst Forsthoff dated 20 May 1950 (Schmitt and Forsthoff 2007, 72). The Morgenthau Plan of August 1944 was a proposal propagated by US Treasury Secretary Henry Morgenthau, who was Jewish, to transform Germany into an agrarian state after the war to preclude the country's remilitarization. Kirchheimer, Neumann, and Herz at the OSS emphatically opposed the plan. Although it was never implemented, it played a key role in the propaganda of Nazi Germany in the final year of the war. To this day, the proposal is one of the subjects of right-wing extremist propaganda in Germany against the Western Allies, combined with antisemitic propaganda.

guardian of the constitution" (Schmitt 1949d, 194). In contrast, he downplayed the role of the parliament. Because of the strong position of the *Bundesverfassungsgericht*, Schmitt felt that what mattered most were the individuals serving on it and the procedure for determining its composition (see Schmitt 1949d, 178 and 193).

These comments echoed Schmitt's earlier criticism of the Weimar *Staatsgerichtshof* (see List of German Courts), which he had called an "arena of the pluralistic system" (Schmitt 1931b, 153). They also foreshadow his later criticism of the judiciary in the Federal Republic allegedly dominating political decision-making. He also used his explanations of the constitution for his own ends, stating that it was unconstitutional for the state to massively cut civil service pensions (see Schmitt 1949d, 180). Schmitt's series of articles did not claim to be scholarly legal writing but was designed to provide an overview for readers without a legal background. Overall, he presented himself as a well-informed and factual legal scholar who viewed the Basic Law from the perspective of a comparison with the Weimar Constitution.

Schmitt's entries in his postwar notebook *Glossarium* clearly reject the Basic Law. While he was working on the series of articles, he noted, "Do they still not understand that a Basic Law is today in itself something much more vile than an organizational statute?"[14] On the occasion of the fifth anniversary of the failed assassination attempt on Hitler that took place on 20 July, 1944, he wrote, "When reading the Bonn Basic Law, I am imbued with the amusement of an omniscient old man."[15] A letter to his wife from this time also evidences that Schmitt was not serious about his factual description. He wrote her that his second article about the Basic Law in *Eisenbahnerzeitung* had been faulted for "still being too critical." He closed this passage sighing, "It is difficult to strike the right tone"[16] when writing about the new German constitution.

Readers of the *Glossarium* can trace how Schmitt gradually switched back into a combative mode, all his self-pity notwithstanding. He formulated slogans of German resistance against the victorious powers of the world war and their alleged German lackeys. "We are vanquished, thrown to the ground, subjugated, quartered, and trampled." Yet the German *Volk* was not yet annihilated: "We are occupied, but not conquered. Only he can conquer who knows his prey better than it knows itself."[17] Resistance began with standing up for oneself intellectually. According to Schmitt's *Glossarium*, this effort to resist meant refusing to participate in "constantly and repeatedly churning up the garbage can" of history and not responding to accusations of crimes. Those who demanded such answers from the Germans merely wanted to "enjoy their very personal revenge."[18] Instead, Schmitt asserted, a sweeping amnesty was needed because amnesty was one of "the most foundational positions of that which one can call justice."[19] In April 1949, he wrote the following about the goal of his personal efforts: "For myself and my *Volk*, I am seeking abso-

14 *Glossarium* entry of 25 April 1949 (Schmitt 2015, 176).

15 *Glossarium* entry of 20 July 1950 (Schmitt 2015, 196).

16 Letter from Carl Schmitt to Duška Schmitt dated 5 October 1949 (Schmitt and Schmitt 2020, 321).

17 *Glossarium* entry of 14 March 1948 (Schmitt 2015, 86).

18 *Glossarium* entry of 4 December 1947 (Schmitt 2015, 43).

19 *Glossarium* entry of 5 December 1947 (Schmitt 2015, 43).

lution from the crime."[20] This was the complete opposite of what Kirchheimer imagined to be the only practicable way to make Germany a country deserving respect again.

In November 1949, Schmitt intervened in the contentious public debate about dealing with Nazi criminals in the Federal Republic with his article "Amnesty or the Force of Forgetting,"[21] which was published anonymously. It appeared in *Christ und Welt* [Christians and the World], one of the most widely circulated weekly newspapers with a conservative Protestant orientation and a large readership among the ruling CDU. Schmitt's piece was reprinted in multiple conservative and right-wing newspapers in the next two weeks. He superimposed an interpretation of an international civil war over all the events of World War II, claiming that that civil war had not come to an end even after the cessation of military operations. The defining characteristic of civil wars was that the other side was treated like criminals, murderers, saboteurs, and gangsters. Schmitt had thus fabricated an interpretive framework, and he placed the denazification procedures conducted by the Allies—in his words: "Denazification was Cold Civil War" (92)—within this. There were only two ways out of such a confrontation: annihilation of the enemy—which Schmitt insinuated was the aim of the communists while failing to mention his fellow Nazis in this context at all—or the "Force of Forgetting" (92). Schmitt's next step was to lecture his readers on the broad historical context. In his interpretation, the lesson from the history of civil wars, ranging from the Peloponnesian Wars to the English Revolution during the lifetime of Thomas Hobbes, was that the conflicts could be brought to an end only with an amnesty.

Schmitt's idea of amnesty included unmistakable recommendations for the debate about policy concerning the past: "The word amnesty means forgetting, and not only forgetting but also the strict prohibition against rummaging in the past and seeking cause for further acts of revenge and further claims of reparation" (92). He believed amnesty was not an act of compassion toward the defeated, nor was it merely a pardon. Amnesty was "a reciprocal act of forgetting" (93). This, according to Schmitt, was the only way to end the Cold Civil War "in a human way" (93). Schmitt's argument went far beyond the calls for amnesty for Nazi perpetrators promoted in the Federal Republic at the time. He did not simply demand that the surviving victims should allow mercy to prevail. No, they should refrain from addressing their personal suffering in public life, let alone demanding compensation for stolen property or brutal treatment. Moreover, in Schmitt's view, the memory of those murdered by the Nazi regime had to be erased, too. Only then could there be a new form of peace. Schmitt's call for amnesty entailed complete amnesia concerning the atrocities of the past.

2. Evaluating the new West German democracy

After Kirchheimer was forced to acknowledge that his efforts toward targeted denazification in the Western zones had come to nothing, his latest attempts to obtain a permanent position at a university so that he could leave the State Department also failed.

20 *Glossarium* entry of 4 April 1949 (Schmitt 2015, 173).

21 See Schmitt (1949a). The following page numbers refer to this text.

All he managed to acquire was temporary teaching appointments at two colleges near Washington.

His main responsibility at the State Department was to prepare internal reports which were published only in exceptional cases. In September 1949, he completed his first longer analysis of the Federal Republic. Co-authored with Arnold Price, a State Department expert on European geography, it provided an interpretation of the results of the election to the first German Bundestag on 14 August 1949 (see Kirchheimer and Price 1949). It was the basis for Kirchheimer's essay "The Composition of the German Bundestag," written in the autumn of 1949 and first published in the journal *Western Political Quarterly* in 1950.[22] It is certainly worth reading this article because it shows two things with respect to a comparison with Schmitt. First, that Kirchheimer, too, viewed the events in the newly established Federal Republic through the lens of comparison with the Weimar Republic. And, second, that in contrast to Schmitt, who merely describes the normativity of the constitution, he prepared empirical studies of politics in the Federal Republic of Germany.

Kirchheimer embedded his analysis of the composition of the parliament in a benign assessment of the election result, which was shockingly disappointing for the Social Democrats. If we did not know from contemporary witnesses' letters and memoirs how disappointed Kirchheimer was about Chancellor Konrad Adenauer's victory because he had to acknowledge that the opportunities for democratic socialism in Germany had been laid to rest, we would read part of his election analysis as an approving commentary. Kirchheimer left no doubt that the Bundestag "can justifiably be called a truly representative body" (177). Turnout was 78 percent, only slightly below the average during the Weimar Republic. Schmitt, incidentally, belonged to the small minority who heeded the call of extreme right-wing groups to boycott the election. Kirchheimer's analysis of the composition of the Bundestag aimed to answer five questions: To what extent did the Bundestag indicate a break with the past in terms of its members? What was the role of leading politicians from the Weimar Republic in the new Bundestag? To what extent were former members of the NSDAP represented in the parliament? What business interests were represented in the Bundestag? What role did ethnic German refugees from the East play in the parliament?

In answering these questions, Kirchheimer arrived at some findings that seemed to surprise him. For example, despite the appearance of some of its leading politicians and although two-thirds of its members had already been politically active during the Weimar Republic, the new Bundestag was "no gerontocracy" (178). Concerning the Nazi past of Bundestag members, Kirchheimer stated laconically, "there are fairly large gaps in the information available about them, even though the biographical accounts have been furnished by the members themselves" (180). A small percentage had been employed by the Nazi regime's state apparatus, a larger percentage had held leading positions in the business and academic communities during the Third Reich. The number of those active in the resistance against the Nazi regime was significantly lower. And only 6.8 percent of the members had returned to Germany from exile. Kirchheimer mentioned the high percentages of civil service employees and the members' high educational status as a feature

22 See Kirchheimer (1950a). The following page numbers refer to this text.

already present during the Weimar Republic. "[N]either political and social upheaval nor changes in the electoral system" (183) had been able to change this pattern.

Kirchheimer found surprising continuities from the Weimar Republic in the election outcome, too. Neither the interim phase of the Nazi regime nor the new electoral law had left "any decisive imprint" (190) on the composition of the new Bundestag. Its image was "clearly one of restoration and return to old institutional patterns" (190). How was this to be explained? "[W]hat are the reasons for this *Wiederkehr des Gleichen*?" (return of the same, 191). Kirchheimer considered this to be an expression of younger Germans' lack of interest in politics and a "deep-seated skepticism toward parliamentarism" (191), which reminded him of the crisis at the end of the Weimar Republic.

He was optimistic that because of the political parties comprising the Bundestag, it would presumably establish itself more as a working parliament than as a stage for extremists' radical speeches, and it would therefore be possible to overcome people's reservations against parliamentarism. At the same time, Kirchheimer identified a transition to a parliament composed of interest group representatives. He thereby disagreed with the demand voiced by Dolf Sternberger, one of the founders of political science at German universities, which was widely discussed by the West German public, that the members of the Bundestag were to be as independent as possible from political interest groups (see Sternberger 1950). Kirchheimer called such a return to the epoch of dignitaries as politicians unrealistic and also doubted that members of the Bundestag who were not bound to interest groups would automatically be more independent in their political judgment.

Only when Kirchheimer had conducted an empirical analysis of affiliations with interest groups did his critical view of the early Federal Republic become apparent, for he arrived at the finding that business interests had the strongest representation in the Bundestag: "business stands out with 9.9 per cent" (189). They were followed by interest groups representing ethnic German refugees from the East and the agricultural sector. The representatives of trade union interests and members of smaller professional organizations only ranked "rather low" (189). Kirchheimer described the Bundestag faction of the CDU/CSU, which elected Chancellor Adenauer, as dominated by the interests of agriculture and the major industrial trade associations. Their power corresponded "more to their economic power than to their numerical strength in the country" (192); thus, at the end of his essay, he in effect refuted his statement quoted above that the Bundestag was "a truly representative body" (177).

Reading Kirchheimer's analyses of the early Federal Republic from Schmitt's perspective, one can perceive them as attempts to explore the opportunities of stabilizing the country politically. Like Schmitt, Kirchheimer asked pointed questions about the continuities from the Weimar Republic. However, he emphasized the discontinuities which gave him reason to hope that a policy of social integration would prevent a repeat of the conflict-ridden Weimar era, whereas it seemed obvious to Schmitt that they would develop into a civil war scenario. This was one more reason why Kirchheimer, in contrast to Schmitt, fully acknowledged the legitimacy of the new Federal Republic of Germany.

3. Meeting face to face in Plettenberg

As described in Chapter 14, it was Schmitt who first took the initiative to make contact with Kirchheimer after World War II when he was transferred to Nuremberg by Ossip K. Flechtheim in 1947, asking the latter to convey his best regards to Kirchheimer. There is no indication that Kirchheimer responded. Two years later, Schmitt was able to get hold of a State Department postal address of Kirchheimer's in Germany; how he did so cannot be reconstructed on the basis of the archival material. He may have received it from Werner Weber, Rudolf Smend's colleague in Göttingen, who was close to Schmitt. Schmitt's papers document that he sent a copy of his essay on Francisco de Vitoria to Kirchheimer on 22 November 1949.[23]

A few days later, on November 27, Kirchheimer took Schmitt by surprise and stopped by at his home in Plettenberg for a visit that lasted two and a half hours. The secondary literature mentions very few trustworthy sources concerning this visit and its consequences for their relationship. The famous anecdote, first mentioned by Alfons Söllner (see Söllner 1996, 114), that Schmitt had asked Kirchheimer: "Are you coming as a friend or an enemy?" when he turned up at his doorstep has never been confirmed by the sources, and even Söllner himself has called it into question as a cleverly contrived allegory.[24] In order to shed light on the matter, additional sources have to be taken into account. These include the exchange of letters between Kirchheimer and Schmitt, letters which both of them wrote to third parties, and my interview with Wilhelm Hennis in 2009, with whom Kirchheimer had spoken extensively about his personal motive to visit Schmitt and to stay in contact with him occasionally.

How did Kirchheimer's visit come about? The documents in the archives do not help answer this question. As far as can be reconstructed today on the basis of the additional archival material mentioned above, the visit was a private undertaking during Kirchheimer's third trip to Europe for the State Department. His 1949 trip was longer than those in the previous two years. This time, he served as a consultant to the US High Commission for Germany (HICOG) in Frankfurt from early October 1949 to mid-January 1950, and he used his time in Europe to visit friends in France, the Federal Republic of Germany, and West Berlin. Posing as a private citizen, he visited East Berlin, the newly designated capital of the GDR.[25] His task during this trip to Germany was to prepare a study on the "Structure of present trade union organizations"[26] in the Federal Republic of Germany for HICOG's Office of Labor Affairs. Some of the trade unionists he visited were acquaintances from before 1933; others he met for the first time. Kirchheimer continued to be a member of the SPD and established a number of new contacts among younger

---

23 List by Carl Schmitt about mailing complimentary copies. I obtained this information thanks to Gerd Giesler. This refers to the essay by Schmitt (1949b) which he later integrated into his *Nomos of the Earth*.

24 Alfons Söllner in a conversation with the author, 21 April 2021.

25 Hanna Kirchheimer-Grossman in a conversation with the author, 14 September 2021.

26 Otto Kirchheimer, Curriculum Vitae (1965). Private collection of Hanna Kirchheimer-Grossman (Arlington).

Social Democrats. The daily *Tägliche Rundschau*, which was published in East Berlin, reported on 18 November 1949 on its front page that "a certain O. Kirchheimer" had been in the American sector of the city for some days and was supplying the West Berlin SPD with money from the US.[27] It can no longer be ascertained today to what extent this description was accurate, but it is not entirely implausible.

What exactly is documented about this visit? There are three sources about it from Schmitt. The first is a letter from 29 November 1949 to his wife Duška, who at the time was undergoing medical treatment in Heidelberg. Schmitt reported to her:

> Day before yesterday, Sunday midday, a big yellow American car drove up, with 'USA' in bold letters. I thought I was going to be picked up [for interrogation or the like] once again. Anima opened the door. Who was there? I don't think, dear Duška, that you would guess. It was Otto Kirchheimer. Fat, but otherwise unchanged. We had a good conversation for 2 $\frac{1}{2}$ hours, then he drove on to Düsseldorf. He works for the State Department in Washington and just wanted to see how I was doing. He was not satisfied with *Ex Captivitate [Salus]* because there was no explanation of what I did in 1933. I gave him the essay on [Francisco de] Vitoria. I enjoyed his visit. Incidentally, I don't believe it is very meaningful. It was simply a stirring of human interest in my fate, nothing more. But it was that, and in that sense, it was still nicer than the typical behavior of the German colleagues.[28]

The second source is a letter by Schmitt to Ernst Rudolf Huber, his former assistant professor in Bonn. It is dated two weeks later and reads:

> Do you remember Otto Kirchheimer? He is now with the State Department in Washington. When he visited me two weeks ago, he mentioned the conference of the German Association of Professors of Public Law. We agreed that an outburst of intellectual freedom and dégagé thinking as sublime as the one we experienced in 1930/32 is hardly to be expected again.[29]

Huber wrote back to Schmitt shortly afterwards and reminded him of their political differences:

> Of course, I remember Otto Kirchheimer well. You may remember that we walked through the Tiergarten [park in Berlin] with him in November 1932, on the day of the Berlin transport strike that had been undertaken jointly by the Nazis and the communists. That remains a day of memorable topicality, leading to the abyss, to be precise.[30]

---

27 "Amerikaner halten Westberliner Parteien aus!", *Tägliche Rundschau*, 18 November 1949, p. 1.

28 Letter from Carl Schmitt to Duška Schmitt dated 29 November 1949. Carl Schmitt Papers, RW 265–29926/46.

29 Letter from Carl Schmitt to Ernst Rudolf Huber dated 10 December 1949. In: Schmitt and Huber (2014, 355).

30 Letter from Ernst Rudolf Huber to Carl Schmitt dated 14 December 1949. In: Schmitt and Huber (2014, 356).

The third source from Schmitt is an entry his *Glossarium*. He did not mention the visit itself, but he did mention a few weeks later that he and Kirchheimer had also talked about Ernst Friesenhahn, who had been a student of Schmitt's in Bonn at the same time as Kirchheimer and was now Dean of the Faculty of Law in Bonn. Schmitt noted that he was outraged when he learned from Kirchheimer that Friesenhahn "would not bring himself to visit a person like me [Schmitt]."[31] How trustworthy are the sources provided by Schmitt? His letter to his wife Duška obviously contradicts an entry in his personal mailing list about complimentary copies. According to this entry, he had sent Kirchheimer his essay on Francisco de Vitoria, but according to the letter to his wife, Kirchheimer already had had the opportunity to read *Ex captivitate salus* before he arrived in Plettenberg. The most probable explanation for this contradiction is that Schmitt had made a mistake when recording what he had sent to Kirchheimer in his personal mailing list.

Based on what Schmitt reported to his wife and to Huber, it is difficult to establish Kirchheimer's intention in visiting Schmitt. Was it primarily "a stirring of human interest" in the fate of his doctoral advisor, as the latter reported to his wife? Was Kirchheimer's main interest to find out what Schmitt thought about his own important role in establishing the Third Reich? Did he want to take up the opportunity to discuss this issue directly and in person with Schmitt? Was he primarily concerned with confronting Schmitt with his complete failure to grapple with his role in the Nazi regime in *Ex captivitate salus*, which Kirchheimer had already had a chance to read before he arrived in Plettenberg? Kirchheimer was not the only one to interpret Schmitt's book as proof of his stubbornness und unwillingness to reflect upon his actions.[32] Or did Kirchheimer have a different motive that had less to do with Schmitt and more to do with himself? There are two sources about the visit from Kirchheimer's side, one direct and one secondary.

The direct source is a letter Kirchheimer wrote almost ten years later. In 1958, he was asked by Arvid Brødersen about his personal relationship with Schmitt. Kirchheimer had known Brødersen, who had studied sociology in Berlin, since 1929; they later became colleagues at the New School for Social Research. Kirchheimer's reply was: "I neither saw C. S. in the period between 1932 and 1949 nor maintained any relations with him at all. After the war, when I was in Germany for the US State Department, I spoke with C. S. twice. 1949 and 1953."[33] He also told Brødersen:

> As early as 1949, when he tried to justify his behavior after 1933, I told him that the authority for his actions could only be his conscience. I have held this view from the beginning, especially in 1947 when I heard in Germany that C.S. was sent to a camp. I still think today that nobody should be held criminally or pseudocriminally responsible for their writings or their intellectual production. To a writer, the authority is the

---

31 *Glossarium*, comment regarding the entry of 4 August 1949. Stenographic addition by Schmitt dated 23 November 1949 (Schmitt 2015, 198). The date 23 November is incorrect; the addition must have been made after 27 November 1949. I am not blaming the editors—Schmitt's stenographic notes are extremely difficult to decipher.

32 For the critical comments by Ernst Niekisch, see van Laak (1993, 78).

33 Letter from Otto Kirchheimer to Arvid Brødersen dated 2 March 1958. Otto Kirchheimer Papers, Series 2, Box 1, Folder 25.

> reaction of the audience and their own conscience. The question of employment sanctioned and paid for by the state is of course a different matter.[34]

According to this letter, the two did talk about the brochure Kirchheimer had written in 1935 and which had been illegally distributed under a pseudonym and in the guise of a series of publications by Schmitt that was widely read in the German Reich. Kirchheimer added to Brødersen on this topic that on this occasion Schmitt "told me, [that he] knew that I was probably the author."[35]

The second source about the visit does provide an answer to the question about Kirchheimer's motive. In the 1960s, Kirchheimer spoke with Wilhelm Hennis a few times about his visit to Schmitt. Hennis was a student of Rudolf Smend's and, in those days, a member of the Social Democratic Party. Hennis recounted Kirchheimer's reports about his visit in an interview with the author more than fifty years later in 2009.[36] Concerning the external circumstances—it was an unannounced visit; the big car; the military uniform[37]—the information provided by Hennis basically corroboarates what Schmitt had written to his wife. He reported that Kirchheimer had read Schmitt's *Ex captivitate salus* and was outraged by Schmitt's unwillingness to grapple self-critically with his personal responsibility for the Nazi regime's policies. According to Hennis, whose memories of his conversations with Kirchheimer were permeated with highly interpretive elements, Kirchheimer considered his visit to Plettenberg first and foremost a sign of "*stolze Selbstbehauptung*" ("proud self-affirmation")[38] vis-à-vis Schmitt. By stopping by in Plettenberg on his way to Düsseldorf, Kirchheimer wanted to demonstrate to Schmitt the extent to which the political tide had turned, appearing in the uniform of a member of the American occupying forces and a big car driven by a chauffeur. Hennis's interpretation of this was that Kirchheimer wanted to show that he, who in 1933 had been one of the people Schmitt had wanted to see driven out of Germany once and for all in his Nazi propaganda writing, had succeeded in surviving, and in a dual sense: as a Jew and as a leftist. He had weathered being persecuted by the Nazi regime and had now come back to his home country as a US citizen and in an important position serving the State Department. Hennis had talked about this with Kirchheimer several times, and I find his interpretation convincing.

Kirchheimer's visit made waves. There was "continuing reserve" (Wiggershaus 1995, 470) toward Kirchheimer from the core group of the Frankfurt School after he had mentioned his visit to Theodor W. Adorno. Perhaps—but this is pure speculation—the visit

---

34 Letter from Otto Kirchheimer to Arvid Brødersen dated 2 March 1958. Otto Kirchheimer Papers, Series 2, Box 1, Folder 25.

35 Letter from Otto Kirchheimer to Arvid Brødersen dated 2 March 1958. Otto Kirchheimer Papers, Series 2, Box 1, Folder 25.

36 Wilhelm Hennis in a conversation with the author, 26 September 2009.

37 The fact that Kirchheimer appeared in an American uniform was also reported by Ernst Hüsmert, the administrator of Schmitt's estate, see e-mail from Reinhard Mehring to the author dated 10 March 2019. Hanna Kirchheimer-Grossman insisted that her father disliked any military uniform and that he never had to wear a uniform while he was serving with the State Department (Hanna Kirchheimer-Grossman in a conversation with the author, 14 September 2021).

38 Hennis used this expression in the conversation with the author, 26 September 2009

was one of the reasons why Horkheimer and Adorno did not make the logical decision to ask Kirchheimer to be involved in reestablishing the Institut für Sozialforschung (IfS), which was preparing to relocate in Frankfurt. Schmitt, on the other hand, immediately wrote to his wife and Huber and proudly told his circle of friends and followers about it. The visit did not change his attitude toward returning émigrés. In June 1949, he had written the following about the philosopher Karl Löwith in a letter to Hans Paeschke, the editor of the journal *Merkur*: "the émigrés are unpredictable and mostly potentially deranged in the moral sense."[39] After Kirchheimer's visit, Schmitt also expressed his outrage about the way he felt he was being treated again. The conservative legal scholar Friedrich A. von der Heydte, whom Schmitt had dismissed at Cologne University in the spring of 1933 because he was a student of Hans Kelsen's, had criticized Schmitt's return to the public eye in the Federal Republic of Germany. On this occasion, Schmitt wrote in a letter to Ernst Forsthoff four days after Kirchheimer's visit: "Never in the 12 years of the Hitler period was such a heinous and spiteful act committed against a Jewish colleague."[40] A few weeks after Kirchheimer's visit, on 12 January 1950, Schmitt entered the following in his *Glossarium*:

> When we began to disagree, the Jews sub-introduced.[41] Today, these people who had sub-introduced themselves are experiencing restoration with colossal claims for restitution and repayments. But still, the sub-introduced are even worse than the returning émigrés who relish their revenge. They should be ashamed of accepting the dollar (Schmitt 2015, 221).

Two months later, on 17 March 1950, he commented about returning émigrés: "Those who did emigrate are declaring those who did not to be enemies of the country" (Schmitt 2015, 226). Schmitt's militant hatred of Jews and émigrés was clearly still as virulent as ever.

4. Schmitt's return to the public eye

Schmitt had been banned from publishing under Allied occupation, but that ban was automatically lifted with the founding of the Federal Republic of Germany in 1949. He immediately contacted various publishers and, in the span of only a few months, he was able to present four monographs to the public under his own name: *Die Lage der europäischen Rechtswissenschaft* [The situation of European legal scholarship], *The Nomos of the Earth*, "A Pan-European Interpretation of Donoso Cortés," and *Ex captivitate salus*. He had already written the first three during the final years of the war. The following almost forty years stand in stark contrast to this flood of publications, as there is nothing really new in Schmitt's postwar oeuvre after 1950. Virtually without exception, Schmitt merely took

39 Quoted in van Laak (1993, 149).

40 Letter from Carl Schmitt to Ernst Forsthoff dated 1 December 1949. (Schmitt and Forsthoff 2007, 59).

41 The verb "*subintroduzieren*," which Schmitt uses here, does not exist in German. As Schmitt used it in this particular context, it can be assumed to mean: "to come in and assume a leading position in place of us."

up and pursued motifs from Weimar and deliberations from the Nazi period and interpreted and commented on his own work.

The booklet *Die Lage der europäischen Rechtswissenschaft* was based on lectures he had repeatedly given abroad in 1943 and 1944 and reworked in 1950. He continued to believe that the liberal concept of the law was unraveling, but now for a different reason than in his other writing after 1933. Because of the war, legislative procedures had been expedited in all European states, and new laws dealt with more issues, so legal scholars were stripped of opportunities to provide input. The task of guiding the economy had further accelerated legislative processes in the modern interventionist state. In light of this, Schmitt spoke of a "motorized legislature" (Schmitt 1950b, 404). He asserted that the law was "transforming itself into a means of planning, and the administrative act into an act of guidance" (Schmitt 1950b, 407). What options were there in this situation, which Schmitt felt was critical for legal thought? One way out was to draw on Romantic legal theorist Friedrich Karl Savigny and his early nineteenth century doctrine of sources of the law. The latter stated that "law as a concrete order" was not set out intentionally—as it was later also understood in the positivism criticized by Schmitt—but arose in an unintentional development. Only then was it recognized as such by professional legal experts, who proceeded to shape it into systematic forms. Schmitt believed that Savigny's doctrine was particularly topical because it formulated an antithesis to the mechanization of the law. Turning to Savigny would enable legal scholarship to "distance itself" (Schmitt 1950b, 414) in a reflective manner from the legality of the state based on laws.

This text can be read as a modification of his concrete-order thinking at the beginning of the Nazi regime. Despite all his talk of a concrete order, Schmitt had been unable (or unwilling) to provide a substantial criterion for distinguishing an order that was concrete in the positive sense from one that was merely factually concrete. By referring back to Savigny's doctrine of sources of the law, Schmitt overcame this shortcoming using institutional means: it was legal scholarship that decided what was the concrete order, and thus the law, in an order that had arisen in an unintentional development. For a law professor who had played a decisive role in legitimizing the destruction of the rule of law in Germany after 1933 to invoke legal scholarship as the guardian of a European awareness of the law, this "bordered on chutzpah" (Neumann 2015, 507).[42]

Schmitt's studies on Shakespeare became his main work in the 1950s, even if he only finalized a few of them, and he did make at least some new points. Schmitt was a passionate theatergoer, and as a *Preußischer Staatsrat* (Member of the Prussian State Council) during the Nazi period, he had had the privilege of a box of his own in the Theater am Gendarmenmarkt, one of Berlin's major theaters at the time, where he had watched the renowned productions of *Hamlet* and *King Lear* multiple times (see Mehring 2021, 241–253). Even in *The Nomos of the Earth*, Schmitt had explained the representations of characters in Shakespeare's dramas with the advent of the "great men" (Schmitt 1950d,

42 The version of his lecture published in 1950 includes a final section that unequivocally no longer bears the signum of the years 1943/44. In that section, legal scholarship is described as the "final refuge of awareness of the law" (420), the validity of the principles of "recognition of the individual based on mutual respect" (422), and "due process of law without which no justice exists" (423).

144) to neutralize the religious conflicts of medieval Europe in the sixteenth and seventeenth centuries. In 1952, he wrote a brief preface to British literary critic Lilian Winstanley's book *Hamlet and the Scottish Succession*, which his daughter Anima translated into German. In his preface to the book, which had first been published in Great Britain in 1921, he self-confidently ignored the criticisms of Winstanley's interpretation that historians had raised at the time (see Höfele 2014, 15–25). Instead, he followed the author's argument and concluded that a play became a tragedy only if there was an "urgent historical presence" (Schmitt 1952a, 168) at the core of its plot. Shakespeare had used Hamlet's character to highlight the constellation of the conflict around the contemporary king, James I, and the audience of the day was fully aware of this. To explain this "theatricalization of one's own historical being" in Shakespeare's play to his readers, Schmitt mentioned the events of the "Night of the Long Knives" familiar to the Berlin theater audience of the summer of 1934.[43]

Four years later, Schmitt published the book *Hamlet or Hecuba: The Intrusion of Time into the Play*. Even the subtitle revealed the author's rejection of the theory of the autonomy of art. In Homer's *Iliad*, Hecuba was the wife of Priam, the last Trojan king, and became Ulysses's slave after her husband and sons had died. Because of her fate, she is considered to be the embodiment of the worst that can happen to a woman in war. In Shakespeare's *Hamlet*, her fate is the subject of a play within the play, and an actor portraying Hecuba must weep when declaiming the death of Priam. Hamlet wonders whether he should weep, too, but does not.[44] Yet Hecuba's fate and the question of empathy with the suffering of others played no role at all in Schmitt's interpretation, the title he selected for the book notwithstanding. Instead, he deciphered two intrusions of time in the play. One was tabooing the complicity of Mary, Queen of Scots, in the murder of her husband. Schmitt believed that in the piece, the murderous mother was to "be left exclusively to her own conscience. Strange revenge drama!" (Schmitt 1956, 14) The second intrusion was Shakespeare's transformation of the figure of the avenger into a melancholic inhibited by unceasing reflection. Schmitt's book made an argument both for the political stakes of art and the continued mythic foundation of politics.[45]

Hamlet or Hecuba was not met with applause from the experts in Shakespeare studies. Schmitt was furious about negative reviews and responded to two critical newspaper reviews of his book with the absurd accusation that the authors had criminalized him as a "disturber of the peace" and an "aggressor" (Schmitt 1957b, 138). Philosopher Hans-Georg Gadamer argued in his critique that he was sympathetic to Schmitt's idea of examining—from the perspective of a historian—how the relationships between the characters in a play were interwoven with the personal and political constellations of the time of its

43 Schmitt used this comment to set the stage for his interpretation of his essay "Der Führer schützt das Recht" [The *Führer* is protecting the law], which he launched soon thereafter and according to which this essay had been a covert and courageous criticism of the murders committed on 30 June 1934, which readers of a later era would be unable to recognize.

44 During the German Empire, the quote from Shakespeare became a well-known saying, but in slightly altered form. "*That's Hecuba to me*," said Chancellor of the Reich Otto von Bismarck in a famous speech before the Reichstag in 1887, seeking to express that he was entirely indifferent about the independence of Bulgaria, which was the subject of fierce public discussion at the time.

45 See Meierhenrich and Simons (2016b, 44–46).

creation. But Schmitt had "underestimated the difficulty of this task" and thereby succumbed to a "false historicism" (Gadamer 1965, 519).

Most contemporary interpretations of Schmitt's book discuss intertextual aspects and Schmitt's theory of tragedy, his aesthetics of reception, or his brief remarks on Walter Benjamin's theory of baroque tragedy.[46] Schmitt has rightly been criticized for developing a criterion of the tragic in his book that only modern tragedy, not classical tragedy, can fulfill (see Heller 2019). In addition, I think Schmitt's interpretation of the play should be understood mainly as an update to suit his own ends. I see an intrusion of urgent historical presence into Schmitt's particular interpretation of the play in two senses. For one thing, his interpretation of the accepted taboo restated his argument for societally agreed amnesia regarding the murders during the Nazi regime. And for another, he made the character of Hamlet a symbol of the European intellectual characterized by the imbalance of thinking and acting and by paralysis through introspection. Hamlet, alias James I, was born "literally from the womb immersed in the schisms of his era" (Schmitt 1956, 27)—no wonder that he became so duplicitous and learned how to deceive his enemies. Schmitt attempted to place himself in the proximity of the threatened intellectual "great reader and writer of books" (Schmitt 1956, 28).

Another *Preußischer Staatsrat*, Gustaf Gründgens, played Hamlet in the theater productions that Schmitt had attended in Berlin. Gründgens enjoyed even greater success in the role of Mephisto in Goethe's *Faust*.[47] What if Schmitt had declared Gründgens's inimitably diabolical performance of Mephisto to be the archetype of the modern intellectual instead of Hamlet? He would have mapped out a much more accurate self-portrait of his own political role using Mephistopheles as a figure from Shakespeare's plays (see Mehring 2021, 242).

Until his *Theory of the Partisan* in 1963, Schmitt did not publish any work as significant as his earlier oeuvre. As he aged, he was increasingly concerned with the "proper" way of reading his work (see van Laak 1993, 67–71). An important element of this self-referential nature of his late oeuvre was the tirelessness with which Schmitt "discovered" the destinies of thinkers from European intellectual history in whom he sought to see the "tragedy" of his existence reflected—among them Niccolò Machiavelli, Thomas Hobbes, Benito Cereno, Donoso Cortés, and Alexis de Tocqueville. Schmitt made literary and mythical figures such as Hamlet and Epimetheus into historical "archetypes" of his own destiny, too. He orchestrated the reception of his works, sending out a large number of copies of his writing and organizing reviews to be written by people in his circles.

There is one exception to be found among his more aesthetic writings and his comments written in retrospect, namely his 1952 piece *Rechtsstaatlicher Verfassungsvollzug* [Execution of the constitution under the *Rechtsstaat*]. Published under his name the same year in the form of an independent brochure, the text is based on a legal opinion he had prepared for the Buderus-Röchling steelworks (see Schmitt 1952b). The company wanted to take legal action before the *Staatsgerichtshof* of the *Land* Hesse (see List of German Courts) to prevent its nationalization under Article 41 of the Constitution of Hesse. In

46 See Höfele (2014), Pan (2016), and the contributions in Telos No. 153 (Winter 2010).

47 Klaus Mann, a son of Nobel laureate Thomas Mann, based his most famous novel *Mephisto*, which he wrote in his Amsterdam exile in 1936, on the career of Gustaf Gründgens.

a sense, working on the legal opinion "reinvigorated" (van Laak 1993, 137) Schmitt again, partly because he could link up to one of the legal topics he had worked on previously, and partly because the elite of constitutional law specialists was involved in the proceedings before that court and later the Federal Constitutional Court. Writing to Ernst Jünger, Schmitt described his own role in the proceedings as bringing "something new to the attention of the judges in great haste after things had been picked apart for two years."[48] In his legal opinion, he did not dispute the legality of Article 41 of the Hessian Constitution but, rather, the legality of the specific instance of nationalization. Schmitt asked whether that article made it legal to directly dispossess property and stated that it did not. Nationalization required a law adopted by the parliament, he asserted, but such a law did not exist, for which reason this instance of nationalization was null and void from the outset. The judges at the *Staatsgerichtshof* of the *Land* Hesse came to a different ruling. They acknowledged the nationalization of some of the companies in the Buderus group but established considerable obstacles for its implementation. In practice, only some parts of the group were nationalized because of transactions within it as well as demands for large amounts of compensation. And overall, Schmitt was pleased to see his opinion prevail.[49] For, in 1965, the government of the *Land* Hesse transferred its own parts of the group back to the private owners. Ultimately, this legal battle became a milestone in German jurisprudence as an important victory of the conservatives in public law—even without Schmitt's help.

5. Kirchheimer as a political scientist

With their decidedly empirical orientation, Otto Kirchheimer's works from the 1950s stood in stark contrast to Schmitt's legal normativism and his interpretations founded in the history of theater. Kirchheimer wrote a number of essays on the transformation of political orders in modern industrial societies. The geographical focus of these studies was on the newly established Federal Republic of Germany as well as on other Western European democracies, the US, and the German Democratic Republic. He was particularly interested in the changes to the party systems, the changing role of the parliamentary opposition, the influence of trade associations and interest groups, the strengthening of the bureaucracy and the executive, and the political attitudes and expectations of citizens in modern democracies.

Some of these works were nothing less than counterpoints to Schmitt; others piqued Schmitt's interest. Among the studies Schmitt read immediately was Kirchheimer's essay "The Waning of Opposition in Parliamentary Regimes,"[50] which was based on lectures he gave in the US and Europe in 1956. It was published in 1957 in *Social Research*, the journal of the New School for Social Research. The German translation by Gurland was published the same year in the *Archiv für Rechts- und Sozialphilosophie* (ARSP)—none other than the

48 Letter from Carl Schmitt to Ernst Jünger dated 24 March 1952 (Schmitt and Jünger 1999, 254).

49 Letter from Carl Schmitt to Ernst Forsthoff dated 2 August 1965 (Schmitt and Forsthoff 2007, 212).

50 Kirchheimer (1957a). The following page numbers refer to this text.

peer-reviewed academic journal of philosophy that Schmitt's younger students had established as a counterweight to the public law journal *Archiv des Öffentlichen Rechts*, which was published by students of Smend's.

First, Kirchheimer offered his readers a typology of various forms of political opposition in parliamentary democracies. The three types, which he described in detail using examples from various European countries, were "classical parliamentary opposition" (392), "opposition of principle" (392), and, as a third "counterconcept" (392), the waning of the opposition, which he observed in the majority of democracies of the day. When sketching out this third type, Kirchheimer drew on the description of constellations of coalitions in Austria and Italy after World War II. He considered Austria to be a model of the "elimination of major political opposition through government by party cartel" (300). He described extensively how it came about that the socialist and the conservative parties, which were roughly the same size, managed to agree on a system of carefully negotiated cooperation, thereby representing almost 90 percent of the electoral votes in total, and he listed the large number of details laid down in their coalition agreements. Kirchheimer used sharp words in his criticism of the Austrian model. Because of the pacts between the two major coalition parties, there was now an "absence of the opposition's control function" (305). He described Italy as a case in which the "opposition of principle," the Italian Communist Party (PCI), was so strong that the other parties had formed a kind of defensive cartel that had led to serious distortions of parliamentary representation.

Besides the forces favoring such cartels, such as the specific party constellations in Austria or Italy, Kirchheimer considered the societal factor driving such coalitions to be the "emergence of a substantial new middle class" (311) in the modern industrialized countries. This continually growing class consisted of skilled manual workers, mid-level civil servants, and employees in very similar economic and social psychological circumstances. Their consumer expectations of constant increases in their material standard of living as well as the services they expected of the state were generally the same. Almost all the political parties considered the new middle class to be important in terms of electoral strategy and oriented their actions and their programs toward this target group which, for its part, expected the political community to quickly reward their electoral votes. These expectations in turn made it completely unattractive for any parliamentary party to take an oppositional role because it could satisfy the needs of its voters much more effectively by being part of the government.

The references Kirchheimer provided for his diagnosis of society were current works by sociologists David Riesman, Helmut Schelsky, and Siegfried Landshut. Unlike them, however, he did not assume that the trend toward a uniform middle-class society would encompass everyone to the same extent. Culturally speaking, modern society may have become a mass society, but socially, it continued to be a class society. Kirchheimer countered the hypotheses of the end of class society by stating that severe poverty and "*classes dirigeantes*" (313) continued to exist—yet they had hardly any opportunities for political representation; he mentioned some kinds of pensioners, low-level employees, and small business owners as those who had suffered most from the societal changes. These were the sections of society most likely to vote for the extreme right-wing "opposition of principle" parties. Overall, however, Kirchheimer believed that to the extent that the political power of a party was measured by the satisfaction of the groups it represented, a parliamentary opposition designed for the long term had lost its projects for the future.

Kirchheimer used a number of metaphors for the "Waning of Opposition" in the title of his essay. The qualifying nouns "desiccation" (300), "erosion" (300), and "vanishing" (300) are to be found in a single paragraph and were intended to designate an irreversible trend. Kirchheimer's conclusion was clear: the cartel-like coalitions in Austria and Italy would not remain exceptions specific to those countries; instead, grand coalitions were to be expected soon in other European countries, too. The trend was the same in almost all postwar democracies: freezing or fencing in any political opposition whose aim was beyond the framework of the existing societal order. This diagnosis was the greatest imaginable contrast to Carl Schmitt's efforts to conjure up political disintegration and civil wars as the inevitable result of granting rights to the opposition and permitting a plurality of parties.

A second focus of Kirchheimer's research was political parties. In comparative political science, a subdiscipline of political science, he is still known to this day for his studies on party typology. His diagnoses are part of the canon of political science, too, and the research literature unanimously credits him with coining the term "catch-all party."[51] Kirchheimer's writing on this subject is an amalgamation of personal observations, theoretical assumptions about changes in the social structure of Western societies, empirical sociology of parties, deliberations from the economic theory of politics, and numerous case studies he learned about from reading various European newspapers and traveling to various Western European countries.

Kirchheimer's interest in questions of party typology had its roots in the Weimar Republic and was sparked by the condition of the Social Democratic Party of Germany (SPD) at the time. The term "catch-all party" occurs for the first time in Kirchheimer's publications in his 1954 essay "Notes on the Political Scene in Western Germany."[52] However, he did not yet use it consistently and as a fixed designation for this type of party, but only sporadically and metaphorically. He coined it in his effort to describe a transformation in the target group orientation of the Christian Democratic Union of Germany (CDU) and the SPD in the electoral campaign of 1953. He called the CDU a "conservative catch-all party" (262) because it had succeeded in winning votes even beyond the traditionally conservative milieus. At this point, he still considered it an open question whether "the SPD [would] develop into a catch-all mass party rather than a democratic working-class party" (263).

Kirchheimer's innovative use of the term can be traced back further, to the intelligence reports about the GDR he prepared for the State Department in 1950. In a report dated 24 May 1950, he wrote the following: "The National Front has become the catch-all organization for political activities emanating from East Germany which are designed

51 See, among a number of other authors, Sartori (1976, 138), Krouwel (2003, 24), Allen (2009, 636), Mair (2013, 82), and Llanque (2021).

52 See Kirchheimer (1954a). The following page numbers refer to this text.

to appeal to a non-communist public all over Germany."[53] Here, the term "catch-all" appeared in an entirely different context; after all, the parties and mass organizations in the GDR had been forced to amalgamate into the *Nationale Front*. In a later memorandum for the State Department about the Bundestag election of 6 September 1953, he wrote "catch-all All-German bloc (BHE),"[54] and in his later analysis of the outcome of the election, "[w]ith the specter of a socialist government fading and potential conflicts within the CDU enhanced by its expansion into a middle class catch-all, centrifugal tendencies may be expected to arise."[55] As for the performance of the Free Democratic Party (FDP), he added, "The FDP has managed to avoid major losses, but the general obliteration of denominational lines robbed the party of any chance of serving as a Catch-all for the Protestant vote." [56] As he used the term, Kirchheimer could apparently envisage multiple possible "catch-alls" for various distinct groups of voters. Since he first used the term "catch-all" in these documents, it should be safe to assume that the wording "catch-all party" came about during coffee break discussions among Kirchheimer's group at the State Department. Kirchheimer developed his new party typology from these initial conceptual exercises within a few years. His deliberations culminated in the 1960s in the essay "The Transformation of the Western European Party Systems" (see Kirchheimer 1966),[57] which was published posthumously.

6. At a distance: More correspondence and another meeting

Kirchheimer's surprise visit to Plettenberg in November 1949 did not revive the close relationship from the late 1920s. Schmitt's estate includes a reprint of Kirchheimer's essay "The Act, the Offense, and Double Jeopardy" (Kirchheimer 1949a), which had been published in the prestigious *Yale Law Journal*, with a personal dedication "with best recom-

---

53 Otto Kirchheimer, *The State of East Germany 1949–50*. Intelligence Report 5230 dated 24 May 1950, p. 12. National Archives at College Park, Maryland. General Records of the Department of State. Record Group 59. Intelligence Reports, 1941–1961 (National Archives Microfilm Publication M 1221).

54 Block der Heimatvertriebenen und Entrechteten (BHE, League of Expellees and Those Deprived of Rights). Otto Kirchheimer, *The West German Election Campaign*. Intelligence Report 6378 dated 13 August 1953, p. 2. National Archives at College Park, Maryland. General Records of the Department of State. Record Group 59. Intelligence Reports, 1941–1961 (National Archives Microfilm Publication M 1221).

55 Otto Kirchheimer, *The West German Bundestag Elections of 1953*. Intelligence Report 6426 dated 6 October 1953, p. 4. National Archives at College Park, Maryland. General Records of the Department of State. Record Group 59. Intelligence Reports, 1941–1961 (National Archives Microfilm Publication M 1221).

56 Otto Kirchheimer, *The West German Bundestag Elections of 1953*. Intelligence Report 6426 dated 6 October 1953, p. 5. National Archives at College Park, Maryland. General Records of the Department of State. Record Group 59. Intelligence Reports, 1941–1961 (National Archives Microfilm Publication M 1221).

57 For a detailed reconstruction of the hypothesis of the catch-all party in Kirchheimer's oeuvre, see Buchstein (2020b, 113–137) and Llanque (2021).

mendations."[58] It is no longer possible to determine whether Kirchheimer gave Schmitt the reprint during his visit or sent it later.

The next documented contact between the two of them is Schmitt's mailing of his booklet *Die Lage der europäischen Rechtswissenschaft* to Kirchheimer in March 1950.[59] He did not get a response. In November 1951, Schmitt sent him a catalog of his publisher's featuring his works.[60] Again, there was no response. In the spring of 1952, Schmitt took the initiative again and mailed him a copy of the legal opinion he had prepared for the Buderus-Röchling steelworks, [61] which sought to take action against its partial expropriation by the government of the *Land* Hesse. He knew that Kirchheimer was definitely interested in this topic since he had published a book and a few articles on the subject of expropriation during the Weimar Republic in which he had contradicted Schmitt. We can now no longer ascertain to what extent Schmitt was also aware that Kirchheimer had argued for far-reaching nationalizations in the memoranda he had prepared for the OSS and was sympathetic to the socialist 1946 Hessian Constitution.

This time, Kirchheimer responded in a letter dated 4 May 1952, his first response to Schmitt two and a half years after his visit in November 1949. He first expressed his condolences on the death of Schmitt's wife Duška. She had passed away about eighteen months earlier so these condolences indicate that the two had had no personal contact since then. In his letter, Kirchheimer went on to thank Schmitt for sending the legal opinion to Washington. However, he commented critically that he could not "fully agree with your differentiation between expropriation through law that was still permissible and [...] expropriation through reinterpretation of the constitution, which was not permissible."[62] This, however, was the main point of Schmitt's line of argument to avoid expropriation, which did not prevail in the trial. Kirchheimer also very briefly commented on Schmitt's book *The Nomos of the Earth*. Since sales of the book had stalled in the autumn of 1952, friends of Schmitt's bought the remaining copies so that he could disseminate them free of charge. He mailed complimentary copies specifically to the US to become better known there (see van Laak 1993, 55). Kirchheimer had apparently also received one of these copies. The only mention of this in his letter to Schmitt dated 4 May 1952 was the vague comment that he had read it "with great pleasure." [63] He closed his letter with the prospect of exchanging views about these subjects in a personal conversation if the opportunity arose.

Schmitt responded shortly afterwards, in June, seeking to discuss the role of the *Bundesverfassungsgericht* in the legal and political system of the newly established Federal Republic of Germany with Kirchheimer.[64] Kirchheimer responded three months later, on 8 September, writing just a few lines. In the meantime, the court in Hesse had rejected

58 Carl Schmitt Papers, RW 265–25658.

59 Mailed on 19 March 1950. List by Carl Schmitt about mailing complimentary copies. I obtained this information thanks to Gerd Giesler.

60 Mailed on 16 November 1951.

61 Mailed on 3 April 1952.

62 Letter from Otto Kirchheimer to Carl Schmitt dated 4 May 1952. Carl Schmitt Papers, RW 265–7598.

63 Letter from Otto Kirchheimer to Carl Schmitt dated 4 May 1952. Carl Schmitt Papers, RW 265–7598.

64 This letter has been lost. The contents can be reconstructed from Kirchheimer's letter dated 8 September 1952.

the claims of the Buderus-Röchling steelworks. Kirchheimer commented favorably on the decision. Regarding the *Bundesverfassungsgericht* in Karlsruhe, he wrote to Schmitt: "sometimes I take a look at the decisions from Karlsruhe; since the legislature did not give much thought to the inherent limits of constitutional jurisdiction when delineating them, the court has to take care not to reduce its entire jurisdiction to absurdity."[65] This statement covers the concern already discussed in the consultations of the *Parlamentarische Rat* (Constitutional Council) that the newly established court would dominate politics in the Federal Republic in the future. This concern also referred to Schmitt's diagnosis of a juridification of politics. In 1953, he noted in his *Glossarium* that the *Bundesverfassungsgericht* was in an unresolvable dilemma. It would either have to avoid all important decisions, thereby calling its own right to exist into question, or become "a breeding ground for apocryphal acts of sovereignty."[66] Schmitt believed the court had opted for the latter. In his letter, Kirchheimer was more positive about the *Bundesverfassungsgericht*. He also called himself "*schreibfaul*" (lazy about writing, i.e., a poor correspondent) and that he preferred reading Marcel Proust and Graham Greene over memoranda from the State Department. He concluded his letter responding to Schmitt's suggestion that they meet in person by agreeing that might be possible in late 1952 or early 1953, although he would be traveling extensively in Europe.

Kirchheimer's longest postwar letter to Schmitt was dated 27 November 1952.[67] It, too, is a response to a letter from Schmitt which has been lost. Schmitt's letter was presumably from October or early November 1952, since in his response to it, Kirchheimer mentioned Schmitt's disquiet about a comment in a devastating critique by Golo Mann of *The Nomos of the Earth* which had been published in the October issue of the magazine *Der Monat*. Schmitt, who in his letters and diaries regularly made disparaging remarks about Golo Mann's father, the "emigrant" Thomas Mann, and his family, once more felt he was being persecuted. Kirchheimer wrote him about Mann's critique, "you need not be particularly unsettled by it—apart from the footnote." In this footnote, Golo Mann had retracted his previous characterization of Schmitt as a "Nazi crown jurist," replacing it with his assessment that Schmitt had interceded on behalf of the Nazi regime coming "from the outside," which he considered no less disgraceful.[68] So once again, it posed the question to what extent Schmitt had been a staunch Nazi in his innermost being while frenetically supporting the Nazi regime; this matter was relevant to the reception of Schmitt.

---

65 Letter from Otto Kirchheimer to Carl Schmitt dated 8 September 1952. Carl Schmitt Papers, RW 265–7599.

66 *Glossarium* entry of 5 February 1953 (Schmitt 2015: 291).

67 Handwritten letter from Otto Kirchheimer to Carl Schmitt dated 27 November 1952. Carl Schmitt Papers, RW 265–7600. The following quotes are from this letter.

68 As already mentioned, the term "crown jurist of the Third Reich" was coined by Waldemar Gurian, following up on Kirchheimer's characterization of Schmitt (see Chapter 7, p. 211–212). The entire text of Golo Mann's footnote reads as follows: "In issue 45 of this magazine, I called Carl Schmitt a 'Nazi crown jurist.' Although Schmitt at times sought to serve Nazism with his technique of thinking, he was too much of an outsider, which is why this characterization is not quite fitting; I hereby retract it for this reason. G.M." (Mann 1952b, 89). This statement by Golo Mann is to be found as an aside in his comments on historian Ludwig Dehio's hypotheses about European hegemonic struggle as the cause of World War II in the June 1952 issue of *Der Monat* (Mann 1952a, 329).

Instead, Kirchheimer suggested that Schmitt regard Mann's review as "an approach for a fruitful criticism—the relationship of your thinking to the question of historical reality."

Kirchheimer took Mann's criticism as an occasion to look back on his essay "Remarks on Carl Schmitt's 'Legality and Legitimacy,'" which he had co-authored with Nathan Leites exactly twenty years earlier.[69] He wrote to Schmitt:

> You will remember that even in my co-authored essay from 1932 on legality & legitimacy, I tried to confront the conceptual realism with the actual tendencies of institutional development; that does not meet the internal consistency of your train of thought, but it may well shift the perspective somewhat.[70]

Kirchheimer directed his criticism at some of Schmitt's students, too:

> When reading [Werner] Weber's little book [[71]], it became clear to me again that critical engagement with the conceptual structure of constitutional theory, which Weber took on board in toto, is overdue; what a pity that there doesn't seem to be anyone in Germany who is taking on such fruitful work; although German constitutional theory, to the extent that it exists intellectually, relies completely on your body of thought, it would benefit more from it if it complemented the act of receiving ideas with critical reception.

Concerning Ernst Forsthoff, Kirchheimer let Schmitt know that he had read Forsthoff's paper on the position of the political parties in terms of constitutional law but had serious doubts that Forsthoff did justice to the subject in the twentieth century, in light of his skepticism with regard to political parties.[72]

In all these points, Kirchheimer was fundamentally concerned with the relationship between Schmitt's general theoretical approach and the empirical "question of historical reality,"[73] which he criticized as being disconnected from one another. He had written the letter by hand on American Christmas-themed paper, embellished with some kitschy Renaissance-style angels. It animated Schmitt to note "Kirchheimer!" on it; he apparently found it as inappropriate as it was typical of Kirchheimer's behavior to turn a letter of Christmas greetings into a critical statement on political theory.

A year before Schmitt's 65th birthday in the summer of 1953, the editors of a planned *Festschrift* invited Kirchheimer to contribute a piece. He rejected the request, as did Ernst Friesenhahn.[74] Five years later, he explained his decision to Arvid Brødersen by noting that he sought to avoid the appearance of contributing to publicly enhancing Schmitt's

---

69 See Chapter 6, p. 151–157.

70 This quote and the following ones are from the handwritten letter from Otto Kirchheimer to Carl Schmitt dated 21 November 1952. Carl Schmitt Papers, RW 265–7600.

71 This refers to Weber (1951).

72 In this article, Forsthoff calls for party office and parliamentary mandate to be made incompatible, among other things, see Forsthoff (1950, 23–25).

73 Handwritten letter from Otto Kirchheimer to Carl Schmitt dated 21 November 1952. Carl Schmitt Papers, RW 265–7600.

74 Letter from Otto Kirchheimer to Ernst Friesenhahn dated 15 November 1952. Otto Kirchheimer Papers, Series 2, Box 2, Folder 61.

reputation among German constitutional theorists.[75] Meanwhile, in late 1952, Schmitt had sent him the German edition of Lilian Winstanley's book *Hamlet and the Scottish Succession* for Christmas in 1952.[76] Kirchheimer thanked him on 28 January 1953 without going into Schmitt's peculiar interpretation of the theme of Hamlet or his remark about what became known as the "Night of the Long Knives" in 1934. In his lost letter to Kirchheimer, Schmitt had obviously suggested another personal meeting with Kirchheimer, and the latter responded positively to this but without making any concrete suggestion.[77]

Kirchheimer contacted Schmitt again in February 1953 from the residence of Richard Schmid, President of Stuttgart's *Oberlandesgericht*. He told him that he would be in his vicinity for professional reasons, presumably in April or May, and would give him a telephone call to arrange a meeting, should the occasion arise.[78] Schmitt responded to him immediately but this letter has been lost, too. On 28 March, Kirchheimer suggested to Schmitt that they meet in Düsseldorf or that he visit him in late May.[79] They ultimately met for a few hours in Cologne in June 1953.[80] Schmitt was accompanied by his daughter Anima. No documents about this encounter seem to have survived. It was the last time they met in person. Shortly after the meeting, on 1 July 1953, Kirchheimer sent Schmitt a letter for his 65th birthday. He told him that he had enjoyed meeting Schmitt and his daughter "recently in Cologne." He added: "Too bad we're both on opposite sides of the pond."[81] In my view, the friendly statements in this letter by Kirchheimer are not to be taken literally but should be interpreted as platitudes because the exchange of letters between the two came to an end for five years after this. Although Schmitt mentioned to journalist Winfried Martini in September 1953 that he wanted to ask Kirchheimer about the whereabouts of sociologist Heinz Otto Ziegler, who he—Schmitt—thought had emigrated to the United States (see Burkhardt 2013, 123),[82] nothing is to be found in the relevant archives relating to this question. The only contact between the two was to continue sending each other copies of reprints, albeit rarely.

In his response to this birthday letter, Schmitt sent Kirchheimer a copy of the first bibliography of his own work, which Belgian sociologist Piet Thomissen had compiled.[83] The next envelope that Schmitt received from Kirchheimer was labeled "printed matter": the typescript of a book review by Kirchheimer about politics and the constitution in the

75 Letter from Otto Kirchheimer to Arvid Brødersen dated 2 March 1958. Otto Kirchheimer Papers, Series 2, Box 1, Folder 25.

76 This letter has been lost. The book is in Kirchheimer's estate in Albany.

77 Letter from Otto Kirchheimer to Carl Schmitt postmarked 28 January 1953. Carl Schmitt Papers, RW 265–7601.

78 Letter from Otto Kirchheimer to Carl Schmitt [no date; presumably February 1953]. Carl Schmitt Papers, RW 265–7593.

79 Letter from Otto Kirchheimer to Carl Schmitt dated 28 March 1953. Carl Schmitt Papers, RW 265–7594.

80 The exact date could not be determined.

81 ("*Schade, dass der grosse Teich doch eben ein sehr wirkungsvoller Graben ist.*") Letter from Otto Kirchheimer to Carl Schmitt dated 1 July 1953. Carl Schmitt Papers, RW 265–7602.

82 Schmitt was obviously not aware that Ziegler had died in military action in May 1944.

83 Mailed on 15 July 1953. List by Carl Schmitt about mailing complimentary copies

history of the United States,[84] with no accompanying card or comment.[85] According to the surviving material in the archives, Kirchheimer sent Schmitt a reprint of his essay "Politische Justiz" [Political justice] in 1955 (Kirchheimer 1955b) and Schmitt sent him a copy of his book *Hamlet or Hecuba* in 1956 (Schmitt 1956). It seems that no letters were attached to these mailings, and apparently, neither side was particularly inspired to arrange another personal meeting after the one in Cologne.

7. Kirchheimer as a professor of political science in the US

Kirchheimer's contacts with his former colleagues from the Institut für Sozialforschung were complicated, which had negative impacts on his prospects for a professorship in Frankfurt. He remained close friends with Gurland, Neumann, and Marcuse; the families vacationed together, and up until his death, Kirchheimer regularly welcomed Marcuse to stay over at his place in Silver Spring.[86] However, his relationship with Max Horkheimer, Theodor W. Adorno, and Friedrich Pollock, who had returned to Germany, was more problematic. None of these three attempted to hire Kirchheimer at the reestablished Institut für Sozialforschung in Frankfurt. Pollock and Kirchheimer had never been friends.[87] Adorno considered him politically suspect because he had visited Schmitt; their correspondence dwindled to almost nothing after this.[88] Horkheimer and Kirchheimer had become downright hostile; from Kirchheimer's perspective, this was because he had been treated poorly by Horkheimer in New York.

The intensity of these hostilities on Kirchheimer's part is illustrated by an episode about "the chest in the basement" (Wiggershaus 1995, 534) with a complete set of copies of the *Zeitschrift für Sozialforschung* that later became one of the well-known anecdotes about the Frankfurt School.[89] Copies of the journal were not automatically made available to new staff members in Frankfurt but were kept in a locked chest in the basement (see Habermas 1980, 415). Kirchheimer played a key role in making all staff members aware of this old journal and making the articles on the early critical theory published in it accessible to them.[90] He was angry about Horkheimer's ban on student protests against an upcoming visit by Chancellor Konrad Adenauer to Frankfurt University and sought to keep the memory of critical theory's radical political past alive. Kirchheimer asked Wilhelm Hennis, who was supposed to begin working as an assistant professor for Carlo

84 Published a few months later, see Kirchheimer (1954d).

85 Carl Schmitt Papers, RW 265–7605.

86 Peter Kirchheimer in a conversation with the author, 27 April 2023.

87 Leo Löwenthal recounted this in a conversation with the author on 5 October 1988.

88 All that is to be found in Adorno's papers is a brief exchange of letters: in 1954, Kirchheimer asked him, in English, to send the manuscript of Adorno's lecture at the Deutsche Soziologentag (Conference of the German Sociological Association) in Heidelberg, and Adorno sent it to him, writing "what are you up to?" in the cover letter. Theodor W. Adorno Papers, Aa 1, 11 (K1).

89 See, among others, Wiggershaus (1995, 544), Dahrendorf (1989, 878), Albrecht et al. (1999, 264), Specter (2010, 31), and Link (2022, 256).

90 The following description is based on a conversation with Wilhelm Hennis, 26 September 2009. This episode is also reported almost verbatim in Schlak (2008, 47–49).

Schmid in Frankfurt in 1953, to purchase copies of the journal still available in a used book store in Paris and to display them in the library. Word soon got around in Frankfurt about the existence and availability of the journal. Thus, members of the younger generation at the Institut für Sozialforschung, including Horkheimer's assistant Alfred Schmidt and new staff member Jürgen Habermas, obtained easy access to the key works from the early days of critical theory. According to Hennis, Kirchheimer was particularly amused to watch Habermas criticize Horkheimer's political and philosophical positions in the early 1960s, armed with the older writing of the Frankfurt School. Against this background, it is hardly surprising that Kirchheimer's contact with Horkheimer remained sporadic from then on; it was limited to "obligatory" letters with best wishes on milestone birthdays and other absolutely necessary communications. His contribution to the 1955 *Festschrift* on the occasion of Horkheimer's 65th birthday was his essay "Politics and Justice," which had previously been published in a slightly different version in *Social Research*, in a translation by Gurland (see Kirchheimer 1955a and 1955b).

In the US, on the other hand, when a temporary position became available at the Graduate Faculty of the New School for Social Research in New York in 1954, Kirchheimer suddenly had reason to hope that he would be able to finally leave the State Department for a university. Alvin Johnson, President of the New School for Social Research, which had been founded in 1919, had established the Graduate Faculty in 1933 as the University in Exile for academic refugees from Europe. Most of its core group were veterans of practical politics from former democracies in Europe, and their research interests were oriented toward political practice. The German staff was recruited mainly from three institutions: Frankfurt University, the Kiel Institute for the World Economy, and the Deutsche Hochschule für Politik (German Academy for Politics) in Berlin. Prominent German sociologists and social democratic intellectuals such as Hans Speier, Albert Salomon, Arnold Brecht, Eduard Heimann, Hans Simons, and Frieda Wunderlich had found academic refuge there right from the outset.[91]

Kirchheimer was already familiar with the New School. He had received a research stipend from the Graduate Faculty from March to July 1942 to work on "Contemporary Legal Trends," and this had enabled him to keep afloat financially for several months. It also helped that his friend John H. Herz had been invited to the Graduate Faculty as a visiting professor in 1953. The position of full professor of political science had remained vacant, and it was Herz who proposed hiring Kirchheimer as his successor.[92] Karl Loewenstein also put in a good word for Kirchheimer with Dean Hans Staudinger.[93] The latter liked the idea, and the Faculty decided in March 1954 to offer Kirchheimer the position for comparative government for the academic year 1954/55. Among the reasons given for selecting him, Kirchheimer's employment at the State Department as Chief of the Central European Branch was especially important. The New School was hoping for connections to

91 On the history of the Graduate Faculty at the New School for Social Research, see Friedlander (2019).

92 John H. Herz in a conversation with the author, 15 November 1985.

93 Letter from Karl Loewenstein to Otto Kirchheimer dated 13 June 1955. Otto Kirchheimer Papers, Series 2, Box 1, Folder 104.

political practitioners and, in particular, to potential funders in the capital, Washington. So, it appears they were not very familiar with Kirchheimer after all. Kirchheimer was able to take a leave of absence from the State Department for his tenure at the New School. The faculty's decision explicitly stated that it "should be pointed out to Dr. Kirchheimer that this appointment is only a temporary one."[94] Kirchheimer accepted the offer gratefully and without hesitation. Another plus was that a number of colleagues he already knew from his Berlin days during the Weimar Republic were at the New School.

Kirchheimer's professional prospects improved once again in spring 1955. The position of full professor of political science at the Graduate Faculty was still vacant. The New School had initially selected Gerhard Leibholz, a judge at the *Bundesverfassungsgericht* who also taught at Göttingen University, for this position at the urging of Arnold Brecht and Hans Staudinger. In light of the disputes around Leibholz at the *Bundesverfassungsgericht*, his supporters at the New School anticipated that he would welcome the opportunity to accept their offer. Whether the New School was poorly informed about Leibholz's personal plans or that they had unrealistic expectations about how attractive a chair in New York would be for an established German professor, the New School was surprised when he rejected the offer and felt pressured to rapidly fill the position, which had been vacant for some time. Staudinger proposed to the faculty to offer it to Kirchheimer. Only a week later, the Faculty decided "unanimously to recommend to the Administration and the Board the appointment of Dr. Otto Kirchheimer as Full Professor in the Department of Political Science."[95] The position was to be limited to two years initially but would then—provided there were no serious reasons against this—become a tenured position.

At the age of 50, Otto Kirchheimer finally felt he had achieved a goal he had been pursuing ever since he had planned to begin his *habilitation* in 1932 in Germany: a proper position as an academic at a university. He was formally welcomed as a new member of the Graduate Faculty in November 1955 alongside the philosopher Hans Jonas.[96] Writing to Smend, he proudly described his position as "successor of Arnold Brecht's."[97] The professorship was for comparative government. Kirchheimer also taught criminology and the political system of the US at the New School from 1955 to 1961.

8. Criticism of Schmittianism in German legal thought

Schmitt never publicly articulated his criticism of the Basic Law in much detail. People knew about it from conversations but had no written text to draw on. The actual criticism of the Basic Law and its interpretation by the *Bundesverfassungsgericht* came from the circle of his students who had subsequently become professors. Kirchheimer engaged with

---

94 Minutes, Executive Faculty Meeting of 3 March 1954. New School for Social Research: New School Institutional Collections. Graduate Faculty, Minutes. NS.02.17.02, Box 1, unprocessed collection.

95 Minutes, Executive Faculty Meeting of 18 May 1955. New School for Social Research: New School Institutional Collections. Graduate Faculty, Minutes. NS.02.17.02, Box 1, unprocessed collection.

96 Minutes, Executive Faculty Meeting of November 1955. New School for Social Research: New School Institutional Collections. Graduate Faculty, Minutes. NS.02.17.02, Box 1, unprocessed collection.

97 Letter from Otto Kirchheimer to Rudolf Smend dated 7 January 1956. Rudolf Smend Papers, correspondence with Kirchheimer.

them multiple times. He articulated his criticism of Schmitt and his students not only in conversations with German friends and in letters but also in shorter publications. He chose book reviews as his way to intervene in the German debates on constitutional theory. The first one was a review of Werner Weber's 1951 book *Spannungen und Kräfte im westdeutschen Verfassungssystem* [Tensions and forces in the West German constitutional system] for the *American Political Science Review*. In 1948, Weber had accepted an appointment to a chair of public law in Göttingen. Rudolf Smend had preferred his former student Kirchheimer for this position instead and expressed his concern to Kirchheimer about his new colleague because of his close ties with Schmitt. It was clear from Kirchheimer's response that he shared this concern.[98] He also told Smend about the review he intended to write about Weber's book. He summarized the essence of the review as follows: "I do not think it is very productive to deal with the current circumstances in Germany using Carl Schmitt's scarcely modified conceptual framework from the years 28–32."[99] Moreover, he agreed with Smend that "[Hermann] Heller [was] more productive than C.S. in the long run."[100] In April 1952, he asked Smend to hand the text of his then completed review over to Weber "with a proper remark"[101] so that, in all fairness, Weber would not find out about it only indirectly once it had been published in the United States.

The review of Weber's book was a frontal attack.[102] Kirchheimer probably sensed that Schmitt was full of praise for Weber's fundamental criticism of the West German constitution.[103] In his introduction, Kirchheimer called Weber an "intelligent and lucid writer" (220), only to add that he remained "hopelessly caught" in the "conceptual framework erected by Carl Schmitt in the late twenties and the early thirties" (220) in his efforts to analyze the political system of the new Federal Republic of Germany. Readers of the book encountered all of Schmitt's "old clichés" (220): the incontestable authority of the state, a strong and neutral executive branch, an elite at the head of state administration, and criticism of pluralism of political parties as well as of the political influence of interest groups. Kirchheimer commented sarcastically that, in hindsight, the Weimar Constitution suddenly came off as not all that bad in Weber's tract, but this served only to pave the way for an all the more vehement attack on the newly established system of the Federal Republic of Germany.

Kirchheimer countered Weber's charge of the Basic Law's misguided perfectionism by pointing out that Weber had blocked his own ability to realistically assess the functioning of the West German system because of his Schmittian conceptual framework. Weber's "complete acceptance of Carl Schmitt's conceptual framework and scale of values has stood in the way of a dispassionate analysis of the interplay between constitu-

---

98 Letter from Otto Kirchheimer to Rudolf Smend dated 9 June 1951. Rudolf Smend Papers, correspondence with Kirchheimer.

99 Letter from Otto Kirchheimer to Rudolf Smend dated 1 February 1952. Rudolf Smend Papers, correspondence with Kirchheimer.

100 Letter from Otto Kirchheimer to Rudolf Smend dated 25 October 1951. Otto Kirchheimer Papers, Series 2, Box 2, Folder 19.

101 Letter from Otto Kirchheimer to Rudolf Smend dated 1 April 1952. Rudolf Smend Papers, correspondence with Kirchheimer.

102 See Kirchheimer (1952b). The following page numbers refer to this text.

103 Letter from Carl Schmitt to Ernst Forsthoff dated 4 March 1952 (Schmitt and Forsthoff 2007, 86–87).

tional order and the social reality of the Bonn state" (220). Neither his excessive criticism of the status of the political parties in the Basic Law nor of the federal order laid down in the Basic Law was convincing, Kirchheimer claimed. Contrary to Weber's reiteration of the old clichés, Kirchheimer described the political parties and interest groups as the primary agencies of political integration whose legitimacy rested "in their ability to channel the political and social energies of their clientele of unions, economic associations or churches into political action" (221). For this reason, it was to be considered positive that they were mentioned explicitly in the Basic Law.

Three years later, Kirchheimer attacked Ernst Forsthoff, the most prominent Schmittian in Germany. He did so in his essay "Parteistruktur und Massendemokratie in Europa" [Party structure and mass democracy in Europe], published in Smend's journal *Archiv des öffentlichen Rechts* in 1954.[104] In a letter to Forsthoff, Schmitt had praised and thanked him for his work on parties.[105] Kirchheimer took a different position. He bluntly rejected Forsthoff's proposal to shield parliamentary party groups from the influence of their party leadership by making it legally incompatible to belong to a parliament and simultaneously hold a party office (see Forsthoff 1950). This proposal was based on the "mistaken assumption" (241) that it was only the parliamentary party group, not the entire party, that was tasked to design policy. If differences between authorities of the parliamentary party groups and the parties themselves arose in current-day parliamentary democracies, this was a clear indication of deeper social and political discrepancies. These would be resolved through splits between parliamentary party groups and the headquarters of political parties and the establishment of new ones, which did not require restrictions by the state.

Kirchheimer's review of the book *Wirtschaftsverwaltungsrecht* [Administrative law concerning trade and industry], published the same year by Ernst Rudolf Huber, a third student from Schmitt's circle in Bonn and later Nazi theorist, was slightly more moderate.[106] This may have been because Kirchheimer knew that Huber had distanced himself from Schmitt's antisemitic activities during the Third Reich. However, Kirchheimer diagnosed a certain Schmittian intellectual legacy in Huber's writing, too, describing his "continuing belief in the possibility of a neutral state with the functions of an arbitrator" (267) and criticizing, as a consequence of this, the anti-trade-union tendency of such convictions. At the same time, Kirchheimer saw the fact that the Basic Law was indeterminate in terms of economic policy, which Huber championed, as somewhat problematic for the author since he refrained from working through the various, and partially competing, imperatives of constitutional law in the context of their "proving themselves over the course of history" (268). Instead, Huber took refuge in conceptual arguments without asking himself "to what extent any correspondence [existed] between the conceptual schema and social reality" (269).

In the summer of 1956, Kirchheimer again picked a fight, this time with a professor of the younger generation, Joseph H. Kaiser, who was one of the Schmittians at the University of Tübingen and later became a confidant of Schmitt's and one of the administra-

104 Kirchheimer (1954b). The following page numbers refer to this text.

105 Letter from Carl Schmitt to Ernst Forsthoff dated 14 August 1950 (Schmitt and Forsthoff 2007, 76).

106 Kirchheimer (1954c). The following page numbers refer to this text.

tors of his estate. His discussion of Kaiser's book *Die Repräsentation organisierter Interessen* [The representation of organized interests] was an uncompromising attack.[107] It started with a critique of Kaiser's methodological approach. Instead of seeking orientation in "empirical social research, which [was] being conducted in the Anglo-American and increasingly also in the Romance cultural region" (271), and instead of limiting his work, as Huber did, to legally classifying social phenomena according to the constitutional order as objectively as possible, Kirchheimer alleged that Kaiser wove "rich material from the history of ideas and contemporary history into a predetermined conceptual schema, persistently adopting Carl Schmitt's ways of thinking and forms of presentation" (271). Kaiser's argument, he stated, was based on a "concept of the political rooted in the antinomy of state and society" (271) that, when discussing the role of societal interest groups, inevitably arrived at the hypothesis that pressure groups undermined the character of the state institutions.

Toward the end of his critique of Kaiser, Kirchheimer provided insight into his own skeptical view of the role of the individual in modern society. If Kaiser was calling for a strong state because it alone could protect individuals from excessive claims by intermediary institutions, Kirchheimer had a more "ambivalent" view (275). States as well as interest groups could "easily and almost unnoticeably" (274) make service provision and protection become intertwined, as well as harassment and oppression. The best protection of the individual, he claimed, consisted in exploiting organizational rivalries and spaces between institutions. It was hardly surprising in the current "lull of the postfascist age and in the neighborhood of the Bolshevist sphere" (275) that many people sought to elude being organized by the state and in interest groups. This was the reason why Kirchheimer countered Kaiser's argument for a strong state with the point "that the individual, the state, and the societal apparatus were to work together sufficiently" (275).

Kirchheimer had attacked the four most prominent avowed Schmittians at German universities in the 1950s: Forsthoff, Huber, Weber, and Kaiser. He used the same blueprint for his interpretation and criticism for all of them, namely denouncing them all for propounding a false theory of the strong state and pointing out the lack of empirical evidence for their deliberations. Kirchheimer forced all four into the model of the Weimar controversies. His accusation that they had all remained stuck in an outdated pattern of thinking, both mentally and in their arguments, was ultimately aimed at Schmitt. His criticisms also had another subtext. Whereas he regarded the four scholars he criticized as more or less uncritical epigones of Schmitt's, he considered himself capable of critically receiving Schmitt's oeuvre with incomparably more independence.

9. Conclusion: The new constellation

By the mid-1950s, Schmitt and Kirchheimer's relationship had become established in a new constellation. The tables had turned. Schmitt had not succeeded in returning to a professorship, whereas Kirchheimer had finally realized his dream—which he had had at least since 1932—of becoming an academic at a university. Schmitt, conversely, tended

107 See Kirchheimer (1956a). The following page numbers refer to this text.

to his "invisible college" from his base in Plettenberg. Within a short space of time, Kirchheimer succeeded in establishing a reputation in the US as a well-informed and sharp-witted political scientist.

It cannot be stated with certainty today why Kirchheimer paid Schmitt a visit at his home in Plettenberg in November 1949. If we follow Wilhelm Hennis—and I see no reason to doubt his key statements—then Kirchheimer had wanted to demonstrate to Schmitt that the political tide had turned. He also wanted to demonstrate to Schmitt that he, who had been among those driven out of Germany with Schmitt's applause in 1933, had managed to survive.

Even though no friendship or any kind of more intimate professional relationship resulted from this visit, they stayed in contact afterwards and even met again once more in Cologne in June 1953. After Kirchheimer's birthday letter to Schmitt shortly after their meeting in Cologne, they stopped writing each other letters for the next five years; their contact was reduced to sending each other reprints, and even that only rarely.

Their correspondence—which was sparse overall compared with Schmitt's communications with others—does not show indications of a close personal bond but was characterized by distant politeness and occasionally a critical comment of Kirchheimer's toward Schmitt or his followers. The original initiative to resume personal contact in 1947 had been Schmitt's, when he asked Ossip K. Flechtheim about Kirchheimer and also asked him to pass on his best regards to him. Kirchheimer was already aware of Schmitt's detention at this time via Flechtheim and presumably also due to his contacts with Karl Loewenstein, Rudolf Smend, Carlo Schmid, and Richard Schmid. The next documented contact is dated 22 November 1949 when Schmitt sent Kirchheimer a copy of his essay on Francisco de Vitoria (or, more likely, *Ex captivitate salus*). Schmitt was eager to provide Kirchheimer with his writing that linked up with subjects they had both worked on during the Weimar Republic. Kirchheimer did not take up Schmitt's offers to conduct a debate with one another. Whereas Schmitt proposed discussing subjects like property rights, expropriation, or constitutional courts, Kirchheimer stuck to pleasantries and did not allow himself to be drawn into any in-depth discussions. The fact that there are no handwritten comments or the like in his copies of these works, which are among his papers, also raises doubts as to whether he even found Schmitt's *Nomos of the Earth* or his interpretations of Hamlet interesting enough to read them attentively.

In most of his letters, Kirchheimer responded only briefly to Schmitt—with one exception: his letter of November 1952, in which he addressed their methodological differences once again. He reminded Schmitt of the essay he had co-authored with Nathan Leites in 1932/1933 and in which he had confronted Schmitt's conceptual realism with empirical evidence regarding actual tendencies of institutional developments in Western democracies. Schmitt had not responded to this criticism twenty years earlier. And he did not respond to Kirchheimer this time, either. Only in his *Glossarium* are some notes to be found on this matter, and he considered the label "conceptual realism," which Kirchheimer had intended to be an accusation, to be an honorary title. He proudly noted "conceptual realism—as a prerequisite of jurisprudence,"[108] followed a few lines later by vi-

108 *Glossarium* entry of 24 April 1947 (Schmitt 2015, 14).

cious antisemitic invective.[109] A year later, he countered "positivist illusionism" by stating that it meant "doing without any conceptual realism, which, however, [constituted] the essence of legal thinking."[110] At the root of this notion of Schmitt's was apparently the idea that it was only the use of concepts itself that created reality. He considered conceptual realism to be a creative practice because he noted at one point in his *Glossarium*: "merely mentioning a word determines the atmosphere."[111] Schmitt positioned non-creative, positivist, and "Jewish" thinking as the opposing view to creative conceptual realism.

Kirchheimer increasingly became convinced that his efforts to achieve targeted denazification in the Western zones had failed. Nevertheless, he did not see the newly established Federal Republic of Germany as having any problems of legitimacy; he considered only the GDR, which had been founded using dictatorial means, to be illegitimate. He became all the more interested in empirical questions of the new German democracy: election campaigns, election laws, election results, government formations, party formation, parliamentary politics, and government policy. Reading Kirchheimer's analyses from Schmitt's perspective, we can interpret them as attempts to explore future opportunities for politically stabilizing the Federal Republic of Germany. Kirchheimer identified certain continuities from the Weimar Republic, for example in the top politicians and the election results. Yet the discontinuities—and the Basic Law played a key role here—seemed more prominent to him. These discontinuities were the empirical basis for his hope that the conflictual Weimar times, which in Schmitt's imagination would rapidly develop into a civil war scenario, could be avoided this time. Here, Kirchheimer was convinced of a policy of social and political integration that was legitimized in the theory of integration put forward by Smend and his students.

Schmitt was of the opinion that the newly established Federal Republic of Germany was not worthy of recognition at all. To him, it was a badly updated version of a weak Weimar Republic. In his *Glossarium*, he formulated slogans of German resistance against the victorious Allied powers and the founding figures of the Federal Republic of Germany; such resistance, he asserted, would first of all have to focus on Germany's spiritual self-assertion. He relished commenting sarcastically on rumors that some deliberations from his *Constitutional Theory* had indirectly been taken up in the Basic Law. Moreover, he thought that the Federal Republic of Germany was doomed to failure and would soon go to ruin because of the potential for political conflict. To Schmitt, the struggle for German self-assertion included the refusal to even begin to deal with the Germans' war crimes. He demanded that the victims refrain from mentioning their personal suffering in public life and even went one step further: remembrance of those murdered by the Nazi regime had to be obliterated, too; only then could there be peace. What Schmitt meant as he called for amnesty was complete amnesia concerning the past.

109 *"Juden bleiben immer Juden. Während der Kommunist sich bessern und ändern kann."* (Jews will always be Jews. Whereas the communist can better himself and change). *"Der assimilierte Jude ist der wahre Feind."* (The assimilated Jew is the real enemy). *Glossarium* entries of 25 April 1947 (Schmitt 2015, 14).

110 *Glossarium* entry of 2 March 1948 (Schmitt 2015, 14).

111 *Glossarium* entry of 7 October 1947 (Schmitt 2015, 21).

A rift had developed among Schmitt's students from his days in Bonn concerning his dedicated work for the Nazi regime, and this rift left its mark on the atmosphere among West German scholars of public law. On one side was the larger group of his students with the prominent figures Ernst-Rudolf Huber, Werner Weber, and Ernst Forsthoff. All three continued to correspond intensely with Schmitt and had a major influence on public law in the early stages of the Federal Republic of Germany—and Schmitt made frequent and lively positive comments in this regard. On the other side were Ernst Friesenhahn and Gerhard Schiedermair, who distanced themselves from Schmitt clearly and in public, not just in private. A second rift among the West German public law scholars of the day was related to the existence of two competing schools around Schmitt and Smend. Vehement and complex struggles for positions took place between these two schools in the first two decades of the Federal Republic of Germany.[112] Kirchheimer took a clear and public position with respect to both fronts. He was in close personal contact with Friesenhahn, not least about the activities of Schmitt and his circle, and also with Smend, and supported the latter's younger generation of students. By contrast, he was not in direct contact with Weber, Huber, or Forsthoff. At the same time, he openly attacked Schmittianism in German public law in multiple publications and did not mince words. The four leading Schmittians were the targets of his criticism: Werner Weber, Ernst Forsthoff, Ernst-Rudolf Huber, and Joseph Kaiser. Kirchheimer criticized the lack of any empirical basis for their claims and accused them of continuing to promote a false theory of the strong state. He portrayed them as uncritical epigones of Schmitt's who were unable to receive Schmitt's oeuvre critically, selectively, and independently.

Not only their different roles during the years 1933 to 1945, but also the differences in how they dealt with the Nazi past created a deep rift between Kirchheimer and Schmitt which could not be papered over by friendly pleasantries in the forms of address in their letters. After Kirchheimer's visit, Schmitt had written his wife Duška that Kirchheimer was not satisfied with Ex captivitate salus because there was no explanation of what Schmitt had done in 1933. Wilhelm Hennis later used stronger words in his conversation with the author: Kirchheimer had been outraged by Schmitt's unwillingness to grapple self-critically with his own responsibility for the Nazi regime's policies.[113] Nevertheless, Kirchheimer did not decide to cut off contact to him completely. None of the surviving sources explain why he continued to respond to him and to keep in contact—and it is important to bear in mind that their postwar contact was by no means close. Perhaps it was another manifestation of what Hennis considered to be Kirchheimer's motive to visit Schmitt at his home in the first place: his way of expressing "proud self-affirmation" ("*stolze Selbstbehauptung*") vis-à-vis Schmitt.

---

112 On the competition between these two schools in the 1950s and 1960s, see Günther (2004).

113 Wilhelm Hennis recounted this in a conversation with the author on 26 September 2009.

Chapter 16: Juridification and Political Justice (1957-1961)

Kirchheimer and Schmitt worked on different subjects during the second half of the 1950s. Schmitt continued to dabble in interpreting Shakespeare (see Schmitt 1956) but also spoke out sporadically on other matters. He again challenged the *Bundesverfassungsgericht* and the juridification of politics that he alleged resulted from its establishment. In his short polemic piece "The Tyranny of Values," he frontally attacked the supposed "value-philosophical foundation" (Schmitt 1960, 4) of the court's jurisprudence, which was influenced by Smend's theory of integration.[1] He also focused on compiling a selection of his essays on constitutional law from the previous three decades. When commenting on his older works, he repeatedly referred to Kirchheimer. At the same time, Kirchheimer published several book reviews and articles providing overviews of the state of West German political science, initially for a US readership. After that, he authored essays on political parties, elections, parliamentarism, and opposition in Germany, France, Italy, and the US. Yet working on his book *Political Justice* was his top priority.

Kirchheimer briefly mentioned his view of Shakespeare in this book, too, which was completely different from Schmitt's ideas in *Hamlet or Hecuba*.[2] Kirchheimer was interested most in Shakespeare's concept of mercy. Concerning the mercy granted to protagonist Angelo, he quoted from *Measure for Measure*: "the quality of mercy is not strain'd; it droppeth as the gentle rain from heaven" and contrasted it with *The Merchant of Venice* (see Kirchheimer 1961a, 392–395). He argued that punishment and mercy followed different principles. Whereas the punishment meted out by a court complied with the law and balanced the public interest, culpability, and remorse, the act of mercy appeared to be arbitrary. Mercy seemed to be the positive version of decisionism for a delinquent, so to speak.

After Kirchheimer and Schmitt stopped corresponding in July 1953, their only direct contact over the next five years consisted of sending each other copies of a few of their essays. In 1955, Kirchheimer sent Schmitt a reprint of his essay "Politische Justiz" [Political justice] (see Kirchheimer 1955b), Schmitt sent a copy of his 1956 book *Hamlet or Hecuba*,

1 See Zeitlin (2020).

2 See Chapter 15, p. 396–397.

and Kirchheimer replied by sending reprints of his essay on the vanishing of political opposition (see Kirchheimer 1957c).[3] In 1957, Kirchheimer briefly returned to directly attacking Schmitt himself, not only his students, in his publications. At the same time, Schmitt, too, began to quote his former doctoral student again; he did so with exquisite politeness and purported to agree with him. Their disagreements subsequently deepened in their correspondence in 1958.

1. Debating each other in public again

In 1957, Kirchheimer seized the opportunity to once again grapple explicitly with Schmitt. His essay "The Political Scene in West Germany,"[4] an omnibus review of thirteen new scholarly publications on politics and the law in Germany, was published in the fall issue of *World Politics*. One key topic was the appropriate interpretation of Article 20 of the Basic Law, which mandated the welfare state. As he presented the controversy between the protagonists of the day, Werner Weber, a student of Schmitt's, and Wolfgang Abendroth, a disciple of Heller's, Kirchheimer clearly sympathized with Abendroth's concept of the "social *Rechtsstaat*."[5] He used the controversy to find fault with Schmitt's multiple criticisms of the *Bundesverfassungsgericht* in the Federal Republic. Kirchheimer considered the establishment of the *Bundesverfassungsgericht* one of the most important "postwar innovation[s] that enhances popular acceptance of the new order" (354). He praised the court for its jurisdiction to date because it was guided by considerations strictly based on the legal framework "instead of concealing political reasoning behind legal exegesis" (355). Kirchheimer placed his hopes for the foreseeable future in an expansion of the court's jurisdiction concerning fundamental rights as well as its judicious actions being "habit-forming" (355) in a positive sense for the political culture of the Federal Republic of Germany.

However, Kirchheimer also used the review to attack Schmitt more profoundly and by name. He did so indirectly, by commenting on a *habilitation* dissertation by young Swiss legal scholar Peter Schneider on Carl Schmitt titled *Ausnahmezustand und Norm* [State of emergency and norm], which had not yet been published as a book (see Schneider 1957). There was a backstory to Peter Schneider's study that played out in the midst of the convoluted struggles between the two competing schools of constitutional law in the Federal Republic of Germany around Smend and Schmitt mentioned in the previous chapter. From his base in Plettenberg, Schmitt was in frequent contact with his former students who held professorships in law. Smend, too, remained active once he was emeritus, continuing to teach his seminar in Göttingen from 1952 until 1964. He regularly invited leading legal experts and political scientists, including Kirchheimer, to speak as guests. Kirchheimer, in turn, tried to familiarize his students with Smend's theory of

3 The books and reprints can be found in the papers of Kirchheimer and Schmitt, with no notes or letters attached.

4 Kirchheimer (1957b). The following page numbers refer to this text.

5 On the controversy at the time, see Stolleis (2012, 281–285).

integration.[6] He intensified his contacts with legal experts of his generation such as Ulrich Scheuner, a former student of Smend's in the late 1920s and a law professor in Bonn from 1951 on, and Ernst Friesenhahn, formerly an assistant of Schmitt's. Both Scheuner and Friesenhahn were part of the growing Smend school of constitutional theory. Smend also introduced Kirchheimer to a new and young generation of German students. Among them were Wilhelm Hennis, Konrad Hesse, Horst Ehmke, and Peter von Oertzen, who were all close to the Social Democratic Party and soon enjoyed successful careers as legal scholars, political scientists, and politicians. In 1962, Kirchheimer wrote a piece for Smend's *Festschrift*, an honor granted "only to the closest circle of students and friends" (Günther 2004, 162).

Peter Schneider, who was Swiss, had originally intended to complete his *habilitation* dissertation under Carlo Schmid at the University of Tübingen. When Schmid could no longer exercise this function because he had been elected to the Bundestag, Ernst Friesenhahn, who was at the University of Bonn, stepped in. One of his colleagues in Bonn was Hans Schneider, a close friend of Schmitt's, who informed Schmitt in detail about the work and the ongoing *habilitation* procedure. Friesenhahn reported in a letter to Kirchheimer that "Hans Schneider [...] had caused him [Peter Schneider] difficulties because of the *habilitation* dissertation."[7] It was only in 1955, after some complications, that the procedure could be completed successfully. Schneider's book was the first detailed and systematic study of Schmitt's theories in Germany.

To this day, the book stands out because not only does it interpret Schmitt's various changes of positions as opportunistic, but it also reconstructs a fundamental position throughout his works. Schneider identified this fundamental position as the "total negation" (Schneider 1957, 121) of the principles of the bourgeois *Rechtsstaat*, that is, a state order whose goal was to secure its citizens' personal liberties. Liberty in Schmitt's sense did not exist for the individual, only for the collective. All his criticisms notwithstanding, Schneider acknowledged the abundant food for thought that Schmitt, the "admirable storyteller" and "first-rate jurist" (Schneider 1957, 20), offered. Critics at the time praised Schneider's book as a great success because he had attempted to follow and understand every aspect of Schmitt's multifaceted theories.[8] It was not until 1964 that constitutional theorist Hasso Hofmann achieved the same high level of critical examination in *Legitimität gegen Legalität* [Legitimacy against legality].

Schmitt was far from amused by the publication of Schneider's monograph. He thanked its author for sending him the, in his words, "vivisection or more precisely (if I may coin such a term): arcanoscopy."[9] He used stronger words in a letter to Ernst Jünger: "The Swiss Peter Schneider, who had published a big fat book about my legal theory (a youth, from Zurich, lacking destiny like a sleeping infant, about me, an old

6 For example, the reading list for his seminar "The Political Institutions of Divided Germany" (1962/63) at Columbia University included Smend's *Verfassung und Verfassungsrecht* from 1928, recommended by Kirchheimer with the comment "creative theory of meaning of constitution." Minutes of the Faculty of Political Science 1957–62. Special Collection, Columbia University Archives.

7 Letter from Ernst Friesenhahn to Otto Kirchheimer dated 4 February 1955. Otto Kirchheimer Papers, Series 2, Box 1, Folder 55.

8 See Sontheimer (1957) and Ehmke (1959).

9 Quoted in Schneider and Gremmels (1994, 227).

man laden with destiny) claims that I am a 'storyteller,' which is all wrong."[10] From 1955 on, Schmitt noted in his *Glossarium* his thoughts on "dogcatcher" Schneider and the risk that he would "actually snatch [him]," on his opponents' goal of "de-Schmittianizing" German legal scholarship, and on the fact that a book written by an author from neutral Switzerland necessarily "lack[ed] [...] destiny." He was also outraged that the book by a Swiss national could even be published in Germany,[11] and he wrote to Forsthoff that he could "hardly believe"[12] Schneider was appointed to a chair in law in Mainz.

After the book had been published, Kirchheimer received a copy of the proofs from Schneider, at Friesenhahn's recommendation. After he had finished reading, he wrote Friesenhahn that the book "stuck too closely to the material" initially, but that after a good sixty pages, "the book becomes first-rate."[13] He hoped it would be widely read in Germany. Kirchheimer was full of praise for Schneider's book in his essay "The Political Scene in West Germany."[14] He saw it as indication that the culture in the legal sciences in Germany was finally changing for the better. Schneider's book was "one of the most encouraging signs on the German intellectual horizon" (348). "In its long-range literary impact," Kirchheimer continued, "it may be presumed to overshadow much of present-day writing" (348). With admirable energy, Schneider had dared to draw a coherent and comprehensive picture of Schmitt's theory based on his many works, thus providing the background for a detailed analysis of discrepancies in Schmitt's theories and imprecisions in his concepts. Kirchheimer agreed with Schneider's basic interpretive hypothesis that in Schmitt's theory, the bourgeois *Rechtsstaat* was the "eternal enemy" (349).

Kirchheimer's review of Schneider's book included almost the entire list of his own key criticisms of Schmitt in compact form:

> The lack of any clear-cut criteria for differentiating between *nomos* and violence; the discrepancy between the traditional liberal concepts of classical international law and the decisive rejection of an *artfremd*[15] and disintegrating liberalism as part of the domestic constitutional order; the brooding omnipresence of the people's constituent power and its incapacity to act as a constituted organ; the indeterminate character of the values underlying concrete decisions; and the conjunction of a relativistic openness to a variety of historical interpretations with an ever-present negation of the rule of law (348).

Schneider was delighted with Kirchheimer's praise. He had been afraid that his harsh criticism of Schmitt had jeopardized his opportunities for any academic job in law in

---

10 Letter from Carl Schmitt to Ernst Jünger dated 7 May 1957 (Schmitt and Jünger 1999, 330).

11 *Glossarium* entries of 11 November 1955; 17 March 1957; 30 March 1957 (Schmitt 2015, 329, 357, 358).

12 Letter from Carl Schmitt to Ernst Forsthoff dated 1 August 1956 (Schmitt and Forsthoff 2007, 127).

13 Letter from Otto Kirchheimer to Ernst Friesenhahn dated 10 March 1957. Otto Kirchheimer Papers, Series 2, Box 1, Folder 61.

14 Kirchheimer (1957b). The following page numbers refer to this text.

15 Here, Kirchheimer used the German word *artfremd* specifically as a term from Schmitt's vocabulary, meaning foreign/alien to the *Volk*, in an exclusionary and antisemitic sense; see Translator's Preface.

Germany.[16] Horst Ehmke was less enthusiastic about the impact of Schneider's book. He had also read and praised Schneider's interpretation of Schmitt. But he expressed his fear to Kirchheimer as follows: "I am not at all sure whether in our situation, the book might have the effect of being propaganda for Carl Schmitt."[17]

Kirchheimer's writing about the political changes in France amounted to a second public confrontation with Schmitt.[18] In his article "France from the Fourth to the Fifth Republic,"[19] Kirchheimer studied the reasons for the collapse of the constitution of the Fourth Republic, which he had criticized back in 1947. At the same time, he attempted to provide an explanation why and how Charles de Gaulle had succeeded so easily in having a new constitution enter into force with him as president. Kirchheimer considered the deeper causes to be general social factors "transcending the French scene" (396). He interpreted France's transition to the Fifth Republic as the expression of a general social transition in which the citizens' desire for political participation was being eroded and the position of the executive branch within the political system was becoming stronger. After World War II, France experienced an industrial modernization that undermined the parliament and the political parties. In the course of this development, the French bureaucracy gradually increased its independence from both legislative control and the leadership of the executive. So Kirchheimer interpreted the malaise in French domestic politics not as a fundamental structural weakness of parliamentarism but, rather, as the result of a modernization process of industrial society unfolding in all Western democracies.

Kirchheimer called de Gaulle a "representative of authoritarian technocracy, rather than of plebeian totalitarianism" (401). He criticized the constitution of the Fifth Republic, which had been drafted under de Gaulle's direction and accepted by referendum in September 1958. The president's power was not sufficiently limited, he asserted, and the prime minister, the cabinet, and the National Assembly were too weak. Yet he stopped short of calling the new French system a presidential dictatorship. Despite its anti-parliamentarian thrust, there was in the constitution "little evidence that the presidency has been intentionally construed to serve as a springboard for the assumption of Caesarean-Napoleonic or modern totalitarian dictatorship" (405). If de Gaulle had wanted that, he would have codified that the president would be elected directly, not indirectly by an electoral college of parliamentarians and municipal leaders—there is no indication in Kirchheimer's text that he had anticipated the direct election of the president introduced by de Gaulle in a referendum in 1962. Kirchheimer was skeptical whether switching course toward a stronger presidential constitution would solve France's social and political problems in the future. He claimed the new constitutional model was "construed from the viewpoint of administrative efficiency rather than from a careful consideration of the supporting political structure" (425). Kirchheimer was of the opinion that the unclear allocation of competencies between the president and the prime minister was

16 Letter from Peter Schneider to Otto Kirchheimer dated 9 February 1957. Otto Kirchheimer Papers, Series 2, Box 2, Folder 1.

17 Letter from Horst Ehmke to Otto Kirchheimer dated 19 May 1957. Horst Ehmke Papers, No. 504.

18 On Kirchheimer's writings on France, see Schale (2011).

19 See Kirchheimer (1958a). The following page numbers refer to this text.

symptomatic of this weakness and provoked the conflicts that erupted later in the era of "cohabitation."

A revised version of the above-mentioned article was published a year later in the German journal *Außenpolitik* [Foreign policy] under the title *In Frankreich regiert ein einziger Mann* [France is ruled by just a single man]. Even the somewhat pointed wording of the title indicated that it was not only an abridged translation into German, but that Kirchheimer expressed his criticism of the political events in France more explicitly. He interpreted de Gaulle's constitutional policy as a kind of incarnation of Carl Schmitt's theory of the presidential dictatorship: "Constituted and constituting authorities to exert power were unified in the same transitional government, embodied in the person of General de Gaulle" (430). Kirchheimer knew that Schmitt had praised the new French constitution as France's authoritarian salvation and considered himself to be one of the sources of inspiration for de Gaulle's constitution. He also knew that it was West German confidants of Schmitt's such as Armin Mohler who welcomed France's transition to a state led by a "strong man" who did not have to have much regard for the parliament, in the sense of a "commissarial dictatorship." In his 1963 *Theory of the Partisan*, Schmitt once again expressed his sympathy for how General de Gaulle had trumped republican legitimacy by means of the traditional and national legitimacy that de Gaulle had claimed (see Schmitt 1963a, 83–84).

Whereas Kirchheimer had again begun attacking Schmitt in his publications, Schmitt took the opposite strategy. He remained exceptionally friendly and quoted Kirchheimer exclusively in positive ways. In so doing, he cherry-picked hypotheses or wording of Kirchheimer's that he believed he could use to support his own position. The platform for reintroducing Kirchheimer to the circle of authors Schmitt considered worthy of being quoted was the edition of his *Verfassungsrechtliche Aufsätze aus den Jahren 1924–54* [Essays on constitutional law from the years 1924–54], which was Schmitt's seventieth birthday present to himself. Schmitt added appendices, some of them extensive, to his old essays. In the comments on his Weimar essays, he was keen to present his justifications of the presidential system as a defense of the republic and to assert his loyalty to the constitution (see Schmitt 1958b, 345–350 and 449–451). Schmitt even turned the tables, accusing the Social Democrats of the Weimar period of having destabilized the republic with what he called their dogmatic actions. The Weimar Social Democrats had succumbed to the "chimera of the 51% majority" according to which "the social order [could be] transformed *uni actu*, as if with a magic wand" (Schmitt 1958b, 346) by a law, Schmitt asserted, quoting Kirchheimer. The SPD, Schmitt claimed, "must have had to recognize the chimera as such as early as 1932" (Schmitt 1958b, 346). Schmitt quoted Kirchheimer's words from the modified German version of the 1957 essay "The Waning of Opposition in Parliamentary Regimes" correctly. But he put them in the wrong context; in contrast to what Schmitt suggested, Kirchheimer's comment had referred to a socialist opposition of principle and not to a social democratic party like the Weimar SPD which had been prepared to form coalitions with the bourgeois parties even to the point of self-denial (see Kirchheimer 1957c, 369).

Schmitt made pointed comments on the Federal Republic of Germany. He criticized a tendency toward juridification in a state founded on the supremacy of the judiciary, the dissolution of the bourgeois concept of property, and the welfare state. In his critique of the Federal Republic of Germany, he again referred to Kirchheimer's essay on the waning of opposition. He used Kirchheimer's phrase of the "chimera" once more, this time to present the article on expropriation of property in the Hessian Constitution of 1946 as absurd (see Schmitt 1958b, 488). He contrasted the Federal Republic of Germany with the Weimar Republic, and his assessment of both political systems was dismal. In the Weimar Republic, German society had been torn apart by radical political alternatives in every election in 1932/33. Now, in 1958, it was the opposite scenario: Schmitt believed there was a lack of real alternatives and, consequently, a lack of real opposition—above all, however, a lack of real politics. He quoted Kirchheimer's wording, the "desiccation" of opposition, multiple times (see Schmitt 1958b, 262, 346, and 366). However, he drew a different conclusion to Kirchheimer, who was concerned that in the absence of a hard-hitting political opposition, citizens would be alienated from politics, and this would result in the disintegration of society over time. To Schmitt, in contrast, Kirchheimer's hypothesis served as evidence of his own diagnosis that the German state had become incapable of acting. In his view, there was no longer any state at all in Germany in the true meaning of the term. Instead, "party cartels and a system of a limited opposition" (Schmitt 1958b, 366) had taken over. Schmitt recommended that Rüdiger Altmann, a publicist of the rightist authoritarian wing of the CDU and subseqeuntly an advisor to Chancellor Ludwig Erhardt, should read Kirchheimer's article on opposition as a forward-looking warning of a flaccid and apolitical Federal Republic of Germany.[20]

2. Resuming correspondence in 1958

Between the summer of 1953 and the summer of 1958, Kirchheimer and Schmitt occasionally sent each other reprints of some of their publications, but no correspondence from that time has survived. The three letters they exchanged in 1958 provide evidence of some of their disagreements.

The first of these three letters is dated more than five years after the last one Kirchheimer had written to Schmitt. He wrote him a few hurried lines on the letterhead of a Copenhagen hotel on the occasion of his seventieth birthday on 11 July 1958, but two weeks late, with belated birthday greetings to Schmitt and congratulations on his daughter's fairly recent wedding, which he had found out about from Werner Weber in Göttingen while he was visiting Smend. He closed his letter with the words: "I would have liked to express my wishes to you in person [...]. I hope to be able to be in Europe again on your 75th; simply to make good on what I missed this time."[21]

A few months before Kirchheimer wrote the above letter, he had again refused to contribute to a *Festschrift* for Schmitt.[22] Schmitt was aware of Kirchheimer's refusal. Nevertheless, he immediately picked up the ball after returning from a trip to Spain to visit his daughter. He responded by sending Kirchheimer a long letter on 6 August 1958 in which

20 Letter from Carl Schmitt to Rüdiger Altmann dated 31 March 1961; quoted in Burckhardt (2013, 155).

21 Letter from Otto Kirchheimer to Carl Schmitt dated 25 July 1958. Carl Schmitt Papers, RW 265–7603.

22 As reported by Kirchheimer in a letter to Arvid Brødersen dated 2 March 1958. Otto Kirchheimer Papers, Series 2, Box 1, Folder 25.

he expressed how pleased he was to be in touch again. He reported in detail about his work from the previous year on the edition of his anthology *Verfassungsrechtliche Aufsätze aus den Jahren 1924–1954* and informed Kirchheimer that he had "been reminded of [him] a lot" [23] while commenting on his own articles from the years 1929 to 1932. He emphasized to Kirchheimer that his writing from the Weimar period was as topical as ever: "[b]ecause of the events in recent months (de Gaulle in France—do you still remember René Capitant? […]—and the referendum in the Federal Republic) the old essays from 1932 have become more current than my new remarks from 1957."[24] He thus hinted to Kirchheimer in passing that the new presidential constitution in France was—via Capitant—inspired by his Weimar book *Der Hüter der Verfassung* [The Guardian of the Constitution].[25] He encouraged Kirchheimer to comment on his (Schmitt's) remarks and referred to the index of names that Kirchheimer could use to easily find "the passages that [he] may be most interested in personally." Schmitt also mentioned that he had a visitor from New York in his hometown Plettenberg:

> I had a visit for some months in the summer from a young student from New York, George Schwab, Columbia University, with whom I had very good conversations and whom I found very friendly. If you ever have the opportunity to speak to him—his teacher is Herbert A. Deane—Public Law and Government, Columbia Univ.—the author of the book on H. J. Laski—I would be interested in your impression of him.

Schmitt closed the letter with his hope for "good conversations about our old topics."

Kirchheimer answered this letter, but only briefly, one month later on 4 September. He brusquely rejected the claim that Schmitt's writing from the Weimar period was still as topical as ever: "I do not believe in the repetition of similar situations; too many qualitative changes have taken place." [26] In a slightly ironic tone, he conveyed to Schmitt his "anticipating thanks for enriching the treasures in my library." And he also told him: "Too bad I didn't see you, but I'm sure we can make good on that later." Three months later, Schmitt sent him a copy of his *Verfassungsrechtliche Aufsätze* with the handwritten dedication: "For Otto Kirchheimer as a Christmas greeting from Carl Schmitt, 20/12/58."[27] This time, Kirchheimer did not comment on the book to Schmitt. Instead, he responded by sending him a Christmas letter with a reprint of his essay "France from the Fourth

---

23 This and the following quotes are from a letter from Carl Schmitt to Otto Kirchheimer dated 6 August 1958. Otto Kirchheimer Papers, Series 2, Box 2, Folder 12.

24 In his letters to Ernst Jünger (Jünger and Schmitt 1999, 353) and Reinhart Koselleck (Schmitt and Koselleck 2019, 164), Schmitt sought to impress the hypothesis on them that his writing from 1930 to 1932 was even more topical in 1958 than at the time.

25 Schmitt stated as much in his letter to Ernst Jünger dated 26 August 1958 (Schmitt and Jünger 1958, 353). Schmitt disseminated this version very widely until it was taken up in a journal, in a 1961 article by Arnulf Baring (see Baring 1961, 103). On the construction and rebuttal of this legend, see Neumann (2015, 539–545).

26 This and the following quotes are from a letter from Otto Kirchheimer to Carl Schmitt dated 4 September 1958. Carl Schmitt Papers, RW 265–7604.

27 Carl Schmitt, *Verfassungsrechtliche Aufsätze*. Otto Kirchheimer Papers, Series 8–9, Box 5.

to the Fifth Republic" (see Kirchheimer 1958a).[28] This was an assessment of the events in France diametrically opposed to Schmitt's praise of the strong presidential system. Schmitt was not amused by what he received from Kirchheimer for another reason, too. He complained in his *Glossarium* that Kirchheimer had penned his Christmas greetings on the back of a print by Renaissance artist Hans Baldung Grien showing two wild horses attacking each other. Schmitt found it "ghastly as a Christmas greeting." And he continued: "Kirchheimer surely does not know what he is doing; nor does he know what he is inflicting on me with this. Nonetheless, I prefer him to *Nathan der Weise* (Nathan the Wise) and *Satan der Leise* (Satan the Quiet)."[29]

Even though Kirchheimer did not share his views on *Verfassungsrechtliche Aufsätze* with Schmitt, he had read the book even before he received the copy from Schmitt. He expressed his sharp criticism in a letter to Ernst Friesenhahn:

> I have Schmitt's essays here. The man is intransigent and the relation between the reality of the concept and responsibility is as unclear to him today as it was 30 years ago. But I fear that the evil lies deeper than the harm that the most brilliant German political thinker since Max Weber could cause. It lies in the entire German attitude that is never willing to take stock of how political and conceptual formulations correspond to reality—I know, it resulted in my having the opposite tendency of not asking deeply enough about the values and going home reassured when I established what the political-sociological equation of a concept and [a] legal construct look like, but the former deplorable custom is simply much more at home in Germany.[30]

He did not mention in the letter how Schmitt had used his (Kirchheimer's) work on political opposition to suit his own purposes and how he had misquoted him about the Weimar Social Democratic Party.

It is striking that from then on, Kirchheimer no longer invested his time and energy in grappling with Schmitt and his students. When the sociologist Otto Stammer asked him in January 1959 whether he could review Jürgen Fijalkowski's book on Schmitt, which had just been published at Stammer's Institute for Political Research at Freie Universität Berlin,[31] he declined to do so: "Please forgive me, but there are many reasons why I would not like to deal with Carl Schmitt academically."[32] He did not write any more reviews on new works by Forsthoff, Weber, Huber, or Kaiser, either. In the meantime, in February

---

28 With the dedication "With best wishes and thanks for the book! Your OK." On the first page, Schmitt underlined the wording "friends and foes" in his copy. Carl Schmitt Papers, RW 265–25656.

29 *Glossarium* entry of 30 December 1958 (Schmitt 2015, 378–379). The play *Nathan the Wise* by Gotthold Ephraim Lessing, written in 1779, is the most famous plea of eighteenth-century Germany for religious tolerance between Christians, Muslims, and Jews.

30 Letter from Otto Kirchheimer to Ernst Friesenhahn dated 15 December 1958. Otto Kirchheimer Papers, Series 2, Box 2, Folder 61.

31 See Fijalkowski (1958). Otto Stammer was the founder of political sociology in Germany in the 1950s, see Buchstein (1992, 296–313).

32 Letter from Otto Kirchheimer to Otto Stammer dated 29 December 1959. Otto Kirchheimer Papers, Series 2, Box 2, Folder 24.

1959, Schmitt advised young leftist scholar Jürgen Seifert, who had contacted him, to read Kirchheimer's *Weimar—and What Then?*[33]

Nonetheless, correspondence between Kirchheimer and Schmitt did not break off entirely. It did, however, take on the nature of smaller jabs, as the subjects of the works they exchanged reveal. In May 1959, Kirchheimer sent Schmitt a reprint of his essay "The Administration of Justice and the Concept of Legality in East Germany," which includes a defense of the *Rechtsstaat* in the face of those condemning it as a solely bourgeois institution (Kirchheimer 1959b).[34] Shortly afterwards, Schmitt recommended the "exciting article about the concept of legality in East Germany" by Kirchheimer to his young admirer Ernst-Wolfgang Böckenförde.[35] In turn, Schmitt mailed Kirchheimer a copy of the brochure *The Tyranny of Values* (see Schmitt 1960), which had been published with a small print run of 200 private copies, in March 1960. He added the dedication "for Otto Kirchheimer, C.S.—20/3/60."[36] Schmitt placed a very particular greeting to Kirchheimer on page 15 of the booklet. It reads: "Back then, in 1920, it was possible […] to suppress writings." In the copy with the dedication, Schmitt wrote by hand "especially *Fest-Schriften*" beside the word *Schriften* (writings). Schmitt was alluding to the fact that the publisher Kohlhammer had decided against publishing the *Festschrift* on the occasion of Schmitt's seventieth birthday, which was already being printed, at the last minute. The publishing house Duncker & Humblot, which generally published Schmitt's work, then produced an opulent volume with the support of affluent backers. With his handwritten addition for Kirchheimer, Schmitt once again stylized himself as the victim of malicious persecutors.

Kirchheimer's riposte could not be found wanting. In the spring of 1961, he sent him "with best compliments" a reprint of his article "German Democracy in the 1950's" (Kirchheimer 1961b). If Schmitt read it, then he surely stumbled over the passages where Kirchheimer used sharp words to criticize what the West Germans "elegantly call[ed] the *unbewältigte Vergangenheit*" (the past with which they have not come to terms) as "little demand for self-criticism" and "collective lack of memory" (Kirchheimer 1961b, 486).[37]

3. Schmitt on political justice

Kirchheimer's main intellectual project in the 1950s was his book on political justice. The backstory to the book and its subject goes back to the Weimar Republic and is closely linked to ideas from Schmitt's *Constitutional Theory*. Schmitt had devoted an entire section to the subject of political justice in his 1928 magnum opus as part of his criticism of the bourgeois *Rechtsstaat*. Schmitt defined "political justice" as the result of the impossible attempt to settle all political conflicts "via a formal judicial procedure" (Schmitt 1928b, 176). Thus, the term "political justice" in this context has nothing to do with the normative

33 As reported by Jürgen Seifert, see Seifert (1996, 118).

34 Carl Schmitt Papers, RW 265–25663.

35 Letter from Carl Schmitt to Ernst-Wolfgang Böckenförde dated 8 May 1959 (Schmitt and Böckenförde 2022, 199).

36 Carl Schmitt, *Die Tyrannei der Werte*. Otto Kirchheimer Papers, Series 8–9, Box 5.

37 The reprint is in Landesarchiv Nordrhein-Westfalen, Papers of Carl Schmitt, RW 265–25657.

question of the moral quality of judicial decisions, but is rather focused solely on the procedural administration of justice.

In his book, Schmitt first described the historical conditions for the emergence of the liberal *Rechtsstaat*, then its key substantive components and institutional organizational principles. He asserted that the fully realized ideal of the *Rechtsstaat* "culminates in the conformity of the entire state life to general judicial forms" (Schmitt 1928b, 176). The *Rechtsstaat* must provide for a procedure for every type of disagreement and dispute between citizens and the state as well as between the various state institutions. The main prerequisite for these procedures to succeed was valid general norms. Some of the disagreements and disputes, however, lacked such a norm laid down in advance. Schmitt recognized a systematic gap in the theory of the bourgeois *Rechtsstaat* at this point. Some disagreements and disputes were so strong "that the political distinctiveness of such cases" (Schmitt 1928b, 176) was inevitable, and it was also inevitable that judges' decisions in these cases were political decisions. That, he thought, constituted "the actual problem of political justice" (Schmitt 1928b, 176) as a component of the *Rechtsstaat* that was as necessary as it was contrary to the system.

Schmitt's concept of political justice was not a negative polemical one. He did not consider political justice to be the abuse of judicial procedures out of political calculus or the camouflaging of political purposes by means of a judicial facade. Quite the contrary. The political was already inherent to the matter in dispute. Because of its political character, "a special procedure or order is provided for special types of genuine legal disputes" (Schmitt 1928b, 177). This occurred less in realms of private law but more often in criminal matters or in disagreements under public law.

Schmitt lists the six most important examples of political justice in his book (see Schmitt 1928b, 176–180). The list begins with high treason and treason against a home country. The second example is ministerial and presidential indictments. Next comes genuine constitutional disputes decided by constitutional courts; in other words: Schmitt thought that all decisions made by constitutional courts were a form of political justice. The fourth example is doubts and differences of opinion over the constitutionality of statutes and decrees by special courts. The fifth example is taken from France and the United States and their special treatment of governmental acts in the area of adjudication. Schmitt's last example is judicial reviews of important elections by constitutional courts or special electoral review commissions (see Schmitt 1928b, 180).

Schmitt claimed that in all the examples listed, purely judicial decisions following the ideal of the *Rechtsstaat* were not possible. The actions of the courts always took place in a political decision-making arena whose sphere of influence therefore had to be, and was indeed, the subject of political decisions in advance. According to Schmitt, the ideal of the bourgeois *Rechtsstaat* was a chimera; and political justice was a symptom of the fact that the perfect *Rechtsstaat* could never be attained. It is clear that the intent behind this hypothesis from his *Constitutional Theory* was critical: Schmitt considered it to be a devastating blow right to the heart of the liberal theory of the *Rechtsstaat*.

Schmitt remained true to this line of criticism throughout the next five years. In 1929, he had criticized proposals to confer the final decision concerning disputes in constitutional law to special courts, which would make them guardians of the Weimar Constitution. He believed that giving this role to a supreme court would ultimately bring about "unrestrained expansion of the justice system" and would transform the courts into political authorities: "Politics would not be juridified; instead, the justice system would be politicized" (Schmitt 1929a, 98). Schmitt thought the judicial institutions should not even attempt to adjudicate social conflicts but should leave this to a strong government and the President of the Reich. In his 1931 book *The Guardian of the Constitution*, in which he promoted the presidential dictatorship, he quoted the well-known statement by Guizot that in the event of such juridifications, "politics had nothing to gain and the justice system had everything to lose" (Schmitt 1931b, 35).

When Hitler's government took power in 1933, Schmitt abandoned his position that the justice system should exercise political restraint. It should suffice here to mention his *Fünf Leitsätze für die Rechtspraxis* [Five guiding principles for legal practice] from the summer of 1933, analyzed in more detail in Chapter 7, in which he supported an openly political justice system not bound by any laws (see Schmitt 1933h, 54–56). After 1945, he again made a complete about-face, complaining bitterly in his prison notebooks that the victors were exercising political justice over the Third Reich actors. In *Ex captivitate salus*, he condemned what he alleged was a transformation of "the means and methods of the judiciary into means and methods of annihilation" (Schmitt 1950a, 48) by the victorious powers. To his mind, the Nuremberg Trials were a prime example of "political trials."[38] In his *Glossarium*, Schmitt packaged his fundamental rejection of political trials in a mythological narrative. Each of the three peoples with a formative influence on current-day European thought had had just a single truly great man: the Greeks Socrates, the Jews Jesus Christ, and the Romans Julius Caesar. "Each of these three people killed its greatest man; but only the people with the strongest sense of law and legal procedure, the Romans, did not kill their Julius Caesar by judicial means, but murdered him."[39] In other words, respect for the integrity of the law dictated the act of murder.

After the founding of the Federal Republic of Germany, Schmitt continued his line of criticism of the juridification of politics from the Weimar era. He was the major voice operating behind the scenes against a constitutional court and an alleged "judiciality" of politics.[40] In the spring of 1961, Schmitt used pseudonyms for a number of letters to the editor as part of a campaign against the *Bundesverfassungsgericht*, which he accused of "political justice" because it had put an end to the efforts of Chancellor Konrad Adenauer to create a government television station. He accused the court of grossly overstepping its authority with this ruling. The consequence, according to Schmitt, was that the decision whether or not to establish a television station for government propaganda was to be left exclusively to the political leadership. Taking up "Bonn is not Weimar," a common saying at the time, he stated, "Bonn is Karlsruhe," Karlsruhe being the seat of the *Bundesverfassungsgericht*.[41]

An astounding parallel to what Kirchheimer wrote in his book *Political Justice* is to be found in one of Schmitt's comments on his own Weimar writing in *Verfassungsrechtliche*

38 *Glossarium* entry of 6 May 1948 (Schmitt 2015, 110).

39 *Glossarium* entry of 10 April 1957 (Schmitt 2015, 359).

40 See Chapter 15, p. 386.

41 Carl Schmitt, letters to the editor of *Deutsche Zeitung*, 17 April and 8 May 1961; quoted in Giesler (2016, 33 and 34).

Aufsätze. About the political role of the Weimar *Reichsgericht* (see List of German Courts), he wrote in retrospect in 1958: "The extent to which the means and methods of the judicial process change its object would require a fundamental scholarly examination" (Schmitt 1958b, 109). Elsewhere, he discussed the concern of President of the Reich Hindenburg in 1932 about the "newly established weapon, the threat of political proceedings before the *Staatsgerichtshof*" (Schmitt 1958b, 350)[42] against him in the event that the state of emergency would be extended. It was only this threat, Schmitt claimed, that had convinced Hindenburg to appoint Hitler Chancellor of the Reich. Regardless of the historical soundness of Schmitt's comments on his own work—the questions he raised were right at the center of Kirchheimer's studies on political justice. Kirchheimer had been working on the subject from 1952 on, and he published his first programmatic essay on it in a German publication in 1955 (see Kirchheimer 1955b). There is no doubt that Schmitt was aware of this essay when he wrote the comments on his own work mentioned above in 1958.

4. The backstory to Kirchheimer's book

Of course, Kirchheimer could not have known about Schmitt's notes in his *Glossarium*, but he was familiar with his *Constitutional Theory* and his *Guardian of the Constitution*. He had also addressed political justice in his Weimar writing himself, albeit in a different way to Schmitt. Back in 1929, in one of his earliest comments on judicial policy,[43] Kirchheimer had criticized the decisions of the *Reichsgericht* on the occasion of the 50th anniversary of its establishment in Leipzig as a faithful reflection of the ideas and views of Germany's ruling classes. He had faulted both a politically one-sided practice of ruling against the left and the court's defense against the law of expropriation codified in the constitution as politically motivated perversion of justice. Kirchheimer's understanding of political justice during the Weimar Republic was synonymous with the accusation of class justice going back to socialist Karl Liebknecht during the German Empire; this was also raised by authors during the Weimar Republic, for instance, his fellow lawyer Ernst Fraenkel.[44]

Kirchheimer had also used the term "political justice" in his 1935 article "State Structure and Law in the Third Reich" in connection with his attacks against Schmitt legitimizing the changes in Nazi criminal law. Kirchheimer saw this as the construction of "*political* cases" (Kirchheimer 1935a, 153)[45] of criminal law for the purpose of persecuting political opponents. The procedural guarantees and the independence of the judiciary in Nazi criminal law had already been replaced by executive orders, with Schmitt's blessing, in the fight against political opponents: "It is thus perfectly legitimate to conclude that political justice in Germany is primarily administered by policemen who punish" (Kirchheimer 1935a, 153). Only if it happened to be opportune for the police or the Gestapo did they bother to hand their cases to the courts.

42 He made the same argument on page 450 of his book.

43 Discussed in Chapter 3, p. 72–74.

44 See Fraenkel (1927) and Deutsche Liga (1927).

45 Emphasis by Otto Kirchheimer.

It is striking against this background that Schmitt and Kirchheimer evaluated political justice in very different ways in the postwar period. Both agreed on the terminological level that political justice was understood as making political use of the opportunities given by judicial proceedings. Unlike the English term, the German term "*Politische Justiz*" does not allow any other meaning. However, Schmitt and Kirchheimer strongly disagreed in their evaluation of the phenomenon of political justice. Schmitt had switched to the mode of complaining about a political justice system he claimed was hostile to him. In contrast, Kirchheimer's usage of the term took two steps in the opposite direction: changing course from the mode of exposing to one of soberly describing, and in the context of his theory of liberal democracy, he even took on a normative, potentially positive understanding of the term. He developed this concept extensively in his 1961 book *Political Justice*.

The editorial backstory of the book dates back to 1952,[46] when the Rockefeller Foundation, under the direction of Franz L. Neumann, established a program to support Legal and Political Philosophy (LAPP). The program was to provide a response to positivism, which was dominant in US political science. Alternatives to analytical philosophy, which was also dominant in the US, were to be strengthened as well. The Rockefeller Foundation devoted 1.7 million US dollars to political theory and the history of political ideas as part of LAPP over ten years. Neumann encouraged Kirchheimer to apply for funding for his book project on political justice, and after Kirchheimer had done so a number of times, it was the eighteenth project to be funded. Book projects by Herbert Marcuse, Henry A. Kissinger, Leo Strauss, Eric Voegelin, and Hannah Arendt received funding, too.

Kirchheimer first applied to the Rockefeller Foundation for support for the book in February 1954. He stated that the research goal was to develop the "legal and socio-political aspects of political justice."[47] In the first project proposal, he envisaged that the book would have six chapters; the English version of the book later had eleven, the German one twelve. He did not yet plan for a theoretical chapter on the concept of political justice or for chapters on political justice in the GDR, asylum law, or the Nuremberg Trials. In his first proposal, he structured the book according to the actors in court proceedings; the analysis was to be centered around the roles of prosecutors, defense attorneys, defendants, and judges in political proceedings. Even in this initial phase of his considerations, Kirchheimer circumscribed the term in two ways. For one thing, in contrast to his previous use of the term, he no longer considered it to have a purely exposing function. And for another, like Schmitt in his *Constitutional Theory*, he distinguished it from an understanding of the justice system as a neutral instance free of politics, following legal positivism.

According to his project proposal, Kirchheimer's conceptualization of his work was to go beyond existing individual studies examining the question of individual guilt in the cases against Ethel and Julius Rosenberg as well as Nicola Sacco and Bartolomeo Vanzetti. By contrast, he intended to analyze the social and political causes of initiating court proceedings in political controversies: "Most of the studies on these subjects

46 On its editorial history, see also Klingsporn and Wilke (2019, 33–40 and 2020, 704–726).

47 Project proposal for "Political Justice" of 23 February 1954, Archive of the Rockefeller Foundation, RF RG1.2 Series 200, U.S.-Social Science Box 539, Folder 4614.

have been written by either Nazis or Communists. Law reviews, as a rule, have not dealt with the subject as a whole."[48] In the interview with the foundation, Kirchheimer praised the study *Ritual of Liquidation: The Case of the Moscow Trials* (see Bernaut and Leites 1954) by Elsa Bernaut and his former Weimar co-author Nathan Leites, which had just been published, as the work going farthest in this direction at that point in time. Seeking to make statements about the values, attitudes, and relationships within the party leadership under Stalin, Leites and Bernaut had used a literary and psychoanalytical approach in their analysis in order to determine the feelings and the attitudes of the accused and their defense lawyers toward Bolshevism.[49]

Kirchheimer's approach to his research was more comprehensive. He wanted to study both the commonalities and the differences in the practice of political justice in the USSR, the GDR, under the Nazi regime, and in democratic *Rechtsstaaten*. His starting point was a concept of political justice that could be used equally in all political systems. Only comparative legal research and an analysis of the roles of those involved in the proceedings would be able to demonstrate the functional differences between political justice in the various forms of regimes.

In April 1954, the Rockefeller Foundation informed Kirchheimer that it could not fund the project because he did not have a professional position at a university.[50] He undertook another attempt to finance his project in November 1954.[51] His application was supported by a peer-reviewed report by Herbert Marcuse. In this report, Marcuse praised Kirchheimer as "one of the most gifted and original scholars in the field of political science and political philosophy."[52] In 1955, Kirchheimer's first programmatic essay "Politische Justiz" [Political justice] was published in the *Festschrift* on the occasion of Max Horkheimer's sixtieth birthday (see Kirchheimer 1955b).

Once the New School for Social Research had formally offered Kirchheimer a full-time position, he had fulfilled the prerequisites for the grant. But now it was the New School that foiled his plans. It was not willing to give him a leave of absence right at the beginning of his employment there.[53] After some back and forth, Kirchheimer finally received the research grant he had desired for so long covering the period from June 1957 to December 1958.[54] Kirchheimer prepared the English-language text of *Political Justice* in

48 Project proposal for "Political Justice" of 23 February 1954, Archive of the Rockefeller Foundation, RF RG1.2 Series 200, U.S.-Social Science, Box 539, Folder 4614.

49 Minutes of the interview with Otto Kirchheimer by Kenneth W. Thompson of 23 February 1954, Archive of the Rockefeller Foundation, RF RG1.2 Series 200, U.S.-Social Sciences, Box 539, Folder 4614.

50 Letter from Kenneth Thompson to Otto Kirchheimer dated 8 April 1954. Archive of the Rockefeller Foundation, RF RG1.2 Series 200, U.S.-Social Sciences, Box 539, Folder 4615.

51 Minutes of the interview with Otto Kirchheimer by Kenneth W. Thompson of 13 November 1954, Archive of the Rockefeller Foundation, RF RG1.2 Series 200, U.S.-Social Sciences, Box 539, Folder 4614.

52 Letter from Herbert Marcuse to John Stewart dated 20 December 1954. Archive of the Rockefeller Foundation, RF RG1.2 Series 200, U.S.-Social Sciences, Box 539, Folder 4614.

53 Letter from Hans Simons to John Stewart dated 14 July 1955. Archive of the Rockefeller Foundation, RF RG1.2 Series 200, U.S.-Social Sciences, Box 539, Folder 4615.

54 Letter from Hans Simons to Flora Rhind dated 15 January 1957. Archive of the Rockefeller Foundation, RF RG1.2 Series 200, U.S.-Social Sciences, Box 539, Folder 4615.

close cooperation with Arkadij Gurland. After spending three years in Germany, which were a professional failure, Gurland was living in New York again and needed to find paid work.[55] His first letters with comments on the manuscript date back to December 1957. He criticized using the word "justice" in English for the German word "*Justiz*" and, even at this early stage, pointed out that the term could be confused with the normative moral quality of being just. He advised Kirchheimer to use "administration of justice" when writing for an English-language readership,[56] but Kirchheimer did not follow this advice. Had he done so, readers would not have misunderstood what the main subject of his book was, and he would not have had to deal with the fallout.

In late 1958, Kirchheimer's grant was extended through September 1960.[57] Besides Gurland, who worked with Kirchheimer on an ongoing basis, Karl Loewenstein now also influenced the book. In an expert review of the manuscript for Princeton University Press in May 1960, he praised the chapters on the Nuremberg Trials and on the current-day problems of granting asylum but also criticized the book for still being too unsystematic.[58] At Loewenstein's suggestion, Kirchheimer explained and theoretically developed his use of the term "political justice" in an introductory chapter. Kirchheimer finalized the manuscript before beginning a semester as a Fulbright Visiting Professor at the University of Freiburg in May 1961 at the invitation of Konrad Hesse and Horst Ehmke, who had both been students of Smend's.[59] A preprint of one chapter was published in German in the *Festschrift* for Rudolf Smend on the occasion of his eightieth birthday. The editors apparently paid such meticulous attention to ensure that no traces of Schmitt could be found in the *Festschrift* that they even made suggestions regarding the language used. Co-editor Ulrich Scheuner wrote a concerned letter on this matter to Kirchheimer. He asked him to replace "*Freund-Feind-Beziehung*" ("friend-enemy relationship"), which he had used once, with "*Kontrastbeziehung*" ("contrasting relationship").[60] Kirchheimer claimed this request was "for reasons of academic politics"[61] in a letter to his translator Gurland. He honored the request. He was apparently able to do so easily not least because six years previously, he had made the same linguistic change in the German-language version of his essay "Politische Justiz" for the *Festschrift* for Max Horkheimer. Whereas in the English version of *Politics and Justice* published in 1955, the term "enemies" was used most of the time and "adversaries" occasionally (see Kirchheimer 1955a, 410, 411, 412, and 413), they were both translated as "*Gegner*" (adversaries or opponents) in the German-language version (see Kirchheimer 1955b, 106, 107, and 109).

55 On Gurland's postwar biography, see Buchstein (2010).

56 Letter from Arkadij Gurland to Otto Kirchheimer dated 31 December 1957. Otto Kirchheimer Papers, Series 2, Box 1, Folder 68.

57 Letter from Kenneth Thompson to Otto Kirchheimer dated 19 February 1959, Archive of the Rockefeller Foundation, RF RG1.2 Series 200, U.S.-Social Sciences, Box 539, Folder 4615.

58 Review Reader B [Karl Loewenstein] for Princeton University Press. Otto Kirchheimer Papers, Series 3, Box 2, Folder 64.

59 Letter from Otto Kirchheimer to Horst Ehmke dated 26 January 1961, Horst Ehmke Papers, No. 504.

60 Letter from Ulrich Scheuner to Otto Kirchheimer dated 25 August 1961. Otto Kirchheimer Papers, Series 2, Box 2, Folder 7.

61 Letter from Otto Kirchheimer to Arkadij Gurland dated 30 August 1961. Arkadij Gurland Papers, NA 5.

Such considerations were irrelevant for the English version of the book. Gurland suggested using "foe" instead of "enemy" in the opening sentence "for reasons of style."[62] This time, Kirchheimer followed his advice (see Kirchheimer 1961a, 3). That he considered "enemy" and "foe" interchangeable in their semantic meaning[63] is evident from the fact that he used "foe" in some places in the book and "enemy" in others without any systematic order.[64] How little the English version followed Schmitt's vocabulary is also apparent from Kirchheimer's use of alternate terms such as "hostile groups," "opponents," and "adversaries" in other places.

5. The ambivalences of political justice

Kirchheimer's book *Political Justice* has the subtitle *The Use of Legal Procedures for Political Ends*. This was suggested by editor Judy Walton of Princeton University Press,[65] who wanted to add it to indicate to American readers what the book was about. Kirchheimer dedicated the book "To the Past, Present, and Future Victims of Political Justice" (Kirchheimer 1961a, v). There were more than two decades between his use of the term "political justice" for criminal justice against political opponents and his understanding of "political justice" as a broader term for any use of the judicial process for the purpose of gaining (or upholding, or enlarging) or limiting (or destroying) political power. The political justice he referred to now pointed well beyond political criminal justice. His book[66] uses examples dating back to European antiquity and right up to the time of writing to provide a seemingly encyclopedic overview of all kinds of political proceedings. Most of the examples are from the US, France, the UK, Russia/the Soviet Union or from Germany during the Weimar Republic, the Third Reich, the GDR, and the Federal Republic of Germany at the time.

The book consists of three main sections with a total of twelve chapters. In the first section, Kirchheimer portrays the emergence of the category of a specifically political offense and devotes a chapter each to the basic forms of political proceedings against individuals and the measures that could be used against political organizations. In the second section, three chapters look at the institutions and actors of political justice: the

62 Letter from Arkadij Gurland to Otto Kirchheimer dated 5 May 1960. Otto Kirchheimer Papers, Series 2, Box 1, Folder 68.

63 "Foe" is of Germanic origin, related to the word "feud." "Enemy" has a Latin source (*inimicus* = not friend). To some native speakers, "foe" sounds more old-fashioned and literary. In their translations of Schmitt's late work, George Schwab and Gary L. Ulmen later created the semantic distinction between enemy, i.e., a legitimate opponent whom one fights according to recognized rules and whom one does not discriminate against as a criminal, and foe, i.e., a lawless opponent whom one must fight to the death (see Schwab 1987 and Ulmen 1987). However, this distinction has not prevailed, even among Schmitt scholars.

64 See, for example, Kirchheimer (1961a, 16, 234, 421).

65 Letter from Judy Walton to Otto Kirchheimer dated 17 January 1961. Otto Kirchheimer Papers, Series 2, Box 2, Folder 65.

66 See Kirchheimer (1961a). The following page numbers refer to this book.

public prosecutor's office, the courts, the judges, the defense attorneys, and the defendants. The next two chapters examine specific constellations of political justice: justice in the GDR in the 1950s and the problem of proceedings against former political and military elites by successor regimes, such as with the Nuremberg Trials. In the third and final part of the book, Kirchheimer explores two phenomena he considered to be closely linked to political proceedings, first, asylum law, understood as an opportunity to escape from politically motivated criminal prosecution by fleeing and emigrating and, second, the institution of mercy, understood as lessening or erasing the punishment for political offenses ex post.

Kirchheimer begins his initial clarifications of terms in the first chapter with the words "Every political regime has its foes or in due time creates them" (3). He then goes on to define politics as a degree of intensity, as did Schmitt. Something was deemed to be political "if it [was] thought to relate in a particularly intensive way to the interests of the community" (25). Political justice was just one area among many in the political struggle for power. Consequently, he commented on political events and the impacts of political justice using categories such as victory and defeat. At the same time, his understanding of politics included an appeal to the public—whose support was essential for victory in political competitions. Kirchheimer understood political proceedings conducted in the public eye to be events with a theatrical dimension and used vocabulary from the sphere of the theater: stage, audience, director, drama, and script. Political proceedings conducted in public, he asserted, had the primary function of legitimation but also to restrict the scope for political action. They could serve as a resource for the production of political legitimacy—or not, as the case may be. Thus, political trials were "marked by a creative element of risk and unpredictability, which distinguishe[d] them from an administrative command performance" (Kirchheimer 1968b, 98).

Viewed against the background of Kirchheimer's political biography and his lifelong debate with Schmitt, four topics in the book stand out: first, Kirchheimer's cursory retrospect of the Weimar Republic; second, the subjects of fleeing, asylum, and amnesty; third, his integration of political justice into a theory of the democratic *Rechtsstaat*; and fourth, his assessment of the Nuremberg Trials in retrospect, which is the subject of the next section of this chapter.

Various examples from the Weimar Republic are presented in multiple chapters of *Political Justice*, yet they do not provide many surprises. Kirchheimer devotes a considerable amount of space to the unsuccessful libel suits that President of the Reich Friedrich Ebert brought against the right-wing press and reminds readers of the right-wing leanings of most of the Weimar judges. The only new addition is a detailed description of the failure of the Weimar courts in combating Nazi and other right-wing extremist perpetrators of violence and insurgents.

In Chapters IX and X, Kirchheimer gives new weight to the subjects of fleeing, asylum, and amnesty. He addresses intent in the context of amnesties, namely that amnesties were "intended to efface the memory and possible consequences" (418). However, Kirchheimer first poses the question of the legitimacy of such intentions before coming to an affirmative answer. After all, the victims might consider amnesties to be illegitimate at times. For the case of Germany after 1945, he saw mostly "former rightists and former National Socialists" (414) rallying around the battle cry for a comprehensive amnesty. In particular, he praised the Bundestag for successfully resisting manifold pressures, and for how it had done so (see 412–415). Schmitt, of course, had the opposite view, calling amnesty the "most foundational position of that which one can call justice"[67] in his *Glossarium*.

Concerning the problem of asylum, Kirchheimer saw a fundamental difference between the nineteenth and twentieth centuries, as had Hannah Arendt before him (see Arendt 1951, 267–271). Whereas a few vulnerable political rebels had become political refugees in the nineteenth century, fleeing from persecution that targeted certain groups grew into a mass phenomenon in the twentieth century. The majority of reasons for fleeing no longer had any connection to political justice. The refugees were not persecuted because they were accused of breaking the law but because they belonged to an unwelcome social, ethnic, or religious group. The prototype of the asylum seeker in the twentieth century was "just one of hundreds of thousands or millions threatened by a policy directed against a social stratum or an ethnic group" (354). Against this background, Kirchheimer advocated expanding the category of political refugee in asylum law to include all actual or suspected victims of racist, national, religious, or political persecution. At the same time, however, he was against a legally binding obligation of states to admit victims of political persecution. He accepted the right of every state to grant or refuse political asylum. His rationale referred to Hugo Grotius, who claimed a state "would come to an end if the right to emigrate were given to every man" (365).

Kirchheimer's examples of political refugees from the Nazi regime being denied asylum in France and Switzerland make for sobering reading. The Soviet Union comes off even worse in historical comparison. It granted asylum solely according to the criterion "how serviceable an individual was to the party machine" (358). Its practice flagrantly contradicted the conferences on asylum law—propaganda events masterminded by communists—which Kirchheimer and his wife had attended in their Paris exile in 1936. He described the US as an ambiguous case whose practice "the lucky winners extol and the losers vilipend" (360). Kirchheimer highly praises the "generous provisions" of the Basic Law of the Federal Republic of Germany that "go beyond the established practice of international law" (356). Such an article of the Basic Law had been possible only under the immediate impression of the relentless persecution of political opponents in the Nazi regime.

Examining how Kirchheimer embedded the phenomenon of political justice in his theory of the democratic *Rechtsstaat*, the third topic in the book, is certainly fruitful. Like Schmitt—but without mentioning his name—Kirchheimer contradicted how Western democracies officially presented themselves, namely that political justice could not exist there. He argued that political justice was unavoidable even in well-functioning *Rechtsstaaten*—and that this was not necessarily all negative. He had formulated this hypothesis in his first project proposal to the Rockefeller Foundation back in 1954. He understood political justice as one of multiple modalities for dealing with the real or imagined principal opponents of a political regime. In *Rechtsstaaten*, an internal contradiction resulted from this, namely between the legal means and the political goal to be achieved. This fusion of politics and opposition is reminiscent of Schmitt at first glance.

67 *Glossarium* entry of 5 December 1947 (Schmitt 2015, 43).

As described above, Schmitt had in fact devoted an entire section to political justice in the *Rechtsstaat* in his *Constitutional Theory*. However, Kirchheimer only addressed two of Schmitt's six main examples of political justice: high treason and indictments of ministers or other high officials, and unlike Schmitt, he wrote about asylum law and amnesty law.

Kirchheimer viewed politics and justice as placed along a continuum. Political actors made use of the opportunities afforded by judicial proceedings—be it at the initiative of state agencies or of members of the opposition who wanted to publicly put the state in a bad light. Yet in liberal *Rechtsstaaten*, those in power could not prevent members of the opposition from gaining control over the justice apparatus, too. They could contrive suits concerning libel or perjury. Groups excluded from political power could thus influence their fellow citizens' political views. Political proceedings were fraught with risk for both sides. If those in power took control of the justice apparatus, this created uncertainties in liberal democracies. Since the development of policies had to take a "detour" (421) via legal procedures and their inflexible rules, it was impossible for political intentions to prevail unchanged. That meant wasting time and limiting the methods that could be used, which implied risks for both sides. This was the specific judicial space, the uncertainty of the outcomes of political trials, in liberal democracies. Consequently, political justice in democratic *Rechtsstaaten* might even have "benefits" (429) by taking on a positive function civilizing the conflict. Then, following Smend, it performed the function of integrating society via conflicts.

Kirchheimer supported his deliberations with the historical hypothesis that using legal procedures politically only made sense with the emergence of the bourgeois *Rechtsstaat*. Limiting state action in procedural terms, or juridification, did not mean the end of political justice but was, rather, its prerequisite. He identified the Federal Republic of Germany and the US as being at opposite ends of the spectrum, Germany as a militant democracy able to repress principal political opposition by banning political parties or depriving citizens of fundamental rights and the US as a country with virtually no legal restrictions to principal political opposition. Kirchheimer's ideal was a liberal practice like what he saw in the UK. There, the government strictly limited itself to repressing unlawful acts while granting generous freedom to propaganda and political organization. He placed France and Italy in between, too, with their administrative discrimination against the communists. Seeking to depoliticize political justice, Kirchheimer believed, was in the best case an optical illusion, in the worst self-deception. In a democratic *Rechtsstaat*, what mattered was not to abolish political justice, but to reduce it to a tolerable level. The alternative to political justice could be worse, for example, if a regime acted more arbitrarily and perhaps violently if it had no recourse to the courts. Like Schmitt, Kirchheimer held the view that political justice could not be overcome. Unlike Schmitt, however, he did not think that this fact was an absolute breach with the system of the principles of the *Rechtsstaat* but an opportunity for its recognition by society.

Nonetheless, the tone of Kirchheimer's book is melancholy in many places. He was of the opinion that political justice united both a repressive and a civilizing element within itself, and the two were inextricably linked to one another. This tension could not be unmade; at best it could be recognized in its ambiguity. Political justice could never bring about complete justice.

6. In dialogue with Hannah Arendt

A fourth key topic of Kirchheimer's *Political Justice* is the Nuremberg War Crimes Trials addressed in Chapter VIII, titled "Trial by Fiat of the Successor Regime." Looking back twenty years later, Kirchheimer considered these trials to be proceedings without any prior models or precedents. Whatever pressure there was did not come from external organized groups but from the situation itself: It was the pressure of "the charnel houses, the millions who had lost their families, husbands, or homes" (340). Kirchheimer was analytically correct in calling these proceedings "successor trial[s]" (324) in hindsight. But he did not conclude from this that they had been illegitimate. On the contrary. He praised them as an achievement of civilization and an example of how transitional justice could succeed. The paradigmatic significance and accomplishment of the trials had been undermined by the dissensions among the wartime partners. The wartime coalition had broken apart "before the ink on the Nuremberg judgement had time to dry" (324) and the opportunity to lay a firm foundation for a new world order was wasted.

Kirchheimer reviewed four main objections of the defense attorneys against the Nuremberg prosecution, rejecting one after the other. Incidentally, Schmitt, too, had independently put forward these four objections, using identical or similar wording, in his letters, anonymous articles, or diary notes,[68] most of which Kirchheimer could not have been aware of. The first objection was that the defendants could plead that they had simply followed the law. Using a line of argument following the philosophers of natural law, Gustav Radbruch and Lon L. Fuller, Kirchheimer countered that the defendants "in those patently exceptional cases" (328) such as the mass murder of Jews and Poles were not permitted to plead that they had implemented existing laws or that their behavior had been legal. He argued that the value of legal certainty "is not strong enough" against the principle that "intentional violation of minimum standards deprives an enactment of the claim to legal validity" (328).

A second objection was following superior orders: the perpetrators had been bound to specific and binding orders. Kirchheimer responded that there had indeed often been superior orders at the lower levels of the military, and that refusing to follow orders had cost soldiers their lives. "[I]n every case, there will come a point when the illusion that one's own influence can arrest more general developments will be dispelled. At this moment there arises the conflict of open resistance or silent withdrawal" (331). Kirchheimer explains that "active resistance will always remain a highly personal decision" (331) and concludes that "active resistance to the oppressor is [...] an illusory yardstick" (331). In his view, the "legitimate yardstick" is "withdrawing" from "significant participation" (331) in the regime. Kirchheimer states that this kind of behavior was possible in Nazi Germany, citing the example that some judges avoided appointments to *Sondergerichte* (see List of

68 See Chapters 13 and 14.

German Courts) without facing reprisals. At the higher level of military command, relations were in fact "more like relations within what might be described as a power elite, and should be judged in these terms" (330). The defendants in Nuremberg belonged to this power elite who had decided not to withdraw from their positions.

The third objection of the defense concerned the alleged bias of the court because it had been appointed by the victorious powers. Kirchheimer did not consider this a special feature. In all political trials conducted by the judges of the successor's regime, "the judges are in a certain sense the victor's judges." (352). He forcefully contradicted the assertion that a German court would have arrived at different rulings in the first postwar years. German judges—provided they had not been Nazis—may have emphasized different points and selected different procedures, yet they would by no means have arrived at rulings more beneficial to the defendants.

A fourth objection concerned the accusation of *tu quoque*: judgment was passed on acts that the Allies had also committed themselves. Kirchheimer stated that this accusation made clear that political proceedings like these concerned not only the past but that they always also concerned the future: "In laying bare the roots of iniquity in the previous regime's conduct, it simultaneously seizes the opportunity to convert the trial into a cornerstone of the new order" (336). In the case of the Nuremberg Trials, the accusation of *tu quoque* was absurd and merely showed to what small degree those who propounded it had understood the Nazi regime's crimes: "Of those misdeeds which we call offenses against the human condition, no comparable practices of any state of the world, whether represented on the bench or not, could serve in exculpation or mitigation" (338).

This last question marked the point at which Kirchheimer again clearly acknowledges his commitment to universal norms. Elsewhere in his oeuvre, he usually refers to the historical variability of norms or to the political instrumentalization of moral norms. But confronted with the evidence of mass annihilation and mass enslavement presented in the Nuremberg Trials, he reverts to a position akin to natural law:

> And in wading through the evidence on mass annihilation and mass enslavement, those fact situations which we have since come to describe as genocide have established signs, imprecise as they might be, that the most atrocious offenses against the human condition lie beyond the pale of what may be considered contingent and fortuitous political action, judgment on which may change from regime to regime (341).

The "lasting contribution" of the Nuremberg Trials was their transformation into "the concerns of the human condition, the survival of mankind in both its universality and diversity" (341). Kirchheimer thought this also raised the question of how to deal with the principle of "individual responsibility" (319) if the entire state had become an *état criminel*.

It is precisely these questions about the link between social norms, the law, and personal responsibility that Hannah Arendt poses in her coverage of the trial against Adolf Eichmann published in *The New Yorker* in 1961. Arendt had read *Political Justice* immediately after its publication and had included it in the later version of her reporting, which was published as a book in 1963. Arendt and Kirchheimer were acquainted from their time in exile in Paris but had never developed a closer personal relationship. This remained the case even though they met privately a few times in New York.[69] Academically, however, they highly respected each other. Arendt, too, was particularly interested in the eighth chapter, as evidenced by what she underlined in her copy of Kirchheimer's book.[70]

In her own book, *Eichmann in Jerusalem*, Arendt built on Kirchheimer's deliberations on the theatrical dimension of political justice and the "irreducible risk" (Arendt 1963, 208) of political trials. Moreover, she referred explicitly to his analyses of the political context of the Nuremberg Trials (see Arendt 1963, 127, 256, 257, and 266). She also quoted a key idea from his work for her book about distinguishing between guilt and innocence. She followed Kirchheimer both in her argument that active resistance to a totalitarian regime was an illusory and wrongful normative standard and in her definition of the appropriate yardstick:

> [T]he only possible way to live in the Third Reich and not act as a Nazi was not to appear at all; 'Withdrawal from significant participation in public life' was indeed the only criterion by which one might have measured individual guilt, as Otto Kirchheimer recently remarked in his *Political Justice* (Arendt 1963, 127).[71]

Kirchheimer had added the following in his book to the parts Arendt had quoted:

> A large body of experience teaches us that many men show a fatal proclivity toward pushing themselves, or allowing themselves to be pushed, into positions where they know in advance the honors and rewards will entail corresponding entanglement and responsibility (Kirchheimer 1961a, 331).

Although he did not mention Schmitt by name in this passage, readers in the know were certainly aware that he was referring not only to jurists in general but to *Preußischer Staatsrat* Carl Schmitt too.

Kirchheimer, for his part, was just as keen to read Arendt's book. Referring also to his wife Anne and his son Peter, he wrote to her after reading the first edition of 1963: "my family of non-professionals is reading it also with great interest and seem to share my enthusiasm." He told her he agreed with her analysis on the whole: "I agreed with about two thirds of what you said and my disagreements are minor and on the legal rather than a political or moral level."[72] His affirmative statement is remarkable considering the vehement criticism Arendt's book received after its publication in the US, in particular with regard to her discussion of the role of the *Judenräte* (Jewish Councils) in the concentration camps. Kirchheimer's former superior at the OSS, Robert M. W. Kempner, was among the first harsh critics in *Aufbau* (see Kempner 1963), followed by the vast majority of Jewish readers (see Renz 2021, 50–131). Along with Arendt's friend sociologist Joseph Maier

---

69 As reported by Peter Kirchheimer in a conversation with the author on 4 May 2023.

70 See Arendt's copy of Otto Kirchheimer, *Political Justice*. Bard College, Hannah Arendt Collection, Call #: KF310.P65 K56.

71 See Kirchheimer (1961a, 331).

72 Letter from Otto Kirchheimer to Hannah Arendt dated 15 May 1963. Otto Kirchheimer Papers, Series 2, Box 1, Folder 11.

(see Maier 1963), who had shared an office with Kirchheimer at the Institute of Social Research in New York, Kirchheimer was among the few Jewish authors who mostly agreed with her view of the Eichmann trial. In his letter to her, he added a list with a number of corrections concerning technical legal aspects of her book. Arendt took them on board in the revised second edition of her book in 1964. Later, in the German edition of his book *Political Justice*, Kirchheimer shared the main point of her criticism of the Israeli government, that it had turned the trial into a "courtroom drama" in order to insrumentalize it "as a pivotal point of demonstrative affirmation of the national state idea in the face of the ongoing external threat to the state's existence" (Kirchheimer 1965c, 44).

Carl Schmitt's readings of Hannah Arendt are worth mentioning in this context, too. He had read and commented extensively on Arendt's books from the 1950s to the 1970s. After the publication of Arendt's *Origins of Totalitarianism* (see Arendt 1951), he enthusiastically called it an interesting and important book.[73] He cited it in private letters and in his *Glossarium* in support of his own retrospective theory concerning the difference between a total state and a totalitarian movement like the Nazis. In particular, her distinction between "real Nazis" and "outstanding intellectuals" fitted into his retrospective self-description. He liked her account of the limited role intellectuals could play within totalitarian Nazi Germany. He also read her book as an indirect justification for his retrospective pose as a victim of the Nazi system. Moreover, he read it as an explanation as to why his career was interrupted in 1936: the more totalitarian a political system becomes, the less space it leaves open for truly original and independent intellectuals like himself.

Schmitt read Arendt's report *Eichmann in Jerusalem* shortly after its publication, and this time, he reacted quite differently. He wrote to legal scholars Roman Schnur and Ernst Forsthoff in October and November 1963, respectively, that reading her book had been an "upsetting" experience that had left him "sick for several weeks."[74] He explained his emotional reaction to Forsthoff as follows: "because once again my legal opinion from August 1945 came to my mind, especially its conclusion."[75] The legal opinion he was referring to was the one for Friedrich Flick.[76] In this, he had argued that Flick could not be held responsible for any significant crime. In Schmitt's view, the same applied to those who served as legal scholars for the Nazi system. Only a very small number of elite Nazis were to be held responsible. Arendt argued from a diametrically opposing point of view when she explained that those who lacked specific influence, such as the Nazi intellectuals, had a "greater realm of freedom" (Suuronen 2022, 19) in their decision whether or not to support the regime. Instead of choosing silence or emigration, all Nazi intellectuals had willingly chosen to support the regime with their writing: "For politics is not like the nursery; in politics obedience and support are the same" (Arendt 1963, 279). Her argument was an echo of her reception of Kirchheimer's criteria to distinguish between guilt

---

73 On Schmitt's apologetic reading of Arendt's *Origins*, see Herberg-Rothe (2004) and Suuronen (2022).

74 Letter from Carl Schmitt to Roman Schnur dated 24 October 1963 (Schmitt and Schnur 2023, 542) and letter from Carl Schmitt to Ernst Forsthoff dated 18 November 1963 (Schmitt and Forsthoff 2007, 198).

75 Letter from Carl Schmitt to Ernst Forsthoff dated 18 November 1963 (Schmitt and Forsthoff 2007, 199).

76 See Chapter 13, p. 337–339.

and innocence mentioned above. After having read Arendt's book on Eichmann, Schmitt finally realized that Arendt's work could not be utilized in an apologetic manner as he had previously thought.[77]

In his letter to Roman Schnur, Schmitt also mentioned Kirchheimer. He told Schnur: "It is interesting that she seeks advice from O. Kirchheimer, who is truly anything but a lawyer."[78] He went so far as to suspect that Kirchheimer was the source of a "squirt of poison" against him, namely Arendt's note that Dieter Wechtenbruch, the assistant of Robert Servatius, Eichmann's lawyer in the Jerusalem trial, was "a disciple of Carl Schmitt" (Arendt 1963, 145), which Schmitt correctly denied. Arendt did not indicate in any of her publications that she got advice from Kirchheimer for her book about the Eichmann trial and, to this day, there is no mention of his supportive role in the secondary literature on Arendt. Nevertheless, Schmitt knew about it. The source of his information is unknown to me. His knowledge about this is a remarkable example of Schmitt's interest in Kirchheimer's activities.

Kirchheimer's book *Political Justice* was published in November 1961. One of its first reviews was by Robert M. W. Kempner, Kirchheimer's superior as they prepared for the Nuremberg Trials.[79] It was published in the February issue of *Aufbau*, the German-Jewish monthly published in New York. Kempner paid particular attention to the chapter about the Nuremberg Trials. He praised Kirchheimer's "scholarly analysis of the Nazi objections against the Nuremberg Trials" as a "necessary [...] clarification." Kempner placed the study in the tradition of the Deutsche Liga für Menschenrechte (German League for Human Rights) he had co-founded in Weimar. *Political Justice* was the scholarly continuation of the position paper of the Deutsche Liga für Menschenrechte titled *Acht Jahre Politische Justiz* [Eight years of political justice], (see Deutsche Liga 1927) which had demonstrated how "anti-democratic forces within the justice system can undermine democracy, let political murderers go free, and defame supporters of democracy by means of 'rulings'" (Kempner 1962, 8). As much as this praise may have pleased Kirchheimer, it also contributed to superficial readers believing more firmly that Kirchheimer's intent was to use the term "political justice" solely in a pejorative sense. The chapter on the Nuremberg Trials has had the most favorable response in the further debate in the English-speaking world to this day. Kirchheimer is credited with having conceptualized "transitional trials" for the first time in this chapter, and his work in this field of research is still seen as groundbreaking today.[80]

Kirchheimer sent Schmitt a copy of the book immediately upon its publication. He added a brief formal dedication "With compliments, your OK."[81] Schmitt is known to add handwritten comments in the books he read; they are very few and far between in this one. He apparently did not read the sections on the legal proceedings relating to the

77 On Schmitt's reading of *Eichmann in Jerusalem*, see Graf (2021), Suuronen (2022), and Plaetzer (2022).

78 Letter from Carl Schmitt to Roman Schnur dated 24 October 1963 (Schmitt and Schnur 2023, 542).

79 See Chapter 13, p. 358–360.

80 "The credit for conceptualizing transitional trials [...] must go to Otto Kirchheimer" (Priemel 2016, 7).

81 Schmitt's copy of Otto Kirchheimer, *Political Justice*. Carl Schmitt Papers, RW 0265–25665.

Nazi crimes at all. But he did read the two-page appendix on the Roman Empire and the Christians, underlining some parts. Schmitt left his few marks in the parts of the book that he believed pointed to his own concept of politics as the friend-enemy distinction. Most of his markings are in the three-page section headed "The Informer: Enemy from Within" in Chapter VI "The Defendant, His Lawyer, and the Court." Moreover, Schmitt could not suppress his urge to add a comment underneath Kirchheimer's dedication "To the Past, Present, and Future Victims of Political Justice": he added a handwritten note: "I do not compare the victims (to whom I—past, present and future—belong), I only compare the judges, C.S."[82] Schmitt did not respond directly to Kirchheimer. As we shall see, his reaction to the book the following year was overshadowed by a new conflict between them.[83]

7. Kirchheimer as a professor at the New School for Social Research

By publishing *Political Justice*, Kirchheimer had fulfilled the formal prerequisites for a tenured position as a professor at Columbia University. Moreover, sales of the book were good; Princeton University Press had already sold 1,100 copies by October 1962, in just under a year.[84] The book was reviewed in all the important journals and major newspapers and received wide praise. A number of reviewers criticized it, however, as "heavy of language" and "too long and meandering," and its title as "misunderstandable."[85] All the same, Kirchheimer was disappointed by this resonance because of some critical voices. His anger was due most of all to the misunderstanding in some reviews that he had used the term "political justice" in a pejorative sense in his book.[86] Four months after its publication, he wrote to Gurland: "As my reviews except for one isolated leftwinger show complete lack of understanding and often bad will to match—I am somehow *angewiesen* [dependent] on the German translation."[87] The "leftwinger" was Henry (Heinz) Paechter (see Paechter 1962). He and Kirchheimer knew each other as activists of the left wing of the Weimar SPD. Paechter was able to escape from Nazi Germany to New York. He was a socialist intellectual and frequently taught German history classes at the New School for Social Research. In 1944, he co-authored the dictionary *Nazi-Deutsch. A Glossary of Contemporary German Usage*, which has been used for this book.

When appointing Kirchheimer, the New School for Social Research had high expectations. Hoping he could draw on his work at the State Department, the President of the

---

82 Schmitt's copy of Otto Kirchheimer, *Political Justice* (page 5). Carl Schmitt Papers, RW 0265–25665. Schmitt used exactly the same wording once again in a letter to Ingeborg Maus in May 1982; see Mehring (2013, 443).

83 See Chapter 17.

84 Letter from Herbert Bailey to Otto Kirchheimer dated 26 October 1962. Otto Kirchheimer Papers, Series 3, Box 2, Folder 51.

85 On the book's reception in the US, see Klingsporn and Wilke (2020, 750–754).

86 On this misunderstanding in the book's reception, see Klingsporn and Wilke (2020, 755–758).

87 Letter from Otto Kirchheimer to Arkadij Gurland dated 5 March 1962. Otto Kirchheimer Papers, Series 2, Box 1, Folder 68.

Graduate Faculty immediately appointed him Chairman of the Committee for Coordinating Research.[88] However, Kirchheimer had not become a successful academic manager. On the contrary. Besides his university teaching, he was primarily interested in continuing to work on his book on political justice. His modest activities to obtain financing were limited to funds from the State of New York for research at the New School on a subject close to his heart ever since he taught classes as Visiting Professor at Howard University in Washington in 1952 and 1953, namely "discrimination against Puerto Ricans and Negroes."[89] Among his colleagues at the New School, he particularly liked to talk to faculty members of the younger generation.[90]

Kirchheimer continued to try to obtain a position at a university in West Germany. He rejected Fritz Bauer's offer in the summer of 1957 to accept a full position as editor-in-chief of the social democratic theory journal *Die Neue Gesellschaft*.[91] Instead, he relied on his good contacts to Carlo Schmid, Hermann Heller's former assistant. In the spring of 1953, immediately after assuming his position as professor of political science at the University of Frankfurt, Schmid had begun his efforts to establish a second chair, to which he wanted to appoint Kirchheimer. Apparently, he received no support at all from Horkheimer or Adorno (see Weber 1996, 511–522). When it seemed that the efforts in Frankfurt would not bear fruit, Schmid interceded for Kirchheimer to be appointed at the universities in Cologne and Bonn but was unsuccessful because of the resistance of the relevant departments there.[92] In a letter to Fritz Bauer, Kirchheimer spoke with resignation of the "fata morgana of a German professorship."[93] In 1960, Gert von Eynern informed him that Kirchheimer—as well as Ossip K. Flechtheim—was under discussion for a chair for domestic policy at the Otto-Suhr-Institut, the political science department at the Freie Universität Berlin.[94] Eynern was a social democrat who had been active in the resistance movement against Hitler and became director of the Otto-Suhr-Institut in 1959. Yet this plan petered out, too. In a letter to Horst Ehmke, Kirchheimer confided he felt that German universities viewed him as "a kind of straying dog that people permit to wander in their front yards if need be."[95]

The situation in the US was different. He received tenure at the New School for Social Research in 1957. Kirchheimer's way of dealing with people changed during his time

88 Minutes, Executive Faculty Meeting of 9 May 1956. New School for Social Research: New School Institutional Collections. Graduate Faculty, Minutes. NS.02.17.02, Box 1, unprocessed collection.

89 Minutes, Executive Faculty Meeting of 16 January 1957. New School for Social Research: New School Institutional Collections. Graduate Faculty, Minutes. NS.02.17.02, Box 1, unprocessed collection.

90 Arthur J. Vidich, who had joined the New School in 1960, in a conversation with the author on April 4, 1995.

91 The reason he gave was: "working with a party is possible; living off a party is not very desirable." Letter from Otto Kirchheimer to Fritz Bauer dated 22 August 1957. Otto Kirchheimer Papers, Series 2, Box 1, Folder 17. Bauer later played a prominent role in the capture of Adolf Eichmann.

92 See letter from Carlo Schmid to Paul Luchtenberg dated 11 June 1958. Carlo Schmid Papers, Reg. Nr. 678.

93 Letter from Otto Kirchheimer to Fritz Bauer dated 22 August 1957. Otto Kirchheimer Papers, Series 2, Box 1, Folder 17.

94 Letter from Gert von Eynern to Otto Kirchheimer dated 31 January 1960. Otto Kirchheimer Papers, Series 2, Box 2, Folder 38.

95 Letter from Otto Kirchheimer to Horst Ehmke dated 26 January 1961. Horst Ehmke Papers, No. 504.

at the New School. Herbert Marcuse wrote in a confidential letter of recommendation for Kirchheimer to the Rockefeller Foundation: "I felt that he was hampered by linguistic difficulties and a psychological failure to adjust to the American scene." Yet Marcuse saw improvement in his newer works: "His recent work shows that this is no longer the case: the last two or three articles I have seen are well written and presented in a fashion fully comprehensible to an American audience."[96] Erich Hula wrote the same, almost verbatim, to the foundation, and added: "I think that he has finally arrived at the rare combination of American empirical method and German sociological learning and theory."[97] In John H. Herz's view, Kirchheimer had changed in a positive way in terms of everyday things as well during his time at the New School. Although he still was no friend of the American way of life, he became more open when dealing with his fellow human beings and more sympathetic to everyday political matters in the US (see Herz 1989, 16).

The years at the Graduate Faculty of the New School for Social Research had been a productive period for Kirchheimer. He completed his book *Political Justice*, and he published a number of essays and book reviews placed prominently in renowned journals as well as regular book reviews in the *Washington Post*. His position at Columbia University further enhanced his reputation in the field of political science in the US. One indicator of this was his co-optation as the successor of Leo Strauss on the editorial board of the *American Political Science Review* (APSR) (see Kettler 2006, 535), the journal published by the American Political Science Association (APSA). Kirchheimer had arrived at a new and prestigious high point of his academic career.

8. Conclusion: A Smendian solution to a Schmittian problem

The front lines between Kirchheimer and Schmitt remained the same in the years 1957 to 1961. At most, the range of main subjects and the means of discussion had changed. In addition to the renewed discussion about the end of the Weimar Republic, subjects now included the new presidential regime in France, the role of the opposition in the Federal Republic of Germany, the welfare state, constitutional jurisdiction, and political justice. Schmitt stuck to his old positions. He taught his young follower Ernst-Wolfgang Böckenförde the following historical lesson about 1933 and the years that followed: "*Jeder anständige Deutsche* (every decent German) who wasn't a communist or a Marxist joined in at the time."[98] Not only did this mean that Schmitt still felt that resistance had not been a viable option, but also that he excluded liberals and social democrats as well as conservatives who had in fact resisted from the circle of those he considered "decent Germans."

Although Kirchheimer and Schmitt corresponded sporadically in 1958, they never met again in person after their encounter in Cologne in 1953. Schmitt proposed that

---

96 Letter from Herbert Marcuse to Kenneth Thompson dated 26 November 1958. Archive of the Rockefeller Foundation, RF RG1.2 Series 200, U.S.-Social Sciences, Box 539, Folder 4615.

97 Letter from Erich Hula to Kenneth Thompson dated 24 December 1958, quoted in Müller (2010, 395).

98 Letter from Carl Schmitt to Ernst-Wolfgang Böckenförde dated 13 April 1961 (Schmitt and Böckenförde 2022, 276).

they engage in more in-depth debates but did not receive the response he desired. Kirchheimer brusquely rejected Schmitt's claim that his (Schmitt's) writing from the Weimar period was still as topical as ever. He informed him that he did not believe in the repetition of similar situations because too many qualitative changes had taken place since then. After 1958, Kirchheimer stopped publicly commenting on Schmitt's work altogether. Their correspondence also petered out and was limited to taunting each other by sending offprints of their new works.

In his public confrontation with Schmitt, Kirchheimer revived his penchant to go on the offense, repeatedly attacking Schmitt head-on again in his publications. In his review of Schneider's book in 1957, he once again summarized all of his substantive and methodological objections to Schmitt in a succinct form: the ever-present negation of *Rechtsstaat*; the discrepancy between the traditional liberal concept of classical international law and the rejection of liberalism as part of the domestic constitutional order; the omnipresence of the people's constituent power and its incapacity to act as a constituted organ; the indeterminate character of the values underlying specific decisions; and the lack of any clear-cut criteria for differentiating between violence and nomos. In a letter to Friesenhahn a year later, Kirchheimer called Schmitt intransigent and repeated his methodological critique of conceptual realism.[99] Kirchheimer thought that Schmitt was still not willing to take stock of how political and conceptual formulations corresponded to social reality.

In contrast, Schmitt chose a different strategy in his public debate with Kirchheimer. He remained friendly and quoted Kirchheimer several times and only positively. He selected short sentences or formulations from Kirchheimer and used them to support his own position. As with Schmitt's "art of quoting" during the Weimar Republic in his *Legality and Legitimacy*,[100] these benevolently worded citations were taken out of context, thus distorting their meaning and conveying inaccuracies.

Kirchheimer had dedicated the book *Political Justice* to the past, present, and future victims of political justice. Schmitt saw himself as a victim of Jewish-American political justice and responded to Kirchheimer's dedication sarcastically that he did not compare the victims but only the judges. The chapter on the Nuremberg Trials can be seen as a full rebuttal of Schmitt's objections from his letters, anonymous newspaper articles, and his *Glossarium*, although it is not possible that Kirchheimer could not have read these.

Nonetheless, Schmitt's influence on Kirchheimer's understanding of political justice, inspiring the subject of the book *Political Justice*, is unmistakable. Just like Schmitt in his *Constitutional Theory*, Kirchheimer stated that political justice was a phenomenon that had never been and could never be abolished even in the best *Rechtsstaat*. But unlike Schmitt, this statement did not mean that Kirchheimer believed that the idea of the *Rechtsstaat* was flawed from the outset and should be rejected. Neither Schmitt nor Kirchheimer considered "political justice" to be a negative concept. The term did not mean the camouflage of political purposes with a judicial facade and the misuse of judicial procedures for political reasons. The political was always inherent in a dispute. The two of them agreed up to

99 Letter from Otto Kirchheimer to Ernst Friesenhahn dated 15 December 1958. Otto Kirchheimer Papers, Series 2, Box 2, Folder 61.

100 See Chapter 5, p. 129.

this point. The divergences began with the adaptation of political justice in the theory of the bourgeois *Rechtsstaat*.

According to Schmitt, the existence of political justice proved that the theory of the bourgeois *Rechtsstaat* widened a systematic gap. The *Rechtsstaat* had to provide a procedure for any kind of disagreement and dispute between citizens and the state and between the various state organs. The prerequisite for the success of these procedures was valid and general standards. However, some of the disagreements and disputes lacked such a predetermined norm. In such cases, the judicial decision inevitably became a political decision. Schmitt saw the systematic gap in the theory of the bourgeois *Rechtsstaat* at this point. Some disagreements and disputes were so prominent that the political specificity of such cases was inescapable. That, according to Schmitt, created the actual problem of political justice as an element that went against the systematic logic of the *Rechtsstaat*.

In contrast, in his theory of *Rechtsstaat*, Kirchheimer conceptualized political and judicial processes as being located on a continuum. Again and again, politics made use of the judicial institutions—and not only on the initiative of state authorities, but often also by oppositional citizens who wanted to publicly criticize the state in such a trial. Political trials were risky for both sides. If those in government wanted to make use of the judiciary, this resulted in imponderables. The detour via the law meant wasting time and limiting the methods that could be used against the political opponent. Political trials in a *Rechtsstaat* created a specific judicial space, namely the fact that results were not predetermined. Political opponents of the government could be acquitted and members of the opposition could attract the attention of the public for their political concerns in political trials. In modern democracies, political justice could even have some benefits. It might take on a positive function that curbed social conflicts.

Ultimately, Kirchheimer countered the problem of political justice, which Schmitt had exaggerated into an insoluble contradiction, with Smend's basic idea about the function of the judiciary in a democratic *Rechtsstaat*. At the time when Kirchheimer had finished his book, Smend was completely unknown in the English-speaking world, which is why Kirchheimer omitted any reference to him in the English version of his book. He added a reference to the German version from 1965:

> More than three decades ago, Rudolf Smend argued insistently that the constitution made the courts independent of the state leadership and thus expressly freed them from the obligation to serve the state to provide integration; in practice, however, he said subsequently, it could be that the courts serve not only to integrate the legal community, but also to integrate the state (Kirchheimer 1965c, 23).

And he added in a footnote: "In an effort to trace the courts on the tortuous paths of their dual role, I seek to penetrate the perils of the courts' liberation from state leadership, which Smend had emphasized" (Kirchheimer 1965c, 23). In the passage from Smend's *Verfassung und Verfassungsrecht* (see Smend 1928, 207–209) quoted by Kirchheimer, Smend described how the role of the judiciary had changed since the transition from the medieval jurisdictional state to the modern *Rechtsstaat*. Even if the constitution had freed the judicial institutions from the explicit task of integrating the unity of the state, both the legislature and the judiciary functioned as political institutions of social integration. According to Kirchheimer, the value of legal procedures in conveying the legitimacy of a political system depends largely on the degree to which such procedures respect the limitations on political prosecution. That is the freedom of defense, the organizational and intellectual distance between the prosecution and the court, the untrammeled introduction and challenge of testimony, and the degree of insistence upon evidence of concrete past action. Performed in such a way, political justice could even contribute to the integration of society through conflict. In a nutshell, Kirchheimer's *Political Justice* provides a Smendian answer to a Schmittian formulation of the problem.

Chapter 17: The Final Break (1962-1965)

In the spring of 1962, after thirty-six years, all forms of direct contact between Otto Kirchheimer and Carl Schmitt, including sending letters or offprints, came to an end once and for all. The trigger for this was a conflict sparked by the unsuccessful attempt of George Schwab, an American student, to obtain a PhD at Columbia University with a dissertation on Carl Schmitt. This conflict was in the offing as Kirchheimer was preparing to move from the New School for Social Research to Columbia University. After the final break between Kirchheimer and Schmitt, their relationship shifted to second-order observations, i.e., they no longer communicated with each other directly but only via third parties. Neither of them initiated personal contact again before Kirchheimer's sudden death in November 1965. What Schmitt, who survived Kirchheimer by two decades, said about Kirchheimer changed markedly in the 1970s, and this contributed to Kirchheimer subsequently being viewed as a groundbreaking "Left-Schmittian."

1. Kirchheimer as a professor at Columbia University

During his time at the New School for Social Research, Kirchheimer succeeded in renewing his contacts at Columbia University from the early 1940s. After Franz L. Neumann had died in a car crash in Europe in September 1954—Kirchheimer included a moving obituary in his review of the posthumously published collection of Neumann's essays in which he emphasized their common and permanent efforts to come up with "new analyses of the progressive and regressive tendencies in society" (Kirchheimer 1957d, 382)—Kirchheimer's most important contact at Columbia University, where he hoped to obtain a position, was Neumann's student Julian H. Franklin. In March 1960, Dean David B. Truman asked Kirchheimer whether he was interested in working at Columbia's Department for Public Law and Government as a visiting professor the following academic year.[1] This chair had been vacant since Neumann's death.[2]

Kirchheimer continued to pay attention to current events and developments in Germany, not least with a view to obtaining a position there. The Basic Law had not provided for reinstating people who had been persecuted by the Nazi regime for political and "racial" reasons in their previous professional positions—in contrast to Article 131 of the Basic Law, which benefited Nazis who had worked in the civil service. Émigrés who wanted to return to Germany had to take action themselves. This applied to citizenship, too. Kirchheimer was one of those who refused to apply for his German citizenship to be reinstated. His reasoning was that it had been the German state that had robbed him of his German citizenship during the Nazi years, and not his own doing. Consequently, he saw it as the duty of the German state to reinstate his citizenship automatically.[3]

In 1961, Carlo Schmid's efforts to establish a second chair of political science in Frankfurt were successful, and he put Kirchheimer at the top of the short list of candidates for the position. Kirchheimer received the offer from Frankfurt in August 1961.[4] He was delighted and negotiated for more than six months, with extensive correspondence about the following questions: his status as a *Beamter* (civil servant) since he was a US citizen and this status was reserved for German citizens; his later pension entitlements; and the compatibility of two part-time positions, one in Frankfurt and one in New York. Everyone involved on the German side was surprised when he eventually rejected the offer in April 1962.[5] The reasons he gave Schmid and Adolf Arndt were his family situation and his health. "In principle," he wanted to "turn his back" on the US, he wrote, but after consulting with his wife, he had committed to spend longer periods in the US on a regular basis until his son Peter had finished school. He also had to consider how to finance Peter's college tuition; he was planning to attend Columbia, and his tuition would be reduced provided that his father was a professor there. He thanked Schmid for his support and said he was "quite sad about this affair" because he had had high hopes for it.[6] Because of his family situation, he would have had to commute between Frankfurt and New York and his "health would not have withstood commuting for 4 years à la [Carl Joachim] Friedrich."[7] Visibly indignant, Carlo Schmid made no secret of his disappointment.[8] In May 1963, Marxism scholar Iring Fetscher accepted the chair in Frankfurt.

1 Letter from David B. Truman to Otto Kirchheimer dated 4 March 1960. Otto Kirchheimer Papers, Series 2, Box 2, Folder 34.

2 Letter from Richard Herpens to Otto Kirchheimer dated 14 April 1960. Otto Kirchheimer Papers, Series 2, Box 1, Folder 29.

3 Hanna Kirchheimer-Grossman in a conversation with the author on 10 September 2021.

4 Letter from Hessian Ministry of Culture and Education to Otto Kirchheimer dated 29 August 1961. Otto Kirchheimer Papers, Series 2, Box 1, Folder 79.

5 Letter from Otto Kirchheimer to Hessian Ministry of Culture and Education dated 13 April 1962. Otto Kirchheimer Papers, Series 2, Box 1, Folder 79.

6 Letter from Otto Kirchheimer to Carlo Schmid dated 3 May 1962. Otto Kirchheimer Papers, Series 2, Box 2, Folder 10.

7 Letter from Otto Kirchheimer to Adolf Arndt dated 19 May 1962. Otto Kirchheimer Papers, Series 2, Box 1, Folder 12.

8 Letter from Carlo Schmid to Otto Kirchheimer dated 10 May 1962. Carlo Schmid Papers, Reg. No. 756.

On 23 May 1962, the Dean of the Graduate Faculty of the New School of Social Research thanked Kirchheimer during his "last meeting with our Faculty."[9] Kirchheimer took a permanent position at Columbia University, and his arrival as a member of the faculty was announced in late April 1962 along with that of historian Peter Gay.[10] Kirchheimer was now Professor for Government in the Department for Public Law and Government.[11] He was not required to teach undergraduates, only graduate students and doctoral candidates. There are few traces of his work in faculty, department, or various university committee meeting minutes. Yet these do provide evidence that he intensively supported the interests of the students.[12] Kirchheimer also served as his faculty's Adviser for Foreign Political Institutions and Political Theory.[13] He successfully recruited colleagues and friends who were important to him to spend time at Columbia University as visiting scholars and professors. For example, it was thanks to his initiative that his former cellmate Paul Kecskemeti, now of the RAND Corporation, came to the department in 1963 and John H. Herz in 1965 as visiting professors.[14] Arkadij Gurland, who had been appointed professor of political science in Darmstadt, Germany, in 1962 thanks to Kirchheimer's vigorous support, spent a semester at Columbia in 1964, and the two jointly taught the research seminar "Studies in the theory and practice of modern government."[15] In the winter term of 1965, Kirchheimer co-taught this course with Juan Linz,[16] who later became a leading researcher on the collapse of authoritarian regimes in Latin America and on political transformations toward liberal democracies.

Kirchheimer also devoted his efforts to applying for additional research fellowships and other opportunities to spend time in Germany. After completing his first spring term as a visiting professor at Columbia University early, he spent May to August 1961 at the Faculty of Law of the University of Freiburg as a Fulbright Professor.[17] He received funding to serve as a Fulbright professor in 1963, 1964, and 1965, too, at times obtaining a leave

9 Minutes, Executive Faculty Meeting on 23 May 1962. New School for Social Research: New School Institutional Collections. Graduate Faculty, Minutes. NS.02.17.02, Box 1, unprocessed collection.

10 Minutes of 27 April 1962. Minutes of the Faculty of Political Science 1950–1962. Special Collection, Columbia University Archives.

11 Columbia University Bulletin. The Graduate Faculties. Series 62, page 262. Special Collection, Columbia University Archives.

12 David Kettler in a conversation with the author on 17 May 2015.

13 See Columbia University Bulletin. The Graduate Faculties. Series 63, page 255. Special Collection, Columbia University Archives.

14 Columbia University Bulletin. The Graduate Faculties. Series 62, page 262 and minutes of 29 April 1965. Minutes of the Faculty of Political Science 1963–1970. Special Collection, Columbia University Archives.

15 Columbia University Bulletin. The Graduate Faculties. Series 64, number 8, page 277. Special Collection, Columbia University Archives.

16 Columbia University Bulletin. The Graduate Faculties. Series 65, number 3, page 284. Special Collection, Columbia University Archives.

17 Letter from Otto Kirchheimer to Carl Anthon dated 9 October 1960. Otto Kirchheimer Papers, Series 2, Box 1, Folder 62.

of absence from Columbia for an entire semester.[18] He spent the spring semester of 1963 as a visiting professor without teaching responsibilities at the Department of Government at Yale University in New Haven, Connecticut. In the meantime, Herbert Marcuse asked him whether he could imagine spending time at Brandeis University, which had been founded fifteen years before as a non-sectarian university sponsored by the Jewish community.[19] Kirchheimer turned him down politely, preferring to spent the academic year 1964/65 at home at his desk in Silver Spring as a fellow of both the Social Science Research Council (SSRC) and the John Guggenheim Foundation.[20]

Kirchheimer regularly spent time in Germany and attended international conferences there. The conferences where he presented the findings of his research on political parties and, in particular, his deliberations on the catch-all party to a larger academic audience for the first time were held in Europe, too. One of the outcomes of these new and renewed contacts was that Kirchheimer was pleased to serve as one of the official liaisons for the Fulbright programs for academic exchange with the US. He willingly provided formal invitations to scholars who needed them. His home in Silver Spring was open to guests and visitors from Germany, among them top SPD politicians such as Carlo Schmid, Fritz Erler, and Willy Brandt as well as trade union leaders such as Hans Böckler, Willi Richter, Hans Matthöfer, and Otto Brenner. Influential colleagues and friends from his generation, such as Richard Schmid und Otto Stammer, were his guests as well as younger scholars such as Wilhelm Hennis, Jürgen Habermas, Horst Ehmke, Helge Pross, and Peter C. Ludz, who subsequently took leading positions in West German academia.

Kirchheimer's relationship with the core members of the Frankfurt School who had returned to Germany remained troubled. His contact with Jürgen Habermas, twenty-five years his junior, who had been Adorno's assistant at the Frankfurt Institut für Sozialforschung since 1956, developed differently and more positively than that with Horkheimer and Adorno. Reading Helmut Ridder as a student, Habermas had become aware of leftist legal experts from the Weimar Republic such as Hermann Heller, Franz L. Neumann, and Kirchheimer (see Ridder 2005, 373). He had met Kirchheimer in person via Neumann's new partner Helge Pross, who also worked at the institute in Frankfurt (she later became one of the first female professors of sociology in Germany and a pioneer in gender studies). In his 1958 essay "Zum Begriff der politischen Beteiligung" [On the concept of political participation], Habermas drew on writings of Kirchheimer's on the transformation of the liberal *Rechtsstaat* to the social welfare state. Habermas also referred to Kirchheimer's newer works on political parties, parliaments, and the decline of the opposition in Western democracies in his *habilitation* dissertation *The*

18 Kirchheimer's appointment card at Columbia University states "leave without salary" for the autumn semester 1964 and the spring semester 1965. Appointment Card Otto Kirchheimer. File 159/9. Special Collection. Columbia University Archives.

19 Letter from Herbert Marcuse to Otto Kirchheimer dated 14 October 1963. Otto Kirchheimer Papers, Series 2, Box 1, Folder 110.

20 Otto Kirchheimer, Curriculum Vitae (1965). Private collection of Hanna Kirchheimer-Grossman (Arlington).

Structural Transformation of the Public Sphere, and he sent him a copy of the book.[21] In 1961, Kirchheimer visited Habermas in Heidelberg, and when the latter was preparing for his first professional trip to the US in the spring of 1965, he mentioned Kirchheimer as his main liaison for New York in his application for funding (see Müller-Dohm 2014, 177) and visited him in Silver Spring that summer.[22]

2. The conflict over George Schwab's dissertation

The conflict that ended direct contact between Kirchheimer and Schmitt erupted over a dissertation about Schmitt's oeuvre by a doctoral student at Columbia University. Born in Lithuania, the student, George David Schwab, belonged to an interwar generation of young Jews who had been lucky enough to escape from persecution in the Holocaust by fleeing to the United States. In New York, he met Franz L. Neumann and decided to study political science as a graduate student at Columbia University. After Neumann's death in 1954, Herbert L. Deane, professor of political theory, became Schwab's supervisor. Deane had written his dissertation with Neumann and recommended Schwab to write a master's thesis about Schmitt. Then he encouraged him to write his PhD dissertation about Schmitt, too. With Carl Joachim Friedrich acting as liaison, Schwab contacted Schmitt in October 1956. On 22 January 1957, he informed Schmitt of his plan to visit him in Plettenberg. Schmitt agreed. In February, Schwab confirmed his intention to write his dissertation on Schmitt's work during the Weimar Republic and the early Nazi period. He arrived in Plettenberg for a two-month stay in April and came back for another visit in the fall of the same year.

Schmitt was immediately impressed by Schwab, who was twenty-six. The same year, he wrote enthusiastically to Ernst Jünger: "In particular, the young American from Columbia University in New York has taken a room at the nearby Hotel Ostermann for 2 months and goes on long walks with me. The diligence with which he is writing his book on me is exemplary."[23]

Schmitt also revealed in the letter that he was supporting the project because he harbored specific hopes for its reception: his motive was to counter Peter Schneider's book, which Kirchheimer had praised so highly, with a work he himself had not authorized. The study by the American Schwab "humiliates the European Peter Schneider from Zurich, who cautiously avoided talking to me or even seeing me although he explicitly aims to show the 'arcanum' of Carl Schmitt, as he says, in his book."[24] Schmitt was apparently

21 See Habermas (1958, 1962) and letter of thanks from Otto Kirchheimer to Jürgen Habermas dated 6 October 1962. Otto Kirchheimer Papers, Series 2, Box 1, Folder 69. On Habermas's reception of Kirchheimer, see Buchstein (2019b).

22 Jürgen Habermas to the author on 10 March 2018. In a note on the development of the social sciences in the early years of the Federal Republic of Germany, Habermas mentioned Kirchheimer by name as one of those "who have made a big contribution to the dense web of personal and academic ties between here [Germany] and over there [in the US]" (Habermas 1992, 151).

23 Letter from Carl Schmitt to Ernst Jünger dated 29 May 1957 (Schmitt and Jünger 1999, 334). On Schmitt's *faible* for taking long walks with his visitors see Braunfels and Grajcarek (2023).

24 Letter from Carl Schmitt to Ernst Jünger dated 29 May 1957 (Schmitt and Jünger 1999, 334).

firmly convinced that a good book about him and his work could only be written in close cooperation with himself and with his approval.

Consequently, Schmitt made plenty of time for his guest over the following months and years. In his *Glossarium*, he mentioned "delightful conversations" with Schwab about questions of "being human"[25] and expressed his enthusiasm about Schwab agreeing with his legal opinion from 1945 about the war of aggression.[26] He encouraged Schwab to devote special attention to his role at the end of the Weimar Republic. For, during Schwab's stay in Germany, Schmitt was also working on his comments on his *Verfassungrechtliche Aufsätze* [Essays on constitutional law] in which he interpreted himself as a tragically failed savior of the Weimar Republic.[27] Schmitt authorized Schwab to translate texts of his into English. The first one (see Schmitt 1958b, 439), in 1958, was "Der Zugang zum Machthaber, ein zentrales verfassungsrechtliches Problem" [Dialogue on Power and Access to the Holder of Power] (see Schmitt 1947). Schmitt had succeeded in giving Schwab what a number of other younger visitors including Ernst-Wolfgang Böckenförde, Reinhart Koselleck, and Christian Meier had raved about all their lives, namely making them feel that he was truly taking them seriously and that he cared about teaching them. Schwab remained enthusiastic about the long and amicable conversations he had with Schmitt, as he wrote in his memoirs in 2021. In his long life, he had never learned as much from any other person and in such a short period of time about legal and political theory and international relations (see Schwab 2021, 145–158). In the summer of 1958, Schwab visited Schmitt again.

In his long letter to Kirchheimer dated 6 August 1958, Schmitt mentioned that he had a visitor from New York in his hometown Plettenberg. He described him using the following words:

> I had a visit for some months in the summer from a young student from New York, George Schwab, Columbia University, with whom I had very good conversations and whom I found very friendly. If you ever have the opportunity to speak to him—his teacher is Herbert A. Deane—Public Law and Government, Columbia Univ.—the author of the book on H. J. Laski—I would be interested in your impression of him.[28]

Schmitt apparently expected Kirchheimer to also be immediately impressed by the young student and that he would support him. Kirchheimer responded, but only briefly, one month later. Concerning Schmitt's visitor from New York, he wrote: "I do not know Mr. Schwab yet, but will try to get in touch with him when the semester has begun."[29] There is no indication, however, that they actually met at this early stage of Schwab's dissertation. Nor are any letters from Schwab in Kirchheimer's papers. In 1959, Schwab visited

---

25 *Glossarium* entry of 10 June 1957 (Schmitt 2015, 362).

26 *Glossarium* entry of 6 October 1957 (Schmitt 2015, 366).

27 See Chapter 16, p. 420.

28 Letter from Carl Schmitt to Otto Kirchheimer dated 6 August 1958. Otto Kirchheimer Papers, Series 2, Box 2, Folder 12.

29 Letter from Otto Kirchheimer to Carl Schmitt dated 4 September 1958. Carl Schmitt Papers, RW 265–7604.

Schmitt again for two months. The two discussed the hypotheses in his dissertation about Schmitt's work and his impact from 1930 to 1936 (see Schwab 2021, 173).

At the same time, Kirchheimer sounded out his chances for potentially leaving the New School for Social Research for a position at Columbia University. Deane had heard from Franz L. Neumann a few years earlier that Kirchheimer was familiar with Schmitt's work, and after Neumann's death, Deane believed that Kirchheimer was the only person he knew who was in a position to make a fair judgement about Schwab's project. So, he suggested to Schwab in the fall of 1959 that he discuss his work with Kirchheimer. The latter agreed without receiving any detailed information about Schwab's dissertation project from Deane. In late 1959 and the first half of 1960, the two had a few cursory conversations about the topic of the dissertation. Kirchheimer urged Schwab to take notice of the German literature about Schmitt, among them the recently published monographs by Christian Graf von Krockow, Jürgen Fijalkowski, and Peter Schneider.[30] At Deane's request, Kirchheimer joined the five-person dissertation committee at Columbia. Schwab sent his manuscript to Kirchheimer and received an official invitation to speak with him in his office shortly afterwards.

According to Schwab, Kirchheimer informed him at this meeting in late May 1960 "in a friendly manner" that he had "failed to understand Schmitt" and that he had to rewrite parts of the dissertation. Kirchheimer told him that he had made two main mistakes. First, he had failed to realize the extent to which Schmitt had helped pave the way for the destruction of the Weimar Republic and, second, Kirchheimer had stated that Schmitt was "already an anti-Semite during the Weimar period" (Schwab 2021, 175). He also instructed Schwab to include additional publications of Schmitt's from 1932 to 1936 that he had not yet considered and to engage with the relevant new secondary literature on Schmitt and the history of the Weimar Republic. In June, Schwab indignantly reported to Schmitt about the—in his view—disappointing conversation with Kirchheimer as the new member of the dissertation committee.[31] He initially considered submitting a petition to the president of Columbia University to have Kirchheimer removed from the committee for lack of impartiality but abandoned the idea because he realized it had no prospect of success.[32]

Kirchheimer did not take the dispute about Schwab's work lightly, either. He correctly assumed that Schmitt had been involved behind the scenes to make another attempt at political rehabilitation via the US. Kirchheimer reported about the matter one month later to Ernst Friesenhahn, asking him to keep the information to himself:

> One of the first doctoral researchers who arrived at Columbia with a finished thesis was Mr. Schwab. He wanted to enlighten the world in American English about Carl Schmitt's life and works, the young man had sat at CS's feet and had actually let himself be talked into believing that CS had actually always wanted to help the Weimar Constitution be

30 See Fijalkowski (1958), Krockow (1958), and Schneider (1957). On the controversial debate about Schneider's book, see Chapter 16, p. 417–419. Kirchheimer mentioned this later in a letter to Ernst Friesenhahn dated 20 November 1960. Otto Kirchheimer Papers, Series 2, Box 2, Folder 61.

31 Letter from George Schwab to Carl Schmitt dated 11 June 1960. Quoted in Mehring 2020, 506.

32 Letter from George Schwab to Carl Schmitt dated 23 May 1961. Quoted in Mehring 2020, 506.

> protected and correctly applied, a kind of democrat in dire times. He was very discomfited when I announced that I would not accept his stupid scribblings, not even if modified, that as far as I was concerned, he could defend CS lock, stock, and barrel, but only if he emphasized his real doctrines and did not disguise him as a democrat and strict constitutionalist. CS had also made his personal correspondence file available to him, and he came to me with transcripts of letters, including approval from the other side, from Franz Neumann, on legality and legitimacy.[33]

In the summer of 1961, Kirchheimer and Schmitt had another exchange of two letters, as the conflict with Schwab was already smoldering. It was another five-line birthday letter in which Kirchheimer asked where Schmitt would spend the summer. He closed with the friendly phrase: "I would be pleased if there might be the opportunity to see you again."[34] Schmitt responded five weeks later and told him later that such a meeting would be impossible because of his own plans to spend the summer in Spain.[35] Instead of using the cordial form of address "*lieber Herr Kirchheimer*" as in his previous letters, he now opted for the formal salutation "*sehr geehrter Herr Kirchheimer.*" In November 1961, Kirchheimer's book *Political Justice* was published. Kirchheimer sent Schmitt a copy with the formal dedication "with best compliments, your OK."[36] Schmitt did not respond. This was their last direct contact.

In the final months of 1961, Schwab finished revising his doctoral dissertation. The defense in February 1962 ended in uproar. Since it was impossible to find any files on the matter in Columbia University's archive, the only source for this passage is Schwab's memoirs.[37] According to his report on the two-hour dissertation defense, the discussion was initially "boring" until Kirchheimer weighed in. He "mercilessly attacked" him (Schwab 2021, 177) and criticized the dissertation as a whole: Schwab had "failed to understand Schmitt's true role in Weimar," he had "turned Schmitt upside down" and had "written an apology of Schmitt." Kirchheimer rejected the sharp distinction in Schwab's work "between racial theory and Catholic anti-Semitism." He also accused Schwab of misinterpreting the principle of equal opportunity in the constitution. Finally, he criticized Schwab's fundamentally misguided understanding of Article 48 of the Weimar Constitution, as a result of which his codification of the emergency decrees in his dissertation was flawed. Kirchheimer explained in detail how Schmitt had paved the way for the Nazi regime with his theory of the emergency decrees. Schmitt had been "among

---

33 Letter from Otto Kirchheimer to Ernst Friesenhahn dated 20 November 1960. Otto Kirchheimer Papers, Series 2, Box 2, Folder 61. Neumann's letter, dated 7 September 1932, is published in Erd (1985, 79–80).

34 Letter from Otto Kirchheimer to Carl Schmitt dated 4 July 1961. Carl Schmitt Papers, RW 265–7605.

35 Letter from Carl Schmitt to Otto Kirchheimer dated 12 August 1961. Otto Kirchheimer Papers, Series 2, Box 2, Folder 12.

36 Copy of Otto Kirchheimer, *Political Justice*. Carl Schmitt Papers, RW 0265- 25665.

37 These and the following quotes of Kirchheimer's words are to be found in Schwab (1988a, 80–81), (2021, 175–178), and for Schwab's response to questions about this matter from Volker Neumann, see Schwab (1988b, 462). Richter (2001, 222–224) and Hitschler (2011, 19–21) base their descriptions of the defense on Schwab's memoirs, too.

the most prominent gravediggers of Weimar."[38] Schwab insisted in his defense that he had "not encountered any references to anti-Semitism prior to 1933" (Schwab 2021, 157) in Schmitt's work and that Schmitt had enthusiastically attempted to defend the Weimar Republic against both Communists and the Nazis. According to Schwab, Kirchheimer was also incensed that Schwab had briefly discussed his own 1930 article "Weimar—and What Then" in a footnote. Schwab had interpreted Kirchheimer's early Weimar writing as attempts to torpedo the Weimar Constitution. "Of all writings," Schwab later quoted Kirchheimer, "you had to single out those." Obviously, not only Schmitt's works but also some of Kirchheimer's were at stake during the defense.

Kirchheimer pronounced that Schwab's work had remained apologetic throughout. Schwab had not even remotely understood Schmitt's role in the destruction of the Weimar Republic, and moreover, the work included several factually incorrect and polemical attacks against critics of Schmitt. Since Herbert Deane and the other members of the committee had nothing substantive to contribute to the debate, they followed Kirchheimer's negative assessment. And they believed he had good reasons for it. Deane knew Kirchheimer from other discussions and valued his knowledge, academic tolerance, and fair judgment. The members of the committee also accepted Kirchheimer's objection that Schwab had failed to include the critical literature on Schmitt that had already been published in Germany, for example, the books by von Krockow, Fijalkowski, and Schneider. After the defense, Deane informed Schwab that he had failed, calling Kirchheimer the decisive voice because he was an "expert in the field" (Schwab 1988a, 81).

Schwab was stunned. He was personally disappointed by Deane, who had encouraged him time and time again over the past seven years in his work on Schmitt. He immediately reported extensively to Schmitt about the result of the defense in letters and during his next visit to Plettenberg. Enraged, he wrote to Schmitt that Kirchheimer had not accepted his description of Schmitt as a defender of the Weimar constitutional order.[39] He now considered Kirchheimer "an enemy you know" (Schwab 2021, 180) and abandoned any new attempt to obtain a doctorate at Columbia University on the same subject. Instead, he decided to write a new dissertation on neutral countries and nuclear weapons in a case study of neutral Switzerland. Kirchheimer died in 1965 and Schwab noted in his memoir: "With Kirchheimer out of the way, I could now peacefully focus on completing the new dissertation" (Schwab 2021, 195). He successfully finished it in 1968.

Even though Schwab viewed Kirchheimer as an "enemy" after his failed dissertation—even according to his own retrospective reports—there was obviously no ill will on Kirchheimer's part. His reasons were based on his factual objections—which the other committee members agreed with—to Schwab's interpretations. Overall, Kirchheimer had four substantive objections: first, Schwab had misunderstood crucial sections of the Weimar Constitution; second, he had misread Schmitt's role in the final days of the Weimar Republic; third, he had misrepresented Schmitt's antisemitism, thereby downplaying it; and, fourth, he had ignored the state of research in the critical literature

38 As reported by George Schwab in response to a question about this matter from Volker Neumann, see Schwab (1988b, 462).

39 See letter from George Schwab to Carl Schmitt dated 11 March 1962. Quoted in Mehring (2014a, 507–508).

on Schmitt. Concerning his academic standards, Kirchheimer wrote in his letter to Friesenhahn at an early stage of the conflict that he would not necessarily even have objected to a defense of Schmitt "lock, stock, and barrel," but only if Schwab had emphasized Schmitt's "real doctrines and did not disguise him as a democrat and strict constitutionalist"[40] before and after 1933.

However, both Schwab and Schmitt viewed Kirchheimer's substantive objections to Schwab's dissertation as a purely politically motivated attack on them. This was not the first time that Schmitt had taken criticism poorly and personally. Rudolf Smend had called him an "effective pacemaker of the violence-based Nazi system" in a 1960 article about the history of the Berlin Law Faculty. Schmitt felt offended and immediately broke off contact with Smend, whom he had known for more than forty years at the time.[41] Two months after Schwab's defense, Schmitt wrote to Ernst-Wolfgang Böckenförde: "I am sure you know that Kirchheimer didn't let poor George pass,"[42] He held Kirchheimer responsible for Schwab's failure and felt it to be an attack by Kirchheimer *ad hominem* and a stab in the back.[43]

3. Second-order observations

No personal contact between Kirchheimer and Schmitt is documented after Schwab's failed dissertation. Whether it was Schmitt who broke off contact or whether both sides had concluded that they no longer wanted anything to do with one another can no longer be determined today. Both, however, still followed the work and activities of the other. Although Kirchheimer refrained from contacting Schmitt again, he closely monitored the steps Schmitt took after he had rejected Schwab's dissertation. One year after the incident, he reported to Ernst Friesenhahn:

> By the way, our friend Carl Schmitt has managed again to take revenge on me semi-anonymously for not accepting his young man's doctoral dissertation. Signed 'C.S.,' he made an unfriendly comment in a German journal, I think it was '*[Die] politische Meinung*,' by saying more or less that the whole book [*Political Justice*] actually doesn't say anything more than my essay from 1955. I somehow also suspect that he was behind a 10-page polemic in a third-rate American law review.[44]

40 Letter from Otto Kirchheimer to Ernst Friesenhahn dated 20 November 1960. Otto Kirchheimer Papers, Series 2, Box 2, Folder 61.

41 The quote is to be found in Smend (1960, 542). On breaking off contact, see Mehring (2010, 150–152).

42 Letter from Carl Schmitt to Ernst-Wolfgang Böckenförde dated 6 April 1962 (Schmitt and Böckenförde 2022, 321).

43 See Schwab (1980a, 81) and Quaritsch (1999, 72).

44 Letter from Otto Kirchheimer to Ernst Friesenhahn dated 31 March 1963. Otto Kirchheimer Papers, Series 2, Box 2, Folder 61.

The polemic Kirchheimer mentioned was presumably the review of *Political Justice* by the anonymous author C. in the *Modern Law Review*.[45] There is no evidence or even merely any indication that Schmitt was involved in its publication (let alone that he could have been "C."). The case is different regarding the review in the journal *Die Politische Meinung* [The political opinion]. One of its editors was conservative publicist Rüdiger Altmann, who had studied under Schmitt in Berlin as a wounded veteran in the final semesters during World War II and had been in touch with him again from the mid-1950s onward (see van Laak 1993, 262–265). It was presumably via this connection that the journal accepted the review. Signed "C. S.," this has not been listed in the bibliographies of Schmitt's works to date. Not only Kirchheimer's statement (presumably informed by Werner Weber or Rudolf Smend), but equally the review's substance, language, and style support the assumption that it was authored by Schmitt. For example, it was characteristic of Schmitt to approach the reviewed book via the index and to refer to the dedication. The choice of wording is quite typical of Schmitt in multiple places as well.[46]

Schmitt began the review[47] by pointing out that Kirchheimer had dedicated the book to the victims of political justice. Schmitt added that to him "any and all political administration of justice is somehow suspect, in most cases an annoyance and a piece of folly" (94)—which was only partly an accurate description of the intention of Kirchheimer's book, however. Schmitt chafed at Kirchheimer's assessment of the case of Paul Jorns during the Weimar Republic. Prosecutor Jorns had been assigned to investigate the murders of Rosa Luxemburg and Karl Liebknecht. Kirchheimer agreed with the opinion liberal journalist Berthold Jacobs had expressed in 1928 that through his way of investigating the case, Jorns had aided and abetted the murderers and had helped them escape from jail. Jorns responded by filing a libel suit against Jacobs. Yet Jorns lost this case in multiple instances because of the facts presented. According to Kirchheimer, the *Reichsgericht* (see List of German Courts) applied a legal trick to avoid having to acquit the journalist once again. Schmitt contradicted this point, rejecting Kirchheimer's "attack" (94) on the court as "unjustified" (94).

Schmitt nitpicked about two minor errors in the index of names and one piece of incorrect information about a judge at the Nazi *Volksgerichtshof.* He caricatured Kirchheimer's argument in the book as an arbitrary concatenation of examples and names: "On page 26, he quotes the *Bundesgerichtshof*, presents Count Harry von Arnim, only to flash back to Henry VIII and then shift his attention to Hermann Göring" (94). Kirchheimer had "processed a downright improbably copious amount of material with unending diligence" (94). Yet this supposed praise was poisoned inasmuch as he judged him a few lines further down: "Incidentally, in his essay '*Political Justice*' [...] the author stated his concerns in a considerably more concise and concentrated way" (94). The purpose of the book "might be for its author to gain influence on the law clerks in the American Supreme Court and thus on its decisions" (94). In other words, Schmitt insinuated that the German discussion would not benefit at all from the book. But he did recommend

45 Volume 26, 1963, pp. 456–459.

46 Reinhard Mehring (e-mail dated 7 December 2021) and Gerd Giesler (e-mail dated 8 December 2021) also support my claim of Schmitt's authorship.

47 See Schmitt (1962) for this and the following quotes.

two publications about state security and the constitution by other authors to the readers of *Die Politische Meinung*. Schmitt concluded his review with lukewarm praise: "In any case, Kirchheimer's book is interesting and instructive. Even if one does not agree with everything he says" (94).

Schmitt was known for his own particular way of approaching the subjects of the books he reviewed. That is why it is not surprising that he cherry-picked just a few points to comment on. But it is surprising how little he engaged with the concept of political justice and also the wider context of Kirchheimer's argument in his review. This raises the question how deeply Schmitt had even read the book. He is known to have made handwritten comments in the books he read, and as mentioned in the previous chapter, there are only very few comments of his in his copy of this book.[48] Apparently, he had read it superficially at best, and he did not reveal to his readers at which points—potentially including the Nuremberg Trials, asylum law, and Nazi criminal law, for example, none of which he even mentioned—he disagreed with its author. He did not devote a single word to Kirchheimer's critical analysis of the legal system of the GDR, either.

Three years earlier, he had responded quite differently when Kirchheimer had sent him an offprint of his essay on the concept of legality in East Germany. Kirchheimer later included this essay with only a few changes in his book *Political Justice*. In May 1959, Schmitt had written in a letter to Ernst-Wolfgang Böckenförde that Kirchheimer's essay was "exciting" and had urged him to read it.[49] Nothing about this recommendation is to be found in his review of Kirchheimer's book. Schmitt no longer praised Kirchheimer in any of his writing and stopped sending him offprints. Kirchheimer stopped sending Schmitt his publications, too. Schmitt's name was not even on the long list Kirchheimer sent to Günther Busch of Suhrkamp publishing house in 1964 of potential recipients of his book *Politik und Verfassung* [Politics and constitution],[50] which included his famous article "Weimar—and What Then?".

The German edition of *Political Justice* was published in March 1965 (see Kirchheimer 1965c). Arkardij Gurland had prepared the translation, which had taken more than four years because he had had to interrupt his work on it several times. In addition, Kirchheimer had made a number of additions to the text. Overall, the German text was 20 percent longer than the English one. Of course, there was no need to include positions on the fundamental debates among legal scholars in Germany in the American edition of the book. Yet Kirchheimer felt he had to take a position for the German edition. In 1965, the rifts between the two remaining major schools of thought on Weimar constitutional law, those following Schmitt and Smend, were as deep as never before. The publication of the *Festschrift* on the occasion of Schmitt's seventieth birthday in 1958, edited by former students of Schmitt's and legal scholars Ernst Forsthoff, Werner Weber, and Hans Barion, had sparked a new debate on the concept of the constitution and the meth-

48 Copy of Otto Kirchheimer, *Political Justice*. Carl Schmitt Papers, RW 0265, No. 25665.

49 Letter from Carl Schmitt to Ernst-Wolfgang Böckenförde dated 8 May 1959 (Schmitt and Böckenförde 2022, 199).

50 Letters from Otto Kirchheimer to Günther Busch dated 19 and 20 November 1964. Otto Kirchheimer Papers, Series 3, Box 2, Folder 68.

ods of interpreting the constitution in the Federal Republic of Germany.[51] The statism of Schmitt's school was countered by the proponents of Smend's theory of integration, who were simultaneously advocating for opening the field up to Western theories of democracy. Kirchheimer took the conceptual introductory passages of the German version of his book as an opportunity to refer to Smend and his idea about the potentially integrative functions of judicial procedures (see Kirchheimer 1965c, 22–24).[52]

By contrast, Schmitt was not mentioned explicitly even once in the German edition. Once again, however, a few passages read like silent dialogues with Schmitt, for instance, where Kirchheimer contradicted Schmitt's student Roman Schnur's interpretation of the history of French parliamentarism in the late sixteenth century as "large-scale neutralization" or attested that the parliament in Paris had successfully adapted to rapidly changing situations (see Kirchheimer 1965c, 660–661). In some passages of his defense of the Nuremberg Trials, to which he added multiple pages for the German edition, he had German critics of the trials speak, at times using Schmitt's vocabulary (see Kirchheimer 1965c, 473–510). Exercising less restraint than previously in his essays on political justice published in German, Kirchheimer now used the word *Feind* (enemy), a signal word of Schmitt's. The word appeared right at the beginning, in the first two sentences of the book, as well as in many other places (see Kirchheimer 1965c, 21, 206, 207, and 237). However, Kirchheimer never used the word *Feind* to signify enmity between individuals or groups of individuals but, rather, in the sense of a group's fundamental opposition to a political system. The term Kirchheimer used as a synonym for *Feindschaft* (enmity) was *systemfeindlich*, inimical to the system (see Kirchheimer 1965c, 243), thereby diverging from Schmitt's usage of the word *Feind* in a personalizing and existential way.

Schmitt continued to observe Kirchheimer's activities and publications from afar. In the following years, he went one step further, writing disrespectful comments about Kirchheimer in letters to his friends and young admirers. As mentioned above, he had realized that Kirchheimer had given advice to Hannah Arendt for her book on the Eichmann trial.[53] Writing to Roman Schnur, he described Kirchheimer as follows: "a superficially reformed Marxist, a kind of sociologist, a debunker of every non-Marxist ideology, but he is truly not a legal scholar in any sense of European jurisprudence."[54] In 1965, he wrote to Armin Mohler about the publication of the German edition of *Politische Justiz*: "[Kirchheimer's] book about political justice does not address the actual problem."[55] Yet, as in his review for the journal *Die Politische Meinung*, he failed to reveal to Mohler what he thought the actual problem of political justice was. Another thing Schmitt did was try to help Schwab get his rejected dissertation published by an American academic publisher. All of Schwab's attempts failed because of negative expert reviews. Furious, he and Schmitt accused Kirchheimer of pulling strings to prevent the publication.[56] There are no documents in Kirchheimer's papers at the State University of New York at Albany

---

51 See Günther (2004).

52 See Chapter 16, p. 444.

53 See Chapter 16, p. 439.

54 Letter from Carl Schmitt to Roman Schnur dated 24 October 1963 (Schmitt and Schnur 2023, 542).

55 Letter from Carl Schmitt to Armin Mohler dated 26 August 1965 (Schmitt and Mohler 1995, 354).

56 See Richter (2001, 222–224) and Hitschler (2011, 19–21).

that would support this claim. In December 1965, Schmitt complained to Roman Schnur that "O. Kirchheimer [was] going after George Schwab."[57]

In 1968, Schmitt wrote to Forsthoff about the fact that no reputable publisher was willing to publish Schwab's book: "What is being done to me is a disgrace, but I do not want to share the glory of that disgrace with anyone."[58] It was only through Schmitt's personal intervention with his publishing house Duncker & Humblot that the text was published eight years later, in English, in Germany (see Schwab 1970).[59] Even in retrospect more than twenty-five years later, Schwab blamed a "[Kirchheimer's] hostile attitude toward Schmitt" (Schwab 1988a, 81) for the failure of his dissertation. He repeated the unfounded accusation that a negative attitude toward Schmitt in the United States, for which Kirchheimer had been instrumental, was the reason why his manuscript on Schmitt was not accepted by any recognized publisher, in his memoirs in 2021 (see Schwab 2021, 180).

Kirchheimer by no means intended to categorically halt the reception of Schmitt's work in the United States, as Schwab insinuated. In a peer review comment on a manuscript for the *American Political Science Review* sent to its editor Harvey Mansfield two years after the conflict over Schwab's work, Kirchheimer wrote: "Schmitt should be presented to the American Political Science Community and on the basis of the numerous German studies [already] existing." Two approaches were to be given preference: "One may treat Carl Schmitt [...] either by studying his conceptual framework, including questions of logical consistency; or, by relating his concepts to the German political reality of his days."[60] He continued to include Schmitt in his teaching at Columbia University. On the reading list of his syllabus for the seminar "The Political Institutions of Divided Germany" (1962/63), he recommended that the students read Schmitt's *Constitutional Theory*, calling it the "most influential constitutional interpretation on [the] basis of antidemocratic-authoritarian theory."[61]

4. On partisans and political partisanship

Schmitt's gift to himself on the occasion of his seventy-fifth birthday in 1963 was the republication of his books *Dictatorship* and *The Concept of the Political*, both with brief retrospective comments. The only book of Schmitt's after 1950 which was not mostly retrospection was *Theory of the Partisan*, which was also published just in time for his seventy-fifth birthday. This is the only book from Schmitt's late oeuvre that has been received just as widely beyond his circles and still to this day as otherwise only his Weimar writings

57 Letter from Carl Schmitt to Roman Schnur dated 1 December 1965 (Schmitt and Schnur 2023, 575). Kirchheimer had died nine days before.

58 Letter from Carl Schmitt to Ernst Forsthoff dated 22 May 1968 (Schmitt and Forsthoff 2007: 261).

59 A number of reviewers accused the book of aiming to construct the apologetic legend that Schmitt had kept his distance from the Nazi regime, see Richter (2001, 224–226).

60 Letter from Otto Kirchheimer to Harvey Mansfield dated 4 June 1964. Otto Kirchheimer Papers, Series 2, Box 1, Folder 51.

61 Minutes of the Faculty of Political Science 1957–62. Special Collection, Columbia University Archives.

have.[62] *Theory of the Partisan* (see Schmitt 1963a)[63]—subtitled *Intermediate Commentary on the Concept of the Political*—was based on lectures Schmitt had held in Spain in 1962. He revealed in this work how strong his political sympathies for fascist Spain continued to be in an aside celebrating the civil war that followed General Franco's coup as a "war of national liberation" against "the international communist movement" (56).

Schmitt considered the substantive core of the book to be a continuation of his reflections on the concept of the political. He described the partisan as a type of fighter with high political intensity. The partisan's origins lay in the Spanish guerrilla resistance against Napoleon.[64] While the bourgeois took off his uniform in order to trade and make money in peace, the partisan took off his uniform in order to fight all the better. The partisan of the Spanish war fought against the universalizing impulses of the Napoleonic project. In his purest form, the partisan was a creature of agrarian provenance. Partisans were mobile and fast. But despite all their tactical mobility, they maintained their intimate relationship to a specific locality and the soil. Schmitt coined the term "telluric" (20) to describe this feature of the partisan. He gave several more historical and more recent examples of the occurrence of this type of fighter. However, he regarded the *Volkssturm* (a militia of poorly equipped civilian boys and men drafted by the Nazi regime in a last-ditch effort to defend the fatherland), to which he had been conscripted for a few days in early 1945, as a regular military corps (see 38–39) and thus not as partisans.

Schmitt demonstrated how difficult it was for the traditional law of war to deal with the phenomenon of the partisan. To Schmitt, the Prussian military expert Carl von Clausewitz was an outstanding author, the first to theoretically recognize and legitimize partisans in his writings on war. In my view, this assertion of Schmitt's is astounding because partisans had not played a particularly significant role during the war of the Prussians against Napoleon's forces. Yet Clausewitz was an important author to Schmitt inasmuch as he had a major impact on Friedrich Engels's and Vladimir Ilyich Lenin's thoughts on war (see Hohendahl 2012, 532–533). Schmitt revised the terminology about enmity he had previously used in *The Concept of the Political* with respect to the partisan. He now differentiated between three categories: the conventional enemy, the real enemy, and the absolute enemy. The conventional enemy corresponded to cabinet war, Schmitt claimed, which was subject to limits under international law that were so strong that it practically amounted to a duel that did not impact civilians and "could be conceived as a play" (88). It was only partisans who had reestablished war as a serious matter and had made the enemy a real enemy. The next step up was the absolute enemy. Schmitt attributed the theoretical foundation of absolute enmity to Lenin's theory of class struggle, enriched by Stalin's and Chairman Mao Zedong's theories on partisan warfare. The true partisan had not taken the step from the real to the absolute enemy. In this sense, as a "partisan of tradition," (Münkler 1992, 122) he, Schmitt—like the protagonist in the book *Forest Passage* by his friend, right-wing author Ernst Jünger (see Jünger 1951)—was

62 For a well-informed discussion of the general place of Schmitt's book in his oeuvre, see Llanque (1990).

63 The following page numbers refer to this text.

64 On the criticism that Schmitt did not include the irregular troops of the Thirty Years' War or the American Revolutionary War, see Hohendahl (2012, 531).

the last remaining proponent of the idea of real enmity; in Schmitt's view, the potential of the political relied on that idea. He considered leftist revolutionaries Ho Chi-Minh, Fidel Castro, and Che Guevara to be the most prominent authors of the day regarding the transition to the concept of the absolute enemy in the theory of war.[65] Thus, Schmitt made Marxism exclusively responsible for the turn to the concept of the absolute enemy.

Schmitt did not say a word about Hitler, however, as a hatemonger of absolute enmity. Nor—and this is hardly surprising—is there any reflection of Schmitt's own writing from the phase leading up to World War II in which he spoke of the "imminent, immutable, real, and total enmity" that "leads to the ordeal by battle of a total war" (Schmitt 1937a, 485). Nor did he mention the crimes of the *Wehrmacht* in the passages about its battles with partisans in the Soviet Union, Greece, or the Balkans (see 19–29). Rooted in the Prussian military tradition, the German *Wehrmacht* was unprepared for partisan warfare when it had invaded Russia, he claimed. It was only in late 1944 that the Supreme Command of the *Wehrmacht* had issued a guideline for fighting partisans that Schmitt praised as "extraordinary" (39). Before then, he alleged, they had been marauders "handled by the police" (33). Thus, Schmitt, too, continued to spin the postwar German legend of the "clean *Wehrmacht*," namely that it was not involved in perpetrating war crimes or the Holocaust.

As part of a new global order in which customary categories of war were losing relevance, partisans had become key figures of global history. At the end of his book, Schmitt conjured the apocalyptic image of entirely "new types of absolute enmity" (94) in modern technical industrial development. It was not enmity that caused the production of new weapons. Instead, it was the development of war technology that produced a need for new enmities and new concepts of the enemy. The new weapons technology had to be given meaning ex post. Schmitt's fear of the tyranny of technology in a nutshell: "absolute weapons of mass destruction require an absolute enemy" (93).

Schmitt ended his work on partisans with associative predictions: "Interested third parties" (75) would provide them with new weapons and other resources and would presumably instrumentalize them more and more often in the future. Thus, they would become a tool of the aggression of the international communist world revolution, a tool that could be manipulated. They would adapt to new technological circumstances with lightning speed, making them the means of their struggle. Schmitt spoke of the "technical-industrial partisan" (79) who would use the most up-to-date biological, chemical, or atomic weapons of annihilation. Schmitt believed that in light of nuclear weapons, partisan warfare with conventional weapons was the last refuge of real enmity. However, the only partisan Schmitt considered legitimate, namely the nationalist partisan, would be replaced by the urban guerilla fighter and the terrorist in the future. In this sense, Schmitt's *Theory of the Partisan* was a nostalgic book melancholically grieving the loss of the telluric and defensive true partisan and pessimistic about global politics.

In the late 1960s and the 1970s, under the formative influence of the Vietnam War, the guerrilla wars in Latin America, and leftist terrorism in Europe, Schmitt's work on partisans caught the attention also, and particularly, of radical left-wing circles in Italy,

65 On these assessments, see Münkler (1992, 111–141). Against Schmitt, he underlined the transitory character of the partisan as a precursor of a regular army in the theories of the authors mentioned.

France, and Germany.[66] However, the left-wing radicals' sympathy for it disregarded the fact that Schmitt wrote about the counterrevolutionary strategies to defend the colonial rule of Raoul Salan, the infamous founder of the Organisation armée nuit (OAS, Secret Army Organization) who embraced terrorist methods in order to fight the Algerian National Liberation Front (FLN), with the same enthusiasm with which he had written about the political significance of the partisan alongside Mao. Nevertheless, the lawyers representing the German Red Army Faction (RAF) terrorists referred to Schmitt's book in their attempt to have their clients acknowledged in court as parties to a civil war (see Preuß 1989, 146–149). At the turn of the millennium, there was renewed interest in Schmitt's work on partisans. After the attacks of 11 September 2001 and the military response of the US, his associative prognoses on the methods of terrorism at the end of the book were hailed as prescient (see Scheuerman 2020). Among military experts, Schmitt is considered to be one of the first theoreticians of the new asymmetric or hybrid wars (see Münkler 2004).

However, it is less apparent what Schmitt's work added in a systematic sense to the concept of the political and, in particular, to the theory of enmity. To begin with, the lack of figures of the enemy is striking, whereas he did discuss these in his previous works. He did not mention the "total enemy" (see Schmitt 1937a, 481) at all, as he had in 1937. Nor did he discuss the "true enemy" he believed he had identified in the assimilated Jew, as he had noted in his *Glossarium* in 1947.[67] Moreover, it remains unclear what the difference between the real and the absolute enemy was supposed to be, as Schmitt considered both to be partisans. And finally, the question arises whether it is even possible to intensify the friend-enemy dichotomy, which he had first detailed in his *Concept of the Political*. After all, Schmitt had defined enmity as the "ultimate distinction" with the "utmost degree of intensity" (Schmitt 1932a, 26) as early as 1932. In a purely logical view, such a concept of enmity cannot be intensified. Schmitt did not solve this problem in his *Vorwort* (Foreword) of 1963 to the new German edition of *The Concept of the Political*, either, where he once again listed the three different kinds of enemies and emphasized the importance of distinguishing between them precisely (see Schmitt 1963b, 17).

Kirchheimer did not engage with this part of Schmitt's oeuvre. He did not have a copy of the book on partisans in his library. Nonetheless, his writing includes a counterpoint to deliberations of Schmitt's from that work. Schmitt viewed partisans as technically adept and fanatic lone wolves but thought that the partisans of the guerrilla wars in Indochina had fallen into dependence on interested third parties, that is, the communists in the Soviet Union and China.

Kirchheimer, by contrast, did not see them as belonging to such fixed categories. He had sympathized with the student protest movement against the US war in Vietnam from the outset and was in animated exchange with Herbert Marcuse about this. He was exposed to this issue at his own university more directly because of his son's political ac-

66 A radio interview that Joachim Schickel, then a revolutionary Maoist, had conducted with Schmitt on partisans in May 1969 contributed to Schmitt's popularity on the political left in Germany, see Schmitt and Schickel (1969).

67 *Glossarium* entry of 25 September 1947 (Schmitt 2015, 14).

tivism.[68] Kirchheimer's letter to the editor published in the *Washington Post* on 27 March 1965 shows how strongly he disapproved of US military policy. In the letter, he rejected domino theory, which had been used to justify the war. Reminding readers of Spain under Franco during and after World War II, Kirchheimer claimed that historical experience showed that countries often structured their alliances differently than assumed ex ante. He believed there was no reason to be convinced that the partisan units in North Vietnam, which felt they were pressed to form an alliance with China because of the demands of the war, would necessarily take China's side in an open political constellation in the future. Kirchheimer cited the early successes of the policy of détente with the Soviet Union in Europe as an alternative to the war in Indochina. He closed his letter to the editor with the rhetorical question, "is it in the long-range interest of a conservative power to tear up the last shreds of international law under dubious pretexts?" (Kirchheimer 1965d, 654). Schmitt assumed, in Cold War diction, that China and the Soviet Union were pulling the strings behind the partisan battles in Indochina. Kirchheimer, conversely, advocated not underestimating the fact that future political developments were still open to surprises and unpredictable turns.

Whereas Schmitt had not clarified the inconsistencies of his concepts of the enemy in his new foreword to *The Concept of the Political*, it does include a passage worth mentioning on the subject of political justice:

> Such a report [on the impacts of *The Concept of the Political* to date] would have to include the development of the views on political crimes and political asylum and on the justiciability of political acts and decisions concerning political questions by the justice system. It would have to take into account the fundamental question of the judicial process, that is, an examination of the extent to which the judicial process itself, as a process, necessarily changes its material, its object, and transforms them into a different aggregate state (Schmitt 1963b, 13–14).

And Schmitt continued: "All this goes far beyond the framework of a foreword and can only be suggested here as a task." (Schmitt 1963b, 14). He presented the desideratum he had formulated as a subject on which work had only just begun and did not mention that it corresponded astoundingly closely to the substance of Kirchheimer's *Political Justice*, which he did not reference, either. Once again, Schmitt wasted an opportunity to enter into a dialogue with Kirchheimer about the subject at hand.

In 1964, Hasso Hofmann published *Legitimität ntui Legalität* [Legitimacy against legality], which dealt with the development of Schmitt's theories until the 1940s and soon became a "milestone" (Neumann 2021, 11) in Schmitt studies. His general thesis was that there was a certain continuity in all the changes in Schmitt's work: the permanent search for new sources of legitimacy which trump legality. His thesis was an extension of the interpretation by Karl Löwith, his doctoral supervisor. Löwith had called the continuity in Schmitt's approach "occasionalist decisionism" (see Löwith 1935, 32–61). After Schmitt

68 In the spring of 1965, Peter Kirchheimer was one of the campus activists resisting the university's involvement with the Reserve Officer Training Corps (ROTC). Peter Kirchheimer in a conversation with the author on 3 May 2024.

had read Hofmann's book, he wrote on a notepad: "sterile echo of [Karl] Löwith's, [Leo] Strauss's, Kirchheimer's criticism." Schmitt had clearly recognized with this note that all three of those named were in agreement in their criticism of him, despite their other differences. They criticized his rapid adjustments to new political situations as well as his methodological approach. Kirchheimer, however, was the only one who also brought in the perspective of the empirical social sciences. Schmitt ignored these differences among the three authors.

5. Against consumer society

Schmitt concluded his letter to Kirchheimer on 6 August 1958 with a historico-philosophical thought: "perhaps history does not consist of a continuous 'flow', but of quantum-like 'epochs' in which the same situation is repeated again and again until a leap into another 'epoch' is successful."[69] Although Kirchheimer contradicted him—as quoted above—with the words "I do not believe in the repetition of similar situations; too many qualitative changes have taken place," he, too, added a gloomy prognosis: "I do not dare imagine what the general process of dulling people's minds and the limitless ability of the next generation to be manipulated will bring."[70] Although the background to the social theories propounded by Kirchheimer and Schmitt was quite different, they did share—to a certain extent—this culturally pessimistic view.

Schmitt could not and would not reconcile with the social and political realities of the Federal Republic of Germany. His disapproval included the rapid and successful development of the economy celebrated as the "economic miracle" and the country's public culture that focused on private consumption. Time and again, his *Glossarium* entries underlined his rejection of that preoccupation of postwar West German society. Schmitt found it nothing less than repugnant because this development thrust aside the serious nature of the political, and he again took up thoughts and motifs with which he had already railed against "economic rationalism" and "irrational consumption" (Schmitt 1923b, 14) in *Roman Catholicism and Political Form* in 1923. He even turned to the Frankfurt School's critical theory to find allies in renouncing modern consumer society. He quoted Jürgen Habermas, Adorno's assistant at the time: "consumption is the continuation of production by other means," indicating his agreement.[71] A week earlier, he had noted sarcastically: "pure consumer society. I suppose that will become the foundation of happiness."[72]

Schmitt used dramatic-sounding words to express his assessment of the situation, which he considered hopeless, to longstanding confidants, such as the Spanish legal historian Álvaro d'Ors: "Germany's situation today is dreadful. Much worse than most peo-

69 Letter from Carl Schmitt to Otto Kirchheimer dated 6 August 1958. Otto Kirchheimer Papers, Series 2, Box 2, Folder 12. Schmitt used identical wording three weeks later in a letter to Ernst Jünger (Schmitt and Jünger 1999, 353).

70 Both quotations in letter from Otto Kirchheimer to Carl Schmitt dated 4 September 1958. Otto Kirchheimer Papers, Series 2, Box 2, Folder 12. The second quote is similar to Kirchheimer's statements about France after de Gaulle took power (Kirchheimer 1958a, 399 and 1959a, 429).

71 *Glossarium* entry of 25 August 1956 (Schmitt 2015, 352).

72 *Glossarium* entry of 17 August 1957 (Schmitt 2015, 351).

ple suspect because they let themselves be bedazzled by the economic miracle. As an old man, I suffer terribly from this and feel veritable Cassandra depressions." In the same letter, he revealed the extent to which his cultural pessimism was fed by right-wing conservative thoughts and motifs: "Those calling themselves Christian in Germany today are more concerned with remaining anti-fascist and agreeing with the leftist slogans than with the courage of finding themselves."[73] Schmitt assumed that the social and political stability of German postwar society was brittle and could collapse into a new crisis at any moment.

Although the wording is less dramatic, Kirchheimer's writing from the 1960s contains a number of melancholy statements about the mentality of affluent consumer society that was taking hold in the US and before long in West Germany, too He followed and sympathized with the activities of the civil rights movement in the US and supported his daughter Hanna as she protested.[74] He voted for the Democrats in the presidential elections in 1960 and 1964 although he disagreed with Lyndon B. Johnson's foreign policy. And from the early 1960s on, Kirchheimer occasionally intervened in the discussions about day-to-day politics in the US with letters to the editor of the *Washington Post*. All his sympathies for the emerging protest movements notwithstanding, his writing lacked both the cautious optimism he had had with respect to the political culture of the Federal Republic of Germany and the specific tone expressing a sense of a new social beginning that had started to spread at US universities from the early 1960s on.

A fatalistic tone is clearly evident in some of Kirchheimer's later works. For example, in his 1962 article "Expertise and Politics in the Administration," he spoke of the "shadow of general barbarism which threatens us daily" (Kirchheimer 1962c, 372). He ended his last major essay on party research in 1965 by emphasizing the functional gap that was becoming apparent because of the transformation in the political systems of Western democracies and opening up the political space for the future success of populist parties and groups. He finished the article with the prognosis: "we may come to regret the passing—even if it was inevitable—of the class-mass party, as we already regret the passing of other features in yesterday's stage of Western civilization" (Kirchheimer 1966a, 371). His posthumously published essay "The *Rechtsstaat* as Magic Wall" in the *Festschrift* for Herbert Marcuse had the following ending:

> A generation which has lived through Auschwitz and Hiroshima and was indifferent or powerless to prevent them, and which is prepared to see bigger Hiroshimas, has no cause for complacency about its preservation or even enlargement of some orderly forms of living. It may have forgotten the essential: there must be life for life to be worth living (Kirchheimer 1967a, 312).

Kirchheimer's writing from the final years of his life has been interpreted in different ways in the secondary literature because of passages like these. Some writers believe that the resigned, stoic, melancholy, pessimistic, or even fatalistic undercurrent dominated

73 Letter from Carl Schmitt to Álvaro d'Ors dated 12 February 1962 (Schmitt and d'Ors 2004, 200).

74 Hanna Kirchheimer-Grossman in a conversation with the author on 15 April 2019.

in his late oeuvre.[75] Yet there is also another way of reading it, which I, too, subscribe to. In this interpretation, his texts are seen as documents of his search for a theoretical approach sympathetic to the program of Max Horkheimer's early critical theory and with which Kirchheimer, with his criticism of capitalist mass society, came closer in substantive terms to the critical theory of his friend Herbert Marcuse.[76] This reading highlights the potential of Kirchheimer's late works to renew critical theory in terms of political science, and it is also supported by the fact that Kirchheimer frequently quoted Adorno and Habermas in his late writing, besides Marcuse. More important, however, are the convergences in matters of substance. When Kirchheimer wrote about what Adorno called political alienation (see Adorno 1963, 382), he more soberly called it "privatization" (Kirchheimer 1967b, 459). In the last pieces he wrote before his death, this tendency of privatization became the analytical center of his diagnoses of the precarious condition of Western democracies beneath the veneer of superficial stability.[77]

Although there seem to be some parallels in their criticisms of consumer society, it is abundantly obvious that Schmitt and Kirchheimer developed them on the basis of quite different fundamentals and that they were imagining completely different sociopolitical alternatives.

6. Kirchheimer's untimely death

When Kirchheimer began his tenure at Columbia University, he continued to commute between Silver Spring and New York, and in the summer months between the US and Germany. He continued to apply for, and receive, research fellowships. Columbia University granted him a leave of absence as a fellow of the Social Science Research Council for the winter term 1964 and as a fellow of the John Guggenheim Foundation for the summer term 1965.[78] The John Simon Guggenheim Memorial Foundation awarded him a research stipend for the academic year 1965/66 to continue his studies of parliamentarism and parties in Western Europe.[79]

Once his son Peter had completed high school and enrolled at Columbia in the autumn of 1964, Kirchheimer again faced the decision of whether or not to move back to Germany for good. Even his family cannot definitively answer the question whether he actually seriously considered moving to Germany permanently.[80] The subject came up time and again, and his wife Anne did not change her position: she did not want to return to the land of the murderers of most of her family. Kirchheimer was clearly flattered by the unceasing interest in him in Germany. He was particularly comfortable in discussions

75 See Herz and Hula (1969), Perels (1988), Kohlmann (1992), and Schale (2006).

76 See Söllner (1982, 1986), Scheuerman (1994), Heins (2006), and Buchstein (2020c, 2023b and 2024).

77 See Söllner (1982) and Buchstein (2020a).

78 Otto Kirchheimer, Curriculum Vitae (1965). Private collection of Hanna Kirchheimer-Grossman (Arlington).

79 Letter from Gordon N. Ray to Otto Kirchheimer dated 17 March 1964. Otto Kirchheimer Papers, Series 2, Box 1, Folder 3.

80 Peter Kirchheimer and Hanna Kirchheimer-Grossman in a conversation with the author in New York on 8 February 2019.

with younger people in Germany, be they students, doctoral candidates, or young professors.[81] Although he appreciated his American colleagues for their knowledge and specialization, he generally found them boring, whereas he did not tire of praising the more in-depth education of young German academics. For their part, they were—as Harry Pross, a former student of Kirchheimer's who later became an influential professor of journalism in Berlin, describes in his memoirs—"deeply impressed" time and again by Kirchheimer's ability to speak in a polished style, convince his audience of his positions, and do so with a dash of humor (Pross 1993, 159).

There was a parallel to Schmitt here. In their memoirs, the highly talented younger scholars of the law and the humanities who flocked to him in the 1950s and 1960s extolled Schmitt's extraordinary goodwill, his exquisite friendliness, and his ability to mesmerize his younger listeners with his rhetorical brilliance.[82] Schmitt also made an effort to nurture the relationships established in person through meticulously composed letters. In these letters, it was often less the clarity of an argument but more a way of establishing associations between ideas that incessantly promised to reveal secret or veiled realms and connections within the humanities, thus generating a special kind of personal attachment (see van Laak 1998, 216). Jacob Taubes called these letters from Schmitt eagerly awaited "messages in a bottle."[83] As philosopher Odo Marquard interpreted his memories of the many conversations with Schmitt, the old man sought to engage with the younger scholars "in order to be present in their minds, then and in the future, as the person he would have liked to have been" (Marquard 2013, 73).

Kirchheimer presented himself to the West German public as a "guest from abroad" (Kirchheimer 1965b, 96), for example, at the 45th Deutsche Juristentag in 1964. When he had repeatedly mentioned his interest in a permanent position in Germany to Horst Ehmke, Ehmke saw to it that he was offered one at the University of Freiburg. Ehmke and Konrad Hesse were both renowned students of Smend's who had come to the Faculty of Law in Freiburg and who were trying to bring together constitutional lawyers and political scientists who shared their mindset.[84] The University of Mainz also expressed interest in Kirchheimer. Ernst Fraenkel asked him in 1964 whether he wanted to assume the Chair of Political Science which was becoming available.[85] Kirchheimer rejected the offer. He wrote in a letter to Gurland: "Both Friesenhahn and Fränkel [*sic*] asked whether I was interested in Mainz, but I indicated that Frankfurt and Freiburg appear more appropriate."[86] Kirchheimer continued to favor Freiburg, where he had regularly taught during summers as a Fulbright professor from 1961 on, and which was not far from his former hometown Heilbronn in southwestern Germany. In late autumn of 1965, Ehmke, then the responsible Dean, officially offered him an appointment at the University of Freiburg.

81 Peter Kirchheimer in a conversation with the author on 31 January 2019.

82 See Seifert (1996, 116–118), Böckenförde (2011, 359–384), and Dunkhase (2019, 412–414).

83 Letter (undated, probably 1958) from Jacob Taubes to Carl Schmitt (Schmitt and Taubes 2012, 24).

84 See Günther (2004, 224), and Schefold (2012, 198–202).

85 Letter from Ernst Fraenkel to Otto Kirchheimer dated 27 May 1964. Otto Kirchheimer Papers, Series 2, Box 1, Folder 57.

86 Letter from Otto Kirchheimer to Arkadij Gurland dated 2 July 1964. Otto Kirchheimer Papers, Series 2, Box 1, Folder 68.

On 22 November 1965, in the midst of this turbulent life as an internationally renowned professor of political science, Otto Kirchheimer suffered a heart attack on an airplane just before it took off from Washington, DC, to New York. He died a few days after his sixtieth birthday. Kirchheimer had still been full of plans for new scholarly projects. Working with Helge Pross, he had begun to arrange for a German translation of Franz L. Neumann's book *Behemoth* (see Erd 1985, 129).[87] Because he opposed the American war effort against the Vietcong partisans, he wanted his next project to be about the problem of hegemony in international relations (see Herz and Hula 1969, xiii). He had also already made arrangements for a trip to the East, to the German Democratic Republic, the following year to collect materials for a study on that country's legal system.

Otto Kirchheimer was not an observant Jew, but he was committed to his Jewish identity.[88] His ashes were buried alongside those of his parents in the Jewish cemetery of his hometown of Heilbronn, as he had wished.[89] His wife Anne Kirchheimer died in Silver Spring in 2008, almost forty-three years after her husband, at the age of 93.

There are no reports on how Schmitt reacted when he learned that his onetime star student had died. Three years later, Jürgen Seifert, a young leftist assistant professor who had worked with Kirchheimer's friend Gurland in Darmstadt from 1963 on, explored questions about Kirchheimer in a letter to Schmitt. Seifert had been part of the group of young German students who had experienced Schmitt in person at a lecture in the mid-1950s and was immediately fascinated by him (see Seifert 1996, 115). They exchanged a few letters over the next few years and, in 1958, Schmitt made Seifert aware of some older works of Kirchheimer's. As an assistant of Gurland's, Seifert was also involved in translating *Political Justice* into German (see Kirchheimer 1965c, 16). After they had met in person to discuss the translation project, Kirchheimer told him about the conflict at Columbia University over George Schwab's dissertation.

Three years after Kirchheimer had died, Seifert asked Schmitt about him in a letter. Schmitt's response to a question—"did the two of you fall out?"—was brusque and hostile; he wrote: "My postwar relations to Mr. Kirchheimer started with his visit in Plettenberg (27 November 1949) and ended in the summer of 1961, when I found out details about his behavior in Schwab's doctoral procedure. [...] the way Kirchheimer prevented the work from being accepted made me recognize an error I had made in 1927."[90] Schmitt wrote this response after he had consulted with George Schwab (see Mehring 2014a, 687). By this point in time, Schmitt viewed Kirchheimer as a persona non grata who did not even deserve to have received his doctorate from him fifty years earlier. It was only when Jürgen Seifert made a second attempt and announced in another letter that he only wanted

87 The fate of the book in Germany is scandalous. It took until 1977 for a German translation to be published. The book was translated into Hebrew and Spanish as early as 1943 (see Söllner, Wildt, Buchstein and Hayes 2023).

88 "Otto was never an observant Jew [...] but he always identified himself as a German Jew. He strongly objected to name changes or other activities he considered to be a denial of a person's Judaism. At the same time, he often declared that Reform Judaism was not the 'real thing.'" (Kirchheimer-Grossman 2010, 63).

89 Peter Kirchheimer in a conversation with the author on 3 May 2023.

90 Letter from Carl Schmitt to Jürgen Seifert dated 30 September 1968 (as cited in Seifert 1996, 120).

to ask Schmitt about it in a conversation that Schmitt gave a different answer: "A conversation about Otto Kirchheimer in unobjectionable openness would be a true blessing for me. The fact that you would like to make that possible is in itself a reason to be grateful."[91] Yet this conversation did not come about in the next few years because Seifert, who was active on the political left, was immersed in various political and professional activities, and as a result, he lost sight of his contact with Schmitt.

Several years passed before Schmitt began to mention Kirchheimer again in more positive terms. In the meantime, he had discovered that some of Kirchheimer's works had found new resonance on the political left in Latin America, Germany, and Italy. When Ingeborg Maus, a leftist political theorist with close ties to the Frankfurt School, received her doctorate in 1972, Schmitt congratulated her in a letter using the words: "At this moment, I am moved by the memory of Otto Kirchheimer's doctorate [...] and by the joy I felt at the time of encountering dissent and understanding it."[92] The joy Schmitt expressed about Kirchheimer's objections at the time must have been a stirring of emotion occurring quite some time later. At least in 1928, Schmitt had noted a different emotion in his diary, his direct impression of the long evening they spent with Erik Peterson after Kirchheimer's doctoral defense: "Kirchheimer lacks any national sentiment, horrendous."[93] His letter to Maus continued: "even though he [Kirchheimer] was sure that he understood me better than I did myself."[94] I doubt that Kirchheimer would have said the same about Schmitt.

Schmitt continued to exchange letters occasionally with Ingeborg Maus, who had in the meantime finished preparing her dissertation on Schmitt's legal theory for publication (see Maus 1976).[95] In September 1975, he wrote her that he was angry that he had to serve as the last remaining scapegoat: "I am befallen by a kind of senile nostalgia when I remember the many conversations with Kirchheimer, Karl Korsch, and others from the autumn of 1932."[96] In 1976, he reported to Armin Mohler about another posthumous edition of essays by Kirchheimer. He took this as an opportunity to praise his own role in Kirchheimer's academic career and to lionize himself as a truly liberal-minded person:

> A new Otto Kirchheimer volume is being published by Suhrkamp [...], with old material, including an excerpt from Kirchheimer's dissertation, which I accepted in Bonn in 1928; it is apparent: liberalism is a matter of the strong, not of the weak.[97]

Four years later, Schmitt gave his version of the story of his relationship to Kirchheimer an additional twist. In an oft quoted conversation with Rainer Erd in July 1980, Schmitt was apparently able to create the impression that Kirchheimer had endeavored time and again to stay in contact with him after 1945. At the time of the interview, Rainer Erd was a

91 Letter from Carl Schmitt to Jürgen Seifert dated 5 October 1968 (as cited in Seifert 1996, 120).

92 Letter from Carl Schmitt to Ingeborg Maus dated 24 January 1972 (as cited in Mehring 2013, 442).

93 Carl Schmitt, diary entry of 25 February 1928 (Schmitt 2018, 208). See Chapter 2, p. 65.

94 Letter from Carl Schmitt to Ingeborg Maus dated 24 January 1972 (as cited in Mehring 2013, 442).

95 Her book is still one of the best critical discussions of Schmitt's legal theory.

96 Letter from Carl Schmitt to Ingeborg Maus dated 24 September 1975 (as cited in Mehring 2013, 442).

97 Letter from Carl Schmitt to Armin Mohler dated 16 July 1976 (Schmitt and Mohler 1995, 410).

young staffer at the Institut für Sozialforschung who was interested in the history of the Frankfurt School's critical theory. He had visited Schmitt unannounced in the summer of 1980 to ask him about Franz L. Neumann. Schmitt was happy to speak with him but insisted that he refrain from taking notes and recording the conversation (see Erd 1985, 14). From then on, what Erd reported about his conversation with Schmitt has lived on in the form of an intensifying rumor. It has been repeated in conversations, including the assertion that Kirchheimer had visited Schmitt at his home in Plettenberg several times. What Erd reported about his interview with Schmitt was mentioned for the first time in writing by Volker Neumann in a 1981 essay (see Neumann 1981, 239). Since then, it has been quoted routinely whenever the subject of the resumption of personal contact between Kirchheimer and Schmitt has come up.

More than forty years after the interview, Rainer Erd recalled that Schmitt's responses had been as friendly as they had been vague. However, Schmitt had taken great pains to express "a certain esteem for him [Kirchheimer]."[98] Schmitt was ninety-one when he spoke with Erd, and one might attribute his account of Kirchheimer's visit and his efforts to keep in touch to memory loss. Yet this is contradicted by the fact that his mental faculties at the time were described as still very sharp (see Mehring 2014a, 530–533). Erd also saw no indication that Schmitt was confused or afflicted by dementia. It is more likely that it was just another attempt of Schmitt's to control the narrative about himself and Kirchheimer.

7. Conclusion: Becoming Schmitt's friend posthumously

The controversy about Schwab's dissertation made for a turbulent finale of the contacts between Kirchheimer and Schmitt. Even at this grand finale, it was typical of their relationship that they did not confront each other directly about it but again used communication channels via third parties to express their mutual displeasure. In Schmitt's view, Kirchheimer's rejection of Schwab's doctoral dissertation at Columbia University was an attack directed personally against him. He felt that Schwab was a scapegoat who had to suffer from this attack. The fact that Kirchheimer so vehemently rejected Schwab's work can be explained not least by his assumption that Schmitt was pulling strings in an attempt at political rehabilitation in Germany via the United States. Kirchheimer was already aware of Schmitt's strategic intentions throughout their communication in the 1950s, and Schmitt had already informed Kirchheimer about Schwab in a letter in August 1958. The controversy around Schwab's dissertation between May 1960 and February 1962 was fueled by Kirchheimer's suspicion that Schmitt wanted to instrumentalize him in his function as a committee member. In the event that the apologetic interpretation of Schmitt's role at the end of the Weimar Republic had been accepted in a dissertation at Columbia University, Schmitt would have been rehabilitated. And, moreover, with Kirchheimer's active participation—a person whose biography as a Jew and an author of the left made him a person above suspicion. Kirchheimer was not willing to partake in

98 E-mail from Rainer Erd to the author dated 25 March 2021.

this game and took an unapologetic stand against it. Schmitt, in return, communicated Kirchheimer's stance as a personal attack against him to the members of his circle.

Kirchheimer's professional career finally took off in the 1960s. His book *Political Justice* brought him mostly positive reviews[99] as well as the professorship at Columbia University. The revised German-language version of the book was published in autumn 1965, for which he also received a great deal of praise, among other authors from future West German President Gustav Heinemann.[100] He also published a number of articles in prestigious journals and was a regular reviewer for the *Washington Post*. His research on the transformation of opposition in modern democracies and his thesis on the emergence of a new type of political party he called a "catch-all party" put him at the forefront of political science in the US. The last four years before his death marked a brilliant high point in his academic career. In the summer of 1965, he had decided to accept the offer of a professorship in Freiburg, Germany; his unexpected death in November put an unhappy end to this plan in which his friends and colleagues in Germany had placed high hopes. Schmitt also experienced a certain high point in his career in the first half of the 1960s, albeit outside of academia. With his book *Theory of the Partisan*, he had proven once again that he had an ntuittion for upcoming topics: this book received wide attention beyond his own circles in the years that followed, not least a new readership among the younger generation in the radical leftist camp.

Kirchheimer and Schmitt did not discuss Kirchheimer's book *Political Justice* in the 1960s—nor did Kirchheimer take the opportunity to discuss Schmitt's *Theory of the Partisan,* although he was interested in the American warfare against the Vietcong. Schmitt wrote dismissive remarks about *Political Justice* in letters to some of his friends. A few months after the conflict over Schwab's dissertation, he published a disparaging book review that did not bear his full name. It reads as if Schmitt wanted to take revenge on Kirchheimer, and Kirchheimer had easily figured out that it was authored by Schmitt. Schmitt's preface to the new German edition of *The Concept of the Political* would have provided a potential starting point for a new dialogue between the two. He wrote that it would be important to explore to what extent the judicial process itself as a procedure changes its substance and object and transfers them to a different aggregate state. It is striking how his wording corresponded to Kirchheimer's research program in *Political Justice*. However, Schmitt left this connection unmentioned. He presented the subject as still in its infancy and thus missed another opportunity to enter into a critical dialogue with Kirchheimer. The latter, on the other hand, gave no indication in his book about the extent to which it was inspired by Schmitt's considerations on political justice in his *Constitutional Theory*. He, too, refrained from openly discussing his old partner's theory in controversial dialogues with him.

A few years after Kirchheimer's untimely death, Schmitt realized that some of his works had found new resonance on the radical political left from the late 1960s on. At this time, he started to speak more positively again about Kirchheimer, whose writing was also rediscovered by authors on the political left. Schmitt had verbalized rosy memories in his letters to Ingeborg Maus and Armin Mohler in the 1970s, and he certainly did

99 See Klingsporn and Wilke (2019, 61–64).

100 See Klingsporn and Wilke (2019, 66–69).

so again in his conversation with Rainer Erd in the summer of 1980. Erd left Schmitt with the Impression not only that Schmitt had always held Kirchheimer in high esteem, but also that Kirchheimer had continued to keep in touch with Schmitt after 1945 and had visited him multiple times. As early as 1949, Schmitt had reported in a letter to Ernst Rudolf Huber, Kirchheimer's fellow student in Bonn, that he and Kirchheimer had both praised the "outburst of intellectual freedom and dégagé thinking as sublime as the one we experienced in 1930/32."[101] With Erd's later report about his own visit in Plettenberg, the kitschy legend about the great unanimity between Schmitt and his leftist student Kirchheimer, about their reconciliation that had bridged all their political differences and all the crimes of the Nazi period, became popular in left-wing circles, too.

101 Letter from Carl Schmitt to Ernst Rudolf Huber dated 10 December 1949 (Schmitt and Huber 2014, 355).

Conclusion

Chapter 18: Kirchheimer's Strategies for Debating Schmitt

Let us suppose for a moment that the anecdote mentioned in the Introduction were true and Carl Schmitt had actually asked Otto Kirchheimer "Are you coming as a friend or as an enemy?" when he unexpectedly visited him at his house in Plettenberg in November 1949. This question would have confronted Kirchheimer with a dilemma because each of the two possible answers would have validated Schmitt—at Kirchheimer's expense. "Enemy" would have endorsed Schmitt's views on Jews as eternal enemies, unwilling to forgive. Schmitt would have taken "friend" as confirmation that he had been forgiven by a Jew for his Nazi and antisemitic propaganda and activities.[1]

Matthew Specter some years ago asked: "Can a political thinker like Schmitt be both intellectual friend and political enemy?" (Specter 2016, 427). This question is of particular significance for the case of Kirchheimer. Today's avowed left-Schmittians answer this question in the affirmative, and this leads to one of the questions raised in the Introduction: can we consider Kirchheimer to be a forerunner of contemporary left-Schmittianism? If we follow Ellen Kennedy's reading of Kirchheimer's works, this question should be answered with a resounding yes. She claimed not only that Schmitt and Kirchheimer agreed on their theoretical apparatus even after the Weimar Republic but also that Kirchheimer adopted from Schmitt certain concepts, namely enthusiasm for decision-making and a specific logic of argumentation in exposing contradictions.[2] She also argued that they shared a dislike of liberal democracy arising from the methods and concepts with which they were in "convergence" (Kennedy 1987a, 37).

In the Introduction, I defined left-Schmittianism as the transformation of Schmittian concepts or categories into the framework of legal or political theories with emancipatory political intentions. Contemporary left-Schmittians such as Chantal Mouffe view Schmitt as a political "adversary of remarkable intellectual quality" from "whom we could benefit" (Mouffe 1999, 1) and propose to "think with Schmitt against Schmitt" (Mouffe

1 I owe the insight into this dilemma to discussions with Sandra H. Lustig.

2 See Kennedy (1987a) and (1987b).

2005, 14).[3] Mouffe and other left-Schmittians of our time such as Andreas Kalyvas, Giorgio Agamben, and Daniolo Zolo are convinced that ignoring Schmitt's work would deprive contemporary political theory of essential and important insights. They claim that Schmitt offers unique resources for political theory, diamonds in the rough, so to speak. In the Introduction. I listed five of these which can be found in the rich literature by left-Schmittians: Schmitt's critique of universalism in international law, his antagonistic concept of the political, his theory of the exceptional state and sovereignty, his declaration of an irreconcilable antagonism between democracy and liberalism, and his critique of parliamentarism including his reflections on the relationship between homogeneity and democracy. It is striking that in one way or another, all five of these subjects had already been part of Kirchheimer's reception of Schmitt's work during the forty-year period in which he grappled with it. However, his evaluation of Schmitt's treatment of them was less positive than that of contemporary left-Schmittians.

Of course, some of Schmitt's statements, theories, categories, and concepts were important intellectual sources for Kirchheimer. But other intellectual sources—various strands of Marxism, Rudolf Smend's theory of integration, the critical theory of the Frankfurt School, and the empirical social sciences—became even more important for the evolution of his intellectual identity. Over the course of time, Kirchheimer reacted to Schmitt's work and political activities in various ways: with attempts at critical revisions, with harsh rejections, with an unmasking critique of ideology, or with sarcastic comments. In the late 1950s, he lost interest in even reading Schmitt's latest works. Upon closer inspection, we can distinguish five modes of reception. The first four correspond closely to the four phases of reception I distinguished in the Introduction. I call them cherry-picking and reframing, frontal attack, condemning Schmitt as a Nazi propagandist, and deliberate disregard. There is also a fifth type of reception that does not fit into the chronological order of these four phases and where Kirchheimer was undoubtedly motivated and inspired by Schmitt's work to move beyond the Schmittian horizon. I refer to this mode of reception as redirecting Schmitt's ideas beyond their original horizon. In those cases, Kirchheimer used certain statements or questions raised by Schmitt as starting points for new insights, albeit without making any direct reference to Schmitt. Kirchheimer's learning process over the course of time propelled his oeuvre far away from Schmitt's original concepts and theories. For this reason, it is inappropriate to characterize him as a godfather or theoretical patron of contemporary left-Schmittianism.

In the following sections of this chapter, I will first summarize and characterize Kirchheimer's five different modes of reception of Schmitt. Then, in conclusion, I will examine Schmitt's reactions to Kirchheimer's reception of him and turn my attention to the way he instrumentalized Kirchheimer for his own purposes.

3 Mouffe notoriously downplays the extremist nature of Schmitt's political position, calling him a "conservative theorist" who made a "compromise with Nazism" (Mouffe 2005, 4).

1. Cherry-picking and reframing

Cherry-picking and reframing is a strategy of reception that involves taking up distinctive concepts and theorems and placing them within a different socio-theoretical framework, thereby arriving at a different evaluation. Following in the footsteps of Ellen Kennedy, a number of authors have classified Kirchheimer's dissertation about the state theories of Bolshevism and social democracy as the very first extensive articulation of left-Schmittianism in political thought. They claim that the dissertation was "quintessentially left-Schmittian" (Scheuerman 1994, 24) and that Kirchheimer's line of argument "closely follows Schmitt's theorems" (Kohlmann 1992, 505–506). In the revised edition of his Schmitt biography, Reinhard Mehring calls Kirchheimer "the founder [...] [of] a kind of Marxist left-Schmittianism" (Mehring 2022a, 198).[4]

However, upon closer inspection of some of the other sources underlying the dissertation, it appears that these authors took an incomplete view and jumped to conclusions. It is important to remember that Schmitt's work was not the only, not even the primary inspiration for Kirchheimer as a twenty-two-year-old doctoral student. His line of thought was a creative—or eclectic—mixture of theoretical fragments from both Schmitt and Rudolf Smend that he squeezed into the basic framework of Max Adler's socialist theory. Kirchheimer incorporated theoretical ideas and political concepts with their origins in conservative and right-wing political thinking. His dissertation shows particularly well how, at this point in time, he thought Schmitt's concepts and wording could be integrated into a Marxist horizon of thinking following Adler. As discussed in Chapter 2, the Marxist patterns in Kirchheimer's work clearly preceded the Schmittian figures of thought. Because he also incorporated concepts from Smend's theory of integration in his dissertation, the label left-Schmittianism is justified for his dissertation only to a certain degree. In my view, "left-Smendianism" or "Adlerism in constitutional theory" would be just as accurate.

Even from Kirchheimer's dissertation, it becomes apparent how fundamentally he and Schmitt differed in their general patterns of thought. Alfons Söllner and Karsten Olson have called these differences the contrast between an ontological and a historical sociological way of thinking.[5] Schmitt argues ontologically when, for instance, he postulates the friend-enemy dichotomy or claims homogeneity is a prerequisite for the existence of a functioning democracy. Kirchheimer did not simply adopt Schmitt's concepts but rather engaged with them from an empirical sociological perspective and with an intention that is critical of ideology. Once we recognize these two fundamentally different ways of thinking, we see a subtle yet all the more significant shift in our understanding of how Kirchheimer dealt with Schmitt's writing.

This shift can also be illustrated with respect to their normative theories of democracy during the final phase of the Weimar Republic, as discussed in Chapters 5 and 6. Whereas Schmitt repeated his sharp conceptual distinction between democracy and *Rechtsstaat* with all his polemical verve, Kirchheimer's understanding of these two elements had a

---

4 Similar assessments are to be found in the interpretations of Tribe (1987), Scheuerman and Caldwell (2000), and Mehring (2007).

5 See Söllner (1987, 90–92) and Olson (2016, 96–97).

different conceptual structure. Schmitt derived the meaning of democracy from the postulate of equality. Kirchheimer rejected this view of democracy and treated the norms of equality and freedom as mutually dependent. Some contemporary left-Schmittians reject Schmitt's understanding of democracy, too. In their view—following Schmitt—the antagonism between a democratic logic of popular sovereignty and a liberal logic of individual rights is irreconcilable. Kirchheimer explicitly and rightly rejected this conceptual antagonism at the end of the Weimar Republic. His main argument was that democracy aspires to realize both collective autonomy and political equality, both individual freedom and social equality. His considerations about the normative core of modern democratic theory read like forerunners of Jürgen Habermas's ideas about the normative co-originality of democracy and the rule of law.[6]

Another example of Kirchheimer's strategy of cherry-picking and reframing during the Weimar Republic is how he simultaneously radicalized and modified a Schmittian term in *Weimar—and What Then?* In his *Constitutional Theory*, Schmitt had labeled the Weimar Constitution a "dilatory formulaic compromise." Kirchheimer took his analysis and radicalized it into the formulation of a "constitution without decision." In other words, he began his considerations by picking up on the term Schmitt used, then placed it in a different theoretical framework, namely the Marxist theory of class struggle, and arrived at a pointed radicalization of Schmitt's original wording. Other Weimar examples are Kirchheimer's early writing about parliamentarism and democracy. To Schmitt, the ongoing structural change of parliamentarism was proof of the historical demise of parliamentarism; Kirchheimer interpreted the same process in a positive light, as a new phase of mass democracy. In a way, with this kind of reception, Kirchheimer exploited Schmitt's outstanding reputation, borrowing the authority of a prominent constitutional law professor to support his own argument—at least as long as it seemed to fit his purpose. He also exposed Schmitt as a bourgeois ideologist several times during the Weimar Republic. He raised this accusation, for instance, in his reflections on property rights and expropriation and in his critique of Schmitt's interpretation of Article 48 of the Weimar Constitution, where he described Schmitt as a supporter of an authoritarianism prepared to transcend the previous limits of the constitutional order in the interests of the ruling classes.

At the beginning of the new wave of left-Schmittianism, Andreas Kalyvas stated that modern democratic theory still suffered from the "absence of a systematic reflection on institutions, rules, and norms" (Kalyvas 1999b, 111). He called it a "scandal" that constituent power had received "neither the recognition it deserve[d] in contemporary constitutional jurisprudence nor its proper place in our political vocabulary" (Kalyvas 2005, 230). In order to overcome that alleged scandal and other shortcomings, Kalyvas suggested that scholars rediscover Schmitt's work on popular sovereignty in his Weimar writings. In my view, this suggestion leads modern democratic theory up a blind alley: the sovereign people in a Schmittian world cannot act as a unit because any and all institutional mechanisms are lacking. Against Kalyvas's exclusively democratic interpretation, Renato Christi has pointed out that Schmitt does not argue that the

6 See Habermas (1996).

sovereign people is the sole bearer of constituent power; he also affirms the legitimacy of the monarchical principle (see Christi 2011).

In contrast to Schmitt's thinking, Kirchheimer's democratic theory during the Weimar Republic embodied a fundamentally different understanding of the institutions of modern democracies. Schmitt attributed the legally unbound people the role of the actual sovereign against all established instances of political decision-making. He did not declare the sovereign people a *Staatsorgan* (organ of the state) because he saw the positive quality in the people's lack of formal status. In Schmitt's interpretation, the sovereign people did not act as a state organ even in referenda. The logical consequence of this idea is that no form of political organization whatsoever can be a people in the proper sense. Parliament, parties, interest groups, trade unions—they all falsify and thus restrict the original unbound sovereign power of the people. Democracy becomes an existential immediate matter of the people that is prior to any law or constitution. Because the people cannot act as a unit due to the lack of any institutional mechanisms in Schmitt's theory, his theory justifies in advance that specific actors—such as a revolutionary group, the army, or a charismatic leader—must fill this institutional gap. Contrary to Schmitt, Kirchheimer resolved the dichotomy between an unorganized and pre-constitutional people on the one hand and a constitutionally organized and thus restricted people on the other hand in favor of procedures that open up democracy to additional forms of political participation. This opening included not only political parties and interest groups but also an extension of democratic decision-making into the economic sphere.

Kirchheimer's reflections on international law, discussed in Chapter 4, are another example of his strategy of cherry-picking and reframing. In 1928, Kirchheimer was still very much under the spell of the theses on national sovereignty of Schmitt on the one hand and Soviet legal scholar Evgeny A. Korovin on the other, who had emphasized the uncompromising differences between capitalist and socialist countries in international relations. Kirchheimer soon turned away from this position and supported the League of Nations and its basic idea of long-term cooperation and peace among nation-states. Whereas Schmitt and Korovin insisted on the existence of an uncompromising heterogeneity of mutually exclusive legal systems—be they nation-based or class-based—Kirchheimer advocated a dynamic concept of homogeneity on the international level based on positive experiences of cooperation. He transposed Smend's theory of integration to the level of international relations, so to speak. To Kirchheimer, Schmitt's realist view of international politics was founded on a set of unproven ontological assumptions about never-ending struggles between nation-states. Both Kirchheimer and Schmitt saw themselves as anti-imperialists, albeit in notably different ways. Schmitt argued in favor of militant nationalism against the enemy of Anglo-American imperialism. He believed he had unmasked universalistic international law and the prospect of an institutionally secured world peace order as perfidious claims to power by what he considered to be the Anglo-American enemies. He believed the right of every state to wage war at any time must not be restricted. Kirchheimer, conversely, supported restrictions on nation-states attacking neighboring countries or oppressing foreign countries as colonies. He hoped that intelligently limited international law could be used as a weapon against the capitalist imperialism of his age.

Even in his Weimar works, Kirchheimer already argued for expanding the juridification of international politics. His position on international law included a normative argument against the Schmittian critique: every objection raised against the one-sided or selective application of universalistic standards in international law must already presuppose these same standards. Schmitt's hermeneutics of suspicion about universalistic international law smuggles moral-normative commitments into his purportedly "realist" diagnosis of international politics. Kirchheimer maintained this position and developed it even further over the following years. He countered the polemic unmasking put forward by Schmitt and other opponents of the war crimes trials after 1945 that international law was always only the instrument of the powerful with a detailed analysis of the conduct of the trials in his book *Political Justice* in 1961. From the perspective of a historiographer of the critical theory of the Frankfurt School, Kirchheimer can be seen as a forerunner of the political project of constitutionalization of international law advocated by Jürgen Habermas since the mid-1990s (see Habermas 2014), whereas contemporary left-Schmittians, with their critique of cosmopolitan international institutions and international law, oppose such a position (see Zolo 2002; Mouffe 2007; Odysseos and Petito 2007).

2. Frontal attack

Kirchheimer began launching frontal attacks against Schmitt early on, during the Weimar Republic. One way he did this was to capitalize on doctrinal shifts and internal contradictions in Schmitt's work. Kirchheimer presented Schmitt as a witness against himself. One example of this strategy is Kirchheimer's criticism of Schmitt's extensive interpretation of Article 48 to justify the presidential dictatorship at the end of the Weimar Republic, as discussed in Chapter 5. Kirchheimer was happy to remind his readers in 1931 and 1932 of Schmitt's works four years earlier in which he had promoted a strict regulation of emergency powers—a position that Kirchheimer and other defenders of the republic also supported. He developed an argument that continues to be relevant to this day for a critical discussion of attempts in contemporary political thought to once again turn Schmitt's theory of emergency power into a fundamental critique of modern constitutional democracies that discredits the *Rechtsstaat*. It should be borne in mind that Schmitt's famous first sentence in his *Political Theology*—"Sovereign is he who decides on the exception" (Schmitt 1922, 5)—turns the original intention of the theory of sovereignty on its head. Following Hermann Heller (see Heller 1927), Kirchheimer argued that the starting point for the doctrine of sovereignty was the problem of the creation of positive rights and not their cancellation in a state of emergency. It was a new theory about the legitimate source of positive law which subsequently led to the theory of popular sovereignty. Schmitt, in contrast, switched the doctrine of sovereignty from the production of laws to a strong executive's power to act, with its greatest moment dawning in the state of emergency.

In contemporary political theory, Giorgio Agamben reverses the sovereignty doctrine in his fundamental critique of sovereignty in a similar way (see Agamben 2003). He interprets the most extreme phenomenon of the state establishing lawless zones in the form of concentration camps as a manifestation of sovereign power. He states that intentionally creating a permanent state of exception has become an essential practice in liberal democracies and that it is impossible to return to the rule of law. Agamben recently made this argument again in his critique of the modest measures taken by the Italian government against the SARS-CoV-2 virus in 2020 (see Agamben 2020). His extension of the semantic meaning of the state of exception leaves us—as William Scheuerman aptly put it—"with nothing more than the deeply mysterious suggestion [...] to 'halt the machine'" (Scheuerman 2020, 294) by ceaselessly trying to interrupt "the working of the machine" (Agamben 2003, 87) rather than by preserving the rule of law. Whereas Agamben argues that "the task at hand is not to bring the state of exception back within its spatially and temporally defined boundaries" (Agamben 2003, 87), Kirchheimer was among those authors who advocated specific legal restraints on the executive for clearly defined situations of emergency; Clinton Rossiter in the US and Ernst Fraenkel in Germany argued in a similar way a few years later (see Rossiter 1949; Fraenkel 1964).

Even during the Weimar Republic, Kirchheimer went a considerable step further in his frontal attacks by declaring Schmitt's positions to be plain nonsense. An early example is the critical remarks in his dissertation on Schmitt's celebration of political myths. To Schmitt, fascist myths functioned as what Ernesto Laclau would later call a "floating signifier" (see Finchelstein 2022, 106). Kirchheimer declared Schmitt's theory of the myth to be a provision for unjustified glorifications of pre-logical irrationalism that recurred to French ethnologist Lucien Lévy-Bruhl, who had argued that such a mythical consciousness belonged to the emotional and spiritual world of "primitive peoples."[7]

Another example is Kirchheimer's repeated criticism of Schmitt's eulogies about the President's impartiality, as discussed in Chapter 5. Kirchheimer took it for granted that every person who holds such an important office has his own political agenda. He blamed Schmitt for being—or pretending to be—naïve in denying this simple fact of political life. Schmitt's political activities during the final crisis of the Weimar Republic and his book *Legality and Legitimacy* in particular provoked Kirchheimer to lay out his frontal attack more extensively. He added another facet to the critical debate that explicitly aimed at questions of methodology, as discussed in Chapter 6. He explicitly raised the question about the status of the empirical social sciences for legal and political theory, coining the term "conceptual realism" to describe Schmitt's way of dealing with political and legal concepts. Kirchheimer used this label to demonstrate that Schmitt derived his knowledge of reality solely from his assertions about the internal logic of a basic idea inherent in the concept in question; in the process, empirical evidence about functional processes (and their problems) became irrelevant. Schmitt understood concepts as the embodiment of principles free of contradictions. Conceptual realism does not accept pragmatic justifications for political institutions. Judging by the doctrinal purity of Schmittian concepts, real political institutions that fulfill diverse, sometimes conflicting, and well-founded functions in practice were set up for failure. Kirchheimer demonstrated this line of Schmitt's thought with regard to parliamentarism, democracy, and constitutionalism. At the same time, he confronted Schmitt's theses with positive counterex-

---

7 Kirchheimer (1928a, 4), see also Lévy-Bruhl (1922, 94–97).

amples from the practice of other Western European democracies and accused him of ignoring the simple facts of political life in modern democracies.

Based on his critique of Schmitt's method, Kirchheimer rejected the way in which Schmitt insisted on the prerequisite of homogeneity for democracy. Following Schmitt's understanding of homogeneity would take democratic theory in a completely wrong direction, he asserted. His concept of social homogeneity was altogether different from Schmitt's concept of the homogeneity of the people. Schmitt's concept was one of the substantive homogeneity of a collective although he did not specify what "substantive" meant. As discussed in Chapter 6, Kirchheimer contradicted Schmitt and argued that all modern democracies accommodate a heterogeneous population by necessity. From class-divided England and France to multiethnic and multilingual Belgium and Switzerland, many heterogenous democracies had flourished. He also noted that there appears to be a global trend in modern societies toward ever increasing heterogeneity. Kirchheimer stated that policies of social justice and political integration were required in order to prevent this heterogeneity from leading to the disintegration of society. Unlike Schmitt, Kirchheimer argued for a certain degree of homogeneity of social living conditions as the basis for a stable democratic state. Contrary to the revolutionary romanticism among a number of contemporary left-Schmittians, Kirchheimer advocated for a reformist program to achieve social justice.

3. Condemning Schmitt as a Nazi propagandist

After Hitler came to power in 1933, Kirchheimer's reception of Schmitt changed drastically, and he mainly condemned him as a Nazi propagandist from then on. Kirchheimer singled him out as the most intelligent of the legal theorists in Nazi Germany. From his exile in London and Paris, he reported in detail about Schmitt's current writing and activities to his international—albeit small—readership and continued to follow Schmitt's activities closely after moving to New York in 1937. In his comments on Schmitt's 1934 programmatic essay "On the Three Types of Juristic Thought" in his co-authored book of 1939, *Punishment and Social Structure*, Kirchheimer argued that Schmitt's theory of concrete-order thinking was the most influential legitimation of the Nazi regime in terms of legal theory, thus making an implicit statement on the question of continuity in Schmitt's work before and after 1933.

However, what is particularly noteworthy here is that when writing in exile, Kirchheimer seemed to show little particular interest in the question of continuity or discontinuity in Schmitt's work. One would expect Kirchheimer to have dealt with this subject more extensively, if only because of his earlier personal relationship with Schmitt and because he had occasionally referred to some of Schmitt's concepts in his own writing during the Weimar Republic. But only rarely did he indicate that Schmitt's seamless transition after 1933 may have been based on his Weimar theories of the state of emergency or presidential dictatorship. For the most part, he treated Schmitt's works from 1933 to 1945 strictly separately from his Weimar writings. In an article for an English audience in the fall of 1933, Kirchheimer even distinguished between Schmitt "the political theorist" and Schmitt "the Nazi partisan," (Kirchheimer 1933c, 534) implying that Schmitt had become a Nazi propagandist and was no longer to be taken seriously as a thinker in the fields of legal and political theory. To this day, the secondary literature on Schmitt is almost as obsessed with the question of whether Schmitt's Weimar theories and his commitment to the Nazi regime are internally consistent as it is in disagreement about this (see Jestaedt 2018, 408). The key question is: To what extent did Schmitt's marriage to Nazism arise inherently from core elements of his theory, at least in part? Or, in other words: To what extent did his Nazi-era publications build on his Weimar writings about the liberal quest to take judicial and administrative actors under control?

Hasso Hofmann has correctly pointed out that all left-Schmittians are united in negating such continuity (see Hofmann 1995, xi). Hofmann himself, in his groundbreaking doctoral dissertation about Schmitt's work, argued that there was "a certain continuity in all changes" (Hofmann 1995: xv) in Schmitt's writing. The continuity is the permanent, never-ending search for new sources of legitimacy which always surpass existing legality. Hofmann's book—first published in 1964—is an extension of how his doctoral supervisor Karl Löwith read Schmitt in 1935. At the time, Löwith had found a temporary safe haven in his exile in Rome and took the time to analyze one of Schmitt's recent works in detail. Löwith interpreted Schmitt's efforts to make the revised 1933 edition of *The Concept of the Political* more appealing to the rulers of the regime as proof of Schmitt's "occasionalist decisionism" (see Löwith 1935, 32–61). It is striking that Kirchheimer, in his Paris and New York exile, came up with a parallel version to Löwith's interpretation of continuity in changes, albeit in a Marxist version of a critique of ideology. The main continuity he identified in Schmitt's changes was the ideological function of Schmitt's writing: in his theories, Schmitt had always represented the interests of the powerful economic and social groups in society. Schmitt had done so during the Weimar Republic and had continued his militant partisanship for the ruling classes with his new doctrines after Hitler came to power. Kirchheimer's assessment can be read as an echo of Hermann Heller's angry interjection during the trial *Prussia vs. Reich* in November 1932 before the *Staatsgerichtshof* (see List of German Courts), when he called Schmitt's way of constructing his arguments in defense of the coup against the Prussian government pure "situational jurisprudence."[8]

Kirchheimer's view of Schmitt after 1938 is documented only in a few scattered statements in articles and reports he wrote at the Institute of Social Research (ISR) about criminal law and the legal order and political system of Nazi Germany, as discussed in Chapters 9 and 10. In these articles, he emphasized the continuities of the Weimar Republic and National Socialism in terms of their social bases. He argued that both systems were founded on certain compromises in the distribution of power and influence between the heads of the industrial and the agricultural monopolies, the state bureaucracy, and the military. During the Weimar Republic, working class organizations had had the opportunity to participate in these negotiations, too. At the end of the republic, the struggles for power escalated into a situation akin to civil war. Kirchheimer interpreted the year 1933 as the takeover of state control by one civil war party in the interest of the industrial and agricultural monopolies, thereby wresting their power posi-

8 Interjection from Heller, 17 October 1932, quoted in the trial's stenographic transcript (Brecht 1933, 469).

tions from the working class. At the ISR, Kirchheimer openly contradicted Pollock's and Horkheimer's theory of stable state capitalism. Conversely, he emphasized the fragility of the Nazi regime. In his view, Hitler had not established a stable political order but, rather, a system of constant rivalries between different power groups. The future stability of the regime depended exclusively on successful imperialist policies. In light of this, Kirchheimer evaluated Schmitt's work during the Nazi era from the perspective of Marxist critique of ideology. In my view, his analytical approach in this phase of his academic career made too many concessions to a simple functionalist version of Marxist critique of ideology.

After Schmitt had turned his main interest to international law in 1937, Kirchheimer gave his reception a new twist. He was of the opinion that Schmitt's *Großraum* (literally: large space, inherently linked to geopolitics and Nazi Germany's expansionist policies; see Glossary) theory had become the most important ideological soundtrack to Nazi Germany's warring imperialism. Kirchheimer did not believe that Schmitt's new thoughts on international law provided intellectual stimulation to seek out further and fruitful alternative ideas. Instead, as discussed in Chapters 11 and 12, he viewed Schmitt's *Großraum* theory as the product of a jurist who was aligning his work with the Nazi regime's situational political needs, lock, stock, and barrel. He accused Schmitt of even being willing to pay the high price of accepting basic and obvious theoretical contradictions in order to keep his official positions in Nazi Germany. Again, Kirchheimer no longer treated Schmitt as a theorist to be taken seriously, but as a propagandist.

The lack of theoretical precision that Kirchheimer accused Schmitt of can also be found, but to a lesser degree, in some of his own interpretations of Schmitt. In 1943/44, Kirchheimer emphasized the economic and technical organizational aspects of Schmitt's *Großraum* theory. He asserted that Schmitt's theory had a rational core inasmuch as it soberly articulated changes both in the domestic German and in the international capitalist system. Owing to the high level of cartelization, monopolization, electrification, and rationalization of German industry, transitioning to the economy of the *Großraum* had become imperative for the ruling classes. Kirchheimer viewed Schmitt's theory as being in line with the major trends in German monopoly capitalist society. It is striking, however, that in his interpretation of Schmitt's theory of the *Großraum*, he ignored its differences to purely *völkisch* and *Rasse*-based *Lebensraum* theories of Nazi authors such as Werner Best and Reinhard Höhn.

4. Deliberate disregard

With the beginning of the postwar period, Kirchheimer's strategy in his reception of Schmitt's work changed again, now becoming deliberate disregard. It should be noted here that during his exile in Paris, Kirchheimer had turned his main research focus not only toward Nazi Germany but also toward Western democracies, in particular, France and the United States. Whereas traces of Marxist critique of ideology can be found in some of his works from this period,[9] they are overshadowed by his detailed empirical

9 See Kirchheimer (1958a), (1959b), and (1962c).

studies in which he addressed what he had previously described as Schmitt's primary methodological shortcoming: the empirical deficit of Schmitt's works. Kirchheimer implemented empirical analyses of political institutions and political processes and transformed his methodological approach from normative legal and political theory to a political theory that takes up the findings of empirical political science.

After 1945, Kirchheimer did not discuss any of Schmitt's activities or articles from the Nazi era in public at all; nor did he ever discuss any of Schmitt's postwar articles or books such as *Ex captivitate salus*, *The Nomos of the Earth*, *Hamlet or Hecuba*, or *Theory of the Partisan* in his writing. He treated Schmitt as if he were an author of the Weimar Republic only. This focus was a statement: he did not consider the theoretical content of any of Schmitt's works during the Nazi regime to be intellectually valuable enough to merit renewed critical examination. He apparently came to the same conclusion about Schmitt's postwar works, and alluded only briefly to Schmitt's *Nomos of the Earth* in a sentence summarizing his criticism of Schmitt (and already quoted above on page 418) in 1957:

> The lack of any clear-cut criteria for differentiating between *nomos* and violence; the discrepancy between the traditional liberal concepts of classical international law and the decisive rejection of an *artfremd*[10] and disintegrating liberalism as part of the domestic constitutional order; the brooding omnipresence of the people's constituent power and its incapacity to act as a constituted organ; the indeterminate character of the values underlying concrete decisions; and the conjunction of a relativistic openness to a variety of historical interpretations with an ever-present negation of the rule of law (Kirchheimer 1957b, 348).

Part of Kirchheimer's reception of Schmitt was his veritable campaign against the renaissance of Schmitt's Weimar writings in legal thought in the early phase of the Federal Republic of Germany. As discussed in Chapter 15, he attacked the four most prominent avowed Schmittians at German universities, Ernst Forsthoff, Werner Weber, Ernst Rudolf Huber, and Joseph H. Kaiser— Kirchheimer and the first three had studied at the same time under Schmitt in Bonn—in a number of publications between 1951 and 1956. Using the same blueprint against all four of them for his interpretation and criticism, he attacked their support of a strong state. He rejected in particular their criticism of both the influence of political parties and the pluralism of political interest groups. Kirchheimer underpinned his criticism by accusing these academics of lacking empirical evidence, of using anti-pluralist reasoning, and of being stuck in Schmitt's outdated thought patterns and authoritarian attitudes from the Weimar era. Ultimately, these accusations were aimed at Schmitt himself. But his criticisms had an additional subtext: whereas he portrayed the four scholars as being more or less uncritical epigones of Schmitt, he considered himself as being capable of analyzing Schmitt's oeuvre in a critical and independent way.

In 1958, Kirchheimer and Schmitt discussed their differences briefly in an exchange of letters. Schmitt had sent Kirchheimer a copy of his collection of essays,

10 Here, Kirchheimer used the German word *artfremd* specifically as a term from Schmitt's vocabulary, meaning foreign/alien to the *Volk*, in an exclusionary and antisemitic sense; see Translator's Preface.

Verfassungsrechtliche Aufsätze, stating proudly that he thought that his writing from the Weimar period was still as topical as ever. In his response, Kirchheimer brusquely rejected Schmitt's claim, explaining that he did not believe in the repetition of similar situations: too many qualitative changes had taken place in modern societies since the Weimar Republic. After this exchange, he stopped commenting on Schmitt's work in his publications in any explicit way.

5. Redirecting Schmitt's ideas beyond their original horizon

There is an additional component to how Kirchheimer dealt with Schmitt's writing that I have not yet discussed in more detail: some of Schmitt's publications served as a starting point for Kirchheimer to redirect Schmitt's ideas or concepts beyond the Schmittian horizon. Both Kirchheimer and Schmitt had an unmistakable sense of the context of *Lage* for the analysis of political and legal institutions. As explained in the Introduction to this book, the German metaphor *Lage* has a double meaning in the work of both Schmitt and Kirchheimer, simultaneously designating the bound aspect of a situation and its potential for change.

Kirchheimer did take up some of Schmitt's questions, ideas, and concepts as starting points for his own academic work—without attacking Schmitt explicitly or making any other reference to him but still provoking him. Four examples illustrate this: first, Kirchheimer's research on antisemitism at the Institute of Social Research in New York; second, his preparations for the Nuremberg Trials at the OSS; third, his research on political parties and interest groups; and finally, his late magnum opus, *Political Justice*.

With respect to the first example, I have argued in the previous chapters that the antisemitic content of Schmitt's work is of greater importance than still assumed by the majority of his interpreters. I agree with the readings of Raphael Gross, Paul Bookbinder, and Nicolaus Sombart, who have all demonstrated how Schmitt was preoccupied with Jews and Jewishness throughout his life. Even during the Weimar Republic, antisemitism was of underlying significance for his criticism of parliamentarism, pluralism, the liberal concept of *Rechtsstaat*, legal positivism, and universal international law. Kirchheimer had been aware of Schmitt's antisemitic attitude during the Weimar Republic. However, antisemitism was not a subject of his academic work prior to his research at the ISR in 1942/43. It is striking to see that the antisemitism analyzed by Kirchheimer and the antisemitism practiced by Schmitt, which are both discussed in Chapter 10, overlapped on a number of issues. The first and particularly notable instance is the fact that Kirchheimer was focused on the role of Catholicism in his research on antisemitism. He was apparently the most suitable among the exiled members of the Frankfurt School to take on this specific aspect of the general topic since he was familiar with the theological debates in Catholic circles when he was a student of Schmitt's in Bonn. Secondly, in his research papers, Kirchheimer entered into the theological debates about Christianity and Judaism by making use of Erik Peterson's work, which he had also become familiar with through Schmitt in Bonn. The third link is Kirchheimer's emphasis on the institutional and ideological elasticity of the Catholic Church. This was a characterization he had obviously drawn from Schmitt's idea of the Catholic Church as a *complexio oppositorum* in his book

Roman Catholicism and Political Form, without mentioning the source, however. A fourth area where Kirchheimer touched on Schmitt's work was his evaluation of the political position of the Catholic Church with respect to the Nazi government. Kirchheimer was skeptical as to the role of the Catholic Church as a potential force of resistance against Nazi policies. This assessment is consistent with Schmitt's retrospective statement after the war that Hitler's government had gained official papal recognition through the Concordat between the Holy See and the German Reich. These four instances of overlap notwithstanding, the differences in the methodological approaches between Kirchheimer and Schmitt are clearly recognizable on this subject, too. In his book *Leviathan*, published in 1938, Schmitt drew on second-hand sources written by antisemites about the Kabbalah to construe a genealogy of the triumph of an international Jewry operating while concealed by masks. Kirchheimer, in contrast, undertook a sociological contextualization of the attitudes toward Jewish people in Europe. To him, Catholic antisemitism was a distinctly modern phenomenon, despite its initially religious sources. It had to be understood as an articulation of negative experiences of capitalist modernization.

The second example of Kirchheimer taking up the productive provocations in Schmitt's work without mentioning his name is his reflections on war crimes trials. Even when Kirchheimer was not aware of any works by Schmitt on a particular subject, it still served as an inspiration for him to redirect some of Schmitt's arguments beyond the latter's original horizon. In some cases, knowing Schmitt's way of thinking was sufficient for him to anticipate his arguments on a certain subject and then counter them with arguments of his own. This can be seen most clearly in his reflections on war crimes trials, as discussed in Chapter 13. Both Kirchheimer and Schmitt were lawyers, and both anticipated the trials against German war criminals. Their anticipation was also that of a prosecutor (or someone identifying with the prosecutorial role) having preempted the defense of a defense lawyer. Schmitt's first professional activity after the war, in the summer of 1945, was to prepare an extensive legal opinion for German industrialist Friedrich Flick, who feared prosecution at one of the Nuremberg Trials. Kirchheimer was not aware of this and never had the opportunity to read that legal opinion (it was not published until 1994). His recommendations for the best prosecution strategy, which he had written at the OSS the year before, read like responses to Schmitt's two main arguments in Flick's defense. Familiar with Schmitt's way of thinking, Kirchheimer accurately anticipated the defendants' main line of defense in the Nuremberg Trials in other cases, too. His knowledge helped him to identify the prosecution's main weak point against the defendants, which was the claim of merely obeying orders from superiors in order to shift responsibility away from themselves; Schmitt had indeed done this in his legal opinion for Flick.

Two main recommendations for the prosecution strategy illustrate how Kirchheimer anticipated what arguments Schmitt would use. The first recommendation in Kirchheimer's legal opinions at the OSS was a preemptive response to the claim he expected that the killings and other brutal deeds had not been unlawful under the Nazi regime, namely proposing "the principle 'selective retroactivity'" (Kirchheimer 1945, 523). According to this principle, all the laws, amnesties, and policy measures that protected Nazis from the consequences of their crimes were to be specifically rescinded. His list of retroactive rescissions of Nazi laws included *Rasse*-based legislation as well as the laws to suppress political opposition. Only if these laws lost their validity retroactively would it be possible to prosecute the members of the official repressive agencies such as the *Volksgerichtshof* and the *Militärgerichte* (see List of German Courts) under criminal law. In his deliberations on selective retroactive revision of a defunct regime's legislation, Kirchheimer also referred to precedents from various countries, going into international law in more depth, examining how the US had dealt with the Southern states, and providing more recent examples from France, Denmark, and Italy. He also discussed the question whether the Nazi regime was constitutional. Kirchheimer followed the hypothesis that after 1933, Schmitt, too, had repeatedly proclaimed a revolutionary break with the order of the Weimar Constitution. A similar break had occurred with the defeat of Nazi Germany, thus giving the victorious Allies the political legitimacy to act legally.

The second recommendation deals with the problem of personal responsibility for atrocities. John H. Herz and Kirchheimer expected the strategy of the defense to be to declare that the accused had merely executed orders given by a few high-ranking superiors. After all, the Nazi legal doctrine of the *Führerprinzip* stated that superiors were responsible for the acts of their subordinates. Their counterstrategy stated that, while all authority and power was theoretically vested in Hitler, as the *Führer*, a considerable amount of discretionary power was still delegated to the group of *Unterführer* (see Glossary). Rather than being tools with no will of their own merely carrying out orders issued by Hitler and other top officials of the Nazi hierarchy, they were considered in Nazi legal writing to be active participants in the Nazi project. As such, they were responsible for formulating broad policies within the sphere of their particular jurisdiction. Quoting from a number of legal documents by Schmitt's former student and trusted colleague Ernst Rudolf Huber, Schmitt's protégé Hans Frank, and other Nazi legal scholars, Kirchheimer and Herz concluded that:

> The more such policies involved a political aspect, the freer they were from any form of legal restraint, and the less likelihood was there that any specific orders would be handed down from the policy making leaders to their executory subordinates (Kirchheimer and Herz 1945, 464).

Kirchheimer and Herz concluded from the Nazi legal literature that while superiors bore criminal responsibility for all crimes committed under their leadership, *Unterführer* bore criminal responsibility, too. They also concluded from Nazi doctrine that any person who had joined the SS (see Glossary) or the NSDAP voluntarily and had been in a superior position must be assumed to have had full knowledge of the practices and functions of the organization and could therefore not avoid sharing responsibility for certain criminal acts.

Another—and third—example of Kirchheimer taking up the productive provocations in Schmitt's work without mentioning his name at all is in his works after World War II about the state of Western democracies, which are discussed in Chapters 14 and 15. From Schmitt's perspective, Kirchheimer's articles on this subject can be interpreted as an attempt to fathom the future opportunities of Western democracies. With respect to the newly founded Federal Republic of Germany, Kirchheimer identified certain continuities with the Weimar Republic, for instance, in the top politicians and in the election results. More important for him, however, were the discontinuities in which the Basic Law and the *Bundesverfassungsgericht* (see List of German Courts) played important roles for the stability and the future development of the Federal Republic of Germany. From these discontinuities, Kirchheimer became confident that a Weimar civil war scenario could be avoided, whereas Schmitt thought that the Federal Republic of Germany was nothing more than an updated version of the weak Weimar Republic and also lacked sovereignty. Nevertheless, several of the issues discussed by Schmitt in his Weimar writings remained on Kirchheimer's research agenda. Most of Kirchheimer's works on modern democracies seem to be motivated by the inspiration to contradict Schmitt's Weimar writings—yet without mentioning Schmitt's name. The most important issues Kirchheimer covered in these contributions include the theory and practice of the *Rechtsstaat*; the developments of parliamentary democracies; the role of political parties and interest groups; presidential democracy; the role of political opposition; constitutional courts; and the legitimate limits to constitutional changes—all of which Schmitt had also addressed.

In my view, Kirchheimer's research agenda in the 1950s and 1960s is best understood as an empirically based counterprogram to Schmitt's mixture of apocalyptic warnings and condescending malice in his critique of modern liberal democracy. To give another example of this in addition to others discussed in the previous chapters: contrary to Schmitt's reiterations of the old clichés on political parties as state-destroying organizations, Kirchheimer described them as primary agencies of political integration whose legitimacy lies in their ability to channel the political and social energies of their clientele into political action. He put his political hopes for democratic stability in a policy of social and political integration, thereby taking up the theory of integration by Rudolf Smend. In contrast to Smend's view, however, Kirchheimer was worried that integration might go too far. He expressed this concern in a number of articles beginning in the second half of the 1950s. This concern again brought him closer to the critical theory of the Frankfurt School and its philosophical critique of "total integration" (Adorno and Horkheimer 1944, x). Unlike Adorno and Horkheimer, Kirchheimer founded his critique of overly far-reaching integration on empirical findings. On the basis of his analysis of the cartel-like formation of political party coalitions in Austria and Italy in the 1950s, he predicted the spread of grand coalitions in other European countries, too. According to his analyses, almost all Western democracies shared the same trend toward catch-all parties and a freezing or enclosing of any political opposition that demanded politics going beyond the given social order of welfare state capitalism. This diagnosis runs completely counter to Schmitt's evocations of political disintegration and civil wars as inevitable consequences of political opposition rights, active interest groups, and party pluralism in modern mass democracies.

My fourth and final example of Kirchheimer redirecting Schmitt's ideas or concepts beyond the Schmittian horizon is the overarching subject of Kirchheimer's 1961 book *Political Justice*, discussed in Chapter 16. The initial question of the book closely follows considerations laid out by Schmitt in his *Constitutional Theory* of 1928. Yet again, Kirchheimer did not mention Schmitt in this context and arrived at completely different conclusions. Kirchheimer began his book *Political Justice* with a definition of the political as the degree of intensity of conflicts. Whereas Schmitt's ultimate aim was to unleash the friend-enemy dichotomy of the political, Kirchheimer sought ways to restrain it. He may well have agreed with Adorno's verdict in *Minima Moralia* that Schmitt's friend-enemy dichotomy was the expression of a "consciousness that [makes] its own regression to the behavior pattern of the child, which either likes things or fears them" (Adorno 1951, 141). Beyond this psychological interpretation, in Kirchheimer's treatment of the administration of justice, the true art of dealing with political conflicts was not to intensify them but to allay them. In Schmitt's dismissive terms, Kirchheimer wanted to depoliticize political conflicts to a certain degree without neutralizing them altogether. Kirchheimer's reflections on the judicial procedures of political justice can be read as an alternative strategy for dealing with political conflicts.

Political Justice presented a set of examples of institutional means within the judicial system that provides opportunities for the articulation of adversary conflicts—and, at the same time, the opportunity to overcome these conflicts peacefully. At first glance, we might see similarities between Kirchheimer's position and the left-Schmittianism of Chantal Mouffe. In a number of books and articles, she argues that political identities consist of a certain type of we/they relations which can easily turn into a friend/enemy relationship.[11] In order to find an understanding of the friend/enemy distinction that is compatible with democratic pluralism, Mouffe transforms the alleged Schmittian diamond of antagonistic politics into a domesticated version of agonistic conflicts. She states that the mitigation of the potential antagonism that exists in social relations cannot be accomplished by transcending the we/they relationship, but only by constructing the we/they relationship in a different way. Conflicts are no longer to take the form of an antagonism—i.e., a fight between enemies—but that of an agonism, i.e., a dispute between adversaries. Adversaries view each other as belonging to the same political unit and sharing a common symbolic space. The conflicts between them are supposed to take place within the shared symbolic space of vibrant pluralistic liberal democracy within which they do not aim to destroy their opponents. Mouffe has rightfully been criticized because her theory of agonistic conflicts dilutes Schmitt's theory and his militant impulses beyond recognition.[12] Another criticism that has been made is that her assertion of never-ending agonism flirts with the regressive ontology of the Schmittian concept of the political which amounts to an essentialization of conflict. In my view, the main shortcoming of her reception of Schmitt is that she fails to answer the question as to which institutional constellations in particular facilitate the transformation from Schmittian antagonistic conflicts to pluralist agonistic ones.

Kirchheimer took a different approach to this question in his work on the administration of justice. To him, the ambivalence of political justice was not proof of the insurmountable paradoxes of the *Rechtsstaat* but offered the opportunity for political integration through the legal system. And again, he chose a Smendian framework. This led him to an understanding of political justice as a potential instrument for political integration in a democratic *Rechtsstaat*, provided that it is practiced reasonably and fairly. Not only did Kirchheimer direct the reader's attention to the institutional infrastructure

11 See Mouffe (1999), (2000), (2005), and (2023).

12 There is a vast body of secondary literature on Mouffe's revisionist reception of Schmitt. For a brief overview, see Specter (2016).

that is essential for processes that overcome the escalation of political conflicts, but he also constructed an example of how to accomplish this goal. His reflections offer a striking illustration of his integrative understanding of "the political" and his concern for the problem of how to institutionalize the articulation of political conflicts.

The story of the relationship between Kirchheimer and Schmitt spans almost forty years. Again and again, both authors kept a keen eye on the theories, concepts, and political partisanships in each other's essays and books. Through the lens of Kirchheimer's reception of Schmitt's works, the latter appeared as a willful destroyer of the Weimar Republic and as an eternal opponent of the democratic *Rechtsstaat* who attempted to delegitimize it in various ways throughout his life. From the beginning, Kirchheimer had no illusions about the fact that Schmitt's intellectual interventions were never simply armchair debates but inevitably had specific political goals. Despite the courteous tone of their interactions from 1928 to 1932 and from 1949 to1961, Kirchheimer viewed Schmitt not only as a political enemy but also as an intellectual enemy.

Through Schmitt's lens, on the other hand, the young Kirchheimer of the Weimar Republic was an author who deserved praise as a leftist radical whose writing served as testimony of the seriousness of the socialist threat, as discussed in Chapters 3 and 5. During the Nazi regime, Schmitt completely ignored Kirchheimer's works both from the Weimar era and later. After World War II, he quoted him again occasionally and praised his Weimar writings in particular—at least until the clash in 1961/62 that resulted in the final rupture of their personal relationship. At no point in his reception of Kirchheimer's work did Schmitt did take Kirchheimer's intellectual development seriously. Instead, as shown in Chapters 15 and 16, he never stopped treating him like a caricature of a left-wing enemy of the Weimar Constitution and as a Jew unable to fully understand his own, superior, way of concrete-order thinking.

The controversies with Schmitt were not constitutive of Kirchheimer's identity as a political scientist and legal scholar. Starting with his dissertation, he took a critical stance toward Schmitt's concepts and theories and integrated multiple quite different works by other authors and schools of thought into his own theoretical aspirations. His reception of Schmitt's work evolved from cherry-picking and reframing to frontal attacks, to condemning Schmitt as a Nazi propagandist, and to deliberate disregard of Schmitt's influence on his own work when he redirected Schmitt's ideas and concepts beyond the original Schmittian horizon, without attacking Schmitt or even making reference to him.

Overall, Schmitt's statements, theories, categories and concepts were obviously an important intellectual source for Kirchheimer throughout his life—even if primarily to contradict Schmitt, however. But other intellectual sources became more important than Schmitt for the evolution of his intellectual identity. These sources include various strands of Marxism, Rudolf Smend's theory of integration, the critical theory of the Frankfurt School, and the empirical social sciences, all of which he received creatively. Nevertheless, Kirchheimer's reception of Schmitt is also an illustration of the eminent difficulties in attempting to separate Schmitt's theoretical concepts and impulses from the overarching context of his legal and political thought. Kirchheimer experienced a learning process that led him far away from Schmitt's original concepts and theories. As discussed in Chapter 3, he had already realized during the Weimar Republic that his original intention, namely to make productive use of Schmitt's key concepts in order to fill some gaps in left-wing political thought, was a lost cause. He also had a sharp eye on the risks of Schmitt's concepts, albeit transformed by authors (like himself) from the left: exalted metaphors of the state of emergency; polemics against discussion, compromise, and mediation; the enthusiasm for decision-making; disregarding findings in the empirical social sciences that did not fit his theory; and, above all, at the methodological level, a style of theorizing that he criticized as conceptual realism. Kirchheimer had already dealt with these five diamonds proposed by contemporary left-Schmittians as he grappled with Schmitt throughout his life. When he tried to polish them, they crumbled in his hands. Clearly, these diamonds were not forever.

Now, this statement is not intended to be an absolute verdict against all attempts to make productive use of Schmitt's work. Unlike those authors who primarily warn of Schmitt as a dangerous mind who is not worth reading, and also unlike those authors who dilute Schmitt's concepts beyond recognition, Kirchheimer was right to take parts of Schmitt's work seriously as diagnoses of problems. But in responding to these problems, he completely broke with Schmitt's theories and concepts. With his selective and sovereignly independent approach to Schmitt's work in his articles on political parties, the vanishing of opposition in Western democracies, and in his book *Political Justice*, as discussed in Chapters 15 and 16, Kirchheimer is a role model for how to deal appropriately with the productive elements in Schmitt's writings today. I see such a selective and highly independent approach to Schmitt's work in Andrew Arato's democratic reinterpretation of the role of the *pouvoir constituant* in constitutional theory (see Arato 2011), in Andreas Kalyvas's brilliant reconstruction of the origins of the *ius publicum Europaeum* from the colonial encounter between Europe and the non-European world (see Kalyvas 2018), or in Benjamin A. Schupmann's liberal theory of militant democracy (see Schupmann 2024). Schmitt's concepts, however, are no necessary condition for their arguments.

When Kirchheimer broke with Schmitt's theories and concepts, his work also moved well beyond Schmitt's theoretical horizon. The fact that he completely abandoned Schmitt's concepts and theories makes him a non-Schmittian, if not an anti-Schmittian. Thus, in my view, Kirchheimer does not fit the role of a godfather or theoretical patron of contemporary left-Schmittianism. If anyone wishes to assign him that role nonetheless, then this means formulating a paradox: Otto Kirchheimer—the person more familiar with the work, activities, and person of Schmitt than anyone else on the political left—was the first left-Schmittian who was no longer a Schmittian at all.

6. Conclusion: Defining Legacies

Schmitt ignored Kirchheimer's barrage of criticism most of the time, at least in his publications. But it does not take much imagination to assume that these criticisms became a subject of their numerous discussions during the Weimar Republic. Nonetheless, whenever Schmitt felt compelled to make public statements about Kirchheimer during that time, he chose to praise him highly. Similar to Kirchheimer's strategy of cherry-picking and reframing, Schmitt placed Kirchheimer's theses in the context of his own frames of reference. He praised Kirchheimer's book on expropriation as particularly instructive because it served him as additional evidence of the threat of socialism. He recommended

Kirchheimer's book *Weimar—and What Then?* even more strongly. In Schmitt's eyes, its critical analysis of the constitution was evidence of the socialist movement's political vitality and determination to fight. His assent to parts of Kirchheimer's *Legality and Legitimacy* was also aimed at declaring the crisis in the republic to be unstoppable. He did not even shy away from misquoting Kirchheimer, a breach of academic standards. Schmitt's reception of Kirchheimer during the Weimar Republic followed one general pattern: he made him a witness from the left supporting his own criticism of the republic, someone who delivered additional credibility to his own warnings about political instability and the outbreak of a civil war.

Not surprisingly, the way they received each other's works changed completely between 1933 and 1945. Whereas Kirchheimer continued to deal with Schmitt, the latter did not say a single word in public about his former partner in dialogue. Schmitt made sure not to mention the name Otto Kirchheimer even once in *Positionen und Begriffe*, the collection of his essays that was published in 1940 and included twenty articles from the Weimar period. Schmitt carefully avoided quoting any of Kirchheimer's works, old or new. In none of his publications, letters, or diary entries did he respond to any of Kirchheimer's exile writing, the sole exception being in 1935, when he asked the Gestapo to go after those responsible for the camouflage brochure.

Here, I would like to point out a remarkable parallel between Schmitt's retrospective analysis of the Nazi regime and some of Kirchheimer's works written in exile. In his responses to Robert M. W. Kempner's questions during his detention in Nuremberg in 1947, Schmitt emphasized a fundamental abnormality of the state organization and the legislative process in Nazi Germany. He presented to Kempner the outline of a structural model of the Nazi regime according to which Hitler and his inner circle had disempowered the various ministries and had created a superministerial political structure from scratch. Its personnel was recruited from three pillars of the regime: the party, the military, and the state. Hitler's orders and commands were implemented through the controlling power of this political structure. Schmitt described his own role in this system as an outsider of the exclusive circles of loyalists that constituted the institutional core of the regime. His model of polycratic personalism sketched out for Kempner has striking similarities to Kirchheimer's structural model of the Nazi regime as a polycracy. However, what had mattered most to Kirchheimer was the conflict dynamics between the main actors, whereas Schmitt wanted his new model to absolve himself of any political responsibility.

After the founding of the Federal Republic of Germany in May 1949, Schmitt was pleased to witness some of his former students obtain prestigious professorships. He did not comment on Kirchheimer's criticism of the renaissance of Schmittianism in German legal thought. He did not even complain about it in letters to him. Instead, he began to quote Kirchheimer again, putting him in a positive light. Similar to his "art of quoting" during the Weimar Republic,[13] some of his benevolently worded citations again contained inaccuracies and distortions of meaning. He also invited Kirchheimer to discuss

13 As discussed in more detail in Chapter 5, p. 129, Schmitt had misquoted Kirchheimer in his book *Legality and Legitimacy* (see Schmitt 1932h, 14), thereby changing the meaning of Kirchheimer's statement. Schmitt had added a general diagnosis in line with a theory of decline to Kirchheimer's state-

his (Schmitt's) Weimar writings once more. In his two surviving postwar letters to Kirchheimer, he also proposed new subjects such as de Gaulle's coup d'état in France, the role of the opposition in the Federal Republic of Germany, the welfare state, and constitutional jurisdiction. And, in fact, Kirchheimer did discuss all of these subjects in his works—but he did so without direct contact with Schmitt and without referring to him in any of his publications.

Schmitt's furious reaction to Kirchheimer's role in the rejection of George Schwab's doctoral dissertation at Columbia University in 1962 changed how he received Kirchheimer from then on. As discussed in Chapter 17, Kirchheimer had raised a number of substantial objections to Schwab's manuscript: Schwab had ignored the state of research in the critical literature on Schmitt, he had misunderstood crucial sections of the Weimar Constitution, he had misread Schmitt's role in the final days of the Weimar Republic, and he had misrepresented Schmitt's antisemitism. The doctoral committee agreed with this overall assessment of the weaknesses and factual errors in Schwab's manuscript. As Kirchheimer had made clear in his letter to Ernst Friesenhahn at the time, quoted in Chapter 17, his own critical stance toward Schmitt notwithstanding, he would have accepted Schwab's dissertation if it had met general academic standards.

As I understand it, Kirchheimer's insistence on the factual errors in Schwab's failed doctoral project was also part of a broader issue: it was about Schmitt's future reception in the research community in the United States and in Germany. Up to this point in time, Schmitt was largely unknown in the American academic field of legal and political theory (see Richter 2001). Consequently, the first accepted academic monograph to emerge from one of the most respected American universities would have a formative effect on Schmitt's image in the English-speaking world. And that, in turn, would probably have repercussions on the German reception of Schmitt. Kirchheimer rightly suspected that Schmitt had tried to instrumentalize Schwab, whose dissertation Schmitt intended to be a clever move to further his own political rehabilitation in Germany via the United States. Kirchheimer was not willing to partake in this game and took an outspoken stand against it. Schmitt, in turn, viewed Kirchheimer's rejection of Schwab's dissertation as an attack directed personally against himself, disregarding the manuscript's shortcomings. Not only had Kirchheimer thwarted his plans for rehabilitation, he had also broken the loyalty Schmitt may well have expected from him as his former doctoral student. Therefore, he saw Kirchheimer's rejection of Schwab's dissertation as an act of treason and deceit. This would have confirmed ex post his view of Kirchheimer as one of his archenemies.

After Schwab's failed dissertation, Schmitt bitterly complained about Kirchheimer in letters to members of his circle, as documented in Chapter 17. Schmitt wrote his last letter to Kirchheimer in August 1961. After that, Kirchheimer sent Schmitt a copy of *Political Justice* in November 1961. He did not contact Schmitt after Schwab's failed dissertation, nor did Schmitt contact him again. Schmitt spoke only negatively about Kirchheimer for the next few years. He even called his former doctoral student to be "not a legal scholar in any sense of European jurisprudence." [14]

---

ment about a negative development of the Weimar Republic that could still be remedied; there was no such diagnosis of unstoppable decline in Kirchheimer's original article.

14 Letter from Carl Schmitt to Roman Schnur dated 24 October 1963 (Schmitt and Schnur 2023, 542).

But both still followed the work and activities of the other after the final break. Schmitt even wrote a malicious critique of *Political Justice*. It was published anonymously (and has been unknown to Schmitt researchers to date, see Chapter 17). Nevertheless, Kirchheimer soon found out that Schmitt was the author. In contrast to Schmitt's allegations and suspicions, he by no means intended to categorically halt the reception of Schmitt's work in the United States or to practice *damnatio memoriae*. On the contrary, he even continued to include original texts from Schmitt in his teaching at Columbia University right up until his death in November 1965.

As also mentioned before, there is no record of how Schmitt reacted when he learned of Kirchheimer's untimely death. Several years passed before Schmitt began to mention him again in positive terms. The *Lage* had changed again in the 1970s in some Western democracies, and Schmitt had discovered that some of his older and more recent works had begun to find new resonance on the political left in Italy, France, and Germany. In light of this, Schmitt opted for a positive reception of Kirchheimer. That decision of Schmitt's was to become the start of the legend of an enduring friendly relationship between Kirchheimer and Schmitt. Kirchheimer's visit to Schmitt's home in Plettenberg in November 1949 and their meeting in Cologne in June 1953 became crucial elements of this narrative. Since the early 1980s, a number of authors have claimed—without any documented evidence—that Kirchheimer was the one who contacted Schmitt first after 1945 and that he visited Schmitt multiple times. This narrative has developed a life of its own and can now be found in almost all scholarly contributions that deal with Schmitt and Kirchheimer, regardless of their author's position on the political spectrum.[15]

As far as I have been able to reconstruct the history of this narrative, Schmitt himself was its original source. In his conversation with Rainer Erd in July 1980, Schmitt was apparently able to create the impression that it was Kirchheimer who repeatedly wanted to stay in contact with him after 1945 and who visited him several times. Articulating to Erd his great esteem for Kirchheimer was part of Schmitt's attempt to create and control the narrative. It would not have been Schmitt's first such attempt. Reinhard Mehring has shown how he attempted to direct and influence the reception of his books and essays as early as the 1920s (see Mehring 2018, 123–125). After 1945, Schmitt became a virtuoso in presenting his oeuvre in ways that served his own political interests with respect to his past. He drew on his early contacts and exchanges with Leo Strauss and Walter Benjamin as evidence that he had always held Jewish scholars in high esteem, and even shrewdly insinuated that he had influenced their works.[16] A number of Schmitt's students, biographers, and editors followed the self-interpretations and legends he propagated. Once these efforts of his had succeeded, and a few years after Kirchheimer's death, he began to include him in his gallery of deceased alibi witnesses and circulate the legend that the driving force behind their contacts after 1945 had been Kirchheimer.

This story later coagulated into a narrative that overshadowed their enduring enmity, remaining vague in its details but leaving the impression of a friendly relationship between Kirchheimer and Schmitt that had overcome all personal and political turbulences,

15 The literature is listed in the Introduction.

16 See Gross (2000, 12 and 346), Mehring (2014b, 137–152), Bredekamp (2016), Palmier (2019, 411–417), and Suuronen (2022, 5–6).

even the Shoah. This narrative ultimately turned young and devoted admirers of Schmitt into champions of the legend of, in late Reinhart Koselleck's words, a "good friendship"[17] with Kirchheimer that, he claimed, proved that in fact, Schmitt had nothing at all against Jews. Schmitt's narrative, and such parroting of it, resurrect Kirchheimer as a posthumous witness for Schmitt.

Schmitt himself might well have summed it up as follows with a variation on the well-known introductory sentence of his book *Political Theology*: sovereign is he who successfully instrumentalizes the dead for his own purposes.

17 Schmitt and Koselleck (2019, 377). On Koselleck's reluctance to face Schmitt's antisemitism, see Lethen (2020).

Appendix

Abbreviations

AAC	Academic Assistance Council
AJC	American Jewish Committee
APSA	American Political Science Association
APSR	*American Political Science Review*
BNSDJ	Bund Nationalsozialistischer Deutscher Juristen (Association of National Socialist German Legal Professionals)
cf.	compare
CARA	Council for At-Risk Academics
CDU	Christlich Demokratische Union
CES	Central European Section of the Research and Analysis Branch of OSS
CID	Central Information Division
CCS	Combined Chiefs of Staff
COI	Coordinator of Information
CSU	Christlich Soziale Union
DAAD	Deutscher Akademischer Austauschdienst (German Academic Exchange Service)
DDP	Deutsche Demokratische Partei
DeWG	Gesellschaft für Europäische Wirtschaftsplanung und Großraumwirtschaft
DHfP	Deutsche Hochschule für Politik
DJZ	*Deutsche Juristenzeitung*
DNVP	Deutschnationale Volkspartei
DVP	Deutsche Volkspartei
EC	Emergency Committee in Aid of Displaced German/Foreign Scholars
ENS	École Normale Supérieure
FBI	(US) Federal Bureau of Investigation
FRG	Federal Republic of Germany
GDR	German Democratic Republic

GF	Graduate Faculty of the New School for Social Research
HICOG	US High Commission for Germany
IfS	Institut für Sozialforschung
IMT	International Military Tribunal
IRIS	Interim Research and Intelligence Service Office of Research and Intelligence (ORI), which was then incorporated in the Office of Intelligence Coordination and Liaison (OCL) and was renamed the Office of Intelligence Research (OIR)
ISR	Institute of Social Research
JCS	Joint Chiefs of Staff
JLC	Jewish Labor Committee
KPD	Kommunistische Partei Deutschlands (Communist Party of Germany)
KPO	Kommunistische Partei / Opposition
LSE	London School of Economics
NSDAP	Nationalsozialistische Deutsche Arbeiterpartei (National Socialist German Workers‘ Party)
NSRB	Nationalsozialistischer Rechtswahrerbund (see Glossary)
OCL	Office of Intelligence Coordination and Liaison
OIR	Office of Intelligence Research
OMGUS	Office of Military Government, United States
ORI	Office of Research and Intelligence
OSS	Office of Strategic Services
R&A	Research & Analysis Branch
RAF	Rote Armee Fraktion (Red Army Faction)
ROTC	Reserve Officer Training Corps
SA	Sturmabteilung (see Glossary)
SAP	Sozialistische Arbeiterpartei (Socialist Workers‘ Party)
SD	SS-Sicherheitsdienst
SIRES	Société Internationale de Recherches Sociales
SPSL	Society for the Protection of Science and Learning
SPD	Sozialdemokratische Partei Deutschlands (Social Democratic Party)
SS	Schutzstaffel (see Glossary)
UK	United Kingdom
US	United States
USSR	Union of Soviet Socialist Republics
ZfS	*Zeitschrift für Sozialforschung*
ZÖR	*Zeitschrift für öffentliches Recht*

List of German Courts

The court system in the Weimar Republic was similar to that in the *Kaiserreich* (the German Empire). It was based on the German Empire's *Gerichtsverfassungsgesetz* (Courts Constitution Act) of 1870. The court of first instance was the *Amtsgericht*, next was the *Landgericht*, followed by the *Oberlandesgericht*. The highest German court dealing with criminal and civil cases was the *Reichsgericht*. The basic structure of the courts (including their personnel) was left unchanged after the democratic revolution of 1918/19 leading to the Weimar Republic. However, some new courts for specific areas of the law were added to the system (e.g., labor courts and finance and revenue courts). A new *Staatsgerichtshof* (as part of the *Reichsgericht*) which dealt with cases between various state entities was established in 1921. Hitler's government also kept the system in place, as well as most of the personnel, of course with the exception of Jews and leftists. It added new courts, for instance *Erbgesundheitsgerichte*, *Militärgerichte*, and *Sondergerichte* such as the *Volksgerichtshof*, all of which were abolished after 1945. The Federal Republic of Germany established the first constitutional courts in German history, both at the level of the *Länder* and at the national level.

The following list includes only the courts in the Weimar Republic, the Third Reich, and the Federal Republic of Germany mentioned in this book.

Amtsgericht: Court of first instance for criminal and civil cases.

Arbeitsgericht: Labor court.

Bundesgerichtshof: Federal Court of Justice, highest court of criminal and civil jurisdiction, established in 1950 and domiciled in Karlsruhe.

Bundesverfassungsgericht: Constitutional Court of the Federal Republic of Germany, established in 1951 and domiciled in Karlsruhe.

Erbgesundheitsgericht: Literally "hereditary health courts" and also known as genetic health courts or sterilization courts, these courts were established on 1 January 1934 on the basis of the *Gesetz zur Verhütung erbkranken Nachwuchses* (Law for the Prevention of

Hereditarily Diseased Offspring) of 14 July 1933, also known as the sterilization law. A tool for implementing the Nazis' *Rassepolitik*, these courts ordered compulsory sterilizations of mentally and physically disabled persons, patients of psychiatric hospitals and nursing homes, alcoholics, and others considered undesirable, including political opponents.

Kammergericht: The name of the *Oberlandesgericht* in the state of Prussia (historically) and the *Land* Berlin (today).

Landgericht: Court of first or second instance (depending on the subject matter of the case) for criminal and civil cases.

Militärgerichte: Military courts, established in May 1933.

Oberlandesgericht: Court of second or third instance (depending on the subject matter of the case) for criminal and civil cases.

Reichsarbeitsgericht: Founded in 1926 and domiciled in Leipzig, the *Reichsarbeitsgericht* was the highest labor court of appeals.

Reichsfinanzhof: Founded in 1918 to handle tax law and domiciled in Munich.

Reichsgericht: Established in 1879 and domiciled in Leipzig, the *Reichsgericht* was the highest court of appeals in criminal and civil law in the German Reich up until 1945.

Sondergerichte were special courts in the Third Reich for political and particularly serious crimes, feared for their swift and severe rulings that could not be appealed.

Staatsgerichtshof: Founded in 1921 and domiciled in Leipzig, the *Staatsgerichtshof* decided on cases between the *Länder* and the Reich as well as those delineating the competencies of the President of the Reich. It was not a permanent court but was convened as needed. The President of the *Reichsgericht* also served as President of the *Staatsgerichtshof.*

Staatsgerichtshof of the *Land* Hesse: The constitutional court of the *Land* Hesse, established in 1948.

Verwaltungsgericht: Administrative court. Even though Article 107 of the Weimar Constitution provided for a national administrative court, administrative courts existed only on the level of the *Länder*. The *Bundesverwaltungsgericht* (Federal Administrative Court) of the Federal Republic of Germany was established in 1952.

Volksgerichtshof: Dissatisfied with the rulings of the *Reichsgericht* on high treason, Hitler ordered the establishment of the *Volksgerichtshof* in Berlin in April 1934 as a special national court for cases concerning high treason and later, during World War II, also subversion of the war effort. The court worked as a commission that was appointed ad hoc by Hitler and was composed of two professional judges and a majority of laymen who had already proven their trustworthiness to the Nazi regime. Its decisions could not be contested. The court served as an instrument of terror used by the Nazi regime against political opponents and passed more than 5,200 death sentences.

Glossary

By Sandra H. Lustig

This Glossary contains Nazi German terms that appear in this book and are key to understanding what Schmitt and some other authors were talking about during the Nazi period. I have taken definitions from various sources; some come from Nazis themselves (see Schmitz-Berning 2007), others from dictionaries prepared at the time to explain Nazi German terms in English (see Paechter et al. 1944 and Neuburger 1944). In some cases, I provide further explanations to illuminate connotations and emotional resonance.

I used the following sources, listed in chronological order of publication, to prepare this Glossary:

Nazi-Deutsch. A Glossary of Contemporary German Usage, with Appendices on Government, Military and Economic Institutions, published in 1944 by Heinz Paechter et al., is based on a range of contemporary sources, including Nazi materials, and it states explicitly: "Except in the case of military ranks, the translations are explanatory and are not intended to approximate American or British equivalents" (3). The introduction to the appendix of terms relating to *Weltanschauung* ("way of seeing the world," philosophy) states: "Words listed here are not new but have a new meaning in the context of Nazi philosophy. The translations try to explain these concepts and to evoke their connotations and emotional background rather than to render slavishly the glittering and often deliberately equivocal meanings of basically untranslatable words" (110).

The German-English Dictionary of German Administrative Terms is a Civil Affairs Guide, War Department Pamphlet No. 31–169, 3 July 1944, prepared by Dr. Otto Neuburger of the Library of Congress for the US military "for the use of Civil Affairs Officers in the field" (Neuburger 1944, no page number); it initially had restricted circulation. Neuburger served as deputy director of the Munich employment office and was a co-author of the *Philo-Lexikon* (a manual of Jewish knowledge) before fleeing Nazi Germany for the US.

Victor Klemperer's *LTI—Notizbuch eines Philologen* (1947; English edition: *The Language of the Third Reich, LTI: Lingua Tertii Imperii, A Philologist's Notebook*) is a collection of the author's observations on *Lingua Tertii Imperii* (Latin for: Language of the Third Reich), which he abbreviated to LTI, likely a play on the Nazis' proclivity for abbreviations. Originally Jewish, Klemperer had converted to Protestantism, and he managed to survive the Nazi period in Germany thanks to his wife Eva, who was an "Aryan" German. His notes capture the flavor and the emotional impact of many Nazi German terms.

Encyclopedia of the Third Reich (1976) was authored by Louis L. Snyder, a historian and scholar of Germany, who also predicted Hitler's rise to power in his 1932 book *Hitlerism: The Iron Fist in Germany*.

The Encyclopedia of the Third Reich (1985, English translation edited by Amy Hackett 1991), edited by historian Christian Zentner and historian and Germanist Friedemann Bedürftig, is a massive two-volume work with more than 3,000 entries and twenty-seven major articles and overviews.

NS-Deutsch: "Selbstverständliche" Begriffe und Schlagwörter aus der Zeit des Nationalsozialismus [NS German: "Self-evident" terms and jargon from the period of National Socialism] by Karl-Heinz Brackmann and Renate Birkenhauer (1988) documents and explains "everyday language, figures of speech, informal—simply "self-evident"—expressions [and] the vocabulary of code words and expressions of administrative language used internally, which also required no explanation at the time" (5). The more than 2,000 entries draw on the knowledge of many older translators and others who had lived through the Nazi period.

In *Vokabular des Nationalsozialismus* [Vocabulary of National Socialism], Cornelia Schmitz-Berning (2007) uses Nazi sources to explain terms from Nazi vocabulary, thus spelling out how the Nazis themselves defined and used such words. Drawing on more than 450 sources, including speeches, newspaper articles, scholarly publications, and compilations of countless individual texts of various types covering a wide range of subjects, Schmitz-Berning frequently cites various editions of the *Duden* German dictionary and *Meyer's Lexikon* [Meyer's encyclopedia] to show how the definition of a word changed from before the Nazi era to during and after this period.

* * *

Art: Biologistic term signifying belonging versus non-belonging and used as a rationale for discrimination, subjugation, and genocide. "[K]indred, type, race, ways of life" (Paechter et al. 1944, 18); "orig.: way of life, typical character, national culture. now: the biological type, the racial character hallowed by blood" (Paechter et al. 1944, 111). *Art* was also defined as "the state of being characterized by *Blut* and *Rasse*" (Brackmann/Birkenhauer 1988, 24). "*Art* [...] had gained an additional meaning under National Socialism besides its meaning 'inborn uniqueness, nature, descent, way of being.' The biological term *Art* (species) for a unit (below the genus) that encompasses organisms with a large number of the same morphological and physiological characteristics had been narrowed in *Rassenkunde* [the science of *Rassen*] and limited to the 'particular uniformity of a *Volk* contingent on and determined by *Rasse*.' This meaning had become part of everyday language through the popularization of *Rassenkunde*" (Schmitz-Berning 2007, 63–64). The concept "narrowed in this way provided the foundation of the Nuremberg Laws; it also provided the pseudoscientific rationale for the persecution of Jews" (Zentner and Bedürftig 1985, 895).

artfremd, fremdartig: *Fremdartig* is an example of a German word with a common meaning—alien, foreign, or unfamiliar—which was given an additional meaning in the right-wing political discourse during the Weimar Republic and in Nazi German, where it also combined the exclusionary meanings of *fremd* and *Art* (see entries in this Glossary). *Fremdartig* could legitimately be interpreted with or without the additional meaning. *Artfremd*, however, has a meaning in biology—not of the same species—and, in Nazi German, its meaning was solely biologistic: foreign/alien to the (German) *Volk*, in an exclusionary and antisemitic sense.

artgleich, gleichartig, homogen; Artgleichheit, Gleichartigkeit, Homogenität: Although these terms might seem synonymous at first glance, they actually express quite different concepts. *Homogen* can be translated directly as homogeneous and may refer to uniform social living conditions, religion, or ethnicity, for instance. *Artgleich* and *gleichartig* are composed of the same roots: *Art* and *gleich* (see entries in this Glossary). *Gleichartig* is an example of a German word with a common meaning—equal, of the same kind, or very similar—which was given an additional meaning in Nazi German, where it combined the exclusionary meanings of *gleich* and *Art*. *Gleichartig* could legitimately be interpreted with or without the additional meaning. *Artgleich*, however, has a meaning in biology—of the same species—and, in Nazi German, its meaning was solely biologistic and exclusionary: belonging to the (German) *Volk*, in an exclusionary and antisemitic sense. The nouns *Artgleichheit*, *Gleichartigkeit*, and *Homogenität* mean the condition of being *artgleich*, the condition of being *gleichartig*, and homogeneity. On Schmitt's changing use of these terms over time, see Chapter 10.

Blut: Biologistic term signifying belonging versus non-belonging and used as a rationale for discrimination, subjugation, and genocide. "*Blut*: a) racially characterized genetic constitution of a *Volk*; also: synonym of: *Rasse*; b) mythically elevated symbol: *the mystique of Nordic Blut*" (Schmitz-Berning 2007, 109); "blood: the 'Myth of the 20th Century', carrier of race quality, bond between fellow-countrymen, basis of national policy. [...] Many compounds indicate the magic blend of death symbols with totemistic sacrifice symbols. The race community is held together by common origin and common destiny" (Paechter et al. 1944, 112). (Paechter was likely referring to Alfred Rosenberg (1930): *Der Mythus des 20. Jahrhunderts*.) "[M]elodramatic for: ancestry, belonging to a *Rasse* [...] a propaganda orator [...] defined: 'yes, *Blut* is material, but not material in the wrong, materialistic sense, but in the sense of *Heimat* [homeland] and soil and *Rasse*-based heritage" (Brackmann/Birkenhauer 1988, 39).

Blutsgemeinschaft: see ***Gemeinschaft***.

Entartung, entartet: "a) biologically: degeneration of the *Art*, the *Rasse* due to changes to the genetic constitution the main cause of which is assumed to be mixing of *Rassen*. b) Decline of culture and morals among peoples, families, and individuals as a consequence of biological *Entartung*, more rarely as a consequence of the immorality of a people" (Schmitz-Berning 2007, 178). "[N]ot corresponding to the official ideas of the Nordic *Rasse* and *Wesensart* (nature, character, spirit, mentality)" (Brackmann/Birkenhauer 1988, 63). "[In the Nazi state, the assumption of] biological and cultural *Entartung* legitimized laws and state measures that involved the most serious interventions in the lives of the people affected and ultimately the annihilation of entire groups of the population" (Schmitz-Berning 2007, 181).

ewig: "a) Emphatically: lasting forever; b) Overly elevated in religious terms: timeless like the divine" (Schmitz-Berning 2007, 220). *Ewig* "is an attribute reserved exclusively for the divine; by calling something [*ewig*], I elevate it to the sphere of the religious" (Klemperer 1947, 114). "*Ostmark* [and] *ewig* "are entirely neutral words *per se* [...] And yet in the context of the LTI they are decidedly Nazi words belonging to a special linguistic register [...] *Ostmark* in place of *Österreich* {Austria}: this represents the link with tradition, reverence for the ancestors, who, rightly or wrongly, are invoked, whose legacy one claims to honour and whose testament one professes to fulfil. *Ewig* points in the same direction; we are links in a chain stretching back into the misty past, one which is supposed to pass through us into the distant future, we always were and always will be. *Ewig* is simply the most forceful special case amongst the numerical superlatives, which are themselves but a special case in the midst of the ubiquitous LTI superlatives" (Klemperer 1947, 272).

fremd, Fremde/r: As an adjective, *fremd* generally means foreign, alien, or unfamiliar; as a noun, *Fremde*/r generally means a person with those characteristics. Alone or as part of a compound word, *fremd* indicated non-belonging and exclusion from the German *Volk* in Nazi German.

fremdartig: see ***artfremd***.

Führer: The literal translation of *Führer* is simply leader. The Nazi neologism *der Führer* (the *Führer*), with the definite article, was used as a title for Adolf Hitler and as a term characterizing him. *Der Führer* no longer meant any political leader, potentially even one elected democratically, as it had before, but solely Adolf Hitler, who was vested not only with absolute political power but with virtually mystical powers, as if he were the leader of a cult. "As it was used in quasi-religious contexts, Hitler's sobriquet *Führer* was idealized and filled with bathos, and Goebbels eventually characterized it as a 'hallowed concept of state'" (Schmitz-Berning 2007, 244). Although only Hitler was called *der Führer*, those in leadership positions in various areas were called *Unterführer* (sub-*Führer*) and had their own areas of responsibility with decision-making power.

Führerprinzip: *Führer* principle. "[A]uthoritarian principle [totalitarian principle of absolute leadership by one; foundation stone of the Nazi system]" (Neuburger 1944, 34). "Antiparliamentarian organizational principle of the Third Reich according to which

Hitler ruled not within the framework of a constitution, but as the alleged personification of the will of the *Volk*" (Schmitz-Berning 2007, 245). According to the *Führerprinzip*, the *Führer* headed the Nazi party and the state, both of which were organized hierarchically. The *Führer* alone held decision-making power, and his authoritative decisions, correct by definition, were to be obeyed by all. The *Führerprinzip* was a general and authoritarian structural principle of the Nazi regime, and it applied at all levels of government and society, including courts and businesses. After World War II, many Nazis invoked the *Führerprinzip* to claim that they themselves were not responsible for any decisions made or crimes committed during the Nazi period and therefore should not be punished for them because they were merely following orders issued by the *Führer*.

Führerstaat: *Führer* state. State organized according to the *Führerprinzip* with an authoritarian government, with the NSDAP as the only political organization, and with a sidelined parliament (see Brackmann/Birkenhauer 1988, 78).

Gefolgschaft: "Term used by Hitler to describe those who were governed by fidelity to the Fuehrer" (Snyder 1976, 110). It "denotes the relationship of those led to the *Führer*, which is based on loyalty, as well as the followers/devotees/subordinates bound in submission/allegiance to the *Führer* because of the relationship based on loyalty [...]. In the Third Reich, the term was often used in an excessive and pseudoreligious way" (Schmitz-Berning 2007, 252–253). A *Gefolgschaft* is "a *Gemeinschaft* that follows the will of a leader; its foundation was the conviction that every genuine *Führertum* [leadership in the sense of the *Führer*] is sent by destiny and that the order to act in *Gefolgschaft* is issued by higher powers to those being led" (Brackmann/Birkenhauer 1988, 80). "But what bothered me every day afresh, never to be entirely erased by other trains of thought [...] was the word *Gefolgschaft*. The whole emotional mendacity of Nazism, the whole mortal sin of deliberately twisting things founded on reason into the realm of the emotions, and deliberate distortion for the sake of sentimental mystification [...]. What does a perfect group of followers do? It doesn't think, and doesn't even feel any more—it follows" (Klemperer 1947, 244 and 252).

Gemeinschaft: Biologistic term signifying belonging versus non-belonging and used as a rationale for discrimination, subjugation, and genocide. *Gemeinschaft* usually means community. In Nazi German, however, it was a "term describing the connectedness of all comrades of the *Volk* with each other and with the *Führer*, the connectedness arising from the experience on the frontlines of the First World War, founded on *Blut* and *Rasse*, and supported by the Nazi worldview" (Schmitz-Berning 2007, 261). It was a "fundamental concept underlying the worldview of Nazi thought and action; described the commonality of *Blut* and soil, culminating in the concept of the *Volksgemeinschaft*" (Brackmann/Birkenhauer 1988, 82). "Structures favorable to the Nazi outlook are presented as indivisible organic units. Community, hence, is conceived, not as a mutual relation between individuals, but as a pre-existing unity of a race or people. Volks[gemeinschaft]: The commonweal of the race, bent on its own perfection or thriving (the usual translation 'community of the people' is a misleading adaptation to liberalistic thinking)" (Paechter et al. 1944, 115). ***Blutsgemeinschaft***: "the German *Volk* as an indivisible *Rasse*-based unit" (Brackmann/Birkenhauer 1988, 39).

gesund: *Gesund* generally means healthy or wholesome, but it had an additional meaning in Nazi German: "corresponding to the norm of the NSDAP" (Schmitz-Berning 2007, 269). "A person is *gesund* if he is like the *Gemeinschaft* wants him to be" (Schmitz-Berning 2007, 270).

gesundes Volksempfinden: "[A]ssessment of a matter in accordance with the Nazi *Volksgemeinschaft* governed by the will of the *Führer*" (Schmitz-Berning 2007, 270). Related to this: ***gesundes Rechtsempfinden***: "The focus and goal/purpose of this system was *Rechtsempfinden* [legal feeling, feelings about justice]; there was never any mention of *Rechtsdenken* [legal thinking, reasoning on justice], nor of a feeling of justice on its own but, rather, always of 'a *gesund* feeling about justice.' And *gesund* meant whatever was in line with the will and the benefit of the party. [...] [T]his *gesund* feeling was used to motivate the plundering of Jewish property" (Klemperer 1947, 251).[1]

gleich, Gleiche/r: As an adjective, *gleich* generally means equal, similar, or alike; as a noun, *Gleiche/r* means a person with those characteristics. Alone or as part of a compound word, *gleich* indicated belonging to the German *Volk*, in an exclusionary and antisemitic sense, in Nazi German.

gleichartig, Gleichartigkeit: see ***artgleich***.

Gleichschaltung: "the alignment of associations, organizations, political parties, and, ultimately, every individual citizen toward the goals of NS [National Socialist] policy. The pattern [of *Gleichschaltung*] was always the same: under pressure by NSDAP members, the executive committee of a professional organization was restructured and National Socialists were admitted. They 'purged' (*säuberte*) the committee and led the association under the umbrella of the party. Where this was not possible, the NS leadership resorted to force: for example, on May 2, 1933, the buildings and offices of the trade unions were occupied by SA and police personnel, their records impounded, and their assets confiscated [...] [A]fter the summer of 1934 there was scarcely a German who was not connected to the party in some way, whether through profession or job, position, or organization" (Zentner and Bedürftig 1985, 940–942). It goes without saying—or perhaps it should be stated explicitly—that this process of "totalitarianiz[ing]" (Neuburger 1944, 40) implied dismissal or incarceration of all Jews and leftists, i.e., their destitution and/or persecution. Where the process of *Gleichschaltung* proved difficult, the Nazis closed down the organization and replaced it with one of their own. Having witnessed *Gleichschaltung* by the state, countless private organizations, down to small sports clubs and choral societies, excluded "non-Aryans" from their ranks and aligned their activities with Nazi ideology.

Großraum: see ***Raum***.

---

1 My translation.

homogen, Homogenität: see ***artgleich***.

Der Jude: "[C]ontemptuous epithet for individuals of Jewish descent or those given the same status" (Schmitz-Berning 2007, 328). A Jew was not to be referred to simply as, for instance, "Hans Kelsen," but as "the Jew Hans Kelsen." "When using the epithet 'Jew,' the Nazis took for granted that because of the constant agitation against Jews [...], because of the stereotypical use of the term in negative contexts, all the revived old and new prejudices were linked to the term 'Jew' as defining characteristics so that 'Jew,' without any additional adjective, could serve as a slur" (Schmitz-Berning 2007, 328). "[T]otalitarian anti-Semitism rejects the Jews because of their race, not because of their individual qualities or because of individual experiences with Jews. The totalitarian concept of races excludes exceptions, denies the possibility that Jews might be different under different conditions and does not admit of any discrimination between good and bad Jews. Totalitarian language does not permit the expression of such a picture but must evoke the image of a type, 'The Jew'. It always speaks of all Jews in the singular, as a collective entity. This counter-image (the 'Gegentyp') is held up to hatred; all its components participate in 'Jewishness'. Similarly, Nazi authors like to speak of 'der Engländer, der Franzose' when speaking of the English or the French nation" (Paechter et al. 1944, 8).

Lebensgesetz: "Lebensform, Lebensgesetz [...] morphological form, biological destiny (Ludwig Klages' philosophy)" (Paechter et al. 1944, 119) (see ***Schicksal***).

Lebensraum: see ***Raum***.

Leistungsraum: see ***Raum***.

Rasse: Biologistic term signifying belonging versus non-belonging and used as a rationale for discrimination, subjugation, and genocide. "Central keyword of Nazism; stands for 'the cornerstone of the Nazi worldview,' the ideology of Aryans being of the highest value and of Jews being subhuman, and for 'hygiene in terms of *Rasse* with its Nordic breeding goal'" (Schmitz-Berning 2007, 481). "Rasse is defined strictly biologically and implies a fateful determinism. It is a unit or personality with a will, a soul etc. of its own which is expressed in the will of the Führer and independent of the wills, souls etc. of the members. Rasse is no classifying term; it stands for a collective entity. No cultural or political denominations enter into its definition" (Paechter et al. 1944, 120). "[T]ypical physical characteristics were linked to moral values and emotional-spiritual behaviors; this enabled not only (a) straightforward labeling of higher and lesser *Rassen*, it also appeared logical to (b) regain the ideal Nordic type from the contemporary jumble of *Rassen* through biological and emotional-spiritual breeding, and to (c) condemn the mixing of *Rassen* with *Fremdblütige* [people of foreign/alien *Blut*] because they allegedly weakened the defensive power of the German *Volk*" (Brackmann/Birkenhauer 1988, 149). "With [the Nuremberg Laws,] racialism, hitherto distinguished mostly by vagueness, became for the first time the legal basis of a modern nation. It was responsible eventually for Hitler's Final Solution of the Jewish question" (Snyder 1976, 278).

Raum, Großraum, Lebensraum, Leistungsraum: A direct translation of *Raum* is space. As used in Nazi German, it was understood not only in the geographical sense, but was inherently linked to geopolitics: "space, range of space, area, scale, width [important term in connection with geopolitics and history, as many forces of history are defined as functions of space or as being graphically represented as space; the following words are only a selection from a very large group of catchwords of geopolitical concepts]: entwehrter Raum—disarmed area; Erdraum—earth area; Grossraum—large space; historischer Raum—historic sphere; Lebensraum—living or vital space" (Neuburger 1944, 75, brackets in the original). From this follows that ***Großraum*** did not merely mean "large space" but, rather, "the political order that Germany was striving to achieve in Europe and in which the *Lebensraum* of multiple *Völker* were to be consolidated to form a unit under German leadership; *Großraumpolitik* was the justification for Hitler's policy of conquest" (Brackmann/Birkenhauer 1988, 91). ***Lebensraum*** did not simply mean "living space" (its direct translation); it was an "NS keyword to legitimize a violent *Rasse*-based expansion of the Third Reich to the east" (Schmitz-Berning 2007, 375). It denoted "territory of expansion [slogan adopted by Nazis to carry out the so-called German destiny]" (Neuburger 1944, 56, brackets in the original) and "1. space necessary for autarchic economy of an increasing population (cf. Hans Grimm's bestseller 'Volk ohne Raum'). 2. space assigned to a people by racial destiny" (Paechter et al. 1944, 119). ***Leistungsraum***: Schmitt coined this term to denote the space required for a country's desired economic performance, thereby providing an economic justification for expansionism. He wrote about *Leistungsraum*, not *Lebensraum*.

Rechtswahrer: "[T]he *Rechtswahrer* [preserver/defender of the law] has replaced the 'jurist.' The focus of his application of the law is not to secure the application of a provision, but to secure the *Volksgemeinschaft*" (Brackmann/Birkenhauer 1988, 152). "An Alliance of Defenders of the Law {*Ein Bund der Rechtswahrer*} sounds incomparably more solemn than an Association of Lawyers {*eine Vereinigung der Rechtsanwälte*}, someone who discharges their duties {*Amtswalter*} sounds incomparably more impressive than an official {*Beamter*} or functionary {*Funktionär*}, and if I were to read '*Amtswaltung* {office for discharging duties}' rather than '*Verwaltung* {administration}' above an office door, then the atmosphere is nothing short of sacred" (Klemperer 1947, 245).

Schicksal: "Schicksal is a basic notion of the teleological NS Weltanschauung. Every 'Gestalt', 'Rasse', 'Typus' has a destiny of its own to fulfil. [...] The concept of Schicksal as set forth by Spengler and accepted by most Nazi authors is close to the antique 'Moira', meaning the pre-destined, unavoidable power represented by race, history, or 'essence', which drives a man or a people to the fulfilment of an assigned end. Göring makes the distinction between Schicksal—the inescapable destiny to which even the Gods are subject—and Vorsehung—the hand of God" (Paechter et al. 1944, 123).

Schutzstaffel (SS): The literal translation of Schutzstaffel is "protective squadron." It was the "name of the organization of the NSDAP established in 1925 to protect Hitler and systematically expanded by Himmler from 1929 on as an instrument of the Nazi regime for domination and terror" (Schmitz-Berning 2007, 590).

Sturmabteilung (SA): The SA was the "uniformed and armed political storm troops of the NSDAP" (Schmitz-Berning 2007, 551). "The Sturmabteilung, or SA, is the organization which the world remembers as the 'Brown Shirts' or Storm Troops—the gangsters of the early days of Nazi terrorism. Since it was the first of the organizations created by the Nazis as instruments to effectuate their illegal objectives, the SA occupied a place of peculiar importance in the scheme of the conspirators. Unlike some of the other organizations, the functions of the SA were not fixed or static. The SA was an agency adapted to many designs and purposes, and its role in the conspiracy changed from time to time [in] various phases toward the final objective—abrogation of the Versailles Treaty and acquisition of the territory of other peoples and nations. If the conspiracy is likened to a pattern, with its various parts fitting together like the pieces of a jig-saw puzzle, the piece representing the SA would be found to constitute the essential link in the pattern" (Office of United States Chief of Counsel for Prosecution of Axis Criminality 1946, 133–134).

Volk (plural: ***Völker***): Biologistic term signifying belonging versus non-belonging and used as a rationale for discrimination, subjugation, and genocide. "A nature-like community of common lineage, history, language and culture which is characterized by *Rasse* and common *Volksboden* (the soil currently or historically settled by Germans, Schmitz-Berning 2007, 649) and requires strong leadership as well as education and Nazi indoctrination at all times" (Schmitz-Berning 2007, 642); "[T]he racial nation: the central concept of home propaganda. It blends many meanings: 1) people, as against respectability and nobility; 2) the natural, original folk (archaic, as in folklore) as against bourgeois decadence; 3) nation as a unity; 4) race as against class; 5) the traditional, given Gestalt of a fate determined by blood and soil. The will of the folk is revealed only to the Führer and never depends upon the will or agreement of individuals" (Paechter et al. 1944, 124). The "entirety of the German nation as a *Blutsgemeinschaft* in terms of politics, *Rasse*, culture, and *Schicksal*—as distinguished from the neighboring *Völker*" (Brackmann/Birkenhauer 1988, 193).

Volksgemeinschaft: see ***Gemeinschaft***.

völkisch: One of the adjectives based on *Volk* (besides *volkhaft* and *volklich*, which were used far less often), *völkisch* is surely one of the most difficult words to translate, and there is certainly no single word in English that even comes close. "[T]he word denotes something more [than national]: the eagerness to cultivate the features typical of the nation and at the same time eliminate the material and spiritual influences of other peoples" (Snyder 1976, 362). During the imperial era and the Weimar Republic, the *völkisch* "movement advocated race mysticism, pseudobiology, and anti-Semitism. Its literature emphasized the idea of recasting history as a primeval battle between the blond Nordic hero and the Jew. The *voelkisch* movement provided the historical roots and constituted the organizational as well as the ideological starting point of National Socialism." (Snyder 1976, 362) "Germanization of the word 'national' meaning: determined by *Rasse* and *Blut*; positive antonym of 'international' (usually in combination with 'Jewish'), which was uttered in contempt" (Brackmann/Birkenhauer 1988, 194). "[R]acial, national, nationalistic; in the true tradition of the type, stock, race, people" (Neuburger 1944, 117). "National, emphasizing the values embodied in the *Rasse* and the character of the *Volk*" (Schmitz-Berning 2007, 645, quoting Volks-Brockhaus encyclopedia 1940). "In NS usage it acquired first and foremost the meanings of antisemitic, chauvinist-nationalistic, and true to blood and species (blut- und artgemäss)" (Zentner and Bedürftig 1985, 1001).

Sources and Bibliography

Archival Sources

Theodor W. Adorno Papers. Institut für Sozialforschung, Frankfurt. Theodor W. Adorno Archiv, Nachlass Theodor W. Adorno.

Hannah Arendt Collection. Bard College, Hannah Arendt Collection, Marginalia.

Bundesarchiv Berlin-Lichterfelde, Akten des Auswärtigen Amtes. Politisches Archiv, RZ 214, R 99744 (69. Ausbürgerungsliste, Ausbürgerungsakte betreffend Otto Kirchheimer).

Bundesarchiv Berlin-Lichterfelde, Akten des Justizministeriums, R 3001/6322, Akten des Justizministeriums betreffend Dr. Otto Kirchheimer.

Bundesarchiv Berlin-Lichterfelde, Akten des Sicherheitsdienstes des Reichsführers SS, Akte Staatsrat Carl Schmitt, R 58/854.

Columbia University, Columbia University Archives. Columbia University Bulletin 1958–1965

Deutsche Bibliothek Frankfurt am Main. Abteilung Exilliteratur. Akten der American Guilt of German Cultural Freedom.

Horst Ehmke Papers. Friedrich-Ebert-Stiftung, Archiv für soziale Demokratie, Horst Ehmke Depositum, Nr. 504.

Henry W. Ehrmann Papers. State University of New York, University at Albany, M.E. Grenander Special Collections & Archives, Papers of Henry W. Ehrmann.

Ernst Fraenkel Papers. Bundesarchiv Koblenz. Nachlass Ernst Fraenkel. N 1274.

Arkadij Gurland Papers. Universitätsbibliothek Frankfurt, Archivzentrum, Nachlass Arkadij Gurland, Na 5.

John H. Herz Papers. State University of New York, University at Albany, M.E. Grenander Special Collections & Archives, Papers of John H. Herz

Max Horkheimer Papers. Universitätsbibliothek Frankfurt, Archivzentrum, Nachlass Max Horkheimer, Na 1.

Humboldt-Universität Berlin, Universitätsarchiv, Akten der Juristischen Fakultät 1945–1948.

Hanna Kirchheimer-Grossman, Private Family Collection (Arlington).

Otto Kirchheimer Papers. State University of New York, University at Albany, M.E. Grenander Special Collections & Archives, Papers of Otto Kirchheimer.

Leo Löwenthal Papers. Universitätsbibliothek Frankfurt, Archivzentrum, Nachlass Leo Löwenthal, Na 4.

Karl Loewenstein Papers. Amherst College, Archives and Special Collections. Karl Loewenstein (1891–1973) Papers.

Herbert Marcuse Papers. Universitätsbibliothek Frankfurt, Archivzentrum, Nachlass Herbert Marcuse, Na 3.

National Archives at College Park, Maryland. General Records of the Department of State. Record Group 59. Intelligence Reports, 1941–1961.

National Archives at College Park, Maryland. Records of the Federal Bureau of Investigation, U.S. Department of Justice, F. B. I., Files on Subject Otto Kirchheimer.

National Archives at College Park, Maryland. Military Records Division. Records of the Office for Strategic Services (OSS).

New School for Social Research, New School Institutional Collections. Graduate Faculty, Minutes.

New York Public Library, New York City. Emergency Committee in Aid of Displaced German /Foreign Scholars, I, A Grantees, 1933–46.

Rudolf Smend Papers. Staats- und Universitätsbibliothek Göttingen, Nachlass Rudolf Smend, Cod. Ms. R. Smend A 441.

Carlo Schmid Papers. Archiv der Sozialen Demokratie, Nachlass Carlo Schmid, Reg. Nr. 678.

Carl Schmitt Papers. Landesarchiv Nordrhein-Westfalen, Nachlass Carl Schmitt, RW 472.

Rheinische-Friedrich-Wilhelms-Universität Bonn, Archiv der Juristischen Fakultät, Prüfungsakten. Prüfungsakte Otto Kirchheimer, Promotion 1927/28, No. 500–524.

Rockefeller Foundation Center, Rockefeller Foundation Records, RF RG1.2 Series 200 U.S.-Social Sciences, Box 539, Folder 4615.

Personal Sources
(interviews, conversations, conference discussions, letters, or e-Mail exchanges)

Johannes Agnoli, Andrew Arato, Reinhard Bendix, Ernst-Wolfgang Böckenförde, Karl Dietrich Bracher, Angelo Bolaffi, Alexander von Brünneck, Horst Ehmke, Dieter Emig, Rainer Erd, Henry W. Erdmann, Jürgen Fijalkowski, Ossip K. Flechtheim, Lili Flechtheim-Faktor, Julian Franklin, Peter Gay, Gerd Giesler, Jürgen Habermas, Wilhelm Hennis, John H. Herz, Raul Hilberg, Albert O. Hirschman, David Kettler, Hanna Kirchheimer-Grossman, Peter Kirchheimer, Leo Löwenthal, Richard Löwenthal, Peter Marcuse, Reinhard Mehring, Wolfgang Mommsen, Ludmilla Müller, Michael Neumann, Volker Neumann, Joachim Perels, Ulrich K. Preuß, Frank Schale, Jürgen Seifert, Alfons Söllner, Kurt Sontheimer, Nicolaus Sombart, Winfried Steffani, Susanne Suhr, Jacob Taubes, Arthur J. Vidich. Herta Zerna, Gilbert Ziebura.

Bibliography

Abendroth, Wolfgang (1978). *Ein Leben in der Arbeiterbewegung*. Frankfurt am Main: Suhrkamp.

Abromeit, John (2011). *Max Horkheimer and the Foundations of the Frankfurt School*. New York: Cambridge University Press.

Adorno, Theodor W. (1951). *Minima Moralia: Reflections from Dangered Life*. London: Verso, 2020.

Adorno, Theodor W. (1963). "Der Begriff der politischen Bildung." In Theodor W. Adorno, *Vorträge 1949–1968*, edited by Michael Schwarz, 377–386. Berlin: Suhrkamp, 2019.

Adorno, Theodor W., Else Frenkel-Brunswik, Daniel J. Levinsonn, and Nevitt R. Sanford (1951). *The Authoritarian Personality*, edited by Theodor W. Adorno, Else Frenkel-Brunswik, Daniel J. Levinsonn, and Nevitt R. Sanford. New York: W. W. Norton & Co.

Adorno, Theodor W., and Max Horkheimer (2004). *Briefwechsel 1938–1944*, edited by Christoph Gödde, and Henri Lonitz. Frankfurt/M.: Suhrkamp.

Adler, Max (1922). *Die Staatsauffassung des Marxismus*. Wien: Verlag der Wiener Volksbuchhandlung.

Adler, Max (1926). *Politische oder soziale Demokratie: ein Beitrag zur sozialistischen Erziehung*. Berlin: E. Laubesche Buchhandlung.

Agamben, Giorgio (2003). *State of Exception: Homo sacer II*. Chicago: University of Chicago Press.

Agamben, Giorgio (2014). *Leviathans Rätsel*, edited by Friedrich Hermanni. Tübingen: Mohr Siebeck.

Agamben, Giorgio (2015). Stasis. Civil War as a Political Paradigm. Stanford: Stanford University Press.

Agamben, Giorgio (2020). *A che punto siamo? L'epidimia come politica*. Rome: Quodlibet.

Akten Papen (1989). *Das Kabinett von Papen: 1. Juni bis 3. Dezember 1932*, edited by Karl-Heinz Minuth. München: Oldenbourg Wissenschaftsverlag.

Albrecht, Clemens et al. (1999). Die intellektuelle Gründung der Bundesrepublik: Eine Wirkungsgeschichte der Frankfurter Schule. Frankfurt am Main: Campus.

Allen, Christopher S. (2009). "›Empty Nets‹: Social Democracy and the ›Catch-all Party Thesis‹." *Party Politics* 15, no. 5 (2009): 635–653.

Ananiadis, Grigoris (1999). "Carl Schmitt and Max Adler: The Irreconciability of Politics and Democracy." In *The Challenge of Carl Schmitt*, edited by Chantal Mouffe, 118–127. London: Verso.

Anschel, Eugene (1990). *The World of a German Jew* (private printing).

Anter, Andreas (2016). *Max Weber und die Staatsrechtslehre*. Tübingen: Mohr Siebeck.

Appelius, Claudia (2003). *Die schönste Stadt der Welt: Deutsch-jüdische Flüchtlinge in New York*. Essen: Klartext Verlag.

Arato, Andrew (2011). "Multi-track Constitutionalism Beyond Schmitt." *Constellations* 18 (3): 324–352.

Arendt, Hannah (1951). *The Origins of Totalitarianism*. Orlando: Harcourt, Brace & Co.

Arendt, Hannah (1955). *Elemente und Ursprünge totaler Herrschaft*. München: Piper, 1986.

Arendt, Hannah (1963). *Eichmann in Jerusalem: A Report on the Banality of Evil*. New York: Viking Press.

Arndt, Hans Joachim (1985). "Politische Lageanalyse." In *Pipers Wörterbuch zur Politik*, Band 1: *Politikwissenschaft. Theorien – Methoden – Begriffe*, edited by Dieter Nohlen and Rainer Olaf Schulze, 754–757. München: Piper.

Auer, Stefan (2015). "Carl Schmitt in the Kremelin: The Ukraine Crisis and the Return of Geopolics." *International Affairs* 91 (2015): 953–968.

Aufricht, Ernst Josef (1969). *Erzähle, damit du dein Recht erweist*. München: Deutscher Taschenbuch-Verlag.

Badia, Gilbert (1998). "Deutsche und österreichische Emigranten in Frankreich 1933–1942." In *Zweimal verjagt: Die deutschsprachige Emigration und der Fluchtweg Frankreich-Lateinamerika 1933–1945*, edited by Anne Saint Sauveur-Henn, 16–33. Berlin: Metropol.

Badia, Gilbert (2002). "Frankreichs Haltung gegenüber den deutschsprachigen Emigranten zwischen 1933 und 1940." In *Fluchtziel Paris. Die deutschsprachige Emigration 1933–1940*, edited by Anne Saint Sauveur-Henn, 29–40. Berlin: Metropol.

Balakrishnan, Gopal (2000). *The Enemy: An Intellectual Portrait of Carl Schmitt*. London: Verso.

Balakrishnan, Gopal (2011). "The Geopolitics of Separation." *New Left Review* 68 (March/April 2011): 57–72.

Baldwin, Neil (2001). *Henry Ford and the Jews: The Mass Production of Hate*. New York: Public Affairs.

Ball, Hugo (1924). "Carl Schmitts Politische Theologie." *Hochland* 2 (1923/24): 265–286.

Balke, Friedrich (2016). "Carl Schmitt and Modernity." In *The Oxford Handbook of Carl Schmitt*, edited by Jens Meierhenrich, and Oliver Simons, 629–656. Oxford: Oxford University Press.

Baring, Arnulf (1961). "Hüter der Verfassung." *Deutsches Verwaltungsblatt* 76 (1961): 101–108.

Barth, Bernd-Rainer (2010). "Neumann, Hilde, geb. Rosenfeld." In *Wer war wer in der DDR?*, edited by Bernd-Rainer Barth, 945–946. Berlin: Fischer.

Bavaj, Riccardo (2005). *Von links gegen Weimar: Linkes antiparlamentarisches Denken in der Weimarer Republik*. Bonn: Dietz.

Bavaj, Riccardo (2007). "Otto Kirchheimers Parlamentarismuskritik in der Weimarer Republik: Ein Fall von ›Linksschmittianismus‹?" *Vierteljahreshefte für Zeitgeschichte* 55 (2007): 33–51.

Bendersky, Joseph W. (1983). *Carl Schmitt. Theorist for the Reich*. Princeton: Princeton University Press.

Bendersky, Joseph W. (1987). "Carl Schmitt at Nuremberg." *Telos* 72 (Summer 1987): 91–96.

Bendersky, Joseph W. (2004). "Introduction." In Carl Schmitt, *On the Three Types of Juristic Thought*, translated by Joseph W. Bendersky 1–39. Westport: Praeger Publishers.

Bendersky, Joseph W. (2007). "Carl Schmitt's path to Nuremberg: A Sixty-Year Reassessment." *Telos* 139 (Fall 2007): 6–34.

Bendersky, Joseph W. (2016). "Schmitt's Diaries." In *The Oxford Handbook of Carl Schmitt*, edited by Jens Meierhenrich, and Oliver Simons, 117–146. Oxford: Oxford University Press.

Benhabib, Seyla (2012). "Carl Schmitt's Critique of Kant: Sovereignty and International Law." *Political Theory* 40, no. 6 (2012): 688–713.

Berdjajew, Nikolai (1924). *Das neue Mittelalter: Betrachtungen über das Schicksal Russlands und Europas*. Darmstadt: Otto Reichl, 1927.

Bernaut, Elsa, and Nathan Leites (1954). *The Ritual of Liquidation. The Case of the Moscow Trials*. Glencoe: The Free Press.

Bernstein, Richard (2011). "The Aporias of Carl Schmitt." *Constellations* 18, no. 3 (2011): 403–430.

Benoist, Alain de (2007). *Carl Schmitt und der Krieg*. Berlin: Edition Junge Freiheit.

Berthold, Lutz (1999). *Carl Schmitt und der Staatnotstandsplan am Ende der Weimarer Republik*. Berlin: Duncker & Humblot.

Beveridge, William (1959). *A Defense of Free Learning*. Oxford: Oxford University Press.

Butha, Nehal (2005). "The Antinomies of Transformative Occupation" *European Journal for International Law* 721: 731–749.

Blanco, Jody and Ivonne del Valle (2014). Carl Schmitt and the Early Modern World. *Politica común* 5 (Special Issue).

Blasius, Dirk (2001). *Carl Schmitt: Preußischer Staatsrat in Hitlers Reich*. Göttingen: Vandenhoeck & Ruprecht.

Blasius, Dirk (2005). *Weimars Ende. Bürgerkrieg und Politik 1930–1933*. Göttingen: Vandenhoeck & Ruprecht.

Blasius, Dirk (2009). *Carl Schmitt und der 30. Januar 1933*. Frankfurt am Main: Peter Lang.

Blasius, Dirk (2021). "Carl Schmitt, April 1993: Der Reichsbegriff im Völkerrecht." *Der Staat* 60, no. 3 (2021): 455–472.

Blau, Joachim (1980). *Sozialdemokratische Staatslehre in der Weimarer Republik: Darstellung und Untersuchung der staatstheoretischen Konzeptionen von Hermann Heller, Ernst Fraenkel und Otto Kirchheimer*. Marburg: Verlag Arbeiterbewegung und Gesellschaftswissenschaft.

Blindow, Felix (1999). *Carl Schmitts Reichsordnung: Strategie für einen europäischen Großraum*. Berlin: Akademie-Verlag

Böckenförde, Ernst-Wolfgang (1988). "Aussprache zu dem Referat von Joseph H. Kaiser." In *Complexio Oppositorum: Über Carl Schmitt*, edited by Helmut Quaritsch. 331–338. Berlin: Duncker & Humblot.

Böckenförde, Ernst-Wolfgang (2011). "Biographisches Interview mit Dieter Gosewinkel". Ernst-Wolfgang Böckenförde, *Wissenschaft, Politik, Verfassungsgericht. Aufsätze und biographisches Interview mit Dieter Gosewinkel*. 307–485. Berlin: Suhrkamp.

Bolsinger, Eckard (2001). *The Autonomy of the Political. Carl Schmitt's and Lenin's Political Realism*. Westport: Greenwood Press.

Bookbinder, Paul (1991). "Carl Schmitt, Der Leviathan, and the Jews." *International Social Science Review* 66, no. 3 (1991): 99–109.

Borkenau, Franz (1932). "Zur Soziologie des Faschismus." *Archiv für Sozialwissenschaft und Sozialpolitik* 68, no. 5 (1932): 513–547.

Borkenau, Franz (1934). *Der Übergang vom feudalen zum bürgerlichen Weltbild*. Paris: Felix Alcan.

Boudon, Louis (1932). *Government by Judiciary*. New York: W. Godwin.

Bracher, Karl Dietrich (1955). *Die Auflösung der Weimarer Republik*. Villingen: Ring Verlag.

Brackmann, Karl-Heinz, Renate Birkenhauer (1988). *NS-Deutsch. "Selbstverständliche" Begriffe und Schlagwörter aus der Zeit des Nationalsozialismus*. Straelen/Niederrhein: Straelener Manuskripte Verlag.

Braunbuch (1933). *Braunbuch über Reichstagsbrand und Hitler-Terror*. Basel: Universum-Bücherei.

Braunfels, Wenzel M. and Johannes Grajcarek (2023). *Herrliche Einsamkeit. Carl Schmitts Wanderungen in und um Plettenberg*. Plettenberger Miniaturen 16. Berlin: Duncker & Humblot.

Brecht, Arnold (1933). *Preußen contra Reich vor dem Staatsgerichtshof.* Edited by Arnold Brecht. Berlin: J.H.W. Dietz Verlag.

Bredekamp, Horst (2012). "Die Brüder und Nachkommen des Leviathans." In *Die Bilder des Leviathans: Eine Deutungsgeschichte,* edited by Philip Manow, Friedbert Rüb, and Dagmar Simon. 45–76. Baden-Baden: Nomos

Bredekamp, Horst (2016). *Der Behemoth: Metamorphosen des Anti-Leviathan*. Berlin: Duncker & Humblot.

Breuer, Stefan (2012). *Carl Schmitt im Kontext: Intellektuellenpolitik in der Weimarer Republik*. Berlin: Akademie Verlag.

Bristler, Eduard (1938). *Die Völkerrechtslehre des Nationalsozialismus*. Zürich: Europa-Verlag. (Eduard Bristler is the pseudonym of John H. Herz).

Brown, Chris (2007). "From Humanized War to Humaniatarian Intervention: Carl Schmitt's Critique of the Just War Tradition." In *The International Thought of Carl Schmitt: Terror, War, and the Crises of the Global Order*, edited by Louiza Odysseos and Fabio Petito, 56–69. London: Routledge.

Brumlik, Micha (2020). *Antisemitismus*. Stuttgart: Reclam-Verlag.

Buchstein, Hubertus (1986). "Otto-Kirchheimer-Konferenz." *Internationale wissenschaftliche Korrespondenz zur Geschichte der deutschen Arbeiterbewegung* 22 (1986): 98–103.

Buchstein, Hubertus (1992). *Politikwissenschaft und Demokratie*. Baden-Baden: Nomos.

Buchstein, Hubertus (2000). *Öffentliche und geheime Stimmabgabe*. Baden-Baden: Nomos.

Buchstein, Hubertus (2002). "Democracy's Secret: Carl Schmitt and the Criticism of Secret Voting in Germany." *Finnish Yearbook of Political Thought* 6 (2002): 107–125.

Buchstein, Hubertus (2003). "A Heroic Reconciliation of Freedom and Power: On the Tension between Democratic and Social Theory in the Late Work of Franz L. Neumann." *Constellations* 10, no. 2 (2003): 228–246.

Buchstein, Hubertus (2010). "From Critical Theory to Political Science: A.R.L. Gurland's Project of Critical Political Science in Postwar Germany." *Redescriptions* 14, no. 1 (2010): 55–82.

Buchstein, Hubertus (2017a). "Einleitung." In Otto Kirchheimer, *Gesammelte Schriften, Band 1: Recht und Politik in der Weimarer Republik,* edited by Hubertus Buchstein, 15–126. Baden-Baden: Nomos.

Buchstein, Hubertus (2017b). "The Eschenburg Controversy in German Political Science." *Redescriptions* 20, no. 1 (2017): 85–115.

Buchstein, Hubertus (2018a). "A Forgotten Chapter of the Frankfurt School's Research on Antisemitism: Otto Kirchheimer's ‚The Policy of the Catholic Church Toward the Jews' (1943)." *Antisemitism Studies* 2, no. 2 (2018): 254–296.

Buchstein, Hubertus (2018b). "Einleitung." In Otto Kirchheimer, *Gesammelte Schriften, Band 2: Faschismus, Demokratie und Kapitalismus*, edited by Hubertus Buchstein and Henning Hochstein, 7–116. Baden-Baden, Nomos.

Buchstein, Hubertus (2019a). "Einleitung." In Otto Kirchheimer, *Gesammelte Schriften, Band 3: Kriminologische Schriften*, edited by Hubertus Buchstein and Lisa Klingsporn, 7–82. Baden-Baden, Nomos.

Buchstein, Hubertus (2019b). "Otto Kirchheimer." In *The Cambridge Habermas Lexicon*, edited by Amy Allen and Eduardo Mendieta, 593–594. Cambridge: Cambridge University Press.

Buchstein, Hubertus (2020a). "Otto Kirchheimer and the Frankfurt School: Failed Collaborations in the Search of a Critical Theory of Politics." *New German Critique* 47, no. 2 (2020): 81–106.

Buchstein, Hubertus (2020b). "Einleitung." In Otto Kirchheimer, *Gesammelte Schriften, Band 5: Politische Systeme im Nachkriegseuropa*, edited by Henning Hochstein and Frank Schale, 7–176. Baden-Baden: Nomos.

Buchstein, Hubertus (2020c). "Das Werk von Otto Kirchheimer und seine Gegenwartsbedeutung: Ein Gespräch mit Helmut König." *Zeitschrift für Politische Theorie* 10, no. 2 (2020): 255–270.

Buchstein, Hubertus (2021a). "The Godfather of Left-Schmittianism? Otto Kirchheimer and his Relationship to Carl Schmitt after 1945." *Redescriptions* 24, no. 1 (1921): 4–26.

Buchstein, Hubertus (2021b). "Entry Otto Kirchheimer." In *Encyclopedia of the Philosophy of Law and Social Philosophy*, edited by Mortimer Sellers and Stephan Kirste, 178–180. Dordrecht: Springer.

Buchstein, Hubertus (2021c). "Five Stages of Antisemitic Political Thought: Review Essay about Carl Schmitt, Gesammelte Schriften 1933–1936." *Redescriptions* 24, no. 2 (2021): 172–177.

Buchstein, Hubertus (2022). "Zwei Versionen des Anti-Imperialismus." In *Gesellschaft und Politik verstehen: Frank Nullmeier zum 65. Geburtstag*, edited by Martin Nonhoff et al., 111–126. Frankfurt am Main: Campus.

Buchstein, Hubertus (2023a). "Being A Master of Metaphors. Habermas and the New Structural Transformation of the Public Sphere." *Constellations* 30 (1): 48– 54.

Buchstein, Hubertus (2023b). "Interdisciplinary Materialism and the Task of Critical Institutionalism." *Constellations* 30 (4): 414–418.

Buchstein, Hubertus (2024). "From the Critique of Power to Critical Institutionalism." In *Critical Theory and the Political*, edited by Anastasia Marinopoulou. Manchester: Manchester University Press (forthcoming).

Buchstein, Hubertus, Dieter Emig, and Dieter Zimmermann (1991). "Zur Einführung." In A.R.L. Gurland, *Sozialdemokratische Kampfpositionen 1925–53*, edited by Hubertus Buchstein, Dieter Emig, and Dieter Zimmermann, 9–43. Baden-Baden: Nomos.

Buchstein, Hubertus and Dirk Jörke (2023). "Herrmann Hellers Wille zur Demokratie." In Hermann Heller, *Kämpfen für die Demokratie. Kleine politische Schriften*: 149–187. Hamburg: Europäische Verlagsanstalt.

Buchstein, Hubertus and Frank Schale (2014). "Introduction to Otto Kirchheimer: Elite-Consent-Control in the Western Political System (1964)." *Redescriptions* 17, no. 2 (2014):140–148.

Buchstein, Hubertus and Gerlinde Schloer (1983). *Politische Theorie in der Kritischen Theorie nach 1950: Franz L. Neumann*. Berlin: FU Berlin (= Occasional Paper 10, Fachbereich Politische Wissenschaft, Freie Universität Berlin).

Bumke, Christian (2002). "Eigentum – Paradigma für ein dem Gemeinwohl verpflichtetes Rechtsinstitut." In *Gemeinwohl und Gemeinsinn im Recht*, edited by Herfried Münkler and Karsten Fischer, 179–230. Berlin: Akademie Verlag.

Burkhardt, Kai (2013). *Carl Schmitt und die Öffentlichkeit: Briefwechsel mit Journalisten, Publizisten und Verlegern aus den Jahren 1923 bis 1983*, edited by Kai Burkhardt. Berlin: Duncker & Humblot.

Burnham, James (1941). *The Managerial Revolution or What is Happening in the World Now?* New York: Day.

Caldwell, Peter C. (2005). "Controversies over Carl Schmitt." *The Journal of Modern History* 77 (June 2005): 357–387.

Christi, Renato (1998) *Carl Schmitt and Authoritarian Liberalism*. Cardiff: University of Wales.

Christi, Renato (2011). "Schmitt on Constituent Power and the Monarchical Principle." *Constellations* 18 (3): 352–364.

Clausewitz, Carl von (1832). *Vom Kriege*, 18th edition. Bonn: Dümmlers Verlag, 1973.

c.m. (1947). "Staatsrechtler Carl Schmitt unter Anklage." *Der Aufbau* (New York), May 2, 1947.

Cohen, Jean (2004). "Whose Sovereingnty? Empire versus International Law." *Ethics & International Affairs* 18, no. 3 (2004): 1–24.

Cohen, Jean (2007). "The Role of International Law in Post-Conflict Constitution Making. Toward a *Jus Post Bellum* for 'Interim Occupations'" *New York Law School Law Review* 51: 497–532.

Dahrendorf, Ralf (1989). "Zeitgenosse Habermas." *Merkur* 43, no. 484 (1989): 478–487.

Denkschrift (1935). *Entwicklungstendenzen im deutschen Strafvollzug: Denkschrift des Parteivorstandes der Sozialdemokratischen Partei Deutschlands*. Prag.

Deutsche Briefe 1934–1938: Ein Blatt der katholischen Emigration, Two Volumes (reprint), edited by Heinz Hürten. Mainz: Matthias-Grünewald-Verlag, 1969.

Deutsche Liga für Menschenrechte (1927). *Acht Jahre politische Justiz*. Berlin: Hensel.

Dewey, John (1927). *The Public and its Problems*. Athens: Ohio University Press.

Doerr, Karin (2002). "Nazi-Deutsch: An Ideological Language of Exclusion, Domination, and Annihilation." In Robert Michael and Karin Doerr, *Nazi-Deutsch/Nazi German: an English lexicon of the language of the Third Reich*, 27–46. Westport, CT: Greenwood Press.

Dubiel, Helmut and Alfons Söllner (1981). "Die Nationalsozialismusforschung des Instituts für Sozialforschung." In *Wirtschaft, Recht und Staat im Nationalsozialismus. Analysen des Instituts für Sozialforschung*, edited by Helmut Dubiel and Alfons Söllner, 7–32. Frankfurt am Main: Suhrkamp.

Dunkhase, Jan Eike (2019). "Asymmetrische Korrespondenz." In Carl Schmitt and Reinhart Koselleck, *Der Briefwechsel 1953–1983*, edited by Jan Eike Dunkhase, 409–429. Berlin: Suhrkamp.

Egener, David (2013). "Zur Stellung des Antisemitismus im Denken Carl Schmitts." *Vierteljahreshefte für Zeitgeschichte* 61 (3), 345–361.

Ehmke, Horst (1959). "Rezension: Peter Schneider: Ausnahmezustand und Norm." *Zeitschrift für die gesamte Staatswissenschaft* 115 (1959): 187–192.

Eisfeld, Rainer (2013). *Ausgebürgert und doch angebräunt: Deutsche Politikwissenschaft 1920–1945*. Baden-Baden: Nomos.

Elden, Stuart (2010). „Reading Schmitt's geopolitically: Nomos, Territory, and Grossraum." *Radical Philosophy* 161, 18–26.

Engels, Friedrich and Karl Kautsky (1887). "Juristen-Sozialismus." In *Marx-Engels Werke (MEW), Band 21*, 491–509. Berlin: Dietz, 1979.

Erd, Rainer (1985). *Reform und Resignation: Gespräche über Franz L. Neumann*, edited by Rainer Erd. Frankfurt am Main: Suhrkamp.

Eschenburg, Theodor (1956). *Herrschaft der Verbände?* Stuttgart: Deutsche Verlagsanstalt.

Fabian, Ruth (1981). "Zur Integration der deutschen Emigranten in Frankreich 1933–1945." In *Leben im Exil: Probleme der Integration deutscher Flüchtlinge im Ausland 1933–1945*, edited by Wolfgang Frühwald and Wolfgang Schieder, 200–206. Hamburg: Hoffmann und Campe.

Fabian, Ruth and Corinna Coulmas (1982). *Die deutsche Emigration in Frankreich nach 1933*. München: K. G. Saur.

Fassbender, Bodo (2002). "Stories of War and Peace: On Writing the History of International Law in the Third Reich and After." *European Journal of International Law* 13, no. 2 (April 2002): 479–512.

Feidel-Mertz, Hildegard (1972). *Zur Ideologie der Arbeiterbildung*. Frankfurt am Main: Europäische Verlagsanstalt.

Feldman, Egal (2001). *Catholics and Jews in Twentieth-Century America*. Chicago: University of Illinois Press.

Fest, Joachim C. (1973). *Hitler: Eine Biographie*. Berlin: Propyläen Verlag.

Fijalkowski, Jürgen (1958). Die Wendung zum Führerstaat: Ideologische Komponenten in der politischen Philosophie Carl Schmitts. Köln and Opladen: Westdeutscher Verlag.

Finchelstein, Federico (2022). *Fascist Mythologies. The History and Politics of Unreason in Borges, Freud, and Schmitt*. New York: Columbia University Press.

Fischer, Dana Davis (1974). *National Socialist Germany and International Law*. Ann Arbor: University Michigan Microfilms.

Flechtheim, Ossip K. (1936). "Die Völkerrechtstheorie von Korovin und Paschukanis." In Ossip K. Flechtheim, *Von Hegel zu Kelsen: Rechtstheoretische Aufsätze*, 55–80. Berlin: Duncker & Humblot, 1963.

Flechtheim, Ossip K. (1938). "Totalitäre Völkerrechtswissenschaft." In Ossip K. Flechtheim, *Von Hegel zu Kelsen: Rechtstheoretische Aufsätze*, 81–88. Berlin: Duncker & Humblot, 1963.

Forsthoff, Ernst (1941). "Besprechung von Schmitt Der Leviathan in der Staatslehre von Thomas Hobbes." *Zeitschrift für deutsche Kulturphilosophie, Neue Folge* 7 (1941): 206–214.

Forsthoff, Ernst (1950). "Zur verfassungsrechtlichen Stellung und inneren Ordnung der Parteien." In *Die politischen Parteien im Verfassungsrecht*, edited by Ernst Forsthoff and Karl Loewenstein, 5–24. Tübingen: Mohr.

Fraenkel, Ernst (1927). "Zur Soziologie der Klassenjustiz." In Ernst Fraenkel, *Gesammelte Schriften, Band 1: Recht und Politik in der Weimarer Republik*, edited by Hubertus Buchstein, 177–212. Baden-Baden: Nomos, 1999.

Fraenkel, Ernst (1930). "Antifaschistische Aufklärungsarbeit." In Ernst Fraenkel, *Gesammelte Schriften, Band 1: Recht und Politik in der Weimarer Republik*, edited by Hubertus Buchstein, 410–417. Baden-Baden: Nomos, 1999.

Fraenkel, Ernst (1932a). "Chronik Die Justiz – Januar 1932." In Ernst Fraenkel, *Gesammelte Schriften, Band 1: Recht und Politik in der Weimarer Republik*, edited by Hubertus Buchstein, 558–563. Baden-Baden: Nomos, 1999.

Fraenkel, Ernst (1932b). "Verfassungsreform und Sozialdemokratie." In Ernst Fraenkel, *Gesammelte Schriften, Band 1: Recht und Politik in der Weimarer Republik*, edited by Hubertus Buchstein, 516–530. Baden-Baden: Nomos, 1999.

Fraenkel, Ernst (1932c). "Um die Verfassung." In Ernst Fraenkel, *Gesammelte Schriften, Band 1: Recht und Politik in der Weimarer Republik*, edited by Hubertus Buchstein, 496–509. Baden-Baden: Nomos, 1999.

Fraenkel, Ernst (1941). *The Dual State: A Contribution to the Theory of Dictatorship.* New edition. Oxford: Oxford University Press, 2017.

Fraenkel, Ernst (1957). "Otto Suhr zum Gedächtnis." In Ernst Fraenkel, *Gesammelte Schriften, Band 6: Internationale Politik und Völkerrecht, Politikwissenschaft und Hochschulpolitik*, edited by Hubertus Buchstein, and Gert Lutterbeck, 374–393. Baden-Baden: Nomos, 2011.

Fraenkel, Ernst (1964). "Martial Law und Staatsnotstand in England und den USA." In Ernst Fraenkel, *Gesammelte Schriften, Band* 4: Amerikastudien. edited by Hubertus Buchstein, and Rainer Kühn, 893–920. Baden-Baden: Nomos, 2000.

Fraenkel, Ernst (1999). *Ernst Fraenkel: Gesammelte Schriften, Band 1 – Recht und Politik in der Weimarer Republik*, edited by Hubertus Buchstein. Baden-Baden: Nomos.

Frank, Tibor (2009). *Double Exile: Migrations of Jewish-Hungarian Professionals through Germany to the United States*, 1919–1949. Oxford. Oxford University Press.

Franke, Julia (2000). *Paris – neue Heimat? Jüdische Emigration aus Deutschland 1933–1939.* Berlin: Duncker & Humblot.

Friedlander, Judith (2019). *A Light in Dark Times: The New School for Social Research and its University in Exile*. New York: Columbia University Press.

Friedrich, Carl Joachim (1931). "Die Entwicklung des amerikanischen öffentlichen Rechts nach dem Kriege." *Jahrbuch des öffentlichen Rechts* 20 (1932): 393–429.

Friesenhahn, Ernst (1950). *Staatsrechtslehrer und Verfassung*. Krefeld: Scherpe.

Friesenhahn, Ernst (1967). "Otto Kirchheimer." In: *Verhandlungen des 46. Deutschen Juristentages 1966 in Essen, Band II – Sitzungsberichte*. C66–C68. München: H. Beck.

Frings, Manfred S. (1997). *The Mind of Max Scheler*. Milwaukee: Marquette University Press.

Fritsch, Theodor (1944). *Handbuch der Judenfrage.* 49th edition. Leipzig: Hammer Verlag.

Fuchshuber, Thorsten (2019). *Rackets: Kritische Theorie der Bandenherrschaft.* Freiburg and Wien: ça ira-Verlag.

Fuldt, Werner (1979). *Walter Benjamin: Biographie*. München: Hanser.

Gadamer, Hans Georg (1965). *Truth and Method*. London: Bloomsbury, 2013.

Gerlach, Hellmut von (1932). "Schleicher und sein Stahlhelm." *Die Weltbühne* 28, no. 2 (September 1932): 342–345.

Giesler, Gerd (2016). *Das Fernseh-Urteil des Bundesverfassungsgerichts 1961: Eine Leserbrief-Kampagne aus dem Hause Carl Schmitt in der Deutsche Zeitung*, edited by Gerd Giesler. Berlin: Duncker & Humblot.

Gowan, Peter (2004). "American Lebensraum." *New Left Review* 30, no. 2 (Nov/Dec 2004): 155–164.

Grabowsky, Adolf (1928). "Der Primat der Außenpolitik: Ein Nachwort." *Zeitschrift für Politik* 17 (1928): 527–542.

Grabowsky, Adolf (1932). *Politik*. Wien: Industrieverlag Spaeth & Linde.

Graf, Sinja (2021). "Carl Schmitt reads Hannah Arendt's Eichmann in Jerusalem: Archival Perspectives on Convergences and Divergences" *American Journal of Political Science* 66 (4), 918–931.

Gramm, Hanns (1956). *The Oberlaender Trust 1931–1953*. Philadelphia: Carl Schurz Memorial Foundation.

Grau, Richard (1932). "Die Diktaturgewalt des Reichspräsidenten." In *Handbuch des Staatrechts II*, edited by Gerhard Anschütz and Richard Thoma, 274–295. Tübingen: Mohr Siebeck, 1998.

Gross, Raphael (1994). "Politische Polykratie 1936: Die legendenumwobene SD-Akte Carl Schmitt." *Tel Aviver Jahrbuch für deutsche Geschichte* 23 (1994): 115–143.

Gross, Raphael (2000). *Carl Schmitt und die Juden: Eine deutsche Rechtslehre*. Frankfurt am Main: Suhrkamp.

Gross, Raphael (2016). "The 'True Enemy': Antisemitism in Carl Schmitt's Life and Work." In *The Oxford Handbook of Carl Schmitt*, edited by Jens Meierhenrich and Oliver Simons, 96–116. Oxford: Oxford University Press.

Gruchmann, Lothar (1962). *Nationalsozialistische Großraumordnung: Die Konstruktion einer deutschen Monroe-Doktrin*. Stuttgart: Deutsche Verlags-Anstalt.

Gruchmann, Lothar (2001). *Justiz im Dritten Reich 1933–1940*. 3rd edition. München: Oldenbourg Wissenschaftsverlag.

Günther, Frieder (2004). *Denken vom Staat her: Die bundesdeutsche Staatsrechtslehre zwischen Dezision und Integration 1949–1970*. München: Oldenbourg Wissenschaftsverlag.

Gurland, Arkadij (1930a). *Marxismus und Diktatur*. Leipzig: Leipziger Buchdruckerei.

Gurland, Arkadij (1930b). "Book review: Otto Kirchheimer ›Weimar ... und was dann?‹." *Die Bücherwarte* 5, no.9 (1930): 135–137.

Gurland, Arkadij (1931). *Das Heute der proletarischen Revolution: Hemmnisse und Wandlungen im Klassenkampf*. Berlin: Laubsche Verlagsbuchhandlung.

Gurland, Arkadij, Otto Kirchheimer, and Franz L. Neumann (1943). *The Fate of Small Business in Nazi Germany*. Reprint. Honolulu: University of the Pacific, 2003.

Gurian, Waldemar (1929). "Der Faschismus." *Heiliges Feuer* 16 (1928/29): 507–518.

Gusy, Christoph (1997). *Die Weimarer Reichsverfassung*. Tübingen: Mohr Siebeck.

Habermas, Jürgen (1958). "Zum Begriff der politischen Beteiligung." In Jürgen Habermas, *Kultur und Kritik*, 11–57. Frankfurt am Main: Suhrkamp, 1973.

Habermas, Jürgen (1962). *The Structural Transformation of the Public Sphere*. Cambridge: MIT Press, 1992.

Habermas, Jürgen (1980). "Die Frankfurter Schule in New York." In Jürgen Habermas, *Philosophisch-politische Profile*, 411–425. Frankfurt am Main: Suhrkamp, 1984.

Habermas, Jürgen (1981). *The Theory of Communicative Action: Volume 2*. Boston: Beacon Press.

Habermas, Jürgen (1986). "Bemerkungen zur Entwicklungsgeschichte des Horkheimerschen Werkes." In *Max Horkheimer heute: Werk und Wirkung*, edited by Alfred Schmidt and Norbert Altwicker, 163–179. Frankfurt am Main: Fischer Taschenbuchverlag.

Habermas, Jürgen (1987). "Die Schrecken der Autonomie: Carl Schmitt auf Englisch." In Jürgen Habermas, *Eine Art Schadensabwicklung*, 101–115. Frankfurt am Main: Suhrkamp.

Habermas, Jürgen (1992). *Autonomy and Solidarity: Interviews with Jürgen Habermas*. London: Verso.

Habermas, Jürgen (1996). *Between Facts and Norms: Contributions to a Discourse Theory of Law and Democracy.* Cambridge: MIT Press.

Habermas, Jürgen (2001). *The Inclusion of the Other*. Cambridge: MIT Press.

Habermas, Jürgen (2014). "Plea for a Constitutionalization of International Law." *Philosophy & Social Criticism* 4 (1), 5–12.

Habermas, Jürgen (2022a). "Der neue Historikerstreit." In Saul Friedländer et al., *Ein Verbrechen ohne Namen: Anmerkungen zum neuen Streit über den Holocaust*, 7–13. München: Beck.

Habermas, Jürgen (2022b). *Der neue Strukturwandel der Öffentlichkeit und die deliberative Politik*. Berlin: Suhrkamp.

Hammer, Espen (2005). *Adorno and the Political*. London: Routledge.

Hardt, Michael and Antonio Negri (2004). *Multitude: War and Democracy in the Age of Empire.* New York: Penguin Press.

Hauriou, Maurice (1923). *Précis de droit constitutionnel*. Paris: Libr. de la Société du Recueil Sirey.

Heckel, Johannes (1932). "Diktatur, Notverordnungsrecht, Verfassungsnotstand." *Archiv des Öffentlichen Rechts.* N.F. 22, no. 3 (1932): 257–338.

Heil, Susanne (1996). *Gefährliche Beziehungen: Walter Benjamin und Carl Schmitt*, edited by Susanne Heil. Stuttgart: Metzler.

Heimes, Claus (2004). *Antipositivistische Staatslehre: Eric Voegelin und Carl Schmitt zwischen Wissenschaft und Ideologie* (Occasional Paper XLII des Voegelin Archivs). München: Eric-Voegelin-Archiv, Ludwig-Maximilians-Universität München.

Heins, Volker (2006). "Zur Frage der Souveränität." In *Schlüsseltexte der Kritischen Theorie*, edited by Axel Honneth, 272–274. Wiesbaden: VS Verlag für Sozialwisssenschaften.

Heller, Hermann (1924). "Grundrechte und Grundpflichten der Deutschen." In Hermann Heller, *Gesammelte Schriften, Band 2: Staatslehre als politische Wissenschaft*, edited by Christoph Müller, 281–318. Tübingen: Mohr, 1992.

Heller, Hermann (1926). "Die politischen Ideenkreise der Gegenwart." In Hermann Heller, *Gesammelte Schriften, Band 1: Orientierung und Entscheidung*, edited by Christoph Müller, 267–413. Tübingen: Mohr, 1992.

Heller, Hermann (1927). *Sovereignty: A Contribution to the Theory of Public and International Law*. Oxford: Oxford University Press, 2019.

Heller, Hermann (1928). "Politische Demokratie und soziale Homogenität." In Hermann Heller, *Gesammelte Schriften, Band 2: Staatslehre als politische Wissenschaft*, edited by Christoph Müller, 421–433. Tübingen: Mohr, 1992.

Heller, Hermann (1929). "Europa und der Faschismus." In Hermann Heller, *Gesammelte Schriften, Band 2: Staatslehre als politische Wissenschaft*, edited by Christoph Müller, 463–609. Tübingen: Mohr, 1992.

Heller, Hermann (1930). "Freiheit und Form in der Reichsverfassung." In Hermann Heller, *Gesammelte Schriften, Band 2: Staatslehre als politische Wissenschaft*, edited by Christoph Müller, 371–377. Tübingen: Mohr, 1992.

Heller, Hermann (1932). "Ziele und Grenzen einer deutschen Verfassungsreform." In Hermann Heller, *Gesammelte Schriften, Band 2: Staatslehre als politische Wissenschaft*, edited by Christoph Müller, 411–417. Tübingen: Mohr, 1992.

Heller, Hermann (1933a). "Autoritärer Liberalismus." In Hermann Heller, *Gesammelte Schriften, Band 2: Staatslehre als politische Wissenschaft*, edited by Christoph Müller, 643–653. Tübingen: Mohr, 1992.

Heller, Hermann (1933b). "Political Science." In: Hermann Heller, *Gesammelte Schriften, Band 3: Recht, Staat, Macht*, edited by Christoph Müller, 45–75. Tübingen: Mohr, 1992.

Heller, Jonas (2019). "Der Zwang zur Tragödie. Zur Selbstdurchbrechung des Politischen bei Carl Schmitt." *Deutsche Zeitschrift für Philosophie* 67, no. 6 (2019): 952–972.

Henkel, Michael (2005). *Positivismuskritik und autoritärer Staat*. Occasional Paper XXXVI Eric-Voegelin-Archiv. München: Ludwig-Maximilians-Universität.

Henkel, Michael (2011). *Hermann Hellers Theorie der Politik und des Staates: die Geburt der Politikwissenschaft aus dem Geiste der Soziologie*. Tübingen: Mohr Siebeck.

Herberg-Rothe, Andreas (2004). "Hannah Arendt und Carl Schmitt: Vermittlung von Freund und Feind." *Der Staat* (43), 35–55.

Herbert, Ulrich (1996). *Best: Biographische Studien über Radikalismus, Weltanschauung und Vernunft 1903–1989*. Bonn: Diez

Herf, Jeffrey (2008). *The Jewish Enemy: Nazi Propaganda During World War II and the Holocaust*. Cambridge: Harvard University Press.

Hertweck, Frank and Dimitrios Kisoudis (2010). *„Solange das Imperium da ist" – Carl Schmitt im Gespräch 1971*, edited by Frank Hertweck and Dimitrios Kisoudis. Berlin: Duncker & Humblot.

Herz, John H. (1948). "The Fiasco of Denazification in Germany." *Political Science Quarterly* 63, no. 4 (1948): 231–242.

Herz, John H. (1989). "Otto Kirchheimer – Leben und Werk." In *Verfassungsstaat, Souveränität, Pluralismus: Otto Kirchheimer zum Gedächtnis*, edited by Wolfgang Luthardt and Alfons Söllner, 11–23. Opladen: VS Verlag für Sozialwissenschaften.

Herz, John H. (1984). *Vom Überleben: Wie ein Weltbild entstand. Autobiographie*. Düsseldorf: Droste.

Herz, John H. (1992). "Looking at Carl Schmitt from the Vantage Point of the 1990s." *Interpretations. A Journal of Political Philosophy* 10, no. 3 (1992): 307–314.

Herz, John H. and Erich Hula (1969). "Otto Kirchheimer – An Introduction to his Life and Work." In *Politics, Law, and Social Change. Selected Essays by Otto Kirchheimer*, edited by Fredric S. Burin and Kurt L. Shell, ix–xxxviii. New York and London: Columbia University Press.

Heydecker, Joe J. and Johannes Leeb (2020). *Der Nürnberger Prozess. Neuausgabe*. Köln: Kiepenheuer & Witsch.

Hilberg, Raul (1990). *Die Vernichtung der europäischen Juden. 3 Bände.* 2nd edition. Frankfurt am Main: Fischer Taschenbuch-Verlag.

Hilberg, Raul (2002). "Die bleibende Bedeutung des *Behemoth.*" In *Kritische Theorie der Politik: Franz L. Neumann*, edited by Mattias Iser and David Strecker, 83–94. Baden-Baden: Nomos.

Hirschfeld, Gerhard (1985). "Die Emigration deutscher Wissenschaftler nach Großbritannien 1933–1945." In *Großbritannien als Gast- und Exilland für Deutsche im 19. und 20. Jahrhundert*, edited by Gottfried Niedhart, 117–140. Bochum: Studienverlag Dr. N. Brockmeyer.

Hitler, Adolf (1925). *Hitler, Mein Kampf. Eine kritische Edition*, edited by Christian Hartmann, Thomas Vordermayer, Othmar Plöckinger, and Roman Töppel. Volume 1. München and Berlin: Institut für Zeitgeschichte, 2017.

Hitschler, Daniel (2011). *Zwischen Liberalismus und Existentialismus. Carl Schmitt im englischsprachigen Schrifttum*. Baden-Baden: Nomos.

Hochstein, Henning and Frank Schale (2021). "Einleitung." In Otto Kirchheimer, *Gesammelte Schriften, Band 6: Politische Analysen für das OSS und Department of State*, edited by Henning Hochstein und Frank Schale, 7–153. Baden-Baden: Nomos.

Hofmann, Hasso (1988). "Die deutsche Rechtswissenschaft im Kampf gegen den jüdischen Geist." In *Geschichte und Kultur des Judentums*, edited by Karlheinz Müller and Klaus Wittstadt, 223–240. Würzburg: Schönigh.

Hofmann, Hasso (1995). *Legitimität gegen Legalität: Der Weg der politischen Philosophie Carl Schmitts.* 3rd edition. Berlin: Duncker & Humblot.

Höfele, Andreas (2014). *Der Einbruch der Zeit: Carl Schmitt liest Hamlet*. München: Beck.

Hohendahl, Peter Uwe (2011). "Revolutionary War and Absolute Enmity: Rereading Schmitt's Theory of the Partisan." *Constellations* 18, no. 4 (2011): 529–544.

Holmes, Stephen (1993). *The Anatomy of Antiliberalism*. Boston: Harvard University Press.

Honigsheim, Paul (1960). "Reminiscences of the Durkheim School." In *Emile Durkheim, 1858–1917: A Collection of Essays*, edited by Kurt H. Wolff, 309–314. Columbus: Ohio State University Press.

Honneth, Axel (1987). "Critical Theory." In *Social Theory Today*, edited by Anthony Giddens and Jonathan Turner, 347–82. Cambridge: Polity.

Hooker, William (2009). *Carl Schmitt's International Thought: Order and Orientation*. Cambridge: Cambridge University Press.

Horkheimer, Max (1930). "Anfänge der bürgerlichen Geschichtsphilosophie" In Max Horkheimer, *Gesammelte Schriften, Band 2: Philosophische Frühschriften 1922–1932*, edited by Gunzelin Schmid Noerr, 177–268. Frankfurt am Main: S. Fischer, 1987.

Horkheimer, Max (1931). "Die gegenwärtige Lage der Sozialphilosophie und die Aufgabe des Instituts für Sozialforschung." In Max Horkheimer, *Gesammelte Schriften, Band 3: Schriften 1931–1936*, edited by Alfred Schmidt, 20–35. Frankfurt am Main: S. Fischer, 1988.

Horkheimer, Max (1935). "Studien über Autorität und Familie." In Max Horkheimer, *Gesammelte Schriften, Band 3: Schriften 1931–1936*, edited by Alfred Schmidt, 329–335. Frankfurt am Main: S. Fischer, 1988

Horkheimer, Max (1937a). "Traditional and Critical Theory." In Horkheimer, Max, *Critical Theory: Selected Essays*, translated by Matthew J. O'Connell, 188–243. New York: Continuum, 2002.

Horkheimer, Max (1937b). "Postscript. Philosophy and Critical Theory." In *Critical Theory: Selected Essays*, translated by Matthew J. O'Connell, 244–252. New York: Continuum, 2002.

Horkheimer, Max (1938a). "Memorandum zu Idee, Aktivität und Programm des Instituts für Sozialforschung." In Max Horkheimer, *Gesammelte Schriften, Band 12: Nachgelassene Schriften 1931–1949*, edited by Gunzelin Schmid Noerr, 131–164. Frankfurt am Main: S. Fischer, 1985.

Horkheimer, Max (1939a). "The Jews and Europe." In *Critical Theory and Society: A Reader*, edited by Stephen Eric Bronner and Douglas MacKay Kellner, 77–94. New York: Routledge, 1989.

Horkheimer, Max (1939b). "Die Rackets und der Geist." In Max Horkheimer, *Gesammelte Schriften, Band 12: Nachgelassene Schriften 1931–1949*, edited by Gunzelin Schmid Noerr, 287–292. Frankfurt am Main: S. Fischer, 1985.

Horkheimer, Max (1940). "The Authoritarian State." In *The Essential Frankfurt School Reader*, edited by Andrew Arato and Eike Gebhardt, 95–117. New York: Continuum, 1993.

Horkheimer, Max (1941). "The End of Reason." *Studies in Philosophy and Social Science* 9 (1941): 366–388.

Horkheimer, Max (1943a). "Zur Soziologie der Klassenverhältnisse." In Max Horkheimer, *Gesammelte Schriften, Band 12: Nachgelassene Schriften 1931–1949*, edited by Gunzelin Schmid Noerr, 75–104. Frankfurt am Main: S. Fischer, 1985.

Horkheimer, Max (1943b). "Plan des Forschungsprojekts über Antisemitismus." In Max Horkheimer, *Gesammelte Schriften, Band 12: Nachgelassene Schriften 1931–1949*, edited by Gunzelin Schmid Noerr, 167–171. Frankfurt am Main: S. Fischer, 1985.

Horkheimer, Max (1995). *Gesammelte Schriften, Band 16: Briefwechsel 1937–1940*, edited by Gunzelin Schmid Noerr. Frankfurt am Main: S. Fischer.

Horkheimer, Max (1996). *Gesammelte Schriften, Band 17: Briefwechsel 1941–1948*, edited by Gunzelin Schmid Noerr. Frankfurt am Main: S. Fischer.

Horkheimer, Max and Theodor W. Adorno (1944). *Dialectic of Enlightenment*, translated by John Cumming. New York: Continuum, 1972.

Huber, Ernst Rudolf (1931a). *Das Deutsche Reich als Wirtschaftsstaat*. Tübingen: Mohr.

Huber, Ernst Rudolf (1931b). "Eine Auseinandersetzung mit einem Theoretiker des Sozialismus über Sozialpolitik." *Der Ring* 4, no. 9 (1931): 161–163 (Pseudonym Friedrich Schreyer).

Huber, Ernst Rudolf (1932). "Die Stellung der Geschäftsregierung in den deutschen Ländern." *Deutsche Juristen-Zeitung* 37 (1932): Spalte 194–199.

Huber, Ernst Rudolf (1941). "Positionen und Begriffe. Eine Auseinandersetzung mit Carl Schmitt." *Zeitschrift für die gesamte Staatswissenschaft* 101, no. 1 (1941): 1–44.

Huber, Ernst Rudolf (1980). "Verfassungswirklichkeit und Verfassungswert im Staatsdenken der Weimarer Zeit." In: *Arbeiten zur Rechtsgeschichte: Festschrift für Gustaf Klemens Schmelzeisen*, edited by Hans-Wolf Thümmel, 126–141. Stuttgart: Klett-Cotta.

Huber, Ernst Rudolf (1984). *Deutsche Verfassungsgeschichte seit 1789, Band VII: Ausbau, Schutz und Untergang der Weimarer Republik*. Stuttgart: Kohlhammer.

Huber, Ernst Rudolf (1988). "Carl Schmitt in der Reichskrise der Weimarer Endzeit." In *Complexio Oppositorum: Über Carl Schmitt*, edited by Helmut Quaritsch, 33–70. Berlin: Duncker & Humblot.

Huber, Ernst Rudolf (1993). *Deutsche Verfassungsgeschichte seit 1789, Band VI: Die Weimarer Reichsverfassung. Revidierter Nachdruck der 1. Auflage*. Stuttgart: Kohlhammer.

Hürten, Heinz (1972). *Waldemar Gurian: Ein Zeuge der Krise unserer Welt in der ersten Hälfte des 20. Jahrhunderts*. Mainz: Matthias-Grünewald-Verlag.

Intelmann, Peter (1996). *Franz L. Neumann: Chancen und Dilemma des politischen Reformismus*. Baden-Baden: Nomos.

ISR (1941). "Institute of Social Research: Research Project on Antisemitism." *Studies in Philosophy and Social Science* 9 (1941): 124–143.

ISR (1944). "Institute of Social Research: Project on Antisemitism: Report to the American Jewish Committee on the First Year of the Project Ending March 15, 1944." (Universität Frankfurt, Archivzentrum, Nachlaß Max Horkheimer, IX, 121a).

Jäckel, Eberhard and Axel Kuhn (1980). *Hitler. Sämtliche Aufzeichnungen 1905–1924.*, edited by Eberhard Jäckel and Axel Kuhn. Stuttgart: Deutsche Verlagsanstalt.

Jäger, Lorenz (2017). *Walter Benjamin: Das Leben eines Unvollendeten*. Berlin: Rowohlt.

Jänicke, Martin (1969). "Die ‚Abgründige Wissenschaft vom Leviathan': Zur Hobbes-Deutung Carl Schmitts im Dritten Reich." *Zeitschrift für Politik* 16, no. 3 (1969): 401–415.

Jansen, Peter-Erwin (2000). *Das Utopische soll Funken schlagen ... Zum hundertsten Geburtstag von Leo Löwenthal*, edited by Peter-Erwin Jansen. Frankfurt am Main: Klostermann.

Jay, Martin (1973). *The Dialectical Imagination: A History of the Frankfurt School and the Institute of Social Research 1923–1950*. Berkeley: University of California Press.

Jay, Martin (1987) "Reconciling the Irreconcilable? A Rejoinder to Kennedy." *Telos* 71 (Spring 1987): 65–80.

Jeffreys-Jones, Rhodri (1989). *The CIA and American Democracy*. New Haven: Yale University Press.

Jellinek, Georg (1908). *Die socialethische Bedeutung von Recht: Unrecht und Strafe*. Second edition. Berlin: O. Häring.

Jestaedt, Matthias (2018). "Carl Schmitt." In *Staatrechtslehrer des 20. Jahrhunderts*, edited by Peter Häberle, Michael Kilian, and Heinrich Wolff. 2nd edition, 391–414. Berlin: de Gruyter.

Jünger, Ernst (1951). *Der Waldgang*. Stuttgart: Klett-Cotta 2021.

Kahn-Freund, Otto (1932). "Der Funktionswandel des Arbeitsrechts." In *Arbeitsrecht und Politik: Quellentexte 1918–1933*, edited by Thilo Ramm, 211–246. Neuwied: Luchterhand, 1966.

Kahn-Freund, Otto (1981). "Autobiographische Erinnerungen an die Weimarer Republik: Ein Gespräch mit Wolfgang Luthardt." *Kritische Justiz* 13, no. 2 (1981):183–200.

Kalyvas, Andreas (1999a). "Carl Schmitt and the Three Moments of Democracy." *Cardozo Law Review* 21, no 5 (1999): 1525–1565.

Kalyvas, Andreas (1999b). "Who's afraid of Carl Schmitt?" *Philosophy & Social Criticism* 25, no. 5 (1999): 87–125.

Kalyvas, Andreas (2005). "Popular Sovereignty, Democracy and Constituent Power." *Constellations* 12 (2), 223–44.

Kalyvas, Andreas (2009). *Democracy and the Politics of the Extraordinary: Max Weber, Carl Schmitt and Hannah Arendt.* Cambridge: Cambridge University Press.

Kalyvas, Andreas (2018). "Carl Schmitt's postcolonial Imagination." *Constellations* 25 (1), 35–53.

Kambas, Chryssoula (1983). *Walter Benjamin im Exil: Zum Verhältnis von Literaturpolitik und Ästhetik.* Tübingen: Niemeyer.

Katz, Jacob (1989). *Vom Vorurteil bis zur Vernichtung: Der Antisemitismus 1700–1933.* München: Beck.

Katz, Barry M. (1989). *Foreign Intelligence. Research and Analysis in the Office of Strategic Services 1942–1945.* Cambridge: Harvard University Press.

Kelly, Duncan (2002). „Rethinking Franz Neumann's Route to *Behemoth*". *History of Political Thought* 23 (3), 458–496

Kelsen, Hans (1928). *Das Problem der Souveränität und die Theorie des Völkerrechts.* 2nd edition. Tübingen: Mohr.

Kemmerer, Alexandra (2008). "Kontroverses zur Weimarer Verfassung." *Frankfurter Allgemeine Zeitung*, January 30, 2009: 13.

Kempner, Robert M. W. (1962). "Die Justiz als politische Waffe." *Aufbau*, February 2, 1962: 8.

Kempner, Robert M. W. (1963). "Der missverstandene Adolf Eichmann." *Aufbau*, April 14, 1963: 8.

Kennedy, Ellen (1986). "Carl Schmitt und die Frankfurter Schule." *Geschichte und Gesellschaft* 12: 380–419.

Kennedy, Ellen (1987a). "Carl Schmitt and the Frankfurt School." *Telos* 71 (Spring 1987): 37–66.

Kennedy, Ellen (1987b). "Carl Schmitt and Frankfurt School: A Re-Joinder." *Telos* 73 (Fall 1987): 101–116.

Kennedy, Ellen (2000). "Note on the Text and Translation." In Carl Schmitt, *The Crisis of Parliamentary Democracy*, translated by Ellen Kennedy, ix–xi. Cambridge: MIT Press.

Kennedy, Ellen (2004). *Constitutional Failure: Carl Schmitt in Weimar.* Durham: Duke University Press.

Kennedy, Ellen (2011). "Emergency Government Within the Bounds of the Constitution." *Constellations* 18: 284–298.

Kehr, Eckard (1928). "Englandhaß und Weltpolitik: Eine Studie über die innerpolitischen und sozialen Grundlagen der deutschen Außenpolitik um die Jahrhundertwende." *Zeitschrift für Politik* 17 (1928): 500–526.

Kersting, Wolfgang (2009). "Carl Schmitt und Thomas Hobbes." In *Der Hobbes-Kristall: Carl Schmitts Hobbes-Interpretation in der Diskussion*, edited by Rüdiger Voigt, 95–122. Stuttgart: Franz Steiner.

Kervégan, Jean Francois (1999). "Carl Schmitt and ‚World Unitiy'." In *The Challenge of Carl Schmitt*, edited by Chantal Mouffe, 54–74. London: Verso.

Kervégan, Jean Francois (2019). *Was tun mit Carl Schmitt?* Tübingen: Mohr Siebeck.

Keßler, Mario (2002). *Exil und Nach-Exil. Vertriebene Intellektuelle im 20. Jahrhundert.* Hamburg: VSA-Verlag.

Keßler, Mario (2007). *Ossip K. Flechtheim: Politischer Wissenschaftler und Zukunftsdenker.* Köln und Weimar: Böhlau.

Kettler, David (2006). "The Political Theory Question in Political Science: 1956–1967." *American Political Science Review* 100, no. 4 (2006): 532–538.

Kettler, David and Thomas Wheatland (2019). *Learning from Franz L. Neumann*. New York: Anthem.

Kießling, Wolfgang (1980). *Exil in Lateinamerika. Kunst und Literatur im antifaschistischen Exil 1933–1945, Band 4*. Leipzig: Reclam.

Kirchheimer, Otto (1928a). "The Socialist and Bolshevik Theory of the State." In *Politics. Law, and Social Change: Selected Essays of Otto Kirchheimer*, edited by Fredric S. Burin and Kurt S. Shell, 3–21. New York: Columbia University Press, 1969.

Kirchheimer, Otto (1928b). "Bedeutungswandel des Parlamentarismus." In Otto Kirchheimer, *Gesammelte Schriften, Band 1: Recht und Politik in der Weimarer Republik*, edited by Hubertus Buchstein, 157–162. Baden-Baden: Nomos, 2017.

Kirchheimer, Otto (1929a). "50 Jahre Deutsches Reichsgericht." In Otto Kirchheimer, *Gesammelte Schriften, Band 1: Recht und Politik in der Weimarer Republik*, edited by Hubertus Buchstein, 187–191. Baden-Baden: Nomos, 2017.

Kirchheimer, Otto (1929b). "Verfassungswirklichkeit und politische Zukunft der Arbeiterklasse: Zum Verfassungstag." Otto Kirchheimer, *Gesammelte Schriften, Band 1: Recht und Politik in der Weimarer Republik*, edited by Hubertus Buchstein, 179–186. Baden-Baden: Nomos, 2017.

Kirchheimer, Otto (1929c). "Die Englische Arbeiterbewegung." In Otto Kirchheimer, *Gesammelte Schriften, Band 1: Recht und Politik in der Weimarer Republik*, edited by Hubertus Buchstein, 192–195. Baden-Baden: Nomos, 2017.

Kirchheimer, Otto (1930a). "Review: Eugene Korovine: Das Völkerrecht der Übergangszeit." In Otto Kirchheimer, *Gesammelte Schriften, Band 1: Recht und Politik in der Weimarer Republik*, edited by Hubertus Buchstein, 323–327. Baden-Baden: Nomos, 2017.

Kirchheimer, Otto (1930b). "Eigentumsgarantie in Reichsverfassung und Rechtsprechung." In Otto Kirchheimer, *Gesammelte Schriften, Band 1: Recht und Politik in der Weimarer Republik*, edited by Hubertus Buchstein, 333–348. Baden-Baden: Nomos, 2017.

Kirchheimer, Otto (1930c). "Artikel 48 – der falsche Weg." In Otto Kirchheimer, *Gesammelte Schriften, Band 1: Recht und Politik in der Weimarer Republik*, edited by Hubertus Buchstein, 202–205. Baden-Baden: Nomos, 2017.

Kirchheimer, Otto (1930d). "Artikel 48 und die Wandlungen des Verfassungssystems." In Otto Kirchheimer, *Gesammelte Schriften, Band 1: Recht und Politik in der Weimarer Republik*, edited by Hubertus Buchstein, 349–354. Baden-Baden: Nomos, 2017.

Kirchheimer, Otto (1930e). "Weimar—and What Then? An Analysis of a Constitution." In *Politics, Law, and Social Change: Selected Essays by Otto Kirchheimer*, edited by Fredric S. Burin and Kurt L. Shell, 33–74. New York: Columbia University Press, 1969.

Kirchheimer, Otto (1930f). "Review: Carl Tannert, Die Fehlgestalt des Volksentscheids." In Otto Kirchheimer, *Gesammelte Schriften, Band 1: Recht und Politik in der Weimarer Republik*, edited by Hubertus Buchstein, 196–198. Baden-Baden: Nomos, 2017.

Kirchheimer, Otto (1930g). "Reichsgericht und Enteignung." In Otto Kirchheimer, *Gesammelte Schriften, Band 1: Recht und Politik in der Weimarer Republik*, edited by Hubertus Buchstein, 251–263. Baden-Baden: Nomos, 2017.

Kirchheimer, Otto (1930h). "The Limits of Expropriation." In *Social Democracy and the Rule of Law*, edited by Keith Tribe, 85–129. Abington: Routledge, 2020.

Kirchheimer, Otto (1930i). "Weimar… und was dann?" In Otto Kirchheimer, *Gesammelte Schriften, Band 1: Recht und Politik in der Weimarer Republik*, edited by Hubertus Buchstein, 209–250. Baden-Baden: Nomos, 2017.

Kirchheimer, Otto (1932a). "Legality and Legitimacy." In *The Rule of Law under Siege*, edited by William E. Scheuerman, 44–63. Berkeley: University of California Press, 1996.

Kirchheimer Otto (1932b). "Review: Georg Schwarzenberger, Die Kreuger-Anleihen." In Otto Kirchheimer, *Gesammelte Schriften, Band 1: Recht und Politik in der Weimarer Republik*, edited by Hubertus Buchstein, 373–376. Baden-Baden: Nomos, 2017.

Kirchheimer, Otto (1932c). "Die staatsrechtlichen Probleme der Reichstagsauflösung." In Otto Kirchheimer, *Gesammelte Schriften, Band 1: Recht und Politik in der Weimarer Republik*, edited by Hubertus Buchstein, 396–407. Baden-Baden: Nomos, 2017.

Kirchheimer, Otto (1932d). "Die Verfassungslehre des Preußen-Konflikts." In Otto Kirchheimer, *Gesammelte Schriften, Band 1: Recht und Politik in der Weimarer Republik*, edited by Hubertus Buchstein, 408–424. Baden-Baden: Nomos, 2017.

Kirchheimer, Otto (1932e). "Constitutional Reaction in 1932." In *Politics, Law, and Social Change: Selected Essays by Otto Kirchheimer*, edited by Fredric S. Burin and Kurt L. Shell, 75–87. New York: Columbia University Press, 1969.

Kirchheimer, Otto (1932f). "Die Verfassungsreform." In Otto Kirchheimer, *Gesammelte Schriften, Band 1: Recht und Politik in der Weimarer Republik*, edited by Hubertus Buchstein, 443–457. Baden-Baden: Nomos, 2017.

Kirchheimer, Otto (1932g). "Legalität und Legitimität." In Otto Kirchheimer, *Gesammelte Schriften, Band 1: Recht und Politik in der Weimarer Republik*, edited by Hubertus Buchstein, 376–395, Baden-Baden: Nomos, 2017.

Kirchheimer, Otto (1932h). "Verfassungsreaktion 1932." In Otto Kirchheimer, *Gesammelte Schriften, Band 1: Recht und Politik in der Weimarer Republik*, edited by Hubertus Buchstein, 429–442. Baden-Baden: Nomos, 2017.

Kirchheimer, Otto (1932i). "Review: Curcio Malaparte, Der Staatsstreich." In Otto Kirchheimer, *Gesammelte Schriften, Band 1: Recht und Politik in der Weimarer Republik*, edited by Hubertus Buchstein, 369–372. Baden-Baden: Nomos, 2017.

Kirchheimer, Otto (1933a). "Review: Adolf Grabowsky, Politik." In Otto Kirchheimer, *Gesammelte Schriften, Band 1: Recht und Politik in der Weimarer Republik*, edited by Hubertus Buchstein, 511–514. Baden-Baden: Nomos, 2017.

Kirchheimer, Otto (1933b). "Marxismus, Diktatur und die Organisationsform des Proletariats." In Otto Kirchheimer, *Gesammelte Schriften, Band 1: Recht und Politik in der Weimarer Republik*, edited by Hubertus Buchstein, 515–526. Baden-Baden: Nomos, 2017.

Kirchheimer, Otto (1933c). "The Growth and the Decay of the Weimar Constitution." In Otto Kirchheimer, *Gesammelte Schriften, Band 1: Recht und Politik in der Weimarer Republik*, edited by Hubertus Buchstein, 527–534. Baden-Baden: Nomos, 2017.

Kirchheimer, Otto (1933d). "Constitutional Reform and Social Democracy." In *Social Democracy and the Rule of Law*, edited by Keith Tribe, 179–193. Abington: Routledge, 2020.

Kirchheimer, Otto (1934a). "Zur Geschichte des Obersten Gerichtshofes der Vereinigten Staaten von Amerika." In Otto Kirchheimer, *Gesammelte Schriften, Band 2: Faschismus, Demokratie und Kapitalismus*, edited by Hubertus Buchstein and Henning Hochstein, 117–131. Baden-Baden: Nomos, 2018.

Kirchheimer, Otto (1934b). "Anmerkungen zur Theorie der nationalen Souveränität in Deutschland und Frankreich." In Otto Kirchheimer, *Gesammelte Schriften, Band 2: Faschismus, Demokratie und Kapitalismus*, edited by Hubertus Buchstein and Henning Hochstein, 132–151. Baden-Baden: Nomos, 2018.

Kirchheimer, Otto (1935a). "State Structure and Law in the Third Reich." In *The Rule of Law under Siege: Selected Essays of Franz L. Neumann and Otto Kirchheimer*, edited by William Scheuerman, 142–172. Berkeley: University of California Press, 1996.

Kirchheimer, Otto (1936a). "Die wirtschaftliche Betätigung der französischen Gemeinden und die Rechtsprechung des Conseil d´Ètat." In Otto Kirchheimer, *Gesammelte Schriften, Band 2: Faschismus, Demokratie und Kapitalismus*, edited by Hubertus Buchstein and Henning Hochstein, 182–190. Baden-Baden: Nomos, 2018.

Kirchheimer, Otto (1936b). "Anmerkungen zur Kriminalstatistik des Nachkriegsfrankreichs." In Otto Kirchheimer, *Gesammelte Schriften, Band 3: Kriminologische Schriften*, edited by Hubertus Buchstein and Lisa Klingsporn, 83–116. Baden-Baden: Nomos, 2019.

Kirchheimer, Otto (1938a). "Recent Trends in German Treatment of Juveline Delinquency." In Otto Kirchheimer, *Gesammelte Schriften, Band 3: Kriminologische Schriften*, edited by Hubertus Buchstein and Lisa Klingsporn, 117–125. Baden-Baden: Nomos, 2019.

Kirchheimer, Otto (1938b). "Sammelrezension zur nationalsozialistischen Völkerrechtstheorie." In Otto Kirchheimer, *Gesammelte Schriften, Band 2: Faschismus, Demokratie und Kapitalismus*, edited by Hubertus Buchstein and Henning Hochstein, 200–202. Baden-Baden: Nomos, 2018.

Kirchheimer, Otto (1939a). "Zu Max Horkheimers ‚Die Juden und Europa'." In Otto Kirchheimer, *Gesammelte Schriften, Band 2: Faschismus, Demokratie und Kapitalismus*, edited by Hubertus Buchstein and Henning Hochstein, 203–212. Baden-Baden: Nomos, 2018.

Kirchheimer, Otto (1940a). "Thesen zu Max Horkheimers Manuskript 'Autoritärer Staat." In Otto Kirchheimer, *Gesammelte Schriften, Band 2: Faschismus, Demokratie und Kapitalismus*, edited by Hubertus Buchstein and Henning Hochstein, 213–215. Baden-Baden: Nomos, 2018.

Kirchheimer, Otto (1940b). "Criminal Law in National-Socialist Germany." *Studies in Philosophy and Social Science* 8 (1940): 444–463.

Kirchheimer, Otto (1940c). "Funktionswandel und Konzentrationstendenzen im Bankgewerbe." In Otto Kirchheimer, *Gesammelte Schriften, Band 2: Faschismus, Demokratie und Kapitalismus*, edited by Hubertus Buchstein and Henning Hochstein, 218–220. Baden-Baden: Nomos, 2018.

Kirchheimer, Otto (1941a). "Changes in the Structure of Political Compromise." *Studies in Philosophy and Social Science* 9 (1941): 264–289.

Kirchheimer, Otto (1941b). "Book Review on Writings about the German Trade Relations." *Studies in Philosophy and Social Science* 9 (1941): 361–364.

Kirchheimer, Otto (1941c). "Book Review, Hermann Mannhein: War and Crime." In Otto Kirchheimer, *Gesammelte Schriften, Band 3: Kriminologische Schriften*, edited by Hubertus Buchstein and Lisa Klingsporn, 427–428. Baden-Baden: Nomos, 2019.

Kirchheimer, Otto (1941d). "The Legal Order of National Socialism." *Studies in Philosophy and Social Science* 9 (1941): 456–475.

Kirchheimer, Otto (1942). "Review: Leopold Schwarzschild: World in Trance.". In Otto Kirchheimer, *Gesammelte Schriften, Band 2: Faschismus, Demokratie und Kapitalismus*, edited by Hubertus Buchstein and Henning Hochstein, 331–332. Baden-Baden: Nomos, 2018.

Kirchheimer, Otto (1943). "The Policy of the Catholic Church Toward the Jews." In Otto Kirchheimer, *Gesammelte Schriften, Band 2: Faschismus, Demokratie und Kapitalismus*, edited by Hubertus Buchstein and Henning Hochstein, 498–518. Baden-Baden: Nomos, 2018.

Kirchheimer, Otto (1944a). "The Abrogation of Nazi Laws in the Early Period of MG." In *Secret Reports on Nazi Germany: The Frankfurt School Contribution to the War Effort*, edited by Raffaele Laudani, 229–252. Princeton: Princeton University Press, 2013.

Kirchheimer, Otto (1944b). "General Principles of Administration and Civil Service in Germany." In *Secret Reports on Nazi Germany: The Frankfurt School Contribution to the War Effort*, edited by Raffaele Laudani, 301–317. Princeton: Princeton University Press, 2013.

Kirchheimer, Otto (1944c). "Administration of Criminal Justice under Military Government." In *Secret Reports on Nazi Germany: The Frankfurt School Contribution to the War Effort*, edited by Raffaele Laudani, 318–344. Princeton: Princeton University Press, 2013.

Kirchheimer, Otto (1944d). "Civil Affairs Handbook Germany. Section 3: Legal Affairs." In Otto Kirchheimer, *Gesammelte Schriften, Band 6: Politische Analysen für das OSS und Department of State*, edited by Henning Hochstein, Frank Schale, and Lisa Klingsporn, 245–339. Baden-Baden: Nomos, 2022.

Kirchheimer, Otto (1944e). "In Quest of Sovereignty." In *Politics, Law, and Social Change: Selected Essays by Otto Kirchheimer*, edited by Frederic S. Burin and Kurt L. Shell, 160–196. New York: Columbia University Press, 1969.

Kirchheimer, Otto (1945). "Nazi Plans for Dominating Germany and Europe: Domestic Crimes." In *Secret Reports on Nazi Germany: The Frankfurt School Contribution to the War Effort*, edited by Raffaele Laudani, 522–556. Princeton: Princeton University Press, 2013.

Kirchheimer, Otto (1946a). "The General Character of a Future German Government." In Otto Kirchheimer, *Gesammelte Schriften, Band 6: Politische Analysen für das OSS und Department of State*, edited by Henning Hochstein, Frank Schale, and Lisa Klingsporn, 340–349. Baden-Baden: Nomos, 2022.

Kirchheimer, Otto (1946b). "Current Political Tendencies in Germany Bearing on Its Future Governmental Structure." In Otto Kirchheimer, *Gesammelte Schriften, Band 6: Politische Analysen für das OSS und Department of State*, edited by Henning Hochstein, Frank Schale, and Lisa Klingsporn, 350–356. Baden-Baden: Nomos, 2022.

Kirchheimer, Otto (1947). "A Constitution for the Fourth Republic." In Otto Kirchheimer, *Gesammelte Schriften, Band 6: Politische Analysen für das OSS und Department of State*, edited by Henning Hochstein, Frank Schale, and Lisa Klingsporn, 357–425. Baden-Baden: Nomos, 2022.

Kirchheimer, Otto (1949a). "The Act, the Offense and Double Jeopardy." *Yale Law Journal* 58 (4), 513–544.

Kirchheimer, Otto (1949b). "Book Review: Andrei Y. Yyshinski: The Law of the Soviet State." In Otto Kirchheimer, *Gesammelte Schriften, Band 4: Politische Justiz und Wandel der Rechtsstaatlichkeit*, edited by Lisa Klingsporn, Merete Peetz, and Christiane Wilke, 839–841. Baden-Baden: Nomos, 2019.

Kirchheimer, Otto (1950a). "The Composition of the German Bundestag." In Otto Kirchheimer, *Gesammelte Schriften, Band 5: Politische Systeme im Nachkriegseuropa*, edited by Hubertus Buchstein and Moritz Langfeldt, 177–193. Baden-Baden: Nomos, 2020.

Kirchheimer, Otto (1952a). "Book Review: Werner Weber: Spannungen und Kräfte im westdeutschen Verfassungssystem." In Otto Kirchheimer, *Gesammelte Schriften, Band 5: Politische Systeme im Nachkriegseuropa*, edited by Hubertus Buchstein and Moritz Langfeldt, 220–230. Baden-Baden: Nomos, 2020.

Kirchheimer, Otto (1954a). "Notes on the Political Scene in Western Germany." In Otto Kirchheimer, *Gesammelte Schriften, Band 5: Politische Systeme im Nachkriegseuropa*, edited by Hubertus Buchstein and Moritz Langfeldt, 252–265. Baden-Baden: Nomos, 2020.

Kirchheimer, Otto (1954b). "Parteistruktur und Massendemokratie in Europa." In Otto Kirchheimer, *Gesammelte Schriften, Band 5: Politische Systeme im Nachkriegseuropa*, edited by Hubertus Buchstein and Moritz Langfeldt, 223–248. Baden-Baden: Nomos, 2020.

Kirchheimer, Otto (1954c). "Book Review: Ernst Rudolf Huber: Wirtschaftsverwaltungsrecht." In Otto Kirchheimer, *Gesammelte Schriften, Band 5: Politische Systeme im Nachkriegseuropa*, edited by Hubertus Buchstein, and Moritz Langfeldt, 266–270. Baden-Baden: Nomos, 2020.

Kirchheimer, Otto (1954d). "Book Review: William W. Crosskey – Politics and the Constitution in the History of the United States." *Politische Literatur* 3 (1954): 902–905.

Kirchheimer, Otto (1955a). "Politics and Justice." In *Politics, Law, and Social Change: Selected Essays by Otto Kirchheimer*, edited by Fredric S. Burin and Kurt L. Shell, 408–417. New York: Columbia University Press, 1969.

Kirchheimer, Otto (1955b). "Politische Justiz." In Otto Kirchheimer, *Gesammelte Schriften, Band 4: Politische Justiz und Wandel der Rechtsstaatlichkeit*, edited by Lisa Klingsporn, Merete Peetz, and Christiane Wilke, 102–134. Baden-Baden: Nomos, 2019.

Kirchheimer, Otto (1956a). "Book Review: Joseph H. Kaiser – Die Repräsentation organisierter Interessen." In Otto Kirchheimer, *Gesammelte Schriften, Band 5: Politische Systeme im Nachkriegseuropa*, edited by Hubertus Buchstein and Moritz Langfeldt, 271–275. Baden-Baden: Nomos, 2020.

Kirchheimer, Otto (1957a). "The Waning of Opposition in Parliamentary Regimes." In *Politics, Law, and Social Change: Selected Essays by Otto Kirchheimer*, edited by Fredric S. Burin and Kurt L. Shell, 319–345. New York: Columbia University Press, 1969.

Kirchheimer, Otto (1957b). "The Political Scene in West Germany." In Otto Kirchheimer, *Gesammelte Schriften, Band 5: Politische Systeme im Nachkriegseuropa*, edited by Hubertus Buchstein and Moritz Langfeldt, 342–356. Baden-Baden: Nomos, 2020.

Kirchheimer, Otto (1957c). "Wandlungen der politischen Opposition." In Otto Kirchheimer, *Gesammelte Schriften, Band 5: Politische Systeme im Nachkriegseuropa*, edited by Hubertus Buchstein and Moritz Langfeldt, 361–388. Baden-Baden: Nomos, 2020.

Kirchheimer, Otto (1957d). "Franz L. Neumann: An Appreciation." *Dissent* 4, no. 4 (1957): 382–386.

Kirchheimer, Otto (1958a). "France from the Fourth to the Fifth Republic." In Otto Kirchheimer, *Gesammelte Schriften, Band 5: Politische Systeme im Nachkriegseuropa*, edited by Hubertus Buchstein and Moritz Langfeldt, 396–426. Baden-Baden: Nomos, 2020.

Kirchheimer, Otto (1959a). "In Frankreich regiert ein einziger Mann." In Otto Kirchheimer, *Gesammelte Schriften, Band 5: Politische Systeme im Nachkriegseuropa*, edited by Hubertus Buchstein and Moritz Langfeldt, 427–438. Baden-Baden: Nomos, 2020.

Kirchheimer, Otto (1959b). "The Administration of Justice and the Concept of Legality in East Germany." *Yale Law Journal* 68 (4): 705–749.

Kirchheimer, Otto (1961a). *Political Justice: The Use of Legal Procedures for Political Ends.* Princeton: Princeton University Press.

Kirchheimer, Otto (1961b). "German Democracy in the 1950's." In Otto Kirchheimer, *Gesammelte Schriften, Band 5: Politische Systeme im Nachkriegseuropa*, edited by Hubertus Buchstein and Moritz Langfeldt, 475–488. Baden-Baden: Nomos, 2020.

Kirchheimer, Otto (1962a). "Heroes in Hitler's Ghettos." *Washington Post*, September 9, 1962: 6.

Kirchheimer, Otto (1962b). "How German Jews Fare." *Washington Post*, November 4, 1962: 9.

Kirchheimer, Otto (1962c). "Expertise and Politics in the Administration." In *Politics, Law, and Social Change: Selected Essays by Otto Kirchheimer*, edited by Fredric S. Burin and Kurt L. Shell, 372–382. New York: Columbia University Press, 1969.

Kirchheimer, Otto (1964). "Vorbemerkung zu ‚Politik und Verfassung'." In Otto Kirchheimer, *Gesammelte Schriften, Band 5: Politische Systeme im Nachkriegseuropa*, edited by Hubertus Buchstein and Moritz Langfeldt, 501. Baden-Baden: Nomos, 2020.

Kirchheimer, Otto (1965a). "Book Review Guenter Lewy, The Catholic Church and Nazi Germany." *Dissent* 12, no. 4, (1965): 92–95.

Kirchheimer, Otto (1965b). "Book Review Gordon Zahn, German Catholics and Hitler's Wars. Manuscript, 3 pages." State University of New York, University at Albany, Special Collections & Archives, Otto Kirchheimer Papers, Series 4, Box 3, Folder 22.

Kirchheimer, Otto (1965c). *Politische Justiz. Verwendung juristischer Verfahrensmöglichkeiten zu politischen Zwecken.* 5th, revised edition. Hamburg: Europäische Verlagsanstalt, 2020.

Kirchheimer, Otto (1965d). "Bombing in Viet-Nam." *Washington Post*, March 27, 1965: 27.

Kirchheimer, Otto (1966). "The Transformation of the Western European Party Systems." In *Politics, Law, and Social Change: Selected Essays by Otto Kirchheimer*, edited by Fredric S. Burin and Kurt L. Shell, 346–371. New York: Columbia University Press, 1969.

Kirchheimer, Otto (1967a). "The *Rechtsstaat* as Magic Wall." In *The Critical Spirit: Essays in the Honor of Herbert Marcuse*, edited by Kurt H. Wolff and Barrington Moore, Jr., 287–312. Boston: Beacon Press, 1969.

Kirchheimer, Otto (1967b). "Private Man and Society." In *Politics, Law, and Social Change: Selected Essays by Otto Kirchheimer*, edited by Fredric S. Burin and Kurt L. Shell, 453–477. New York: Columbia University Press, 1969.

Kirchheimer, Otto (1968a). "Die Justiz in der Weimarer Republik." In Otto Kirchheimer, *Gesammelte Schriften, Band 1: Recht und Politik in der Weimarer Republik*, edited by Hubertus Buchstein, 549–558. Baden-Baden: Nomos, 2017.

Kirchheimer, Otto (1968b). "Political Justice." In *International Encyclopedia of the Social Sciences,Volume 12*, edited by David L. Sills, 246–249. New York: Macmillan.

Kirchheimer, Otto, Hans G. Gerth, and Arkadij Gurland (1947). "Bureaucratic Trends in German Society: A Study of Obstacles to Democratization in Trade Unions, Political Parties, and Public Administration." *OMGUS Research Branch ICD*, Special Report No. 4. Washington D.C.

Kirchheimer, Otto and John H. Herz (1943). "The ‚Statement of Atrocities' of the Moscow Tripartite Conference." In *Secret Reports on Nazi Germany: The Frankfurt School Contribution to the War Effort*, edited by Raffaele Laudani, 451–456. Princeton: Princeton University Press, 2013.

Kirchheimer, Otto and John H. Herz (1945). "Leadership Principle and Criminal Responsibility." In *Secret Reports on Nazi Germany: The Frankfurt School Contribution to the War Effort*, edited by Raffaele Laudani, 464–475. Princeton: Princeton University Press, 2013.

Kirchheimer, Otto and Nathan Leites (1933). "Remarks on Carl Schmitt's *Legality and Legitimacy*." In *The Rule of Law under Siege: Selected Essays of Franz L. Neumann and Otto Kirchheimer*, edited by William E. Scheuerman, 64–99. Berkeley: University of California Press, 1996.

Kirchheimer, Otto and Arnold Price (1949). "Analysis and Effects of the Elections in Western Germany." *U.S. Department of State Bulletin, Volume 21*, 565–573.

Kirchheimer, Otto and Georg Rusche (1939). *Punishment and Social Structure*. New York: Columbia University Press.

Kirchheimer-Grossman, Hanna (2010). *The Kirchheimer Family: From Baden to the Wider World* (private printing).

Klein, Max (2022). "Der unbestimmte Ort des Eigentums." *Politische Vierteljahresschrift* 63, no. 1 (2022): 25–52.

Klemperer, Victor (1947). *The Language of the Third Reich. LTI: Lingua Terti Imperii. A Philologist's Notebook.* Translated by Martin Brady. London: Bloomsbury Publishing. 2021.

Klingsporn, Lisa (2024). *Kampf und Kompromiss. Otto Kirchheimers Analysen von Politik und Recht in der Rezeption*. Baden-Baden: Nomos.

Klingsporn, Lisa and Christiane Wilke (2019). "Einleitung." In Otto Kirchheimer, *Gesammelte Schriften, Band 4: Politische Justiz und Wandel der Rechtsstaatlichkeit*, edited by Lisa Klingsporn, Merete Peetz, and Christiane Wilke, 7–95. Baden-Baden: Nomos.

Klingsporn, Lisa and Christiane Wilke (2020). "Nachwort." In Otto Kirchheimer, *Politische Justiz*, edited by Lisa Klingsporn, Merete Peetz, and Christiane Wilke. New edition., 700–766. Hamburg: Europäische Verlagsanstalt, 2020.

Koellreutter, Otto (1932). "Parteien und Verfassung im heutigen Deutschland." In *Festgabe für Richard Schmidt zu seinem 70. Geburtstag 19. Januar 1932*, edited by Hans Gmelin and Otto Koellreutter, 107–118. Leipzig: C. L. Hirschfeld.

Koellreutter, Otto (1938). "Leviathan und totaler Staat." *Reichsverwaltungsblatt* 59, no. 38, September 17, 1939: 803–807.

Kohlmann, Ulrich (1992). "Der Denkweg Otto Kirchheimers." *Archiv für Rechts- und Sozialphilosophie* 78, no. 4 (1992): 502–515.

Koenen, Andreas (1995). *Der Fall Carl Schmitt: Sein Aufstieg zum „Kronjuristen des Dritten Reiches"*. Darmstadt: Wissenschaftliche Buchgesellschaft.

König, Helmut (2016). *Elemente des Antisemitismus: Kommentare und Interpretationen zu einem Kapitel der Dialektik der Aufklärung*. Weilerswist: Velbrück.

Koestler, Arthur (1954). *Die Geheimschrift: Bericht eines Lebens 1932 bis 1940*. Wien: Verlag Kurt Desch.

Korioth, Stefan (1990). *Integration und Bundesstaat: Ein Beitrag zur Staats- und Verfassungslehre Rudolf Smends*. Berlin: Duncker & Humblot.

Korovine, Eugene A. (1929). *Das Völkerrecht der Übergangszeit: Grundlagen der völkerrechtlichen Beziehungen der Union der Sowjetrepubliken*. Berlin: Rothschild.

Korsch, Karl (1932). "Review of Carl Schmitt, Der Hüter der Verfassung." *Zeitschrift für Sozialforschung* 1, no. 1–2 (1932): 204–205.

Koskenniemi, Martti (2001). *The Gentle Civilizer of Nations: The Rise and Fall of International Law 1870–1960*. Cambridge: Cambridge University Press.

Koskenniemi, Martti (2016). "Carl Schmitt and International Law." In *The Oxford Handbook of Carl Schmitt*, edited by Jens Meierhenrich and Oliver Simons, 592–611. Oxford: Oxford University Press.

Kotowski, Elke-Vera (2011). *Aufbau: Sprachrohr, Heimat, Mythos. Geschichte(n) einer deutsch-jüdischen Zeitung aus New York*, edited by Elke-Vera Kotowski and Hermann Simon. Berlin: Hentrich & Hentrich.

Kraus, Hans-Christoph (1998). "Anmerkungen zur Begriffs- und Thesenbildung bei Carl Schmitt." In *Jahrbuch Politisches Denken 1998*, edited by Karl Graf Ballestrem, Volker Gerhardt, Henning Ottmann, and Martyn P. Thompson, 161–176. Stuttgart: Metzler.

Krockow, Christian Graf von (1958). *Die Entscheidung. Eine Untersuchung über Ernst Jünger, Carl Schmitt und Martin Heidegger.* Stuttgart: Deutsche Verlags-Anstalt.

Krohn, Claus-Dieter (1987). *Wissenschaft im Exil: Deutsche Sozial- und Wirtschaftswissenschaftler in den USA und die New School for Social Research*. Frankfurt am Main: Campus.

Kronlund, Anna (2022). *US Congress' Powers under Debate. Separation of Powers and Parliamentary Politics in Times of War and Crisis.* Baden-Baden: Nomos.

Krouwel, André (2003). "Otto Kirchheimer and the ‚catch-all-party'." *West European Politics* 26, no. 2 (2003): 23–40.

Kunz, Josef L. (1940). "Germany's Lebensraum." American Journal of International Law 34 (1), 170–176.

Ladwig-Winters, Simone (2007). *Anwalt ohne Recht: Das Schicksal jüdischer Rechtsanwälte in Berlin nach 1933*. Berlin: beb.ra-Verlag.

Ladwig-Winters, Simone (2009). *Ernst Fraenkel. Ein politisches Leben*. Frankfurt am Main: Campus.

Lahusen, Benjamin (2022). *Der Dienstbetrieb ist nicht gestört: Die Deutschen und ihre Justiz 1943–1948*. München: C.H. Beck.

Landois, Leonard (2007). *Konterrevolution von links: Das Staats- und Gesellschaftsverständnis der ‚68er' und dessen Quellen bei Carl Schmitt*. Baden-Baden: Nomos.

Lang, Markus (2007). *Karl Loewenstein: Transatlantischer Denker der Politik*. Stuttgart: Steiner.

Langkau-Alex, Ursula (1989). "Deutsche Emigrationspresse." In *Exilliteratur 1933–1945*, edited by Wulf Koepke and Michael Winkler, 169–15. Darmstadt: Wissenschaftliche Buchgesellschaft.

Langkau-Alex, Ursula (2005a). *Deutsche Volksfront 1932–1939: Zwischen Berlin, Paris, Prag und Moskau, 1. Band.* Berlin: Akademie-Verlag.

Langkau-Alex, Ursula (2005b). *Deutsche Volksfront 1932–1939: Zwischen Berlin, Paris, Prag und Moskau. 2. Band.* Berlin: Akademie-Verlag.

Laudani, Raffaele (2013). *Secret Reports on Nazi Germany: The Frankfurt School Contribution to the War Effort*, edited by Raffaele Laudani. Princeton: Princeton UP.

Lethen, Helmut (2018). *Die Staatsräte: Elite im Dritten Reich*. Berlin: Rowohlt.

Lethen, Helmut (2020): „Armer Allende!" *Zeitschrift für Ideengeschichte* 14 (1): 116–119.

Lévy, René and Hartwig Zander (1994). "Introduction." In Georg Rusche and Otto Kirchheimer, *Peine et structure sociale: histoire et «théorie critique» du régime penal*, 9–82. Paris: Cerf.

Lewis, David (2020). *Russia's New Authoritarianism: Putin and the Politics of Order*. Edinburgh: Edinburgh University Press.

Lewis, Sinclair (1935). *It Can't Happen Here*. New York: Doubleday.

Lévy-Bruhl, Lucien (1922). *La mentalité primitive*. Paris: Alcan.

Lietzmann, Hans (1988). "Vater der Verfassungsväter? Carl Schmitt und die Verfassungsgründung in der Bundesrepublik Deutschland." In *Carl Schmitt und die Liberalismuskritik*, edited by Klaus Hansen and Hans Lietzmann, 107–119. Opladen: Leske + Budrich.

Lietzmann, Hans (1999). *Politikwissenschaft im ›Zeitalter der Diktaturen‹: Die Entwicklung der Totalitarismustheorie Carl Joachim Friedrichs*. Opladen: VS Verlag für Sozialwissenschaften.

Lilla, Mark (2016). *The Shipwrecked Mind: On Political Reaction*. New York: New York Review of Books.

Link, Fabian (2022). *Demokratisierung nach Auschwitz. Eine Geschichte der westdeutschen Sozialwissenschaften in der Nachkriegszeit.* Göttingen: Wallstein.

Llanque, Marcus (1990). "Ein Träger des Politischen nach dem Ende der Staatlichkeit: Der Partisan in Carl Schmitts politischer Theorie." In *Der Partisan: Theorie, Strategie, Gestalt*, edited by Herfried Münkler, 61–80. Opladen: Westdeutscher Verlag.

Llanque, Marcus (2011). "Otto Kirchheimer und die sozialistische Staatslehre." In *Kritische Verfassungspolitologie. Das Statsverständnis von Otto Kirchheimer,* edited by Robert van Oyen and Frank Schale, 69–86. Baden-Baden: Nomos.

Llanque, Marcus (2021). "Otto Kirchheimer und der Wandel von der Integrationspartei zur Allerweltspartei als Problem der politischen Bindung." In *Parteiendemokratie. Theorie und Praxis*, edited by Detlef Lehnert, 241–266. Berlin: Metropol.

Loewenstein, Karl (2023). *Des Lebens Überfluß. Erinnerungen eines ausgewanderten Juristen*, edited by Oliver Lepsius, Robert van Ooyen, and Frank Schale. Tübingen: Mohr Siebeck.

Löwith, Karl (1935). "Der okkasionelle Dezisionismus von Carl Schmitt." In Karl Löwith, *Sämtliche Schriften, Band 8: Heidegger – Denker in dürftiger Zeit*, 32–71. Stuttgart: Metzler, 1984

Lokatis, Siegfried (1992). *Hanseatische Verlagsanstalt: Buchmarketing im ›Dritten Reich‹*. Frankfurt am Main: Buchhändler-Vereinigung.

Lübbe, Hermann (1988). "Carl Schmitt liberal rezipiert." In *Complexio Oppositorum: Über Carl Schmitt*, edited by Helmut Quaritsch, 427–440. Berlin: Duncker & Humblot.

Lukács, Georg (1923). *History and Class Consciousness*. Cambridge: MIT Press, 1971.

Lukács, Georg (1928). "Book Review: Carl Schmitt, Politische Romantik." *Archiv für die Geschichte des Sozialismus und der Arbeiterbewegung* 13 (1928): 307–308.

Lustig, Sandra (2006). "'The Germans will never forgive the Jews for Auschwitz'. When things go wrong in the Jewish Space: The case of the Walser-Bubis Debate. In Sandra Lustig and Ian Leveson (eds.), *Turning the Kaleidoscope. Perspectives on European Jewry*, 205–222. New York/Oxford: Berghahn Books.

Luthardt, Wolfgang (1976). "Einleitung zu ›Otto Kirchheimer: Staatsgefüge und Recht des dritten Reiches‹." *Kritische Justiz* 7, no. 1 (1976): 33–39.

Luthardt, Wolfgang and Alfons Söllner (1989). *Verfassungsstaat, Souveränität, Pluralismus: Otto Kirchheimer zum Gedächtnis*, edited by Wolfgang Luthardt and Alfons Söllner. Opladen: Westdeutscher Verlag.

Maier, Joseph (1963). "Was hat Hannah Arendt eigentlich gesagt?" *Aufbau*, December 20, 1963: 17–18.

Mair, Peter (2013). *Ruling the Void: The Hollowing of Western Democracy*. London: Verso.

Malkopoulou, Anthoula (2023). "Hermann Heller on Democratic Self-Defense." *Hermann Hellers demokratischer Konstitutionalismus*, edited by Verena Frick and Oliver W. Lembcke, 185–202. Wiesbaden: Springer VS.

Mann, Golo (1952a) "Historische Selbstbesinnung." *Der Monat* 4 (1952): 328–329.

Mann, Golo (1952b). "Carl Schmitt und die schlechte Juristerei." *Der Monat* 5 (1952): 89–93.

Manow, Philip (2022). *Nehmen, Teilen, Weiden: Carl Schmitts politische Ökonomien*. Konstanz: Konstanz University Press.

Marchal, Kai and Carl K.Y. Shaw (2017). *Carl Schmitt and Leo Strauss in the Chinese-Speaking World*, edited by Kai Marchal and Carl K.Y. Shaw. Lanham: Lexington Books.

Marcuse, Herbert (1934). "The Struggle against Liberalism in the Totalitarian View of the State." In Herbert Marcuse, *Negotiations: Essays in Critical Theory*, 1–30. London: MayFlyBooks, 2009.

Marquard, Odo (2013). *Endzeitphilosophisches*. Stuttgart: Reclam.

Marquardt-Bigman, Petra (1995). *Amerikanische Geheimdienstanalysen über Deutschland 1942–1949*. München: Oldenbourg Wissenschaftsverlag.

Maschke, Günter (1995a). "Zum ‚Leviathan' von Carl Schmitt." In Carl Schmitt, *Der Leviathan in der Staatslehre des Thomas Hobbes: Sinn und Fehlschlag eines politischen Symbols*, 179–244. Stuttgart: Klett-Cotta, 1995.

Maschke, Günter (1995b). "Editor's Notes." In Carl Schmitt, *Staat, Großraum, Nomos: Aufsätze aus den Jahren 1916–1969*, edited by Günter Maschke. Berlin: Duncker & Humblot.

Maschke, Günter (2005). "Anmerkungen des Herausgebers." In Carl Schmitt, *Frieden oder Pazifismus? Arbeiten zum Völkerrecht und zur Internationalen Politik 1924–1978*, edited by Günter Maschke, 219–239. Berlin: Duncker& Humblot, 2005.

Maschke, Günter (2011). "Erkenne die Lage!" *Sezession* 42 (June 2011): 18–22.

Massing, Paul W. (1949). *Rehearsal for Destruction: A Study of Political Antisemitism in Imperial Germany*. New York: Harper.

Mastnak, Tomaz (2015). „Hobbes in Kiel: From Ferdinand Tönnies to Carl Schmitt". *History of European Ideas* 41, 966–991.

Matthias, Erich and Werner Link (1968). *Mit dem Gesicht nach Deutschland: Eine Dokumentation der sozialdemokratischen Emigration*, edited by Erich Matthias and Werner Link. Düsseldorf: Droste.

Maus, Ingeborg (1976). *Bürgerliche Rechtstheorie und Faschismus: Zur sozialen Funktion und aktuellen Wirkung der Theorie Carl Schmitts*. München: Wilhelm Fink Verlag.

Mayer, Hans (1988). *Ein Deutscher auf Widerruf: Erinnerungen, Band 1*. Frankfurt am Main: Suhrkamp.

McCormick, John (1994). "Fear, Technology, and the State: Carl Schmitt, Leo Strauss, and the Revival of Hobbes in Weimar and National Socialist Germany." *Political Theory* 22, no. 4 (1994): 619–652.

McCormick, John (1998). "Political Theory and Political Theology: The Second Wave of Carl Schmitt in English." *Political Theory* 26, no. 6 (1998): 830–854.

McCormick, John (2004). "Introduction." In Carl Schmitt, *Legality and Legitimacy*, edited by John McCormick, xiii–xliii. Durham & London: Duke University Press.

McCormick, John (2016). "Teaching in Vain: Carl Schmitt, Thomas Hobbes, and the Theory of the Sovereign State." In *The Oxford Handbook of Carl Schmitt*, edited by Jens Meierhenrich and Oliver Simons, 269–290. Oxford: Oxford University Press.

Mehring, Reinhard (2006). "Begriffssoziologie, Begriffsgeschichte, Begriffspolitik." In *Politische Ideengeschichte im 20. Jahrhundert*, edited by Harald Bluhm and Jürgen Gebhardt, 31–50. Baden-Baden: Nomos.

Mehring, Reinhard (2007). "Otto Kirchheimer und der Linksschmittianismus." In *Der Staat des Dezisionismus: Carl Schmitt in der internationalen Debatte*, edited by Rüdiger Voigt, 60–82. Baden-Baden: Nomos.

Mehring, Reinhard (2010). "Zum Ende der Beziehung." In Carl Schmitt and Rudolf Smend, *Auf der gefahrenvollen Straße des öffentlichen Rechts: Briefwechsel Carl Schmitt – Rudolf Smend, 1921–1961*, edited by Reinhard Mehring, 149–153. Berlin: Duncker & Humblot.

Mehring, Reinhard (2013). "Kant gegen Schmitt: Ingeborg Maus über Volkssouveränität." *Der Staat* 52, no. 3 (2013): 435–454.

Mehring, Reinhard (2014a). *Carl Schmitt: A Biography*. Cambridge: Polity.

Mehring (2014b). *Kriegstechniker des Begriffs: Biographische Studien zu Carl Schmitt*. Tübingen: Mohr Siebeck.

Mehring, Reinhard (2017). *Carl Schmitt: Denker im Widerstreit*. Freiburg and München: Alber.

Mehring, Reinhard (2018). *Vom Umgang mit Carl Schmitt*. Baden-Baden: Nomos.

Mehring, Reinhard (2021). *Carl Schmitts Gegenrevolution*. Hamburg: Europäische Verlagsanstalt.

Mehring, Reinhard (2022). *Carl Schmitt: Aufstieg und Fall*. 2nd, revised edition. München: Beck.

Mehring, Reinhard (2023). „'Beschleuniger wider Willen'? Carl Schmitts Hobbes im Spätwerk. Eine entstehungsgeschichtliche Klärung nach neuen Quellen. *Der Staat* 61, 461–483.

Meier, Christian (1988). "Zu Carl Schmitts Begriffsbildung." In *Complexio Oppositorum: Über Carl Schmitt*, edited by Helmut Quaritsch, 535–557 and 605–607. Berlin: Duncker & Humblot.

Meier, Heinrich (1994). *Die Lehre Carl Schmitts: vier Kapitel zur Unterscheidung Politischer Theologie und Politischer Philosophie*. Stuttgart: Metzler.

Meier, Heinrich (1995). *Carl Schmitt and Leo Strauss: The Hidden Dialogue*. Chicago: University of Chicago Press.

Meier, Heinrich (1996). *Die Denkbewegung von Leo Strauss*. Stuttgart: Metzler.

Meier, Heinrich (1998). *The Lesson of Carl Schmitt*. Chicago: University of Chicago Press.

Meierhenrich, Jens and Oliver Simons (2016a). *The Oxford Handbook of Carl Schmitt*, edited by Jens Meierhenrich and Oliver Simons. Oxford: Oxford University Press.

Meierhenrich, Jens and Oliver Simons (2016b). "A Fanatic of Order in an Epoch of Confusing Turmoil." In *The Oxford Handbook of Carl Schmitt*, edited by Jens Meierhenrich and Oliver Simons, 3–70. Oxford: Oxford University Press.

Meinel, Florian (2011). *Der Jurist in der industriellen Gesellschaft: Ernst Forsthoff und seine Zeit*. Berlin: de Gruyter.

Meyer N.N. 1942: "Art. Schmitt, Carl." In *Meyers Lexikon. 8. Auflage*, 1175–1176. Leipzig.

Meyer, Hans (2018). "Ernst Friesenhahn." In *Staatsrechtlehrer des 20. Jahrhunderts: Deutschland – Österreich – Schweiz*, edited by Peter Häberle, Michael Kilian, and Heinrich Wolff. 2nd edition, 693–708. Berlin: de Gruyter.

Meyer, Thomas (2023). *Hannah Arendt. Die Biografie*. München: Piper.

Mezger, Edmund (1934). *Kriminalpolitik auf kriminologischer Grundlage*. Stuttgart: Enke.

Michael, Robert and Karin Doerr (2002). *Nazi-Deutsch/Nazi German: An English Lexicon of the Language of the Third Reich*. Westport, CT: Greenwood Press.

Mieder, Wolfgang (2002). "Foreword." In Robert Michael and Karin Doerr, *Nazi-Deutsch/Nazi German: an English lexicon of the language of the Third Reich*, xv-xvii. Westport, CT: Greenwood Press.

Middendorf, Stephanie (2022). *Macht der Ausnahme: Reichsfinanzministerium und Staatlichkeit 1919–1945*. Berlin: de Gruyter.

Minca, Claudio and Rory Rowan (2016). *On Schmitt and Space*. London: Routledge.

Möller, Horst (1984). *Exodus der Kultur: Schriftsteller, Wissenschaftler und Künstler in der Emigration nach 1933*. München: Beck.

Morris, Leslie (2002). "Foreword." In Robert Michael and Karin Doerr, *Nazi-Deutsch/Nazi German: an English lexicon of the language of the Third Reich*, xi-xiv. Westport, CT: Greenwood Press.

Mouffe, Chantal (1999). "Introduction: Schmitt's Challenge." In *The Challenge of Carl Schmitt*, edited by Chantal Mouffe, 1–6. London: Verso.

Mouffe, Chantal (2000). *The Democratic Paradox*. London: Verso.

Mouffe, Chantal (2005). *On the Political*. London: Routledge.

Mouffe, Chantal (2007). "Carl Schmitt's Warnings on the Dangers of an Unipolar World." In *The International Thought of Carl Schmitt: Terror, War, and the Crises of the Global Order*, edited by Louiza Odyssecs and Fabio Petito, 137–153. London: Routledge.

Mouffe, Chantal (2022). *Towards a Green Democratic Revolution*. London: Routledge.

Müller, Ernst and Falko Schmieder (2016). *Begriffsgeschichte und historische Semantik*. Berlin: Suhrkamp.

Müller, Jan Werner (2003). *A Dangerous Mind: Carl Schmitt in Post-War European Thought*. New Haven: Yale University Press.

Müller, Jan Werner (2007). "Mit Schmitt gegen Schmitt und gegen die liberale Weltordnung." In *Der Staat des Dezisionismus: Carl Schmitt in der internationalen Debatte*, edited by Rüdiger Voigt, 201–217. Baden-Baden: Nomos.

Muth, Henrich (1971). "Carl Schmitt in der deutschen Innenpolitik des Sommers 1932." In *Beiträge zur Geschichte der Weimarer Republik* (Historische Zeitschrift, Beiheft N.F. 1), edited by Theodor Schieder, 75–147. München:

Müller, Ingo (1987). *Furchtbare Juristen: die unbewältigte Vergangenheit unserer Justiz*. München: Kindler.

Müller, Tim B. (2010). *Krieger und Gelehrte: Herbert Marcuse und die Denksysteme im Kalten Krieg*. Hamburg: Hamburger Edition.

Müller-Doohm, Stefan (2011). *Adorno: Eine Biographie*. 2nd edition. Berlin: Suhrkamp.

Müller-Doohm, Stefan (2014). *Jürgen Habermas: Eine Biographie*. Berlin: Suhrkamp.

Münkler, Herfried (1984). "Carl Schmitt und Thomas Hobbes." *Neue Politische Literatur* 29, 352–356.

Münkler, Herfried (1992). *Gewalt und Ordnung: Bilder des Krieges im politischen Denken*. Frankfurt am Main: Fischer.

Münkler, Herfried (2004). The New Wars. Cambridge: Polity Press, 2004.

Neuburger, Otto (1944). *Civil Affairs Guide: German-English Dictionary of German Administrative Terms*, Pamphlet No. 31–169. Washington, D.C.: War Department.

Neumann, Franz L. (1930). "Die soziale Bedeutung der Grundrechte in der Weimarer Verfassung." In Franz L. Neumann, *Wirtschaft, Staat, Demokratie: Aufsätze 1930–1954*, edited by Alfons Söllner, 57–75. Frankfurt am Main: Suhrkamp, 1978.

Neumann, Franz L. (1931). "Über die Voraussetzungen und den Rechtsbegriff einer Wirtschaftsverfassung." In Franz L. Neumann, *Wirtschaft, Staat, Demokratie: Aufsätze 1930–1954*, edited by Alfons Söllner, 76–102. Frankfurt am Main: Suhrkamp, 1978.

Neumann, Franz L. (1932). *Koalitionsfreiheit und Reichsverfassung*. Berlin: Carl Heymanns.

Neumann, Franz L. (1933). *Das gesamte Pressenotrecht vom 4. Februar 1933. Mit Anhang: Das Pressenotrecht vom 28. Februar 1928*. Berlin.

Neumann, Franz L. (1935). *The Rule of Law: Political Theory and the Legal System in Modern Society*. Dover: Berg Publishers, 1986.

Neumann, Franz L. (1937). "The Change in the Function of Law in Modern Society." In Franz L. Neumann, *The Democratic and the Authoritarian State*, 22–68. New York: Free Press, 1957.

Neumann, Franz L. (1940). "Types of Natural Law." *Studies in Philosophy and Social Science* 8 (1940): 338–361.

Neumann, Franz L. (1944a). *Behemoth: The Structure and Practice of National Socialism 1933–1944*. Second edition. New York: Oxford University Press.

Neumann, Franz L. (1944b). "Problems Concerning the Treatment of War Criminals." In *Secret Reports on Nazi Germany: The Frankfurt School Contribution to the War Effort*, edited by Raffaele Laudani, 457–463. Princeton: Princeton University Press, 2013.

Neumann, Sigmund (1932). *Die deutschen Parteien: Wesen und Wandel nach dem Kriege*. Berlin: Junker und Dünnhaupt.

Neumann, Volker (1981). "Verfassungstheorien politischer Antipoden: Otto Kirchheimer und Carl Schmitt." *Kritische Justiz* 14, no. 3 (1981): 235–254.

Neumann, Volker (1983). "Carl Schmitt und die Linke." *Die Zeit*, July 8, 1983: 32.

Neumann, Volker (1988). "Die Wirklichkeit im Lichte der Idee." In *Complexio Oppositorum: Über Carl Schmitt*, edited by Helmut Quaritsch, 557–575. Berlin: Duncker & Humblot.

Neumann, Volker (2015). *Carl Schmitt als Jurist*. Tübingen: Mohr Siebeck.

Neumann, Volker (2021). "Nachruf auf Hasso Hofmann." *Jahrbuch Politisches Denken* 31, 11–14.

Nipperdey, Thomas and Reinhard Rürup (1972). "Antisemitismus." In *Geschichtliche Grundbegriffe: Historisches Lexikon zur politisch-sozialen Sprache in Deutschland, Band 1*, 129–153. Stuttgart: Klett.

Noack, Paul (1993). *Carl Schmitt: Eine Biographie*. Berlin: Propyläen.

Nunan, Timothy (2011). "Translator's Introduction." In Carl Schmitt, *Writings on War*, edited and translated by Timothy Nunan, 1–26. Malden and Cambridge: Polity Press, 2011.

Oberlercher, Reinhold (1993). "Zerlegung der Lage." In *Politische Lageanalyse: Festschrift für Hans-Joachim Arndt*, edited by Volker Beismann and Markus J. Klein, 234–240. Bruchsal: San Cascino Verlag.

Odysseos, Louiza (2007). "Carl Schmitt on the Spaceless Universalism of Cosmopolitanism and the War on Terror." In *The International Thought of Carl Schmitt: Terror, War, and the Crises of the Global Order*, edited by Louiza Odysseos and Fabio Petito, 124–143. London: Routledge.

Odysseos, Louiza and Fabio Petito (2007). *The International Thought of Carl Schmitt: Terror, War, and the Crises of the Global Order*, edited by Louiza Odysseos and Fabio Petito. London: Routledge.

Offe, Claus (1969). "Politische Herrschaft und Klassenstrukturen." In *Politikwissenschaft*, edited by Gisela Kress and Dieter Senghaas, 155–189. Frankfurt am Main: Europäische Verlags-Anstalt.

Offe, Claus (1985). *Disorganized Capitalism*. Cambridge: Polity Press.

Olbrich, Josef (2001). *Geschichte der Erwachsenenbildung in Deutschland*. Opladen: Leske + Budrich.

Olechowski, Thomas (2020). *Hans Kelsen: Biographie eines Rechtswissenschaftlers*. Tübingen: Mohr Siebeck.

Olson, Karsten (2016). "Historical-Sociology vs. Ontology: The Role of Economy in Kirchheimer and Schmitt's Essay ‚Legality and Legitimacy'." *History of the Human Sciences* 29, no. 2 (2016): 96–112.

Ooyen, Robert Chr. van and Frank Schale 2011: *Kritische Verfassungspolitologie. Das Statsverständnis von Otto Kirchheimer*, edited by Robert van Oyen and Frank Schale. Baden-Baden: Nomos.

Paechter, Henry (1962). "Politics and the Rule of Law." *Dissent* 9 (Summer 1962): 298–300.

Paechter, Heinz in association with Bertha Hellman, Hedwig Paechter, Karl O. Paetel (1944). *Nazi-Deutsch. A Glossary of Contemporary German Usage, with Appendices on Government, Military and Economic Institutions*. New York: Frederick Ungar Publishing Co.

Palmier, Jean-Michel (2019). *Walter Benjamin*. Berlin: Suhrkamp.

Pan, David (2016). "Tragedy as Exception in Carl Schmitt's Hamlet or Hecuba." In *The Oxford Handbook of Carl Schmitt*, edited by Jens Meierhenrich and Oliver Simons, 731–750. Oxford: Oxford University Press.

Papcke, Sven (1993). *Deutsche Soziologie im Exil: Gegenwartsdiagnose und Epochenkritik 1933–45*. Frankfurt am Main: Campus.

Pauer-Studer, Herlinde (2014). "Rechtfertigungen des Unrechts. Das Rechtsdenken des Nationalsozialismus." In *Rechtfertigungen des Unrechts: Das Rechtsdenken des Nationalsozialismus in Originaltexten*, edited by Herlinde Pauer-Studer and Julian Fink, 15–39. Frankfurt am Main: Suhrkamp.

Perels, Joachim (1988). "Otto Kirchheimer: Demokratischer Marxist und Verfassungstheoretiker." In *Streitbare Juristen: Eine andere Tradition*, edited by Redaktion Kritische Justiz, 401–414. Baden-Baden: Nomos.

Perels, Joachim (1989). "Weimarer Demokratie und gesellschaftliche Machtverhältnisse: Zur Methode der Verfassungsanalysen Otto Kirchheimers." In *Verfassungsstaat, Souveränität, Pluralismus: Otto Kirchheimer zum Gedächtnis*, edited by Wolfgang Luthardt and Alfons Söllner, 57–66. Opladen: Westdeutscher Verlag.

Perels, Joachim (2002). "Franz L. Neumanns Beitrag zur Konzipierung der Nürnberger Prozesse." In *Kritische Theorie der Politik: Franz L. Neumann*, edited by Mattias Iser and David Strecker, 83–94. Baden-Baden: Nomos.

Perry, Jonathan (2014). "From Frankfurt to Westermann: Forced Labor and the Early Development of Finley's Thought." *American Journal of Philology* 135, no. 2 (2014): 221–241.

Pfabigan, Alfred (1982). *Max Adler: Eine politische Biographie*. Frankfurt am Main: Campus.

Plaetzer, Niklas (2022). "Eichmann in Plettenberg. Carl Schmitt Reads Hannah Arendt." *Modern Intellectual History* 19 (1), 1–28.

Preuß, Ulrich K. (1973). *Legalität und Pluralismus*. Frankfurt am Main: Suhrkamp.

Preuß, Ulrich K. (1978). "Strategien staatsbürgerlicher Diskriminierung." *In 3. Internationales Russel-Tribunal: Zur Situation der Menschenrechte in der Bundesrepublik Deutschland, Band 1*, 78–96. Berlin: Rotbuch-Verlag.

Preuß, Ulrich K. (1987). "Carl Schmitt und die Frankfurter Schule: Liberalismuskritik im 20. Jahrhundert." *Geschichte und Gesellschaft* 13, no. 3 (1987): 400–418.

Preuß, Ulrich K. (1989). "Politische Justiz im demokratischen Verfassungsstaat." In *Verfassungsstaat, Souveränität, Pluralismus: Otto Kirchheimer zum Gedächtnis*, edited by Wolfgang Luthardt and Alfons Söllner, 129–152. Opladen: Westdeutscher Verlag.

Preuß, Ulrich K. (2016). "Carl Schmitt and the Weimar Constitution." In *The Oxford Handbook of Carl Schmitt*, edited by Jens Meierhenrich and Oliver Simons, 471–490. Oxford: Oxford University Press.

Priemel, Kim Christian (2016). *The Betrayal: The Nuremberg Trials and German Divergence*. Oxford: Oxford University Press.

Priester, Karin (2015). "Erkenne die Lage! Über die rechtspopulistische Versuchung des bundesdeutschen Konservatismus." *Indes* no. 3 (2015): 84–92.

Pross, Harry (1993). *Memoiren eines Inländers*. München: Artemis & Winkler Verlag.

Puglierin, Jana (2011). *John H. Herz: Leben und Denken zwischen Idealismus und Realismus, Deutschland und Amerika*. Berlin: Duncker & Humblot.

Pyta, Wolfram and Gabriel Seiberth (1999). "Die Staatskrise der Weimarer Republik im Spiegel des Tagesbuchs von Carl Schmitt." *Der Staat* 38, no. 3 and 4 (1999): 423–448 and 594–610.

Quaritsch, Helmut (1994). "Nachwort." In Carl Schmitt, *Das international-rechtliche Verbrechen des Angriffskrieges und der Grundsatz ‚nullum crimen, nulla poena sine lege'*, edited by Helmut Quaritzsch, 125–250. Berlin: Duncker & Humblot.

Quaritsch, Helmut (1995). *Positionen und Begriffe Carl Schmitts*. 3rd edition. Berlin: Duncker & Humblot.

Quaritsch, Helmut (2000). "Carl Schmitt im Nürnberger Justizgefängnis." In *Carl Schmitt, Antworten in Nürnberg*, edited by Helmut Quaritsch, 11–50. Berlin: Duncker & Humblot.

Quaritsch, Helmut (2018). *Positionen und Begriffe Carl Schmitts*. 5th edition. Berlin: Duncker & Humblot.

Rand Corporation (1988). *Remembering Nathan Leites, An Appreciation: Recollections of Some Friends, Colleagues, and Students*. Santa Monica: Rand Corporation.

Rasch, William (2000). "A Just War: Or Just a War? Schmitt, Habermas, and the cosmopolitan orthodoxy." *Cardozo Law Review* 21: 1665–1684.

Raulet, Gérard (2000). "Unfall der Republik oder strukturelles Problem? Überlegungen zum antiparlamentarischen Denken in der Weimarer Republik." In *Intellektuelle im Nationalsozialismus*, edited by Wolfgang Bialas and Manfred Gangl, 50–67. Frankfurt am Main: Peter Lang.

Reinthal, Angela and Matthias Hurst (2022). *Eine Welt voll Bildern. Carl Schmitt und der Film der Weimarer Republik*. Berlin: Duncker & Humblot.

Renner, Karl (1929). *Die Rechtsinstitute des Privatrechts und ihre soziale Funktion: Ein Beitrag zur Kritik des bürgerlichen Rechts*. 2nd edition. Tübingen: Mohr.

Rensmann, Lars (2017). *The Politics of Unreason: The Frankfurt School and the Origins of Modern Antisemitism*. Albany: SUNY Press.

Renz, Werner (2021). *Ad Hannah Arendt Eichmann in Jerusalem*. Hamburg: Europäische Verlagsanstalt.

Richter, Emanuel (2001). "Der falsche Prophet: Carl Schmitt in den USA." In *Mythos Staat: Carl Schmitts Staatsverständnis*, edited by Rüdiger Voigt, 215–265. Baden-Baden: Nomos.

Ridder, Helmut (1977). "Vom Wendekreis der Grundrechte." *Leviathan* 5, no. 4 (1977): 467–521.

Ridder, Helmut (2005). "Stationen im Leben eines Juristen: Helmut Ridder im Gespräch mit Joachim Perels." *Neue Politische Literatur* 50 (2005): 365–382.

Rose, Paul (2002). "Foreword." In Robert Michael and Karin Doerr, *Nazi-Deutsch/Nazi-German: an English lexicon of the language of the Third Reich*, vii-ix. Westport, CT: Greenwood Press.

Rossiter, Clinton (1949). *Constitutional Dictatorship: Crisis Government in Modern Democracies*. Princeton: Princeton University Press

Roth, Klaus (2005). "Carl Schmitt – Ein Verfassungsfreund? Seine Stellung Zur Weimarer Republik in der Phase der relativen Stabilisierung (1924–29)." *Zeitschrift für Politik* 52, no. 2 (2005): 141–56.

Rottleuthner, Hubert (1983). "Substantieller Dezisionismus. Zur Funktion der Rechtsphilosophie im Nationalsozialismus." In: *Recht, Rechtsphilosophie und Nationalsozialismus*, edited by Hubert Rottleuthner, 20–35. Wiesbaden: Steiner.

Rückert, Joachim (1993). "Franz L. Neumann. Ein Jurist mit Prinzipien." In *Der Einfluss deutscher Emigranten auf die Rechtsentwicklung in den USA und Deutschland*, edited by Marcus Lutter, Ernst C. Stiefel, and Michael H. Höflich, 437–474. Tübingen: Mohr Siebeck.

Rüthers, Bernd (1990). *Carl Schmitt im Dritten Reich*. 2nd edition. München: Beck.

Rumpf, Helmut (1972). *Carl Schmitt und Thomas Hobbes: Ideelle Beziehungen und aktuelle Bedeutung mit einer Abhandlung über die Frühschriften Carl Schmitts*. Berlin: Duncker & Humblot.

Sabrow, Martin (1998). *Die verdrängte Verschwörung: Der Rathenau-Mord und die deutsche Gegenrevolution*. Göttingen: Wallstein.

Salter, Michael (1999). "Neo-Fascist Legal Theory on Trial: An Interpretation of Carl Schmitt's Defense at Nuremberg from the Perspective of Franz Neumann's Critical Theory of Law." *Res Publica* 5 (1999): 161–194.

Salter, Michael (2007). *Nazi War Crimes, US Intelligence and Selective Prosecution at Nuremberg*. New York: Routledge.

Salzborn, Samuel 2009: "Leviathan und Behemoth. Staat und Mytos bei Thomas Hobbes und Carl Schmitt." In *Der Hobbes-Kristall: Carl Schmitts Hobbes-Interpretation in der Diskussion*, edited by Rüdiger Voigt, 143–164. Stuttgart: Steiner.

Sander, Hans-Dietrich (1993). "Das Gastmahl des Leviathan." In *Politische Lageanalyse: Festschrift für Hans-Joachim Arndt*, edited by Volker Beismann and Markus J. Klein, 259–276. Bruchsal: San Cascino Verlag.

Sartori, Giovanni (1976). *Parties and Party Systems*. Cambridge: Cambridge University Press.

Scelle, Georges (1938). "Préface." In Eduard Bristler, *Die Völkerrechtslehre des Nationalsozialismus*, 5–15. Zürich: Europa-Verlag.

Schäffer, Albert (1987) "Ein Trauma. Carl Schmitt und die Frankfurter Schule." *Frankfurter Allgemeine Zeitung*, February 18, 1987, 14.

Schale, Frank (2003). "Otto Kirchheimer: Linkssozialistische Analysen in der Weimarer Republik." In *Linke Juristen in der Weimarer Republik*, edited by Manfred Gangl, 276–290. Frankfurt am Main: Peter Lang.

Schale, Frank (2006). *Zwischen Engagement und Skepsis: Eine Studie zu den Schriften von Otto Kirchheimer*. Baden-Baden: Nomos.

Schale, Frank (2011). "Abschied vom Ideal: Otto Kirchheimers Frankreich-Schriften." In *Deutsche Frankreich-Bücher aus der Zwischenkriegszeit*, edited by Alfons Söllner, 293–317. Baden-Baden: Nomos.

Schale, Frank, Lisa Klingsporn, and Hubertus Buchstein (2018). "Otto Kirchheimer: Capitalist State, Political Parties, and Political Justice." In *SAGE-Handbook of Frankfurt School Critical Theory*, edited by Beverly Best, Werner Bonefeld, and Chris O'Kane, 105–122. London: Sage.

Schefold, Dian (2012). *Bewahrung der Demokratie: Ausgewählte Aufsätze*. Berlin: Berliner Wissenschaftsverlag.

Schelsky, Helmut (1938). "Die Totalität des Staates bei Hobbes." *Archiv für Rechts- und Sozialphilosophie* 31 (1937/38): 176–201.

Schelsky, Helmut (1979). *Die Hoffnung Blochs: Kritik der marxistischen Existenzphilosophie eines Jugendbewegten*. Stuttgart: Klett-Cotta.

Scheuerman, William E. (1994). *Between the Norm and the Exception: The Frankfurt School and the Rule of Law*. Cambridge: MIT Press.

Scheuerman, William E. (1996). "Introduction." In *The Rule of Law under Siege: Selected Essays of Franz L. Neumann and Otto Kirchheimer*, edited by William E. Scheuerman, 1–27. Berkeley: University of California Press.

Scheuerman, William E. (2000). "Introduction." In *The Rule of Law under Siege: Selected Essays of Franz L. Neumann and Otto Kirchheimer*, edited by William E. Scheuerman, 9–28. Berkeley: University of California Press.

Scheuerman, William E. (2006). "Carl Schmitt and the Road to Abu Graib." *Constellations* 13, no. 1 (2006): 108–124.

Scheuerman, William E. (2020). *The End of Law: Carl Schmitt in the Twenty-First Century*. 2nd Edition. London: Rowman & Littlefield Publishers.

Scheuerman, William E. and Peter C. Caldwell (2000). "Introduction." In *From Liberal Democracy to Fascism: Legal and Political Thought of the Weimar Republic*, edited by William E. Scheuerman and Peter C. Caldwell, 1–19. Boston: Brill.

Schickel, Joachim (1970). *Guerrilleros, Partisanen, Theorie und Praxis*. München: Hanser.

Schiller, Dieter et al. (1981). *Kunst und Literatur im antifaschistischen Exil 1933–1945, Band 7: Exil in Frankreich*. Leipzig: Reclam.

Schlak, Stephan (2008). *Wilhelm Hennis: Szenen einer Ideengeschichte der Bundesrepublik*. München: Beck.

Schmid, Carlo (1979). *Erinnerungen*. Bern und München: Scherz.

Schmid, Richard (1948). "Denazification: A German Critique." *American Perspective* 2, no. 5 (1948): 231–242.

Schmitt, Carl (1919). *Political Romanticism*. Chicago: MIT Press, 1986.

Schmitt, Carl (1921). *Dictatorship: From the Origin of the Modern Concept of Sovereignty to Proletarian Class Struggle*. Cambridge: Polity, 2014.

Schmitt, Carl (1922). *Political Theology: Four Chapters on the Concept of Sovereignty*. Chicago: University of Chicago Press, 2005.

Schmitt, Carl (1923a). *The Crisis of Parliamentary Democracy*. Cambridge: MIT Press, 2000.

Schmitt, Carl (1923b). *Roman Catholicism and Political Form*. Westport: Greenwood Press, 1988.

Schmitt, Carl (1924a). "Die Kernfrage des Völkerbundes." In *Carl Schmitt: Frieden oder Pazifismus? Arbeiten zum Völkerrecht und zur Internationalen Politik 1924–1978*, edited by Günter Maschke, 1–26. Berlin: Duncker & Humblot, 2005.

Schmitt, Carl (1924b). "Reichstagsauflösungen." In Carl Schmitt, *Verfassungsrechtliche Aufsätze aus den Jahren 1924–1954*, 13–26. Berlin: Duncker & Humblot, 1958.

Schmitt, Carl (1924c). "Appendix: The Dictatorship of the President of the Reich according to Article 48 of the Weimar Constitution." In Carl Schmitt, *Dictatorship: From the Origin of the Modern Concept of Sovereignty to Proletarian Class Struggle*, 180–226. Cambridge: Polity, 2014.

Schmitt, Carl (1925a). "Die Rheinlande als Objekt internationaler Politik." In Carl Schmitt, *Frieden oder Pazifismus? Arbeiten zum Völkerrecht und zur Internationalen Politik 1924–1978*, edited by Günter Maschke, 26–50. Berlin: Duncker & Humblot, 2005.

Schmitt, Carl (1925b). "Der Status Quo und der Friede." In Carl Schmitt, *Frieden oder Pazifismus? Arbeiten zum Völkerrecht und zur Internationalen Politik 1924–1978*, edited by Günter Maschke, 51–72. Berlin: Duncker & Humblot, 2005.

Schmitt, Carl (1926a). "Die Kernfrage des Völkerbundes." In Carl Schmitt, *Frieden oder Pazifismus? Arbeiten zum Völkerrecht und zur Internationalen Politik 1924–1978*, edited by Günter Maschke, 73–194. Berlin: Duncker & Humblot, 2005.

Schmitt, Carl (1926b). "Preface of the Second Edition." In Carl Schmitt, *The Crisis of Parliamentary Democracy*, 1–17. Cambridge: MIT Press, 2000.

Schmitt, Carl (1926c). "Das Ausführungsgesetz zu Artikel 48." In Carl Schmitt, *Staat, Großraum, Nomos: Aufsätze aus den Jahren 1916–1969*, edited by Günter Maschke, 38–42. Berlin: Duncker& Humblot, 1995.

Schmitt, Carl (1926d). *Unabhängigkeit der Richter, Gleichheit vor dem Gesetz und Gewährleistung des Privateigentums nach der Weimarer Verfassung*. Berlin: Duncker & Humblot.

Schmitt, Carl (1926e). *Die geistesgeschichtliche Lage des heutigen Parlamentarismus*. Berlin: Duncker & Humblot.

Schmitt, Carl (1927a). "Der Begriff des Politischen." In Carl Schmitt, *Positionen und Begriffe im Kampf mit Weimar – Genf – Versailles 1923–1939*, 67–74. Hamburg: Hanseatische Verlagsanstalt, 1995.

Schmitt, Carl (1927b). "Der Völkerbund und Europa." In Carl Schmitt, *Frieden oder Pazifismus? Arbeiten zum Völkerrecht und zur Internationalen Politik 1924–1978*, edited by Günter Maschke, 240–248. Berlin: Duncker & Humblot, 2005.

Schmitt, Carl (1928a). "Völkerrechtliche Probleme im Rheingebiet." In Carl Schmitt, *Frieden oder Pazifismus? Arbeiten zum Völkerrecht und zur Internationalen Politik 1924–1978*, edited by Günter Maschke, 255–273. Berlin: Duncker & Humblot, 2005.

Schmitt, Carl (1928b). *Constitutional Theory*. Durham and London: Duke University Press, 2008.

Schmitt, Carl (1928c). *Verfassungslehre*. Berlin: Duncker & Humblot,1983.

Schmitt, Carl (1928d). "Der bürgerliche Rechtsstaat." *In* Carl Schmitt, *Staat, Großraum, Nomos. Aufsätze aus den Jahren 1916–1969*, edited by Günter Maschke, 44–50. Berlin: Duncker & Humblot, 1995.

Schmitt, Carl (1929a). "Das Reichsgericht als Hüter der Verfassung." In Carl Schmitt, *Verfassungsrechtliche Aufsätze aus den Jahren 1924–1954*, 63–100. Berlin: Duncker & Humblot, 1958.

Schmitt, Carl (1929b). "Wesen und Werden des faschistischen Staates." In Carl Schmitt, *Positionen und Begriffe im Kampf mit Weimar – Genf – Versailles 1923–1939*, 109–115. Hamburg: Hanseatische Verlagsanstalt, 1940.

Schmitt, Carl (1929c). "Die Auflösung des Enteignungsbegriffs." In Carl Schmitt, *Verfassungsrechtliche Aufsätze aus den Jahren 1924–1954*, 110–118. Berlin: Duncker & Humblot, 1958.

Schmitt, Carl (1930a). "Die politische Lage der entmilitarisierten Rheinlande." In Carl Schmitt, *Frieden oder Pazifismus? Arbeiten zum Völkerrecht und zur Internationalen Politik 1924–1978*, edited by Günter Maschke, 274–280. Berlin: Duncker & Humblot, 2005.

Schmitt, Carl (1930b). "Der Völkerbund und das politische Problem der Friedenssicherung." In Carl Schmitt, *Frieden oder Pazifismus? Arbeiten zum Völkerrecht und zur In-*

ternationalen Politik 1924–1978, edited by Günter Maschke, 281–332. Berlin: Duncker & Humblot, 2005.

Schmitt, Carl (1930c). "Hugo Preuß. Sein Staatsbegriff und seine Stellung in der deutschen Staatsrechtslehre. Tübingen." In Carl Schmitt, *Der Hüter der Verfassung*. 5th editon, 161–184. Berlin: Duncker & Humblot, 2016.

Schmitt, Carl (1930d). "Einberufung und Vertagung des Reichstages nach Art. 24 Reichsverfassung." In *Deutsche Juristenzeitung* 35 (1930): 1285–1289.

Schmitt, Carl (1931a). "Der Völkerbund." In Carl Schmitt, *Frieden oder Pazifismus? Arbeiten zum Völkerrecht und zur Internationalen Politik 1924–1978*, edited by Günter Maschke, 222–249. Berlin: Duncker & Humblot, 2005.

Schmitt, Carl (1931b). *Der Hüter der Verfassung*. 5th edition. Berlin: Duncker & Humblot, 2016.

Schmitt, Carl (1931c). "Die staatsrechtliche Bedeutung der Notverordnung, insbesondere ihre Rechtsgültigkeit." In Carl Schmitt, *Verfassungsrechtliche Aufsätze aus den Jahren 1924–1954*, 233–260. Berlin: Duncker & Humblot, 1958.

Schmitt, Carl (1931d). "Freiheitsrechte und institutionelle Garantien." In Carl Schmitt, *Verfassungsrechtliche Aufsätze aus den Jahren 1924–1954*, 140–171. Berlin: Duncker& Humblot, 1958.

Schmitt, Carl (1931e). "Verfassungsstaat und Staatsnotstand: Rundfunkvortrag, 4. November 1931." In *Schmittiana, Band 8: Beiträge zu Leben und Werk Carl Schmitts*, edited by Piet Tommissen, 11–15. Berlin: Duncker & Humblot, 2003.

Schmitt, Carl (1932a). *The Concept of the Political*. Chicago: University of Chicago Press, 1996.

Schmitt, Carl (1932b). "USA und die völkerrechtlichen Formen des modernen Imperialismus." In Carl Schmitt, *Frieden oder Pazifismus? Arbeiten zum Völkerrecht und zur Internationalen Politik 1924–1978*, edited by Günter Maschke, 349–377. Berlin: Duncker & Humblot, 2005.

Schmitt, Carl (1932c). *Legality and Legitimacy*. Durham and London: Duke University Press, 2004.

Schmitt, Carl (1932d). "Grundrechte und Grundpflichten." In Carl Schmitt, *Verfassungsrechtliche Aufsätze aus den Jahren 1924–1954*, 181–231. Berlin: Duncker & Humblot, 1958.

Schmitt, Carl (1932e). "Starker Staat und gesunde Wirtschaft." In Carl Schmitt, *Staat, Großraum, Nomos. Aufsätze aus den Jahren 1916–1969*, edited by Günter Maschke, 71–85. Berlin: Duncker & Humblot, 1995.

Schmitt, Carl (1932f). "Die Verfassungsgemäßheit der Bestellung eines Reichskommissars für das Land Preußen." *Deutsche Juristen-Zeitung* 37 (1932): 557–564.

Schmitt, Carl (1932g). "Konstruktive Verfassungsprobleme." In Carl Schmitt, *Staat, Großraum, Nomos. Aufsätze aus den Jahren 1916–1969*, edited by Günter Maschke, 55–64. Berlin: Duncker & Humblot, 1995.

Schmitt, Carl (1932h). *Legalität und Legitimität*. Berlin: Duncker & Humblot, 1958.

Schmitt, Carl (1933a). "Das Gesetz zur Behebung der Not von Volk und Reich." *Deutsche Juristen-Zeitung* 38 (1933): 455–458.

Schmitt, Carl (1933b). "Das gute Recht der deutschen Revolution." In Carl Schmitt, *Gesammelte Schriften 1933–1936: mit ergänzenden Beiträgen aus der Zeit des Zweiten Weltkriegs*, 28–32. Berlin: Duncker & Humblot, 2021.

Schmitt, Carl (1933c). "Die deutschen Intellektuellen." In Carl Schmitt, *Gesammelte Schriften 1933–1936: mit ergänzenden Beiträgen aus der Zeit des Zweiten Weltkriegs*, 32–35. Berlin: Duncker & Humblot, 2021.

Schmitt, Carl (1933d). *Staat, Bewegung, Volk: Die Dreigliederung der politischen Einheit.* Hamburg: Hanseatische Verlagsanstalt.

Schmitt, Carl (1933e). "Der Staat des 20. Jahrhunderts." In Carl Schmitt, *Gesammelte Schriften 1933–1936 mit ergänzenden Beiträgen aus der Zeit des Zweiten Weltkriegs*, 36–38. Berlin: Duncker & Humblot, 2021.

Schmitt, Carl (1933f). "Der Neubau des Staats- und Verwaltungsrechts." In: *Deutscher Juristentag* 1933, 242–252. Berlin: Duncker & Humblot.

Schmitt, Carl (1933g). "Das Reichsstatthaltergesetz." In Carl Schmitt, *Gesammelte Schriften 1933–1936: mit ergänzenden Beiträgen aus der Zeit des Zweiten Weltkriegs*, 7–27. Berlin: Duncker & Humblot, 2021.

Schmitt, Carl (1933h). "Neue Leitsätze für die Rechtspraxis." In Carl Schmitt, *Gesammelte Schriften 1933–1936: mit ergänzenden Beiträgen aus der Zeit des Zweiten Weltkriegs*, 54–56. Berlin: Duncker & Humblot, 2021.

Schmitt, Carl (1933i). *Der Begriff des Politischen.* Hamburg: Hanseatische Verlagsanstalt.

Schmitt, Carl (1933j). "Weiterentwicklung des totalen Staats in Deutschland." In Carl Schmitt, *Positionen und Begriffe im Kampf mit Weimar– Genf– Versailles 1923–1939*, 185–190. Hamburg: Hanseatische Verlagsanstalt, 1940.

Schmitt, Carl (1933k). "Der Neubau des Staats- und Verwaltungsrechts." In Carl Schmitt, *Gesammelte Schriften 1933–1936: mit ergänzenden Beiträgen aus der Zeit des Zweiten Weltkriegs*, 57–69. Berlin: Duncker & Humblot, 2021.

Schmitt, Carl (1933l). "Reich – Staat – Bund." In Carl Schmitt, *Positionen und Begriffe im Kampf mit Weimar – Genf – Versailles 1923–1939*, 190–198. Berlin: Duncker & Humblot, 1941.

Schmitt, Carl (1934a). "Nationalsozialismus und Völkerrecht." In Carl Schmitt, *Frieden oder Pazifismus? Arbeiten zum Völkerrecht und zur Internationalen Politik 1924–1978*, edited by Günter Maschke, 391–423. Berlin: Duncker & Humblot, 2005.

Schmitt Carl (1934b). "Sowjet-Union und Genfer Völkerbund." In Carl Schmitt, *Frieden oder Pazifismus? Arbeiten zum Völkerrecht und zur Internationalen Politik 1924–1978*, edited by Günter Maschke, 424–435. Berlin: Duncker & Humblot, 2005.

Schmitt, Carl (1934c). "Ein Jahr nationalsozialistischer Verfassungsstaat." *Deutsches Recht* 4 (1934): 27-–30.

Schmitt, Carl (1934d). "Nationalsozialismus und Rechtsstaat." *Juristische Wochenschrift* 63 (1934): 712–718.

Schmitt, Carl (1934e). "Der Führer schützt das Recht." In Carl Schmitt, *Positionen und Begriffe im Kampf mit Weimar – Genf – Versailles 1923–1939*, 199–203. Hamburg: Hanseatische Verlagsanstalt, 1940.

Schmitt, Carl (1934f). *Staatsgefüge und Zusammenbruch des zweiten Reiches: Der Sieg des Bürgers über den Soldaten*, edited by Günter Maschke. Berlin: Duncker & Humblot, 2011.

Schmitt, Carl (1934g). *On the three Types of Juristic Thought.* Westport and London: Praeger, 2004.

Schmitt, Carl (1934h). "Nationalsozialistisches Rechtsdenken." In Carl Schmitt, *Gesammelte Schriften 1933–1936: mit ergänzenden Beiträgen aus der Zeit des Zweiten Weltkriegs*, 156–164. Berlin: Duncker & Humblot, 2021.

Schmitt, Carl (1934i). "Preface to the Second Edition." In Carl Schmitt, *Political Theology: Four Chapters on the Concept of Sovereignty*, edited by George Schwab, 1–4. Chicago: University of Chicago Press, 2005.

Schmitt, Carl (1935a). "Die Verfassung der Freiheit." In Carl Schmitt, *Gesammelte Schriften 1933–1936: mit ergänzenden Beiträgen aus der Zeit des Zweiten Weltkriegs*, 282–284. Berlin: Duncker & Humblot, 2021.

Schmitt, Carl (1935b). "Der Rechtsstaat." In Carl Schmitt, *Staat, Großraum, Nomos. Aufsätze aus den Jahren 1916–1969*, edited by Günter Maschke, 108–117. Berlin: Duncker & Humblot, 1995.

Schmitt, Carl (1935c). "Die nationalsozialistische Gesetzgebung und der Vorbehalt des ‚ordre public' im Internationalen Privatrecht." In Carl Schmitt, *Gesammelte Schriften 1933–1936: mit ergänzenden Beiträgen aus der Zeit des Zweiten Weltkriegs*, 369–385. Berlin: Duncker & Humblot, 2021.

Schmitt, Carl (1936a). "Über die neuen Aufgaben der Verfassungsgeschichte." In Carl Schmitt, *Gesammelte Schriften 1933–1936: mit ergänzenden Beiträgen aus der Zeit des Zweiten Weltkriegs*, 408–413. Berlin: Duncker & Humblot, 2021.

Schmitt, Carl (1936b). "Stellungnahme der Wissenschaftlichen Abteilung des Nationalsozialistischen Rechtswahrerbundes." In Carl Schmitt, *Gesammelte Schriften 1933–1936: mit ergänzenden Beiträgen aus der Zeit des Zweiten Weltkriegs*, 431–481. Berlin: Duncker & Humblot, 2021.

Schmitt, Carl (1936c). "Das Judentum in der Rechtswissenschaft. Eröffnungsvortrag." In Carl Schmitt, *Gesammelte Schriften 1933–1936: mit ergänzenden Beiträgen aus der Zeit des Zweiten Weltkriegs*, 482–485. Berlin: Duncker & Humblot, 2021.

Schmitt, Carl (1936d). "Das Judentum in der Rechtswissenschaft: Schlußwort." In Carl Schmitt, *Gesammelte Schriften 1933–1936: mit ergänzenden Beiträgen aus der Zeit des Zweiten Weltkriegs*, 486–491. Berlin: Duncker& Humblot, 2021.

Schmitt, Carl (1936e). "Faschistische und nationalsozialistische Rechtswissenschaft." In Carl Schmitt, *Gesammelte Schriften 1933–1936: mit ergänzenden Beiträgen aus der Zeit des Zweiten Weltkriegs*, 333–334. Berlin: Duncker & Humblot, 2021.

Schmitt, Carl (1936f). "Die Grundzüge des nationalsozialistischen Staates." In Carl Schmitt, *Gesammelte Schriften 1933–1936: mit ergänzenden Beiträgen aus der Zeit des Zweiten Weltkriegs*, 335–349. Berlin: Duncker & Humblot, 2021.

Schmitt, Carl (1936g). "Die Bedeutung des Rheinbundes für die spätere deutsche Verfassungsgeschichte." In Carl Schmitt, *Gesammelte Schriften 1933–1936: mit ergänzenden Beiträgen aus der Zeit des Zweiten Weltkriegs*, 362–363. Berlin: Duncker & Humblot, 2021.

Schmitt, Carl (1937a). "Totaler Feind, totaler Krieg, totaler Staat." In Carl Schmitt, *Frieden oder Pazifismus? Arbeiten zum Völkerrecht und zur Internationalen Politik 1924–1978*, edited by Günter Maschke, 481–485. Berlin: Duncker & Humblot, 2005.

Schmitt, Carl (1937b). "The Turn of the Discriminating Concept of War." In Carl Schmitt, *Writings on War*, edited by Timothy Nunan, 30–74. Cambridge: Polity Press, 2011.

Schmitt, Carl (1937c). "The State as Mechanism in Hobbes and Descartes." In Carl Schmitt, *The Leviathan in the State Theory of Thomas Hobbes,* 91–103. Chicago: University of Chicago Press, 2008.

Schmitt, Carl (1938a). *The Leviathan in the State Theory of Thomas Hobbes*. Chicago: University of Chicago Press, 2008.

Schmitt, Carl (1938b). "Grußbotschaft an die Hobbes-Gesellschaft in Kiel, 5. April 1938." In Carl Schmitt, *Gesammelte Schriften 1933–1936: mit ergänzenden Beiträgen aus der Zeit des Zweiten Weltkriegs*, 495. Berlin: Duncker & Humblot, 2021.

Schmitt, Carl (1938c). "Völkerrechtliche Neutralität und völkische Totalität." In Carl Schmitt, *Frieden oder Pazifismus? Arbeiten zum Völkerrecht und zur Internationalen Politik 1924–1978*, edited by Günter Maschke, 617–622. Berlin: Duncker & Humblot, 2005.

Schmitt, Carl (1938d). *Der Leviathan in der Staatslehre des Thomas Hobbes*. Stuttgart: Klett-Cotta 1982.

Schmitt, Carl (1939a). *Völkerrechtliche Großraumordnung mit Interventionsverbot für raumfremde Mächte.* Berlin: Hanseatische Verlagsanstalt.

Schmitt, Carl (1939b). "Über die zwei großen 'Dualismen' des heutigen Rechtssystems." In Carl Schmitt, *Positionen und Begriffe im Kampf mit Weimar – Genf- Versailles 1923–1939*. 261–271. Hamburg: Hanseatische Verlagsanstalt, 1940.

Schmitt, Carl (1940). *Positionen und Begriffe im Kampf mit Weimar – Genf– Versailles 1923–1939*. Hamburg: Hanseatische Verlagsanstalt.

Schmitt, Carl (1941a). "Preliminary Remark of the 1941 edition of The Großraum Order of International Law." In Carl Schmitt, *Writings on War*, edited by Timothy Nunan, 76–77. Cambridge: Polity Press, 2011.

Schmitt, Carl (1941b). "Staatliche Souveränität und freies Meer Totalität." In Carl Schmitt, *Frieden oder Pazifismus? Arbeiten zum Völkerrecht und zur Internationalen Politik 1924–1978*, edited by Günter Maschke, 401–430. Berlin: Duncker & Humblot, 2005.

Schmitt, Carl (1941c). "Das Meer gegen das Land." In Carl Schmitt, *Frieden oder Pazifismus? Arbeiten zum Völkerrecht und zur Internationalen Politik 1924–1978*, edited by Günter Maschke, 395–400. Berlin: Duncker & Humblot, 2005.

Schmitt, Carl (1941d). "Völkerrechtliche Großraumordnung mit Interventionsverbot für raumfremde Mächte. Text der vierten Auflage". In Carl Schmitt, *Staat, Großraum, Nomos: Aufsätze aus den Jahren 1916–1969*, edited by Günter Maschke, 269–371. Berlin: Duncker & Humblot, 1995.

Schmitt, Carl (1941e). "The Großraum Order of International Law with a Ban on Intervention from Spatially Foreign Powers." In *Carl Schmitt, Writings on War*, edited by Timothy Nunan, 75–124. Cambridge: Polity Press, 2011.

Schmitt, Carl (1942a). *Land and Sea: A World-Historical Meditation*. Candor, NY: Telos Press, 2015.

Schmitt, Carl (1942b). "Beschleuniger wider Willen." In Carl Schmitt, *Staat, Großraum, Nomos: Aufsätze aus den Jahren 1916–1969*, edited by Günter Maschke, 431–436. Berlin: Duncker & Humblot, 1995.

Schmitt, Carl (1942c). "Die Formung des französischen Geistes durch die Legisten." In Carl Schmitt, *Staat, Großraum, Nomos: Aufsätze aus den Jahren 1916–1969*, edited by Günter Maschke, 184–210. Berlin: Duncker & Humblot, 1995.

Schmitt, Carl (1943). "Behemoth, Leviathan und Greif." In Carl Schmitt, *Gesammelte Schriften 1933–1936: mit ergänzenden Beiträgen aus der Zeit des Zweiten Weltkriegs*, 520–527. Berlin: Duncker & Humblot, 2021.

Schmitt, Carl (1944). "Die letzte globale Linie." In Carl Schmitt, *Staat, Großraum, Nomos: Aufsätze aus den Jahren 1916–1969*, edited by Günter Maschke, 444–452. Berlin: Duncker & Humblot, 1995.

Schmitt, Carl (1945). "The International Crime of the War on Aggression and the Principle ‚Nullum crimen, nulla poena sine lege." In Carl Schmitt, *Writings on War*, edited by Timothy Nunan, 125–197. Cambridge: Polity Press, 2011.

Schmitt, Carl (1947). "Der Zugang zum Machthaber, ein zentrales verfassungsrechtliches Problem" In Carl Schmitt, *Verfassungsrechtliche Aufsätze aus den Jahren 1924–1954*, 430–4437. Berlin: Duncker & Humblot, 1958.

Schmitt, Carl (1948). "Völkerrecht: Ein juristisches Repetitorium." In Carl Schmitt, *Frieden oder Pazifismus? Arbeiten zum Völkerrecht und zur Internationalen Politik 1924–1978*, edited by Günter Maschke, 701–798. Berlin: Duncker & Humblot.

Schmitt, Carl (1949a). "Amnesty or the Force of Forgetting." In Carl Schmitt, *The Tyranny of Values and Other Texts*, 91–93. New York: Telos Press, 2018.

Schmitt, Carl (1949b). "Francisco de Vitoria und die Geschichte seines Ruhms." *Die neue Ordnung* 3 (1949): 289–313.

Schmitt, Carl (1949c). "Gegenwartsfragen der Verfassung." In *Carl Schmitt und die Liberalismuskritik*, edited by Klaus Hansen and Hans Lietzmann, 171–174. Opladen: Leske + Budrich, 1988.

Schmitt, Carl (1949d). "Das Grundgesetz der Bundesrepublik Deutschland." In *Carl Schmitt und die Liberalismuskritik*, edited by Klaus Hansen and Hans Lietzmann, 175–194. Opladen: Leske–Budrich, 1988.

Schmitt, Carl (1950a). *Ex captivitate salus: Experiences 1945–47*, edited by Andreas Kalyvas. Cambridge: Polity Press, 2017.

Schmitt, Carl (1950b). "Die Lage der europäischen Rechtswissenschaft (1943/44)." In Carl Schmitt, *Verfassungsrechtliche Aufsätze aus den Jahren 1924–1954*, 386–426. Berlin: Duncker & Humblot, 1958.

Schmitt, Carl (1950c). "A Pan-European Interpretation of Donoso Cortés." *Telos* 125 (2002): 100–112.

Schmitt, Carl (1950d). *The Nomos of the Earth*. New York: Telos Press, 2006

Schmitt, Carl (1950e). *Der Nomos der Erde*. Berlin: Duncker & Humblot.

Schmitt, Carl (1952a). "Foreword of the German Edition of Lilian Winstanley's Hamlet and the Scottish Succession." *Telos* 153 (2010): 164–177.

Schmitt, Carl (1952b). "Rechtsstaatlicher Verfassungsvollzug." In Carl Schmitt, *Verfassungsrechtliche Aufsätze aus den Jahren 1924–1954*, 452–486. Berlin: Duncker & Humblot, 1958.

Schmitt, Carl (1956). *Hamlet or Hecuba: The Intrusion of the Time into the Play*. New York: Telos Press, 2009.

Schmitt, Carl (1957). "What have I done?" In Carl Schmitt, *The Tyranny of Values and Other Texts*, 136–143. New York: Telos Press, 2018.

Schmitt, Carl (1958a). "Foreword of the Spanish edition (of Ex captivitate salus)." In Carl Schmitt, *Ex captivitate salus*, 75–78. Cambridge: Polity Press, 2017.

Schmitt, Carl (1958b). *Verfassungsrechtliche Aufsätze aus den Jahren 1924–1954*. Berlin: Duncker & Humblot.

Schmitt, Carl (1960). "The Tyranny of Values." In *Carl Schmitt: The Tyranny of Values and Other Texts*, 26–44. New York: Telos Press, 2018.

Schmitt, Carl (1962). "Buchbesprechung Otto Kirchheimer, Political Justice." In *Die Politische Meinung*, August 1962: 94.

Schmitt, Carl (1963a). *Theory of the Partisan*. New York: Telos Press, 2007.

Schmitt, Carl (1963b). "Vorwort (zur Neuauflage von 1963)." In Carl Schmitt, *Der Begriff des Politischen*, 9–19. Berlin: Duncker & Humblot, 1979.

Schmitt, Carl (1970). "Gespräch über den Partisanen." In Joachim Schickel, *Gespräche mit Carl Schmitt*, 9–59. Berlin: Merve.

Schmitt, Carl (1975). "Ein Rundfunkgespräch vom 1. Februar 1933." *Eclectica* 5, no 2 (1975): 113–124.

Schmitt, Carl (1987). "Interrogation by Robert Kempner." *Telos* 72 (Summer 1987): 97–129.

Schmitt, Carl (1995). *Staat, Großraum, Nomos: Arbeiten aus den Jahren 1916–1969*, edited by Günter Maschke. Berlin: Duncker & Humblot.

Schmitt, Carl (2000). *Antworten in Nürnberg*, edited by Helmut Quaritsch. Berlin: Duncker & Humblot.

Schmitt, Carl (2005). *Frieden oder Pazifismus? Arbeiten zum Völkerrecht und zur internationalen Politik 1924–1978*, edited by Günter Maschke. Berlin: Duncker & Humblot.

Schmitt, Carl (2010). *Tagebücher 1930 bis 1934*, edited by Wolfgang Schuller and Gerd Giesler. Berlin: Duncker & Humblot.

Schmitt, Carl (2015). *Glossarium: Aufzeichnungen aus den Jahre 1947 bis 1958*, edited by Gerd Giesler and Martin Tielke, 2nd, corrected edition. Berlin: Duncker & Humblot.

Schmitt, Carl (2018). *Tagebücher 1925 bis 1929*, edited by Martin Tielke and Gerd Giesler. Berlin: Duncker & Humblot.

Schmitt, Carl (2019). *Der Begriff des Politischen: Synoptische Darstellung der Texte*, edited by Marco Walter. Berlin: Duncker & Humblot.

Schmitt, Carl (2021). *Gesammelte Schriften 1933–1936: mit ergänzenden Beiträgen aus der Zeit des Zweiten Weltkriegs*, edited by Alexander Schmitz and Marcel Lepper. Berlin: Duncker & Humblot.

Schmitt, Carl and Hans Blumenberg (2021). *Briefwechsel 1971–1978 und weitere Materialien*, edited by Alexander Schmitz. Berlin: Suhrkamp.

Schmitt, Carl and Ernst-Wolfgang Böckenförde (2022). *Welch gütiges Schicksal: Briefwechsel 1953–1984*, edited by Reinhard Mehring. Baden-Baden: Nomos.

Schmitt, Carl and Luis Cabral de Moncada (1997). *Briefwechsel 1943–1973*, edited by Erik Jayme. Heidelberg: C.F. Müller.

Schmitt, Carl and Álvaro d'Ors (2004). *Briefwechsel*, edited by Montserat Herrero. Berlin: Duncker & Humblot.

Schmitt, Carl and Ernst Forsthoff (2007). *Briefwechsel 1926–1974*, edited by Dorothee Mußgnug. Berlin: Akademie Verlag.

Schmitt, Carl and Ernst Rudolf Huber (2014). *Briefwechsel 1926–1981: Mit ergänzenden Materialien*, edited by Ewald Grothe. Berlin: Duncker & Humblot.

Schmitt, Carl and Ernst Jünger (1999). *Briefwechsel 1930–1983*, edited by Helmuth Kiesel. Stuttgart: Klett-Cotta.

Schmitt, Carl and Reinhart Koselleck (2019). *Der Briefwechsel: 1953–1983 und weitere Materialien*, edited by Jan Eike Dunkhase. Berlin: Suhrkamp.

Schmitt, Carl and Armin Mohler (1995). *Briefwechsel mit einem seiner Schüler*, edited by Armin Mohler. Berlin: Duncker & Humblot.

Schmitt, Carl and Joachim Schickel (1969). "Gespräch über den Partisanen." In Joachim Schickel, *Gespräche mit Carl Schmitt*, 9–30. Berlin: Merve Verlag, 1993.

Schmitt, Carl and Duška Schmitt (2020). *Briefwechsel 1923–1950*, edited by Martin Tielke. Berlin: Duncker & Humblot.

Schmitt, Carl and Rudolf Smend (2010). *Auf der gefahrenvollen Straße des öffentlichen Rechts: Briefwechsel 1921–61*, edited by Reinhard Mehring. Berlin: Duncker & Humblot.

Schmitt, Carl and Roman Schnur (2023). *Briefwechsel 1951 bis 1983*, edited by Martin Tielke. Berlin: Duncker & Humblot.

Schmitt, Carl and Jacob Taubes (2012). *Briefwechsel*, edited by Herbert Kopp-Oberstebrink, Thorsten Palzhoff, and Martin Treml. Paderborn: Fink.

Schmitt, Carl and Ferdinand Tönnies (2016). "Korrespondenz 1924–1930." In *Schmittiana, Neue Folge, Band 3*, edited by Carl-Schmitt-Gesellschaft, 92–116. Berlin: Duncker & Humblot.

Schmitt, Carl and Eric Voegelin (2014). "Korrespondenz." In *Schmittiana, Neue Folge, Band 2*, edited by Carl-Schmitt-Gesellschaft, 183–199. Berlin: Duncker & Humblot.

Schmitz, Alexander and Marcel Lepper (2021). "Nachwort." In Hans Blumenberg and Carl Schmitt, *Briefwechsel 1971–1978*, edited by Alexander Schmitz and Marcel Lepper, 252–395. Berlin: Suhrkamp.

Schmoeckel, Mathias (1994). *Die Großraumtheorie: Ein Beitrag zur Geschichte der Völkerrechtswissenschaft im Dritten Reich, insbesondere der Kriegszeit*. Berlin: Duncker & Humblot.

Schneider, Hans (1993). "Der Entscheidungsvorgang." In *Politische Lageanalyse: Festschrift für Hans-Joachim Arndt*, edited by Volker Beismann and Markus J. Klein, 277–281. Bruchsal: San Cascino Verlag.

Schneider, Peter (1957). *Ausnahmezustand und Norm: Eine Studie zur Rechtslehre von Carl Schmitt.* Stuttgart: Deutsche Verlags-Anstalt.

Schneider, Peter and Heinrich Gremmels (1994). "Eine Diskussion über Carl Schmitt." In *Schmittiana: Beiträge zu Leben und Werk Carl Schmitts, Band IV*, edited by Piet Tommissen, 227–248. Berlin: Duncker & Humblot.

Schuller, Wolfgang (2008). "Carl Schmitt in Leipzig." *Recht und Politik* 44, no. 1(2008): 35–48.

Schulze, Hagen (1977). *Otto Braun oder Preußens demokratische Sendung*. Frankfurt am Main: Propyläen.

Schulze-Fielitz, Helmuth (2018). "Rudolf Smend." In *Staatrechtslehrer des 20. Jahrhunderts*, edited by Peter Häberle, Michael Kilian, and Heinrich Wolff, 2nd edition, 318–333. Berlin: de Gruyter.

Schumpeter, Joseph A. (1919). "Zur Soziologie des Imperialismus." *Archiv für Sozialwissenschaft und Sozialpolitik* 46, no. 1–2 (1919): 1–39.

Schupmann, Benjamin A. (2024). *Democracy despite Itself: Liberal Constitutionalism and Militant Democracy.* Oxford: Oxford University Press (forthcoming).

Schwab, George D. (1970). The Challenge of the Exception: An Introduction of the Political Ideas of Carl Schmitt between 1921 and 1936. Berlin: Duncker & Humblot.

Schwab, George D. (1980). "Schmitt Scholarship." *Canadian Journal of Political and Social Theory* 4, no. 2 (1980): 149–155.

Schwab, George D. (1987). "Enemy or Foe: A Conflict of Modern Politics." *Telos* 72 (Summer 1987): 194–201.

Schwab, George D. (1988a). "Carl Schmitt through a Glass Darkly." *Eclectica* 17, no. 71/72 (1988): 70–87.

Schwab, George D. (1988b). "Progress of Schmitt Studies in the English-Speaking World." In *Complexio Oppositorum: Über Carl Schmitt*, edited by Helmut Quaritsch, 447–459 and 461–464 (Aussprache). Berlin: Duncker & Humblot.

Schwab, George D. (1990). "Carl Schmitt through a Glass Darkly." In Schmittiana 3 (1990, third edition), 70–87.

Schwab, George D. (1992). "Carl Schmitt Hysteria in the United States: The Case of Bill Scheuermann." *Telos* 92 (1992): 99–107.

Schwab, George D. (1996). "Introduction." In Carl Schmitt, *The Leviathan in the State Theory of Thomas Hobbes*. New Edition, XXXI–LIII. Chicago: University of Chicago Press, 2008.

Schwab, George D. (1996). "Introduction." In Carl Schmitt, *The Concept of the Political*. XXXI–LII. University of Chicago Press, 2008.

Schwab, George D. (2021). *Odyssey of a Child Survivor: From Lavitia through the Camps to the United States* (private printing).

Schwarzenberger, Georg (1932). *Die Kreuger-Anleihen: Ein Beitrag zur Auslegung der internationalen Anleihe- und Monopolverträge sowie von der Lehre zum Staatsbankrott*. München und Leipzig: Duncker & Humblot.

Schwarzschild, Leopold (1934). *Ende der Illusionen*. Amsterdam: Querido.

Seiberth, Gabriel (2001). *Anwalt des Reiches: Carl Schmitt und der Prozess Preußen contra Reich vor dem Staatsgerichtshof.* Berlin: Duncker & Humblot.

Seiberth, Gabriel (2002). "Legalität oder Legitimität? ‚Preußenschlag' und Staatsnotstand als juristische Herausforderung für Carl Schmitt in der Reichskrise der Weimarer Endzeit." In *Schmittiana: Beiträge zu Leben und Werk Carl Schmitts, Band VII*, edited by Piet Tommissen, 132–164. Berlin: Duncker & Humblot.

Seifert, Jürgen (1985). "Theoretiker der Gegenrevolution. Carl Schmitt 1888–1985." *Kritische Justiz* 18, no. 2 (1985): 193–200.

Seifert, Jürgen (1996). "Unterwegs zur Ebene über den Gegensatz." In *Schmittiana: Beiträge zu Leben und Werk Carl Schmitts, Band V*, edited by Piet Tommissen, 109–150. Berlin: Duncker & Humblot.

Simard, Augustin (2022). "L'exception… et après? Exception, État autoritaire et exécutif selon Otto Kirchheimer." In *De la dictature à l'état d'exception*, edited by Marie Goupy and Yann Rivière. Rome: Éditions de l'École française de Rome.

Simard, Augustin (2023). *Le droit et son autre. Otto Kirchheimer et la critique de l'État de droit*. Montréal: Les Presses de l'Université de Montréal.

Sinzheimer, Hugo (1938). *Jüdische Klassiker der deutschen Rechtswissenschaft*. Amsterdam: Hertzberger.

Slomp, Gabriella (2006). "Carl Schmitt's Five Arguments Against the Idea of Just War." *Cambridge Review of International Affairs* 19, no. 3 (2006): 435–447.

Slomp, Gabriella (2009). *Carl Schmitt and the Politics of Hostility, Violence and Terror*. London: Palgrave Macmillan.

Smend, Rudolf (1919). "Die Verschiebung der konstitutionellen Ordnung durch die Verhältniswahl." In Rudolf Smend, *Staatsrechtliche Abhandlungen und andere Aufsätze*. 2nd edition, 60–67. Berlin: Duncker & Humblot, 1968.

Smend, Rudolf (1923). "Die politische Gewalt im Verfassungsstaat und das Problem der Staatsform." In Rudolf Smend, *Staatsrechtliche Abhandlungen und andere Aufsätze*. 2nd edition, 68–88. Berlin: Duncker & Humblot, 1968.

Smend, Rudolf (1928). "Verfassung und Verfassungsrecht." In Rudolf Smend, *Staatsrechtliche Abhandlungen und andere Aufsätze*. 2nd edition, 119–276. Berlin: Duncker & Humblot, 1968.

Smend, Rudolf (1931). "Les actes de gouvernement en Allemagne." *Annuaire de l'institut international de droit public* 2(1931): 192–232.

Smend, Rudolf (1933). "Bürger und Bourgeois im deutschen Staatsrecht." In Rudolf Smend, *Staatsrechtliche Abhandlungen und andere Aufsätze*. 2nd edition, 309–326. Berlin: Duncker & Humblot, 1968.

Smend, Rudolf (1949). "Staatsrechtslehrer-Vereinigung neu gegründet." *Göttinger Universitätsanzeiger* 4, no. 21 (1949): 15–17.

Smend, Rudolf (1960). "Zur Geschichte der Berliner Juristenfakultät im 20. Jahrhundert." In Rudolf Smend, *Staatsrechtliche Abhandlungen und andere Aufsätze*. 2nd edition, 527–546. Berlin: Duncker & Humblot, 1968.

Smith, Richard Harris (1973). *OSS: The Secret History of America's First Central Intelligence Agency*. New York: Dell.

Snyder, Louis L. (1932). *Hitlerism: The Iron Fist in Germany*. New York: Mohawk Press.

Söllner, Alfons (1979). *Geschichte und Herrschaft: Studien zur materialistischen Sozialwissenschaft 1929–1942*. Frankfurt am Main: Suhrkamp.

Söllner, Alfons (1982). "Politische Dialektik der Aufklärung: Zum Spätwerk von Franz Neumann und Otto Kirchheimer." In *Sozialforschung als Kritik*, edited by Wolfgang Bonß and Axel Honneth, 281–327. Frankfurt am Main: Suhrkamp.

Söllner, Alfons (1983). "Linke Schüler der konservativen Revolution? Zur politischen Theorie Neumanns, Kirchheimers und Marcuses am Ende der Weimarer Republik." *Leviathan* 11, no. 2 (1983): 214–232.

Söllner, Alfons (1986a). *Zur Archäologie der Demokratie in Deutschland, Band 1: Analysen von politischen Emigranten im amerikanischen Geheimdienst: 1943–1945*. Frankfurt am Main: Europäische Verlagsanstalt.

Söllner, Alfons (1986b). *Zur Archäologie der Demokratie in Deutschland, Band 2: Analysen von politischen Emigranten im amerikanischen Außenministerium, 1946–1949*. Frankfurt am Main: Europäische Verlagsanstalt.

Söllner, Alfons (1987). "Beyond Carl Schmitt: Political Theory in the Frankfurt School." *Telos* 71 (Spring 1987): 81–108.

Söllner, Alfons (1996). *Deutsche Politikwissenschaftler in der Emigration*. Wiesbaden: Westdeutscher Verlag.

Söllner, Alfons (2018). *Political Scholar: Zur Intellektuellengeschichte des 20. Jahrhunderts*. Hamburg: Europäische Verlags-Anstalt.

Söllner, Alfons, Michael Wildt, Hubertus Buchstein and Peter Hayes (2023). *ad Franz L. Neumann Behemoth*. Hamburg: Europäische Verlagsanstalt.

Sombart, Nicolaus (1997). *Die deutschen Männer und ihre Feinde: Carl Schmitt – ein deutsches Schicksal zwischen Männerbund und Matriachatsmythos*. Frankfurt am Main: Fischer.

Sontheimer, Kurt (1957). "Vom Rechtsstaat zum totalen Staat." In *Schmittiana: Beiträge zu Leben und Werk Carl Schmitts, Band IV*, edited by Piet Tommissen, 243–248. Berlin: Duncker & Humblot, 1994.

Sontheimer, Kurt (1962). *Antidemokratisches Denken in der Weimarer Republik*. München: Nymphenburger Verlagsanstalt.

Sorel, Georges (1906). *Reflections on Violence*. Cambridge: Cambridge University Press, 1999.

Später, Jörg (2017). *Siegfried Kracauer: Eine Biographie*. Berlin: Suhrkamp.

Specter, Matthew G. (2010). *Habermas: An Intellectual Biography*. New York: Cambridge University Press.

Specter, Matthew G. (2016). "What's ‚Left' in Schmitt? From Aversion to Appropriaton in Contemporary Political Theory." In *The Oxford Handbook of Carl Schmitt*, edited by Jens Meierhenrich and Oliver Simons, 426–455. Oxford: Oxford University Press.

Speier, Hans (1932). "Rezension: Carl Schmitt, Der Begriff des Politischen." *Zeitschrift für Sozialforschung* 1 (1932): 203–204.

Sperber, Manès (1982). *Bis man mir Scherben auf die Augen legt: All das Vergangene*. München: dtv.

Steinle, Stephanie (2004). "Georg Schwarzenberger (1908–1991)." In *Jurists unprooted: German-speaking émigré lawyers in twentieth century Britain*, edited by Jack Beatson and Reinhard Zimmermann, 663–680. Oxford: Oxford University Press.

Sternberger, Dolf (1950). "Berufspolitiker und Politikerberufe." *Die Gegenwart* 5 (1950): 9–11.

Stiefel, Ernst C. and Frank Mecklenburg (1991). *Deutsche Juristen im amerikanischen Exil (1933–1950)*. Tübingen: Mohr Siebeck.

Stirk, Peter (1999). "Carl Schmitt's Völkerrechtliche Großraumordnung." *History of Political Thought* 20, no. 2 (1999): 257–274.

Stirk, Peter (2005). *Carl Schmitt: Crown Jurist of the Third Reich on Pre-emptive War, Military Occupation, and World Empire*. Lewiston: Edwin Mellen Press.

Stolleis, Michael (1999). *Geschichte des öffentlichen Rechts in Deutschland, Dritter Band: Staats- und Verwaltungsrechtswissenschaft in Republik und Diktatur 1914–1945*. München: Beck.

Stolleis, Michael (2012). *Geschichte des öffentlichen Rechts in Deutschland, Vierter Band: Staats- und Verwaltungsrechtswissenschaft in West und Ost 1945–1990*. München: Beck.

Strauss, Leo (1932). "Notes on Carl Schmitt, The Concept of the Political." In Carl Schmitt, *The Concept of the Political*. New Edition, 81–107. Chicago: The University of Chicago Press, 1996.

Streeck, Wolfgang (2016). *Buying Time: The Delayed Crisis of Democratic Capitalism*. London: Verso.

Strong, Tracy B. (1996). "Dimensions of the New Debate around Carl Schmitt." In Carl Schmitt, *The Concept of the Political*. New Edition, ix–xxvii. Chicago: The University of Chicago Press, 1996.

Strong, Tracy B. (2008). "Foreword." In Carl Schmitt, *The Concept of the Political*, vii–xxvii. Chicago: The University of Chicago Press, 2008.

Suuronen, Ville (2022). "Carl Schmitt's Confrontation with the Work of Hannah Arendt." *Global Intellectual History* 47 (5): 270–305.

Taylor, Telford (1993). *The Anatomy of the Nuremberg Trials: A Personal Memoir*. London: Bloomsbury.

Teschke, Benno G. (2011a). "Fatal Attraction: A Critique of Carl Schmitt's International Political and Legal Theory." *International Theory* 3, no. 2 (2011): 179–227.

Teschke, Benno G. (2011b). *The Fetish of Geopolitics: A Reply to Gopal Balakrishnan. New Left Review* 69 (2011): 81–100.

Teschke, Benno G. (2016). "Carl Schmitt's Concepts of War: A Categorial Failure." In *The Oxford Handbook of Carl Schmitt*, edited by Jens Meierhenrich and Oliver Simons, 367–400. Oxford: Oxford University Press.

Thoma, Richard (1925). "On the Ideology of Parliamentarism" In Carl Schmitt, *The Crisis of Parliamentary Democracy*, 77–84. Cambridge: MIT Press 1985.

Tielke, Martin (2007). *Der stille Bürgerkrieg: Ernst Jünger und Carl Schmitt im Dritten Reich*. Berlin: Landt-Verlag.

Tielke, Martin (2018). "Einführung." In Carl Schmitt, *Tagebücher 1925 bis 1929*, edited by Martin Tielke and Gerd Giesler, ix–xxxviii. Berlin: Duncker & Humblot.

Tielke, Martin (2020). "Einführung." In Carl Schmitt and Duška Schmitt, *Briefwechsel 1923–1950*, edited by Martin Tielke, 15–45. Berlin: Duncker & Humblot.

Tielke, Martin, and Gerd Giesler (2020). "Die Tagebücher Carl Schmitts." *In Jahrbuch für internationale Germanistik* 53, no. 2 (2020): 29–60.

Tilitzki, Christian (1994). "Carl Schmitt an der Handels-Hochschule 1928–1933." In *Schmittiana: Beiträge zu Leben und Werk Carl Schmitts, Band IV*, edited by Piet Tommissen, 157–201, Berlin: Duncker & Humblot, 1994.

Tilitzki, Christian (1998). "Die Vortragsreisen Carl Schmitts während des Zweiten Weltkriegs In *Schmittiana: Beiträge zu Leben und Werk Carl Schmitts, Band VI*, edited by Piet Tommissen, 191–269. Berlin: Duncker & Humblot, 1998.

Tocqueville, Alexis de (1835). *Democracy in America*, translated by James Schleifer. Indianapolis: Liberty Fund 2010.

Tommissen, Piet (1990). "Bemerkungen zum Verhör Carl Schmitts durch Ossip K. Flechtheim." In *Schmittiana: Beiträge zu Leben und Werk Carl Schmitts, Band II*, edited by Piet Tommissen, 142–148. Berlin: Duncker & Humblot.

Tönnies, Ferdinand (1925). *Thomas Hobbes: Leben und Werk*. 3rd edition. Stuttgart: Fr. Frommann.

Trefz, Bernhard (1997). *Jugendbewegung und Juden in Deutschland: Eine historische Untersuchung mit besonderer Berücksichtigung des Deutsch-Jüdischen Wanderbundes ›Kameraden‹*. Frankfurt am Main: Peter Lang.

Tribe, Keith (1987). "Introduction." In *Social Democracy and the Rule of Law: Otto Kirchheimer und Franz Neumann*, edited by Keith Tribe, 1–12. London: Allen & Unwin, 2000.

Tribe, Keith (2020). *Social Democracy and the Rule of Law: Otto Kirchheimer und Franz Neumann*, edited by Keith Tribe. New York: Routledge.

Troy, Thomas F. (1981). *Donovan and the CIA: A History of the Establishment of the Central Intelligence Agency*. Frederick: Aletheia Books.

Turner, Stephen (2011). "Schmitt, *Telos*, the Collapse of the Weimar Constitution, and the Bad Conscience of the Left." In *A Journal of no Illusions: Telos, Paul Piccone, and the Americanization of Critical Theory*, edited by Timothy W. Luke and Ben Agger, 113–140. New York: Telos Press.

Ulmen, Gary (1987). "Return of the Foe." *Telos* 72 (Summer 1987): 187–193.

United States Holocaust Memorial Museum, *Holocaust Encyclopedia*, https://encyclopedia.ushmm.org/content/en/article/indoctrinating-youth.

Vagst, D. (1990). "International Law in the Third Reich." *American Journal of International Law* 84, no. 3 (1990): 661–704.

van Laak, Dirk (1993). *Gespräche in der Sicherheit des Schweigens: Carl Schmitt in der politischen Geistesgeschichte der frühen Bundesrepublik*. Berlin: Akademie Verlag.

van Laak, Dirk (1998). "Buchbesprechung Carl Schmitt, Briefwechsel mit einem seiner Schüler" In *Jahrbuch Politisches Denken 1998*, edited by Karl Graf Ballestrem, Volker Gerhardt, Henning Ottmann, and Martyn P. Thompson, 212–218. Stuttgart: Metzler.

van Roon, Gert (1967). *Neuordnung im Widerstand: Der Kreisauer Kreis innerhalb der deutschen Widerstandsbewegung*. München: Oldenbourg.

Vinx, Lars (2015). *The Guardian of the Constitution: Hans Kelsen and Carl Schmitt on the Limits of Constitutional Law.* Cambridge: Cambridge University Press.

Voegelin, Eric (1931). "Die Verfassungslehre von Carl Schmitt: Versuch einer konstruktiven. Analyse ihrer staatstheoretischen Prinzipien." *Zeitschrift für Öffentliches Recht* 11, no. 1 (1931): 89–109.

Voigt, Rüdiger (2009). *Der Hobbes-Kristall: Carl Schmitts Hobbes-Interpretation in der Diskussion*, edited by Rüdiger Voigt. Stuttgart: Steiner.

Vorländer, Karl (1926). *Von Machiavelli bis Lenin: Neuzeitliche Staats- und Gesellschaftstheorien*. Leipzig: Quelle & Meyer.

Vormeier, Barbara (1981). *Exil in Frankreich: Die deutschen Emigranten 1933–1944*: Braunschweig: Westermann.

Vormeier, Barbara (2002). "Die Schaffung des internationalen Flüchtlingsstatus und die Rolle der Pariser Asylrechts- und Flüchtlingskonferenz." In *Fluchtziel Paris: Die deutschsprachige Emigration 1933–1940*, edited by Anne Sauveur-Henn, 41–50. Berlin: Metropol.

Walter, Marco (2018). "Kommentar." In Carl Schmitt, *Der Begriff des Politischen: Synoptische Darstellung der Texte*, edited by Marco Walter, 279–304. Berlin: Duncker & Humblot.

Walz, Rainer (1995). "Der vormoderne Antisemitismus: Religiöser Fanatismus oder Rassenwahn?" *Historische Zeitschrift* 260, no. 3 (1995): 719–748.

Weber, Petra (1996). *Carlo Schmid: Eine Biographie*. München: Beck.

Weber, Werner (1949). "Rezension Repetitorien." *Neue Juristische Wochenschrift*, 2, no. 21 (1949): 819–821.

Weber, Werner (1951). *Spannungen und Kräfte im westdeutschen Verfassungssystem*. Stuttgart: Friedrich Vorwerk Verlag.

Weil, Felix J. (1951). "The Cold War Against the Jews in Argentina." *The Chicago Jewish Forum* 10, no. 1 (1951): 11.

Wendenburg, Helge (1984). *Die Debatte um die Verfassungsgerichtsbarkeit und der Methodenstreit der Staatrechtslehre in der Weimarer Republik*. Göttingen: Schwartz.

Werner, Wouter G. (2010). "The Changing Face of Enmity: Carl Schmitt's International Theory and the Evolution of the Legal System of War." *International Theory* 2, no. 3 (2010): 531–573.

Wheatland, Thomas (2009). *The Frankfurt School in Exile*. Minneapolis: University of Minnesota Press.

Wieland, Claus-Dietrich (1987). "Carl Schmitt in Nürnberg (1947)." *1999: Zeitschrift für Sozialgeschichte des 20. und 21. Jahrhunderts* 2 (1987): 96–106.

Wiggershaus, Rolf (1995). *The Frankfurt School: Its History, Theories, and Political Significance*. Cambridge: MIT Press.

Wiggershaus, Rolf (2010). *Die Frankfurter Schule*. Reinbek: Rowohlt.

Wildt, Michael (2023). "Franz Neumann und die NS-Forschung". In Alfons Söllner, Michael Wildt, Hubertus Buchstein, and Peter Hayes, *ad Franz Neumann Behemoth*, 63–110. Hamburg: Europäische Verlagsanstalt.

Willms, Bernard (1982). *Die deutsche Nation: Theorie – Lage – Zukunft*. Köln: Hohenheim Verlag.

Winkler, Heinrich August (2001). *Der lange Weg nach Westen: Deutsche Geschichte vom ‚Dritten Reich' bis zur Wiedervereinigung*. München: Beck.

Wollenberg, Jörg (eds.) (1991). *Von der Hoffnung aller Deutschen. Wie die BRD entstand 1945–1949*. Köln: Pahl-Rugenstein.

Worrell, Mark P. (2008). *Dialectic of Solidarity: Labor, Antisemitism, and the Frankfurt School*. Leiden: Brill.

Wulf, Joseph (1963). *Aus dem Lexikon der Mörder: "Sonderbehandlung" und verwandte Worte in nationalsozialistischen Dokumenten*. Gütersloh: Mohn.

Zeitlin, Samuel Garret (2015). "Translator's Note." In Carl Schmitt, *Land and Sea. A World-Historical Meditation*. Candor, NY: Telos Press Publishing.

Zeitlin, Samuel Garret (2020). "Indirection and the Rhetoric of Tyranny: Carl Schmitt's The Tyranny of Values 1960–1967." *Modern Intellectual History* 18, 1–24.

Ziege, Eva-Maria (2009). *Antisemitismus und Gesellschaftstheorie: Die Frankfurter Schule im amerikanischen Exil*. Frankfurt am Main: Suhrkamp.

Zolo, Danielo (2007). "The Re-emerging Notion of Empire and the Influence of Carl Schmitt's Thought." In *The International Thought of Carl Schmitt: Terror, War, and the Crises of the Global Order*, edited by Louiza Odysseos, and Fabio Petito, 154–165. London: Routledge.

Zwarg, Robert (2017). *Die Kritische Theorie in Amerika: Das Nachleben einer Tradition*. Göttingen: Vandenhoeck und Ruprecht.

Index of Names

J

K

N

O

P

Q

R